Springer Proceedings in Mathematics

Volume 4

For further volumes:
http://www.springer.com/series/8806

Springer Proceedings in Mathematics

The book series will feature volumes of selected contributions from workshops and conferences in all areas of current research activity in mathematics. Besides an overall evaluation, at the hands of the publisher, of the interest, scientific quality, and timeliness of each proposal, every individual contribution will be refereed to standards comparable to those of leading mathematics journals. It is hoped that this series will thus propose to the research community well-edited and authoritative reports on newest developments in the most interesting and promising areas of mathematical research today.

Jaroslav Fořt • Jiří Fürst • Jan Halama
Raphaèle Herbin • Florence Hubert
Editors

Finite Volumes for Complex Applications VI
Problems & Perspectives

FVCA 6, International Symposium,
Prague, June 6-10, 2011, Volume 1

Editors
Jaroslav Fořt
Czech Technical University
Faculty of Mechanical Engineering
Karlovo náměstí 13
Prague
Czech Republic
Jaroslav.Fort@fs.cvut.cz

Jiří Fürst
Czech Technical University
Faculty of Mechanical Engineering
Karlovo náměstí 13
Prague
Czech Republic
Jiri.Furst@fs.cvut.cz

Jan Halama
Czech Technical University
Faculty of Mechanical Engineering
Karlovo náměstí 13
Prague
Czech Republic
Jan.Halama@fs.cvut.cz

Raphaèle Herbin
Université Aix-Marseille
LATP
Laboratoire d'Analyse
Probabilités et T
rue Joliot Curie 39
13453 Marseille
France
Raphaele.Herbin@latp.univ-mrs.fr

Florence Hubert
Université Aix-Marseille
LATP
Laboratoire d'Analyse
Probabilités et T
rue Joliot Curie 39
13453 Marseille
France
Florence.Hubert@latp.univ-mrs.fr

ISSN 2190-5614   e-ISSN 2190-5622
ISBN 978-3-642-20670-2   e-ISBN 978-3-642-20671-9
DOI 10.1007/978-3-642-20671-9
Springer Heidelberg Dordrecht London New York

Library of Congress Control Number: 2011930263

© Springer-Verlag Berlin Heidelberg 2011

This work is subject to copyright. All rights are reserved, whether the whole or part of the material is concerned, specifically the rights of translation, reprinting, reuse of illustrations, recitation, broadcasting, reproduction on microfilm or in any other way, and storage in data banks. Duplication of this publication or parts thereof is permitted only under the provisions of the German Copyright Law of September 9, 1965, in its current version, and permission for use must always be obtained from Springer. Violations are liable to prosecution under the German Copyright Law.

The use of general descriptive names, registered names, trademarks, etc. in this publication does not imply, even in the absence of a specific statement, that such names are exempt from the relevant protective laws and regulations and therefore free for general use.

Cover design: deblik, Berlin

Printed on acid-free paper

Springer is part of Springer Science+Business Media (www.springer.com)

Editors Preface

The sixth International Symposium on Finite Volumes for Complex Applications, held in Prague (Czech Republic, June 2011) follows the series of symposiums held successively in Rouen (France, 1996), Duisburg (Germany, 1999), Porquerolles (France, 2002), Marrakech (Morocco, 2005), Aussois (France, 2008).

The sixth symposium, similarly to the previous ones, gives the opportunity of a large and critical discussion about the various aspects of finite volumes and related methods: mathematical results, numerical techniques, but also validations via industrial applications and comparisons with experimental test results.

This book tries to assemble the recent advances in both the finite volume method itself (theoretical aspects of the methods, new or improved algorithms, numerical implementation problems, benchmark problems and efficient solvers) as well as its application in complex problems in industry, environmental sciences, medicine and other fields of technology, so as to bring together the academic world and the industrial world. The topics of the proceedings reflect this wide range of perspectives and include: advanced schemes and methods (complex grid topology, higher order methods, efficient implementation), convergence and stability analysis, global error analysis, limits of methods, purely multidimensional difficulties, non homogeneous systems with stiff source terms, complex geometries and adaptivity, complexity, efficiency and large computations, chaotic problems (turbulence, ignition, mixing, . . .), new fields of application, comparisons with experimental results. The proceedings also include the results to a benchmark on three–dimensional anisotropic and heterogeneous diffusion problems, which was designed to test some 16 different schemes, among which finite volume methods, finite element methods, discontinuous Galerkin methods, mimetic methods and discrete gradient schemes. A new feature of this benchmark is the comparison of various iterative solvers on the matrices resulting from the different schemes.

Of course, the success of the symposium crucially depends on the quality of the contributions. Therefore we would like to express many thank all the authors of regular papers, who provided high quality papers on the above mentioned wide range of subjects, or contributed to the 3D anisotropic diffusion benchmark. The level the contributions was ensured by the Scientific Committee members, who organized the reviewing process of each paper. We express our gratitude to members of the Scientific Committee as well as to many other reviewers.

The symposium could not have been organized without the local support of Czech Technical University, Faculty of Mechanical Engineering and financial support of French contributors: CMLA ENS Cachan, IFP Energies nouvelles, IRSN, LATP Université Aix Marseille I, MOMAS group, Université Paris XIII, Université Paris Est Marne la Vallée, Université Pierre et Marie Curie.

Finally we would like to thank Springer Verlag Editor's team for their cooperation in the proceedings preparation, conference secretary T. Němcová and all others, who ensured logistic and communication before and during the conference.

Jaroslav Fořt
Jiří Fürst
Jan Halama
Raphaèle Herbin
Florence Hubert

Organization

Committees

Organizing Committee:

Jaroslav Fořt
Jiří Fürst
Jan Halama
Rémi Abgrall
Fayssal Benkhaldoun

Robert Eymard
Jean-Michel Ghidaglia
Jean-Marc Hérard
Raphaèle Herbin
Martin Vohralík

Scientific Committee:

Rémi Abgrall
Brahim Amaziane
Fayssal Benkhaldoun
Vít Dolejší
François Dubois
Denys Dutykh
Robert Eymard
Jaroslav Fořt
Jürgen Fuhrmann
Jiří Fürst
Thierry Gallouet
Jean-Michel Ghidaglia

Hervé Guillard
Jan Halama
Khaled Hassouni
Jean-Marc Hérard
Raphaèle Herbin
Florence Hubert
Raytcho Lazarov
Karol Mikula
Mario Ohlberger
Frédéric Pascal
Martin Vohralík

Contents

Vol. 1

Part I Regular Papers

Volume-Agglomeration Coarse Grid In Schwarz Algorithm 3
H. Alcin, O. Allain, and A. Dervieux

A comparison between the meshless and the finite volume methods for shallow water flows... 13
Yasser Alhuri, Fayssal Benkhaldoun, Imad Elmahi, Driss Ouazar, Mohammed Seaïd, and Ahmed Taik

Time Compactness Tools for Discretized Evolution Equations and Applications to Degenerate Parabolic PDEs 21
Boris Andreianov

Penalty Methods for the Hyperbolic System Modelling the Wall-Plasma Interaction in a Tokamak 31
Philippe Angot, Thomas Auphan, and Olivier Guès

A Spectacular Vector Penalty-Projection Method for Darcy and Navier-Stokes Problems 39
Philippe Angot, Jean-Paul Caltagirone, and Pierre Fabrie

Numerical Front Propagation Using Kinematical Conservation Laws 49
K.R. Arun, M. Lukáčová-Medviďová, and P. Prasad

Preservation of the Discrete Geostrophic Equilibrium in Shallow Water Flows.. 59
E. Audusse, R. Klein, D.D. Nguyen, and S. Vater

**Arbitrary order nodal mimetic discretizations
of elliptic problems on polygonal meshes** 69
Lourenço Beirão da Veiga, Konstantin Lipnikov,
and Gianmarco Manzini

**Adaptive cell-centered finite volume method for non-
homogeneous diffusion problems: Application to transport
in porous media** 79
Fayssal Benkhaldoun, Amadou Mahamane, and Mohammed Seaïd

**A Generalized Rusanov method for Saint-Venant Equations
with Variable Horizontal Density** 89
Fayssal Benkhaldoun, Kamel Mohamed, and Mohammed Seaïd

Hydrostatic Upwind Schemes for Shallow–Water Equations 97
Christophe Berthon and Françoise Foucher

**Finite Volumes Asymptotic Preserving Schemes for Systems
of Conservation Laws with Stiff Source Terms** 107
C. Berthon and R. Turpault

Development of DDFV Methods for the Euler Equations 117
Christophe Berthon, Yves Coudière, and Vivien Desveaux

**Comparison of Explicit and Implicit Time Advancing
in the Simulation of a 2D Sediment Transport Problem** 125
M. Bilanceri, F. Beux, I. Elmahi, H. Guillard, and M.V. Salvetti

**Numerical Simulation of the Flow in a Turbopump Inducer
in Non-Cavitating and Cavitating Conditions** 135
M. Bilanceri, F. Beux, and M.V. Salvetti

On Some High Resolution Schemes for Stably Stratified Fluid Flows ... 145
Tomáš Bodnár and Luděk Beneš

**Convergence Analysis of the Upwind Finite Volume Scheme
for General Transport Problems** 155
Franck Boyer

**A Low Degree Non–Conforming Approximation of the Steady
Stokes Problem with an Eddy Viscosity** 165
F. Boyer, F. Dardalhon, C. Lapuerta, and J.-C. Latché

**Some Abstract Error Estimates of a Finite Volume
Scheme for the Wave Equation on General Nonconforming
Multidimensional Spatial Meshes** 175
Abdallah Bradji

**A Convergent Finite Volume Scheme for Two-Phase Flows
in Porous Media with Discontinuous Capillary Pressure Field** 185
K. Brenner, C. Cancès, and D. Hilhorst

Contents

Uncertainty Quantification for a Clarifier–Thickener Model with Random Feed 195
Raimund Bürger, Ilja Kröker, and Christian Rohde

Asymptotic preserving schemes in the quasi-neutral limit for the drift-diffusion system 205
Chainais-Hillairet Claire and Vignal Marie-Hélène

A Posteriori Error Estimates for Unsteady Convection–Diffusion–Reaction Problems and the Finite Volume Method 215
Nancy Chalhoub, Alexandre Ern, Tony Sayah, and Martin Vohralík

Large Time-Step Numerical Scheme for the Seven-Equation Model of Compressible Two-Phase Flows 225
Christophe Chalons, Frédéric Coquel, Samuel Kokh, and Nicole Spillane

Asymptotic Behavior of the Scharfetter–Gummel Scheme for the Drift-Diffusion Model 235
Marianne CHATARD

A Finite Volume Solver for Radiation Hydrodynamics in the Non Equilibrium Diffusion Limit 245
D. Chauveheid, J.-M. Ghidaglia, and M. Peybernes

An Extension of the MAC Scheme to some Unstructured Meshes 253
Eric Chénier, Robert Eymard, and Raphaèle Herbin

Multi-dimensional Optimal Order Detection (MOOD) — a Very High-Order Finite Volume Scheme for Conservation Laws on Unstructured Meshes 263
S. Clain, S. Diot, and R. Loubère

A Relaxation Approach for Simulating Fluid Flows in a Nozzle 273
Frédéric Coquel, Khaled Saleh, and Nicolas Seguin

A CeVeFE DDFV scheme for discontinuous anisotropic permeability tensors 283
Yves Coudière, Florence Hubert, and Gianmarco Manzini

Multi-Water-Bag Model And Method Of Moments For The Vlasov Equation 293
Anaïs Crestetto and Philippe Helluy

Comparison of Upwind and Centered Schemes for Low Mach Number Flows 303
Thu–Huyen DAO, Michael NDJINGA, and Frédéric MAGOULES

On the Godunov Scheme Applied to the Variable Cross-Section Linear Wave Equation 313
Stéphane Dellacherie and Pascal Omnes

Towards stabilization of cell-centered Lagrangian methods for compressible gas dynamics 323
Bruno Després and Emmanuel Labourasse

Hybrid Finite Volume Discretization of Linear Elasticity Models on General Meshes 331
Daniele A. Di Pietro, Robert Eymard, Simon Lemaire, and Roland Masson

An A Posteriori Error Estimator for a Finite Volume Discretization of the Two-phase Flow 341
Daniele A. Di Pietro, Martin Vohralík, and Carole Widmer

A Two-Dimensional Relaxation Scheme for the Hybrid Modelling of Two-Phase Flows 351
Kateryna Dorogan, Jean-Marc Hérard, and Jean-Pierre Minier

Finite Volume Method for Well-Driven Groundwater Flow 361
Milan Dotlić, Dragan Vidović, Milan Dimkić, Milenko Pušić, and Jovana Radanović

Adaptive Reduced Basis Methods for Nonlinear Convection–Diffusion Equations 369
Martin Drohmann, Bernard Haasdonk, and Mario Ohlberger

Adaptive Time-Space Algorithms for the Simulation of Multi-scale Reaction Waves 379
Max Duarte, Marc Massot, Stéphane Descombes, and Thierry Dumont

Dispersive wave runup on non-uniform shores 389
Denys Dutykh, Theodoros Katsaounis, and Dimitrios Mitsotakis

MAC Schemes on Triangular Meshes 399
Robert Eymard, Jürgen Fuhrmann, and Alexander Linke

Multiphase Flow in Porous Media Using the VAG Scheme 409
Robert Eymard, Cindy Guichard, Raphaèle Herbin, and Roland Masson

Grid Orientation Effect and MultiPoint Flux Approximation 419
Robert Eymard, Cindy Guichard, and Roland Masson

Gradient Schemes for Image Processing 429
Robert Eymard, Angela Handlovičová, Raphaèle Herbin, Karol Mikula, and Olga Stašová

Gradient Scheme Approximations for Diffusion Problems 439
Robert Eymard and Raphaèle Herbin

Cartesian Grid Method for the Compressible Euler Equations M. Asif Farooq and B. Müller	449
Compressible Stokes Problem with General EOS A. Fettah and T. Gallouët	457
Asymptotic Preserving Finite Volumes Discretization For Non-Linear Moment Model On Unstructured Meshes Emmanuel Franck, Christophe Buet, and Bruno Després	467
Mass Conservative Coupling Between Fluid Flow and Solute Transport Jürgen Fuhrmann, Alexander Linke, and Hartmut Langmach	475
Large Eddy Simulation of the Stable Boundary Layer Vladimír Fuka and Josef Brechler	485
3D Unsteady Flow Simulation with the Use of the ALE Method Petr Furmánek, Jiří Fürst, and Karel Kozel	495
FVM-FEM Coupling and its Application to Turbomachinery J. Fořt, J. Fürst, J. Halama, K. Kozel, P. Louda, P. Sváček, Z. Šimka, P. Pánek, and M. Hajsman	505
Charge Transport in Semiconductors and a Finite Volume Scheme Klaus Gärtner	513
Playing with Burgers's Equation T. Gallouët, R. Herbin, J.-C. Latché, and T.T. Nguyen	523
On Discrete Sobolev–Poincaré Inequalities for Voronoi Finite Volume Approximations Annegret Glitzky and Jens A. Griepentrog	533
A Simple Second Order Cartesian Scheme for Compressible Flows Y. Gorsse, A. Iollo, and L. Weynans	543
Efficient Implementation of High Order Reconstruction in Finite Volume Methods Florian Haider, Pierre Brenner, Bernard Courbet, and Jean-Pierre Croisille	553
A Well-Balanced Scheme For Two-Fluid Flows In Variable Cross-Section ducts Philippe Helluy and Jonathan Jung	561
Discretization of the viscous dissipation term with the MAC scheme F. Babik, R. Herbin, W. Kheriji, and J.-C. Latché	571
A Sharp Contact Discontinuity Scheme for Multimaterial Models Angelo Iollo, Thomas Milcent, and Haysam Telib	581

Numerical Simulation of Viscous and Viscoelastic Fluids Flow by Finite Volume Method 589
Radka Keslerová and Karel Kozel

An Aggregation Based Algebraic Multigrid Method Applied to Convection-Diffusion Operators 597
Sana Khelifi, Namane Méchitoua, Frank Hülsemann, and Frédéric Magoulès

Stabilized DDFV Schemes For The Incompressible Navier-Stokes Equations 605
Stella Krell

Higher-Order Reconstruction: From Finite Volumes to Discontinuous Galerkin 613
Václav Kučera

Flux-Based Approach for Conservative Remap of Multi-Material Quantities in 2D Arbitrary Lagrangian-Eulerian Simulations 623
Milan Kucharik and Mikhail Shashkov

Optimized Riemann Solver to Compute the Drift-Flux Model 633
Anela Kumbaro and Michaël Ndjinga

Finite Volume Schemes for Solving Nonlinear Partial Differential Equations in Financial Mathematics 643
Pavol Kútik and Karol Mikula

Monotonicity Conditions in the Mimetic Finite Difference Method 653
Konstantin Lipnikov, Gianmarco Manzini, and Daniil Svyatskiy

Discrete Duality Finite Volume Method Applied to Linear Elasticity 663
Benjamin Martin and Frédéric Pascal

Model Adaptation for Hyperbolic Systems with Relaxation 673
Hélène Mathis and Nicolas Seguin

Inflow-Implicit/Outflow-Explicit Scheme for Solving Advection Equations 683
Karol Mikula and Mario Ohlberger

4D Numerical Schemes for Cell Image Segmentation and Tracking 693
K. Mikula, N. Peyriéras, M. Remešíková, and M. Smíšek

Rhie-Chow interpolation for low Mach number flow computation allowing small time steps 703
Yann Moguen, Tarik Kousksou, Pascal Bruel, Jan Vierendeels, and Erik Dick

| Study and Approximation of IMPES Stability: the CFL Criteria | 713 |
C. Preux and F. McKee

Numerical Solution of 2D and 3D Atmospheric Boundary
Layer Stratified Flows 723
Jiří Šimonek, Karel Kozel, and Zbyněk Jaňour

On The Numerical Validation Study of Stratified Flow Over
2D–Hill Test Case 731
Sládek Ivo, Kozel Karel, and Janour Zbynek

A Multipoint Flux Approximation Finite Volume Scheme
for Solving Anisotropic Reaction–Diffusion Systems in 3D 741
Pavel Strachota and Michal Beneš

Higher Order Chimera Grid Interface for Transonic
Turbomachinery Applications 751
Petr Straka

Application of Nonlinear Monotone Finite Volume Schemes
to Advection-Diffusion Problems 761
Yuri Vassilevski, Alexander Danilov, Ivan Kapyrin,
and Kirill Nikitin

Scale-selective Time Integration for Long-Wave Linear Acoustics 771
Stefan Vater, Rupert Klein, and Omar M. Knio

Nonlocal Second Order Vehicular Traffic Flow Models
And Lagrange-Remap Finite Volumes 781
Florian De Vuyst, Valeria Ricci, and Francesco Salvarani

Unsteady Numerical Simulation of the Turbulent Flow around
an Exhaust Valve 791
M. Žaloudek, H. Deconinck, and J. Fořt

Vol. 2

Part II Invited Papers

Lowest order methods for diffusive problems on general
meshes: A unified approach to definition and implementation 803
Daniele A. Di Pietro and Jean-Marc Gratien

A Unified Framework for a posteriori Error Estimation
in Elliptic and Parabolic Problems with Application
to Finite Volumes 821
Alexandre Ern and Martin Vohralík

Staggered discretizations, pressure correction schemes
and all speed barotropic flows 839
L. Gastaldo, R. Herbin, W. Kheriji, C. Lapuerta, and J.-C. Latché

ALE Method for Simulations of Laser-Produced Plasmas 857
Liska R., Kuchařík M., Limpouch J., Renner O., Váchal P.,
Bednárik L., and Velechovský J.

A two-dimensional finite volume solution of dam-break
hydraulics over erodible sediment beds 875
Fayssal Benkhaldoun, Imad Elmahi, Saïda Sari,
and Mohammed Seaïd

Part III Benchmark Papers

3D Benchmark on Discretization Schemes for Anisotropic
Diffusion Problems on General Grids 895
Robert Eymard, Gérard Henry, Raphaèle Herbin, Florence Hubert,
Robert Klöfkorn, and Gianmarco Manzini

Benchmark 3D: a linear finite element solver............................. 931
Hanen Amor, Marc Bourgeois, and Gregory Mathieu

Benchmark 3D: a version of the DDFV scheme with cell/vertex
unknowns on general meshes 937
Boris Andreianov, Florence Hubert, and Stella Krell

Benchmark 3D: Symmetric Weighted Interior Penalty
Discontinuous Galerkin Scheme... 949
Peter Bastian

Benchmark 3D: A Mimetic Finite Difference Method 961
Peter Bastian, Olaf Ippisch, and Sven Marnach

Benchmark 3D: A Composite Hexahedral Mixed Finite Element 969
Ibtihel Ben Gharbia, Jérôme Jaffré, N. Suresh Kumar,
and Jean E. Roberts

Benchmark 3D: CeVeFE-DDFV, a discrete duality scheme
with cell/vertex/face+edge unknowns 977
Yves Coudière, Florence Hubert, and Gianmarco Manzini

Benchmark 3D: The Cell-Centered Finite Volume Method
Using Least Squares Vertex Reconstruction ("Diamond Scheme") 985
Yves Coudière and Gianmarco Manzini

Benchmark 3D: A Monotone Nonlinear Finite Volume Method
for Diffusion Equations on Polyhedral Meshes 993
Alexander Danilov and Yuri Vassilevski

Benchmark 3D: the SUSHI Scheme... 1005
Robert Eymard, Thierry Gallouët, and Raphaèle Herbin

Benchmark 3D: the VAG scheme 1013
Robert Eymard, Cindy Guichard, and Raphaèle Herbin

**Benchmark 3D: The Compact Discontinuous
Galerkin 2 Scheme** 1023
Robert Klöfkorn

**Benchmark 3D: Mimetic Finite Difference Method
for Generalized Polyhedral Meshes** 1035
Konstantin Lipnikov and Gianmarco Manzini

**Benchmark 3D: CeVe-DDFV, a Discrete Duality Scheme
with Cell/Vertex Unknowns** 1043
Yves Coudière and Charles Pierre

**Benchmark 3D: A multipoint flux mixed finite element method
on general hexahedra** 1055
Mary F. Wheeler, Guangri Xue, and Ivan Yotov

Part I
Regular Papers

Volume-Agglomeration Coarse Grid In Schwarz Algorithm

H. Alcin, O. Allain, and A. Dervieux

Abstract The use of volume-agglomeration for introducing one or several levels of coarse grids in an Additive Schwarz multi-domain algorithm is revisited. The purpose is to build an algorithm applicable to elliptic and convective models. The sub-domain solver is ILU. We rely on algebraic coupling between the coarse grid and the Schwarz preconditioner. The Deflation Method and the Balancing Domain Decomposition (BDD) Method are experimented for a coarse grid as well as domain-by-domain coarse gridding. Standard coarse grids are built with the characteristic functions of the sub-domains. We also consider the building of a set of smooth basis functions (analog to smoothed-aggregation methods). The test problem is the Poisson problem with a discontinuous coefficicent. The two options are compared for the standpoint of coarse-grid consistency and for the gain in scability of the global Schwarz iteration.

Keywords domain decomposition, coarse grid
MSC2010: 65F04, 65F05

1 Volume agglomeration in MG and DDM

The idea of Volume Agglomeration is directly inspired by the multi-grid idea, but inside the context of Finite-Volume Method. In this paper the finite-volume partition considered is built as the dual of triangles, Fig. 1, right. In order to

---

H. Alcin and A. Dervieux
INRIA, B.P. 93, 06902 Sophia-Antipolis, France, e-mail: Hubert.Alcin@inria.fr, Alain.Dervieux@inria.fr

O. Allain
LEMMA, Les Algorithmes (Le Thales A), 2000 route des Lucioles, 06410 BIOT, France, e-mail: olivier.allain@lemma-ing.com

Fig. 1 Left: finite-Volume partition built as dual of a triangulation. Right: Greedy Algorithm for finite-volume cell agglomeration: four fine cells (left) are grouped into a coarse cell

build a coarser grid, it is possible to build coarse cells by sticking together neighboring cells for example with a greedy algorithm, Fig. 1, left. The coarser grid is *a priori* unstructured as is the fine one. By the magic of FVM, a consistent coarse discretisation of a divergence-based first-order PDE is directly available. Indeed, we can consider that the new unknown is constant over the coarse cell and it remains to apply a Godunov quadrature of the fluxes between any couple of two coarse cells. Elliptic PDE can also be addressed in similar although more complicated way.

As a result, consistent linear and non-linear coarse grid approximations are built using the agglomeration principle. Linear and nonlinear MG have been derived, in contrast with AMG algorithms. This method extends to Discontinuous Galerkin approximations [13]. The extension of Agglomeration MG to multi-processor parallel computing, however, are less easily achieved, as compared to Domain Decomposition Methods.

The many works on multi-level methods *à la* Bramble-Pasciak-Xu [2] has drawn attention to the question of basis smoothness. Indeed, the underlying basis function in volume-agglomeration is a characteristic function equal to zero or one. In [10], the agglomeration basis is extended to H^1 consistent ones in an analog way to smoothed-aggregation. In [4], a Bramble-Pasciak-Xu algorithm is built on these bases for an optimal design application.

While MG appeared, at least for a while, as the best CFD solution algorithm, Domain Decomposition methods (DDM) were seen as a new star for computational Structural Dynamics due to matrix stiffness issues. Domain decomposition methods assume the partition of the computational domain into sub-domains and assume that representative sub-problems on sub-domains can be rather easily computed and help convergence towards global problem's solution. An ideal DDM should be weakly scalable, that is, when it produces in some time with p processors a result on a given mesh, the result on a two times larger mesh should be produced in the same time with $2p$ processors. In Schwarz DDM, The set of local problems preconditions the global loop. Boundary conditions for each sub-domain problem are fetched in neighboring domains. The resulting iterative solver generally involves a Krylov iteration and is often refered as Newton-Krylov-Schwarz. It has been shown by S. Brenner [3] that the resulting algorithm is not scalable, unless a extension called coarse grid is added. In [3], the coarse grid correction is computed on a particular coarser mesh, embedded into the main mesh. The advantage of this approach is to produce a convergent coarse mesh solution. However the coarse mesh option is not practical in many cases, in particular for arbitrary unstructured meshes. As a result, it was tried later to build a coarse basis using other principles. An option is to look for a few global eigenvectors of the operator, see for example [15]. For CPU cost reasons, these eigenvectors should not be exactly computed but only approximated. In a recent study [11, 12], it is proposed to compute eigenvectors of the local Dirichlet-to-Neumann operator, which can be computed in parallel on each subdomain. The evaluation of eigenvectors is difficult when the matrix has a dominent Jordan behaviour (as for convection dominent models, the privilegiated domain of finite-volume methods). In the proposed study, we try to build a convergent coarse mesh basis for an arbitrary unstructured fine mesh. It has been observed that coarser meshes for unstructured meshes are elegantly build with volume-agglomeration. In this study, we follow this track, define a convergent basis and examine how it behaves as a coarse grid preconditioner. The test problem we concentrate on is inspired by a pressure-correction phase in Navier-Stokes (see for example [6]), and expresses as a Neumann problem with strongly discontinuous coefficient and writes:

$$-\nabla^* \frac{1}{\rho} \nabla p = RHS \text{ in } \Omega \qquad \frac{\partial p}{\partial n} = 0 \text{ on } \partial\Omega \qquad p(0) = 0.$$

in which the well-posedness is fixed with a Dirichlet condition on one cell.

1.1 Basic Additive exact and ILU Schwarz algorithm

Our discrete model relies on a vertex-centered formulation expressed on a triangulation. Let us assume that the computational domain Ω is split into two sub-domains, Ω_1 and Ω_2, with an intersection $\overline{\Omega_1} \cap \overline{\Omega_2}$ with a thickness of at least one layer of elements. The *Additive Schwarz* algorithm is written in terms of preconditioning, as $M^{-1} = \sum_{i=1}^{2} A_{|\Omega_i}^{-1}$ where $A_{|\Omega_i}^{-1}$ holds for the Dirichlet problem on sub-domain Ω_i. The preconditioner M^{-1} can be used in a Krylov subspace method. In this paper, in order to keep some generality in our algorithms, we use GMRES, also used in [15]. In the *Additive Schwarz-ILU* version, the exact solution of the Dirichlet on each sub-domain is replaced by the less costly Incomplete Lower Upper (ILU) approximate solution.

1.2 Algebraic Coarse grid

As shown by S. Brenner [3], the combination $M^{-1} = A_0^{-1} + \sum_{i=1}^{N} A_{|\Omega_i}^{-1}$ of the Additive Schwarz method with a coarse grid A_0^{-1} reduces the complexity to an essentially scalable one. Two methods have been proposed in the literature for introducing a coarse grid in an *algebraic* manner. Both rely on the following ingredients:

- $A_h u = f_h$ is the linear system to solve in V, fine-grid approximation space.
- $V_0 \subset V$ coarse approximation space. $V_0 = [\Phi_1 \cdots \Phi_N]$.
- Z an extension operator from V_0 in V and Z^T a restriction operator from V in V_0.
- $Z^T A_h Z u_H = Z^T f_h$ is the coarse system.

The Deflation Method (DM) has been introduced by Nicolaides [14] and is used by many authors. Saad *et al.* [15] encapsulates it into a Conjugate Gradient. Aubry *et al.* [1] apply it to a pressure Poisson equation. In DM, the projection operator is defined as:

$$P_D = I_n - A_h Z (Z^T A_h Z)^{-1} Z^T \text{ avec } A_h \in R^{n \times n} \text{ et } Z \in R^{n \times N}$$

The DM algorithm consists in solving first the coarse system $Z^T A_h Z u_H = Z^T f_h$, then the projected system $P_D A_h \check{u} = P_D f_h$ in order to get finally $u = (I_n - P_D^T) u + P_D^T u = Z (Z^T A_h Z)^{-1} Z^T f_h + P_D^T \check{u}$. The Balancing Domain Decomposition has been introduced by J. Mandel [9] and applied to a complex system in [7]. In [16] a formulation close to DM is proposed. It consists in replacing the preconditioner M^{-1} (ex.: global ILU, Schwarz, or Schwarz-ILU) by:

$$P_B = P_D^T M^{-1} P_D + Z(Z^T A_h Z)^{-1} Z^T.$$

1.3 Smooth and non-smooth coarse grid

The coarse grid is then defined by set of basis functions. A central question is the smoothness of these functions. According to Galerkin-MG, smooth enough functions provide consistent coarse-grid solutions. Conversely, DDM methods preferably use the characteristic functions of the sub-domains, $\Phi_i(x_j) = 1 \text{ si } x_j \in \Omega_i$. In the case of P^1 finite-elements, for example, the typical basis function corresponds to setting to 1 all degrees of freedom in sub-domain. According to [10], the coarse system

$$U^H(x) = \Sigma_i U_i \Phi_i(x) \; ; \quad \int \nabla U^H \nabla \Phi_i = \int f \Phi_i \quad \forall i$$

produces a solution U^H which does not converge towards the continous solution U when H tends to 0.

In order to build a better basis, we need to introduce a hierarchical coarsening process from the fine grid to a coarse grid which will support the preconditioner. Level j is made of N_j macro-cells C_{jk}, i.e. $\mathscr{G}_j = \cup_{k=1}^{N_j} C_{jk}$. Transfer operators are defined between successive levels (from coarse to fine):

$$P_i^j : \mathscr{G}_i \to \mathscr{G}_j \qquad P_i^j(u)(C_{k'i}) = u(C_{kj}) \text{ with } C_{k'i} \subset C_{kj}$$

Following [10] we introduce the smoothing operator:

$$(L_k u)_i = \sum_{j \in \mathcal{N}(i) \cup \{i\}} \text{meas}(j) \, u_j / \{\sum_{j \in \mathcal{N}(i) \cup \{i\}} \text{meas}(j)\}$$

where $\mathcal{N}(i)$ holds for the set of cells which are direct neigbors of cell i. The smoothing is applied at each level between the coarse level k defining the characteristic basis and the finest level.

$$\Psi_k = (L_1 P_1^2 L_2 \cdots P_{p-2}^{p-1} L_{p-1} P_{p-1}^p) \Phi_k.$$

The resulting smooth basis function is compared with the characteristic one in Fig. 2.

The inconsistency of the characteristic basis and the convergence of this new smooth basis is illustrated by the solution of a Poisson equation with a *sin* function as exact solution, Fig. 3.

1.4 Three-level Algorithm

Because the local solver is not an exact one but an ILU solver, computing with a larger number of nodes in each sub-domain leads to a degradation of the convergence. It is then interesting to add a coarse grid on each sub-domain. This principle has been investigated in [8], where the authors use a non-smoothed aggregated basis.

Fig. 2 Left: characteristic coarse grid basis function. Right: smooth coarse grid basis function

Fig. 3 Accuracy of the coarse grid approximation for a Poisson problem with a *sin* function (of amplitude 2.) as exact solution. Left: coarse grid solution with the characteristic basis (amplitude is 0.06). Right: coarse grid solution with a smooth basis (amplitude is 1.8)

Our proposition is to build sub-domain bases which are consistent with the Dirichlet condition of the Schwarz interface condition. To satisfy this, the Dirichlet condition is introduced in each smoothing step of the smooth basis function building process.

The *global algorithm* is made of a GMRES iteration preconditioned by the P_B operator combining a global coarse system with sub-domain preconditioners. The latter ones combine the local medium basis and the local ILU solver.

2 Numerical evaluation

We present some performance evaluations for the proposed algorithm. In all cases the conjugate gradient is used as fixed-point. The test case is a Neumann problem with discontinuous coefficient as in Section 2.1. The computational domain is a square. The coefficient takes two values with a ratio 100., on two regions separated by the diagonal of the domain. The right-hand side is a *sin* function. In the sequel, convergence is always measured for a division of the residual by 10^{20}. Convergence at this level were problematic with DM and the results are presented for BDD.

We recall first how behaves the *original Schwarz method* with one layer overlapping when the number of domains is fixed but the number of nodes increased. We compare in Table 1 a 2D calculation with two domains and 400 nodes with the analog computation with two domains and 10,000 nodes, which correspond to a h ratio of 5. We observe (Table 1) that the convergence of a Schwarz-ILU is four times slower on the finer mesh. We also observe that the convergence of the Schwarz algorithm with exact sub-domain solution is also degraded by a factor 2.6, a loss which may be explained by the thinner overlapping.

We continue with the study of the impact of choosing a *smooth basis* for the two-level Additive Schwarz ILU method. We observe that the scalability again does not hold, but it is nearly attained for the smooth basis option. It is rather bad for the characteristic basis. The rest of the paper uses only the smooth basis.

Table 1 Additive Schwarz method

# sub-domains	# cells	Local solver	# Iterations
2	400	ILU	55
2	400	Direct	28
2	10,000	ILU	221
2	10,000	Direct	74

Table 2 Scalability of the two-level AS-ILU method

Cells	10K	20K	47K	94K
Domains	12	28	66	142
Cells/domain	833	714	712	661
Char. basis	480	546	750	810
Smooth basis	400	391	444	491

The impact of the *medium grid* is examined in a third series of experiment is performed on a mesh of 40,000 cells, with 4 sub-domains and a total of 64 medium basis function (8 per sub-domain). In Table 3, we observe that without a coarse grid,

Table 3 Convergence of the different preconditioners (40,000 cells)

Type of preconditioner M^{-1}	# sub-domains	Iterations
Global ILU	1	348
Schwarz-ILU	4	431
Schwarz-ILU+coarse-grid	4	334
Three-level	4	264
Three-level	16	164

the Schwarz-ILU solver is 20% slower than the global (1-sub-domain) ILU solver (in terms of iteration count for 20 decades), the Schwarz-ILU with coarse-grid is slightly faster and the three level is 30% faster.

The *speedup* is measured for a given problem, set on a mesh of 40,000 cells. We compare the iteration count between a 4-sub-domain computation and a 16-sub-domain one. The coarse system solution with 16 unknowns is not parallel, but its cost is very small. Using four times more processors turn into a 6.4 smaller number of iterations before obtaining the solution (Table 3).

For a *scalability* measure, the mesh is taken finer and the number of sub-domain increased accordingly. We compare a 40,000-cell computation on 4 processors with a 160,000-cell on with 16 processors. We would like to mention that the Schwarz method with exact sub-domain resolution is far from being scalable: in Table 4, increase in iteration count is 40%. These bad news were announced by Table 1. We

Table 4 Scalability for the Schwarz, two-level Schwarz and three level Schwarz-ILU

Method	# cells	# sub-domains	# medium basis funct	Iterations
Schwarz	40,000	4		320
Schwarz	160,000	16		451
two-level Schwarz	40,000	4		130
two-level Schwarz	160,000	16		212
Three level	40,000	4	64	164
Three level	160,000	16	256	176

turn the combination of the Schwarz method with our smooth coarse grid. Exact solution is again performed on each sub-domain. Convergence becomes at least twice better. However, passing from 40,000 cells with 4 sub-domains to 160,000 cells with 16 sub-domains increases the iteration count by 60%, Table 4. We have checked that results with characteristic coarse grid are worse. In order to perform the analog comparison for the proposed three-level method (smooth coarse grid, smooth medium grid, ILU), we specify a four times higher number of medium-grid basis functions for the computation with four times higher number of cells (and sub-domains). Scalability in iterations is nearly satisfied, with 7% loss, Table 4.

3 Concluding remarks

We have proposed a three-level algorithm for solving a linear system with a Schwarz method. The basis functions are independent of the system to solve and building them is not computationally expensive. The coarse grid solution is obtained after one iteration and yields a good initial solution. A few preliminary results show that the proposed method appears to be suitable for a pressure-projection system. The CPU cost (measured on a 2.6GHz workstation) for the heaviest example is of $0.05nS$ for the coarse factorization, $660nS$ ($20nS$ per processor) for the coarse system assembly while the Schwarz preconditioner cost is $124\mu S$. Further measures and applications to convection-diffusion models are in progress, as well as the introduction into a compressible Navier-Stokes model, [5].

References

1. R. Aubry, F. Mut , R. Lohner , J. R. Cebral, Deflated preconditioned conjugate gradient solvers for the Pressure-Poisson equation, Journal of Computational Physics, 227:24, 10196-10208, 2008.
2. J. Bramble, J. Pasciak, and J. Xu, Parallel multilevel preconditioners, Math. Comput., 55(191):122, 1990.
3. S. Brenner, Two-level additive schwarz preconditioners for plate elements, Numerische Mathematik, 72:4, 1994.
4. F. Courty, A. Dervieux, Multilevel functional Preconditioning for shape optimisation, IJCFD, 20:7, 481-490, 2006.
5. B. Koobus, S. Camarri, M.V. Salvetti, S. Wornom, A. Dervieux, Parallel simulation of three-dimensional flows: application to turbulent wakes and two-phase compressible flows, Advances in Engineering Software, 38, 328-337, 2007.
6. A.-C. Lesage, O. Allain and A. Dervieux, On Level Set modelling of Bi-fluid capillary flow, Int. J. Numer. Methods in Fluids, 53:8, 1297-1314, 2007.
7. P. Le Tallec, J. Mandel, M. Vidrascu, Balancing Domain Decomposition for Plates, Eigth International Symposium on Domain Decomposition Methods for Partial Differential Equations, Penn State, October 1993, Contemporary Mathematics, 180, AMS, Providence, 1994, 515-524.
8. P.T . Lin, M. Sala, J.N. Shadi, R S. Tuminaro, Performance of Fully-Coupled Algebraic Multilevel Domain Decomposition Preconditioners for Incompressible Flow and Transport.
9. J. Mandel, Balancing domain decomposition, Comm. Numer. Methods Engrg., 9, 233-241, 1993.
10. N. Marco, B. Koobus, A. Dervieux, An additive multilevel preconditioning method and its application to unstructured meshes, INRIA research report 2310, 1994 and Journal of Scientific Computing, 12:3, 233-251, 1997.
11. F. Nataf, H. Xiang, V. Dolean, A two level domain decomposition preconditioner based on local Dirichlet-to-Neumann maps, C. R. Mathématiques, 348:21-22, 1163-1167, 2010.
12. F. Nataf, H. Xiang, V. Dolean, N. Spillane, A coarse space construction based on local Dirichlet to Neumann maps, to appear in SIAM J. Sci Comput., 2011.
13. C.R. Nastase, D. J. Mavriplis, High-order discontinuous Galerkin methods using an hp-multigrid approach, Journal of Computational Physics 213:1, 330-357, 2006.
14. R. A. Nicolaides, Deflation of conjugate gradients with applications to boundary value problem, SIAM J.Numer.Anal., 24, 355-365, 1987.

15. Y. Saad, M. Yeung, J. Erhel, and F. Guyomarc'H, A deflated version of the conjugate gradient algorithm, SIAM J. Sci. Comput., 21:5, 1909-1926, 2000.
16. C. Vuik , R. Nabben A comparison of deflation and the balancing preconditionner, SIAM J. Sci. Comput., 27:5, 1742-1759, 2006.

The paper is in final form and no similar paper has been or is being submitted elsewhere.

A comparison between the meshless and the finite volume methods for shallow water flows

Yasser Alhuri, Fayssal Benkhaldoun, Imad Elmahi, Driss Ouazar, Mohammed Seaïd, and Ahmed Taik

Abstract A numerical comparison is presented between a meshless method and a finite volume method for solving the shallow water equations. The meshless method uses the multiquadric radial basis functions whereas a modified Roe reconstruction is used in the finite volume method. The obtained results using both methods are compared to experimental measurements.

Keywords Meshless method, shallow water equations, finite volume method, radial basis functions, numerical simulation
MSC2010: 15A09, 65N08, 65F10

1 Introduction

Finite volume method have been widely used to solve shallow water flows due to their conservation properties and the ability to handle complex geometries. Recently, meshless methods using radial basis functions have attracted many researcher in mechanical engineering as well as in computational fluid dynamics. Application of

---

Yasser Alhuri and Ahmed Taik
Dept. Mathematics, UFR-MASI FST, Hassan II University Mohammedia, Morocco

Fayssal Benkhaldoun
LAGA, Université Paris 13, 99 Av J.B. Clement, 93430 Villetaneuse, France

Imad Elmahi
ENSAO Complex Universitaire, B.P. 669, 60000 Oujda, Morocco

Driss Ouazar
Dept. Genie Civil, LASH EMI, Mohammed V University Rabat, Morocco

Mohammed Seaïd
School of Engineering and Computing Sciences, University of Durham, Durham DH1 3LE, UK

meshless methods for numerical solution of shallow water equations has already been addressed in many references in the literature, see for example [9] and further references are therein. However, to our best knowledge, no comparison between the meshless method and the finite volume method is available in the literature for shallow water flows. The aim of the present work is therefore, to perform a comparative study between the meshless and the finite volume methods for solving the shallow water equations rearranged in a conservative form as

$$\frac{\partial \mathbf{W}}{\partial t} + \frac{\partial \mathbf{F(W)}}{\partial x} + \frac{\partial \mathbf{G(W)}}{\partial y} = \mathbf{0}, \tag{1}$$

where $\mathbf{W}$ is the vector of conserved variables, $\mathbf{F}$ and $\mathbf{G}$ are the tensor fluxes

$$\mathbf{W} = \begin{pmatrix} h \\ hu \\ hv \end{pmatrix}, \quad \mathbf{F(W)} = \begin{pmatrix} hu \\ hu^2 + \tfrac{1}{2}gh^2 \\ huv \end{pmatrix}, \quad \mathbf{G(W)} = \begin{pmatrix} hv \\ huv \\ hv^2 + \tfrac{1}{2}gh^2 \end{pmatrix},$$

where g is the gravitational acceleration, $h(t, x, y)$ is the water depth, $u(t, x, y)$ and $v(t, x, y)$ are the depth-averaged velocities in the x- and y-direction, respectively. Note that the equations (1) has to be solved in a bounded spatial domain Ω with smooth boundary Γ, equipped with given boundary and initial conditions. It is well known that the system (1) is strictly hyperbolic with real and distinct eigenvalues.

The basic references for the present finite volume method are [1, 2]. In [1], a description of the overall structure of the finite volume method is presented. In particular, the discretization of the gradient fluxes using the sign matrix of the Jacobian is described in details for both scalar equations and hyperbolic systems of conservation laws with source terms. In [2], the implementation of the finite volume scheme on unstructured grids is analyzed and applied to pollutant transport by shallow water flows. This implementation involves an original treatment of the flux derivatives coupled with the source term in unstructured meshes. The current work aims to compare this finite volume method to a meshless method using the multiquadric radial basis. The numerical results are presented for the shallow water flow in a backward facing step. This test problem has been experimentally investigated in [6]. The obtained results using the meshless and the finite volume methods are compared against the measurements from [6].

2 Solution procedures

In this section we briefly describe the two methods used in solve the shallow water equations (1). Further details on the formulation and implementation of these techniques can be found in the cited references.

2.1 A meshless method

The principal idea of the radial basis interpolation is to interpolate a finite series of an unknown function $f(\mathbf{X})$ at N distinct points $\mathbf{X}_j$ on Ω by the following expansion

$$f(\mathbf{X}) \simeq \sum_{j=1}^{N} \alpha_j \varphi(\|\mathbf{X} - \mathbf{X}_j\|), \qquad (2)$$

Here α_j's are the unknown coefficients to be calculated at each time step, and $\varphi(\|\mathbf{X} - \mathbf{X}_j\|)$ is the radial basis function, $X_j \in \mathbb{R}^n$, $j = 1, 2, \ldots, N$, $\|\mathbf{X} - \mathbf{X}_j\| = r_j$, $r_j = \sqrt{(x_i - x_j)^2 + (y_i - y_j)^2}$ is the Euclidean distance and $\mathbf{X} = (x, y)$, $\mathbf{X}_j = (x_j, y_j)$. Since multiquadrics ($MQ$) are infinitely smooth functions, they are often chosen as the trial function for φ, i.e.,

$$\varphi(r_j) = \sqrt{r_j^2 + c^2} = \sqrt{(x - x_j)^2 + (y - y_j)^2 + c^2},$$

where $c \neq 0$ is the shape parameter controlling the fitting of a smooth surface to the data, see for instance [7].

The application of collocation radial basis functions to a system (1) and its boundary conditions start by first selecting a set of $(x_1, y_1), \ldots, (x_b, y_b)$ boundary and $(x_{b+1}, y_{b+1}), \ldots, (x_{d+b}, y_{d+b=N})$ domain nodes. The unknown solution of the problem at each time t can be determined under the form

$$\Phi(X, t) = \sum_{j=1}^{N} \alpha_j(t) \varphi(r_j), \qquad (3)$$

where $X = (x, y)^T$ and $\{\alpha_j\}$ are unknown coefficients to be determined. To solve the two-dimensional time-dependent differential equations given by (1), the time explicit forward difference scheme is used, then

$$\Phi_i^{n+1} = \Phi_i^n - \Delta t \left(\frac{\partial G_i^n}{\partial x} + \frac{\partial F_i^n}{\partial y} \right), \qquad (4)$$

where Δt is the time step, Φ_i^{n+1} is the numerical solution vector at points $X_i = (x_i, y_i)$ in time $n\Delta t$. The values of the interpolate Φ^n are given by the following MQ radial basis function

$$\Phi^n(X_i, t) = \sum_{j=1}^{N} \alpha_j^n(t) \sqrt{r_{ij}^2 + c^2}, \qquad (5)$$

where $r_{ij} = \sqrt{(x_i - x_j)^2 + (y_i - y_j)^2 + c^2}$, which are collocating with a set of data $(x_i, y_i)_{i=1}^{N}$ over the domain $\Omega \subset \mathbb{R}^2$, and forms a system of N linear algebraic equations in N unknowns

$$\begin{pmatrix} \Phi_1^n \\ \Phi_2^n \\ \vdots \\ \Phi_N^n \end{pmatrix} = \begin{pmatrix} \varphi(r_{11}) & \varphi(r_{12}) & \cdots & \varphi(r_{1N}) \\ \varphi(r_{21}) & \varphi(r_{22}) & \cdots & \varphi(r_{2N}) \\ \vdots & \vdots & \ddots & \vdots \\ \varphi(r_{N1}) & \varphi(r_{N2}) & \cdots & \varphi(r_{NN}) \end{pmatrix} \begin{pmatrix} \alpha_1^n \\ \alpha_2^n \\ \vdots \\ \alpha_N^n \end{pmatrix}, \quad (6)$$

The numerical scheme (4) gives a system of N linear equations in N unknowns can be expressed in matrix form

$$\vec{\Phi} = A\vec{\alpha}, \quad (7)$$

where $A = [\varphi_j(x_i, y_i)]$ is a $N \times N$ matrix, $\vec{\alpha} = [\alpha_j^n]$ and $\vec{\Phi} = [\Phi_j^n]$ are N vectors. Note that for $n = 0$ the coefficients $\{\alpha_j^0\}$ can be found using the initial conditions. Hence the solution Φ_1^0 is well-determined and it will be used as initial condition for the scheme (4). The numerical values of the unknown spatial derivatives of $\Phi^n(x_i, y_i)$ is approximated using the multiquadric radial basis functions as

$$\frac{\partial^m \Phi^n}{\partial x^m}(x_i, y_i, t) = \sum_{j=1}^N \alpha_j^n(t) \frac{\partial^m \varphi_j}{\partial x^m}(x_i, y_i),$$

$$\frac{\partial^m \Phi^n}{\partial y^m}(x_i, y_i, t) = \sum_{j=1}^N \alpha_j^n(t) \frac{\partial^m \varphi_j}{\partial y^m}(x_i, y_i),$$

where $m = 1, 2$. Thus, at each time step n, the numerical solution of the vector $\Phi(x_i, y_i, t)_{i=b+1}^{b+d}$ at the interior points are computed by substituting the Φ and its spatial derivatives into equation (4). The boundary values $\Phi(x_i, y_i, t)_{i=1}^b$ are given by boundary conditions.

Finally, the numerical solution is obtained by solving the system of N linear equations

$$\vec{\Phi}^{n+1} = (A - \Delta t A_L) \vec{\alpha}^n, \quad (8)$$

where $A_L = [L(\varphi^n(r_{ij}))]$ is an $N \times N$ matrix coefficient of which are defined by

$$L(\varphi^n(r_{ij})) = \left(\frac{\partial G(\varphi^n(r_{ij}))}{\partial x} + \frac{\partial F(\varphi^n(r_{ij}))}{\partial y} \right).$$

Hence, the unknown coefficients vector $[\alpha_j^n]$ can be determined using Gaussian elimination or Gmres methods.

2.2 A finite volume method

Discretizing the computational domain in control volumes, the finite volume method applied to (1) results in

$$\mathbf{W}_i^{n+1} = \mathbf{W}_i^n - \frac{\Delta t}{|\mathcal{T}_i|} \sum_{j \in N(i)} \int_{\Gamma_{ij}} \mathscr{F}(\mathbf{W}^n; \mathbf{n}) \, d\sigma, \qquad (9)$$

where $N(i)$ is the set of neighboring triangles of the cell $\mathcal{T}_i$, $\mathbf{W}_i^n$ is an average value of the solution $\mathbf{W}$ in the cell $\mathcal{T}_i$ at time t_n. $|\mathcal{T}_i|$ denotes the area of $\mathcal{T}_i$ and ∂V is the surface surrounding the control volume V. Here, $\mathbf{n} = (n_x, n_y)^T$ denotes the unit outward normal to the surface ∂V, and

$$\mathscr{F}(\mathbf{W}; \mathbf{n}) = \mathbf{F}(\mathbf{W})n_x + \mathbf{G}(\mathbf{W})n_y.$$

Following the formulation in [1], the proposed finite volume scheme consists of a predictor stage and corrector stage. It can be formulated as

$$\mathbf{W}_{ij}^n = \frac{1}{2}\left(\mathbf{W}_i^n + \mathbf{W}_j^n\right) - \frac{1}{2}\,\mathrm{sgn}\left[\nabla\mathscr{F}\left(\overline{\mathbf{W}}_{ij}^n; \mathbf{n}_{ij}\right)\right]\left(\mathbf{W}_j^n - \mathbf{W}_i^n\right),$$

$$\mathbf{W}_i^{n+1} = \mathbf{W}_i^n - \frac{\Delta t}{|\mathcal{T}_i|} \sum_{j \in N(i)} \mathscr{F}\left(\mathbf{W}_{ij}^n; \mathbf{n}_{ij}\right) |\Gamma_{ij}| + \Delta t\, \mathbf{S}_i^n, \qquad (10)$$

with sgn [$\mathbf{A}$] denotes the sign matrix of $\mathbf{A}$, and $\overline{\mathbf{W}}_{ij}^n$ is approximated either by Roe's average state or simply by the mean state

$$\overline{\mathbf{W}}_{ij}^n = \frac{1}{2}\left(\mathbf{W}_i^n + \mathbf{W}_j^n\right). \qquad (11)$$

A detailed formulation for the sign matrix in (10) are given in [1,2] and will not be repeated here.

3 Numerical Results

To validate the results obtained using the meshless and finite volume methods we consider the test problem of flow in a backward facing step. Experimental data for this test problem have been provided in [6] and are used here to compare our numerical results. The domain geometry is depicted in Fig. 1 and for the other involved parameters we refer the reader to [6]. On the upstream and downstream boundaries we used the condition as in [6] *i.e.*

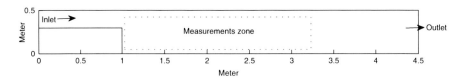

Fig. 1 Schematic domain used in the experimental setup and in our simulations

- On the upstream boundary: The discharge, $Q = 20.2 \, l/s$, is imposed.
- On the downstream boundary: The measured depth, $h = 24.2 \, cm$, is imposed.

In our finite volume simulations we have used an unstructured mesh shown in the left plot of Fig. 2. This mesh contains 5209 triangles and 2779 nodes. In the computations reported herein, the Courant number C is set to 0.8 and the time stepsize Δt is adjusted at each step according to the stability condition

$$\Delta t = C \min_{\Gamma_{ij}} \left(\frac{|\mathcal{T}_i| + |\mathcal{T}_j|}{2 |\Gamma_{ij}| \max_p |(\lambda_p)_{ij}|} \right),$$

where λ_p are the eigenvalues of the system (1), Γ_{ij} is the edge between two triangles $\mathcal{T}_i$ and $\mathcal{T}_j$. For the meshless method we used the node distribution shown in the right plot of Fig. 2. Here we used 229 collocation points uniformly distributed in the computational domain. It should be stressed that the stability condition in the meshless method is

$$\Delta t \leq C \frac{d_{min}}{\max \left(\sqrt{U \pm gh}, \sqrt{V \pm gh} \right)}, \quad (12)$$

where d_{min} is the minimum distance between any two adjacent collocation points and C is the courant number set to 0.1 in our simulations. In our computations the shape coefficient in the multiquadric radial basis functions is selected according to [5, 9]. It has been shown in these references that a near-optimal approximation of the model hydrodynamic can be achieved by using the proposed value

$$c = 0.815 d_{min}.$$

Steady-state solutions are presented for both methods.

Figure 3 illustrates the cross sections of the velocity component u at vertical lines located in $x = 2.03$ and in $x = 2.53$. These two location belong to the zone where measurements have been taken. The agreement between the simulations using the finite volume method and measurements is fairly good. The velocity magnitude and recirculation location are well predicted by the both numerical methods. As expected, a reverse flow is formed near the upper and lower walls and propagates

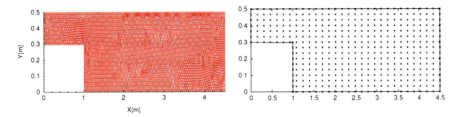

Fig. 2 Finite volume mesh (left plot) and node distribution for meshless method (right plot)

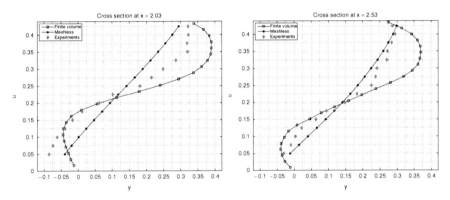

Fig. 3 Comparison results for cross sections of the velocity field u at vertical line $x = 2.03$ (left plot) and at vertical line $x = 2.53$ (right plot)

upstream. However, the location of the recirculation is less accurately predicted by the numerical methods. This may be attributed to the absence of shear stresses from the bed and eddy viscosity in the governing equations (1). It should also be pointed out that the numerical diffusion is more pronounced in the results obtained using the finite volume method than those obtained using the meshless method.

In terms of computational costs for this test problem, the CPU time for the meshless method is about 34 minutes. The considered finite volume method requires more than four times the CPU used for the meshless method. This is a clear indication that the meshless method is more efficient than the finite volume method regardless the number and the distribution of the collocation points in the computational domain.

Acknowledgements The authors would like to thank Prof. Jaime Fe Marqués for providing the experimental data to us.

References

1. F. Benkhaldoun, I. Elmahi, M. Seaïd, "A new finite volume method for flux-gradient and source-term balancing in shallow water equations", Computer Methods in Applied Mechanics and Engineering. **199** pp:49-52 (2010).
2. F. Benkhaldoun, I. Elmahi, M. Seaïd, "Well-balanced finite volume schemes for pollutant transport by shallow water equations on unstructured meshes", J. Comp. Physics. **226** pp:180-203 (2007).
3. T. Belytschko, Y. Krongauz, D. Organ, M. Fleming, P. Krysl, "Meshless methods: an overview and recent developments", Computer methods in applied mechanics and engineering", special issue on Meshless Methods, 139, pp:3-47, (1996).
4. M. Buhamman, Radial basis function: theory and implementations, Cambridge university press, (2003).
5. R. L. Hardy, "Multiquadric equations of topography and other irregular surfaces". J. Geophys. Res, 176, pp:1905-1915, (1971).
6. J. Fe, F. Navarrina, J. Puertas, P. Vellando and D. Ruiz, "Experimental validation of two depth-averaged turbulence models", Int. J. Numer. Meth. Fluids, 60, pp:177-202, (2009).
7. Y. C. Hon, K. F. Cheung, X. Z. Mao and E. J. Kansa, "A Multiquadric solution for the shallow water equations", ASCE J. of hydrodlic engineering", vol.125, No.5, pp:524-533, (1999).
8. P. Roe, "Approximate riemann solvers, parameter vectors and difference schemes", J. Comp. Physics. 43, pp:357-372, (1981)
9. S. M. Wong, Y.C. Hon, M.A. Golberg, "Compactly supported radial basis functions shallow water equations", J. Appl. Sci. Comput, 127, 79-101, (2002).

The paper is in final form and no similar paper has been or is being submitted elsewhere.

Time Compactness Tools for Discretized Evolution Equations and Applications to Degenerate Parabolic PDEs

Boris Andreianov

Abstract We discuss several techniques for proving compactness of sequences of approximate solutions to discretized evolution PDEs. While the well-known Aubin-Simon kind functional-analytic techniques were recently generalized to the discrete setting by Gallouët and Latché [15], here we discuss direct techniques for estimating the time translates of approximate solutions in the space L^1. One important result is the Kruzhkov time compactness lemma. Further, we describe a specific technique that relies upon the order-preservation property. Motivation comes from studying convergence of finite volume discretizations for various classes of nonlinear degenerate parabolic equations. These and other applications are briefly described.

Keywords time translates, Kruzhkov lemma, order-preservation, finite volumes
MSC2010: Primary: 65M12, 46B50, 35K65; Secondary: 46E39, 65M08

1 Introduction

Let us think of evolution equations set on a cylindrical domain $Q := (0, T) \times \Omega \subset \mathbb{R}^+ \times \mathbb{R}^N$. Proving convergence of space-time discretizations of such equations often includes the three following steps: constructing discrete solutions and getting uniform (in appropriate discrete spaces) estimates; extracting a convergent subsequence; writing down a discrete weak formulation (e.g., with discretized test functions) and passing to the limit in the equation in order to infer convergence.

For the first step, obtention of estimates is greatly simplified by preservation, at the discrete level, of the key structure properties of the PDE (such as symmetry, coercivity, monotonicity of the diffusion operators involved; entropy dissipation,

---

Boris Andreianov
CNRS UMR 6623, 16 route de Gray, Besançon, France, e-mail: boris.andreianov@univ-fcomte.fr

for the nonlinear convection operators in the degenerate parabolic case; etc.). For getting discrete *a priori* estimates test functions are often used, as in the continuous case. Therefore, some analogues of integration-by-parts formulas and chain rules are instrumental for the first step. For the examples we give in this paper, "discrete duality" type schemes (mimetic, co-volume, DDFV; see, e.g., [3] and references therein) can be used to guarantee an exact integration-by-parts feature. In contrast, chain rules for derivation in time or in space must be replaced by approximate analogues, often taking the form of convexity inequalities (see, e.g., [4], [3, Sect. 4]).

In this note, we give some insight into convergence proofs for different subclasses of degenerate elliptic-parabolic-hyperbolic PDEs under the general form[1]

$$u = b(v), w = \varphi(v), \quad u_t - \mathrm{div}\left[\mathbf{G}(v) - \mathbf{a_0}(\nabla w)\right] + \psi(v) = f \quad \text{in } Q = (0,T) \times \Omega \tag{1}$$

with $b(\cdot), \varphi(\cdot), \psi(\cdot)$ continuous[2] non-decreasing on $\mathbb{R}$, normalized by zero at zero, with a continuous convection flux $\mathbf{G}(\cdot)$ and with $\mathbf{a_0} : \mathbb{R}^N \to \mathbb{R}^N$ of Leray-Lions type (see e.g. [1,4]; p-laplacian, with $\mathbf{a_0}(\xi) = |\xi|^{p-2}\xi$ is a typical example). For the sake of simplicity, homogeneous Dirichlet boundary condition on $(0,T) \times \partial\Omega$ is taken.

But our main goal is to discuss the second step of the proofs[3], the one of getting compact[4] sequences of discrete solutions. For linear problems, the two latter steps are somewhat trivial; indeed, mere functional-analytic bounds would lead to compactness in a weak topology, which is enough to pass to the limit from the discrete to the continuous weak formulation of the PDE. Thinking of nonlinear problems and passage to the limit in nonlinear terms, bounds in functional spaces can be sufficient when combined with basic compact embeddings; but this requires rather strong bounds involving e.g. some estimates of the derivatives. Regarding evolution PDEs of, say, porous medium type, L^p bounds are available on the space derivatives but not on the time derivatives (those belong to some *negative* Sobolev spaces). In this situation, either compactness in an *ad hoc* strong topology is needed; or the weak compactness coming from uniform boundedness should be combined with some compactification arguments (compensated compactness, Young measures and their reduction, etc.) that exploit in a non-trivial way the particular structure of the PDE in hand (div-curl structure, pseudo-monotonicity, entropy inequalities, etc.).

In this note, we first recall in § 2 the fundamental techniques using only bounds in well-chosen functional spaces (see [2, 9, 11, 17] for the continuous setting; see [12, 15] for the corresponding discrete results). In § 3, we present a collection

---

[1] See [5] and references therein for well-posedness theory of such "triply nonlinear" equations. These are mathematical models for porous media, sedimentation, Stefan problem, etc..
[2] Actually, we assume that either these functions are uniformly continuous, or v is bounded *a priori*.
[3] When the compactification methods strongly utilize a particular structure of the underlying PDE, this step is in fact combined with the third step of passing to the limit.
[4] Throughout the note, "compact" actually signifies "relavively compact".

of complementary techniques for estimation of time translates of families of functions that already possess some estimate of space translates. In § 4, we describe one indirect method for proving compactness and convergence of families of approximate solutions. The method heavily exploits the order-preservation property, required both for the PDE in hand and for the approximation scheme in use. Throughout the note, the exposition is motivated and illustrated by applications to approximation of several cases of problem (1) (different cases requiring different approaches).

2 Functional-analytic approach of Aubin-Lions-Dubinskii-Simon

In the continuous setting, one celebrated result is the Aubin-Lions or Dubinskii lemma ([9] and [11]) and its generalization by Simon [17] (see also Amann [2]). To give an example relevant for the applications we have in mind, let us simply state here that a sequence $(u_h)_h$ bounded in $L^1(0, T; W^{1,1}(\Omega))$ and such that $((u_h)_t)_h$ is bounded in $L^1(0, T; W^{-1,1}(\Omega))$ is relatively compact in $L^1(Q)$, cf. [15]. More generally, compactness comes from an *a priori bound* on u_h in some space $L^p(0, T, X)$ with X compactly embedded in $L^1(\Omega)$ (e.g., $X = W^{1,1}(\Omega)$) while the PDE brings information on boundedness of the time derivatives $(u_h)_t$ in some space $L^q(0, T; Y)$ where Y can be a subspace of distributions on Ω equipped with a rather weak topology (e.g., $Y = W^{-1,1}(\Omega)$). A discrete version of the Aubin-Simon lemma was recently proposed by Gallouët and Latché in [15]; it is based upon a careful reformulation of estimates in terms of "coherent" families $(X_h)_h$, $(Y_h)_h$ of discrete spaces.

A related result taken from Simon [17] and Amann [2] uses a bound on fractional time derivatives of u_h. As it was demonstrated by Emmrich and Thalhammer in [12], this version is quite appropriate in the time-discretized setting. Indeed, time fractional derivatives of order less than $1/2$ exist even for piecewise constant functions. Technically, this method involves an indirect estimation of weighted time translates, under a form $\int_0^T \int_0^T \frac{|u_h(t) - u_h(s)|^p}{|t-s|^{1+\sigma p}} \, ds \, dt$ with some $p \geq 1$ and $\sigma \in (0, 1/2)$.

These results only use bounds in functional spaces and very few of the underlying PDE properties. They offer a very wide spectrum of applications; yet they are difficult to apply on degenerate parabolic problems with non-Lipschitz nonlinearities. The difficulty comes from the fact that non-Lipschitz mappings make bad correspondence between *linear* functional spaces. Yet this difficulty is not a fundamental one; roughly speaking, it is settled by a careful use of translation arguments and of moduli of continuity. This is the object of the next section.

3 Direct estimation of time translates

In this section, the compactness question is studied using the one and only space[5] $L^1(Q)$. By the Fréchet-Kolmogorov compactness criterion in $L^1(Q)$, uniform bounds on space and time translates of u_h are needed. In the setting of the present note, the first ones are readily available. The difficulty lies in estimating the time translates as

$$\forall h \quad \int_0^{T-\delta}\!\!\int_{\Omega} \big|u_h(t+\delta) - u_h(t)\big| \leq \omega(\delta) \quad \text{with } \lim_{\delta \to 0} \omega(\delta) = 0, \qquad (2)$$

$\omega(\cdot)$ being a modulus of continuity, uniform in h. Here are two ways to obtain (2).

A discrete Kruzhkov lemma[6]

Lemma 1 (Kruzhkov [16]). *Assume that the families of functions* $(u_h)_h$, $(F_h^\alpha)_{h,\alpha}$ *are bounded in $L^1(Q)$ and satisfy* $\frac{\partial}{\partial t} u_h = \sum_{|\alpha| \leq m} D^\alpha F_h^\alpha$ *in* $\mathscr{D}'(Q)$. *Assume that u_h can be extended outside Q, and one has*[7]

$$\iint_Q |u_h(t, x+\delta) - u_h(t,x)|\, dx\, dt \leq \omega(\delta), \quad \text{with } \lim_{\delta \to 0} \omega(\delta) = 0, \qquad (3)$$

where $\omega(\cdot)$ does not depend on h. Then $(u_h)_h$ is (relatively) compact in $L^1(Q)$.

Clearly, this is an L^1_{loc} compactness result (one can apply the lemma locally in Q).

For problem (1), the value $m = 1$ is relevant, because an L^1 bound is available for the flux $\big(\mathbf{G}(v) - \mathbf{a_0}(\nabla \varphi(v))\big)$; therefore we limit to this case the discussion of discrete analogues of Lemma 1. To give an idea of discrete versions of the Kruzhkov lemma[8], assume we are given a family of meshes of Ω indexed by their size h and satisfying mild proportionality restrictions (e.g., for the case of two-point flux finite volume schemes as described in [13], one needs for all neighbour volumes K,L, $\text{diam}(K) + \text{diam}(L) \leq \text{const } d_{K,L}$ uniformly in h). Assume that on these meshes, spaces of

---

[5] Working in an h-independent space is an advantage for producing discrete versions of compactness arguments; yet the approach of [15] exhibits a simple and efficient use of h-dependent spaces.

[6] There is a strong relation to the method of § 2. The Kruzhkov lemma allows for general moduli of continuity. E.g., for problem (1) with $\varphi = Id$, the Aubin-Lions-Dubinskii-Simon argument can be used if $b(\cdot)$ is Lipschitz continuous (with $X = W_0^{1,p}(\Omega)$) or Hölder continuous (with a fractional Sobolev space chosen for X), and the Kruzhkov lemma can be used for any continuous $b(\cdot)$.

[7] In practice, space translation estimates of the kind (3) can be obtained via an estimate of some discrete gradients; notice that estimates of kind (3) are stable upon composing $(u_h)_h$ by a function $b(\cdot)$ which is uniformly continuous (as in (1), we mean that $u_h = b(v_h)$).

[8] Here we give a rather heuristic presentation; see [7] and [3] for two precise formulations covering, e.g., the two-point flux finite volume schemes ([13]) and DDFV schemes ([3] and ref. therein).

discrete functions $\mathbb{R}_h$ and discrete fields $(\mathbb{R}^N)_h$ are defined (each element $u_h \in \mathbb{R}_h$ or $\mathbf{F}_h \in (\mathbb{R}^N)_h$ is a piecewise constant on Ω function reconstructed from the degrees of freedom of the discretization method). Assume we are given discrete gradient and discrete divergence operators ∇_h and div_h mapping between these spaces. Thus all discrete objects (functions, fields, gradient, divergence) are naturally lifted to $L^1(Q)$.

Let $(\delta_h)_h$ be the associated time steps, let N_h be the entire part of T/δ_h. Assume that we are given an initial condition b_h^0 and discrete evolution equations under the form

$$\text{for } n \in [1, N_h + 1], \quad \frac{b(v_h^n) - b(v_h^{n-1})}{\delta_h} = \text{div}_h [\mathbf{F}_h^n] + f_h^n \quad \text{in } \mathbb{R}_h, \qquad (4)$$

where families $(((u_h^n)_n)_h$, $(((f_h^n)_n)_h$ (discrete functions) and $((\mathbf{F}_h^n)_n)_h$ (discrete fields) are bounded in $L^1(Q)$. Assume that the discrete gradients $((\nabla_h v_h^n)_n)_h$ are bounded in $L^1(Q)$ and that this bound implies a uniform translation bound in space of the family v_h (this is true, e.g., when discrete Poincaré inequalities can be proved). Under these assumptions, reproducing at the discrete level the proof [16] of Lemma 1 as it is done in [3, 7], one concludes that the family $(b(v_h))_h$ is relatively compact in $L^1(Q)$. Note that, the case $m \geq 2$ would require more work.

A classical technique for the "variational" setting

Following [1], by "variational" we mean a setting where the solution w is an admissible test function in the weak formulation of the PDE; e.g., (1) can be tested with $w = \varphi(v)$. It typically comes along with *a priori* estimates that can be reproduced at the discrete level, provided the discretization is somewhat structure-preserving.[9]

The technique of [1] used, in its finite volume version, e.g., in [4, 13, 14], is to integrate[10] the equation in time from t to $t+\delta$, take $w_h(t+\delta) - w_h(t)$ for test function, then integrate in (t, x). On problem (1), this leads to a uniform estimate

$$\forall h > 0 \quad \int_0^{T-\delta} \int_\Omega \Big(b(v_h)(t+\delta) - b(v_h)(t) \Big) \Big(\varphi(v_h)(t+\delta) - \varphi(v_h)(t) \Big) \leq \omega(\delta). \quad (5)$$

Then Lipschitz continuity of $\varphi \circ b^{-1}$ (resp., of $b \circ \varphi^{-1}$) can be used to infer uniform L^2 time translates of $w_h = \varphi(v_h)$ (resp., of $u_h = b(v_h)$). Yet the L^1 time translates can be obtained in the case $\varphi \circ b^{-1}$ (resp., $b \circ \varphi^{-1}$) is a merely continuous function.

---

[9] Notice that for evolution PDEs governed by accretive in $L^1(\Omega)$ operators, of which (1) is an example, time-implicit discretizations are better suited for structure preservation. Use of numerical schemes in space that possess a kind of discrete duality (mimetic, co-volume, DDFV schemes, etc.) enables getting discrete estimates analogous to the continuous ones. For notions of solution involving some version of chain rule (e.g., entropy, renormalized solutions) orthogonality assumption on the meshes and isotropy assumption on the diffusion operator may be needed, see e.g. [4].

[10] Here, for the sake of simplicity, we stick to the terminology and notation of the continuous case.

• *A technique for L^1 estimates involving non-Lipschitz nonlinearities (see [4])*

Consider the case where $\widetilde{\varphi} := \varphi \circ b^{-1}$ is a uniformly continuous function (moreover, it is non-decreasing). Let π be a concave modulus of continuity for $\varphi \circ b^{-1}$, Π be its inverse, and set $\widetilde{\Pi}(r) = r\, \Pi(r)$. Let $\widetilde{\pi}$ be the inverse of $\widetilde{\Pi}$. Note that $\widetilde{\pi}$ is concave, continuous, and $\widetilde{\pi}(0) = 0$. Set $u^\delta = b(v_h)(t+\delta, x)$ and $u = b(v_h)(t, x)$. We have

$$\int_Q |\widetilde{\varphi}(u^\delta) - \widetilde{\varphi}(u)| = \int_Q \widetilde{\pi} \circ \widetilde{\Pi}\big(|\widetilde{\varphi}(u^\delta) - \widetilde{\varphi}(u)|\big) \le |Q|\,\widetilde{\pi}\Big(\tfrac{1}{|Q|}\int_Q \widetilde{\Pi}\big(|\widetilde{\varphi}(u^\delta) - \widetilde{\varphi}(u)|\big)\Big).$$

Since $|\widetilde{\varphi}(u^\delta) - \widetilde{\varphi}(u)| \le \pi(|u^\delta - u|)$, we have $\Pi(|\widetilde{\varphi}(u^\delta) - \widetilde{\varphi}(u)|) \le |u^\delta - u|$ and

$$\widetilde{\Pi}(|\widetilde{\varphi}(u^\delta) - \widetilde{\varphi}(u)|) = \Pi(|\widetilde{\varphi}(u^\delta) - \widetilde{\varphi}(u)|)|\widetilde{\varphi}(u^\delta) - \widetilde{\varphi}(u)| \le |u^\delta - u|\,|\widetilde{\varphi}(u^\delta) - \widetilde{\varphi}(u)|.$$

Therefore, (5) implies an L^1 estimate of the kind (2) on $w_h = \varphi(v_h)$:

$$\int_Q |w_h(t+\delta) - w_h(t)| \le |Q|\,\widetilde{\pi}\Big(\tfrac{1}{|Q|}\int_Q |u^\delta - u|\,|\widetilde{\varphi}(u^\delta) - \widetilde{\varphi}(u)|\Big) = |Q|\,\widetilde{\pi}\Big(\tfrac{1}{|Q|}\omega(\delta)\Big).$$

• *Use of contraction arguments and absorption terms (see [8])*

Let us mention one more possibility for getting estimates of kind (2) for (1), which takes advantage of the monotonicity of $\psi(\cdot)$. Assume $\varphi = Id$ in (1); to shorten the arguments, assume $f = 0$. Then L^1 translates in time of $u_h = b(v_h)$ can be estimated with every of the two preceding methods, the Kruzhkov lemma and a direct estimation of translates with variational techniques. This makes $(b(v_h))_h$ relatively compact; yet, when $b^{-1}(\cdot)$ is discontinuous, no information on compactness of $(v_h)_h$ is obtained this way. Now, let us use the translation (in time) invariance of the equation and the L^1 contraction property[11] natural for (1). This yields the estimate (see [8])

$$\int_\Omega |b(v_h^\delta) - b(v_h)|(T-\delta) + \int_s^{T-\delta}\int_\Omega |\psi(v_h^\delta) - \psi(v_h)| \le \int_\Omega |b(v_h^\delta)(s) - b(v_h)(s)| \quad (6)$$

for all $s \in (0, T-\delta)$, where $v_h^\delta(t) = v_h(t+\delta)$. Integrating in $s > \alpha > 0$, using the time translation bound for $(b(v_h))_h$ we get an $L^1((\alpha, T) \times \Omega)$ estimate of time translates of $\psi(v_h)$. If $\psi(\cdot)$ is strictly increasing, this is enough for L^1_{loc} compactness of $(v_h)_h$.

Applications to (1) and some other parabolic PDEs

• *Application to a parabolic-hyperbolic PDE (see [4])*

For problem (1) with $b = Id$, provided $L^p(Q)$ estimates of the discrete gradient of $\varphi(v_h)$ are available, space translates of $\varphi(v_h)$ (and the functions $\varphi(v_h)$ themselves)

---

[11]Discrete version of (6) (see [8]) assumes the L^1 contraction property (linked to order preservation via the Crandall-Tartar lemma) is preserved at the discrete level. Estimate (6) is exploited in § 4.

can be estimated uniformly, and an estimate of the form (5) can be obtained. Then the above technique for exploiting (5) assesses the $L^1(Q)$ compactness of $(\varphi(v_h))_h$, which is a first step of the convergence proof for this problem (see [4])[12].

- *Application to an elliptic-parabolic PDE with the structure condition (see [3])*
Assume $\varphi = Id$. Estimate (5) controls the L^1 time translates of $b(v_h)$ similarly to what was described above[13]. If the *structure condition* $\mathbf{G}(v) = \mathbf{F}(b(v))$ is satisfied, compactness of $(b(v_h))_h$ is enough to pass to the limit, see [1] (cf. [10] and § 4).

- *Application to a cross-diffusion system (see [7])*
The following kind of models comes from population dynamics:

$$\begin{cases} u_t - D_1 \Delta u - \text{div}\big((u+v)\nabla u + u\nabla v\big) = u(a_1 - b_1 u - c_1 v), \\ v_t - D_2 \Delta v - \text{div}\big(v\nabla u + (u+v)\nabla v\big) = v(a_2 - b_2 u - c_2 v). \end{cases} \quad (7)$$

Natural estimates for approximate solutions of (7) are L^2 bounds on $\sqrt{1+u+v}\,\nabla u$, $\sqrt{1+u+v}\,\nabla v$; this gives only an $L^{4/3}$ bound on the diffusion fluxes in (7), thus we are not in a variational setting[14]. Therefore for a proof of convergence of finite volume approximations of the kind [13], the Kruzhkov lemma was used in [7][15].

- *Application to convergence of some linearized implicit schemes (see [6])*
In [6], discretization of the simplified version of cardioelectrical bidomain model:

$$\begin{cases} v_t - \text{div}\big[\mathbf{M}_i(\cdot)\nabla u_i\big] + H(v) = I_{ap}(\cdot), \\ v_t + \text{div}\big[\mathbf{M}_e(\cdot)\nabla u_e\big] + H(v) = I_{ap}(\cdot), \end{cases} \quad v = u_i - u_e, \quad (8)$$

was considered; here, the "ionic current" $H(\cdot)$ is a cubic polynomial. This nonlinear reaction term brings an estimate of $vH(v)$ which bounds v in $L^4(Q)$. Time-implicit DDFV discretization of (8) preserves this structure; then the problem falls into the "variational" framework[16] and time translates can be estimated like in [1, 13, 14]. From the practical point of view, it is important to accelerate computations, and to consider a linearized method where the discretization of the reaction term $H(v)$ is not fully implicit. Unfortunately, for theoretical analysis L^4 estimate for v_h is not available any more; only a weaker estimate can be obtained with interpolation arguments. In [6], we applied the Kruzhkov lemma to exploit this weaker estimate[17].

---

[12] For Lipschitz $\varphi(\cdot)$, also the Aubin-Lions-Dubinskii-Simon and Kruzhkov lemmas could be used. For general $\varphi(\cdot)$, the author thinks that neither of these lemmas can replace the direct use of (5).
[13] Alternatively, the Kruzhkov lemma can be used in a straightforward way, see [3].
[14] From the practical point of view, e.g. the first equation cannot be tested with $u(t+\delta)$.
[15] Alternatively, the discrete Aubin-Lions-Dubinskii-Simon lemma (see [15]) could be used here.
[16] Indeed, we have v_h bounded in $L^4(Q)$ and $H(v_h)$ is bounded in $L^{4/3}(Q) = (L^4(Q))^*$.
[17] A discrete Aubin-Lions-Dubinskii-Simon argument could have been applied as well.

4 Advanced use of the underlying PDE features

Often mere functional-analytic bounds are not enough, but additional constraints coming from the particular structure of the approximated PDE may permit an indirect compactness/convergence proof. E.g., for the parabolic-hyperbolic PDE (1) (case $b = Id$) we proved the compactness of $(\varphi(v_h))_h$ in § 3. The two final steps (see [4]; see also [14]) exploit fine PDE tools. First, the Minty argument (see, e.g., [1]) is used for $(\mathbf{a_0}(\nabla_h \varphi(v_h))_h$; second, the "nonlinear weak-* convergence" ([4, 13, 14]) for $(v_h)_h$ is upgraded to strong convergence using entropy inequalities for (1).

Let us show how one very delicate case of (1), see [10], can be treated indirectly.

Compactness from monotone penalization and order-preservation

For getting (6), we already used the order-preservation structure for (1). Its further use, in conjunction with penalization, may lead to the following convergence proof.

- *The structure needed for compactification*

Assume that one can prove *uniqueness* of a solution to a PDE (Eq^0) under study. Assume that (Eq^0) can be embedded "continuously" into a family (Eq^ε) of perturbed PDEs *having the property that* $v_h^{\varepsilon_1} \leq v_h^{\varepsilon_2}$ when $\varepsilon_1 \leq \varepsilon_2$, where $v_h^{\varepsilon_1}$, $v_h^{\varepsilon_2}$ are the associated discrete solutions. Continuity in $\varepsilon \in [-1, 1]$ means, we assume that limits as $\varepsilon \to 0$ (if they exist) of exact solutions v^ε of (Eq^ε) solve the limit equation (Eq^0).

Assume that for $\varepsilon \neq 0$, the corresponding sequence $(v_h^\varepsilon)_h$ is well defined and it converges to an exact solution v^ε of (Eq^ε). Then solutions $(v_h^0)_h$ to the discretized equation (Eq^0) converge a.e., as $h \to 0$, to the unique solution of (Eq^0). Indeed, write

$$v_h^{-1} \leq v_h^{-1/2} \leq \ldots \leq v_h^{-1/m} \leq \ldots \leq v_h^0 \leq \ldots \leq v_h^{1/m} \leq \ldots \leq v_h^{1/2} \leq v_h^1, \qquad (9)$$

and pass to the limit as $h \to 0$ to define $v^{\pm 1/m} := \lim_{h \to 0} v_h^{\pm 1/m}$ (up to extraction of a subsequence) solution to $(Eq^{\pm 1/m})$; then, (9) is inherited at the limit (except that $(v_h^0)_h$ may not have a limit). By monotonicity, we can define $\underline{v} := \lim_{m \to \infty} v^{-1/m}$ and $\overline{v} := \lim_{m \to \infty} v^{1/m}$; furthermore, we have $\underline{v} \leq \liminf_{h \to 0} v_h^0 \leq \limsup_{h \to 0} v_h^0 \leq \overline{v}$. Both $\underline{v}, \overline{v}$ solve (Eq^0). Thus, by uniqueness, $(v_h^0)_h$ converges to $\underline{v} \equiv \overline{v}$ the solution of (Eq^0).

- *Application to an elliptic-parabolic PDE without the structure condition (see [8])*

We assume that $\varphi = Id$, $\psi = 0$ in (1). We have seen that compactness of $(b(v_h))_h$ can be established, e.g., with the Kruzhkov lemma. Under the *structure condition* $\mathbf{G}(v) = \mathbf{F}(b(v))$, this is enough to pass to the limit in the equation. But in general (see [10]) one lacks control of time oscillations of $\mathbf{G}(v_h)$, and the method of [1] fails. Yet it is enough to add penalization term of the form $\psi^\varepsilon(v) = \varepsilon(\arctan v \mp \frac{\pi}{2} \operatorname{sign} \varepsilon)$ to get into the setting where (6) can be exploited to control discrete solutions $(v_h^\varepsilon)_h$ and to pass to the limit, as $h \to 0$, for the ψ^ε-penalized equation (1^ε). The order-preservation assumptions of the above method being fulfilled due to the choice of ψ^ε, we get convergence of $(v_h)_h$ in the cases where uniqueness for (1) can be shown.

Acknowledgements The author thanks E. Emmrich for discussions on the above techniques.

References

1. H.W. Alt and S. Luckhaus. Quasilinear elliptic-parabolic differential equations. *Mat. Z.*, (1983), **183**:311–341.
2. H. Amann. Compact embeddings of vector-valued Sobolev and Besov spaces. *Glasnik Matematički*, (2000), **35**:161–177.
3. B. Andreianov, M. Bendahmane and F. Hubert. On 3D DDFV discretization of gradient and divergence operators. II. Discrete functional analysis tools and applications to degenerate parabolic problems. Preprint HAL, (2011), http://hal.archives-ouvertes.fr/hal-00567342
4. B. Andreianov, M. Bendahmane, and K.H. Karlsen. Discrete duality finite volume schemes for doubly nonlinear degenerate hyperbolic-parabolic equations. *J. Hyp. Diff. Eq.*, (2010), **7**:1–67.
5. B. Andreianov, M. Bendahmane, K.H. Karlsen and S. Ouaro. Well-posedness results for triply nonlinear degenerate parabolic equations. *J. Diff. Eq.*, (2009), **247**(1):277–302.
6. B. Andreianov, M. Bendahmane, K.H. Karlsen and Ch. Pierre. Convergence of Discrete Duality Finite Volume schemes for the macroscopic bidomain model of the heart electric activity. *Netw. Het. Media*, (2011), to appear; available at http://hal.archives-ouvertes.fr/hal-00526047
7. B. Andreianov, M. Bendahmane and R. Ruiz Baier. Analysis of a finite volume method to solve a cross-diffusion population system. *Math. Models Meth. Appl. Sci.*, (2011), to appear.
8. B. Andreianov and P. Wittbold. Convergence of approximate solutions to an elliptic-parabolic equation without the structure condition. Preprint, (2011).
9. J.-P. Aubin. Un théorème de compacité. (French) *C.R. Acad. Sc. Paris*, (1963), **256**:5042–5044.
10. Ph. Bénilan and P. Wittbold. Sur un problème parabolique-elliptique. (French) *M2AN Math. Modelling and Num. Anal.*, (1999), **33**(1):121–127.
11. J.A. Dubinskii. Weak convergence for elliptic and parabolic equations. (Russian) *Math. USSR Sbornik*, (1965), **67**:609–642.
12. E. Emmrich and M. Thalhammer. Doubly nonlinear evolution equations of second order: Existence and fully discrete approximation. *J. Diff. Eq.*, (2011), to appear.
13. R. Eymard, T. Gallouët, and R. Herbin. *Finite Volume Methods*. Handbook of Numerical Analysis, Vol. VII (2000). P. Ciarlet, J.-L. Lions, eds., North-Holland.
14. R. Eymard, T. Gallouët, R. Herbin and A. Michel. Convergence of a finite volume scheme for nonlinear degenerate parabolic equations. *Numer. Math.*, (2002), **92**(1):41–82.
15. T. Gallouët and J.-C. Latché. Compactness of discrete approximate solutions to parabolic PDEs - Application to a turbulence model. *Comm. on Pure and Appl. Anal.*, (2011), to appear.
16. S.N. Kruzhkov. Results on the nature of the continuity of solutions of parabolic equations and some of their applications. *Mat. Zametki (Math. Notes)*, (1969), **6**(1):517-523.
17. J. Simon. Compact sets in the space $L^p(0, T; B)$. *Ann. Mat. Pura ed Appl.*, (1987), **146**:65–96.

The paper is in final form and no similar paper has been or is being submitted elsewhere.

Penalty Methods for the Hyperbolic System Modelling the Wall-Plasma Interaction in a Tokamak

Philippe Angot, Thomas Auphan, and Olivier Guès

Abstract The penalization method is used to take account of obstacles in a tokamak, such as the limiter. We study a non linear hyperbolic system modelling the plasma transport in the area close to the wall. A penalization which cuts the transport term of the momentum is studied. We show numerically that this penalization creates a Dirac measure at the plasma-limiter interface which prevents us from defining the transport term in the usual sense. Hence, a new penalty method is proposed for this hyperbolic system and numerical tests reveal an optimal convergence rate without any spurious boundary layer.

Keywords hyperbolic problem, penalization method, numerical tests
MSC2010: 00B25, 35L04, 65M85

1 Introduction

A tokamak is a machine to study plasmas and the fusion reaction. The plasma at high temperature ($10^8 K$) is confined in a toroïdal chamber thanks to a magnetic field. One of the main goals is to perform controlled fusion with enough efficiency to be a reliable source of energy. But, since the magnetic confinement is not perfect, the plasma is in contact with the wall. In order to preserve the integrity of the wall and to limit the pollution of the plasma, it is crucial to control these interactions.

We study, using a fluid approximation of the plasma, a simplified system of equations governing the plasma transport in the scrape-off layer, parallel to the magnetic field lines. In this paper, after a numerical study of the penalization

---

Philippe Angot, Thomas Auphan, and Olivier Guès
Université de Provence, Laboratoire d'Analyse Topologie et Probabilités, Centre de Mathématiques et Informatique, 39 rue Joliot Curie, 13453 Marseille Cedex 13, France,
e-mail: [angot,tauphan,gues]@cmi.univ-mrs.fr

introduced by Isoardi *et al.* [9], we modify the boundary conditions to ensure the well-posedness of the hyperbolic system and we propose another penalty method which seems to be free of boundary layer.

2 The model hyperbolic problem

In this paper, we consider a very simple model taking only into account the transport in the direction parallel to the magnetic field lines, (see for example [9, 13]). It is a one dimensional 2×2 hyperbolic system of conservation laws for the particle density N and the particle flux Γ, which reads:

$$\begin{cases} \partial_t N + \partial_x \Gamma = S \\ \partial_t \Gamma + \partial_x \left(\dfrac{\Gamma^2}{N} + N \right) = 0 \qquad (t, x) \in \mathbb{R}_*^+ \times]-L, L[\quad (1) \\ \text{Initial conditions: } N(0, .) = N_0 \text{ and } \Gamma(0, .) = \Gamma_0 \end{cases}$$

Here, the boundaries of the domain $x = L$ and $x = -L$ correspond to the "limiters", which are material obstacles for the fluid (see Fig. 1). In the right-hand side, S is a source term.

There is a difficulty with the choice of the boundary conditions for the system (1). From physical arguments, it follows that the domain (namely the scrape-off layer) is basically divided into two regions [13]:

- One region far from the limiter, the pre-sheath, where the plasma is neutral and the Mach number $M = \Gamma/N$ of the plasma satisfies $|M| \leq 1$.
- One region next to the limiter (in a thin layer called the sheath area, whose typical thickness is of the order of $10^{-5}m$), where the electroneutrality hypothesis does not hold and we have $|M| > 1$. More precisely $M > 1$ close to $x = L$ and $M < -1$ close to the boundary $x = -L$.

It could seem natural to prescribe $M = 1$ (resp. $M = -1$) as a boundary condition at $x = L$ (resp. $x = -L$) for the system, since the physical arguments imply that $M = \pm 1$ very close to the obstacle (Bohm criterion). These are exactly the boundary conditions which are chosen in [9]. However, in that case, as the eigenvalues of the Jacobian of the flux function are $M - 1$ and $M + 1$, it follows that at the plasma limiter interface one eigenvalue is 0 (the boundary is characteristic) and the other one is outgoing (it is also true at $x = -L$), and clearly the problem does not satisfy the usual sufficient conditions for well posedness, see [3, 8, 11]: the number of boundary conditions ($= 1$) is not equal to the number of incoming eigenvalues ($= 0$).

In order to test our penalty approach with a well-defined hyperbolic boundary value problem, in Sect. 3, we slightly modify the boundary conditions of the paper [9], and impose $M = 1 - \epsilon$ on $x = L$ and $M = -1 + \epsilon$ on $x = -L$ with a fixed $\epsilon > 0$, which leads to a well-posed hyperbolic problem. In our numerical simulations we use $\epsilon = 0.1$.

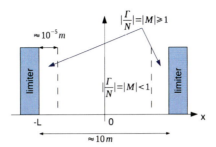

Fig. 1 Schematic representation of the scrape-off layer. The x-axis corresponds to the curvilinear coordinate along a magnetic line close to the wall of the tokamak

The numerical tests presented below, use a finite volume scheme with a second order extension: MUSCL reconstruction with the *minmod* slope limiter and the Heun scheme which is a second order Runge–Kutta TVD time discretization. The finite volume scheme is the VFRoe using the non conservative variables for the linearized Riemann solver [7]; here, the non conservatives variables are N and M. To avoid stability issues, the penalized terms are treated implicitly for the time discretization.

3 Study of penalty methods

3.1 A first penalty method

The following penalty approach has been proposed by Isoardi *et al.* [9]. Let's χ be the characteristic function of the limiter, i.e. $\chi(x) = 1$ if x is in the limiter, and $\chi(x) = 0$ elsewhere, and η the penalization parameter. The penalized system is given by:

$$\begin{cases} \partial_t N + \partial_x \Gamma + \dfrac{\chi}{\eta} N = (1-\chi) S_N & \text{in } \mathbb{R}_*^+ \times \mathbb{R} \\ \partial_t \Gamma + (1-\chi) \partial_x \left(\dfrac{\Gamma^2}{N} + N \right) + \dfrac{\chi}{\eta} (\Gamma - M_0 N) = (1-\chi) S_\Gamma \\ \text{Initial conditions: } N(0,.) = N_0 \text{ and } \Gamma(0,.) = \Gamma_0 \end{cases} \quad (2)$$

M_0 is a function such that, at the plasma-limiter interface we have $|M_0| = 1$. Here, the two components of the unknown are penalized although there is no incoming wave. At least formally, N is forced to converge to 0 inside the limiter when η tends to 0.

The flux of the second equation is cut inside of the limiter, and this causes some troubles from the mathematical point of view. Indeed, this is an hyperbolic system with discontinuous coefficients and the meaning of the term

$$(1-\chi)\partial_x \left(\frac{\Gamma^2}{N} + N \right)$$

is not clear because it can involve the product of a measure with a discontinuous function which has no distributional sense. As a consequence and as a confirmation of this fact, our numerical tests show the existence of a strong singularity at the interface for the numerical discrete solution. Concerning the interpretation of this numerical singularity, it could happen (but we don't have any rigorous proof and this is just an open question) that this system admits generalized solutions in the spirit of Bouchut–James [4] (see also Poupaud–Rascle [10], or Fornet–Guès [5]) such as measure-valued solutions, which can for example exhibit a Dirac measure at the interface, and this generalized solution could be selected by the numerical approximation process.

For the numerical test, we choose S_N and S_Γ so that the following functions define a solution of the boundary value problem:

$$N(t,x) = \exp\left(\frac{-x^2}{0.16(t+1)}\right) \qquad \Gamma(t,x) = \sin\left(\frac{\pi x}{0.8}\right) \exp\left(\frac{-x^2}{0.16(t+1)}\right)$$

These test solutions are regular (at least inside the plasma area) and has no singularity at the plasma-limiter interface. In the Fig. 2, we observe that a peak appears very quickly, then $|M_i^n|$ become very large (about 10^8) in a few points. The same computations are made for two more refined meshes (respectively for

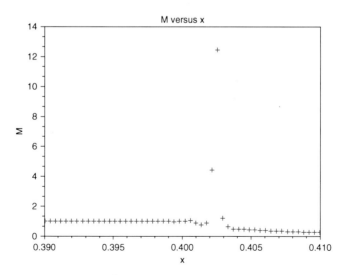

Fig. 2 M versus x with $\eta = 10^{-3}$, a mesh of $J = 1280$ cells using the penalization of Isoardi et al. [9]. The computations are stopped when $\max_{i \in \{1,\dots,J\}}(|M_i^n|) > 10$, which corresponds to the time: $t = 0.008822$. The computational domain was $[0, 0.5]$ and $L = 0.4$ (plasma-limiter interface). At $x = 0$, we impose a symmetry condition

2560 and 10240 cells) and we observe that the peak is nearer and nearer to the plasma limiter interface, when the resolution increases. Besides, when the mesh step decreases, the peak appears earlier and earlier. We stop the computations when $\max_{i \in \{1,\ldots,J\}}(|M_i^n|) > 10$ but similar results are obtained when the stop criterion is $\max_{i \in \{1,\ldots,J\}}(|M_i^n|) > 100$. This leads one to believe that, if the solution converges to a generalized solution of the continuous problem, then this generalized solution must have a singularity supported by the interface (that could be a Dirac measure for example). We notice that the presence of a Dirac measure at the interface is not only a theoretical issue since it has been observed numerically and that the Dirac measure destabilizes numerical schemes. In the following section, we propose a modification of the boundary value problem to obtain a well-posed version.

3.2 A new penalty method for the modified boundary conditions

After the modifications proposed in Sect. 2, the well-posed initial boundary value problem reads:

$$\begin{cases} \partial_t N + \partial_x \Gamma = S \\ \partial_t \Gamma + \partial_x \left(\dfrac{\Gamma^2}{N} + N \right) = 0 \\ M(.,-L) = -1 + \epsilon \text{ and } M(.,L) = 1 - \epsilon \\ N(0,.) = N_0 \text{ and } \Gamma(0,.) = \Gamma_0 \end{cases} \quad (t,x) \in \mathbb{R}_*^+ \times]-L, L[\quad (3)$$

For this problem, the boundary is not characteristic, and the boundary conditions are maximally dissipative. Hence, for compatible initial data, the problem has a unique local in time solution, which is regular enough: at least $\mathscr{C}^1$ is sufficient to perform the asymptotic analysis; see e.g. [3, 12].

To penalize (3), we use a method developed in the semi-linear case by Fornet and Guès [6]. In order to have an homogeneous Dirichlet boundary condition for the theoretical study, the system is reformulated with the unknowns $\tilde{u} = \ln(N)$ and $\tilde{v} = \Gamma/N - M_0$. Although our system is quasi-linear (and not semi-linear), the method can be extended to this case. An interesting feature of the method is that it yields to a convergence result without generation of a boundary layer inside the limiter. Up to now, we don't know if this method can be extended to more general quasi-linear first order hyperbolic system with maximally dissipative conditions.

We assume that M_0 is a constant such that $0 < M_0 < 1$. We denote by χ the characteristic function associated to the limiter, i.e. $\chi(x) = 1$ if the point x is in the limiter.

The new penalized problem reads:

$$\begin{cases} \partial_t N + \partial_x \Gamma = S_N \\ \partial_t \Gamma + \partial_x \left(\dfrac{\Gamma^2}{N} + N \right) + \dfrac{\chi}{\eta} \left(\dfrac{\Gamma}{M_0} - N \right) = S_\Gamma \\ N(0,.) = N_0 \text{ and } \Gamma(0,.) = \Gamma_0 \end{cases} \quad \text{in } \mathbb{R}_*^+ \times \mathbb{R} \quad (4)$$

The formal asymptotic expansion of a continuous solution to (4) with the BKW (Brillouin–Kramers–Wentzel) method does not contain any boundary layer term [1] and this suggests strongly that there is no boundary layer at all in the solution. Notice that the penalization is incomplete: only one field is penalized, which is natural since there is only one boundary condition.

For the numerical tests, we use a regular solution:

$$N(t,x) = \exp\left(\frac{-x^2}{0.16(t+1)}\right) \quad \Gamma(t,x) = M_0 \sin\left(\frac{\pi x}{0.8}\right) \exp\left(\frac{-x^2}{0.16(t+1)}\right)$$

and S_N, S_Γ are well chosen. The spatial domain is $[0, 0.5]$ with a symmetry condition at $x = 0$ and the limiter set corresponds to $x \in [0.4, 0.5]$.

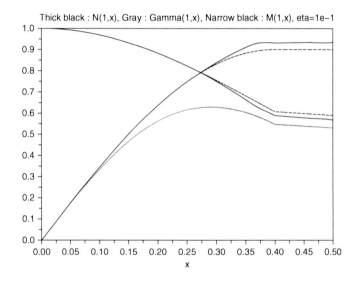

Fig. 3 Plot of N, Γ and M as functions of x (at $t = 1$) with the penalty method free of boundary layer for $\eta = 0.1$. The continuous lines represent the numerical solutions whereas the dashed lines corresponds to the exact solution of the hyperbolic limit problem ($\eta \to 0$). The limiter corresponds to the area $x \in [0.4, 0.5]$. For smaller values of η, for instance for $\eta = 10^{-5}$, the plot is almost the same as the plot of the exact solution (dotted lines)

We analyze the convergence when the penalization parameter η tends to 0 using a uniform spatial mesh of step $\delta x = 10^{-5}$. We calculate the error in L^1 norm for N, $\partial_x N$, Γ and $\partial_x \Gamma$. The goal is to confirm numerically the absence of boundary layer with an optimal rate of convergence as $\mathcal{O}(\eta)$.

One of the main difficulties for the implementation of the penalization, is the choice of a boundary condition at $x = 0.5$ which is necessary for the numerical scheme. As only Γ is penalized, we need a transparent boundary condition for N. For the numerical tests, the boundary condition comes from the asymptotic

Penalty Methods for the Hyperbolic System Modelling the Wall-Plasma Interaction

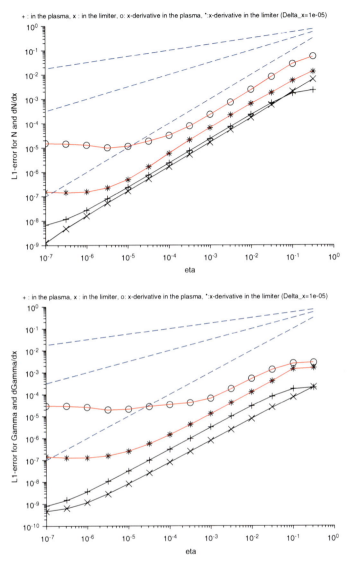

Fig. 4 Errors for N, $\partial_x N$, Γ and $\partial_x \Gamma$ in L^1 norms with the penalization free of boundary layer. The dashed lines represent the curves $\eta^{\frac{1}{4}}$, $\eta^{\frac{1}{2}}$ and η

expansion up to the first order of the BKW analysis. We carry out the computations up to $t = 1$ with an adaptive time step so that the CFL condition is always satisfied. The results are plotted in Fig. 3. In Fig. 4, we observe that the optimal rate of convergence $\mathcal{O}(\eta)$ is reached for the L^1 norms, even for the derivatives. This gives a numerical evidence of the absence of boundary layer. The same numerical results in $\mathcal{O}(\eta)$ are obtained if the penalty term in (4) is replaced by $\frac{\chi}{\eta}\left(\frac{\Gamma}{N} - M_0\right)$, see [2].

When the parameter $\epsilon = 0.01$, i.e. close to a characteristic boundary, the computations show that, for η sufficiently small, $\eta \leq \mathscr{O}(\epsilon)$, the convergence results are similiar; see details in [1].

Acknowledgements This work has been funded by the ANR ESPOIR (Edge Simulation of the Physics Of ITER Relevant turbulent transport)and the *Fédération nationale de Recherche Fusion par Confinement Magnétique* (FR-FCM). We thank Guillaume Chiavassa, Guido Ciraolo and Philippe Ghendrih for fruitful discussions.

References

1. Angot, P., Auphan, P., Guès, O.: An optimal penalty method for the hyperbolic system modelling the edge plasma transport in a tokamak. Preprint in preparation (2011)
2. Auphan, T.: Méthodes de pénalisation pour des systèmes hyperboliques application au transport de plasma en bord de tokamak. Master's thesis, Ecole Centrale Marseille (2010)
3. Benzoni-Gavage, S., Serre, D.: Multidimensional hyperbolic partial differential equations. First-order systems and applications. Oxford Mathematical Monographs. Oxford University Press (2007)
4. Bouchut, F., James, F.: One-dimensional transport equations with discontinuous coefficients. Nonlinear Anal. **32**, 891–933 (1998)
5. Fornet, B.: Small viscosity solution of linear scalar 1-d conservation laws with one discontinuity of the coefficient. Comptes Rendus Mathematique **346**(11-12), 681 – 686 (2008)
6. Fornet, B., Guès, .: Penalization approach of semi-linear symmetric hyperbolic problems with dissipative boundary conditions. Discrete and Continuous Dynamical Systems **23**(3), 827 – 845 (2009)
7. Gallouët, T., Hérard, J.M., Seguin, N.: Some approximate godunov schemes to compute shallow-water equations with topography. Computers and Fluids **32**(4), 479 – 513 (2003)
8. Guès, O.: Problème mixte hyperbolique quasi-linéaire caractéristique. Communications in Partial Differential Equations **15**, 595–654 (1990)
9. Isoardi, L., Chiavassa, G., Ciraolo, G., Haldenwang, P., Serre, E., Ghendrih, P., Sarazin, Y., Schwander, F., Tamain, P.: Penalization modeling of a limiter in the tokamak edge plasma. Journal of Computational Physics **229**(6), 2220 – 2235 (2010)
10. Poupaud, F., Rascle, M.: Measure solutions to the linear multi-dimensional transport equation with non-smooth coefficients. Communications in Partial Differential Equations **22**, 225–267 (1997)
11. Rauch, J.B.: Symmetric positive systems with boundary characteristic of constant multiplicity. Trans. Amer. Math. Soc. **291**(1), 167–187 (1985)
12. Rauch, J.B., Massey, F.J.I.: Differentiability of solutions to hyperbolic initial-boundary value problems. Trans. Amer. Math. Soc. **189**, 303–318 (1974)
13. Tamain, P.: Etude des flux de matière dans le plasma de bord des tokamaks, alimentation, transport et turbulence. Ph.D. thesis, Université de Provence (2007)

The paper is in final form and no similar paper has been or is being submitted elsewhere.

A Spectacular Vector Penalty-Projection Method for Darcy and Navier-Stokes Problems

Philippe Angot, Jean-Paul Caltagirone, and Pierre Fabrie

Abstract We present a new *fast vector penalty-projection method (VPP$_\varepsilon$)*, issued from noticeable improvements of previous works [3,4,7], to efficiently compute the solution of unsteady Navier-Stokes/Brinkman problems governing incompressible multiphase viscous flows. The method is also efficient to solve anisotropic Darcy problems. The key idea of the method is to compute at each time step an accurate and curl-free approximation of the pressure gradient increment in time. This method performs a *two-step approximate divergence-free vector projection* yielding a velocity divergence vanishing as $\mathcal{O}(\varepsilon\,\delta t)$, δt being the time step, with a penalty parameter ε as small as desired until the machine precision, e.g. $\varepsilon = 10^{-14}$, whereas the solution algorithm can be extremely fast and cheap. The method is numerically validated on a benchmark problem for two-phase bubble dynamics where we compare it to the Uzawa augmented Lagrangian (UAL) and scalar incremental projection (SIP) methods. Moreover, a new test case for fluid-structure interaction problems is also investigated. That results in a robust method running faster than usual methods and being able to efficiently compute accurate solutions to sharp test cases whatever the density, viscosity or anisotropic permeability jumps, whereas other methods crash.

Keywords Vector penalty-projection; Penalty method; Splitting method; Multiphase Navier-Stokes/Brinkman; Anisotropic Darcy problem; Incompressible flows
MSC 2010: 35Q30, 35Q35, 65M12, 65M85, 65N12, 65N85, 74F10, 76D05, 76D45, 76M25, 76R10, 76S05, 76T10

Philippe Angot
Aix-Marseille Université, LATP - CMI UMR CNRS 6632, 39 rue F. Joliot Curie, 13453 Marseille Cedex 13 - France, e-mail: angot@cmi.univ-mrs.fr

Jean-Paul Caltagirone
Université de Bordeaux & IPB, IMIB, 16 Av Pey-Berland 33607 Pessac - France,
e-mail: calta@enscbp.fr

Pierre Fabrie
Université de Bordeaux & IPB, IMB UMR CNRS 5251, ENSEIRB-MATMECA,
Talence - France, e-mail: pierre.fabrie@math.u-bordeaux1.fr

1 Introduction to model incompressible multiphase flows

Let $\Omega \subset \mathbb{R}^d$ ($d = 2$ or 3 in practice) be an open bounded and connected domain with a Lipschitz continuous boundary $\Gamma = \partial\Omega$ and $\mathbf{n}$ be the outward unit normal vector on Γ. For $T > 0$, we consider the following unsteady Navier-Stokes/Brinkman problem [9] governing incompressible non-homogeneous or multiphase flows where Dirichlet boundary conditions for the velocity $\mathbf{v}_{|\Gamma} = 0$ on Γ, the volumic force $\mathbf{f}$ and initial data $\mathbf{v}(t = 0) = \mathbf{v}_0$, $\varphi(t = 0) = \varphi_0 \in L^\infty(\Omega)$ with $\varphi_0 \geq 0$ a.e. in Ω, are given. For sake of briefness here, we just focus on the model problem (1-3) where $\mathbf{d}(\mathbf{v}) = (\nabla \mathbf{v} + (\nabla \mathbf{v})^T)/2$, as a part of more complex fluid mechanics problems.

$$\rho\,(\partial_t \mathbf{v} + (\mathbf{v}\cdot\nabla)\mathbf{v}) - 2\nabla\cdot(\mu\,\mathbf{d}(\mathbf{v})) + \mu\,\mathbf{K}^{-1}\mathbf{v} + \nabla p = \mathbf{f} \quad \text{in } \Omega \times (0, T) \quad (1)$$

$$\nabla\cdot\mathbf{v} = 0 \quad \text{in } \Omega \times (0, T) \quad (2)$$

$$\partial_t \varphi + \mathbf{v}\cdot\nabla\varphi = 0 \quad \text{in } \Omega \times (0, T). \quad (3)$$

The permeability tensor $\mathbf{K}$ in the Darcy term is supposed to be symmetric, uniformly positive definite and bounded in Ω. We refer to [1, 9] for the modeling of flows inside complex fluid-porous-solid heterogeneous systems with the Navier-Stokes/Brinkman or Darcy equations. The equation (3) for the positive phase function φ governs the transport by the flow of the interface between two phases, either fluid or solid, respectively in the case of two-phase fluid flows or fluid-structure interaction problems. The force $\mathbf{f}$ may include some volumic forces like the gravity force $\rho\,\mathbf{g}$ as well as the surface tension force to describe the capillarity effects at the phase interfaces Σ. The advection-diffusion equation for the temperature $\mathcal{T}$ is not precised here and we assume some given state laws: $\rho = \rho(\varphi, \mathcal{T})$ and $\mu = \mu(\varphi, \mathcal{T})$ for each phase, where the functions are continuous and positive.

2 The fast vector-penalty projection method (VPP$_\varepsilon$)

2.1 The (VPP$_\varepsilon$) method for multiphase Navier-Stokes/Brinkman

We describe hereafter the two-step vector penalty-projection (VPP$_\varepsilon$) method with a penalty parameter $0 < \varepsilon \ll 1$; see more details in [5]. For φ^0 with $\varphi^0 \geq 0$ a.e. in Ω, $\mathbf{v}^0$ and $p^0 \in L_0^2(\Omega)$ given, the method reads as below with usual notations for the semi-discrete setting in time, $\delta t > 0$ being the time step. For all $n \in \mathbb{N}$ such that $(n+1)\,\delta t \leq T$, find $\tilde{\mathbf{v}}^{n+1}$, $\mathbf{v}^{n+1}$, $p^{n+1} \in L_0^2(\Omega)$, $\varphi^{n+1} \in L^\infty(\Omega)$, such that:

$$\rho^n \left(\frac{\tilde{\mathbf{v}}^{n+1} - \mathbf{v}^n}{\delta t} + (\mathbf{v}^n \cdot \nabla) \tilde{\mathbf{v}}^{n+1} \right) - 2\nabla \cdot \left(\mu^n \, \mathbf{d}(\tilde{\mathbf{v}}^{n+1}) \right) + \mu^n \, \mathbf{K}^{-1} \tilde{\mathbf{v}}^{n+1} + \nabla p^n = \mathbf{f}^n \quad (4)$$

$$\frac{\varepsilon}{\delta t} \rho^n \, \hat{\mathbf{v}}^{n+1} - \nabla \left(\nabla \cdot \hat{\mathbf{v}}^{n+1} \right) = \nabla \left(\nabla \cdot \tilde{\mathbf{v}}^{n+1} \right) \quad (5)$$

$$\mathbf{v}^{n+1} = \tilde{\mathbf{v}}^{n+1} + \hat{\mathbf{v}}^{n+1}, \quad \text{and} \quad \nabla(p^{n+1} - p^n) = -\frac{\rho^n}{\delta t} \hat{\mathbf{v}}^{n+1} \quad (6)$$

$$p^{n+1} = p^n + \phi^{n+1} \quad \text{with } \phi^{n+1} \text{ reconstructed from } \nabla \phi^{n+1} = -\frac{\rho^n}{\delta t} \hat{\mathbf{v}}^{n+1} \quad (7)$$

$$\frac{\varphi^{n+1} - \varphi^n}{\delta t} + \mathbf{v}^{n+1} \cdot \nabla \varphi^n = 0 \quad (8)$$

with: $\tilde{\mathbf{v}}^{n+1}_{|\Gamma} = 0$, or for non homogeneous Dirichlet conditions: $\tilde{\mathbf{v}}^{n+1}_{|\Gamma} = \mathbf{v}^{n+1}_D$, and $\hat{\mathbf{v}}^{n+1} \cdot \mathbf{n}_{|\Gamma} = 0$. Here $\mathbf{v}^n$, p^n are desired to be first-order approximations of the exact velocity and pressure solutions $\mathbf{v}(t_n)$, $p(t_n)$ at time $t_n = n \, \delta t$. Since the end-of-step velocity divergence is not exactly zero, the additional spherical part $\lambda \, \nabla \cdot \mathbf{v} \, \mathbf{I}$ of the Newtonian stress tensor is included within the dynamical pressure gradient ∇p. Once the equations (4-8) have been solved, the advection-diffusion equation of temperature can be solved too for $\mathcal{T}^{n+1}$ and we can find: $\rho^{n+1} = \rho(\varphi^{n+1}, \mathcal{T}^{n+1})$ and $\mu^{n+1} = \mu(\varphi^{n+1}, \mathcal{T}^{n+1})$.

The key feature of our method is to calculate an accurate and curl-free approximation of the momentum vector correction $\rho^n \, \hat{\mathbf{v}}^{n+1}$ in (5). Indeed (5-6) ensures that $\rho^n \, \hat{\mathbf{v}}^{n+1}$ is exactly a gradient which justifies the choice for $\nabla \phi^{n+1} = \nabla(p^{n+1} - p^n)$ since we have:

$$\rho^n \, \hat{\mathbf{v}}^{n+1} = \frac{\delta t}{\varepsilon} \nabla \left(\nabla \cdot \mathbf{v}^{n+1} \right) \Rightarrow \nabla(p^{n+1} - p^n) = -\frac{\rho^n}{\delta t} \hat{\mathbf{v}}^{n+1} = -\frac{1}{\varepsilon} \nabla \left(\nabla \cdot \mathbf{v}^{n+1} \right). \quad (9)$$

The (VPP$_\varepsilon$) method effectively takes advantage of the splitting method proposed in [4] for augmented Lagrangian systems or general saddle-point computations to get a very fast solution of (5); see Theorem 1. When we need the pressure field itself, e.g. to compute stress vectors, it is calculated in an incremental way as an auxiliary step. We propose to reconstruct $\phi^{n+1} = p^{n+1} - p^n$ from its gradient $\nabla \phi^{n+1}$ given in (6) with the following method.

Reconstruction of $\phi^{n+1} = p^{n+1} - p^n$ from its gradient.

By circulating on a suitable path starting at a point on the border where $\phi^{n+1} = 0$ is fixed and going through all the pressure nodes in the mesh, we get with the gradient formula between two neighbour points A and B using the mid-point quadrature:

$$\phi^{n+1}(B) - \phi^{n+1}(A) = \int_A^B \nabla \phi^{n+1} \cdot d\mathbf{l} = -\int_A^B \frac{\rho^n}{\delta t} \hat{\mathbf{v}}^{n+1} \cdot d\mathbf{l} \approx -\frac{\rho^n}{\delta t} |\hat{\mathbf{v}}^{n+1}| h_{AB} \quad (10)$$

with $h_{AB} = $ distance (A, B). The field ϕ^{n+1} is calculated point by point from the boundary and then passing successively by all the pressure nodes. This fast algorithm is performed at each time step to get the pressure field p^{n+1} from the known field p^n. We refer to [5] for more details and validations on the present method.

2.2 The (VPP$_\varepsilon$) method for anisotropic Darcy problems

We present below the fast solution to incompressible Darcy flow problems in porous media with the (VPP$_\varepsilon$) method. The model problem reads in dimensionless form:

$$s \, \partial_t \mathbf{v} + \mu \, \mathbf{K}^{-1} \mathbf{v} + \nabla p = \mathbf{f} \quad \text{in } \Omega \times (0, T) \tag{11}$$

$$\nabla \cdot \mathbf{v} = 0 \quad \text{in } \Omega \times (0, T) \tag{12}$$

$$\mathbf{v} \cdot \mathbf{n} = 0 \quad \text{on } \Gamma \times (0, T) \tag{13}$$

where the viscosity $\mu > 0$ is constant and the permeability tensor $\mathbf{K}$ is supposed to be symmetric, bounded in Ω and uniformly positive definite. The dimensionless stationarity parameter $s > 0$ includes the Darcy number: $Da = K_{ref}/L_{ref}^2$ and thus we have $s \ll 1$ for most practical problems or even $s = 0$ for the steady anisotropic Darcy problem. The equations (11-13) also model flows inside heterogeneous porous-solid systems by letting the permeability tend to zero inside the impermeable media; see also [1, 9] for the analysis and validations of the so-called L^2 volume penalty method.

The (VPP$_\varepsilon$) method with $r = \mathcal{O}(\varepsilon) > 0$ and $0 < \varepsilon \ll 1$ to solve (11-13) reads as follows. For all $n \in \mathbb{N}$ such that $(n+1)\,\delta t \leq T$, find $\tilde{\mathbf{v}}^{n+1}$, $\mathbf{v}^{n+1}$ and p^{n+1} such that:

$$s \frac{\tilde{\mathbf{v}}^{n+1} - \mathbf{v}^n}{\delta t} + \mu \mathbf{K}^{-1} \tilde{\mathbf{v}}^{n+1} - r \nabla \left(\nabla \cdot \tilde{\mathbf{v}}^{n+1} \right) + \nabla p^n = \mathbf{f}^n \tag{14}$$

$$\varepsilon \left(\frac{s}{\delta t} + \mu \mathbf{K}^{-1} \right) \hat{\mathbf{v}}^{n+1} - \nabla \left(\nabla \cdot \hat{\mathbf{v}}^{n+1} \right) = \nabla \left(\nabla \cdot \tilde{\mathbf{v}}^{n+1} \right) \tag{15}$$

$$\mathbf{v}^{n+1} = \tilde{\mathbf{v}}^{n+1} + \hat{\mathbf{v}}^{n+1},$$

$$\text{and} \quad \nabla(p^{n+1} - p^n) = -\left(\frac{s}{\delta t} + \mu \mathbf{K}^{-1} \right) \hat{\mathbf{v}}^{n+1} - r \nabla \left(\nabla \cdot \tilde{\mathbf{v}}^{n+1} \right) \tag{16}$$

$$p^{n+1} = p^n + \phi^{n+1} \quad \text{with } \phi^{n+1} \text{ reconstructed from its gradient } \nabla \phi^{n+1} \tag{17}$$

with the boundary conditions: $\tilde{\mathbf{v}}^{n+1} \cdot \mathbf{n}_{|\Gamma} = 0$ and $\hat{\mathbf{v}}^{n+1} \cdot \mathbf{n}_{|\Gamma} = 0$ on Γ. The space discrete solution to the prediction step (14) is explicit for s and r sufficiently small to invert a perturbation of the Identity matrix with a Neumann asymptotic expansion.

3 On the fast discrete solution to the (VPP$_\varepsilon$) method

The great interest for solving (5) or (15) instead of a usual augmented Lagrangian problem lies in the following result issued from [4] which shows that the method can be ultra-fast and very cheap if $\eta = \varepsilon/\delta t$ is sufficiently small.

Let us now consider any space discretization of our problem. We denote by $B = -div_h$ the $m \times n$ matrix corresponding to the discrete divergence operator, $B^T = grad_h$ the $n \times m$ matrix corresponding to the discrete gradient operator, whereas I denotes the $n \times n$ identity matrix with $n > m$ and D the $n \times n$ diagonal nonsingular matrix containing all the discrete density values of $\rho^n > 0$ a.e. in Ω. Here n is the number of velocity unknowns whereas m is the number of pressure unknowns. Then, the discrete vector penalty-projection problem corresponding to (5) with $\varepsilon = \eta \delta t$ reads:

$$\left(D + \frac{1}{\eta} B^T B\right) \hat{v}_\eta = -\frac{1}{\eta} B^T B \tilde{v}, \quad \text{with} \quad v_\eta = \tilde{v} + \hat{v}_\eta. \tag{18}$$

We proved in [4] the crucial result below due to the *adapted right-hand side* in the correction step (18) which lies in the range of the limit operator $B^T B$. Indeed, (18) can be viewed as a singular perturbation problem with well-suited data in the right-hand side. More precisely, we give in Theorem 1 the zero-order term of the solution $\hat{v}_\eta$ to (18):

$$\hat{v}_\eta = -\frac{1}{\eta} \left(D + \frac{1}{\eta} B^T B\right)^{-1} B^T B \tilde{v} \tag{19}$$

when the penalty parameter η is chosen sufficiently small; see the asymptotic expansion of $\hat{v}_\eta$ and the proof in [4, Theorem 1.1 and Corollary 1.3].

Theorem 1 (Fast solution of the discrete vector penalty-projection). *Let D be an $n \times n$ positive definite diagonal matrix, I the $n \times n$ identity matrix and B an $m \times n$ matrix. If the rows of B are linearly independent, $\text{rank}(B) = m$, then for all η small enough, $0 < \eta < 1/\|S^{-1}\|$ where $S = BD^{-1}B^T$, there exists an $n \times n$ matrix C_1 bounded independently on η such that the solution of the correction step (19) writes for any vector $\tilde{v} \in \mathbb{R}^n$:*

$$\hat{v}_\eta = C_0 \tilde{v} + \eta C_1 \tilde{v} \quad \text{with} \quad C_0 = -D^{-1} B^T S^{-1} B = -D^{-1} B^T (BD^{-1}B^T)^{-1} B. \tag{20}$$

If $\text{rank}(B) = p < m$, there exists a surjective $p \times n$ matrix T such that $B^T B = T^T T$ and a similar result holds replacing B by T.

Hence, for a constant density $\rho > 0$ and choosing now $\eta = \rho\varepsilon/\delta t$, we have: $D = I$, $S = BB^T$ and $C_0 = -B^T S^{-1} B = -B^T (BB^T)^{-1} B$. Moreover, if $\text{rank}(B) = p \leq m \leq n$, the zero-order solution $\hat{v} = C_0 \tilde{v}$ in (20) is the solution of minimal Euclidean norm in $\mathbb{R}^n$ to the linear system: $B \hat{v} = -B \tilde{v}$ by the least-squares method, and the matrix $B^\dagger = B^T (BB^T)^{-1}$ is the Moore-Penrose pseudo-inverse of B such that $C_0 = -B^\dagger B$. Indeed, a singular value decomposition (SVD) or a QR factorization of B yields: $C_0 = -I_0$ where I_0 is the $n \times n$ diagonal *matrix having only* 1 *or* 0 *coefficients, the zero entries in the diagonal being the* $n - p$ *null eigenvalues of the operator* $B^T B$.

Hence, for η small enough, the computational effort required to solve (18) amounts to approximate the matrix C_0 which includes both D and D^{-1} inside non commutative products. Thus, we always use the diagonal preconditioning in the case of a variable density which makes the effective condition number quasi-independent on the density or permeability jumps. We also use the Jacobi preconditioner in the prediction step (4) to cope with the viscosity or permeability jumps as performed in [9]. However, for a constant density when $D = I$, we get $C_0 = -I_0$. This explains why the solution can be obtained with only one iteration of a suitable preconditioned Krylov solver whatever the size of the mesh step or the dimension n; see the numerical results in [4].

4 Numerical validations with discrete operator calculus

The (VPP$_\varepsilon$) method has been implemented with discrete exterior calculus (DEC) methods, see the recent review in [6], for the space discretization of the Navier-Stokes equations on unstructured staggered meshes. The (DEC) methods ensure primary and secondary discrete conservation properties. In particular, the space discretization satisfies for the discrete operators: $\nabla_h \times (\nabla_h \phi) = 0$ and $\nabla_h \cdot (\nabla_h \times \psi) = 0$, which is not usually verified by other methods; see [6]. Hence, the (VPP$_\varepsilon$) method is now validated on unstructured meshes both in 2-D or 3-D.

The structure and solver of the computational code are issued from previous works, originally implemented with a Navier-Stokes finite volume solver on the staggered MAC mesh and using the Uzawa augmented Lagrangian (UAL) method to deal with the divergence-free constraint; see [9]. We refer to [1, 2, 9] and the references therein for the analysis and numerical validations of the fictitious domain model using the so-called L^2 or H^1-penalty methods to take account of obstacles in flow problems with the Navier-Stokes/Brinkman equations. Hence, our approach is essentially Eulerian with a Lagrangian front-tracking of the sharp interfaces accurately reconstructed on the fixed Eulerian mesh, see e.g. [10, 11] and the references therein. Thus we use no Arbitrary Lagrangian-Eulerian (ALE) method, no global remeshing nor moving mesh method.

4.1 Multiphase flows: dispersed two-phase bubble dynamics

The (VPP$_\varepsilon$) method is numerically validated for multiphase incompressible flows by performing with the three methods (UAL), (SIP) and (VPP), the benchmark problem studied in [8] for 2-D bubble dynamics. In that problem, we compute the first test case which considers an initial circular bubble of diameter $0.05\,m$ with density and

 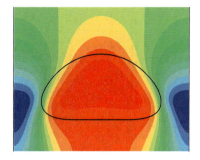

Fig. 1 Benchmark for 2-D bubble dynamics with (VPP$_\varepsilon$) method, $\varepsilon = 10^{-8}$: motion of a circular bubble with surface tension at time $t = 3$ and Re $= 35$ - bubble initial diameter $\varnothing = 0.05$, $\rho_1/\rho_2 = 1000/100 = 10$, $\mu_1/\mu_2 = 10/1 = 10$, domain 0.1×0.2, mesh size 128×256, $\delta t = 0.007143$, circular bubble initially with no motion at height $y = 0.05$. LEFT: isobars and isoline $\varphi = 0.5$ of the phase function at interface. RIGHT: superposition of isoline $\varphi = 0.5$ at interface for (UAL), (SIP), (VPP) and vertical velocity field (in absolute referential)

viscosity ratios equal to 10 which undergoes moderate shape deformation. In this case, the bubble is driven up by the external gravity force $\mathbf{f} = \rho \mathbf{g}$, whereas the surface tension effect on the interface Σ between the two fluid phases is taken into account through the following force balance at the interface Σ:

$$[\![\mathbf{v}]\!]_\Sigma = 0 \text{ and } [\![(-p\mathbf{I} + \mu (\nabla \mathbf{v} + (\nabla \mathbf{v})^T))\cdot \mathbf{n}]\!]_\Sigma = \sigma \kappa \mathbf{n}_{|\Sigma}, \text{ or } \mathbf{f}_{st} = \sigma \kappa \mathbf{n}_{|\Sigma} \delta_\Sigma$$

where $\sigma = 24.5$ is the surface tension coefficient, κ the local curvature of the interface, $\mathbf{n}_{|\Sigma}$ the outward unit normal to the interface and δ_Σ the Dirac measure supported by the interface Σ. The solution of the phase transport (3) is carried out by the so-called *VOF-PLIC* method, *i.e.* the famous *VOF* method using a piecewise linear interface construction proposed in [12] to precisely reconstruct the sharp interface Σ at the isoline $\varphi = 0.5$, with $\varphi^0 = 0$ in Ω_1 and $\varphi^0 = 1$ in Ω_2; see [10, 11].

The results of the three methods (UAL), (SIP) and (VPP) after 420 time iterations are presented in Fig. 1 by superposing the different fields to get a more precise comparison. We observe an excellent agreement both between the three methods and the reference solution in [8]. However, the (VPP) method runs faster.

4.2 A test case for fluid-structure interaction problems

To evaluate the robustness of the (VPP$_\varepsilon$) method with respect to large density or viscosity ratios, we compute the motion of an heavy solid body which freely falls vertically in air with the gravity force $\mathbf{f} = \rho_s \mathbf{g}$. The rigid behaviour of the body is obtained by letting the viscosity μ_s tend to infinity inside the ball in order to penalize the tensor of deformation rate $\mathbf{d}(\mathbf{v})$. This fictitious domain method using a

Fig. 2 ACF11-ball with (VPP$_\varepsilon$) method, $\varepsilon = 10^{-6}$: free fall of a heavy solid body in air at time $t = 0.15$ and Re $= 7358$ - Cylinder diameter $\oslash = 0.05$, $\rho_s = 10^6$, $\rho_f = 1$, $\mu_s = 10^{12}$, $\mu_f = 10^{-5}$, domain 0.1×0.2, mesh size 256×512, $\delta t = 0.0002$, cylinder initially with no motion at height $y = 0.15$. LEFT: isobars and isoline $\varphi = 0.5$ of the phase function at interface. RIGHT: vertical velocity field and horizontal velocity isolines

penalty was studied in [1] (see the references therein) to design a numerical wind-tunnel, then numerically validated in several works, e.g. [11], and also analyzed theoretically in [1, 2] where optimal global error estimates are proved for the H^1 penalty method. Moreover, this fictitious domain method allows us to easily compute the forces applied on the obstacle, see [9]; the error estimate being proved in [1] when the nonlinear convection term is neglected inside the solid obstacle.

The results obtained by the (VPP$_\varepsilon$) method are presented in Fig. 2 at time $t = 0.15\,s$ after 750 time iterations when the ball velocity reaches: $V_b = g\,t = 1.4715\,m/s$. The computation shows that the strain rate tensor inside the ball Ω_s vanishes as $\|\mathbf{d}(\mathbf{v})\|_{L^2(\Omega_s)} = \mathcal{O}(\mu_f/\mu_s)$, i.e. of the order of the machine precision. Hence, the (VPP$_\varepsilon$) method efficiently ensures both the rigidity of the solid body and a velocity divergence vanishing as $\mathcal{O}(\varepsilon\,\delta t)$ [5], whereas it avoids the blocking effect observed with other methods; see e.g. [11].

The (SIP) method crashes after a few time iterations. The (UAL) method is still able to compute the flow with a larger velocity divergence and the computation is far more expensive than with the (VPP$_\varepsilon$) method.

References

1. PH. ANGOT, Analysis of singular perturbations on the Brinkman problem for fictitious domain models of viscous flows, Math. Meth. in the Appl. Sci. ($M2AS$) **22**(16), 1395-1412, 1999.
2. PH. ANGOT, C.-H. BRUNEAU AND P. FABRIE, A penalization method to take into account obstacles in incompressible viscous flows, Nümerische Mathematik **81**(4), 497-520, 1999.
3. PH. ANGOT, J.-P. CALTAGIRONE AND P. FABRIE, Vector penalty-projection methods for the solution of unsteady incompressible flows, in *Finite Volumes for Complex Applications V*, R. Eymard and J.-M. Hérard (Eds), pp. 169-176, ISTE Ltd and J. Wiley & Sons, 2008.
4. PH. ANGOT, J.-P. CALTAGIRONE AND P. FABRIE, A new fast method to compute saddle-points in constrained optimization and applications, Appl. Math. Letters, 2011 (submitted).
5. PH. ANGOT, J.-P. CALTAGIRONE AND P. FABRIE, A fast vector penalty-projection method for incompressible non-homogeneous or multiphase Navier-Stokes problems, Applied Mathematics Letters, 2011 (submitted).
6. J. BLAIR PEROT, Discrete conservation properties of unstructured mesh schemes, Annu. Rev. Fluid Mech. **43**, 299-318, 2011.
7. J.-P. CALTAGIRONE AND J. BREIL, Sur une méthode de projection vectorielle pour la résolution des équations de Navier-Stokes, C. R. Acad. Sci. Paris, IIb **327**, 1179-1184, 1999.
8. S. HYSING, S. TUREK, D. KUZMIN, N. PAROLINI, E. BURMAN, S. GANESAN AND L. TOBISKA, Quantitative benchmark computations of two-dimensional bubble dynamics, Int. J. Numer. Meth. Fluids, **60**, 1259-1288, 2009.
9. K. KHADRA, PH. ANGOT, S. PARNEIX AND J.-P. CALTAGIRONE, Fictitious domain approach for numerical modelling of Navier-Stokes equations, Int. J. Numer. Meth. in Fluids, **34**(8), 651-684, 2000.
10. A. SARTHOU, S. VINCENT, J.-P. CALTAGIRONE AND PH. ANGOT, Eulerian-Lagrangian grid coupling and penalty methods for the simulation of multiphase flows interacting with complex objects, Int. J. Numer. Meth. in Fluids **56**(8), 1093-1099, 2008.
11. S. VINCENT, A. SARTHOU, J.-P. CALTAGIRONE, F. SONILHAC, P. FÉVRIER, C. MIGNOT AND G. PIANET, Augmented Lagrangian and penalty methods for the simulation of two-phase flows interacting with moving solids. Application to hydroplaning flows interacting with real tire tread patterns, J. Comput. Phys. **230**, 956-983, 2011.
12. D.L. YOUNGS, Time-dependent multimaterial flow with large fluid distortion, in *Numerical Methods for Fluid Dynamics*, K.W. Morton, M.J. Baines (Eds.), Academic Press, 1982.

The paper is in final form and no similar paper has been or is being submitted elsewhere.

Numerical Front Propagation Using Kinematical Conservation Laws

K.R. Arun, M. Lukáčová-Medviďová, and P. Prasad

Abstract We use the newly formulated three-dimensional (3-D) kinematical conservation laws (KCL) to study the propagation of a nonlinear wavefront in a polytropic gas in a uniform state at rest. The 3-D KCL forms an under-determined system of six conservation laws with three involutive constraints, to which we add the energy conservation equation of a weakly nonlinear ray theory. The resulting system of seven conservation laws is only weakly hyperbolic and therefore poses a real challenge in the numerical approximation. We implement a central finite volume scheme with a constrained transport technique for the numerical solution of the system of conservation laws. The results of a numerical experiment is presented, which reveals some interesting geometrical features of a nonlinear wavefront.

Keywords kinematical conservation laws, kink, weakly nonlinear ray theory, wavefront, polytropic gas
MSC2010: 35L60, 35L65, 35L67, 35L80

1 Introduction

A curved nonlinear wavefront or a shock front during its evolution develops certain curves of discontinuity, across which the normal to the front and the amplitude

---

K.R. Arun
Institut für Geometrie und Praktische Mathematik, RWTH Aachen, Templergraben 55, D-52056 Aachen, Germany, e-mail: arun@igpm.rwth-aachen.de

M. Lukáčová-Medviďová
Institut für Mathematik, Johannes Gutenberg-Universität Mainz, Staudingerweg 9, D-55099 Mainz, Germany, e-mail: lukacova@mathematik.uni-mainz.de

P. Prasad
Department of Mathematics, Indian Institute of Science, Bangalore - 560012, India,
e-mail: prasad@math.iisc.ernet.in

distribution on it are discontinuous. Some of these curves of discontinuity are called kinks, which are shocks in a corresponding ray coordinate system in which a physically realistic system of conservation laws has been formulated. The conservation form of the system of evolution equations of a surface is called kinematical conservation laws (KCL). The KCL is a pure geometrical result and it does not take into consideration any dynamics of the propagating front. This makes the KCL an incomplete system and additional closure equations derived by considering the dynamical conditions of the propagating front are required for applications. Prasad and collaborators have used the KCL in two dimensions along with some closure equations derived on physical considerations to solve several interesting problems, see the review paper [6] and the references therein. The KCL for a surface evolving in three space dimensions, called 3-D KCL, is a system of six conservation laws with three divergence-free type stationary constraints, all three together are termed as 'geometric solenoidal constraint', see [3]. The analysis of the 3-D KCL system, with the closure equation from a weakly nonlinear ray theory (WNLRT), was done in [3] and it has been shown that the resulting system of conservation laws, the so-called conservation laws of 3-D WNLRT give rise to a weakly hyperbolic system; in the sense that the system has zero as a repeated eigenvalue with multiplicity five, but the associated eigenspace is only four-dimensional.

Despite the 3-D WNLRT being a weakly hyperbolic system, in [1, 2] we have been able to develop efficient numerical approximations for it using simple, but robust central schemes. It is well known that the solution to the Cauchy problem for a weakly hyperbolic system (with deficiency in dimension of the eigenspace by one) typically contains a mode, the so-called 'Jordan mode', which grows linearly in time. However, it has been proved in [1] that when the geometric solenoidal constraint is satisfied initially, the solution to the Cauchy problem for linearised 3-D WNLRT at any time does not exhibit the Jordan mode. Motivated by this, a constrained transport technique has been employed to enforce the geometric solenoidal constraint in the numerical solution of 3-D WNLRT, see [1] for more details.

The aim of the present paper is to give a brief overview of the recent results obtained with 3-D WNLRT and to show its efficacy to model propagating wavefronts. The layout of the paper is as follows. In Sect. 2 we introduce the governing equations of 3-D WNLRT. The numerical approximation and the constrained transport strategy are outlined in Sect.3. In Sect.4 we present the results of a numerical experiment, showing the efficiency and robustness of the present method. Finally, we close this article with some concluding remarks in Sect.5.

2 Governing equations

Consider a one parameter family of surfaces in (x_1, x_2, x_3)-space such that it represents the successive positions of a moving surface Ω_t as time varies. Associated with the family, we have a ray velocity χ at any point (x_1, x_2, x_3) on the surface Ω_t. We consider only the isotropic evolution of Ω_t so that we take χ to be in the direction of the unit normal $\mathbf{n}$ to Ω_t, i.e. $\chi = m\mathbf{n}$, where m is the normal velocity of propagation of Ω_t. Hence, the evolution of Ω_t is governed by

$$\frac{d\mathbf{x}}{dt} = m\mathbf{n}. \tag{1}$$

We introduce a ray coordinate system (ξ_1, ξ_2, t) such that for $t = $ const, we get (ξ_1, ξ_2) as the surface coordinates on Ω_t. Further, $\xi_1 = $ const, $\xi_2 = $ const represent the rays, a two parameter family of curves orthogonal to Ω_t. Let $\mathbf{u}$ and $\mathbf{v}$ be respectively unit tangent vectors to the curves $\xi_2 = $ const and $\xi_1 = $ const on Ω_t. Let $\mathbf{n}$ be a unit normal to Ω_t given by

$$\mathbf{n} = \frac{\mathbf{u} \times \mathbf{v}}{\|\mathbf{u} \times \mathbf{v}\|} \tag{2}$$

so that $(\mathbf{u}, \mathbf{v}, \mathbf{n})$ forms a right handed system. Let an element of distance along a curve $(\xi_2 = $ const, $t = $ const$)$ be $g_1 d\xi_1$. Analogously, denote by $g_2 d\xi_2$, the element of distance along a curve $(\xi_1 = $ const, $t = $ const$)$. The element of distance along a ray $(\xi_1 = $ const, $\xi_2 = $ const$)$ is $m dt$. Based on geometrical considerations we can derive the 3-D KCL [3],

$$(g_1 \mathbf{u})_t - (m\mathbf{n})_{\xi_1} = 0, \tag{3}$$

$$(g_2 \mathbf{v})_t - (m\mathbf{n})_{\xi_2} = 0 \tag{4}$$

subject to the condition

$$(g_1 \mathbf{u})_{\xi_2} - (g_2 \mathbf{v})_{\xi_1} = 0. \tag{5}$$

Note that the constraint (5) is an involution, i.e. if it is satisfied at time $t = 0$, then the equations (3)-(4) imply that it is satisfied for every time. Note that each of the scalar equations in (5) can be written as div$(\mathcal{B}_k) = 0$, where $\mathcal{B}_k := (-g_2 v_k, g_1 u_k), k = 1, 2, 3$. Therefore, the vector constraint (5) has been designated as geometric solenoidal constraint. The 3-D KCL (3)-(4), being a system of six evolution equations in seven unknowns $u_1, u_2, v_1, v_2, m, g_1$ and g_2, is underdetermined. We use the closure equation by considering the energy propagation along the rays of a WNLRT, c.f. [6]. The energy transport equation of WNLRT for a polytropic gas initially at rest and in uniform state can be written in a conservation form [3]

$$\left((m-1)^2 e^{2(m-1)} g_1 g_2 \sin \chi\right)_t = 0, \tag{6}$$

where χ is the angle between the vectors $\mathbf{u}$ and $\mathbf{v}$. The system of equations (3)-(4) and (6), hereafter designated as the conservation laws of 3-D WNLRT, is the complete set of equations describing the evolution of the nonlinear wavefront Ω_t.

Remark 1. It has been proved in [3] that the eigenvalues of 3-D WNLRT are $\lambda_1, \lambda_2(= -\lambda_1), \lambda_3 = \cdots = \lambda_7 = 0$, where λ_1 is given by

$$\lambda_1 = \left\{ \frac{m-1}{2\sin^2\chi} \left(\frac{e_1^2}{g_1^2} - \frac{2e_1 e_2}{g_1 g_2} \cos\chi + \frac{e_2^2}{g_2^2} \right) \right\}^{1/2}. \tag{7}$$

Here, $(e_1, e_2) \in \mathbb{R}^2$ with $e_1^2 + e_2^2 = 1$. Further, there are only four independent eigenvectors for the eigenvalue zero. Note that λ_1 is real for $m > 1$ and purely imaginary for $m < 1$. Hence, the 3-D WNLRT forms a weakly hyperbolic system when $m > 1$. In this article we consider only the case when $m > 1$.

3 Numerical approximation

In this section we present a numerical approximation of the conservation laws of 3-D WNLRT to study evolution of a weakly nonlinear wavefront Ω_t and formation and propagation of kink curves on it. Note that the system of conservation laws of 3-D WNLRT can be recast in the usual divergence form

$$W_t + F_1(W)_{\xi_1} + F_2(W)_{\xi_2} = 0, \tag{8}$$

where the vector of conserved variables W and the flux-vectors $F_1(W)$ and $F_2(W)$ in the ξ_1- and ξ_2-directions respectively, are given by

$$\begin{aligned} W &= \left(g_1 \mathbf{u}, g_2 \mathbf{v}, (m-1)^2 e^{2(m-1)} g_1 g_2 \sin\chi \right)^T, \\ F_1(W) &= (m\mathbf{n}, \mathbf{0}, 0)^T, \quad F_2(W) = (\mathbf{0}, m\mathbf{n}, 0)^T. \end{aligned} \tag{9}$$

In what follows we briefly summarise the central finite volume scheme for (8), first employed in [1].

1. The cell integral averages $\overline{W}_{i,j}$ of the conservative variable W are used in the discretisation of the system of conservation laws (8).
2. A second order TVD Runge-Kutta method [8] is used for time integration. The time-step is chosen to be inversely proportional to the maximum of the nonzero eigenvalue λ_1, c.f. (7), taken over the entire computational domain.
3. A nonlinear iterative solver is employed to recover the values of $\mathbf{u}, \mathbf{v}, g_1, g_2$ and m from the computed values of W.
4. A second order MUSCL reconstruction with a central weighted essentially non-oscillatory (CWENO) limiter [4] is used to reconstruct the variables at the cell interfaces.
5. The Kurganov-Tadmor high resolution flux [5] is used as the numerical flux at a cell interface, for example at a right hand vertical edge

$$\mathscr{F}_{i+\frac{1}{2},j}\left(W_{i,j}^R, W_{i+1,j}^L\right) = \frac{1}{2}\left(F_1\left(W_{i+1,j}^L\right) + F_1\left(W_{i,j}^R\right)\right) - \frac{a_{i+\frac{1}{2},j}}{2}\left(W_{i+1,j}^L - W_{i,j}^R\right), \tag{10}$$

where $W_{i,j}^{L(R)}$ denote respectively the left and right interpolated states. Here, $a_{i+1/2,j}$ is the maximal wave-speed, which can be computed with the help of the maximum of eigenvalues, c.f. [5]. The numerical flux at a horizontal edge can be computed in an analogous manner.

6. In order that the numerical solution satisfy a discrete version of the geometric solenoidal constraint (5), we use a constrained transport algorithm [7]. We employ three potentials $\mathbb{A}_1, \mathbb{A}_2, \mathbb{A}_3$, corresponding to the three components of the vectors $g_1\mathbf{u}$ and $g_2\mathbf{v}$. Note that the geometric solenoidal constraint (5) implies the conditions

$$g_1 u_k = \mathbb{A}_{k\xi_1}, \quad g_2 v_k = \mathbb{A}_{k\xi_2}, \quad k = 1, 2, 3. \tag{11}$$

The use of (11) in the 3-D KCL system (3)-(5) immediately yields the evolution equations

$$\mathbb{A}_{k_t} - m n_k = 0. \tag{12}$$

We numerically solve (12) to get the updated values of the potentials $\mathbb{A}_k$. The resulting values of $\mathbb{A}_k$ are used to suitably discretise (11) to yield the corrected values of $g_1\mathbf{u}$ and $g_2\mathbf{v}$. It is these updated values which satisfy a discrete version of (5), see [1] for more details.

At any time t, we approximate the wavefront Ω_t by a discrete set of points $\mathbf{x}_{i,j}(t) := \mathbf{x}(\xi_{1_i}, \xi_{2_j}, t)$. To get the successive positions of Ω_t, we numerically solve the system of ODEs (1) in the discretised form $d\mathbf{x}_{i,j}(t)/dt = m_{i,j}(t)\mathbf{n}_{i,j}(t)$ where $m_{i,j}(t)$ and $\mathbf{n}_{i,j}(t)$ are the corresponding values of m and $\mathbf{n}$ obtained from $\overline{W}_{i,j}(t)$.

In order to start the algorithm, the conserved variable W has to be initialised at each mesh point. Here, some care has to be taken, so that (11) is satisfied by the initial values. Let us assume that the initial wavefront Ω_0 is given a parametric form $\mathbf{x} = \mathbf{x}_0(\xi_1, \xi_2)$, with some appropriate choice of surface coordinates ξ_1 and ξ_2. The initial values for $g_1\mathbf{u}$ and $g_2\mathbf{v}$ and the potentials $\mathbb{A}_1, \mathbb{A}_2, \mathbb{A}_3$ can be chosen to be

$$g_1\mathbf{u}(\xi_1, \xi_2, 0) = \mathbf{x}_{0\xi_1}(\xi_1, \xi_2), \quad g_2\mathbf{v}(\xi_1, \xi_2, 0) = \mathbf{x}_{0\xi_2}(\xi_1, \xi_2), \tag{13}$$

$$\mathbb{A}_k(\xi_1, \xi_2, 0) = x_k(\xi_1, \xi_2), \quad k = 1, 2, 3. \tag{14}$$

Note that (5) and (11) are satisfied by the above choice of initial values. In the numerical test problem considered here, the normal velocity m on Ω_0 has been assigned a constant value $m_0 = 1.2$. For more details of the numerical scheme and its implementation, we refer the reader to [1].

4 Numerical test problem

We choose initial wavefront Ω_0 in such a way that it is not axisymmetric. The front Ω_0 has a single smooth dip. The initial shape of the wavefront is given by

$$\Omega_0: x_3 = \frac{-\kappa}{1 + \frac{x_1^2}{\alpha^2} + \frac{x_2^2}{\beta^2}}, \tag{15}$$

where the parameter values are set to be $\kappa = 1/2, \alpha = 3/2, \beta = 3$. The ray coordinates (ξ_1, ξ_2) are chosen initially as $\xi_1 = x_1$ and $\xi_2 = x_2$. The computational domain $[-20, 20] \times [-20, 20]$ is divided into 401×401 mesh points. The simulations are done up to $t = 2.0, 6.0, 10.0$. We have set non-reflecting boundary conditions for all the variables.

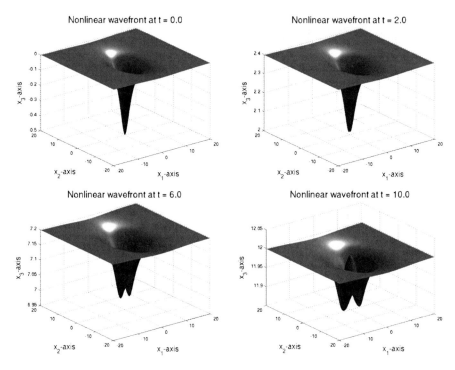

Fig. 1 The successive positions of the nonlinear wavefront Ω_t with an initial smooth dip which is not axisymmetric

In Fig. 1 we plot the initial wavefront Ω_0 and the successive positions of the wavefront Ω_t at times $t = 2.0, 6.0, 10.0$. It can be seen that the wavefront has moved up in the x_3-direction and the dip has spread over a larger area in x_1- and x_2-directions. The lower part of the front moves up leading to a change in shape of the initial front Ω_0. It is very interesting to note that two dips appear in the central part of the wavefront, which are clearly visible at $t = 6.0$ and $t = 10.0$. These two dips are separated by an elevation almost like a wall parallel to the x_2-axis.

To explain the results of convergence of the rays we give in Fig. 2 the slices of the wavefront in $x_2 = 0$ section and $x_1 = 0$ section from time $t = 0.0$ to $t = 10.0$. Due to the particular choice of the parameters α and β in the initial data (15), the section of the front Ω_0 in $x_2 = 0$ plane has a smaller radius of curvature than that of the section in $x_1 = 0$ plane. This results in a stronger convergence of the rays in $x_2 = 0$ plane compared to those in $x_1 = 0$ plane as evident from Fig. 2. In the diagram on the top in Fig. 2, we clearly note a pair of kinks at times $t = 3.0$ onwards in the $x_2 = 0$ section. However, there are no kinks in the bottom diagram in Fig. 2 in $x_1 = 0$ section.

We give now the plots of the normal velocity m in (ξ_1, ξ_2) plane along ξ_1- and ξ_2-directions in Fig. 3. It is observed that m has two shocks in the ξ_1-direction which correspond to the two kinks in the x_1-direction. We have also plotted the numerical values of the divergence of $\mathcal{B}_1$ at time $t = 10.0$ in Fig. 4. It is evident that the

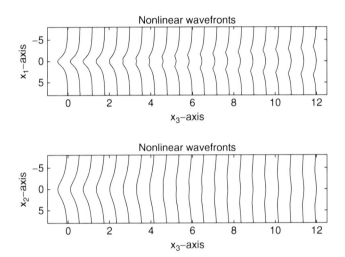

Fig. 2 The sections of the nonlinear wavefront at times $t = 0.0, \ldots, 10.0$ with a time step 0.5. On the top: in $x_2 = 0$ plane. Bottom: in $x_1 = 0$ plane

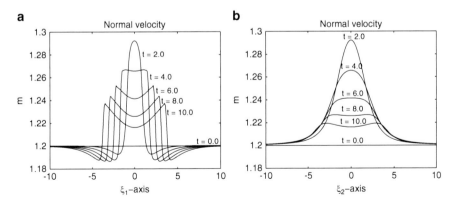

Fig. 3 The time evolution of the normal velocity m. (a): along ξ_1-direction in the section $\xi_2 = 0$. (b): along ξ_2-direction in the section $\xi_1 = 0$

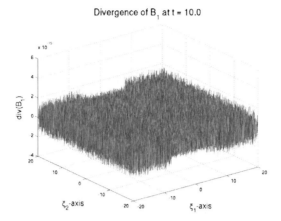

Fig. 4 The divergence of $\mathfrak{B}_1$ at $t = 10.0$. The error is of the order of 10^{-15}

geometric solenoidal condition is satisfied with an error of 10^{-15}. The divergences of $\mathfrak{B}_2$ and $\mathfrak{B}_3$ also show the same trend.

5 Concluding remarks

An efficient central finite volume scheme for the weakly hyperbolic system of conservation laws of 3-D WNLRT has been described and tested. Reconstruction is achieved component-wise and a simple central flux is employed in the numerical flux evaluation. Based on our numerical experiment and the ones reported in [1], it can be concluded that the solenoidal condition is preserved up to machine accuracy if the present finite volume scheme with a constrained transport technique is used. Moreover, none of the solution components exhibits any linearly growing Jordan mode.

Acknowledgements K. R. A. wishes to thank the Alexander von Humboldt Foundation for a postdoctoral fellowship. P. P. is supported by the Department of Atomic Energy, Government of India, under Raja-Ramanna Fellowship Scheme.

References

1. Arun, K.R.: A numerical scheme for three-dimensional front propagation and control of Jordan mode. Tech. rep., Department of Mathematics, Indian Institute of Science, Bangalore (2010)
2. Arun, K.R., Lukáčová-Medviďová, M., Prasad, P., Raghurama Rao, S.V.: An application of 3-D kinematical conservation laws: propagation of a three dimensional wavefront. SIAM. J. Appl. Math. **70**, 2604–2626 (2010)

3. Arun, K.R., Prasad, P.: 3-D kinematical conservation laws (KCL): evolution of a surface in $\mathbb{R}^3$-in particular propagation of a nonlinear wavefront. Wave Motion **46**, 293–311 (2009)
4. Jiang, G.S., Shu, C.W.: Efficient implementation of weighted ENO schemes. J. Comput. Phys. **126**, 202–228 (1996)
5. Kurganov, A., Tadmor, E.: New high-resolution central schemes for nonlinear conservation laws and convection-diffusion equations. J. Comput. Phys. **160**, 241–282 (2000)
6. Prasad, P.: Ray theories for hyperbolic waves, kinematical conservation laws (KCL) and applications. Indian J. Pure Appl. Math. **38**, 467–490 (2007)
7. Ryu, D., Miniati, F., Jones, T.W., Frank, A.: A divergence-free upwind code for multidimensional magnetohydrodynamic flow. Astrophys. J. **509**, 244–255 (1998)
8. Shu, C.W.: Total-variation-diminishing time discretizations. SIAM J. Sci. Stat. Comput. **9**, 1073–1084 (1988)

The paper is in final form and no similar paper has been or is being submitted elsewhere.

Preservation of the Discrete Geostrophic Equilibrium in Shallow Water Flows

E. Audusse, R. Klein, D.D. Nguyen, and S. Vater

Abstract We are interested in the numerical simulation of large scale phenomena in geophysical flows. In these cases, Coriolis forces play an important role and the circulations are often perturbations of the so-called geostrophic equilibrium. Hence, it is essential to design a numerical strategy that preserves a discrete version of this equilibrium. In this article we work on the shallow water equations in a finite volume framework and we propose a first step in this direction by introducing an auxiliary pressure that is in geostrophic equilibrium with the velocity field and that is computed thanks to the solution of an elliptic problem. Then the complete solution is obtained by working on the deviating part of the pressure. Some numerical examples illustrate the improvement through comparisons with classical discretizations.

Keywords Geostrophic Adjustment, Shallow Water Flows, Finite Volume Method, Well-balanced Scheme
MSC2010: 65M08, 76U05, 86A05

1 Introduction

We are interested in the numerical simulation of large scale phenomena in geophysical flows. At these scales, Coriolis forces play an important role and the atmospheric or oceanic circulations are frequently observed near geostrophic equilibrium situations, see for example [11, 12]. For this reason it is essential to design a numerical strategy that preserves a discrete version of this geostrophic

---

E. Audusse and D.D. Nguyen
LAGA, Université Paris Nord, 99 av. J.B. Clement, 93430 Villetaneuse, France,
e-mail: audusse@math.univ-paris13.fr, name2@email.adress

R. Klein and S. Vater
Institut für Mathematik, Freie Universität Berlin, Arnimallee 6, D-14195 Berlin, Germany,
e-mail: rupert.klein@math.fu-berlin.de, stefan.vater@math.fu-berlin.de

equilibrium: if numerical spurious waves are created, they quickly become higher than the physical ones we want to capture. This phenomenon is well known but its solution in the context of finite volume methods is still an open problem. We address this question in this article.

One of the most popular systems that is used to model such quasi-geostrophic flows are the shallow water equations with β-plane approximation

$$h_t + \nabla \cdot (h\bar{u}) = 0, \tag{1}$$

$$(h\bar{u})_t + \nabla \cdot (h\bar{u} \otimes \bar{u}) + \nabla(\frac{gh^2}{2}) = -f\mathbf{e}_z \times (h\bar{u}), \tag{2}$$

The shallow water system is the simplest form of equations of motion that can be used to model Rossby and Kelvin waves in the atmosphere or ocean, and the use of the β-plane approximation allows the model to take into account a non-constant Coriolis parameter f that varies linearly with the latitude without considering a spherical domain. We choose to work in a finite volume framework to discretize the equations because of its ability to deal with complex geometries and its inherent conservation property, see [4, 9]. In this context, the discrete preservation of the geostrophic equilibrium, which is mainly the balance between pressure gradient and Coriolis forces in (2), is a hard touch: the main reason is that the fluxes are upwinded for stability reasons while the source terms are usually discretized in a centered way.

The question of the preservation of non-trivial equilibria in geophysical fluid models has received great attention in the area of numerical modeling in the last decade. Many studies were devoted to the preservation of the so-called hydrostatic and also lake-at-rest equilibria, see [2–4] and references therein. More recently some authors investigated the problem of the geostrophic equilibrium [5, 6, 8, 10]. However, this question is more delicate for two reasons: it is an essentially 2d problem, and it involves a non-zero velocity field. It follows that its solution is still incomplete. In this work we propose a solution to this problem by introducing an auxiliary water depth which is in geostrophic balance with the velocity field and then by working on the deviation between the actual and auxiliary water depths instead of considering the water depth itself. The auxiliary water depth is computed through the solution of a Poisson problem on a dual grid [13].

2 Position of the problem

In this short note we present the method by considering a constant Coriolis parameter. In order to exhibit the importance of the geostrophic equilibrium, we introduce the non-dimensional version of the shallow water equations (1)–(2) written in non-conservative form

$$h_t + \nabla \cdot (h\bar{u}) = 0,$$

$$\bar{u}_t + \bar{u} \cdot \nabla \bar{u} + \frac{1}{\mathbf{Fr}^2}\nabla h + \frac{1}{\mathbf{Ro}}2\mathbf{e}_z \times \bar{u} = 0.$$

Here, h and $\bar{u}$ are the unknown dimensionless depth and velocity fields and

$$\mathbf{Fr} = \frac{\bar{U}}{\sqrt{gH}}, \qquad \mathbf{Ro} = \frac{\bar{U}}{\Omega L}$$

are the Froude and Rossby numbers, respectively, with $\bar{U}$, L and H some characteristic velocity, length and depth for the flow, g the gravity coefficient and Ω the angular velocity of the earth. For large scale phenomena typical values for these numbers are

$$\mathbf{Fr} \approx \mathbf{Ro} \approx \epsilon = 10^{-2},$$

We then expand the unknowns in term of ε

$$h = h_0 + \varepsilon h_1 + \varepsilon^2 h_2 + \ldots, \qquad \bar{u} = \bar{u}_0 + \varepsilon \bar{u}_1 + \varepsilon^2 \bar{u}_2 + \ldots$$

and we keep the leading order terms to exhibit the following stationary state

$$O\left(\epsilon^{-2}\right): \nabla h_0 = 0 \tag{3}$$

$$O\left(\epsilon^{-1}\right): \nabla h_1 + 2e_z \times \bar{u}_0 = 0 \tag{4}$$

$$O\left(\epsilon^{0}\right): \nabla \cdot \bar{u}_0 = 0, \tag{5}$$

This set of equations is called the geostrophic equilibrium. It follows from equation (3) that the water depth is constant at the leading order and from equation (5) that the main part of the velocity field is divergence free. Equation (4) is nothing but the fact that the pressure gradient and the Coriolis term are in balance for leading varying terms h_1 and $\bar{u}_0$. Let's now turn to the numerical point of view. Preservation of the discrete equilibrium (3) is obvious. The divergence free condition (5) is much more delicate to deal with but it has been widely investigated for Stokes or Navier-Stokes equations, mostly in the framework of finite element methods. It is also the subject of a recent work [13], where the authors study the zero Froude number limit of the shallow water equations. In this note we focus on a proper way to preserve the balance in equation (4).

3 The well-balanced finite volume scheme

We choose to discretize the shallow water equations (1)–(2) in a finite volume framework [4, 9]. The reason to consider this particular method is related to its inherent conservation properties that are interesting for geophysical applications and in particular for long time simulations [1]. A second reason is that the finite volume method is also able to deal with sharp fronts that can occur in geophysical applications. We first recall the formulation of the finite volume method and the classical centered discretization of the Coriolis source term. Then, we derive the new well-balanced scheme by introducing an auxiliary pressure that is computed through the solution of a Laplace equation on a dual grid.

System (1)–(2) is a particular case of a 2d conservation law with source term:

$$U_t + (F(U))_x + (G(U))_y = S(U), \qquad (6)$$

in which $U = (h, hu, hv)^T$ and

$$F(U) = \begin{pmatrix} hu \\ hu^2 + \frac{1}{2}gh^2 \\ huv \end{pmatrix}, \ G(U) = \begin{pmatrix} hv \\ huv \\ hv^2 + \frac{1}{2}gh^2 \end{pmatrix}, \ S(U) = \begin{pmatrix} 0 \\ 2\Omega hv \\ -2\Omega hu \end{pmatrix}.$$

In this note we only consider Cartesian grids. Then, the finite volume discretization of equation (6) leads to the computation of approximated solutions $U_{i,j}^n$ through the discrete formula

$$U_{i,j}^{n+1} = U_{i,j}^n - \frac{\delta t}{\delta x}\left(\mathbf{F}_{i+\frac{1}{2},j}^n - \mathbf{F}_{i-\frac{1}{2},j}^n\right) - \frac{\delta t}{\delta y}\left(\mathbf{G}_{i,j+\frac{1}{2}}^n - \mathbf{G}_{i,j-\frac{1}{2}}^n\right) + \delta t\, S_{i,j}^n,$$

where $\mathbf{F}_{i+\frac{1}{2},j}^n$ is a discrete approximation of the flux $F(U)$ along the interface between cells $C_{i,j}$ and $C_{i+1,j}$ that is constructed through a three points formula

$$\mathbf{F}_{i+\frac{1}{2},j}^n = \mathscr{F}\left((h_{i,j}^n, u_{i,j}^n, v_{i,j}^n), (h_{i+1,j}^n, u_{i+1,j}^n, v_{i+1,j}^n)\right). \qquad (7)$$

Here we use the HLL solver [7] to compute these approximations.

The classical discretization of the source term $S_{i,j}^n$ is computed through the centered formula

$$S_{i,j}^n = \begin{pmatrix} 0 \\ -2\Omega_z \times (h_{i,j}^n \overline{u}_{i,j}^n) \end{pmatrix}$$

where $h_{i,j}^n$ denotes the approximated value at time t^n on cell $C_{i,j}$. We will exhibit in the last section that this approach suffers from important drawbacks when we consider applications for small Froude and Rossby numbers.

The main idea of our method to overcome this problem is to introduce an auxiliary water depth h_c that is in balance with Coriolis forces related to the actual velocity field. This idea is an extension of the notion of hydrostatic reconstruction that was introduced in [3] for the Euler equations and in [2] for shallow water flows. Here, h_c will satisfy the equation

$$g\nabla h_c = -2\Omega \times \overline{u}. \qquad (8)$$

In our approach, h_c is discretized as a grid function, which is piecewise bilinear on each grid cell and continuous at the interfaces. The second ingredient of the well-balanced scheme is the representation of the Coriolis forces by the gradient of this quantity. Furthermore, the fluxes in the conservative part of the scheme are modified in the following way: For each cell, we introduce a deviation in the water depth by

$$\Delta h_{i,j}^n = h_{i,j}^n - h_c^n(x_i, y_j)$$

Then, the interface water depths are computed by

$$\widehat{h}_{i+\frac{1}{2},j}^{n,k,x} = \frac{1}{2}\left[h_c^n\left(x_{i+\frac{1}{2}, j+\frac{1}{2}}\right) + h_c^n\left(x_{i+\frac{1}{2}, j-\frac{1}{2}}\right)\right] + \Delta h_{k,j}^n, \quad \text{for} \quad k = i, i+1.$$

and the original three points formula (7) for the flux is replaced by

$$\mathbf{F}_{i+\frac{1}{2},j}^n = \mathscr{F}\left((\widehat{h}_{i+\frac{1}{2},j}^{n,i,x}, u_{i,j}^n, v_{i,j}^n), (\widehat{h}_{i+\frac{1}{2},j}^{n,i+1,x}, u_{i+1,j}^n, v_{i+1,j}^n)\right),$$

If the flow satisfies the geostrophic equilibrium, $\widehat{h}$ and h_c are equal. The consistency of the flux will then provide some numerical balance between the conservative part and the source term that will directly impact the results. Note also that the time step is now related to the interface water depths. Nevertheless, in the numerical applications, the numerical values remain very close for both methods.

It remains to explain the computation of the auxiliary water depth h_c that is the solution of a discrete equivalent of equation (8). We first take the divergence of this equation and then search for the solution of a Poisson equation

$$-\Delta\phi = \nabla \cdot \left(\overline{k} \times \overline{u}\right).$$

Integration of this equation on the dual cell $C_{i+\frac{1}{2}, j+\frac{1}{2}}$ and application of the Gauss theorem leads to

$$\int_{\partial C_{i+\frac{1}{2},j+\frac{1}{2}}} \nabla\phi \cdot \overline{n}\, d\sigma = -\int_{\partial C_{i+\frac{1}{2},j+\frac{1}{2}}} \overline{k}_z \overline{u} \cdot \overline{t}\, d\sigma,$$

where $\overline{n}$ (resp. $\overline{t}$) is a normal (resp. tangential) vector to the interface of the dual cell. We solve this equation by using the technique presented in [13] for the solution of a similar problem. We refer the reader to this article for the details of the method that is in particular proved to provide an inf-sup-stable projection. We finally obtain a linear system with a nine point stencil. The boundary conditions for this auxiliary problem are prescribed by using the fact that the computed pressure (or height) field is equivalent to a stream function for the associated balanced geostrophic flow. For example a rigid wall type boundary condition for the flow translates into a Dirichlet type boundary condition for the stream function. Similar types of equivalences can be used to prescribe other types of boundary conidtions.

4 Numerical results

In order to test the new scheme, we consider a stationary vortex in the square domain $[0, 1] \times [0, 1]$. We consider periodic boundary conditions, and as initial conditions we choose a velocity field of the form

$$\overline{u}_0(r, \theta) = v_\theta(r)\overline{e}_\theta, \quad v_\theta(r) = \varepsilon \left[5r \; \chi\left(r < \frac{1}{5}\right) + (2 - 5r) \chi\left(\frac{1}{5} \leq r < \frac{2}{5}\right) \right],$$

where r is the distance to the center of the domain and χ denotes the characteristic function of a given interval. Some computations show that the vortex is a stationary solution of the shallow water equations (1)-(2), if the initial water depth $h_0(r)$ is a radial solution of the ODE

$$h'_0(r) = \frac{1}{g} \left(2\Omega v_\theta + \frac{v_\theta^2}{r} \right).$$

Note that if we choose a water depth and an angular velocity of order $O(1)$, the Froude and Rossby numbers are of order $O(\epsilon)$. It follows that our interest is for small values of the parameter ϵ.

We first work on a regular grid with 30×30 cells and we consider four Froude resp. Rossby numbers: $0.05, 0.1, 0.5$ and 1. The numerical solution is computed by using both schemes described in the previous section. In order to compare the accuracy of the schemes, we compute the relative L^2 error in the water depth. In Fig. 1 we present the time evolution of this error for the four values of ϵ. It appears that for both schemes the error is increasing with time before reaching a stationary value. More interesting is that for the classical (resp. well-balanced) scheme the error is increasing (resp. decreasing), when the Froude number is decreasing. While the error is of the same order for both schemes when $\epsilon = 1$, for other values of ϵ the well-balanced scheme is always more precise than the classical one.

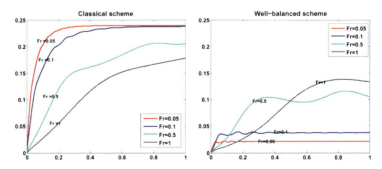

Fig. 1 Error in time for both classical and well-balanced schemes (30×30 grid cells) and for four different Froude numbers

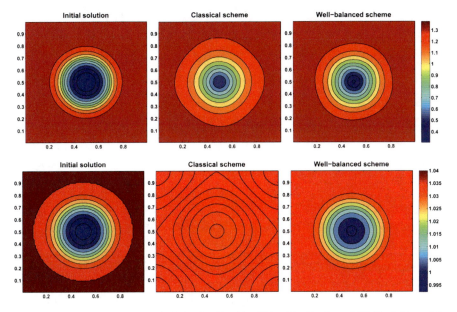

Fig. 2 Contour of the computed fluid depth with 100×100 grid cells. **Fr** $= 0.95$ (Top), **Fr** $= 0.1$ (bottom)

We then consider a finer grid with 100×100 cells and we present the water depth for both schemes and for two values of ϵ: 0.95 (large) and 0.1 (small). In Fig. 2 we present the 2d contour of the water depth. The results look similar and quite close to the initial solution when ϵ is large (top row). But when ϵ is small (bottom row), the classical scheme totally fails to compute the right solution, whereas the water depth computed by the well-balanced scheme stays close to the initial one. In Fig. 3 we give more quantitative results by presenting a cut of the solution along x–axis at $y = 0.5$. These pictures clearly exhibit that the results are very close when ϵ is large, but very different when ϵ is small. In this last case the classical scheme is not able to maintain the vortex, whereas the well-balanced scheme preserves the shape of the free surface. Note that the small diffusion that is observed even for the well-balanced scheme is due to the fact that we consider only first order schemes in this work. We end this short note by some words on the CPU time. We first notice that for the last numerical test case, the time steps are very close for both methods, as it is reported in the table below. We then consider the CPU time for both methods and conclude that it is four times larger for the well-balanced scheme. It is obviously due to the solution of the linear system related to the elliptic problem at each time step. This observation leads to two comments. First, and since the solution of the linear system is only required for the computation of the auxiliary water depth, it is possible to obtain a compromise between accuracy and efficiency of the whole process by considering iterative methods with a small number of iterations. Second we recall that our final objective is to couple the presented process with a numerical

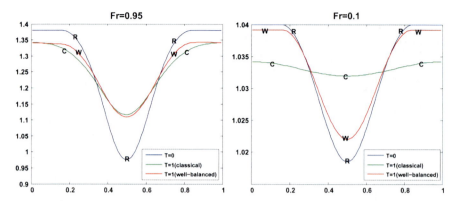

Fig. 3 Fluid depth profiles – cut along the x–axis at $y = 0.5$ with 100×100 grid cells. R line: Initial solution, W line: Well-balanced scheme, C line: Classical scheme

scheme adapted for small Froude number flows and then to generalize the method presented in [13] to rotating flows. Since the technique introduced in [13] already requires for the solution of a related linear system, the additional computational cost of the well-balanced process presented here is very small.

	Classical scheme	Well-balanced scheme
Time Step	9.7702e-005	9.7464e-005
CPU Time	1547 s	5564 s

References

1. E. Audusse, R.Klein, A. Owinoh, Conservative discretization of Coriolis force in a finite volume framework, Journal of Computational Physics, **228**, 2934-2950 (2009).
2. E. Audusse, F. Bouchut, M.O. Bristeau, R. Klein, B. Perthame, A fast and stable well-balanced scheme with hydrostatic reconstruction for shallow water flows, SIAM J. Sci. Comput., **25**, 2050–2065 (2004).
3. N. Botta, R. Klein, S. Langenberg, S. Lützenkirchen, Well-balanced finite volume methods for nearly hydrostatic flows, JCP, **196**, 539–565 (2004).
4. F. Bouchut , Nonlinear stability of finite volume methods for hyperbolic conservation laws and well-balanced schemes for sources, Birkaüser (2004).
5. F. Bouchut, J. Le Sommer, V. Zeitlin, Frontal geostrophic adjustment and nonlinear wave phenomena in one dimensional rotating shallow water. Part 2: high-resolution numerical simulations, J. Fluid Mech., **513**, 35–63 (2004).
6. M.J. Castro, J.A. Lopez, C. Pares, Finite Volume Simulation of the Geostrophic Adjustment in a Rotating Shallow-Water System SIAM J. on Scientific Computing, **31**, 444–477 (2008).
7. A. Harten, P. Lax, B. Van Leer, On upstream differencing and Godunov type schemes for hyperbolic conservation laws, SIAM Review, **25**, 235–261 (1983).

8. A. Kuo, L. Polvani, Time-dependent fully nonlinear geostrophic adjustment, J. Phys. Oceanogr. **27**, 1614–1634 (1997).
9. R.J. LeVeque, Finite Volume Methods for Hyperbolic Problems, Camb. Univ. Press (2002).
10. N. Panktratz, J.R. Natvig, B. Gjevik, S. Noelle, High-order well-balanced finite-volume schemes for barotropic flows. Development and numerical comparisons, Ocean Modelling, **18**, 53–79 (2007).
11. J. Pedlosky, Geophysical Fluid Dynamics, Springer, 2nd edition (1990).
12. G. K. Vallis, Atmospheric and Oceanic Fluid Dynamics: Fundamentals and Large-scale Circulation, Cambridge University Press (2006).
13. S. Vater, R. Klein, Stability of a Cartesian Grid Projection Method for Zero Froude Number Shallow Water Flows, Numerische Mathematik, **113**, 123–161 (2009).

The paper is in final form and no similar paper has been or is being submitted elsewhere.

Arbitrary order nodal mimetic discretizations of elliptic problems on polygonal meshes

Lourenço Beirão da Veiga, Konstantin Lipnikov, and Gianmarco Manzini

Abstract We develop and analyze a new family of mimetic methods on unstructured polygonal meshes for the diffusion problem in primal form. The new nodal MFD formulation that we propose in this work extends the original low-order formulation of [3] to arbitrary orders of accuracy by requiring that the consistency condition holds for polynomials of arbitrary degree $m \geq 1$. An error estimate is presented in a mesh-dependent norm that mimics the energy norm and numerical experiments confirm the convergence rate that is expected from the theory.

Keywords mimetic finite difference, diffusion problems, unstructured mesh, polygonal element
MSC2010: 65N06

1 The nodal mimetic finite difference method

We consider a nodal mimetic discretization of the steady diffusion problem for the scalar solution field u given by

$$\mathrm{div}(\mathbf{K}\nabla u) = f \quad \text{in } \Omega, \tag{1}$$

$$u = g \quad \text{on } \Gamma, \tag{2}$$

L. Beirão da Veiga
Dipartimento di Matematica, Università di Milano, Italy, e-mail: lourenco.beirao@unimi.it

K. Lipnikov
Los Alamos National Laboratory, New Mexico (US), e-mail: lipnikov@lanl.gov

G. Manzini
IMATI-CNR and CeSNA-IUSS, Pavia, Italy, e-mail: Marco.Manzini@imati.cnr.it

where the computational domain Ω is a bounded, open, polygonal subset of $\mathbb{R}^2$ with Lipshitz boundary Γ, the diffusion tensor $\mathbf{K}$ is a 2×2 bounded, measurable, strongly elliptic and symmetric tensor describing the material properties, $f \in L^2(\Omega)$ is the forcing term and $g \in H^{1/2}(\Gamma)$ is the boundary function that defines the non-homogeneous Dirichlet boundary condition.

Let us consider the set of functions $H_g^1(\Omega) = \{v \in H^1(\Omega),\ v_{|\Gamma} = g\}$. Problem (1)-(2) can be restated in the variational form:

find $u \in H_g^1(\Omega)$ such that

$$\int_\Omega \mathbf{K}\nabla u \cdot \nabla v\, dV = \int_\Omega f v\, dV \qquad \forall v \in H_0^1(\Omega). \tag{3}$$

Under the previous assumptions problem (3) is well-posed [5]. The existence and uniqueness of the weak solution follows from continuity and coercivity of the bilinear form in (3).

The mimetic discretizations that are available in the literature for this problem have usually low-order accuracy. The nodal MFD method in [3] uses degrees of freedom associated with the mesh vertices and is derived from a consistency condition, i.e., a discrete integration by parts formula, that is exact for linear polynomials. This method was proven to be first-order accurate in a mesh-dependent H^1-seminorm.

To some extend, the mixed and mixed-hybrid MFD scheme in [4] can be interpreted as a generalization of the $RT_0 - \mathbb{P}_0$ mixed finite element (FE) method to polygonal and polyhedral meshes Similarly, the nodal MFD method in [3] can be viewed as a generalization of the linear Galerkin FE method. The essential difference between FE and MFD methods is that the latter does not use shape functions and operates directly with degrees of freedom. This result in a number of useful consequences such as a family of equivalent MFD methods.

The numerical approximation to (3) is performed on a sequence of polygonal partitions $\{\Omega_h\}_h$ of the domain Ω, which are required to satisfy a suitable set of regularity conditions [4]. For any mesh Ω_h, the subscripted label h is the mesh size and is defined by $h = \max_{P\in\Omega_h} h_P$ where $h_P = \sup_{\mathbf{x},\mathbf{y}\in P} |\mathbf{x}-\mathbf{y}|$ is the diameter of the polygonal cell $P \in \Omega_h$. On a mesh Ω_h, we approximate the scalar fields from $H^1(\Omega)$ through a set of suitable *degrees of freedom* $u_h, v_h \in \mathcal{V}_h$, where $\mathcal{V}_h$ denotes the linear space of the discrete scalar fields. Then, we introduce the bilinear form $\mathcal{A}_h(\cdot,\cdot) : \mathcal{V}_h \times \mathcal{V}_h \to \mathbb{R}$, which approximates the left-hand side of (3), and the bilinear form $(\cdot,\cdot)_h : L^2(\Omega) \times \mathcal{V}_h \to \mathbb{R}$, which approximates the right-hand side of (3). In the nodal mimetic finite difference formulation, the Dirichlet boundary conditions are *essential* and are incorporated through the subspace $\mathcal{V}_{h,g}$ of $\mathcal{V}_h$. The set of discrete scalar fields $\mathcal{V}_{h,g}$ is formed by the elements of $\mathcal{V}_h$ whose degrees of freedom associated with the boundary edges approximate the boundary datum g. We also consider the linear space $\mathcal{V}_{h,g}$, which is obtained by setting $g=0$. Finally, the mimetic finite difference method for (3) reads:

Arbitrary order nodal mimetic discretizations of elliptic problems on polygonal meshes

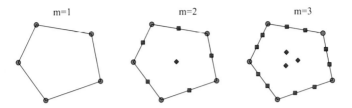

Fig. 1 Degrees of freedom for $m = 1, 2, 3$

Find $u_h \in \mathcal{V}_{h,g}$ such that:

$$\mathcal{A}_h(u_h, v_h) = (f, v_h)_h \qquad \forall v_h \in \mathcal{V}_{h,0}. \tag{4}$$

The well-posedness of the numerical approximation (4) follows from the coercivity and the continuity properties of the bilinear form $\mathcal{A}_h(\cdot, \cdot)$.

Degrees of freedom, norms, and the interpolation operator. Let $\mathcal{V}$ and $\mathcal{F}$ be sets of mesh vertices and edges, respectively. Let m be a positive integer number. A discrete scalar field v_h in $\mathcal{V}_h$ consists of:

(i) one real number v_v per mesh vertex $\mathsf{v} \in \mathcal{V}$;
(ii) $(m-1)$ real numbers $v_{\mathsf{f},i}$ per mesh edge $\mathsf{f} \in \mathcal{F}$, where $i = 1, \ldots, m-1$;
(iii) $m(m-1)/2$ real numbers $v_{\mathsf{P},k,i}$ per mesh cell $\mathsf{P} \in \Omega_h$, where $k = 0, \ldots, m-2$, and $i = 0, \ldots, k$.

The first two sets of degrees of freedom represent *nodal degrees of freedom*. They are associated with the vertices of edge f and with the $m - 1$ internal nodes of the Gauss-Lobatto numerical integration rule of order $2m - 1$. These nodes are defined uniquely and symmetrically on each edge $\mathsf{f} \in \mathcal{F}$ (see formula 25.4.32 and Table 25.6 of [1] for details). The last set is introduced into the nodal MFD scheme to represent k-th order moments of scalar fields over polygons P. We refer to the latter degrees of freedom as the *internal degrees of freedom*. Examples of the degrees of freedom on a generic cell are shown in Fig. 1 for $m = 1, 2, 3$.

Combining the previous definitions allows us to write

$$v_h = \left\{ (v_\mathsf{v})_{\mathsf{v} \in \mathcal{V}}, (v_{\mathsf{f},i})_{\mathsf{f} \in \mathcal{F}, i=1,\ldots,m-1}, (v_{\mathsf{P},k,i})_{\mathsf{P} \in \Omega_h, k=0,\ldots,m-2, i=0,\ldots,k} \right\} \tag{5}$$

for any $v_h \in \mathcal{V}_h$. Therefore, the global approximation space $\mathcal{V}_h$ has dimension $N^\mathcal{V} + N^\mathcal{F}(m-1) + N^\mathcal{P} m(m-1)/2$, where $N^\mathcal{V}$ is the number of mesh vertices, $N^\mathcal{F}$ the number of mesh edges, and $N^\mathcal{P}$ the number of mesh cells.

The sub-set $\mathcal{V}_{h,g}$ is obtained by approximating the datum g by a (globally continuous) piecewise polynomial function g_m of order m, and then enforcing $v_{h,\mathsf{f}} = g_m|_\mathsf{f}$ for every $\mathsf{f} \in \Gamma$. Note that if g is continuous, this can be simply achieved by interpolating g at the Gauss nodes of each edge.

We will find it useful to introduce $\mathcal{V}_{h,P} := \mathcal{V}_h|_P$, the linear space of discrete scalar fields whose members contain only the degrees of freedom associated with the vertices, the edge and the interior of cell P. The local linear space $\mathcal{V}_{h,P}$ has dimension $m_{\mathcal{V}_{h,P}} = N_P^{\mathcal{F}} m + m(m-1)/2$, where $N_P^{\mathcal{F}}$ is the number of the edges forming the boundary ∂P and the second term is the number of internal degrees of freedom.

Let $\bar{v}_{h,P} = \sum_{v \in \partial P} v_v / N_P^{\mathcal{V}}$ be the arithmetic mean of the vertex values of $v_{h,P}$. The mesh-dependent norm $\|v_h\|_{1,h}^2 = \sum_{P \in \Omega_h} \|v_h\|_{1,h,P}^2$ for the elements of $\mathcal{V}_h$ mimics the $H^1(\Omega)$ seminorm when the summation arguments are given by

$$\|v_h\|_{1,h,P}^2 = \sum_{f \in \partial P} h_P \left\| \frac{\partial v_{h,f}}{\partial s} \right\|_{L^2(f)}^2 + \left(v_{P,0,0} - \bar{v}_{h,P}\right)^2 + \sum_{k=1}^{m-2}\sum_{i=0}^{k} |v_{P,k,i}|^2, \quad (6)$$

where $v_{h,f}$ is the polynomial on edge f corresponding to the values $v_{f,i}$, $i = 0, \ldots, m$ and $\partial v_{h,f}/\partial s$ is the directional derivative along f.

For every polygon P of Ω_h and every function $v \in H^1(P) \cap C^0(\overline{P})$, we define the *interpolation operator* v_P^I to the discrete local space $\mathcal{V}_{h,P}$. For the nodal degrees of freedom, we set

$$v_v^I = v(\mathbf{x}_v) \quad \forall v \in \partial P; \quad (7)$$

$$v_{f,i}^I = v(\mathbf{x}_{f,i}) \quad \forall f \in \partial P,\ i = 1, 2, \ldots, m-1. \quad (8)$$

Now, for every cell $P \in \Omega_h$, we consider the set of $m(m-1)/2$ polynomial functions $\varphi_{k,i} : P \to \mathbb{R}$ for $k = 0, \ldots, m-2$ and $i = 0, \ldots, k$ such that $\varphi_{0,0} = 1$ and for every k the set $\{\varphi_{k,i}\}_{i=0,\ldots,k}$ forms an L^2-orthogonal basis for the linear space of the polynomials of degree exactly equal to k on P and orthogonal to the polynomials of degree up to $(k-1)$. Then, the interior degrees of freedom of the interpolated field v^I are given by

$$v_{P,k,i}^I = \frac{1}{|P|} \int_P v \varphi_{k,i}\, dV, \quad k = 0, \ldots, m-2,\ i = 0, 1, \ldots, k. \quad (9)$$

Construction of the mimetic bilinear form $\mathcal{A}_h(\cdot, \cdot)$. The bilinear form $\mathcal{A}_h(\cdot, \cdot)$ is obtained by assemblying the contributions from each polygonal cell

$$\mathcal{A}_h(u_h, v_h) = \sum_{P \in \Omega_h} \mathcal{A}_{h,P}(u_{h,P}, v_{h,P}) \quad \forall u_h, v_h \in \mathcal{V}_h,$$

where $\mathcal{A}_{h,P}(\cdot, \cdot) : \mathcal{V}_{h,P} \times \mathcal{V}_{h,P} \to \mathbb{R}$ is a *local* symmetric bilinear form defined on P. To define it, we proceed throughout the following three steps. In the first step, we introduce the linear vector functional $\mathcal{G}^k : (L^2(P))^2 \to (\mathbb{P}_k(P))^2$, which is the L^2-orthogonal projection on the linear space of two-dimensional vectors of polynomials of degree k on P. For $p \in H^1(\mathsf{P})$ and $\mathbf{K} \in L^\infty(\mathsf{P})$ it holds that $\mathrm{div}(\mathscr{G}^{m-1}(\mathbf{K}\nabla p))$ belongs to $\mathbb{P}_{m-2}(\mathsf{P})$. Therefore, we can express this divergence as the unique linear combination of the polynomial basis functions $\varphi_{k,i}$ of $\mathbb{P}_{m-2}(\mathsf{P})$:

$$\mathrm{div}(\mathscr{G}^{m-1}(\mathbf{K}\nabla p)) = \sum_{k=0}^{m-2}\sum_{i=0}^{k} \alpha_{k,i}\,\varphi_{k,i}, \tag{10}$$

where the coefficients $\alpha_{k,i}$ depend on p. In the second step, we define the "*numerical integration*" formula $\mathscr{I}_\mathsf{P}(v_{h,\mathsf{P}},p)$ as:

$$\mathscr{I}_\mathsf{P}(v_{h,\mathsf{P}},p) = \sum_{k=0}^{m-2}\sum_{i=0}^{k} |\mathsf{P}|\,\alpha_{k,i}\,v_{\mathsf{P},k,i}, \tag{11}$$

where the real numbers $\alpha_{k,i}$ are the coefficients for the polynomial p in summation (10). In the third step, we assume that the symmetric bilinear form $\mathscr{A}_{h,\mathsf{P}}(\cdot,\cdot)$ satisfies the following two conditions:

(S1) *spectral stability*: there exists two positive constants σ_* and σ^* such that for every $v_{h,\mathsf{P}} \in \mathscr{V}_{h,\mathsf{P}}$ there holds:

$$\sigma_*\|v_{h,\mathsf{P}}\|_{1,h,\mathsf{P}}^2 \leq \mathscr{A}_{h,\mathsf{P}}(v_{h,\mathsf{P}},v_{h,\mathsf{P}}) \leq \sigma^*\|v_{h,\mathsf{P}}\|_{1,h,\mathsf{P}}^2;$$

(S2) *local consistency*: for every $v_{h,\mathsf{P}} \in \mathscr{V}_{h,\mathsf{P}}$ and for every $p \in \mathbb{P}_m(\mathsf{P})$ there holds:

$$\mathscr{A}_{h,\mathsf{P}}(v_{h,\mathsf{P}},p_\mathsf{P}^I) = -\mathscr{I}_\mathsf{P}(v_{h,\mathsf{P}},p) + \sum_{\mathsf{f} \in \partial \mathsf{P}} \int_\mathsf{f} v_{h,\mathsf{f}}(s)\,\mathscr{G}^{m-1}(\mathbf{K}\nabla p)\cdot\mathbf{n}_{\mathsf{P},\mathsf{f}}\,ds. \tag{12}$$

As usual, in the mimetic methods, the bilinear form $\mathscr{A}_{h,\mathsf{P}}$ is not unique. Its representing matrix has two terms, a unique consistency term due to (S2) and a non-unique stabilizing term due to (S1). The possibility to vary the stabilizing term is the unique characteristic of the mimetic approach; see also [2].

Discretization of the load term $(\cdot,\cdot)_h$. Let $\mathscr{P}_\mathsf{P}^k : L^2(\mathsf{P}) \to \mathbb{P}_k(\mathsf{P})$ be the L^2-orthogonal projector of scalar functions onto the space of polynomials of degree at most k. We introduce $\hat{f}_\mathsf{P} = \mathscr{P}_\mathsf{P}^{m-2}(f)$, where f is the forcing term in (3). Since $\hat{f}_\mathsf{P} \in \mathbb{P}_{m-2}(\mathsf{P})$, we can write it as a linear combination of the basis functions $\varphi_{k,i}$:

$$\hat{f}_\mathsf{P} = \sum_{k=0}^{m-2}\sum_{i=0}^{k} c_{k,i}\,\varphi_{k,i} \tag{13}$$

using the $(m+1)(m+2)/2$ real coefficients $c_{k,i}$. Then, we define

Table 1 Number of degrees of freedom

	Mesh Family $\mathscr{M}_1$				Mesh Family $\mathscr{M}_2$			
lev	$m=2$	$m=3$	$m=4$	$m=5$	$m=2$	$m=3$	$m=4$	$m=5$
0	441	861	1381	2001	881	1521	2261	3101
1	1681	3321	5361	7801	3361	5841	8721	12001
2	6561	13041	21121	30801	13121	22881	34241	47201
3	25921	51681	83841	122401	51841	90561	135681	187201
4	103041	205761	334081	488001	206081	360321	540161	745601

Table 2 Test Case 1: relative errors and convergence rates on mesh family $\mathscr{M}_1$ (randomized quadrilaterals) for different polynomial order $m = 2, 3, 4, 5$ and non-constant diffusion tensor K

	$m=2$		$m=3$		$m=4$		$m=5$	
lev	Error	Rate	Error	Rate	Error	Rate	Error	Rate
0	$1.529\ 10^{-1}$	--	$9.510\ 10^{-2}$	--	$8.590\ 10^{-3}$	--	$8.237\ 10^{-3}$	--
1	$2.577\ 10^{-2}$	2.60	$1.043\ 10^{-2}$	3.23	$1.496\ 10^{-3}$	2.55	$5.240\ 10^{-4}$	4.03
2	$5.218\ 10^{-3}$	2.29	$1.590\ 10^{-3}$	2.70	$9.476\ 10^{-5}$	3.96	$2.507\ 10^{-5}$	4.37
3	$1.192\ 10^{-3}$	2.19	$1.885\ 10^{-4}$	3.16	$6.620\ 10^{-6}$	3.94	$8.663\ 10^{-7}$	4.98
4	$2.991\ 10^{-4}$	2.06	$2.413\ 10^{-5}$	3.06	$4.199\ 10^{-7}$	4.11	$2.850\ 10^{-8}$	5.09

$$\left(f, v_h\right)_h = \sum_{\mathsf{P}\in\Omega_h} \left(\hat{f}_\mathsf{P}, v_{h,\mathsf{P}}\right)_{h,\mathsf{P}}, \qquad \left(\hat{f}_\mathsf{P}, v_{h,\mathsf{P}}\right)_{h,\mathsf{P}} = |\mathsf{P}| \sum_{k=0}^{m-2} \sum_{i=0}^{k} c_{k,i}\, v_{\mathsf{P},k,i},$$

where we use the coefficients $c_{k,i}$ from (13).

2 Convergence theorem

For simplicity, we consider the homogeneous boundary value problem (3), i.e. $g = 0$.

Theorem 1. *Let $u \in H^{m+1}(\Omega)$ be the solution of the variational problem (3) under assumptions (H1)-(H3). Let $u^I \in \mathscr{V}_h$ be its interpolant defined by (7)-(9). Let u_h be solution of the MFD problem (4) under assumption (HG) and (S1)-(S2). Let us assume that $\mathsf{K}|_\mathsf{P} \in W^{m,\infty}(\mathsf{P})$ for any polygon P. Finally, let mesh Ω_h be shape regular. Then, there exists a positive constant C, which depends only on the shape regularity constants and is independent of h, such that*

$$\|u^I - u_h\|_{1,h} \leq C h^m \|u\|_{H^{m+1}(\Omega)}. \tag{14}$$

Table 3 Test Case 2: relative errors and convergence rates on mesh family $\mathcal{M}_2$ (non-convex cells) for different polynomial order $m = 2, 3, 4, 5$ and non-constant diffusion tensor K

	$m = 2$		$m = 3$		$m = 4$		$m = 5$	
lev	Error	Rate	Error	Rate	Error	Rate	Error	Rate
0	3.007	--	$9.873\ 10^{-1}$	--	$2.059\ 10^{-1}$	--	$1.988\ 10^{-2}$	--
1	$8.081\ 10^{-1}$	1.89	$2.760\ 10^{-1}$	1.84	$1.367\ 10^{-2}$	3.92	$1.016\ 10^{-3}$	4.29
2	$2.071\ 10^{-1}$	1.96	$5.621\ 10^{-2}$	2.29	$7.562\ 10^{-4}$	4.18	$3.924\ 10^{-5}$	4.69
3	$5.303\ 10^{-2}$	1.97	$9.083\ 10^{-3}$	2.63	$4.210\ 10^{-5}$	4.17	$1.351\ 10^{-6}$	4.86
4	$1.348\ 10^{-2}$	1.98	$1.292\ 10^{-3}$	2.81	$2.441\ 10^{-6}$	4.11	$4.472\ 10^{-8}$	4.92

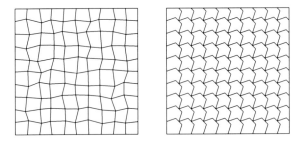

Fig. 2 The first mesh in families $\mathcal{M}_1$ (left) and $\mathcal{M}_2$ (right)

3 Numerical experiments

The numerical experiments presented in this section are aimed to confirm the *a priori* analysis developed in the previous section. To this purpose, we solve the discrete problem (4) on the domain $\Omega =]0, 1[\times]0, 1[$ with the diffusion tensor given by

$$\mathsf{K}(x, y) = \begin{pmatrix} (x+1)^2 + y^2 & -xy \\ -xy & (x+1)^2 \end{pmatrix}.$$

The forcing term f and the Dirichlet boundary condition g are set in accordance with the following exact solution:

- test case 1: $u(x, y) = (x - e^{2(x-1)})(y^2 - e^{3(y-1)})$;
- test case 2: $u(x, y) = e^{-2\pi y} \sin(2\pi x)$.

The performance of the MFD method is investigated by evaluating the rate of convergence on a sequence of refined meshes. Test case 1 is solved using mesh family $\mathcal{M}_1$, where each mesh is formed by randomized quadrilaterals; test case 2 is solved using mesh family $\mathcal{M}_2$, where each mesh is obtained by filling the unit square with a suitably scaled non-convex octagonal reference cell, see the Fig. 2. The meshes are parametrized by the number of partitions in each direction. The

Table 4 Mesh parameters

lev	Mesh Family $\mathcal{M}_1$				Mesh Family $\mathcal{M}_2$			
	$N^{\mathcal{P}}$	$N^{\mathcal{F}}$	$N^{\mathcal{V}}$	h	$N^{\mathcal{P}}$	$N^{\mathcal{F}}$	$N^{\mathcal{V}}$	h
0	100	220	121	$1.922\,10^{-1}$	100	440	341	$1.458\,10^{-1}$
1	400	840	441	$9.705\,10^{-2}$	400	1680	1281	$7.289\,10^{-2}$
2	1600	3280	1681	$4.838\,10^{-2}$	1600	6560	4961	$3.644\,10^{-2}$
3	6400	12960	6561	$2.467\,10^{-2}$	6400	25920	19521	$1.822\,10^{-2}$
4	25600	51520	25921	$1.263\,10^{-2}$	25600	103040	77441	$9.111\,10^{-3}$

starting mesh for both families is built from a 10×10 regular grid, and the refined meshes are obtained by doubling this parameter. The first mesh in each family is shown in Fig. 2. Mesh data for the refinement level lev are reported in Table 4. Here, $N^{\mathcal{P}}$, $N^{\mathcal{F}}$ and $N^{\mathcal{V}}$ are the numbers of mesh elements, edges and vertices, respectively. Table 1 shows the total number of degrees of freedom that are required by the nodal MFD method for $m = 2, \ldots, 5$.

The rate of convergence is measured in the mesh-dependent norm (6). Relative errors and convergence rates are reported in Tables 2-3 and are in good accordance with the theoretical prediction of Theorem 1.

4 Conclusions

In this paper, we presented a new family of nodal arbitrary-order accurate MFD methods for unstructured polygonal meshes. The construction of the method is based on a local consistency condition, i.e., a discrete integration by parts formula, that holds for polynomials of degree m. The arbitrary-order accurate MFD methods use nodal degrees of freedom on mesh edges representing solution values at the quadrature nodes of the Gauss-Lobatto formulas and internal degrees of freedom inside polygons representing solution moments.

Acknowledgements The work of the second author was supported by the Department of Energy (DOE) Advanced Scientific Computing Research (ASCR) program in Applied Mathematics. The work of the third author was partially supported by the Italian MIUR through the program PRIN2008.

References

1. M. Abramowitz and I. A. Stegun. *Handbook of Mathematical Functions with Formulas, Graphs, and Mathematical Tables.* Dover, New York, ninth Dover printing, tenth gpo printing edition, 1964.
2. L. Beirão da Veiga, K. Lipnikov, and G. Manzini. High-order nodal mimetic discretizations of elliptic problems on polygonal meshes. Submitted to SIAM J. Numer. Anal. (also IMATI-CNR Technical Report 32PV10/30/0, 2010).

3. F. Brezzi, A. Buffa, and K. Lipnikov. Mimetic finite differences for elliptic problems. *M2AN Math. Model. Numer. Anal.*, 43(2):277–295, 2009.
4. F. Brezzi, K. Lipnikov, and M. Shashkov. Convergence of the mimetic finite difference method for diffusion problems on polyhedral meshes. *SIAM J. Numer. Anal.*, 43(5):1872–1896, 2005.
5. P. Grisvard. *Elliptic problems in nonsmooth domains*, volume 24 of *Monographs and Studies in Mathematics*. Pitman, Boston, 1985.

The paper is in final form and no similar paper has been or is being submitted elsewhere.

Adaptive cell-centered finite volume method for non-homogeneous diffusion problems: Application to transport in porous media

Fayssal Benkhaldoun, Amadou Mahamane, and Mohammed Seaïd

Abstract We investigate time stepping schemes for the adaptive cell-centered finite volume solution of diffusion equations with heterogeneous diffusion coefficients. The proposed finite volume method uses the cell-centered techniques to discretize the diffusion operators on unstructured grids. Explicit and implicit time integration schemes are used and a comparative study is presented in terms of accuracy and efficiency. Numerical results are presented for a transient diffusion equation with known analytical solution. We also apply these methods to a problem of oil recovery using a two-phase flow problem in porous media.

Keywords Diffusion problems, finite volume, flow in porous media
MSC2010: 76S05, 65N08, 65Y20

1 Introduction

In the last decade, finite volume methods have offered a remarkable level of accuracy and robustness required for solving complex flow problems governed by hyperbolic systems of conservation laws. However, engineering applications often involve coupled hyperbolic and elliptic partial differential equations which have to be solved on complex geometries, thus suggesting the use of the same spatial discretization for both hyperbolic and elliptic equations. As an example of these applications where hyperbolic and elliptic equations coexist we mention

---

Fayssal Benkhaldoun and Amadou Mahamane
LAGA, Université Paris 13, 99 Av J.B. Clement, 93430 Villetaneuse, France,
e-mail: fayssal@math.univ-paris13.fr, mahamane@math.univ-paris13.fr

Mohammed Seaïd
School of Engineering and Computing Sciences, University of Durham, South Road, Durham DH1 3LE, UK, e-mail: m.seaid@durham.ac.uk

the multi-phase flows in porous media, see for example [1, 3, 5]. In practice, the focus is on unstructured meshes where a non-trivial reconstruction scheme is required to have a high-order spatial accuracy. Most of upwind finite volume methods for unstructured grids proposed to date employ a cell-vertex discretization, since it allows a natural definition of the flow gradients: using a dual mesh, a gradient-based reconstruction is applied on the two sides of each interface, where an approximate Riemann solver is finally applied to select the proper upwind contributions. However, solving diffusion equations using the finite volume methods is still a considerable task in the case of unstructured meshes; particularly when these equations have to be solved in conjunction with partial differential equations of hyperbolic type. The emphasis in this work is on the time integration of the resultant system of ordinary differential equations induced from cell-centered finite volume discretization in space variable of the transient diffusion problems. The proposed schemes are the explicit Euler and implicit Euler scheme. These two different methods lead to techniques all of which are occurring in time integration framework since years. Theoretical considerations can provide some ideas concerning stability, convergence rates, restriction on time stepsizes, or qualitative behaviour of the solution, but a complete quantitative analysis is not possible today. Therefore, the only way to make a judgment is to perform numerical tests, at least for some problems which seem to be representative. However, looking into the literature, it seems that there have not been many studies of this type which can give satisfactory answers.

2 Adaptive cell-centered finite volume method

Our main concern in the present study is on the finite volume discretization of the two-dimensional gradient operator $\nabla = \left(\frac{\partial}{\partial x}, \frac{\partial}{\partial y}\right)^T$ resulting from the weak formulation of the diffusion equations. To this end we discretize the spatial domain $\bar{\Omega} = \Omega \cup \partial\Omega$ in conforming triangular elements K_i as $\bar{\Omega} = \cup_{i=1}^{N} K_i$, with N is the total number of elements. Each triangle represents a control volume and the variables are located at the geometric centers of the cells. To discretize the diffusion operators we adapt the so-called cell-centered finite volume method based on a Green-Gauss diamond reconstruction, see for example [5] and further references are therein. Hence, a co-volume, D_σ, is first constructed by connecting the barycentres of the elements that share the edge σ and its endpoints as shown in Fig. 1. Then, the discrete gradient operator ∇_σ is evaluated at an inner edge σ as

$$\nabla_\sigma u_h = \frac{1}{2\text{meas}(D_\sigma)}\left((u_L - u_K)\text{meas}(\sigma)\mathbf{n}_{K,\sigma} + (u_S - u_N)\text{meas}(s_\sigma)\mathbf{n}'_\sigma\right), \quad (1)$$

where u_h is the finite volume discretization of a generic function u, meas(D) denotes the area of the element D, $\mathbf{n}_{K,\sigma}$ denotes the unit outward normal to the surface σ, u_K and u_L are the values of the solution u_h in the elements K and L, respectively. In (1),

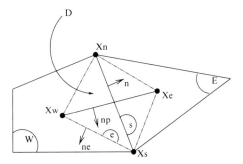

Fig. 1 A generic two-dimensional finite volume and notations

u_S and u_N are the values of the solution u_h at the co-volume nodes approximated by a linear interpolation from the values on the cells sharing the same vertex S and N, respectively. For further details on the formulation and analysis of the considered cell-centered finite volume method we refer to [4,5] among others.

The treatment of boundary conditions in the cell-centered finite volume method is performed using similar techniques as those described in [2,5]. In order to improve the efficiency of the proposed finite volume method, we have performed a mesh adaptation to construct a nearly optimal mesh able to capture the small solution features without relying on extremely fine grid in smooth regions far from steep gradients. In the present work, this goal is achieved by using an error indicator for the gradient of the solution. This indicator requires only information from solution values within a single element at a time and it is easily calculated, see for instance [2].

3 Time stepping schemes

For simplicity in the presentation we consider the transient diffusion problem

$$\frac{\partial u}{\partial t} - \nabla \cdot (\mathbb{K}(\mathbf{x})\nabla u) = f(\mathbf{x},t), \qquad (\mathbf{x},t) \in \Omega \times (0,T], \qquad (2)$$

where Ω is a subset of $\mathbb{R}^2$ with smooth boundary $\partial\Omega$, $(0,T]$ is the time interval, $\mathbb{K}(\mathbf{x})$ is a 2×2 matrix with entries k_{ij}, and $f(\mathbf{x},t)$ is an external force. In the current study the spatial discretization of the diffusion equation (2) is carried out using the cell-centered finite volume method and two time stepping schemes are considered for the time integration.

For the time integration of (2) we discretize the time interval into subintervals $[t_n, t_{n+1}]$ with length Δt, $0 = t_0, t_1, \ldots, t_N = T$ and $t_n = n\Delta t$. We use the notation ω^n to denote the value of a generic function ω at time t_n. Hence, using the forward Euler method, the fully discrete version of the diffusion equation (2) reads

$$u_K^{n+1} = u_K^n + \frac{\Delta t}{\text{meas}(K)} \sum_{\sigma \in \mathscr{E}_K} F_{K,\sigma}\left(u_h^n\right) \text{meas}(\sigma) + \Delta t f_K^n, \qquad \forall\, K \in \mathscr{T}, \quad (3)$$

where $\mathscr{E}_K$ is the set of all edges of the control volume K and $F_{K,\sigma}^n$ are the numerical fluxes reconstructed as

$$F_{K,\sigma}(u) = \mathbb{K}_\sigma \nabla_\sigma u \cdot \mathbf{n}_{K,\sigma}, \qquad (4)$$

with $\mathbb{K}_\sigma$ is an averaged values of the diffusion matrix $\mathbb{K}$ on the edge σ and ∇_σ is the cell-centered finite volume discretization of the gradient operator defined in (1).

An implicit time stepping scheme for (2) is formulated using the backward Euler scheme as

$$u_K^{n+1} = u_K^n + \frac{\Delta t}{\text{meas}(K)} \sum_{\sigma \in \mathscr{E}_K} F_{K,\sigma}\left(u_h^{n+1}\right) \text{meas}(\sigma) + \Delta t f_K^{n+1}, \qquad \forall\, K \in \mathscr{T}. \quad (5)$$

To find the solution u_K^{n+1} from (5) one has to solve, at each time level, a linear system of algebraic equations. In our simulations, the linear system is solved using the preconditioned GMRES solver with a convergence criteria of 10^{-6}, we use the diagonal as a preconditionner.

4 Numerical Results

In this section we examine the accuracy and performance of the proposed time stepping schemes using two test examples. The first example solves a transient diffusion equation with analytical solution that can be used to quantify errors in the time stepping schemes. The second example considers a problem of oil recovery using the equations of two-phase flows in porous media. This last example is used to qualify the considered implicit time stepping scheme for more complicated nonlinear convection-dominated flows. In all the computations reported herein, a two-level refining and fixed CFL numbers are used.

4.1 Accuracy test problem

First we consider the problem of a diffusion problem with manufactured exact solution in a squared domain $\Omega = [-1, 1] \times [-1, 1]$. Here, we solve the transient equation (2) subject to a nonhomogenuous diffusion tensor given by

$$\mathbb{K} = (1 + \alpha x)^2 \begin{pmatrix} 1 & 10^{-2} \\ 10^{-2} & 10^{-6} \end{pmatrix}. \qquad (6)$$

Adaptive cell-centered finite volume method for non-homogeneous diffusion problems 83

The reaction term f is explicitly calculated such that the exact solution of the diffusion problem (2) and (6) is

$$U(x, y, t) = \sin(\pi x) \sin(\pi y) \left(1 - e^{-\lambda t}\right). \tag{7}$$

In our computations $\alpha = \lambda = 0.1$, the initial condition is calculated from the analytical solution (7) and homogeneous Dirichlet boundary conditions are imposed on $\partial \Omega$. To quantify the errors in this test example we consider the L^2-error norm defined as

$$\|e_h\| = \left(\sum_{K \in \mathcal{T}} \text{meas}(K) e_K^2\right)^{\frac{1}{2}},$$

where $e_K(T) = |U_K - u_K|$ with u_K and U_K are respectively, the computed and exact solutions on the control volume K. In the current study, we present numerical results at the transient regime corresponding to the simulation time $T = 1$.

Table 1 Relative error and statistics using the explicit and implicit schemes at the transient time $T = 1$ and different CFL numbers. min Δt is the minimum Δt used in the scheme and GMRES iter refers to the mean number of iterations in the GMRES solver

	Explicit scheme	Implicit scheme			
	CFL = 1	CFL = 5	CFL = 10	CFL = 50	CFL = 100
		Coarse mesh			
min Δt	4.70E-04	2.35E-03	4.70E-03	2.35E-02	4.70E-02
Relative error	1.15E-02	9.65E-03	8.07E-03	1.44E-02	3.49E-02
# time steps	2126	426	213	43	22
CPU time	0.76	0.44	0.32	0.20	0.16
GMRES iter	—	5	8	28	44
# elements	896	896	896	896	896
# nodes	481	481	481	481	481
		Fine mesh			
min Δt	7.32E-06	3.66E-05	7.32E-05	3.66E-04	7.32E-04
Relative error	1.73E-04	1.41E-004	1.097E-04	1.958E-04	5.569E-04
# time steps	136576	27316	13658	2732	1366
CPU time	4814.781	2187.42	1415.296	1981.735	1297.94
GMRES iter	—	2	4	29	42
# elements	57344	57344	57344	57344	57344
# nodes	28929	28929	28929	28929	28929
		Adaptive mesh			
min Δt	8.91E-06	4.46E-05	8.92E-05	4.46E-04	8.91E-04
Relative error	1.83E-03	1.82E-03	1.80E-03	1.918E-03	2.12E-03
# time steps	112120	22383	11166	2192	1070
CPU time	1496.57	595.63	294.846	148.28	124.38
GMRES iter	—	2	3	9	19
# elements	17256	17249	17248	16480	15507
# nodes	8681	8676	8676	8292	7806

To quantify the considered time stepping schemes applied to this example we summarize in Table 1 the results obtained at the transient time $T = 1$. In this table we present the minimum time stepsize, the relative error, the number of time steps required to reach the steady state, the CPU time in seconds, the number of iterations in the GMRES solver, the number of elements and the number of nodes in the considered meshes. A simple inspection of these results shows that the implicit schemes can use larger CFL numbers than those required for a stable explicit scheme. Note that using large CFL numbers in the implicit scheme results in a decrease on the number of time steps needed to reach the steady state and at the same time results in an increase on the number of the iterations in the GMRES solver. As expected the highest accuracy is obtained for both explicit and implicit schemes on the fine mesh but with a large CPU time compared to the results on coarse and adaptive meshes. No remarkable difference is obtained in the relative errors for the implicit and explicit schemes on the coarse and fine meshes. However, using an adaptive mesh the implicit scheme produces smaller errors than the explicit scheme. On the other hand, due to grid adaptation the final mesh at CFL = 100 consists of 15507 cells only compared to the 57344 cells for the fixed fine mesh. This results in a significant reduction of the computational cost. An examination of the CPU times in the tables reveals that, the cell-centered finite volume method on fixed meshes requires more computational work than its adaptive counterpart.

4.2 Transport in porous media

We solve a problem of oil recovery in a two-dimensional porous reservoir. The problem statement is solving a two-phase flow problem in porous media on the computational domain defined by $\bar{\Omega} = \Omega \cup \partial\Omega$ with $\partial\Omega = \Gamma_1 \cup \Gamma_2 \cup \Gamma_3$ as illustrated in Fig. 2. Here, the governing equations consist of the pressure equation

$$\mathbf{q} = -d(u)\mathbb{K}(\mathbf{x})\nabla p, \qquad (\mathbf{x},t) \in \Omega \times (0,T],$$

$$\nabla \cdot \mathbf{q} = 0, \qquad (\mathbf{x},t) \in \Omega \times (0,T], \qquad (8)$$

$$\mathbf{q}\cdot\mathbf{n}\big|_{\Gamma_1} = -1.4, \quad \mathbf{q}\cdot\mathbf{n}\big|_{\Gamma_2} = 0, \quad p\big|_{\Gamma_3} = 0, \qquad t \in (0,T],$$

and the saturation equation

$$\phi(\mathbf{x})\frac{\partial u}{\partial t} - \nabla \cdot \left(b(u)\mathbf{q} - \mathbb{K}(\mathbf{x})a(\mathbf{x})\nabla u\right) = 0, \qquad (\mathbf{x},t) \in \Omega \times (0,T],$$

$$u\big|_{\Gamma_1} = 1, \quad \mathbb{K}\cdot\mathbf{n}\big|_{\Gamma_2} = 0, \quad u\big|_{\Gamma_3} = 0, \qquad t \in (0,T], \qquad (9)$$

$$u(\mathbf{x},0) = u_0(\mathbf{x}), \qquad \mathbf{x} \in \Omega,$$

where p is the pressure, $\mathbf{q}$ the Darcy velocity, u the saturation, $\mathbb{K}$ is the permeability of the medium, ϕ the porosity and $d(u) = k_w(u) + k_o(u)$ is the total mobility with $k_w(u)$ and $k_o(u)$ are the mobility of water and oil, respectively. In (9),

Adaptive cell-centered finite volume method for non-homogeneous diffusion problems 85

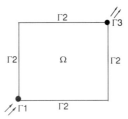

Fig. 2 Computational domain for the example of transport in porous media

$$b(u) = \frac{k_w(u)}{k_w(u) + k_o(u)}, \qquad a(u) = \frac{k_w(u)k_o(u)}{k_w(u) + k_o(u)} p'(u),$$

with $p(u)$ represents the capillary pressure. Note that, in most practical applications, the effects of Darcy velocity dominates the effects of diffusion. As a consequence the saturation equation (9) results in a convection-dominated problem which requires special numerical treatment and many numerical methods from the literature fail to accurately approximate its solution. In addition, since the diffusion in (9) depends on the Darcy velocity, the accuracy on the solution of saturation equation (9) strongly needs an accurate solution of the pressure equation (8). For more details on this model we refer the reader to [1, 3, 5] among others. In our simulations, the permeability $\mathbb{K} = Id$, with Id is the 2×2 identity matrix, the porosity $\phi = 0.2$, the water and oil mobility along with the capillary pressure are given by

$$k_w(u) = \frac{1}{2}u^5, \qquad k_o(u) = \frac{1}{3}(1-u)^3, \qquad p(u) = -\sqrt{\frac{1-u}{u}}.$$

The initial condition u_0 is defined as

$$u_0(\mathbf{x}) = \begin{cases} 1, & \text{if } \mathbf{x} \in \Gamma_1, \\ 0, & \text{elsewhere.} \end{cases}$$

This problem has an interesting structure and will be used to verify the adaptive cell-centered finite volume method namely, to verify that the adaptation methodology is able to compute the right speed of the saturation fronts, and to verify that adaptive refinement is computationally cheaper than fixed mesh for a given level of solution resolution. Based on the conclusions drawn in the previous test example, only results using the implicit time stepping scheme are presented for this example. Figure 3 shows the adapted meshes and plots of the saturation at two different times, namely $t = 0.022$ and $t = 0.048$. The initial mesh contains 3662 cells and a $\Delta t = 2 \times 10^{-5}$ is used in our simulations. At earlier time of the simulation, the front entering the reservoir starts to develop and will be advected later on by the flow at far exit

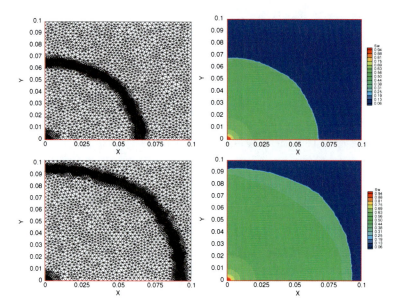

Fig. 3 Adaptive meshes and saturation contours at time $t = 0.022$ (top) and $t = 0.048$ (bottom)

of the reservoir. The interaction between the Darcy pressure and the saturation is detected across the reservoir during the simulation time. It can be clearly seen that the saturation front structures being captured by the cell-centered finite volume method. Another important result is that positions of the saturated waves are not deteriorated by the multiple mesh adaptations. The adaptive cell-centered finite volume method accurately approximates the solution to this problem of two-phase flow in porous media. In addition, the comparison with similar numerical results available in the literature [1] on the same test case is also satisfactory. It should be stressed that, due to grid adaptation the final mesh at times $t = 0.022$ and $t = 0.048$ consists of 4039 and 4947 cells, respectively. This results in a significant reduction of the computational cost compared to a cell-centered finite volume method on fixed meshes.

References

1. M. Afif, B. Amaziane, Convergence of finite volume schemes for degenerate convection-diffusion equation arising in flow in porous media, *Comput. Methods Appl. Mech. Engrg.* (2002) 5265–5286.
2. F. Benkhaldoun, I. Elmahi and M. Seaïd, Well-balanced finite volume schemes for pollutant transport by shallow water equations on unstructured meshes, *J. Comput. Phys.* **226** (2007) 180–203.
3. G. Chavent, J. Jaffré, *Mathematical models and finite element for reservoir simulation*, North-Holland. 1986.

4. Y. Coudière, J.P. Vila, P. Villedieu, Convergence rate of finite volume scheme for a two dimensional convection-diffusion problem, *M2AN*. **33** (1999) 493–516.
5. A. Mahamane, Analyse et estimation d'erreur en volumes finis, Application aux écoulements en milieu poreux et á l'adaptation de maillage, *Dissertation, Université Paris 13*, 2009.

The paper is in final form and no similar paper has been or is being submitted elsewhere.

A Generalized Rusanov method for Saint-Venant Equations with Variable Horizontal Density

Fayssal Benkhaldoun, Kamel Mohamed, and Mohammed Seaïd

Abstract We present a class of finite volume methods for the numerical solution of Saint-Venant equations with variable horizontal density. The model is based on coupling the Saint-Venant equations for the hydraulic variables with a suspended sediment transport equation for the concentration variable. To approximate the numerical solution of the considered models we propose a generalized Rusanov method. The method is simple, accurate and avoids the solution of Riemann problems during the time integration process. Using flux limiters, a second-order accuracy is achieved in the reconstruction of numerical fluxes. The proposed finite volume method is well-balanced, conservative, non-oscillatory and suitable for Saint-Venant equations for which Riemann problems are difficult to solve. The numerical results are presented for two test examples.

Keywords Shallow water equations, variable density, finite volume method
MSC2010: 35L04, 65N08, 76L05

1 Introduction

In this paper we are interested to develop a robust finite volume method for solving Saint-Venant equations with variable horizontal density. The governing equations

---

Fayssal Benkhaldoun
LAGA, Université Paris 13, 99 Av J.B. Clement, 93430 Villetaneuse, France,
e-mail: fayssal@math.univ-paris13.fr

Kamel Mohamed
Department of Computer Science, Faculty of Applied Sciences, University of Taibah, Madinah KSA, e-mail: kamel16@yahoo.com

Mohammed Seaïd
School of Engineering and Computing Sciences, University of Durham, South Road, Durham DH1 3LE, UK, e-mail: m.seaid@durham.ac.uk

can be formulated in a conservative form as

$$\frac{\partial \mathbf{W}}{\partial t} + \frac{\partial \mathbf{F}(\mathbf{W})}{\partial x} = \mathbf{Q}(\mathbf{W}), \tag{1}$$

where $\mathbf{W}$, $\mathbf{F}(\mathbf{W})$ and $\mathbf{Q}(\mathbf{W})$ are vector-valued functions in $\mathbb{R}^3$ given by

$$\mathbf{W} = \begin{pmatrix} \rho h \\ \rho h u \\ \rho_s h c \end{pmatrix}, \quad \mathbf{F}(\mathbf{W}) = \begin{pmatrix} \rho h u \\ \rho h u^2 + \frac{1}{2} g \rho h^2 \\ \rho_s h u c \end{pmatrix}, \quad \mathbf{Q}(\mathbf{W}) = \begin{pmatrix} 0 \\ -g\rho h \frac{\partial Z}{\partial x} \\ 0 \end{pmatrix},$$

where h is the water height above the bottom, u the water velocity, g the acceleration due to gravity, Z the function characterizing the bottom topography and ρ_s the sediment density. For constant density ρ, the equations (1) reduce to the standard Saint-Venant equations. In the current work, we assume that a sediment transport takes place such that the density depends on space and time variables, i.e., $\rho = \rho(x,t)$. This requires an additional equation for its evolution. Here, the equation used to close the system is given by

$$\rho = \rho_w + (\rho_s - \rho_w) c, \tag{2}$$

where ρ_s is the sediment density and c is the depth-averaged concentration of the suspended sediment. Further details on the formulation of the above equations we refer to [3] and further references are therein. It is clear that the system (1) is hyperbolic and the associated eigenvalues λ_k ($k = 1, 2, 3$) are

$$\lambda_1 = u - \sqrt{gh}, \qquad \lambda_2 = u \quad \text{and} \quad \lambda_3 = u + \sqrt{gh}. \tag{3}$$

Note that in the above hydrodynamical model, we have considered only the source terms related to bottom topography while the source terms related to bed friction are neglected. Moreover, the bed-load sediment transport is assumed to be negligible in the considered model compared to the suspended sediment load. It should also be stressed that the transport of suspended sediments involves different physical mechanisms occurring within different time scales according to their time response to the hydrodynamics. In practice, the sediment transport of the bed occurs on a transport time scale much longer than the flow time scale. It is therefore desirable to construct numerical schemes that preserve stability for all time scales. In the current study we propose a modified Rusanov method studied and analyzed in [1] for the numerical solution of conservation laws with source terms. This method is simple, accurate and avoids the solution of Riemann problems during the time integration process. Our main goal is to present a class of numerical methods that are simple, easy to implement, and accurately solves the Saint-Venant equations with variable horizontal density without relying on Riemann solvers or front tracking techniques.

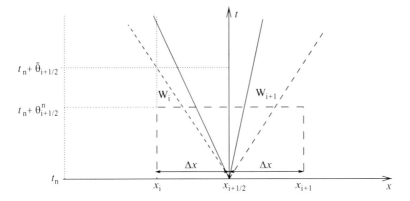

Fig. 1 An illustration of modified Riemann problems in the proposed finite volume method

2 A generalized Rusanov method

To formulate our finite volume method, we discretize the spatial domain into control volumes $[x_{i-1/2}, x_{i+1/2}]$ with uniform size $\Delta x = x_{i+1/2} - x_{i-1/2}$ and we divide the temporal domain into subintervals $[t_n, t_{n+1}]$ with uniform size Δt. Following the standard finite volume formulation, we integrate the equation (1) with respect to time and space over the domain $[t_n, t_{n+1}] \times [x_{i-1/2}, x_{i+1/2}]$ to obtain the following discrete equation

$$\mathbf{W}_i^{n+1} = \mathbf{W}_i^n - \frac{\Delta t}{\Delta x}\left(\mathbf{F}(\mathbf{W}_{i+1/2}^n) - \mathbf{F}(\mathbf{W}_{i-1/2}^n)\right) + \Delta t \mathbf{Q}_i^n, \quad (4)$$

where $\mathbf{W}_i^n$ is the time-space average of the solution $\mathbf{W}$ in the domain $[x_{i-1/2}, x_{i+1/2}]$ at time t_n and $\mathbf{F}(\mathbf{W}_{i\pm 1/2}^n)$ is the numerical flux at $x = x_{i\pm 1/2}$ and time t_n. The spatial discretization of the equation (4) is complete when a numerical construction of the fluxes $\mathbf{F}(\mathbf{W}_{i\pm 1/2}^n)$ is chosen and a discretization of the source term $\mathbf{Q}_i^n$ is performed. In general, the construction of numerical fluxes requires a solution of Riemann problems at the interfaces $x_{i\pm 1/2}$.

In order to avoid these difficulties and reconstruct an approximation of $\mathbf{W}_{i+1/2}^n$, we adapt a finite volume method proposed in [1] for numerical solution of conservation laws with source terms. The key idea is to integrate the equation (1) over a control domain $[t_n, t_n + \theta_{i+1/2}^n] \times [x_i, x_{i+1}]$ containing the point $(t_n, x_{i+1/2})$ as depicted in Fig. 1. It should be stressed that, the integration of the equation (1) over the control domain $[t_n, t_n + \theta_{i+1/2}^n] \times [x_i, x_{i+1}]$ is used only at a predictor stage to construct the intermediate states $\mathbf{W}_{i\pm 1/2}^n$ which will be used in the corrector stage (4). Here, $\mathbf{W}_{i\pm 1/2}^n$ can be viewed as an approximation of the averaged Riemann solution over the control volume $[x_i, x_{i+1}]$ at time $t_n + \theta_{i+1/2}^n$. Thus, the resulting intermediate state is given by

$$\mathbf{W}^n_{i+1/2} = \frac{1}{2}\left(\mathbf{W}^n_i + \mathbf{W}^n_{i+1}\right) - \frac{\theta^n_{i+1/2}}{\Delta x}\left(F(\mathbf{W}^n_{i+1}) - F(\mathbf{W}^n_i)\right) + \theta^n_{i+1/2} Q^n_{i+1/2}, \quad (5)$$

where $Q^n_{i+1/2}$ is an approximation of the averaged source term Q i.e.

$$Q^n_{i+1/2} = \frac{1}{\theta^n_{i+1/2}\Delta x} \int_{t_n}^{t_n+\theta^n_{i+1/2}} \int_{x_i}^{x_{i+1}} Q(\mathbf{W})\, dt\, dx.$$

In order to complete the implementation of the above finite volume method the parameters $\theta^n_{i+1/2}$ and $Q^n_{i+1/2}$ have to be selected. Based on the stability analysis reported in [1] for conservation laws with source terms, the variable $\theta^n_{i+1/2}$ is selected as

$$\theta^n_{i+1/2} = \alpha^n_{i+1/2} \bar{\theta}_{i+1/2}, \qquad \bar{\theta}_{i+1/2} = \frac{\Delta x}{2 S^n_{i+1/2}}, \quad (6)$$

where $\alpha^n_{i+1/2}$ is a positive parameter to be calculated locally and $S^n_{i+1/2}$ is the local Rusanov's velocity defined as

$$S^n_{i+1/2} = \max_{k=1,2,3}\left(\max\left(|\lambda^n_{k,i}|, |\lambda^n_{k,i+1}|\right)\right), \quad (7)$$

with $\lambda^n_{k,i}$ is the kth eigenvalue in (3) evaluated at the solution state $\mathbf{W}^n_i$. Notice that the introduction of the local time step $\theta^n_{i+1/2}$ in the predictor stage (5) is motivated by the fact that $\theta^n_{i+1/2}$ should not be larger than the value $\bar{\theta}_{i+1/2}$ which corresponds to the time required for the fastest wave generated at the interface $x_{i+1/2}$ to leave the cell $[x_i, x_{i+1}]$, compare Fig. 1.

It is clear that by setting $\alpha^n_{i+1/2} = 1$ the proposed finite volume method reduces to the well-established Rusanov method for linear systems of conservation laws, whereas for $\alpha^n_{i+1/2} = \Delta t / \Delta x\, S^n_{i+1/2}$ one recovers the well-known Lax-Wendroff scheme. Another choice of the slopes $\alpha^n_{i+1/2}$ leading to a first-order scheme is $\alpha^n_{i+1/2} = \tilde{\alpha}^n_{i+1/2}$ with

$$\tilde{\alpha}^n_{i+1/2} = \frac{S^n_{i+1/2}}{s^n_{i+1/2}}, \quad (8)$$

where

$$s^n_{i+1/2} = \min_{k=1,2,3}\left(\min\left(|\lambda^n_{k,i}|, |\lambda^n_{k,i+1}|\right)\right). \quad (9)$$

In the current study we incorporate limiters in its reconstruction as

$$\alpha^n_{i+1/2} = \tilde{\alpha}^n_{i+1/2} + \sigma^n_{i+1/2} \Phi\left(r_{i+1/2}\right), \quad (10)$$

where $\tilde{\alpha}^n_{i+1/2}$ is given by (8) and $\Phi_{i+1/2} = \Phi\left(r_{i+1/2}\right)$ is an appropriate limiter which is defined by using a flux limiter function Φ acting on a quantity that measures the ratio $r_{i+1/2}$ of the upwind change to the local change, see for instance [6]. In the present study,

$$\sigma_{i+1/2}^n = \frac{\Delta t}{\Delta x} S_{i+1/2}^n - \frac{S_{i+1/2}^n}{S_{i+1/2}^n},$$

and the ratio of the upwind change is calculated locally as

$$r_{i+1/2} = \frac{W_{i+1-q} - W_{i-q}}{W_{i+1} - W_i}, \qquad q = \text{sgn}\left[\tilde{\alpha}_{i+1/2}^n\right].$$

As a slope limiter function, we consider the Minmod function

$$\Phi(r) = \max(0, \min(1, r)). \tag{11}$$

Note that other slope limiter functions functions from [4, 6] can also apply. The reconstructed slopes (10) are inserted in (6) and the numerical fluxes $\mathbf{W}_{i+1/2}^n$ are computed from (5). Remark that if we set $\Phi = 0$, the spatial discretization (10) reduces to the first-order scheme.

3 Numerical Results

Two test examples are selected to check the accuracy and the performance of the proposed finite volume scheme. As with all explicit time stepping methods the theoretical maximum stable time step Δt is specified according to the Courant-Friedrichs-Lewy condition

$$\Delta t = Cr \frac{\Delta x}{\max_i \left(\left|\alpha_{i+1/2}^n\right|\right)}, \tag{12}$$

where Cr is a constant to be chosen less than unity. In all our simulations, the fixed Courant number $Cr = 0.5$ is used and the time step is varied according to (12).

3.1 Example 1

We consider a density dam-break problem with a single initial discontinuity. The problem consists of solving the equations (1) in a flat channel of length 500 m filled with two liquids with density $\rho = 10 \, kg/m^3$ in the left section and $\rho = 1 \, kg/m^3$ in the right section. Initially, the system is at rest with constant water height $h = 1 \, m$ and $g = 1 \, m/s^2$. In Fig. 2 we display the time evolution of the density, water height, velocity and concentration variables using a mesh with 500 gridpoints. It is clear from these results that at the initial time, the hydrostatic pressure difference at the interface of the two liquids drives a flow of higher density liquid towards the right, pushing the lower density liquid ahead. To conserve mass, the free surface of the lower density liquid rises and a rightward propagating shock-like bore forms. This flow features have been accurately captured by our generalized Rusanov scheme. It should be stressed that the mechanisms of the density dam-break problems are similar to that of the standard dam-break induced by change in free-surface depth, in that a leftward rarefaction, a rightward shock and a contact wave are formed. Similar wave structures also occur in shock tube gas dynamics.

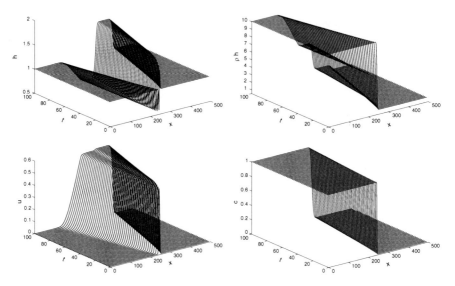

Fig. 2 Numerical results for density dam-break problem with a single initial discontinuity

For the sake of comparison, we present in Fig. 3 the results for the water height at $t = 70$ s obtained using the classical Rusanov method and the proposed method using a mesh with 500 and 1000 gridpoints. We have also included a reference solution obtained using a refined mesh with 100000 gridpoints. As can be seen from this figure, the results obtained using the classical Rusanov method are more diffusive than those obtained using our finite volume method. Similar conclusion can be drawn from other results (not reported here) obtained for the velocity field and sediment concentration.

3.2 Example 2

In this example we solve a density dam-break problem with two initial discontinuities. Here, a flat channel of length 100 m is filled at the left-hand side and right-hand side of the channel with a liquid with density $\rho = 1$ kg/m^3. At the centre of the channel there is a liquid column of density $\rho = 10$ kg/m^3 and width

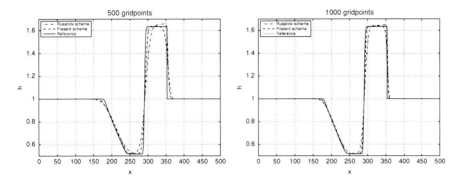

Fig. 3 Comparative results for the water height at $t = 70\ s$ for density dam-break problem with a single initial discontinuity using a mesh with 500 gridpoints (left) and 1000 gridpoints (right)

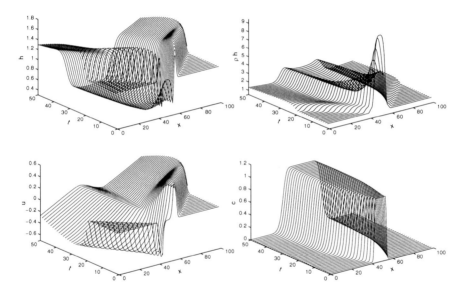

Fig. 4 Numerical results for density dam-break problem with two initial discontinuities

of 1 m. Initially, the system is at rest with constant water height $h = 1\ m$ and $g = 1\ m/s^2$. The computed results are illustrated in Fig. 4 for the t-x phase space. It is evident that the sudden collapse of the denser liquid in the central column causes primary shock waves to be created and propagate as bores in the direction from high to low density. Two outward propagating bores are generated, traveling in opposite directions. Each primary bore decreases in strength with time, which can be seen from the curved shock path. On the other hand, a pair of rarefaction waves travels inward from the interfaces. The rarefaction waves are almost immediately reflected at the center, and then move outward, weakening rapidly. The accuracy of the proposed finite volume is highly achieved in reproducing these physical features.

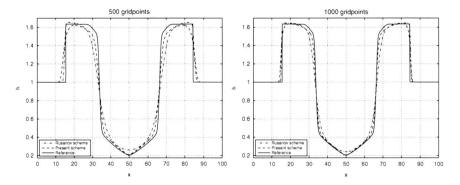

Fig. 5 Comparative results for the water height at $t = 20$ s for density dam-break problem with two initial discontinuities using a mesh with 500 gridpoints (left) and 1000 gridpoints (right)

In Fig. 5 we illustrate a comparison between the results for the water height at $t = 20$ s obtained using the classical Rusanov method and the proposed method using a mesh with 500 and 1000 gridpoints. Again, a reference solution obtained using a refined mesh with 100000 gridpoints is included in this figure. As in the previous test example, an excessive numerical diffusion is detected in the results obtained using the classical Rusanov method. This numerical diffusion has been noticeably reduced in the results obtained using the proposed finite volume method.

References

1. Mohamed, K.: Simulation numérique en volume finis, de problémes d'écoulements multidimensionnels raides, par un schéma de flux á deux pas. Dissertation, University of Paris 13, (2005)
2. Rusanov, V.: Calculation of interaction of non-steady shock waves with obstacles. Comp. Math. Phys. USSR. **1**, 267–279 (1961)
3. Leighton, F.Z. Borthwick, A.G.L. Taylor, P.H.: 1-D numerical modelling of shallow flows with variable horizontal density. International Journal for Numerical Methods in Fluids. **62**, 1209–1231 (2010)
4. Randall, J.L.: Numerical Methods for Conservation Laws, Lectures in Mathematics. ETH Zürich, (1992)
5. Roe, P.: Approximate riemann solvers, parameter vectors and difference schemes. J. Comp. Physics. **43**, 357–372 (1981)
6. Sweby, P.K.: High resolution schemes using flux limiters for hyperbolic conservation laws. SIAM J. Numer. Anal. **21**, 995–1011 (1984)

The paper is in final form and no similar paper has been or is being submitted elsewhere.

Hydrostatic Upwind Schemes for Shallow–Water Equations

Christophe Berthon and Françoise Foucher

Abstract We consider the numerical approximation of the shallow–water equations with non–flat topography. We introduce a new topography discretization that makes all schemes to be well–balanced and robust. At the discrepancy with the well–known hydrostatic reconstruction, the proposed numerical procedure does not involve any cut–off. Moreover, the obtained scheme is able to deal with dry areas. Several numerical benchmarks are performed to assert the interest of the method.

Keywords shallow–water equations, finite volume schemes, source term approximations, well–balanced schemes, positive preserving schemes
MSC2010: 65M12, 76M12, 35L65

1 Introduction

The present work concerns the derivation of finite volume methods to approximate the solutions of the shallow–water equations when involving non–flat topography. The model under interest is given as follows:

$$\begin{cases} \partial_t h + \partial_x hu = 0, \\ \partial_t hu + \partial_x \left(hu^2 + g\frac{h^2}{2} \right) = -gh\partial_x z, \end{cases} \quad (1)$$

where $h > 0$ is the local water–depth and $u \in \mathbb{R}$ is the depth–average velocity. Here, $z : \mathbb{R} \to \mathbb{R}^+$ denotes the topography while $g > 0$ is the gravitational constant.

C. Berthon and F. Foucher
Laboratoire de Mathématiques Jean Leray, UMR 6629, 2 rue de la Houssinière, BP 92208, 44322 Nantes Cedex 3, France, e-mail: christophe.berthon@univ-nantes.fr, francoise.foucher@ec-nantes.fr

To shorten the notations, let us set

$$w = \begin{pmatrix} h \\ hu \end{pmatrix}, \quad f(w) = \begin{pmatrix} hu \\ hu^2 + g\dfrac{h^2}{2} \end{pmatrix},$$

respectively the state vector $\mathbb{R} \times \mathbb{R}^+ \to \Omega$ and the flux function $\Omega \to \mathbb{R}^2$ where Ω stands for the convex space of admissible states:

$$\Omega = \{w \in \mathbb{R}^2;\ h > 0,\ hu \in \mathbb{R}\}.$$

Let us recall that the steady state of the lake at rest, defined by $u = 0$ and $h + z =$ cste, is of primary importance and it must be preserved by the numerical schemes.

The objective is now to derive schemes which preserve positive the water depth and exactly capture the lake at rest. Several approaches have been recently introduced in the literature to approximate the solutions of (1). Most of them propose to consider the discretization of the associated homogeneous system:

$$\partial_t w + \partial_x f(w) = 0, \qquad (2)$$

to suggest a relevant approximation of the topography source term able to restore the expected lake at rest. For instance, we refer to [10] where a suitable correction of the VFRoe scheme is proposed (see also [4, 11, 12]). Other approaches consider systematic corrections to enforce the required *well–balanced* property; i.e. to enforce the exact capture of the lake at rest. The hydrostatic reconstruction [1] (for instance, see also [5, 18]) is certainly one of the most celebrate *well–balanced* technique. However, to be water–depth positive preserving, this approach introduces a cut–off procedure which may involve some failures. Indeed, after the work by Delestre [9], let us consider a small water–depth on a topography with constant slope. Next, let us increase the slope to reduce the water–depth. Considering coarse grids, the hydrostatic reconstruction introduces a wrong behavior since the water–depth increases (see Fig. 4).

In this paper, we derive a new strategy to systematically enforce the well–balanced property. In fact, the proposed scheme is able to deal with vanishing water height without any additional correction and it is proved to be robust. Arguing [2, 15–17], we suggest a relevant upwind form of the topography source term but by involving the free surface.

The paper is organized as follows. In the next section, we introduce a new formulation of (1) and we present the suggested scheme. Section 3 is devoted to some essential properties as consistency and robustness. The last section presents numerical experiments to illustrate the interest of the method. Specifically, the adopted numerical scheme is shown to capture the correct behavior for large topography slopes.

2 Upwind source term discretization

In order to derive our scheme, we need considering a reformulation of system (1) by introducing the free surface $H = h + z$ and a fraction of water $X = \frac{h}{H}$. As usual, one can chose arbitrarily the topography origin (see [6, 13] and references therein), and thus we impose a bottom reference so that $H > 0$. As a consequence, X is well–defined. Involving such notations, the weak solutions of (1) satisfy the following system:

$$\begin{cases} \partial_t h + \partial_x XHu = 0, \\ \partial_t hu + \partial_x \left(X \left(Hu^2 + g\frac{H^2}{2} \right) - \frac{g}{2} hz \right) + gh\partial_x z = 0. \end{cases} \quad (3)$$

Let us remark the following identity:

$$\frac{1}{2}\partial_x (hz) - h\partial_x z = \frac{H^2}{2}\partial_x X,$$

to simplify the discharge equation. Then the smooth solutions of (1) also satisfy:

$$\begin{cases} \partial_t h + \partial_x XHu = 0, \\ \partial_t hu + \partial_x \left(X \left(Hu^2 + g\frac{H^2}{2} \right) \right) - g\frac{H^2}{2}\partial_x X = 0. \end{cases} \quad (4)$$

For the sake of simplicity in the notations, let us set

$$W = (H, Hu)^T,$$

to rewrite (4) as follows:

$$\partial_t w + \partial_x Xf(W) - \begin{pmatrix} 0 \\ g\frac{H^2}{2}\partial_x X \end{pmatrix} = 0.$$

Now, we propose a discretization of (4), but let us note from now on that the discrete form we will obtain for (4) will be consistent with (3). As a consequence, the suggested discretization will be relevant to approximate the weak solutions of (3) or equivalently (1). We suggest to modify any scheme associated with the homogeneous system (2). Hence, let us consider $f^{\Delta x} : \Omega \times \Omega \to \mathbb{R}^2$ a consistent numerical flux function, i.e. $f^{\Delta x}(w, w) = f(w)$. We restrict ourselves to the regular meshes of size Δx such that $\Delta x = x_{i+1/2} - x_{i-1/2}$ for all $i \in \mathbb{Z}$, and we denote the time step by Δt, with $t^{n+1} = t^n + \Delta t$ for all $n \in \mathbb{N}$. The proposed scheme thus reads:

$$w_i^{n+1} = w_i^n - \frac{\Delta t}{\Delta x}\left(X_{i+1/2}^n f^{\Delta x}(W_i^n, W_{i+1}^n) - X_{i-1/2}^n f^{\Delta x}(W_{i-1}^n, W_i^n)\right)$$
$$+ \begin{pmatrix} 0 \\ \frac{\Delta t}{\Delta x}\frac{g}{2} H_{i+1/2}^n H_{i-1/2}^n (X_{i+1/2}^n - X_{i-1/2}^n) \end{pmatrix}, \quad (5)$$

where we have set

$$H_{i+1/2}^n = \begin{cases} H_i^n, & \text{if } f_h^{\Delta x}(W_i^n, W_{i+1}^n) > 0, \\ H_{i+1}^n, & \text{otherwise,} \end{cases} \quad (6)$$

$$X_{i+1/2}^n = \begin{cases} X_i^n, & \text{if } f_h^{\Delta x}(W_i^n, W_{i+1}^n) > 0, \\ X_{i+1}^n, & \text{otherwise,} \end{cases} \quad (7)$$

where $f_h^{\Delta x}$ denotes the first component of the numerical flux function. Here, we have set $W_i^n = X_i^n h_i^n$ where $X_i^n = h_i^n/(h_i^n + Z_i)$.

Before we establish the main properties of the scheme, let us emphasize that the suggested numerical procedure is an extremely easy way to consider the topography from a relevant discretization of the homogeneous shallow–water equation (2). The reader is referred to [15] where similar ideas were introduced.

3 Main properties

First of all, let us remark that the scheme (5) is obviously consistent with (4). Since (4) turns out to be a non conservative formulation of (3), or equivalently (1), after the work by Dal Maso, LeFLoch and Murat [8] (see also [3,7,14]), this may suggest that our approach cannot deal with weak solutions of (1). As a consequence, we first prove the consistency of (5) with (1).

Lemma 1. *Let $(w_i^{n+1})_{i \in \mathbb{Z}}$ be given by (5)-(6)-(7). Then $(w_i^{n+1})_{i \in \mathbb{Z}}$ satisfies in addition:*

$$h_i^{n+1} = h_i^n - \frac{\Delta t}{\Delta x}\left(X_{i+1/2}^n f_h^{\Delta x}(W_i^n, W_{i+1}^n) - X_{i-1/2}^n f_h^{\Delta x}(W_{i-1}^n, W_i^n)\right), \quad (8)$$

$$(hu)_i^{n+1} = (hu)_i^n - \frac{\Delta t}{\Delta x}\left(X_{i+1/2}^n f_{hu}^{\Delta x}(W_i^n, W_{i+1}^n) - X_{i-1/2}^n f_{hu}^{\Delta x}(W_{i-1}^n, W_i^n)\right)$$
$$+ \frac{\Delta t}{\Delta x}\frac{g}{2}\left(h_{i+1/2}^n z_{i+1/2}^n - h_{i-1/2}^n z_{i-1/2}^n\right)$$
$$- \frac{\Delta t}{\Delta x}\frac{g}{2}(h_{i+1/2}^n + h_{i-1/2}^n)(z_{i+1/2}^n + z_{i-1/2}^n), \quad (9)$$

where $h_{i+1/2}^n = H_{i-1/2}^n X_{i-1/2}^n$ and $z_{i+1/2}^n = H_{i-1/2}^n(1 - X_{i-1/2}^n)$. As a consequence, the scheme (5) is consistent with (3) and thus with (1).

We skip the proof of this result since it is just a reformulation of (5).

At this level, the adopted scheme turns out to be relevant when approximating the weak solutions of (1). Moreover, let us note that the above scheme derivation does not need some smoothness argument concerning the topography function z. Now, let us state our main result.

Theorem 1. *Let w_i^n belongs to Ω for all i in $\mathbb{Z}$. Under a suitable CFL like restriction, let us assume that the numerical flux function $f^{\Delta x}$ is Ω-preserving as follows:*

$$w_i^n - \frac{\Delta t}{\Delta x}\left(f^{\Delta x}(w_i^n, w_{i+1}^n) - f^{\Delta x}(w_{i-1}^n, w_i^n)\right) \in \Omega.$$

1. *Assume an additional CFL restriction given by*

$$\frac{\Delta t}{\Delta x}\left(\max(0, f_h^{\Delta x}(W_i^n, W_{i+1}^n)) - \min(0, f_h^{\Delta x}(W_{i-1}^n, W_i^n))\right) < H_i^n,$$

then w_i^{n+1} given by (5) stays in Ω for all i in $\mathbb{Z}$.
2. *The scheme (5) is well–balanced. Assume $u_i^n = 0$ and $h_i^n + z_i = H$ a positive constant, then $u_i^{n+1} = 0$ and $h_i^{n+1} + z_i = H$ for all i in $\mathbb{Z}$.*

We here do not prove the above result but we just establish the preservation of the lake at rest. Let us assume $u_i^n = 0$ and $h_i^n + z_i = H > 0$ for all $i \in \mathbb{Z}$. As a consequence, we have $W_i^n = (H, 0)^T$ for all $i \in \mathbb{Z}$. Arguing the consistency of the numerical flux function $f^{\Delta x}$, we have

$$f^{\Delta x}(W_i^n, W_{i+1}^n) = f(W) = \begin{pmatrix} 0 \\ g\frac{H^2}{2} \end{pmatrix}, \quad \forall i \in \mathbb{Z}.$$

Concerning the water heigh, we immediately deduce $h_i^{n+1} = h_i^n$ for all $i \in \mathbb{Z}$ to obtain $h_i^{n+1} + z_i = h_i^n + z_i = H$. Now, let us rewrite the discretization of the discharge:

$$(hu)_i^{n+1} = (hu)_i^n - \frac{\Delta t}{\Delta x}\left(X_{i+1/2}^n g\frac{H^2}{2} - X_{i-1/2}^n g\frac{H^2}{2}\right)$$

$$+ \frac{\Delta t}{\Delta x}\frac{g}{2} H_{i+1/2}^n H_{i-1/2}^n (X_{i+1/2}^n - X_{i-1/2}^n).$$

Since $H_{i+1/2}^n = H$ for all $i \in \mathbb{Z}$, we get $(hu)_i^{n+1} = 0$. The scheme is thus well–balanced.

4 Numerical results

The numerical experiments are performed on a grid made of 500 cells. Here, the CFL number is systematically fixed to 0.5. To validate the derived numerical scheme, we first propose to consider transcritical flow with shock over a bump. Figure 1 shows a comparison between the classic hydrostatic reconstruction and our scheme. From now on, let us underline that the scheme formulation (5) depends on the choice of the bottom origin. Hence, the comparison is performed by involving three distinct origins. Since the consistency property is not modified, the approximated solutions stay similar. In Fig. 2, we give the approximation obtained when considering a numerical flux function of HLLC type, VFRoe type and Lax–Friedrichs type. In Fig. 3 we present a comparison between first and second order schemes. Here, a MUSCL second order extension has been performed. The second test concerns the known hydrostatic reconstruction failure. Here, the topography is made of a constant slope. As the slope increases, the expected water height must decrease. Because of the cut–off involved into the hydrostatic reconstruction, a wrong water behavior is noted. The same simulation obtained with the hydrostatic upwind scheme gives the required water behavior, see Fig. 4.

The last experiment, presented in Fig. 5, concerns a drain on a non flat bottom in order to simulate dry areas. Once again we obtain an excellent behavior of the derived numerical scheme. As a perspective of the derived 1D technique, we must now propose an extension for 2D unstructured meshes. To address such an issue, arguing the rotational invariance of the model, we should write the 2D formulation

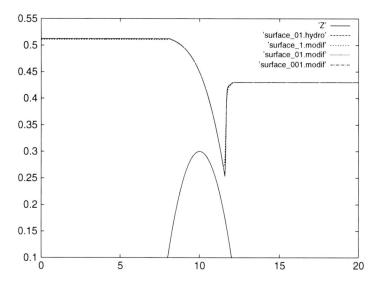

Fig. 1 Comparison hydrostatic reconsturction/hydrostatic upwind involving an arbitrary topography origin so that $\min(z) = 1$, 0.1 and 0.001

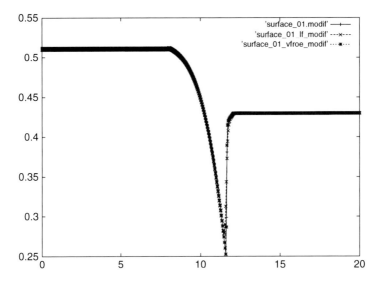

Fig. 2 Comparison between HLLC, VFRoe and Lax–Friedrichs flux functions

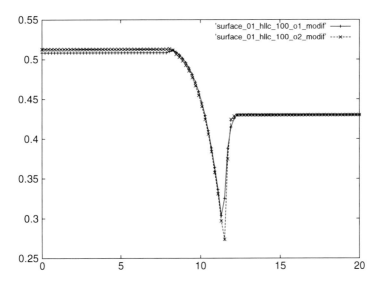

Fig. 3 Comparison first and second order

associated with (4) in the x–direction to obtain the required flux approximation per edge. The interest of such an extension should be an easy topography discretization to obtain 2D well–balanced schemes.

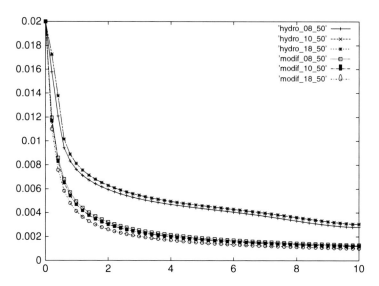

Fig. 4 Failure of the hydrostatic reconstruction. Wrong behavior of h for 8%, 10% and 18% topography slopes. Because of the cut–off the water height increase. Moreover, with 10% and 18% slope, h exactly coincides and thus the obtained approximation is non–physical

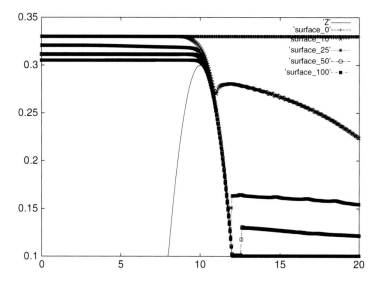

Fig. 5 Drain on a nonflat bottom: Water height at time $t = 0, 1, 10, 25, 50$ and 100

References

1. Audusse E., Bouchut F., Bristeau M.O., Klein R., Perthame B.: A fast and stable well–balanced scheme with hydrostatic reconstruction for shallow water flows, SIAM J.Sci.Comp., **25**, 2050–2065 (2004).

2. Bermudez A., Vazquez–Cendon M.E., Upwind Methods for Hyperbolic Conservation Laws with Source Terms, Computers and Fluids. **23** 1049–1071 (1994).
3. Berthon C., Coquel F.: Nonlinear projection methods for multi–entropies Navier–Stokes systems, Math. Comput., **76**, 1163–1194 (2007).
4. Berthon C., Dubois J., Dubroca B., Nguyen–Bui T.H., Turpault R.: A Free Streaming Contact Preserving Scheme for the M1 Model, Adv. Appl. Math. Mech., **3**, 259–285 (2010).
5. Berthon C., Marche F.: A positive well–balanced VFRoe–ncv scheme for non–homogeneous shallow–water equations, ISCM-EPMESC proceedings, AIP Conference Proceedings, 1495–1500 (2010).
6. Bouchut F.: Non–linear stability of finite volume methods for hyperbolic conservation laws and well–balanced schemes for sources, Frontiers in Mathematics, Birkhauser, 2004.
7. Castro M., LeFloch P., Munoz–Ruiz M.L., Parés C.: Why many theories of shock waves are necessary: convergence error in formally path–consistent schemes, J. Comput. Phys. **227**, 8107-8129 (2008).
8. Dal Maso G., LeFloch P., Murat F., Definition and weak stability of a non conservative product, J. Math. Pures Appl., **74**, 483–548 (1995).
9. Delestre O., Simulation du ruissellement d'eau de pluie sur des surfaces agricoles, PhD Thesis, University of Orléans, http://tel.archives-ouvertes.fr/tel-00531377/en (2010).
10. Gallouët T., Hérard J.M., Seguin N.: Some recent Finite Volume schemes to compute Euler equations using real gas EOS, Int J. Num. Meth. Fluids, **39**, 1073–1138 (2002).
11. Gallouet T., Hérard J.M., Seguin N.: Some approximate Godunov schemes to compute shallow–water equations with topography, Computers and Fluids, **32**, 479–513 (2003).
12. Gallouet T., Hérard J.M., Seguin N.: On the use of some symetrizing variables to deal with vacuum, Calcolo, **40**, 163–194 (2003).
13. Greenberg J.M., Leroux A.Y.: A well–balanced scheme for the numerical processing of source terms in hyperbolic equations, SIAM J. Numer. Anal., **33**, 1–16 (1996).
14. Hou T.Y., LeFloch P.: Why non–conservative schemes converge to wrong solutions: error analysis, Math. Comput., **206**, pp. 497–530 (1994).
15. Jin S.: A steady–state capturing method for hyperbolic systems with geometrical source terms, M2AN Math. Model. Numer. Anal., **35**, 631–645 (2001).
16. Jin S., Wen X.: Two interface–type numerical methods for computing hyperbolic systems with geometrical source terms having concentrations, SIAM J. Sci. Comput., **26**, 2079–2101 (2005).
17. Jin S., Wen X.: An efficient method for computing hyperbolic systems with geometrical source terms having concentrations, Special issue dedicated to the 70th birthday of Professor Zhong-Ci Shi. J. Comput. Math., **22**, 230–249 (2004).
18. Marche F., Bonneton P., Fabrie P., Seguin N.: Evaluation of well–balanced bore–capturing schemes for 2D wetting and drying processes, Int. J. Numer. Meth. Fluids, **53**, 867–894 (2007).

The paper is in final form and no similar paper has been or is being submitted elsewhere.

Finite Volumes Asymptotic Preserving Schemes for Systems of Conservation Laws with Stiff Source Terms

C. Berthon and R. Turpault

Abstract We consider here a numerical technique that allows to build asymptotic-preserving schemes for hyperbolic systems of conservation laws with eventually stiff source terms. The scheme is build in 1D and extended to unstructured 2D meshes. Its behavior is illustrated by numerical experiments on different physical applications.

Keywords systems of conservation laws, stiff, source terms, asymptotic preserving schemes, finite volumes schemes
MSC2010: 65N08, 65Z05

1 Introduction

Our objective is to develop numerical schemes adapted to the resolution of hyperbolic systems of conservation laws with source terms of the form:

$$\partial_t U + \text{div}(\mathbf{F}(U)) = -\gamma R(U), \tag{1}$$

where the state vector $U \in \mathbb{R}^N$ lies in a convex set $\Omega \subset \mathbb{R}^N$. Here, $\gamma \in \mathbb{R}$, which may be a function of U, controls the stiffness of the source term. The function $R(U)$ is supposed to fulfill the compatibility properties required in [2] (see also [10]). In particular, we assume the existence of a constant $n \times N$ matrix Q with rank $n < N$ such that $QR(U) = 0$. It has been showed in [2] that when γ is large, the long-time behavior of such systems degenerates into a nonlinear parabolic system which can be written as:

C. Berthon and R. Turpault
Université de Nantes, Laboratoire de Mathématiques Jean Leray, 2, rue de la Houssinière 44322 Nantes, e-mail: christophe.berthon@univ-nantes.fr, rodolphe.turpault@univ-nantes.fr

$$\partial_t u = -\text{div}(\mathcal{M}(u)\nabla u), \tag{2}$$

where $u = QU$ and $\mathcal{M}(u)$ is a nonlinear diffusion matrix.

Such systems are involved in numerous physical models found for instance in radiotherapy, radiative transfer or fluid dynamics with friction. Typical applications may involve domains where the source term is neglectable (hyperbolic-dominant zones), very stiff (diffusion-dominant zones) or in-between. Therefore, it is crucial to dispose of a numerical scheme able to handle every regime. The construction of such schemes is generally very difficult. Former works (see for instance [1, 6, 8, 9, 12]) usually concentrate on modifying the HLL scheme [16] to adequately include the source term with respect to the physics of a given problem.

In this article, we will propose a generic numerical technique which extends any approximate Riemann solver into an asymptotic preserving scheme for (1). We will first introduce the construction of a finite volumes scheme adapted to the approximation of the solutions of (1) in 1D. This scheme will then be extended for 2D unstructured meshes. Finally, it will be applied on three numerical simulations that will emphasize the relevance of this numerical technique and underline a possibility to improve it.

2 Description of the Scheme

2.1 Construction in 1D

We first show the construction of the numerical technique as it was introduced in [3] and extended in [2]. It consists in a suitable modification of an approximate Riemann solver designed for the transport part of (1) (ie $\gamma = 0$).

Therefore, we start by selecting such a solver. A Riemann problem is thus approximated at each cell interface:

$$\tilde{U}_{\mathcal{R}}\left(\frac{x}{t}; U_L, U_R\right) = \begin{cases} U_L & \text{if } \frac{x}{t} < b^-, \\ \tilde{U}^\star & \text{if } b^- < \frac{x}{t} < b^+, \\ U_R & \text{if } \frac{x}{t} > b^+, \end{cases} \tag{3}$$

where $|b^\pm|$ are chosen to be larger than the fastest wave speed of the problem. For the sake of simplicity in the notations, we will consider in the following that $b^+ = -b^- = b > 0$. Furthermore, $\tilde{U}^\star$ represents the value of the intermediate states and hence generally depends on U_L, U_R and x/t.

As soon as the CFL condition $b\frac{\Delta t}{\Delta x} \leq \frac{1}{2}$ holds, we are considering a juxtaposition of non-interacting approximate Riemann solvers denoted $\tilde{U}^n_{\Delta x}(x, t^n + t)$ for $t \in [0, \Delta t)$. The updated approximated solution at time t^{n+1} is then naturally defined as

follows:
$$\tilde{U}_i^{n+1} = \frac{1}{\Delta x} \int_{x_{i-1/2}}^{x_{i+1/2}} \tilde{U}_{\Delta x}^n(x, t^n + \Delta t) dx. \qquad (4)$$

This scheme can be written in the following usual conservation form:
$$\tilde{U}_i^{n+1} = U_i^n - \frac{\Delta t}{\Delta x}(\mathscr{F}_{i+1/2} - \mathscr{F}_{i-1/2}), \qquad (5)$$

where $\mathscr{F}_{i+1/2}$ denotes the numerical flux at the interface $x_{i+1/2}$. Any (approximate) Riemann solver enter this framework, including for instance Godunov, HLL, HLLC and relaxation schemes. As an example, in the case of the well-known HLL scheme [16], $\tilde{U}^*$ and $\mathscr{F}_{i+1/2}$ are given by:

$$\tilde{U}^{\star,HLL} = \frac{1}{2}(U_L + U_R) - \frac{1}{2b}(F(U_R) - F(U_L)), \qquad (6)$$

$$\mathscr{F}_{i+1/2} = \frac{1}{2}(F(U_i^n) + F(U_{i+1}^n)) - \frac{b}{2}(U_{i+1}^n - U_i^n). \qquad (7)$$

In order to take into account the source term, we now modify the approximate Riemann solver (3) as follows:

$$U_{\mathscr{R}}(\frac{x}{t}; U_L, U_R) = \begin{cases} U_L & \text{if } \frac{x}{t} < -b, \\ U^{\star L} & \text{if } -b < \frac{x}{t} < 0, \\ U^{\star R} & \text{if } 0 < \frac{x}{t} < b, \\ U_R & \text{if } \frac{x}{t} > b, \end{cases} \qquad (8)$$

where we have set:
$$\begin{aligned} U^{\star L} &= \underline{\alpha}\tilde{U}^\star + (\mathbb{I}_d - \underline{\alpha})(U_L - \bar{R}(U_L)), \\ U^{\star R} &= \underline{\alpha}\tilde{U}^\star + (\mathbb{I}_d - \underline{\alpha})(U_R - \bar{R}(U_R)). \end{aligned} \qquad (9)$$

Here, $\underline{\alpha}$, which denotes a $N \times N$ matrix, and $\bar{R}(U)$ are defined by:

$$\underline{\alpha} = \left(\mathbb{I}_d + \frac{\gamma \Delta x}{2b}(\mathbb{I}_d + \underline{\sigma})\right)^{-1}, \quad \bar{R}(U) = (\mathbb{I}_d + \underline{\sigma})^{-1} R(U). \qquad (10)$$

The $N \times N$ matrices $\mathbb{I}_d$ and $\underline{\sigma}$ respectively denote the identity matrix and a parameter matrix to be defined. The updated approximated solution at time t^{n+1} is once again naturally defined:

$$U_i^{n+1} = \frac{1}{\Delta x}\int_{x_{i-1/2}}^{x_{i+1/2}} U_{\Delta x}^n(x, t^n + \Delta t)dx. \tag{11}$$

A straightforward computation leads to:

$$\frac{1}{\Delta t}(U_i^{n+1} - U_i^n) + \frac{1}{\Delta x}(\underline{\alpha}_{i+1/2}\mathscr{F}_{i+1/2} - \underline{\alpha}_{i-1/2}\mathscr{F}_{i-1/2})$$
$$= \frac{1}{\Delta x}(\underline{\alpha}_{i+1/2} - \underline{\alpha}_{i-1/2})F(U_i^n) - \frac{\gamma}{2}(\underline{\alpha}_{i+1/2} + \underline{\alpha}_{i-1/2})R(U_i^n). \tag{12}$$

Observe that whenever $\gamma = 0$, then $\underline{\alpha} = \mathbb{I}_d$ and (12) is nothing but (5).

It was proved in [2] that the scheme (12) is consistant with (1) and preserves Ω as soon as the approximate Riemann solver for the transport part does so. These properties hold for any relevant choice of the parameter matrices $\underline{\sigma}$. These matrices may therefore be chosen to enforce the scheme (12) to be consistant with (2) in the asymptotic regimes. Indeed, an asymptotic analysis of the scheme shows that it is asymptotic preserving if $\underline{\sigma}_{i+1/2}$ is chosen so that the following relation holds:

$$Q(\mathbb{I}_d + \underline{\sigma}_{i+1/2})^{-1} = \frac{1}{b^2}\mathscr{M}_{i+1/2}Q, \tag{13}$$

where $\mathscr{M}_{i+1/2}$ is a discretization of the diffusion matrix $\mathscr{M}(u)$ at the interface $x_{i+1/2}$. One of the edges of this scheme is that it allows to consider applications where γ is a nonlinear function of x and U (see examples in [3] and [4]).

2.2 Extension for 2D unstructured grids

In the case of unstructured grids, the 1D scheme (12) can be extended into the following scheme:

$$U_K^{n+1} = U_K^n - \frac{\Delta t}{|C_K|}\sum_{e\in\partial K}|e|\underline{\alpha}_e\left[\mathscr{F}_e.n_e - F(U_K^n)n_x - G(U_K^n)n_y\right]$$
$$+ \frac{c\Delta t}{|C_K|}\sum_{e\in\partial K}|e|\beta_e b_e(\mathbb{I}_d - \underline{\alpha}_e)\bar{R}(U_K^n), \tag{14}$$

where $|K|$ is the measure of the cell K and $|e|$ is the measure of the interface e.

Furthermore, $\underline{\alpha}_e$ is chosen as:

$$\underline{\alpha}_e = |e|\left(|e|\mathbb{I}_d + \frac{\gamma|K|}{2b}(\mathbb{I}_d + \underline{\sigma})\right)^{-1},$$

Finally, β is set to $1/2$. It is to note that the choices of $\underline{\alpha}$ and β are the simplest admissible ones. However, they are not unique and other expressions may even improve the accuracy of the scheme.

This scheme has successfully been used in the case of cartesian grids in 2D (see for example [5]). In the case of unstructured grids however, in order to enforce the asymptotic preserveness of (14), the choice of $\underline{\sigma}_e$ implies the knowledge of a relevant scheme for the diffusion equation (2). Due to the nonlinear nature of the anisotropy of the diffusion matrix $\mathcal{M}(u)$, the classical two-point scheme (aka FV4, see [15]) lacks of consistance. Therefore, efficient compact schemes have to be considered in order to discretize the diffusion operator. In this framework, we are considering Discrete-Duality Finite Volumes schemes (see for instance [7, 11, 13, 17]). The rich structure of the DDFV schemes can obviously also be used to improve the hyperbolic solvers.

3 Numerical Results

In this section, numerical examples illustrate the behavior of the scheme (12) on three different test-cases. For the sake of simplicity, we used the HLL solver for the transport part.

TC1: Euler with friction
We first consider the 1D isentropic Euler equations with friction. The system reads:

$$\partial_t \rho + \partial_x q = 0,$$

$$\partial_t q + \partial_x \left(\frac{q^2}{\rho} + p(\rho) \right) = -\kappa q,$$

where $\rho > 0$ denotes the density and $q \in \mathbb{R}$ is the fluid momentum. The pressure function $p : \mathbb{R}_+ \to \mathbb{R}_+$ is assumed to be regular enough and to satisfy $p'(\rho) > 0$ in order to ensure the first-order homogeneous associated system to be hyperbolic.

The associated diffusive regime is governed by:

$$\partial_t \rho = \partial_x (p'(\rho) \partial_x \rho). \tag{15}$$

Figure 1 shows the density computed at time $t = 20$. The reference solution is a grid-converged result with a scheme that approximates the diffusion equation (15).

The results of the scheme (12) are in very good agreement with the reference solution even on a coarse grid ($\Delta x = 0.02$). The results of the scheme with $\underline{\sigma} = 0$ are also plotted on Fig. 1. They are representative of what happens with a scheme which is not asymptotic-preserving (although consistant). Indeed, an asymptotic analysis of this scheme shows that it is consistant with a diffusion equation with the wrong diffusion coefficient (see [3]).

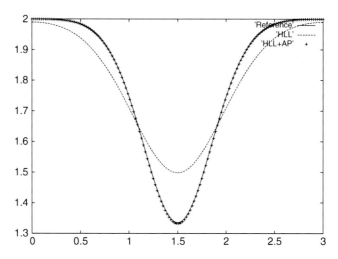

Fig. 1 TC1: computed values of ρ at time $t = 20$. Reference solution (full line) and HLL scheme with (+) or without (dashed line) AP correction

TC2: M1 model for radiative transfer

Now we are interested in the 2D $M1$ model for radiative transfer:

$$\partial_t E + \partial_x \mathbf{F}_x + \partial_y \mathbf{F}_y = c\sigma(aT^4 - E),$$

$$\partial_t \mathbf{F}_x + c^2 \partial_x \mathbf{P}_{xx} + c^2 \partial_y \mathbf{P}_{xy} = -c\sigma \mathbf{F}_x,$$

$$\partial_t \mathbf{F}_y + c^2 \partial_y \mathbf{P}_{xy} + c^2 \partial_y \mathbf{P}_{yy} = -c\sigma \mathbf{F}_y,$$

$$\rho C_v \partial_t T = c\sigma(E - aT^4),$$

where $E, \mathbf{F}$ and $\mathbf{P}$ respectively denote the radiative energy, the radiative flux vector and the radiative pressure tensor. Moreover, T is the material temperature, σ is the opacity, a and c are physical parameters. Finally $\mathbf{P} = \mathbf{P}(\frac{\|\mathbf{F}\|}{cE})$ is a prescribed function (see [14]).

The associated asymptotic regime is described by the so-called equilibrium diffusion equation:

$$\partial_t(\rho C_v T + aT^4) + \mathrm{div}\left(\frac{4acT^3}{3\sigma}\nabla T\right) = 0.$$

In order to obtain a scheme which is consistant with the diffusion operator, unknowns on the triangular mesh were considered at the orthocenter and the classical FV4 scheme (see [15]) was used. Of course, this trick is not valid in general so that other approaches have to be considered as was mentionned in Sect. 2.

Figure 2 shows the results of the scheme (14) on a left-entering Marshak wave inside a square 1m-wide domain with an obstacle. The parameters are $E(t = 0) =$

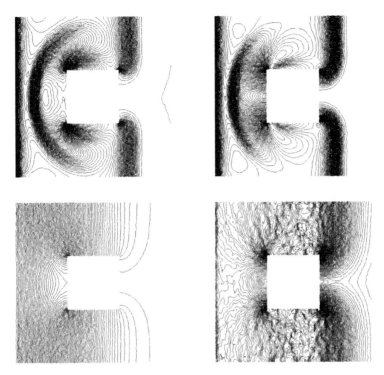

Fig. 2 TC2: Radiative energy (l) and normalized flux (r). Top: $t = 1.e - 8$ and $\sigma = 0$. Bottom: $t = 1.e - 5$ and $\sigma = 10$. Same contours for the energy, same number of contours for the flux (max $\simeq 0.8$ (T) and 0.1 (B)). Triangular mesh with $h \simeq 6.5e - 3$

$a1000^4$, $F(t = 0) = (0,0)$, $T(t = 0) = 1000$, $E_L = a2000^4$, $F_L = (0,0)$ and $T_L = 2000$. Two computations were carried on with $\sigma = 0$ and $\sigma = 10$.

TC3: toy model

For this last application, we consider an interesting toy model that is one of the simplest nontrivial example where the asymptotic regime is described by a system of two equations. It writes:

$$\partial_t \rho + \partial_x q = 0,$$

$$\partial_t q + \partial_x \left(\frac{q^2}{\rho} + p(\rho) \right) = -\kappa q + \sigma f,$$

$$\partial_t e + \partial_x f = 0,$$

$$\partial_t f + \partial_x \chi \left(\frac{f}{e} \right) e = -\sigma f,$$

where $\chi(\xi) = \frac{3+4\xi^2}{5+2\sqrt{4-3\xi^2}}$.

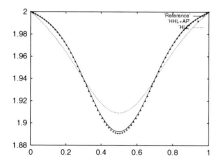 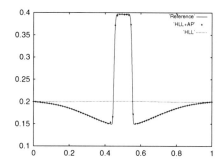

Fig. 3 TC3: computed values of e (l) and ρ (r) at time $t = 50$. Reference solution (full line) and HLL scheme with (+) or without (dashed line) AP correction

The asymptotic regime of this system is given by:

$$\partial_t \rho - \frac{1}{\kappa}\partial_x^2 p(\rho) - \frac{1}{3\kappa}\partial_x^2 e = 0,$$
$$\partial_t e - \frac{1}{3\sigma}\partial_x^2 e = 0. \qquad (16)$$

Figure 3 shows the results of scheme (12) at time $t = 50$ compared with a reference solution for the following test-case: the initial values are $\rho(t = 0) = 0.2$, $q(t = 0) = f(t = 0) = 0$ and $e(t = 0) = 2 - 0.5 \times 1_{[0.45;0.55]}$. The other parameters are $\kappa = 2000$, $\sigma = 1000$, $p(\rho) = 10^{-3}\rho^2$.

With these parameters, the solution is governed by the asymptotic system (16). The results given by the AP preserving scheme (12) are in excellent agreement with the reference solution, even on a very coarse grid (only 80 points where used). It is to note that this test-case is very challenging and that a scheme which does not preserve the asymptotics gives poor results here. As a illustration, the results given by the choice $\sigma = 0$ are also showed on Fig. 3.

References

1. Berthon C., Charrier P., Dubroca B.: An HLLC Scheme to Solve the M_1 Model of Radiative Transfer in Two Space Dimensions, J. Scie. Comput., J. Sci. Comput., **31** 3, 347-389 (2007).
2. Berthon C., LeFloch P., Turpault, R.: Late-time relaxation limits of nonlinear hyperbolic systems. A general framework. (2010) Available via arXiv. http://arxiv.org/abs/1011.3366
3. Berthon C., Turpault, R.: Asymptotic-preseverving HLL schemes. Numerical Methods for Partial Differential Equations, (2010) doi:10.1002/num.20586
4. Berthon C., Turpault, R.: A numerical correction of the $M1$-model in the diffusive limit. NMCF09 proceedings (2009).
5. Berthon C., Dubois J., Dubroca B., Nguyen Bui T.H., Turpault R.: A Free Streaming Contact Preserving Scheme for the M1 Model, Adv. Appl. Math. Mech., 3 (2010), 259-285.

6. Bouchut F., Ounaissa H., Perthame B.: Upwinding of the source term at interfaces for Euler equations with high friction, J. Comput. Math. Appl. **53**, No. 3-4, 361–375 (2007).
7. Boyer F., Hubert F.: Finite volume method for 2D linear and nonlinear elliptic problems with discontinuities, SIAM J. Numer. Anal. **46**, 6, 30323070 (2008).
8. Buet C., Cordier S.: An asymptotic preserving scheme for hydrodynamics radiative transfer models: numerics for radiative transfer, Numer. Math. **108**, 199–221 (2007).
9. Buet C., Després B.: Asymptotic preserving and positive schemes for radiation hydrodynamics, J. Comput. Phys. **215**, 717–740 (2006).
10. Chen G.Q., Levermore C.D., Liu T.P.: Hyperbolic Conservation Laws with Stiff Relaxation Terms and Entropy, Comm. Pure Appl. Math. **47**, 787–830 (1995).
11. Coudière Y., Manzini G.: The discrete duality finite volume method for convection-diffusion problems, SIAM J. Numer. Anal. **47**, 6, 41634192 (2010).
12. Degond P., Deluzet F., Sangam A., Vignal M.H.: An Asymptotic Preserving scheme for the Euler equations in a strong magnetic field, J. Comput. Phys. **228**, 3540–3558 (2009).
13. Domelevo K., Omnes P.: A finite volume method for the Laplace equation on almost arbitrary two-dimensional grids., M2AN Math. Model. Numer. Anal. **39**, no. 6, 12031249 (2005).
14. Dubroca B., Feugeas J.L: Entropic Moment Closure Hierarchy for the Radiative Transfer Equation, C. R. Acad. Sci. Paris, Ser. I, **329**, 915–920 (1999).
15. Eymard R., Gallouët T., Herbin R.: Finite Volume Methods, Handbook of Numerical Analysis, Vol. VII, 713-1020 (2000).
16. Harten A., Lax P.D., Van Leer B.: On upstream differencing and Godunov-type schemes for hyperbolic conservation laws, SIAM Review, **25**, 35–61 (1983).
17. Hermeline F.: Approximation of diffusion operators with discontinuous tensor coefficients on distorted meshes, Comput. Methods Appl. Mech. Engrg., **192**, 1939–1959 (2003).

The paper is in final form and no similar paper has been or is being submitted elsewhere.

Development of DDFV Methods for the Euler Equations

Christophe Berthon, Yves Coudière, and Vivien Desveaux

Abstract We propose to extend some recent gradient reconstruction, the so-called DDFV approaches, to derive accurate finite volume schemes to approximate the weak solutions of the 2D Euler equations. A particular attention is paid on the limitation procedure to enforce the required robustness property. Some numerical experiments are performed to highlight the relevance of the suggested MUSCL–DDFV technique.

Keywords Finite volume methods for hyperbolic problems, Euler equations, DDFV reconstruction, MUSCL reconstruction, Robustness
MSC2010: 65M08, 65N12, 76N99

1 Introduction

This work is devoted to the numerical approximation of the 2–D Euler equations, given as follows:

$$\partial_t \begin{bmatrix} \rho \\ \rho u \\ \rho v \\ E \end{bmatrix} + \partial_x \begin{bmatrix} \rho u \\ \rho u^2 + p \\ \rho uv \\ u(E+p) \end{bmatrix} + \partial_y \begin{bmatrix} \rho v \\ \rho uv \\ \rho v^2 + p \\ v(E+p) \end{bmatrix} = 0, \qquad (1)$$

where $\rho > 0$ denotes the density, $(u, v) \in \mathbb{R}^2$ the velocity vector and $E > 0$ the total energy. For the sake of the simplicity in the presentation, the pressure is given by

---

Christophe Berthon, Yves Coudière, and Vivien Desveaux
Laboratoire de Mathématiques Jean Leray, UMR 6629, 2 rue de la Houssinière - BP 92208 - 44322 Nantes Cedex 3, France, e-mail: Christophe.Berthon@univ-nantes.fr, Yves.Coudiere@univ-nantes.fr, Vivien.Desveaux@univ-nantes.fr

the perfect gas law $p = (\gamma - 1)\left[E - \frac{\rho}{2}(u^2 + v^2)\right]$. The forthcoming developments will easily extend to general pressure laws. To shorten the notations, the system can be rewritten as follows:

$$\partial_t W + \partial_x f(W) + \partial_y g(W) = 0, \qquad (2)$$

where $W = {}^t(\rho, \rho u, \rho v, E) : \mathbb{R}^2 \times \mathbb{R}^+ \to \Omega$ is the unknown state vector and $f(W) : \Omega \to \mathbb{R}^4$ and $g(W) : \Omega \to \mathbb{R}^4$ are the flux functions which find clear definitions. The convex set of admissible states is defined by:

$$\Omega = \left\{ W \in \mathbb{R}^4; \rho > 0, (u, v) \in \mathbb{R}^2, E - \frac{\rho}{2}(u^2 + v^2) > 0 \right\}. \qquad (3)$$

When approximating (1), several strategies have been proposed to increase the accuracy of the numerical solutions among which the most popular is certainly the MUSCL scheme (for instance see [12, 13, 15, 16]). This scheme extends any first–order scheme into a second–order approximation using a piecewise linear reconstruction. In the 2–D case, the main difficulty is to find a technique to reconstruct gradients that can be extended to unstructured meshes (see [4]).

The DDFV (Discrete Duality Finite Volume) method was introduced in the field of elliptic equations in order to reconstruct gradients on distorted meshes (see [1, 6, 9, 10]). The idea of this method is to combine two distinct finite volume schemes on two overlapping meshes: the primal mesh and the dual mesh whose cells are built around the vertices of the primal mesh. This process adds new numerical unknowns at the vertices of the primal mesh, but it will allow to reconstruct very accurate gradients.

It was first proposed to take advantage of the DDFV gradient in order to built second order schemes for the linear convection–diffusion equation in [5]. In this paper, new values of the unknown are built at the midpoint of the interfaces by mean of some averages of the DDFV gradient. The resulting scheme is proved to be of second order in the diffusive regime.

The aim of this work is to extend DDFV–like methods to the case of the Euler equations. As a first step, we have only developed such a method on structured meshes in order to simplify the computation and to check its efficiency. On unstructured meshes, the extension of the DDFV gradient is straightforward. Our reconstruction and limitation procedures generalize although being more technical. Note that the vertices of the primal cells do not coincide with the center of gravity of the dual cells. It might influence the accuracy of the method and some alternatives will be considered in future work.

The paper is organized as follows. In Sect. 2, we introduce the dual mesh and we describe the reconstruction process and the limitation process of our scheme. Section 3 concerns the robustness of our scheme. Indeed, with most of first-order schemes, if a numerical solution is initially valued in Ω, then it remains in Ω. Such a property must be preserved by the second–order accurate scheme. Section 4 is devoted to numerical experiments to illustrate the relevance of DDFV approach when evaluating second–order reconstructions. We give some conclusions and future developments in Sect. 5.

2 Presentation of the scheme

First let us introduce the main notations. We consider a primal mesh composed of rectangular cells

$$K_{i,j} = [x_{i-\frac{1}{2}}, x_{i+\frac{1}{2}}] \times [y_{j-\frac{1}{2}}, y_{j+\frac{1}{2}}], \quad i, j \in \mathbb{Z}. \tag{4}$$

For the sake of simplicity, we will assume that the mesh is uniform, and we enforce $x_{i+\frac{1}{2}} - x_{i-\frac{1}{2}} = y_{j+\frac{1}{2}} - y_{j-\frac{1}{2}} = h$, for all $i, j \in \mathbb{Z}$, where $h > 0$ is fixed.

Let $W_{i,j}^n$ stand for an approximation of the mean value of W on the cell $K_{i,j}$ at time t^n. We denote by $\Delta t > 0$ the time increment. At time $t^{n+1} = t^n + \Delta t$, the updated first–order approximation is given by (see [7, 12, 13]):

$$W_{i,j}^{n+1} = W_{i,j}^n - \frac{\Delta t}{h} \left(F(W_{i,j}^n, W_{i+1,j}^n) - F(W_{i-1,j}^n, W_{i,j}^n) \right.$$

$$\left. + G(W_{i,j}^n, W_{i,j+1}^n) - G(W_{i,j-1}^n, W_{i,j}^n) \right), \tag{5}$$

where $F : \Omega \times \Omega \to \mathbb{R}^4$ and $G : \Omega \times \Omega \to \mathbb{R}^4$ are consistent numerical flux functions. In addition, to avoid some instabilities [12, 13], the time step is restricted according to a CFL–like condition given as follows:

$$\frac{\Delta t}{h} \max_{(i,j) \in \mathbb{Z}^2} \left(\left| \lambda_F^\pm(W_{i,j}^n, W_{i+1,j}^n) \right|, \left| \lambda_G^\pm(W_{i,j}^n, W_{i,j+1}^n) \right| \right) \leq \frac{1}{4}, \tag{6}$$

where $\lambda_\Phi^\pm(W_L, W_R)$ denotes suitable numerical wave velocities associated to the numerical flux function $\Phi(W_L, W_R)$.

2.1 The dual mesh

We denote by $B_{i+\frac{1}{2}, j+\frac{1}{2}} = \left(x_{i+\frac{1}{2}}, y_{j+\frac{1}{2}}\right)$ the vertices of the primal mesh and by $B_{i,j} = (x_i, y_j)$ the center of the primal cell $K_{i,j}$. Around each vertex of the primal mesh $B_{i+\frac{1}{2}, j+\frac{1}{2}}$, we construct a dual cell $K_{i+\frac{1}{2}, j+\frac{1}{2}} = [x_i, x_{i+1}] \times [y_j, y_{j+1}]$. The set of the dual cells $\left(K_{i+\frac{1}{2}, j+\frac{1}{2}}\right)_{i,j \in \mathbb{Z}}$ constitutes a second mesh which we call dual mesh. The centers of the dual cells are the vertices of the primal mesh and conversely.

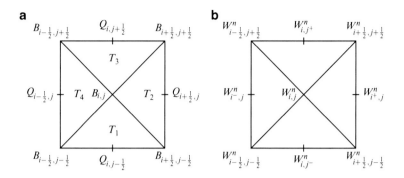

Fig. 1 (Left) Geometry of the cell $K_{i,j}$. (Right) Location of the known states and of the reconstructed states

At time t^n, we assume known approximations $W^n_{i+\frac{1}{2},j+\frac{1}{2}}$ of the mean values of W on cells $K_{i+\frac{1}{2},j+\frac{1}{2}}$. As a consequence, at time t^n, on each primal or dual cell, we know four approximate values at the vertices and one approximate value at the center (see Fig. 1b).

In the sequel, we will deal simultaneously with primal and dual cells. We thus define the set of the indexes of primal and dual cells $\mathbb{S} = \mathbb{Z}^2 \cup (\mathbb{Z} + \frac{1}{2})^2$. The set of primal and dual cells is then $\{K_{i,j}\}_{(i,j)\in\mathbb{S}}$. For $(i,j) \in \mathbb{S}$, we denote by $Q_{i+\frac{1}{2},j} = (x_{i+\frac{1}{2}}, y_j)$, the middle of the interface between the cells $K_{i,j}$ and $K_{i+1,j}$ and by $Q_{i,j+\frac{1}{2}} = (x_i, y_{j+\frac{1}{2}})$, the middle of the interface between the cells $K_{i,j}$ and $K_{i,j+1}$ (see Fig. 1a). On each cell $K_{i,j}$ for $(i,j) \in \mathbb{S}$, we reconstruct values $W^n_{i\pm,j}$ and $W^n_{i,j\pm}$ at points $Q_{i\pm\frac{1}{2},j}$ and $Q_{i,j\pm\frac{1}{2}}$ (see Fig. 1b). Arguing these notations, the second order scheme reads as follows:

$$W^{n+1}_{i,j} = W^n_{i,j} - \frac{\Delta t}{h}\left[F\left(W^n_{i+,j}, W^n_{i+1-,j}\right) - F\left(W^n_{i-1+,j}, W^n_{i-,j}\right) \right.$$
$$\left. + G\left(W^n_{i,j+}, W^n_{i,j+1-}\right) - G\left(W^n_{i,j-1+}, W^n_{i,j-}\right)\right]. \quad (7)$$

We now detail the evaluation of $W^n_{i\pm,j}$ and $W^n_{i,j\pm}$. We recall that both the primal and dual unknowns are solutions of a finite volume scheme. The two schemes are coupled through the gradient reconstruction.

2.2 Gradient reconstruction

As a first step, we perform a gradient reconstruction. To address such an issue, we derive a relevant cell splitting. We consider a primal or dual cell $K_{i,j}$, $(i,j) \in \mathbb{S}$. The cell can be decomposed into four triangles using the four vertices and the center.

We denote by T_1 the bottom triangle and the other ones are denoted by T_2, T_3 and T_4, clockwise (see Fig. 1a).

We define a function $\widehat{W} : K_{i,j} \to \mathbb{R}^4$ piecewise linear on the T_l and which coincides with the approximate values at the four vertices and at the center.

Next, we project each coordinate $\widehat{W}_k$ of $\widehat{W}$ on the space of linear function which takes the value $(W_{i,j})^n_k$ at the point $B_{i,j}$. This means that for all integers $k \in [1,4]$, we seek $\mu_k \in \mathbb{R}^2$ which minimizes the functional $E_k(v) : \mathbb{R}^2 \to \mathbb{R}$ defined by

$$E_k(v) = \int_{K_{i,j}} \left| \widehat{W}_k(X) - \left[\left(W^n_{i,j}\right)_k + v \cdot (X - B_{i,j}) \right] \right|^2 dX. \tag{8}$$

Existence and uniqueness of the minimum are immediate since the functional is strictly convex. The numerical computation of the minimum is quite easy since we only need to compute the Jacobian of E_k and to find its zero. For the sake of simplicity in the notations, we denote by $\mu = {}^t(\mu_1, \mu_2, \mu_3, \mu_4)$, the vector of the solutions of these minimization problems. Hence, we define $\widetilde{W}_\mu(X) : K \to \mathbb{R}^4$ the function whose k–th coordinate is $\left(W^n_{i,j}\right)_k + \mu_k \cdot (X - B_{i,j})$.

2.3 Limitation

We assume that the states $W^n_{i,j}$, $(i,j) \in \mathbb{S}$, are in Ω. Let us remark that the reconstructed function $\widetilde{W}_\mu$ does not necessarily remain in Ω. As a consequence, we have to limit the slopes μ_k. To address such an issue, we propose to substitute the slope μ by $\theta\mu$ where $\theta \in [0,1]$ is a limitation parameter to be fixed according to the required robustness property. To ensure existence and uniqueness of an optimal limited slope, we have to restrict Ω to a close set. We fix a small parameter $\epsilon > 0$ and we define

$$\Omega_\epsilon = \left\{ W \in \mathbb{R}^4; \rho \geq \epsilon, (u,v) \in \mathbb{R}^2, E - \frac{\rho}{2}(u^2 + v^2) \geq \epsilon \right\}. \tag{9}$$

Since we need the values of the reconstructed function only at points $B_{i\pm\frac{1}{2},j}$ and $B_{i,j\pm\frac{1}{2}}$, we require $\widetilde{W}_{\theta\mu}(B_{i\pm\frac{1}{2},j}) \in \Omega_\epsilon$ and $\widetilde{W}_{\theta\mu}(B_{i,j\pm\frac{1}{2}}) \in \Omega_\epsilon$. We thus define the optimal slope limiter by

$$\theta = \max\left\{ t \in [0,1]; \widetilde{W}_{t\mu}(B_{i\pm\frac{1}{2},j}) \in \Omega_\epsilon, \widetilde{W}_{t\mu}(B_{i,j\pm\frac{1}{2}}) \in \Omega_\epsilon \right\}. \tag{10}$$

We emphasize that this set is nonempty since it contains 0. Besides, the maximum is reached because Ω_ϵ is a close set and $t \mapsto \widetilde{W}_{t\mu}(B_{l,m})$ is continuous. Solving for θ requires to find the roots of some quadratic functions (the energy). Finally, the reconstructed states are given by $W^n_{i\pm,j} = \widetilde{W}_{\theta\lambda}(B_{i\pm\frac{1}{2},j})$ and $W^n_{i,j\pm} = \widetilde{W}_{\theta\lambda}(B_{i,j\pm\frac{1}{2}})$.

3 Robustness

We now establish the robustness of the proposed reconstruction. First, let us assume that the directional flux functions F and G are first–order robust on both primal and dual meshes. Indeed, under the CFL condition

$$\frac{\Delta t}{h} \max_{(i,j)\in \mathbb{S}} \left(\left|\lambda_F^\pm(W_{i,j}^n, W_{i+1,j}^n)\right|, \left|\lambda_G^\pm(W_{i,j}^n, W_{i,j+1}^n)\right| \right) \leq \frac{1}{4}, \qquad (11)$$

we assume that the updated states, given by (5) for all pairs (i,j) in $\mathbb{S}$, stay in Ω. Now, let us recall the following statements (for instance see [2, 12]) about robustness of the directional numerical flux functions:

Theorem 1. *Let us consider a robust numerical flux Φ. Assume that W_1, W_2 and W_3 are in Ω. Let W_2^- and W_2^+ be two reconstructed states in Ω such that $W_2 = \frac{W_2^- + W_2^+}{2}$. Assume the CFL condition*

$$\frac{\Delta t}{h} \max \left(|\lambda_\Phi^+(W_1, W_2^-)|, |\lambda_\Phi^\pm(W_2^-, W_2^+)|, |\lambda_\Phi^-(W_2^+, W_3)| \right) \leq \frac{1}{4}. \qquad (12)$$

Then we have $W_2 - \frac{\Delta t}{h} \left(\Phi(W_2^+, W_3) - \Phi(W_1, W_2^+) \right) \in \Omega$.

We assume that the 1D numerical fluxes F and G are robust. In addition, we assume that the states $W_{i,j}^n$, $(i,j) \in \mathbb{S}$ are in Ω, so that the limitation procedure described in Sect. 2.3 ensures that the reconstructed states $W_{i\pm,j}^n$ and $W_{i,j\pm}^n$, $(i,j) \in \mathbb{S}$, remain in Ω. To shorten the notations, we set

$$\Lambda_F = \max_{(i,j)\in\mathbb{S}} \left(|\lambda_F^\pm(W_{i-,j}^n, W_{i+,j}^n)|, |\lambda_F^\pm(W_{i+,j}^n, W_{i+1-,j}^n)| \right),$$

$$\Lambda_G = \max_{(i,j)\in\mathbb{S}} \left(|\lambda_G^\pm(W_{i,j-}^n, W_{i,j+}^n)|, |\lambda_G^\pm(W_{i,j+}^n, W_{i,j+1-}^n)| \right).$$

By applying Theorem 1 we have

$$W_{i,j}^n - \frac{\Delta t}{h} \left[F\left(W_{i+,j}^n, W_{i+1-,j}^n\right) - F\left(W_{i-1+,j}^n, W_{i-,j}^n\right) \right] \in \Omega, \qquad (13)$$

as soon as the CFL restriction $\frac{\Delta t}{h} \Lambda_F \leq \frac{1}{4}$ holds, and we get

$$W_{i,j}^n - \frac{\Delta t}{h} \left[G\left(W_{i,j+}^n, W_{i,j+1-}^n\right) - G\left(W_{i,j-1+}^n, W_{i,j-}^n\right) \right] \in \Omega, \qquad (14)$$

under the CFL condition $\frac{\Delta t}{h} \Lambda_G \leq \frac{1}{4}$.

Considering half sum of (13) and (14), we finally obtain $W_{i,j}^{n+1} \in \Omega$, for all $(i,j) \in \mathbb{S}$ under the CFL condition [12] $\frac{\Delta t}{h} \max(\Lambda_F, \Lambda_G) \leq \frac{1}{8}$. The robustness of the proposed numerical method is thus established.

4 Numerical tests

We have chosen two cases from the collection of 2D Riemann problems proposed by [11], namely configuration 3 (p. 594) and 6 (p. 596). They are called case 1 and case 2. These problems are solved on the square $[0, 1] \times [0, 1]$ divided in four quadrants by lines $x = 1/2$ and $y = 1/2$. The Riemann problems are defined by initial constant states on each quadrant. All four 1D Riemann Problems between quadrants have exactly one wave: four shocks for the case 1 and four contact discontinuities for the case 2. Both cases were computed with primal grids of 200×200 cells which represent about 80,000 cells counting the dual mesh. In order to complete the scheme (7), the adopted numerical flux functions F and G are given by the well–known HLLC approximate Riemann solver (see [3, 8, 14]). The results are displayed for density in Fig. 2. We also provide a comparison with the classical MUSCL scheme on the line $y = x$ and a comparison of the CPU time between the two methods.

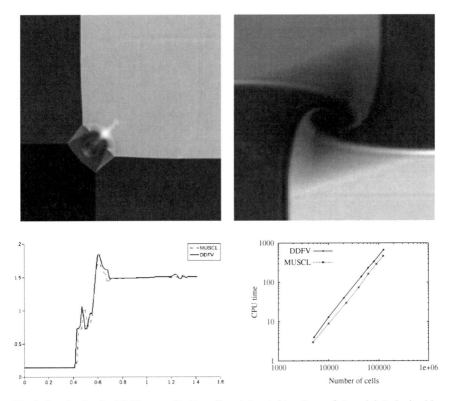

Fig. 2 Results for the 2D Riemann Problem Case 1 (top left) and case 2 (top right) obtained by the derived MUSCL–DDFV scheme. Comparison between the MUSCL–DDFV scheme and the classical MUSCL scheme for case 1: density on the line $y = x$ (bottom left) and CPU time (bottom right)

5 Conclusion

We have presented a second–order robust scheme to approximate the solutions of the 2D Euler equations. The main novelty of this work lies in the gradient reconstruction based on the DDFV methods and the use of two overlapping meshes. We have shown that the method gives good results on structured meshes. Arguing the properties of the DDVF approach, unstructured mesh extensions will be easily obtained.

In order to ensure the robustness, we have enforced that the reconstructed state vectors remain conservative. Another improvement must be performed to propose robust non–conservative reconstructions.

References

1. Andreianov, B., Boyer, F., Hubert, F.: Discrete duality finite volume schemes for Leray-Lions-type elliptic problems on general 2D meshes. Numerical Methods for Partial Differential Equations **23**(1), 145–195 (2007)
2. Berthon, C.: Stability of the MUSCL schemes for the Euler equations. Comm. Math. Sci **3**, 133–158 (2005)
3. Bouchut, F.: Nonlinear stability of finite volume methods for hyperbolic conservation laws and well-balanced schemes for sources. Frontiers in Mathematics. Birkhäuser Verlag, Basel (2004)
4. Buffard, T., Clain, S.: Monoslope and multislope MUSCL methods for unstructured meshes. Journal of Computational Physics **229**(10), 3745–3776 (2010)
5. Coudière, Y., Manzini, G.: The Discrete Duality Finite Volume Method for Convection-diffusion Problems. SIAM Journal on Numerical Analysis **47**(6), 4163–4192 (2010)
6. Domelevo, K., Omnes, P.: A finite volume method for the Laplace equation on almost arbitrary two-dimensional grids. Mathematical Modelling and Numerical Analysis **39**(6), 1203–1249 (2005)
7. Godlewski, E., Raviart, P.A.: Numerical approximation of hyperbolic systems of conservation laws, *Applied Mathematical Sciences*, vol. 118. Springer-Verlag, New York (1996)
8. Harten, A., Lax, P., Van Leer, B.: On upstream differencing and Godunov-type schemes for hyperbolic conservation laws. SIAM review pp. 35–61 (1983)
9. Hermeline, F.: A finite volume method for the approximation of diffusion operators on distorted meshes. Journal of computational Physics **160**(2), 481–499 (2000)
10. Hermeline, F.: Approximation of 2-D and 3-D diffusion operators with variable full tensor coefficients on arbitrary meshes. Computer Methods in Applied Mechanics and Engineering **196**(21-24), 2497–2526 (2007)
11. Kurganov, A., Tadmor, E.: Solution of two-dimensional Riemann problems for gas dynamics without Riemann problem solvers. Numerical Methods for Partial Differential Equations **18**(5), 584–608 (2002)
12. LeVeque, R.: Finite volume methods for hyperbolic problems. Cambridge Univ Pr (2002)
13. Toro, E.: Riemann solvers and numerical methods for fluid dynamics: a practical introduction. Springer Verlag (2009)
14. Toro, E., Spruce, M., Speares, W.: Restoration of the contact surface in the HLL-Riemann solver. Shock waves **4**(1), 25–34 (1994)
15. Van Leer, B.: Towards the ultimate conservative difference scheme. V. A second-order sequel to Godunov's method. Journal of Computational Physics **32**(1), 101–136 (1979)
16. Van Leer, B.: A historical oversight: Vladimir P. Kolgan and his high-resolution scheme. Journal of Computational Physics (2010)

The paper is in final form and no similar paper has been or is being submitted elsewhere.

Comparison of Explicit and Implicit Time Advancing in the Simulation of a 2D Sediment Transport Problem

M. Bilanceri, F. Beux, I. Elmahi, H. Guillard, and M.V. Salvetti

Abstract The simulation of sediment transport, based on the shallow-water equations coupled with Grass model for the sediment transport equation is considered. The aim of the present paper is to investigate the behavior of implicit linearized schemes in this context. A finite-volume method is considered and second-order accuracy in space is obtained through MUSCL reconstruction. A second-order time accurate explicit version of the scheme is obtained through a two step Runge-Kutta method. Implicit linearized schemes of second-order of accuracy in time are derived thanks to a BDF method associated with a Defect Correction technique. The different time-advancing schemes are compared, using a 2D sediment transport problem, with different types of flow/bed interactions. The implicit one largely outperforms the explicit version for slow flow/bed interactions while in the case of fast flow/bed interactions, the CPU time of both time integration schemes are comparable. Thus, the implicit scheme turns out to be a good candidate to simulate flows with sediment transport in practical applications.

Keywords sediment transport, Grass model, linearized implicit time advancing, automatic differentiation
MSC2010: 65-06

M. Bilanceri and M.V. Salvetti
University of Pisa (Italy), e-mail: marco.bilanceri@gmail.com, mv.salvetti@ing.unipi.it

F. Beux
Alta S.p.A., Pisa (Italy), e-mail: f.beux@alta-space.com

I. Elmahi
EMCS, Ensa, Oujda, Complexe Universitaire, Oujda (Morocco), e-mail: ielmahi@ensa.ump.ma

H. Guillard
INRIA, Sophia Antipolis and Laboratoire Jean-Alexandre Dieudonné, University of Nice Sophia-Antipolis, Parc Valrose 06108 Nice Cedex, (France), e-mail: Herve.Guillard@sophia.inria.fr

1 Introduction

A huge amount of work has been done in the last decades to develop numerical methods for the simulation of sediment transport problems (see, e.g., the references in [1, 4]). In this context, the hydrodynamics part is usually modeled through the classical shallow-water equations coupled with an additional equation for the morphodynamical component. The Grass equation [6] is considered herein, which is one of the most popular and simple models. In this context, the treatment of the source terms and of the bed-load fluxes has received the largest attention while time advancing has received much less attention and it is usually carried out by explicit schemes. The focus of the present paper is on the comparison between explicit and implicit schemes in the simulation of a 2D sediment transport problem. We only consider flows over wet areas. The extension to cases in presence of dry areas will the object of further studies. If the interaction of the water flow with the mobile bed is slow, the characteristic time scales of the flow and of the sediment transport can be very different introducing time stiffness in the global problem. Thus, for these cases, it can be advantageous to use implicit schemes. On the other hand, since the considered problems are unsteady, attention must be paid for implicit schemes in the choice of the time step. Another difficulty with implicit schemes is that, in order to avoid the solution of a nonlinear system at each time step, the numerical fluxes must be linearized in time. In order to overcome these difficulties, we use an automatic differentiation tool (Tapenade, [7]). Our starting point was the SRNH numerical scheme, specifically developed and validated for the numerical simulation of sediment transport problems [1]. An implicit version of this scheme is derived herein by computing the Jacobian matrices of the first-order accurate numerical fluxes by the previously mentioned automatic differentiation tool. A defect-correction approach [10] is finally used to obtain second-order accuracy at limited computational costs. The implicit method is compared with the explicit one in a 2D benchmark.

2 Physical model and Numerical Method

The physical model used in this work consists in the well known shallow-water equations coupled with an additional equation to describe the transport of sediment:

$$\frac{\partial \mathbf{W}}{\partial t} + \frac{\partial \mathbf{F}(\mathbf{W})}{\partial x} + \frac{\mathbf{G}(\mathbf{W})}{\partial y} = \mathbf{S}(\mathbf{W}) \quad (1)$$

where x and y are the spatial coordinates, t is the time, and $\mathbf{W}$, $\mathbf{F}(\mathbf{W})$, $\mathbf{G}(\mathbf{W})$ and $\mathbf{S}(\mathbf{W})$ are defined as follows:

$$\begin{cases} \mathbf{W} & = (h, \quad hu, \quad hv, \quad Z \quad)^T \\ \mathbf{F(W)} & = \left(hu, \quad hu^2 + \frac{1}{2}gh^2 + ghZ, \quad huv, \quad \frac{1}{1-p}Q_x \right)^T \\ \mathbf{G(W)} & = \left(hv, \quad hvu, \quad hv^2 + \frac{1}{2}gh^2 + ghZ, \quad \frac{1}{1-p}Q_y \right)^T \\ \mathbf{S(W)} & = \left(0, \quad gZ\frac{\partial h}{\partial x}, \quad gZ\frac{\partial h}{\partial y}, \quad 0 \right)^T \end{cases}$$

(2)

In (2) h is the height of the flow above the bottom Z, g is acceleration of gravity and u and v are the velocity components in the x and y directions. The first three equations of (1) are the standard 2D Shallow Water equations, recast as in [8] in order to avoid the singularity of the Jacobian of the flux function. The last one is the well-known Exner equation for the evolution of the bed level. We restrict our attention to the case in which the sediment transport porosity p is constant and the bed-load sediment transport fluxes Q_x and Q_y are defined by the Grass model:

$$Q_x = Au\left(u^2 + v^2\right)^{\frac{m-1}{2}}, \quad Q_y = Av\left(u^2 + v^2\right)^{\frac{m-1}{2}} \qquad (3)$$

where A and $1 \leq m \leq 4$ are experimental constants depending on the particular problem under consideration. The classical case $m = 3$ is considered here.

The numerical method proposed to discretize in space the system of equations (1)-(2) is a finite-volume approach, applicable to unstructured grids. Namely, it is the SRNH scheme introduced in [11]. A brief summary of the main characteristics of the scheme is given herein, for additional details we refer to [1, 11].

The scheme is composed by a predictor and a corrector stage: in the predictor stage an averaged state $\mathbf{U}_{ij}^n$ is computed, then this predicted state is used in the corrector stage to update the solution. The predictor stage is based on primitive variables projected on the normal and tangential directions with respect to the cell interface, $\mathbf{n}$ and $\boldsymbol{\tau}$. Hence, by introducing the normal and tangential components of the velocity, $u_\mathbf{n}$ and u_τ, it is possible to reformulate the system (1) as follows:

$$\frac{\partial \mathbf{U}}{\partial t} + \mathbf{A_n(U)}\frac{\partial \mathbf{U}}{\partial \mathbf{n}} = 0 \qquad (4)$$

$$\mathbf{U} = \begin{pmatrix} h \\ u_\mathbf{n} \\ u_\tau \\ Z \end{pmatrix}, \quad \mathbf{A_n(U)} = \begin{pmatrix} u_\mathbf{n} & h & 0 & 0 \\ g & u_\mathbf{n} & 0 & g \\ 0 & 0 & u_\mathbf{n} & 0 \\ 0 & A(1-p)^{-1}(3u_\mathbf{n}^2 + u_\tau^2) & 2A(1-p)^{-1}u_\mathbf{n}u_\tau & 0 \end{pmatrix}$$

(5)

Starting from (5) it is possible to introduce a Roe average state $\overline{\mathbf{U}}_{ij}$ and a sign matrix $\text{sgn}\left[\mathbf{A_n}(\overline{\mathbf{U}})\right]$ defined as:

$$\overline{\mathbf{U}}_{ij} = \left(\frac{h_i + h_j}{2}, \frac{u_{\mathbf{n},i}\sqrt{h_i} + u_{\mathbf{n},j}\sqrt{h_j}}{\sqrt{h_i} + \sqrt{h_j}}, \frac{u_{\tau,i}\sqrt{h_i} + u_{\tau,j}\sqrt{h_j}}{\sqrt{h_i} + \sqrt{h_j}}, \frac{Z_i + Z_j}{2} \right)^T \quad (6)$$

$$\text{sgn}\left[\mathbf{A_n}(\overline{\mathbf{U}})\right] = \mathscr{R}(\overline{\mathbf{U}}) \Lambda_{\text{sgn}}(\overline{\mathbf{U}}) \mathscr{R}^{-1}(\overline{\mathbf{U}}) \quad (7)$$

where the elements of the diagonal matrix $\Lambda_{\text{sgn}}(\overline{\mathbf{U}})$ are the sign function of the eigenvalues of $\mathbf{A_n}(\overline{\mathbf{U}})$ and $\mathscr{R}(\overline{\mathbf{U}})$ is the corresponding right-eigenvector matrix.

The explicit SRNH scheme is then formulated as follows:

$$\mathbf{U}_{ij}^n = \frac{1}{2}\left(\mathbf{U}_i^n + \mathbf{U}_j^n\right) - \frac{1}{2}\text{sgn}\left[\mathbf{A_n}(\overline{\mathbf{U}}_{ij})\right]\left(\mathbf{U}_j^n - \mathbf{U}_i^n\right) \quad (8)$$

$$\frac{\mathbf{W}_i^{n+1} - \mathbf{W}_i^n}{\Delta^n t} = -\frac{1}{|V_i|} \sum_{j \in N(i)} \mathscr{F}(\mathbf{W}_{ij}^n, \mathbf{n}_{ij})|\Gamma_{ij}| + \mathbf{S}_i^n \quad (9)$$

where $\mathbf{W}_{ij}^n$ is obtained from $\mathbf{U}_{ij}^n$, $N(i)$ is the set of neighboring cells of the ith cell, $|V_i|$ is the area of the cell, Γ_{ij} is the interface between cell i and j, $\Delta^n t$ is the nth time-step and $\mathscr{F}$ is the analytical flux function. $\mathbf{S}_i^n$ is the discretization of the source term which, in order to satisfy the C-property [2] is defined as follows:

$$\begin{cases} \overline{Z}_{x,i}^n = \frac{1}{2} \dfrac{\sum_{j \in N(i)} \left(Z_{ij}^n\right)^2 n_{x,ij}|\Gamma_{ij}|}{\sum_{j \in N(i)} Z_{ij}^n n_{x,ij}|\Gamma_{ij}|}, \quad \overline{Z}_{y,i}^n = \frac{1}{2} \dfrac{\sum_{j \in N(i)} \left(Z_{ij}^n\right)^2 n_{y,ij}|\Gamma_{ij}|}{\sum_{j \in N(i)} Z_{ij}^n n_{y,ij}|\Gamma_{ij}|} \\ \mathbf{S}_i^n = \left(0, \ g\overline{Z}_{x,i}^n \sum_{j \in N(i)} h_{ij}^n n_{x,ij}|\Gamma_{ij}|, \ g\overline{Z}_{y,i}^n \sum_{j \in N(i)} h_{ij}^n n_{y,ij}|\Gamma_{ij}|, \ 0 \right)^T \end{cases} \quad (10)$$

To switch from an explicit scheme to an implicit one it is sufficient, to compute the quantities $\mathscr{F}_{ij}^{n+1} = \mathscr{F}(\mathbf{W}_{ij}^{n+1}, \mathbf{n}_{ij})$ and $\mathbf{S}_i^{n+1}$ instead of $\mathscr{F}(\mathbf{W}_{ij}^n, \mathbf{n}_{ij})$ and $\mathbf{S}_i^n$. However, from a practical point of view this would require the solution of a large non-linear system of equations at each time step. The computational cost for this operation is in general not affordable in practical applications and generally greatly overcomes any advantage that an implicit scheme could have with respect to its explicit counterpart. A common technique to overcome this difficulty is to linearize the numerical scheme, i.e. to find an approximation of $\mathscr{F}_{ij}^{n+1}$ and $\mathbf{S}_i^{n+1}$ in the form:

$$\Delta^n \mathscr{F}_{ij} \simeq D_{1,ij}\Delta^n \mathbf{W}_i + D_{2,ij}\Delta^n \mathbf{W}_j, \quad \Delta^n \mathbf{S}_i \simeq \sum_{j \in \tilde{N}(i)} D_{3,ij}\Delta^n \mathbf{W}_j \quad (11)$$

where $\Delta^n(\cdot) = (\cdot)^{n+1} - (\cdot)^n$ and $\tilde{N}(i) = N(i) \cup \{i\}$. Using this approximation, the following linear system must be solved at each time step:

$$\frac{\mathbf{W}_i^{n+1} - \mathbf{W}_i^n}{\Delta t} + \frac{1}{|V_i|} \sum_{j \in N(i)} |\Gamma_{ij}| \left(D_{1,ij} \Delta^n \mathbf{W}_i + D_{2,ij} \Delta^n \mathbf{W}_j \right) - \sum_{j \in \bar{N}(i)} D_{3,ij} \Delta^n \mathbf{W}_j$$

$$= -\frac{1}{|V_i|} \sum_{j \in N(i)} \mathscr{F}(\mathbf{W}_{ij}^n, \mathbf{n}_{ij}) |\Gamma_{ij}| + \mathbf{S}_i^n \quad (12)$$

The implicit linearized scheme is completely defined once a suitable definition for the matrices $D_{1,ij}, D_{2,ij}, D_{3,ij}$ is given. If the flux function and the source term are differentiable, a common choice is to use the Jacobian matrices. Nevertheless, it is not always possible nor convenient to exactly compute the Jacobian matrices. In fact, it is not unusual to have some lack of differentiability of the numerical flux functions. Furthermore the explicit scheme (9) is composed by a predictor and a corrector stage and this significantly increases the difficulty in linearizing. This problem has been solved herein by computing through the automatic differentiation software Tapenade [7] the flux Jacobians, which are used to approximate $\mathscr{F}_{ij}^{n+1}$ and $\mathbf{S}_i^{n+1}$, as defined in Eq. (11). Given the source code of a routine which computes the explicit numerical fluxes, the differentiation software generates a new source code which computes the flux Jacobians, and, thus, the derivation and the implementation of their analytical expressions can be avoided.

The extension to second-order accuracy in space can be achieved by using a classical MUSCL technique [9], in which (8) is computed by using extrapolated values at the cell interfaces. The extrapolation is done here as in [3] associated with the Minmod slope limiter. For the explicit scheme, second-order accuracy in time is achieved through a two-step Runge-Kutta scheme. Considering the implicit case, it is possible to obtain a space and time second-order accurate formulation by considering the MUSCL technique for space as previously defined and a second-order backward differentiation formula in time. However, the linearization for the second-order accurate fluxes and source terms and the solution of the resulting linear system implies significant computational costs and memory requirements. Thus, a defect-correction technique [10] is used here, which consists in iteratively solving simpler problems obtained, just considering the same linearization as used for the first-order scheme. Thus defining $\mathscr{W}^0 = \mathbf{W}^n$, the defect-correction iterations write as follows, the unknown being $\Delta^s \mathscr{W}_i$:

$$\frac{(1+2\tau)}{\Delta^n t (1+\tau)} \Delta^s \mathscr{W}_i + \frac{1}{|V_i|} \sum_{j \in N(i)} |\Gamma_{ij}| \left(D_{1,ij} \Delta^s \mathscr{W}_i + D_{2,ij} \Delta^s \mathscr{W}_j \right) - \sum_{j \in \bar{N}(i)} D_{3,ij} \Delta^s \mathscr{W}_j$$

$$= \frac{(1+2\tau)\mathscr{W}_i^s - (1+\tau)^2 \mathbf{W}_i^n + \tau^2 \mathbf{W}_i^{n-1}}{\Delta^n t (1+\tau)} - \frac{1}{|V_i|} \sum_{j \in N(i)} \mathscr{F}\left(\mathscr{W}_{ij}^s, \mathscr{W}_{ji}^s \right) |\Gamma_{ij}| + [\mathbf{S}_2]_i^s$$

$$(13)$$

for $s = 0, \cdots, r - 1$. In (13), $\tau = \frac{\Delta^n t}{\Delta^{n-1} t}$, $D_{1,ij}, D_{2,ij}, D_{3,ij}$ are the matrices of the approximation (11) and the update solution is $\mathbf{W}^{n+1} = \mathscr{W}^r$. It can be shown

[10] that only one defect-correction iteration is theoretically needed to reach a second-order accuracy while few additional iterations (one or two) can improve the robustness.

3 Numerical Experiments

The 2D test case considered herein is a well-known benchmark test, proposed in several papers (see, e.g. [1,5]). It is a sediment transport problem in a square domain Ω of dimensions $1000 \times 1000 \ m^2$ with a non constant bottom relief. The initial bottom topography is defined as follows:

$$Z(0,x) = \sin^2\left(\frac{(x-300)\pi}{200}\right)\sin^2\left(\frac{(y-400)\pi}{200}\right) \text{ if } (x,y) \in Q_h, \ 0 \text{ elsewhere}$$
(14)

where $Q_h = [300, 500] \times [400, 600]$. Given $Z(0, x)$, the remaining initial conditions are $h(0, x, y) = 10 - Z(0, x, y)$, $u(0, x, y) = \frac{10}{h(0,x,y)}$ and $v(0, x, y) = 0$. Considering the boundaries, Dirichlet boundary conditions are imposed at the inlet, while at the outlet characteristic based conditions are used. Finally, free-slip is imposed on the lateral boundaries. The spatial discretization of the computational domain has been carried out by using two different grids: for the first grid GR1, the number of the nodes and the characteristic length of the elements are, respectively, $l_m = 20 \ m$ and $N_c = 2901$. The second grid GR2 is characterized by $l_m = 10 \ m$ and $N_c = 11425$.

Two different values of the parameter A are considered, namely a case with slow interaction between the flow and the bed, $A = 0.001$ and a fast one, $A = 1$. Due to the different time scales for the evolution of the bottom topography, different time intervals have been simulated for the considered cases: the total simulation time is 500 seconds for $A = 1$ and 360000 seconds for $A = 0.001$.

For the slow speed of interaction case, Figure 1a shows a comparison of the results obtained by means of the explicit version of the scheme at CFL $= 0.8$ with those of the implicit one at CFL $= 1000$ both for 1st and 2nd-order accuracy for grid GR2. For the definition of the CFL number we refer to [1]. There is practically no difference between the solutions obtained with the implicit and explicit version of the schemes, while the results obtained at 1st-order of accuracy significantly differ from the 2nd-order ones. Note that the results shown in Fig. 1a for the 2nd-order implicit scheme are computed using only one DeC iteration. By increasing the number of DeC iterations it is possible to further increment, without loosing in accuracy, the CFL number of the 2nd-order implicit scheme (see Fig. 1b). In particular, when 3 DeC iterations are considered it is possible to use a CFL number equal to 10^4 (see also Table 1). As shown in Fig. 1b, similar results can be obtained by considering the grid GR1 instead of the GR2. The profiles of $h + Z$ are shown if Fig. 1c. Slightly larger oscillations are observed for the second-order implicit scheme, but at the first order both schemes gave practically the same results. As

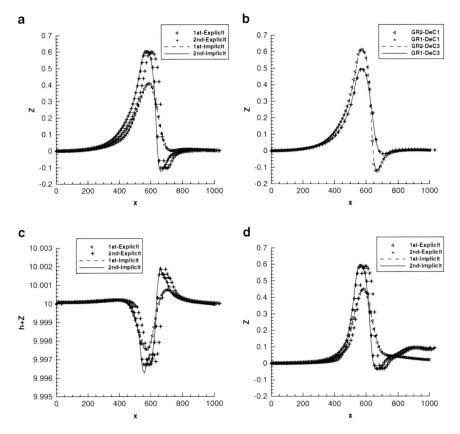

Fig. 1 Comparison of the results given the explicit and implicit schemes; profiles along the line $y = 500$ of: (a) Z for $A = 10^{-3}$ and GR2, (b) Z for implicit schemes and $A = 10^{-3}$, (c) $h + Z$ for $A = 10^{-3}$ GR2, (d) Z for $A = 10^0$ and GR2

for the computational costs, Table 1 shows that already at CFL = 1000 the gain in CPU time obtained with the implicit scheme is large, both for 1st and 2nd-order of accuracy. The CPU gain obtained with the implicit scheme is significantly larger for 2nd-order accuracy. Indeed, when the implicit formulation is used, there are not significant differences, in terms of CPU time, between the 1st and 2nd-order simulations. Instead in the explicit case an important computational cost increase is observed to reach 2nd-order accuracy: the 2nd-order approach is $\simeq 2.4$ times more expensive than the 1st-order one. As a consequence, already at CFL = 1000 using 1 DeC iteration the 2nd-order implicit approach is more than 60 times faster than the explicit one on GR1 and about 30 times faster on GR2. The CPU gain of the 2nd-order implicit approach can be further increased considering 3 DeC iterations and CFL = 10^4. For the fast speed of interaction case, to avoid loss of accuracy the CFL number of the implicit scheme must be lowered down to 1. On the other hand, by increasing the number of DeC iterations, it is possible to increase the maximum

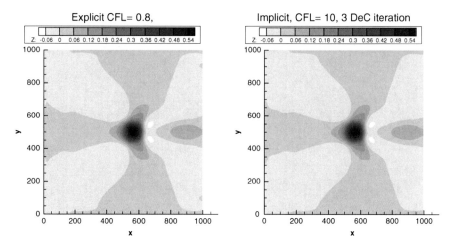

Fig. 2 Comparison of the results for the bed profile of the 2nd-order scheme, $A = 1$, Grid GR2

CFL value by a factor 10. As an example Fig. 2 shows a comparison between the explicit and implicit approach at different CFL values for the grid GR2. Due to the reduced CFL number achievable without loss of accuracy for the implicit scheme in this test case the computational cost of the implicit scheme is larger than for the explicit one, both at first and second order of accuracy, as it is shown in Table 1. Summarizing, in order to avoid loss of accuracy, the CFL number of the implicit scheme must be reduced to a value roughly inversely proportional to the velocity of the interaction between the flow and the bed-load. Also, the increase of the number of DeC iterations allows the maximum CFL number achievable without loosing in accuracy to be increased, and therefore the simulation CPU time is reduced. The implicit code has been found to be computationally more efficient than the explicit one for slow rates of the interaction between the bed and the flow.

Table 1 CPU time required for the considered simulations, comparison between explicit and implicit approach, both at first and second-order of accuracy

Method	$A = 0.001$			$A = 1$		
	GR1	GR2	CFL	GR1	GR2	CFL
Explicit 1st order	12824s	103238s	0.8	21.0s	169.7s	0.8
Explicit 2nd order	30996s	247215s	0.8	52.4s	409.9s	0.8
Implicit 1st order	323.6s	4336s	10^3	191.5s	1541s	10^0
Implicit 2nd order 1 DeC	481.5s	8537s	10^3	198.7s	1582s	10^0
Implicit 2nd order 3 DeC	265.9s	4866s	10^4	74.5s	606.8s	10^1

Acknowledgements This work has been realized in the framework of the EuroMéditerranée $3+3$ network MhyCoF.

References

1. F. Benkhaldoun, S. Sahmim, M. Seaïd. *A two-dimensional finite volume morphodynamic model on unstructured triangular grids.* Int. J. Numer. Meth. Fluids, 63:1296–1327, 2010.
2. A. Bermudez, M.E. Vazquez. *Upwind methods for hyperbolic conservation laws with source terms.* Computers & Fluids, 23(8):1049–1071, 1994.
3. S. Camarri, M.V. Salvetti, B. Koobus, A. Dervieux. *A low-diffusion MUSCL scheme for LES on unstructured grids.* Computers & Fluids, 33:1101–1129, 2004.
4. M.J. Castro Díaz, E.D. Fernández-Nieto, A.M. Ferreiro. *Sediment transport models in shallow water equations and numerical approach by high order finite volume methods.* Computers & Fluids, 37(3):299–316, 2008.
5. M.J. Castro Díaz, E.D. Fernández-Nieto, A.M. Ferreiro, C. Parés. *Two-dimensional sediment transport models in shallow water equations. A second order finite volume approach on unstructured meshes.* Computer Meth. Appl. Mech. Eng., 198:2520–2538, 2009.
6. A.J. Grass. *Sediments transport by waves and currents.* Tech. Rep., SERC London Cent. Mar. Technol., Report No. FL29, 1981.
7. L. Hascoët, V. Pascual. *TAPENADE 2.1 User's Guide.* Tech. Rep. n 300. INRIA, 2004.
8. J. Hudson, P. K. Sweby. *Formulations for Numerically Approximating Hyperbolic Systems Governing Sediment Transport.* J. Sci. Comput., 19:225-252, 2003.
9. B. van Leer. *Towards the ultimate conservative difference scheme V: a second-order sequel to Godunov's method.* J. Comput. Phys., 32(1):101–136, 1979.
10. R. Martin, H. Guillard. *A second order defect correction scheme for unsteady problems.* Computers & Fluids, 25(1):9–27, 1996.
11. S. Sahmim, F. Benkhaldoun, F. Alcrudo. *A sign matrix based scheme for non-homogeneous PDE's with an analysis of the convergence stagnation phenomenon.* J. Comput. Phys., 226(2):1753–1783, 2007.

The paper is in final form and no similar paper has been or is being submitted elsewhere.

Numerical Simulation of the Flow in a Turbopump Inducer in Non-Cavitating and Cavitating Conditions

M. Bilanceri, F. Beux, and M.V. Salvetti

Abstract A numerical methodology for the simulation of cavitating flows in real complex geometries is presented. A homogeneous-flow cavitation model, accounting for thermal effects and active nuclei concentration, which leads to a barotropic state law is adopted. The continuity and momentum equations are discretized through a mixed finite-element/finite-volume approach, applicable to unstructured grids. A robust preconditioned low-diffusive HLL scheme is used to deal with all speed barotropic flows. Second-order accuracy in space is obtained through MUSCL reconstruction. Time advancing is carried out by a second-order implicit linearized formulation together with the Defect Correction technique. The flow in a real 3D inducer for rockets turbopumps is simulated for a wide range of conditions: different flow rates and rotating speeds as well as non-cavitating and cavitating flows are considered. The results obtained with this numerical approach are compared with experimental data.

Keywords cavitating flows, homogeneous flow model, low diffusive HLL scheme, linearized implicit time advancing
MSC2010: 65-06

1 Introduction

A tool for numerical simulation of 3D compressible flows satisfying a barotropic equation of state is presented in this work. In particular, we are interested in simulating cavitating flows through the barotropic homogeneous flow model proposed in [1]. The numerical method used in this work is based on a mixed

---

M. Bilanceri and M.V. Salvetti
University of Pisa (Italy), e-mail: marco.bilanceri@gmail.com, mv.salvetti@ing.unipi.it

F. Beux
Alta S.p.A., Pisa (Italy), e-mail: f.beux@alta-space.com

finite-element/finite-volume spatial discretization on 3D unstructured grids. Viscous fluxes are discretized using P1 finite-elements while for the convective fluxes the LD-HLL scheme [2], a low-diffusive modification of the Rusanov scheme, is adopted. Second-order in space is obtained using a MUSCL reconstruction technique and time-consistent preconditioning is introduced to deal with the low Mach number regime. A linearized implicit time-advancing is associated to a defect-correction technique to obtain a second-order accurate (both in time and space) formulation at a limited computational cost. A non inertial reference frame, rotating at constant angular velocity, is used to account for possible solid-body rotation and the standard $k - \varepsilon$ turbulence model is introduced to capture turbulence effects. The considered numerical tool is used to simulate the flow in a real 3D inducer in both non-cavitating and cavitating conditions.

2 Physical model and numerical method

The physical model considered in this work consists in the standard Navier-Stokes equations for a barotropic flow. Due to the barotropic equation of state (EOS) considered, the energy equation can be discarded since it is decoupled from the mass and momentum balances. Thus, considering a reference frame rotating with constant angular velocity ω, the following system of equations is obtained:

$$\frac{\partial \mathbf{W}}{\partial t} + \frac{\partial}{\partial x_j} F_j(\mathbf{W}) - \frac{\partial}{\partial x_j} \mu V_j(\mathbf{W}, \nabla \mathbf{W}) = \mathbf{S}(\omega, \mathbf{x}, \mathbf{W}) \qquad (1)$$

In Eq. (1) the Einstein notation is used, μ is the molecular viscosity of the fluid, $\mathbf{W} = (\rho, \rho u_1, \rho u_2, \rho u_3)^T$ is the unknown vector, where ρ is the density and u_i the velocity component in the ith direction. $\mathbf{F} = (F_1, F_2, F_3)$ and $\mathbf{V} = (V_1, V_2, V_3)$ are, respectively, the convective fluxes and the diffusive ones (not shown here for sake of brevity). Finally, $\mathbf{S}$ is the source term appearing in a frame of reference rotating with constant speed ω:

$$\begin{cases} \overline{\mathbf{S}} = -(2\omega \wedge \rho \mathbf{u} + \rho \omega \wedge (\omega \wedge \mathbf{x})) \\ \mathbf{S}(\omega, \mathbf{x}, \mathbf{W}) = \left(0, \overline{\mathbf{S}}^T\right)^T \end{cases} \qquad (2)$$

System (1) is completely defined once a suitable constitutive equation $p = p(\rho)$ is introduced. In this work a weakly-compressible liquid at constant temperature T_L is considered as working fluid. The liquid density ρ is allowed to locally fall below the saturation limit $\rho_{Lsat} = \rho_{Lsat}(T_L)$ thus originating cavitation phenomena. A regime-dependent (wet/cavitating) constitutive relation is therefore adopted. As for the wet regime ($\rho \geq \rho_{Lsat}$), a barotropic model of the form

$$p = p_{sat} + \frac{1}{\beta_{sL}} \ln\left(\frac{\rho}{\rho_{Lsat}}\right) \qquad (3)$$

is adopted, $p_{sat} = p_{sat}(T_L)$ and $\beta_{sL} = \beta_{sL}(T_L)$ being the saturation pressure and the liquid isentropic compressibility, respectively. As for the cavitating regime ($\rho < \rho_{Lsat}$), a homogeneous-flow model explicitly accounting for thermal cavitation effects and for the concentration of the active cavitation nuclei in the pure liquid has been adopted [1]:

$$\frac{p}{\rho}\frac{d\rho}{dp} = (1-\alpha)\left[(1-\varepsilon_L)\frac{p}{\rho_{Lsat}a_{Lsat}^2} + \varepsilon_L g^\star \left(\frac{p_c}{p}\right)^\eta\right] + \frac{\alpha}{\gamma_V} \qquad (4)$$

where $g^\star$, η, γ_V and p_c are liquid parameters, a_{Lsat} is the liquid sound speed at saturation, $\alpha = 1 - \rho/\rho_{Lsat}$ and $\varepsilon_L = \varepsilon_L(\alpha, \zeta)$ is a given function (see [1] for its physical interpretation and for more details). The resulting unified barotropic state law for the liquid and for the cavitating mixture only depends on the two parameters T_L and ζ. For instance, for water at $T_L = 293.16 K$, the other parameters involved in (3) and (4) are: $p_{sat} = 2806.82$ Pa, $\rho_{Lsat} = 997.29$ kg/m³, $\beta_{sL} = 5\,10^{-10}$ Pa^{-1}, $g^\star = 1.67$, $\eta = 0.73$, $\gamma_V = 1.28$, $p_c = 2.21\,10^7$ Pa and $a_{Lsat} = 1415$ m/s [6]. Note that despite the model simplifications leading to a unified barotropic state law, the transition between wet and cavitating regimes is extremely abrupt. Indeed, the sound speed falls from values of order 10^3 m/s in the pure liquid down to values of order 0.1 or 1 m/s in the mixture. The corresponding Mach number variation renders this state law very stiff from a numerical viewpoint. As for the definition of the molecular viscosity, a simple model, which is linear in the cavitating regime, is considered:

$$\mu(\rho) = \begin{cases} \mu_L & \text{if } \rho \geq \rho_{Lsat} \\ \mu_v & \text{if } \rho \leq \rho_v \\ \alpha\mu_v + (1-\alpha)\mu_L & \text{otherwise} \end{cases} \qquad (5)$$

in which μ_v and μ_L are the molecular viscosity of the vapor and of the liquid respectively, which, consistently with the assumptions made in the adopted cavitation model, are considered constant and computed at $T = T_L$.

The spatial discretization of the governing equations is based on a mixed finite-element/finite-volume formulation on unstructured grids. Starting from an unstructured tetrahedral grid, a dual finite-volume tessellation is obtained by the rule of medians. The semi-discrete balance applied to cell C_i reads (not accounting for boundary contributions):

$$V_i \frac{d\mathbf{W}_i}{dt} + \sum_{j \in N(i)} \Phi_{ij} + \Upsilon_i = \Omega_i \qquad (6)$$

where $\mathbf{W}_i$ is the semi-discrete unknown associated with C_i, V_i is the cell volume, and $N(i)$ represents the set of neighbors of the ith cell. The numerical discretization of the convective flux crossing the boundary ∂C_{ij} shared by C_i and C_j (positive towards C_j) is denoted Φ_{ij}, while Υ_i and Ω_i are the numerical discretizations for, respectively, the viscous fluxes and the source term. Let us describe, first, the first-order version of the used numerical method. Once defined $\mathbf{n}_{ij} = (n_{ij,1}, n_{ij,2}, n_{ij,3})^T$

as the integral over ∂C_{ij} of the outer unit normal to the cell boundary, it is possible to approximate Φ_{ij} by the following preconditioned flux function:

$$\Phi_{ij} = \frac{n_{ij,k}\left(F_k(\mathbf{W}_i) + F_k(\mathbf{W}_j)\right)}{2} -$$
$$\frac{1}{2}\begin{pmatrix} \lambda_1^p & 0 & 0 & 0 \\ 0 & (\Delta^{32}\lambda^p)\,n_{ij,1}^2 + \lambda_3^p & (\Delta^{32}\lambda^p)\,n_{ij,1}n_{ij,2} & (\Delta^{32}\lambda^p)\,n_{ij,1}n_{ij,3} \\ 0 & (\Delta^{32}\lambda^p)\,n_{ij,2}n_{ij,1} & (\Delta^{32}\lambda^p)\,n_{ij,2}^2 + \lambda_3^p & (\Delta^{32}\lambda^p)\,n_{ij,2}n_{ij,3} \\ 0 & (\Delta^{32}\lambda^p)\,n_{ij,3}n_{ij,1} & (\Delta^{32}\lambda^p)\,n_{ij,3}n_{ij,2} & (\Delta^{32}\lambda^p)\,n_{ij,3}^2 + \lambda_3^p \end{pmatrix}(\mathbf{W}_j - \mathbf{W}_i)$$
(7)

where $\Delta^{32}\lambda^p = \lambda_2^p - \lambda_3^p$, $\lambda_1^p = \theta^{-1}\lambda_{ij}$, $\lambda_2^p = \theta\lambda_{ij}$, $\lambda_3^p = \lambda_{ij}$ and the parameters θ and λ_{ij} are defined as follows:

$$\theta = \theta(M) = \begin{cases} 10^{-6} & \text{if } M \leq 10^{-6} \\ \min(M, 1) & \text{otherwise} \end{cases}, \quad M = \frac{|\tilde{\mathbf{u}}_{ij}|}{\tilde{a}_{ij}}, \quad \lambda_{ij} = \tilde{u}_{ij} + \tilde{a}_{ij} \quad (8)$$

$\tilde{\mathbf{u}}_{ij}$ and $\tilde{a}_{ij}$ being the Roe averages for, respectively, the velocity and the sound of speed. The discretization (7) is the 3D extension of LD-HLL scheme defined in [2] as a low diffusive modification of the Rusanov scheme.

The discretization of the viscous fluxes is instead based on P1 finite-elements in which the test functions are linear functions on the tetrahedral element. The source term is discretized as follows:

$$\Omega_i := \begin{pmatrix} 0 \\ -2\,\omega \wedge \rho_i\mathbf{u}_i + \rho_i\,\mathbf{r}_i \end{pmatrix} \quad \mathbf{r}_i := -\omega \wedge (\omega \wedge \mathbf{g}_i) \quad (9)$$

$\mathbf{g}_i$ being the centroid of the ith cell.

A first-order implicit Euler method can be used for time-advancing. As a consequence, at each time step it is necessary to compute $\mathscr{F}_i^{n+1} = \mathscr{F}(\mathbf{W}_j^{n+1}, j \in \bar{N}(i))$, where $\bar{N}(i) = N(i) \cup \{i\}$ and $\mathscr{F}_i$ is defined as $\mathscr{F}_i = \sum_{j \in N(i)} \Phi_{ij} + \Upsilon_i - \Omega_i$.

In order to avoid the direct solution of large non-linear system of equations at each time step a linearization can be performed finding an approximation of $\mathscr{F}_i^{n+1}$ in the form:

$$\Delta^n \mathscr{F}_i \simeq \sum_{j \in \bar{N}(i)} D_{ij}\,\Delta^n \mathbf{W}_j \quad (10)$$

where $\Delta^n(\cdot) = (\cdot)^{n+1} - (\cdot)^n$. Using this approximation, the following linear system must be solved at each time step:

$$|V_i|\frac{\mathbf{W}_i^{n+1} - \mathbf{W}_i^n}{\Delta t} + \sum_{j \in \bar{N}(i)} D_{ij}\,\Delta^n \mathbf{W}_j = -\mathscr{F}(\mathbf{W}_j^n, j \in \bar{N}(i)) \quad (11)$$

The implicit linearized scheme is completely defined once a suitable definition for the matrices D_{ij} is given. Since viscous and source terms are easily differentiable, the use of the Jacobian matrices has been considered here to compute their contribution to D_{ij}. However the computation of the Jacobian matrices can be more challenging for the convective fluxes. Thus, in this work the approximate linearization developed in [2] for the numerical flux function (7) has been used. Once matrices D_{ij} are given, the first-order numerical method (11) is completely defined.

Since viscous and source terms are already second-order accurate in space, the extension to second-order accuracy in space can be achieved by simply using a classical MUSCL technique [4], in which the convective fluxes are computed by using extrapolated values at the cell interfaces. The second-order accuracy in time is then achieved through the use of a backward differentiation formula. However, the linearization for the second-order accurate fluxes and the solution of the resulting linear system imply significant computational costs and memory requirements. Thus, a defect-correction technique [5] is used here. It consists in iteratively solving simpler problems obtained just considering the same linearization as used for the first-order scheme. The number of DeC iterations r is typically chosen equal to 2. Indeed, it can be shown [5] that only one defect-correction iteration is theoretically needed to reach a second-order accuracy while few additional iterations (one or two) can improve the robustness.

Finally, in order to account for the turbulence effects the RANS approach together with the standard turbulence model $k - \varepsilon$ have been used. For the sake of brevity the additional terms introduced in the system of equation by this model are not shown. We just mention that the convective and viscous turbulent fluxes are discretized using the same methods considered for their laminar counterparts. Similarly, the turbulent source term appearing in the equations for k and ε is discretized using the same approach considered for the source term associated to the rotating frame of reference.

3 Numerical experiments

In this section the numerical tool described in Sect. 1 is applied to the simulations of the flow in a real three blade axial inducer [6]. It is a three blade inducer with a tip blade radius of 81 mm and 2 mm radial clearance between the blade tip and the external case. Experimental data are available for all the numerical simulations described in the following. In particular the pressure jump between two different stations has been measured for a wide range of working conditions: from small to large mass flow rates, non-cavitating and cavitating conditions and different values of the rotational speed ω_z. The results are presented in terms of the mean adimensionalized pressure jump Ψ as a function of the adimensionalized discharge Φ:

$$\Psi = \frac{\Delta P}{\rho_L \omega_z^2 R_T^2} \qquad \Phi = \frac{Q}{\pi R_T^2 \omega_z R_T} \tag{12}$$

where Q is the discharge, R_T is the radius of the tip of blade, ρ_L the density of the liquid and ω_z is the angular velocity. Note that the numerical pressure jump is averaged over one complete revolution of the inducer. A cylindrical computational domain is used, whose external surface is coincident with the inducer case. The inlet is placed 249 mm ahead of the inducer nose and the outlet is placed 409 mm behind. A second computational domain, characterized by a larger streamwise length (the inlet 1120 mm ahead the inducer nose) has also been considered. Two different grids have been generated to discretize the shorter domain: the basic one G1 (1926773 cells) and G2 (3431721 cells) obtained from G1 by refining the region between the blade tip and the external case. In particular, inside the tip clearance region there are 3−4 nodes for the grid G1, while there are 9−10 points for G2. The larger domain has been discretized by grid G1L (2093770 cells), which coincides with G1 in the original domain. The working conditions considered in this work are shown in Table 1, where p_{out} is the outlet pressure of the flow and $\sigma = \dfrac{p - p_{Lsat}}{0.5\rho\omega_z^2 R_T^2}$ is the cavitating number (only shown for cavitating simulations). Note that, except when differently stated, the simulations do not include turbulence effects.

Table 1 Conditions of the numerical simulations and of the experiments

Benchmark	Ind1	Ind2	Ind3	Ind4	Ind5	Ind6
Φ	0.0584	0.0391	0.0185	0.0531	0.0531	0.0531
ω_z (rpm)	1500	1500	1500	3000	3000	3000
p_{out} (Pa)	125000	125000	125000	60000	85000	82500
T (C°)	25°	25°	25°	16.8°	16.8°	16.8°
σ	-	-	-	0.056	0.084	0.077

As shown in Table 1, all the simulations in non cavitating conditions use the same rotational velocity of 1500 rpm. In the $\Phi - \Psi$ plane the experimental curves of the performances of the inducer are roughly independent from the rotational velocity ω_z [6]. As a consequence, validating the numerical tool for a specific rotational velocity and different flow rates should validate the proposed numerical tool for a generic rotational velocity. Table 2 shows the results for the non-cavitating simulations. It clearly appears that the lower is the discharge Φ, the worse are the results. Already with the coarsest grid G1, rather satisfactory results are obtained for intermediate and high discharge values, Ind2 and Ind1, respectively. Furthermore the quantitative agreement is further improved considering the more refined grid G2 for the case Ind2. Conversely, for the low discharge case, Ind3, the simulations with the grid G1 and G2 greatly overestimate the pressure jump by, respectively, 41% and 30%. The magnitude of this error could be ascribed to the backward flow between the inducer blades and the external case. The correct resolution of this flow is of crucial importance for the determination of the performance of an inducer. Since the smaller is the mass flow rate the greater is the backflow, we investigated

Table 2 Pressure jump in non-cavitating conditions

	Experimental Ψ	Numerical Ψ	Error%
G1-Ind1	0.122	0.114	−6.6%
G1-Ind2	0.186	0.204	+9.7%
G2-Ind2	0.186	0.179	−3.8%
G1-Ind3	0.214	0.302	+41%
G2-Ind3	0.214	0.278	+30%
G1L-Ind3	0.214	0.297	+39%
G1L-Ind3-T	0.214	0.239	+12%

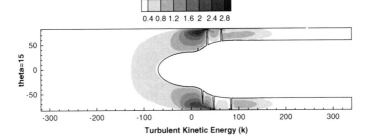

Fig. 1 Cross section of the averaged k field at $\theta = 15°$, simulation G1L-Ind3 (view of the shorter domain)

two possible explanations of this behavior. The first one was that the distance of the inlet from the inducer nose was not large enough to avoid spurious effects on the solution, the second one was that for this case turbulence effects have to be included. The results of the first simulations for the longer computational domain, G1L-Ind3, show that even if there is a small effect, a decrease from 41% to 39%, this is not the source of the error. Instead the results of the simulation G1L-Ind3-T, i.e. the one done considering the RANS model, show that in this case turbulence is a key-issue. Indeed, in this case the error falls down to 12%, less than the error obtained with the refined grid G2 in laminar conditions. As expected the effects of turbulence are particularly important near the gap between the blades and the external case, as it is shown by Fig. 1 by considering the isocontours of k. This strongly affects the backflow and, thus, the pressure jump. This also explains why for larger flow rates, for which the backflow is less important, the effects of turbulence are not so strong and a good agreement with experimental data can be obtained also in laminar simulations.

The mass flow rate for the cavitating cases is large enough to prevent the issues related to the backflow previously described, thus only laminar simulations are considered. The results for the cavitating conditions, reported in Table 3, show that the first grid G1 is not enough refined to correctly describe cavitation for this case. The pressure jump is greatly overestimated. For these conditions the error is related to the underestimation of the cavitating region: the experimental data for $\sigma = 0.056$ show a large cavitating zone and consequently the performance of the

Table 3 Numerical results for the cavitating simulations

	Experimental Ψ	Numerical Ψ	Error%
G1-Ind4	0.105	0.143	+36%
G2-Ind5	0.143	0.130	−8.9%
G2-Ind6	0.137	0.130	−5.0%

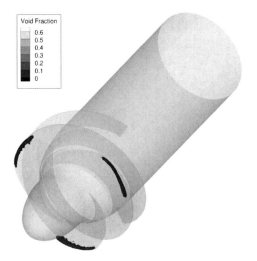

Fig. 2 Isocontours of the cavitating region, $\alpha = 0.005$, for the simulation G2-Ind6

inducer is significantly deteriorated. Instead, in the simulation with grid G1 the extension of the cavitating region is greatly underestimated and, as a consequence, the "numerical" performance of the inducer is similar to the non cavitating case. Grid refinement is particularly effective as shown by the results for the simulations, G2-Ind5 and G2-Ind6. The error in the prediction of the pressure jump is reduced and the extension of the cavitating region, even if it is still underestimated, is closer to the one found in experiments, as it is shown by Fig. 2 which plots the isocontours of the void fraction, corresponding to the cavitating region. Note that when the coarse grid G1 is used the cases Ind5 and Ind6 are not cavitating.

Acknowledgements The support of the European Space Agency under Contract number 20081/06/NL/IA is gratefully acknowledged. The authors also wish to thank the Italian Computer Center CASPUR for having provided computational resources and support.

References

1. L. d'Agostino, E. Rapposelli, C. Pascarella, A. Ciucci *A Modified Bubbly Isenthalpic Model for Numerical Simulation of Cavitating Flows.* 37th AIAA/ASME/SAE/ASEE Joint Propulsion Conference, Salt Lake City, UT, USA, 2001.

2. M. Bilanceri, F. Beux, M.V. Salvetti *An Implicit Low-Diffusive HLL Scheme with Complete Time Linearization: Application to Cavitating Barotropic Flows.* Computer & Fluids, 39(10):1990–2006, 2010.
3. S. Camarri, M.V. Salvetti, B. Koobus, A. Dervieux. *A low-diffusion MUSCL scheme for LES on unstructured grids.* Computers & Fluids, 33:1101–1129, 2004.
4. B. van Leer. *Towards the ultimate conservative difference scheme* V: *a second-order sequel to Godunov's method.* J. Comput. Phys., 32(1):101–136, 1979.
5. R. Martin, H. Guillard. *A second order defect correction scheme for unsteady problems.* Computers & Fluids, 25(1):9–27, 1996.
6. L. Torre, G. Pace, P. Miloro, A. Pasini, A. Cervone, L. d'Agostino, *Flow Instabilities on a Three Bladed Axial Inducer at Variable Tip Clearance.* 13th International Symposium on Transport Phenomena and Dynamics of Rotating Machinery, Honolulu, Hawaii, USA.

The paper is in final form and no similar paper has been or is being submitted elsewhere.

On Some High Resolution Schemes for Stably Stratified Fluid Flows

Tomáš Bodnár and Luděk Beneš

Abstract The aim of this paper is to present some high-resolution numerical methods in the context of the solution of stably stratified flow of incompressible fluid. Two different numerical methods are applied to a simple 2D test case of wall bounded flow and results are compared and discussed in detail with emphasize on the specific features of stratified flows. The two numerical methods are the AUSM finite–volume scheme and the high order compact finite-difference scheme.

Keywords finite–volume, finite–difference, stratification, compact, AUSM
MSC2010: 65M08, 65M06, 76D05, 76D50, 76D33

1 Introduction

The numerical solution of stably stratified fluid flows represents a challenging class of problems in modern CFD. This study was motivated by the air flow in the stably stratified Atmospheric Boundary Layer, where the presence of stratification leads to appearance of gravity waves in the proximity of terrain obstacles. These small–amplitude waves are affecting the flow–field at large distances which is in contrast to the typical non-stratified case, where the flow–field is only affected locally in the close proximity of the obstacle. The wavelength of these waves is governed by the Brunt–Väisälä frequency, i.e depends on the product of the gravity acceleration and the background density gradient.

Tomáš Bodnár
Institute of Thermomechanics, Academy of Sciences of Czech Republic, Dolejškova 5, 182 00 Prague 8, Czech Republic, e-mail: bodnar@marian.fsik.cvut.cz

Luděk Beneš
Department of Tech. Mathematics, Faculty of Mech. Engineering, Czech Technical University in Prague, Karlovo Náměstí13, 121 35 Prague 2 Czech Republic, e-mail: Ludek.Benes@fs.cvut.cz

From the numerical point of view, the simulations of stratified fluid flows are in general more demanding than the solution of similar non-stratified flow cases (see our previous work [10], [4], [1], or [6]). First of all the *model of stratified fluid flow* has to be chosen. Such models are based on variable-density incompressible fluid model including gravity force terms. A simple approximation of such model is developed in Sect. 2. The appearance of the gravity waves in the computational field adds some more constrains on the choice of *numerical scheme and grid*. The limiting factor here is the proper resolution of gravity waves in the whole domain with sufficient number of grid points per wavelength and low amount of numerical dumping to preserve the resolved gravity waves rather than excessively dumping them. Last but not least problem comes with *boundary conditions*. Their proper choice and implementation affects the computational field much strongly than in the non–stratified case.

One of the aims of this paper is to demonstrate that the high-order compact finite-difference schemes offer an interesting alternative to the modern finite-volume discretizations. Beside of the high resolving capabilities of both methods, the compact discretizations have well defined dispersion/diffusion properties and thus can safely be applied to the numerical simulations of wave phenomena. These specific properties of compact discretizations have been successfully used in computational aeroacoustics. This paper is one of the first attempts to use these wave resolving capabilities in the numerical solution of stratified fluid flows.

2 Mathematical Model

Full incompressible model The motion equations describing the flow of incompressible Newtonian fluid could be written in the following general form

$$\frac{\partial u}{\partial x} + \frac{\partial v}{\partial y} + \frac{\partial w}{\partial z} = 0 \tag{1}$$

$$\frac{\partial \rho}{\partial t} + \frac{\partial (\rho u)}{\partial x} + \frac{\partial (\rho v)}{\partial y} + \frac{\partial (\rho w)}{\partial z} = 0 \tag{2}$$

$$\rho \left(\frac{\partial u}{\partial t} + \frac{\partial (u^2)}{\partial x} + \frac{\partial (uv)}{\partial y} + \frac{\partial (uw)}{\partial z} \right) = -\frac{\partial p}{\partial x} + \mu \Delta u \tag{3}$$

$$\rho \left(\frac{\partial v}{\partial t} + \frac{\partial (uv)}{\partial x} + \frac{\partial (v^2)}{\partial y} + \frac{\partial (vw)}{\partial z} \right) = -\frac{\partial p}{\partial y} + \mu \Delta v \tag{4}$$

$$\rho \left(\frac{\partial w}{\partial t} + \frac{\partial (uw)}{\partial x} + \frac{\partial (vw)}{\partial y} + \frac{\partial (w^2)}{\partial z} \right) = -\frac{\partial p}{\partial z} + \mu \Delta w + \rho g \tag{5}$$

The governing system (1)–(5) for unknowns u, p and ρ is sometimes called the Non-homogeneous (incompressible) Navier-Stokes equations.

The small perturbation approximation Now we will assume that the pressure and density fields are perturbation of hydrostatic equilibrium state, i.e.:

$$\rho(x,y,z,t) = \rho_0(z) + \rho'(x,y,z,t) \qquad (6)$$

$$p(x,y,z,t) = p_0(z) + p'(x,y,z,t) \qquad (7)$$

where the background density and pressure fields are linked by the hydrostatic relation:

$$\frac{\partial p_0}{\partial z} = \rho_0 g. \qquad (8)$$

The small perturbation approximation of momentum equations is obtained by introducing the above decomposition of density and pressure into the momentum equations (3), (4) and (5). The density perturbation ρ' is neglected on the left–hand side while on the right–hand side it is retained. On the right–hand side we have removed the hydrostatic pressure using the relation (8) and the fact that according to (7) the horizontal parts of the background pressure gradient are zero.

$$\frac{\partial u}{\partial x} + \frac{\partial v}{\partial y} + \frac{\partial w}{\partial z} = 0 \qquad (9)$$

$$\frac{\partial \rho'}{\partial t} + \frac{\partial (\rho' u)}{\partial x} + \frac{\partial (\rho' v)}{\partial y} + \frac{\partial (\rho' w)}{\partial z} = -w \frac{\partial \rho_0}{\partial z} \qquad (10)$$

$$\frac{\partial u}{\partial t} + \frac{\partial (u^2)}{\partial x} + \frac{\partial (uv)}{\partial y} + \frac{\partial (uw)}{\partial z} = \frac{1}{\rho_0}\left(-\frac{\partial p'}{\partial x} + \mu \Delta u\right) \qquad (11)$$

$$\frac{\partial v}{\partial t} + \frac{\partial (uv)}{\partial x} + \frac{\partial (v^2)}{\partial y} + \frac{\partial (vw)}{\partial z} = \frac{1}{\rho_0}\left(-\frac{\partial p'}{\partial y} + \mu \Delta v\right) \qquad (12)$$

$$\frac{\partial w}{\partial t} + \frac{\partial (uw)}{\partial x} + \frac{\partial (vw)}{\partial y} + \frac{\partial (w^2)}{\partial z} = \frac{1}{\rho_0}\left(-\frac{\partial p'}{\partial z} + \mu \Delta w + \rho' g\right) \qquad (13)$$

This model is in 2D (x–z) version used for all the simulations presented in this work.

3 Numerical Methods

Two different numerical methods were chosen to perform a comparative study allowing for cross-comparison of results. The first method is the AUSM finite–volume scheme. For comparison, the compact finite–difference schemes were implemented.

3.1 AUSM Finite–Volume Scheme

This method has been chosen to represent the modern high resolution finite–volume schemes. This particular scheme was previously used for the simulation of stratified flow and compared successfully with other methods in [2], [3].

Space discretizations For numerical solution the artificial compressibility method in dual time was used. Continuity equation is rewritten in the form (in 2D, x–z plane)

$$\frac{\partial p}{\partial \tau} + \beta^2 \left(\frac{\partial u}{\partial x} + \frac{\partial w}{\partial z} \right) = 0$$

where τ is the artificial time. The equations (9)–(13) rewritten in the 2D conservative form are

$$PW_t + F(W)_x + G(W)_y = S(W).$$

Here $W = [\rho', u, v, p]^T$, $F = F^i - \nu F^\nu$ and $G = G^i - \nu G^\nu$ contain the inviscid fluxes F^i, G^i and viscous fluxes F^ν and G^ν, S is the source term, and $P = diag(1, 1, 1, 0)$. the fluxes and the source term are

$$F^i(W) = [\rho'u, u^2 + p, uw, \beta^2 u]^T, \qquad G^i(W) = [\rho'w, uw, w^2 + p, \beta^2 w]^T, \quad (14)$$

$$F^\nu(W) = [0, u_x, w_x, 0]^T, \quad G^\nu(W) = [0, u_y, w_y, 0]^T, \quad S(W) = [-wd\rho_0/dz, 0, \rho'g, 0]^T.$$

The finite volume AUSM scheme was used for spatial discretizations of the inviscid fluxes:

$$\int_\Omega (F^i_x + G^i_y) dS = \oint_{\partial\Omega} (F^i n_x + G^i n_y) dl \approx \sum_{k=1}^4 \left[u_n \begin{pmatrix} \varrho \\ u \\ w \\ \beta^2 \end{pmatrix} + p \begin{pmatrix} 0 \\ n_x \\ n_y \\ 0 \end{pmatrix} \right]_{L/R} \Delta l_k$$
(15)

where n is normal vector, u_n is normal velocity vector, and $(q)_{L/R}$ are quantities on left/right hand side of the face respectively. These quantities are computed using MUSCL reconstruction with Hemker–Koren limiter.

$$q_R = q_{i+1} - \frac{1}{2}\delta_R \quad q_L = q_i + \frac{1}{2}\delta_L$$

$$\delta_{L/R} = \frac{a_{L/R}(b_{L/R}^2 + 2) + b_{L/R}(2a_{L/R}^2 + 1)}{2a_{L/R}^2 + 2b_{L/R}^2 - a_{L/R}b_{L/R} + 3}$$

$$a_R = q_{i+2} - q_{i+1} \quad a_L = q_{i+1} - q_i \quad b_R = q_{i+1} - q_i \quad b_L = q_i - q_{i-1}$$

The viscous fluxes are discretized in central way on a dual mesh. This scheme is formally of the second order of accuracy in space.

Time integration For the finite–volume AUSM scheme a fully unsteady solver was used. The dual time stepping approach was adopted, so the separate time–discretizations were needed for physical and artificial time. The derivative with respect to the physical time t is discretized by the second order BDF formula,

$$P\frac{3W^{n+1} - 4W^n + W^{n-1}}{2\Delta t} + F_x^{n+1}(W) + G_y^{n+1}(W) = S^{n+1} \quad (16)$$

$$Rez^{n+1}(W) = P(\frac{3}{2\Delta t}W^{n+1} - \frac{2}{\Delta t}W^n + \frac{1}{2\Delta t}W^{n-1}) + F_x^{n+1}(W) + G_y^{n+1}(W) - f^{n+1} - S^{n+1}.$$

Arising system of equations is solved by artificial compressibility method in the dual (artificial) time τ by an explicit 3–stage Runge-Kutta method.

3.2 Compact Finite-Difference Schemes

Here again the artificial compressibility method was used. The solver is limited to steady problems solution, employing the time–marching method.

Space discretizations The spatial discretizations used in this work is directly based on the paper [8], where the class of very high order compact finite difference schemes was introduced and analyzed. The main idea used to construct this family of schemes is that instead of approximating the spatial derivatives ϕ' of certain quantity ϕ explicitly from the neighboring values ϕ_i, the (symmetric) linear combination of neighboring derivatives $(\ldots, \phi'_{i-1}, \phi'_i, \phi'_{i+1}, \ldots)$ is approximated by weighted average of central differences.

The simplest compact finite–difference schemes use the approximation in the form

$$a\phi'_{i-1} + \phi'_i + a\phi'_{i+1} = \alpha_1 \frac{\phi_{i+1} - \phi_{i-1}}{2h} + \alpha_2 \frac{\phi_{i+2} - \phi_{i-2}}{4h} \quad (17)$$

Here $h = x_i - x_{i-1}$ is the spatial step, while a and α_k are the coefficients determining the specific scheme within the family described by (17). It is easy to see that e.g. for $a = 0$, $\alpha_1 = 1$ and $\alpha_2 = 0$, the explicit second order central discretizations is recovered. For the simulations presented here, the following coefficients were used:

$$\alpha_1 = \frac{2}{3}(a+2) \qquad \alpha_2 = \frac{1}{3}(4a-1). \quad (18)$$

This choice of parameters leads to a one–parametric family of formally fourth order accurate schemes. For $a = 0$ the classical explicit fourth order discretizations is recovered, while for $a = 0.25$ the well known Padé scheme is obtained.

The above presented schemes are based on central discretizations in space and thus non-physical oscillations can occur in the numerical approximations. A very efficient algorithm for filtering out these high frequency oscillations was proposed in [8]. The low–pass filter (for the filtered values $\widehat{\phi}_i$) can be formulated in a form very similar to (17):

$$b\widehat{\phi}_{i-1} + \widehat{\phi}_i + b\widehat{\phi}_{i+1} = 2\beta_0\phi_i + \beta_1\frac{\phi_{i+1}+\phi_{i-1}}{2h} + \beta_2\frac{\phi_{i+2}+\phi_{i-2}}{4h} + \ldots \quad (19)$$

The filters of different orders could be obtained for various choices of coefficients. Here the sixth order filter with coefficients

$$\beta_0 = \frac{1}{16}(11+10b); \ \beta_1 = \frac{1}{32}(15+34b); \ \beta_2 = \frac{1}{16}(6b-3); \ \beta_3 = \frac{1}{32}(1-2b) \quad (20)$$

was used. For other filters see e.g. [12]. The parameter $-0.5 < b < 0.5$ is used to fine–tune the filter.[1] More information on the compact space discretizations can be found in [8], [12], [7].

Temporal discretizations The system of governing Partial Differential Equations was discretized in space using the above described finite–difference technique. This leads to a system of Ordinary Differential Equations for time-evolution of grid values of the vector of unknowns W. Resulting system of ODE's can be solved by a suitable time-integration method. In this study we have used the so called Strong Stability Preserving Runge–Kutta methods [9,11]. The three stage second order SSP Runge–Kutta method was used to obtain the results presented here.

4 Numerical Results

Computational domain The 2D computational domain is selected as a part of wall-bounded half space with low smooth cosine-shaped hill. The hill height is $h = 1m$, while the whole domain has dimensions $90 \times 30\,m$. The numerical simulations were performed on a structured, non-orthogonal wall-fitted grid that has 233×117 points with the minimum cell size in the near-wall region $\Delta z = 0.03m$. The grid is smoothly coarsened from the proximity of the hill towards the far field. The maximum growth of consequent cells is 3%.

Boundary conditions On the *inlet* the velocity profile $\boldsymbol{u} = (u(z), 0, 0)$ is prescribed. The horizontal velocity component u is given by $u(z) = U_0(z/H)^{1/r}$ with $U_0 = 1m/s$ and $r = 40$. Density perturbation ρ' is set to zero, while homogeneous Neumann condition is used for pressure. On the *outlet* the homogeneous Neumann condition is prescribed for all velocity components, as well as for the density perturbations. Pressure is set to a constant. On the *wall* the no-slip conditions are

---

[1] In order to distinguish between different finite–difference schemes we use the notation $CX_{aaa}FY_{bbb}$ for compact scheme of order X with parameter $a = $ aaa combined with filter of order Y applied with the dumping parameter $b = $ bbb.

used for velocity. Homogeneous Neumann condition is used for pressure and density perturbation. For *free stream* the homogeneous Neumann condition is used for all quantities including pressure and density perturbations.

The background density field is given by $\rho_0(z) = \rho_w + \gamma z$ with $\rho_w = 1.2 \, kg \cdot m^{-3}$ and $\gamma = -0.01 \, kg \cdot m^{-4}$. The gravity acceleration was set to $g = -50 \, m \cdot s^{-2}$ to test the behavior of the model and numerical method for sufficiently high Brunt–Väisälä frequency (i.e. short wavelength). The Reynolds number was in the range 100–500 (i.e. $\mu = 1 \cdot 10^{-2} - 2 \cdot 10^{-3} kg \cdot m^{-1} \cdot s^{-1}$).

Numerical results The small hill placed at the origin of the coordinate system generates a perturbation in the density field. Due to the buoyancy term in the equation (13), this density perturbation is translated into vertical motion that is superposed to the mean horizontal flow. The gravity waves are best visible in the vertical velocity contours (Fig. 1–4). The same color scale was used in all figures. The results of both schemes are quite close to each other as it is visible from the

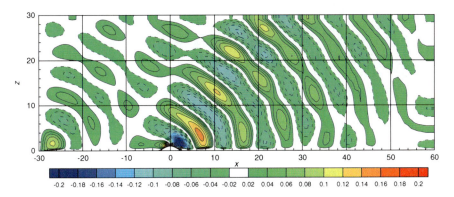

Fig. 1 Vertical velocity contours - $Re = 500$ - Compact scheme $C4_{038}F6_{049}$

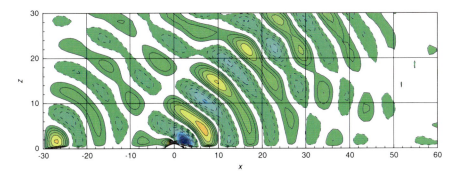

Fig. 2 Vertical velocity contours - $Re = 500$ - AUSM scheme

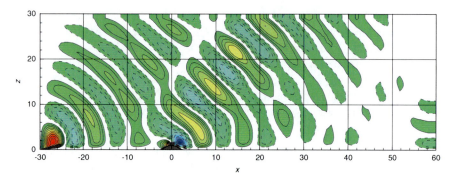

Fig. 3 Vertical velocity contours - $Re = 100$ - Compact scheme $C4_{038}F6_{049}$

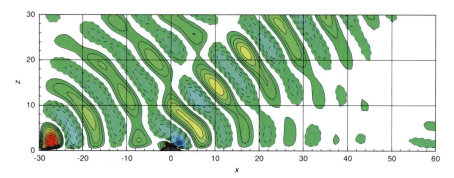

Fig. 4 Vertical velocity contours - $Re = 100$ - AUSM scheme

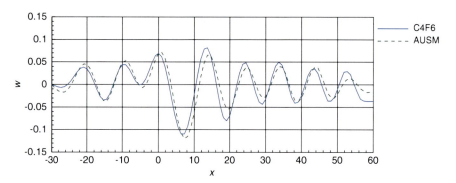

Fig. 5 Vertical velocity profiles comparison for $Re = 500$ in a horizontal cut at the height $z = 10\,m$

vertical velocity profiles shown in the Fig. 5. The basic structure of the results of both numerical methods is very similar. The compact finite difference scheme has clearly an advantage in being able to resolve the gravity waves in the far–field regions where the grid is very coarse.

5 Conclusions and Remarks

The numerical test have demonstrated the ability of both numerical methods to capture the main features of stably stratified flows. The advantage of high order methods was well documented by resolving the gravity waves in the regions of very coarse grid. The numerical simulations have brought some more problems that need to be further explored in detail. From these issues let's mention the the non–physical waves generated and reflected by the artificial boundaries, strong influence of numerical discretization/stabilization techniques on the small amplitude gravity waves, and the grid spacing limits related to Brunt–Väisälä frequency. Some of these problems are discussed in more detail in our recent study [5].

Acknowledgements The financial support for this work was partly provided by the Research Plan *MSM 6840770010* of the Ministry of Education of Czech Republic.

References

1. L. Beneš, T. Bodnár, P. Fraunié, K. Kozel, Numerical modelling of pollution dispersion in 3D atmospheric boundary layer, in: B. Sportisse (Ed.), Air Pollution Modelling and Simulation, Springer Verlag, 2002, pp. 69–78.
2. L. Beneš, J. Fürst, Numerical simulation of stratified flows past a body, in: ENUMATH 2009, Springer, 2009, p. 8p.
3. L. Beneš, J. Fürst, Comparison of the two numerical methods for the stratified flow, in: ICFD 2010 10th Conference on Numerical Methods for Fluid Dynamics, Univ. Reading, 2010, p. 6p.
4. T. Bodnár, L. Beneš, K. Kozel, Numerical simulation of flow over barriers in complex terrain, Il Nuovo Cimento C 31 (5–6) (2008) 619–632.
5. T. Bodnár, L. Beneš, Ph. Fraunié, K. Kozel, Application of Compact Finite-Difference Schemes to Simulations of Stably Stratified Fluid Flows, Preprint submitted to Applied Mathematics and Computation (2011).
6. T. Bodnár, K. Kozel, P. Fraunié, Z. Jaňour, Numerical simulation of flow and pollution dispersion in 3D atmospheric boundary layer, Computing and Visualization in Science 3 (1–2) (2000) 3–8.
7. D. V. Gaitonde, J. S. Shang, J. L. Young, Practical aspects of higher-order numerical schemes for wave propagation phenomena, International Journal for Numerical Methods in Engineering 45 (1999) 1849–1869.
8. S. K. Lele, Compact finite difference schemes with spectral-like resolution, Journal of Computational hysics 103 (1992) 16–42.
9. C. W. Shu, S. Osher, Efficient implementation of essentially non-oscillatory shock-capturing schemes, Journal of Computational Physics 77 (1988) 439–471.
10. I. Sládek, T. Bodnár, K. Kozel, On a numerical study of atmospheric 2D and 3D - flows over a complex topography with forest including pollution dispersion, Journal of Wind Engineering and Industrial Aerodynamics 95 (9–11).
11. R. J. Spiteri, S. J. Ruuth, A new class of optimal high-order strong-stability-preserving time discretization methods, SIAM Journal on Numerical Analysis 40 (2) (2002) 469–491.
12. M. R. Visbal, D. V. Gaitonde, On the use of higher-order finite-difference schemes on curvilinear and deforming meshes, Journal of Computational Physics 181 (2002) 155–185.

The paper is in final form and no similar paper has been or is being submitted elsewhere.

Convergence Analysis of the Upwind Finite Volume Scheme for General Transport Problems

Franck Boyer

Abstract This work is devoted to the convergence analysis of the implicit upwind finite volume scheme for the initial and boundary value problem associated to the linear transport equation in any dimension, on general unstructured meshes. We are particularly concerned with the case where the initial and boundary data are in L^∞ and the advection vector field v has low regularity properties, namely $v \in L^1(]0,T[,(W^{1,1}(\Omega))^d)$, with suitable assumptions on its divergence. We prove strong convergence in $L^\infty(]0,T[,L^p(\Omega))$ with $p < +\infty$, of the approximate solution towards the unique weak solution of the problem as well as the strong convergence of its trace.

Keywords Transport equation, upwind finite volume method, renormalized solutions
MSC2010: 35D30, 35L04, 65M08, 65M12

1 Introduction

Let $d \geq 1$, $\Omega \subset \mathbf{R}^d$ a bounded polygonal (or polyhedral) domain, and $T > 0$ given. We are interested here in the following initial and boundary value problem

$$\begin{cases} \partial_t \rho + \mathrm{div}\,(\rho v) + c\rho = 0, & \text{in }]0,T[\times\Omega, \\ \rho(0,\cdot) = \rho_0, & \text{in } \Omega, \\ \rho = \rho^{in}, & \text{on }]0,T[\times\Gamma, \text{ where } (v\cdot\nu) < 0. \end{cases} \qquad (1)$$

Franck Boyer
Aix-Marseille Université, LATP, FST Saint-Jérôme, Avenue Escradille Normandie-Niemen, 13397 MARSEILLE Cedex 20, FRANCE, e-mail: fboyer@latp.univ-mrs.fr

We consider the following assumptions for the data

$$c \in L^1(]0, T[\times \Omega), \qquad (2)$$

$$v \in L^1(]0, T[, (W^{1,1}(\Omega))^d), \text{ and } (v \cdot \nu) \in L^\alpha(]0, T[\times \Gamma), \text{ for some } \alpha > 1, \quad (3)$$

$$(c + \operatorname{div} v)^- \in L^1(]0, T[, L^\infty(\Omega)), \text{ and } (\operatorname{div} v)^+ \in L^1(]0, T[, L^\infty(\Omega)), \quad (4)$$

where x^+ and x^- stands for the positive and negative parts of any real number x. Associated to the vector field v, we introduce the measure $d\mu_v = (v \cdot \nu) \, dx \, dt$ on $]0, T[\times \Gamma$ and we denote by $d\mu_v^+$ (resp. $d\mu_v^-$) its positive (resp. negative) part in such a way that $|d\mu_v| = d\mu_v^+ + d\mu_v^-$. The support of $d\mu_v^+$ (resp. $d\mu_v^-$) is the outflow (resp. inflow) part of the boundary. In this framework, Problem (1) is well-posed in the following sense.

Theorem 1 (**Existence and uniqueness**). *We assume that assumptions* (2), (3), (4) *hold. For any* $\rho^0 \in L^\infty(\Omega)$ *and* $\rho^{in} \in L^\infty(]0, T[\times \Gamma, d\mu_v^-)$, *there exists a unique weak solution* $(\rho, \gamma(\rho)) \in L^\infty(]0, T[\times \Omega) \times L^\infty(]0, T[\times \Gamma, |d\mu_v|)$ *of* (1) *in the sense that*

$$\int_0^T \int_\Omega \rho(\partial_t \phi + v \cdot \nabla \phi - c\phi) \, dx \, dt - \int_0^T \int_\Gamma \gamma(\rho) \phi (v \cdot \nu) \, dx \, dt$$

$$+ \int_\Omega \rho^0 \phi(0, .) \, dx = 0, \quad \forall \phi \in \mathscr{C}_c^1([0, T[\times \overline{\Omega}), \quad (5)$$

with $\gamma(\rho) = \rho^{in}$, $d\mu_v^-$-*almost everywhere.*
Moreover, $\rho \in \mathscr{C}^0([0, T], L^p(\Omega))$ *for any* $p < +\infty$, *and we have*

$$\|\rho\|_{L^\infty(]0,T[\times\Omega)} \leq \max(\|\rho_0\|_{L^\infty}, \|\rho^{in}\|_{L^\infty}) e^{\int_0^T \|(c+\operatorname{div} v)^-\|_{L^\infty(\Omega)} \, dt}.$$

Finally, the following **renormalization property** *holds: for any smooth function* $\beta : \mathbf{R} \mapsto \mathbf{R}$, *the following system holds in the weak sense*

$$\begin{cases} \partial_t \beta(\rho) + \operatorname{div}(\beta(\rho)v) + c\beta'(\rho)\rho + (\operatorname{div} v)(\beta'(\rho)\rho - \beta(\rho)) = 0, & \text{in }]0, T[\times \Omega, \\ \beta(\rho)(0, .) = \beta(\rho^0), \\ \gamma(\beta(\rho)) = \beta(\gamma(\rho)), & \text{on }]0, T[\times \Gamma. \end{cases}$$

This result originates first from DiPerna-Lions theory [10] in the case when $v \cdot \nu = 0$ on the boundary. The initial and boundary value problem is studied in [4] in the case $c = \operatorname{div} v = 0$ and for a smooth domain Ω, whereas the general case is studied in [6]. Note that the assumptions we consider on the vector field v are almost minimal to prove the well-posedness of the transport problem. In fact, for $v \in L^1(]0, T[, (\operatorname{BV}(\Omega))^d)$, it is known that many of the results of the renormalized solutions theory still hold [1], but we will not consider this case here.

Convergence Analysis of the Upwind Finite Volume

The main aim of this work is to prove, in the above weak regularity framework, the convergence of the upwind finite volume method in a strong sense. The detailed proofs of all the results given here can be found in [5]. To our knowledge, the only available result in this framework is a L^∞ weak-$\star$ convergence result, in the case where $v \cdot \nu = 0$ on $\partial\Omega$, which can be found in [12]. When the transport vector field v is more regular (say Lipschitz continuous) and when $v \cdot \nu = 0$ on the boundary, the upwind finite volume scheme was studied in many references (see for instance [3, 7–9, 13, 14]). In summary, it is known that the convergence rate of the scheme is $\frac{1}{2}$ as soon as the initial data is discontinuous, even for regular meshes, whereas this rate is 1 for smooth data and regular meshes.

2 The finite volume setting

2.1 Notation

We introduce here the main notation we need to define and analyse the finite volume method, following quite closely the notation introduced for instance in [11].

A finite volume mesh of the domain Ω is a set $\mathscr{T} = (K)_{K \in \mathscr{T}}$ of closed connected polygonal subsets of $\mathbf{R}^d$, with disjoint interiors and such that $\overline{\Omega} = \bigcup_{K \in \mathscr{T}} K$. The boundary of each control volume $K \in \mathscr{T}$ can be written as the union of a finite number of edges/faces (we will use the word "edge" even for $d > 2$) which are closed connected sets of dimension $d - 1$. We denote by $\mathscr{E}_K$ the set of the faces/edges of K. We assume that for any K, L such that $K \neq L$ and $K \cap L$ is of co-dimension 1, then $K \cap L \in \mathscr{E}_K \cap \mathscr{E}_L$, in that case the corresponding face is denoted by $K|L$. The set of all the faces in the mesh is denoted by $\mathscr{E}$ and $\mathscr{E}_{bd}$ denote the subset of the faces which are included in the boundary $\partial\Omega$, $\mathscr{E}_{int} = \mathscr{E} \setminus \mathscr{E}_{bd}$ the set of the interior faces.

- For each $K \in \mathscr{T}$, and $\sigma \in \mathscr{E}_K$, we denote by $\nu_{K\sigma}$ the unit outward normal vector to K on σ. If $\sigma = K|L \in \mathscr{E}_{int}$, we we observe that $\nu_{K\sigma} = -\nu_{L\sigma}$.
- We will denote by $|K|$ (resp. $|\sigma|$) the d-dimensional Lebesgue measure of the control volume K (resp. the $d - 1$ dimensional measure of the face σ).
- The diameter of a control volume K shall be denoted by d_K and the size of the mesh is defined by $h_\mathscr{T} = \max_{K \in \mathscr{T}} d_K$.

We will need to measure the regularity of the mesh. To this end, we denote by $\text{reg}(\mathscr{T})$, the smallest positive number such that

$$\|f\|_{L^1(\partial K)} \leq \frac{\text{reg}(\mathscr{T})}{d_K} \|f\|_{W^{1,1}(K)}, \quad \forall K \in \mathscr{T}, \forall f \in W^{1,1}(K). \tag{6}$$

In the convergence results given below we shall assume that $\text{reg}(\mathscr{T})$ remains bounded as $h_\mathscr{T} \to 0$, which amounts to assume that the control volumes do not degenerate when one refines the mesh.

2.2 Definition of the scheme

Let us first define the discretization of the data needed to define the scheme.

- For any $K \in \mathscr{T}, n \in [\![0, N-1]\!]$, we define

$$c_K^n = \frac{1}{\delta t |K|} \int_{t^n}^{t^{n+1}} \int_K c \, dx \, dt, \text{ and } v_{K\sigma}^n = \frac{1}{\delta t |\sigma|} \int_{t^n}^{t^{n+1}} \int_\sigma (v \cdot \nu_{K\sigma}) \, dx \, dt, \forall \sigma \in \mathscr{E}_K.$$

Furthermore, if $\sigma \in \mathscr{E}_{\text{int}}$, with $\sigma = K|L$ we shall use the notation $v_{KL}^n = v_{K\sigma}^n = -v_{L\sigma}^n$, and if $\sigma \in \mathscr{E}_K \cap \mathscr{E}_{\text{bd}}$ we will denote $v_\sigma^n = v_{K\sigma}^n$.
- For any boundary edge $\sigma \in \mathscr{E}_{\text{bd}}$ and any $n \in [\![0, N-1]\!]$, we define

$$\rho_\sigma^{in,n+1} = \frac{1}{\delta t |\sigma|} \int_{t^n}^{t^{n+1}} \int_\sigma \rho^{in} \, dx \, dt. \tag{7}$$

Notice that ρ^{in} is *a priori* only given $d\mu_\nu^-$-almost everywhere so that in this formula we need, in fact, to consider an extension of ρ^{in} in $L^\infty(]0, T[\times \Gamma)$.

The implicit upwind finite volume scheme we consider is the following: Find approximate values $\{\rho_K^n, n \in [\![0, N]\!], K \in \mathscr{T}\}$ such that

$$\begin{cases} |K| \dfrac{\rho_K^{n+1} - \rho_K^n}{\delta t} + \displaystyle\sum_{\sigma \in \mathscr{E}_K \cap \mathscr{E}_{\text{int}}} |\sigma| \left((v_{K\sigma}^n)^+ \rho_K^{n+1} - (v_{K\sigma}^n)^- \rho_L^{n+1} \right) \\ + \displaystyle\sum_{\sigma \in \mathscr{E}_K \cap \mathscr{E}_{\text{bd}}} |\sigma| v_{K\sigma}^n \rho_\sigma^{n+1} + |K| c_K^n \rho_K^{n+1} = 0, \quad \forall n \in [\![0, N-1]\!], \forall K \in \mathscr{T}, \\ \rho_K^0 = \dfrac{1}{|K|} \displaystyle\int_K \rho^0 \, dx, \quad \forall K \in \mathscr{T}, \\ \rho_\sigma^{n+1} = \rho_\sigma^{in,n+1}, \quad \forall n \in [\![0, N-1]\!], \forall \sigma \in \mathscr{E}_{\text{bd}}, \text{ s.t. } v_{K\sigma}^n \leq 0, \\ \rho_\sigma^{n+1} = \rho_K^{n+1}, \quad \forall n \in [\![0, N-1]\!], \forall \sigma \in \mathscr{E}_{\text{bd}}, \text{ s.t. } v_{K\sigma}^n > 0. \end{cases} \tag{8}$$

Note that the boundary data ρ^{in} is only taken into account on the boundary edges such that $v_{K\sigma}^n \leq 0$. Those edges are not necessarily included in the inflow boundary, that is the support of the measure $d\mu_\nu^-$. That's the reason why we need to extend the definition of ρ^{in} to the whole boundary $]0, T[\times \Gamma$.

2.3 Existence and uniqueness

The first result we can prove is the following existence and uniqueness result.

Theorem 2. *Assume that (2),(3) and (4) hold. There exists $\delta t_{\max} > 0$ (depending only on $(c+\mathrm{div}\, v)^-$) such that for any $\delta t \leq \delta t_{\max}$, any mesh $\mathcal{T}$ and any data ρ^{in}, ρ^0, there exists an unique solution to the scheme (8).*

Moreover, the approximate solution is non-negative as soon as the data is. Finally, we have the following a priori bound

$$\max_{\substack{K\in\mathcal{T} \\ n\in[\![0,N]\!]}} |\rho_K^n| \leq \max(\|\rho^0\|_{L^\infty}, \|\rho^{in}\|_{L^\infty}) \exp\left(2\int_0^T \|(c+\mathrm{div}\, v)^-\|_{L^\infty}\, dt\right). \quad (9)$$

In the case of the pure transport problem, that is when $c = -\mathrm{div}\, v$, we find that $\delta t_{\max} = +\infty$. From now on we will denote by $\rho_{\mathcal{T},\delta t}$ (and its trace $(\gamma \rho_{\mathcal{T},\delta t})$) the piecewise constant function build upon the approximate solution as follows

$$\rho_{\mathcal{T},\delta t} = \sum_{n=0}^{N-1}\sum_{K\in\mathcal{T}} \rho_K^{n+1} 1_{]t^n,t^{n+1}[\times K}, \quad \gamma(\rho_{\mathcal{T},\delta t}) = \sum_{n=0}^{N-1}\sum_{\sigma\in\mathcal{E}_{\mathrm{bd}}} \rho_K^{n+1} 1_{]t^n,t^{n+1}[\times\sigma},$$

where, in the last sum, K is the unique control volume in $\mathcal{T}$ such that $\sigma \in \mathcal{E}_K$.

3 Convergence analysis

3.1 Uniform in time strong convergence

The main result of this work is the following

Theorem 3. *Assume that (2), (3) and (4) hold. Let $\mathrm{reg}_{\max} > 0$ be given and consider a family of meshes and time steps, such that $(\delta t, h_{\mathcal{T}}) \to 0$ and satisfying the bound*

$$\max\left(\mathrm{reg}(\mathcal{T}), \max_{K\in\mathcal{T}} \frac{\delta t}{d_K}\right) \leq \mathrm{reg}_{\max}.$$

Then, for any bounded data ρ^0 and ρ^{in}, we have the following convergences

$$\rho_{\mathcal{T},\delta t} \xrightarrow[(\delta t,h_{\mathcal{T}})\to 0]{} \rho, \quad \text{in } L^\infty(]0,T[,L^p(\Omega)),$$

$$\gamma(\rho_{\mathcal{T},\delta t}) \xrightarrow[(\delta t,h_{\mathcal{T}})\to 0]{} \gamma(\rho), \quad \text{in } L^p(]0,T[\times\Gamma, |d\mu_v|), \quad \forall p < +\infty,$$

where $(\rho, \gamma(\rho))$ is the unique solution to (5).

We want to emphasize the fact that the convergence of $\rho_{\mathcal{T},\delta t}$ is uniform in time with values in $L^p(\Omega)$. We describe now the main steps of the proof of this result.

- Using the *a priori* L^∞ bound (9), we can extract a subsequence of $\rho_{\mathscr{T},\delta t}$ (resp. $\gamma(\rho_{\mathscr{T},\delta t})$) which weak-$\star$ converges towards a limit denoted by ρ (resp. g) in $L^\infty(]0,T[\times\Omega)$ (resp. in $L^\infty(]0,T[\times\Gamma)$).
  - We prove that $g = \rho^{in}\,d\mu_\nu^-$-almost-everywhere.
  - We prove that (ρ, g) satisfy the weak formulation of the problem.
  - Since the weak solution $(\rho, \gamma(\rho))$ is unique, we deduce the weak-$\star$ convergence of the whole sequence of approximate solutions.
- For any $\varepsilon > 0$ we construct a smooth function ρ^ε associated to ρ such that
  - $\|\rho^\varepsilon\|_{L^\infty} \leq \|\rho\|_{L^\infty}$;
  - ρ^ε converges to ρ in $\mathscr{C}^0([0,T], L^p(\Omega))$, for any $p < +\infty$;
  - $\gamma(\rho^\varepsilon)$ converges to $\gamma(\rho)$ in $L^p(]0,T[\times\Gamma, |d\mu_\nu|)$ for any $p < +\infty$;
  - ρ^ε solves
  $$\partial_t \rho^\varepsilon + \operatorname{div}(\rho^\varepsilon v) + c\rho^\varepsilon = R_\varepsilon,$$
  where $R_\varepsilon \in L^1(]0,T[\times\Omega)$ tends to 0 in L^1.

  Following [10], this sequence is built by convolution with a mollifier and the regularity assumptions on v and c let us prove the property of the remainder term R_ε (Friedrichs commutator lemma). Nevertheless, since the boundary of Ω is not characteristic for v, the argument has to be adapted (see [2,4,6]).
- We define the discretization of ρ^ε to be

  $$\rho^\varepsilon_{\mathscr{T},\delta t} = \sum_{n=0}^{N-1} \delta t \sum_{K \in \mathscr{T}} \rho^\varepsilon(t^{n+1}, x_K) \mathbf{1}_{]t^n, t^{n+1}[\times K},$$

  where x_K is an arbitrary point in K.
- We use now the triangle inequality to get

  $$\|\rho_{\mathscr{T},\delta t} - \rho\|_{L^\infty(]0,T[,L^2(\Omega))} \leq \|\rho_{\mathscr{T},\delta t} - \rho^\varepsilon_{\mathscr{T},\delta t}\|_{L^\infty(]0,T[,L^2(\Omega))}$$
  $$+ \|\rho^\varepsilon_{\mathscr{T},\delta t} - \rho^\varepsilon\|_{L^\infty(]0,T[,L^2(\Omega))} + \|\rho^\varepsilon - \rho\|_{L^\infty(]0,T[,L^2(\Omega))}. \quad (10)$$

  In this inequality, the last term converges to 0 when $\varepsilon \to 0$ by construction of ρ^ε. The second term can be easily estimated by $C\|\rho^\varepsilon\|_{W^{1,\infty}}(h_{\mathscr{T}} + \delta t)$. Obviously $\|\rho^\varepsilon\|_{W^{1,\infty}}$ blows up when $\varepsilon \to 0$, but if ε is fixed, this term converges to 0 when $(\delta t, h_{\mathscr{T}}) \to 0$.
- Finally, we are led to compare the approximate solution $\rho_{\mathscr{T},\delta t}$ and the projection $\rho^\varepsilon_{\mathscr{T},\delta t}$ of ρ^ε. This can be done by writing the discrete equations satisfied by $\rho^\varepsilon_{\mathscr{T},\delta t}$ in a form which is similar to that of (8) with additional remainder terms in the right-hand side.
  We subtract the two set of equations and we perform an usual $L^\infty(]0,T[, L^2(\Omega))$ estimate of the difference $\rho_{\mathscr{T},\delta t} - \rho^\varepsilon_{\mathscr{T},\delta t}$. It can then be proved that

all the contribution of the remainder terms can be controlled by two types of quantities:
- some of them tend to zero when $\varepsilon \to 0$, independently on δt and $h_\mathcal{T}$.
- some of them tend to zero when $(\delta t, h_\mathcal{T}) \to 0$, as soon as ε is fixed.

The conclusion is then clear. The main tools we use in these estimates are
- The fact that, by the Friedrichs Lemma, we have $\|R_\varepsilon\|_{L^1} \to 0$.
- The weak $L^2(H^1)$ estimate satisfied by the approximate solution which reads

$$\sum_{n=0}^{N-1} \delta t \sum_{\sigma \in \mathcal{E}_{bd}} |\sigma| |v^n_{K\sigma}| (\rho_K^{n+1} - \rho_\sigma^{n+1})^2$$

$$+ \sum_{n=0}^{N-1} \delta t \sum_{\substack{\sigma \in \mathcal{E}_{int} \\ \sigma = K|L}} |\sigma| |v^n_{KL}| (\rho_L^{n+1} - \rho_K^{n+1})^2 \leq M, \quad (11)$$

for some $M > 0$ which only depends on the data. This estimate corresponds in the linear case to the so-called "weak BV estimate" for nonlinear hyperbolic equations (see [11]).
- The density of the set of smooth vector fields in $L^1(]0, T[, (W^{1,1}(\Omega))^d)$.

3.2 Discrete renormalization property and consequences

We now state a result which is a discrete counter-part of the renormalization property given in Theorem 1 for the weak solution of the continuous problem. This kind of result might be important when studying the coupling between problem (1) and some other equations (in the nonhomogeneous incompressible Navier-Stokes system for instance).

Theorem 4. *For any function $\beta : \mathbf{R} \mapsto \mathbf{R}$ which is continuous and piecewise $\mathcal{C}^1$, the approximate solution $\rho_{\mathcal{T},\delta t}$ satisfy the following set of equations*

$$|K| \frac{\beta(\rho_K^{n+1}) - \beta(\rho_K^n)}{\delta t} + \sum_{\sigma \in \mathcal{E}_K \cap \mathcal{E}_{int}} |\sigma| \left(v^{n+}_{K\sigma} \beta(\rho_K^{n+1}) - v^{n-}_{K\sigma} \beta(\rho_L^{n+1}) \right)$$

$$+ \sum_{\sigma \in \mathcal{E}_K \cap \mathcal{E}_{bd}} |\sigma| v^n_{K\sigma} \beta(\rho_\sigma^{n+1}) + |K| c_K^n \beta'(\rho_K^{n+1}) \rho_K^{n+1}$$

$$+ |K| (\operatorname{div} v)_K^n \left(\beta'(\rho_K^{n+1}) \rho_K^{n+1} - \beta(\rho_K^{n+1}) \right) = |K| R_K^{n+1}, \quad \forall n \in [\![0, N-1]\!], \forall K \in \mathcal{T},$$

$$(12)$$

where the remainder term $(R_K^{n+1})_{K\in\mathcal{T}, n\in[\![0,N-1]\!]}$ strongly converges towards zero in $L^1(]0,T[\times\Omega)$, that is

$$\sum_{n=0}^{N-1}\delta t \sum_{K\in\mathcal{T}} |K||R_K^{n+1}| \xrightarrow[(\delta t,h_\mathcal{T})\to 0]{} 0.$$

Furthermore, when β is convex we have $R_K^{n+1} \leq 0$, $\forall K \in \mathcal{T}, \forall n \in [\![0, N-1]\!]$.

Note that this result holds for any arbitrary choice of the value of β' at singular points. We can deduce from Theorem 4 many properties of the approximate solution. For instance, we can prove:

- For any $\alpha \in \mathbf{R} \setminus \{0\}$, we have

$$\sum_{n=0}^{N-1}\delta t \sum_{K\in\mathcal{T}} |K||c_K^n + (\text{div } v)_K^n|\mathbf{1}_{\{\rho_K^{n+1}=\alpha\}} \xrightarrow[(\delta t,h_\mathcal{T})\to 0]{} 0.$$

This is the discrete counter part of the following property of the weak solution of the problem

$$c + \text{div } v = 0, \quad \text{for almost every } (t,x) \text{ in the level set } \{\rho = \alpha\}.$$

- The total numerical dissipation term associated with the upwind discretization (that is the left-hand side term in (11)) tends to 0 when $(\delta t, h_\mathcal{T}) \to 0$. This is an improvement of the weak $L^2(H^1)$ estimate which only says that this quantity is bounded.

Acknowledgements The author wishes to thank T. Gallouët and R. Herbin for many stimulating discussions on the topic of this work.

References

1. Ambrosio, L.: Transport equation and Cauchy problem for BV vector fields. Invent. Math. **158**(2), 227–260 (2004)
2. Blouza, A., Le Dret, H.: An up-to-the-boundary version of Friedrichs's lemma and applications to the linear Koiter shell model. SIAM J. Math. Anal. **33**(4), 877–895 (electronic) (2001)
3. Bouche, D., Ghidaglia, J.M., Pascal, F.: Error estimate and the geometric corrector for the upwind finite volume method applied to the linear advection equation. SIAM J. Numer. Anal. **43**(2), 578–603 (electronic) (2005)
4. Boyer, F.: Trace theorems and spatial continuity properties for the solutions of the transport equation. Differential Integral Equations **18**(8), 891–934 (2005)
5. Boyer, F.: Analysis of the upwind finite volume method for general initial and boundary value transport problems. submitted (2011). http://hal.archives-ouvertes.fr/hal-00559586/fr/
6. Boyer, F., Fabrie, P.: Elements of analysis for the study of some models of incompressible viscous flows. in preparation. Springer-Verlag (2011)

7. Delarue, F., Lagoutière, F.: Probabilistic analysis of the upwind scheme for transport equations. Archive for Rational Mechanics and Analysis **199**, 229–268 (2011)
8. Desprès, B.: Convergence of non-linear finite volume schemes for linear transport. In: Notes from the XIth Jacques-Louis Lions Hispano-French School on Numerical Simulation in Physics and Engineering (Spanish), pp. 219–239. Grupo Anal. Teor. Numer. Modelos Cienc. Exp. Univ. Cádiz, Cádiz (2004)
9. Desprès, B.: Lax theorem and finite volume schemes. Math. Comp. **73**(247), 1203–1234 (electronic) (2004)
10. DiPerna, R., Lions, P.L.: Ordinary differential equations, transport theory and Sobolev spaces. Invent. Math. **98**(3), 511–547 (1989)
11. Eymard, R., Gallouët, T., Herbin, R.: Finite volume methods. In: Handbook of numerical analysis, Vol. VII, Handb. Numer. Anal., VII, pp. 713–1020. North-Holland, Amsterdam (2000)
12. Fettah, A.: Analyse de modèles en mécanique des fluides compressibles. Ph.D. thesis, Université de Provence (2011)
13. Merlet, B.: L^∞- and L^2-error estimates for a finite volume approximation of linear advection. SIAM J. Numer. Anal. **46**(1), 124–150 (2007/08)
14. Merlet, B., Vovelle, J.: Error estimate for finite volume scheme. Numer. Math. **106**(1), 129–155 (2007)

The paper is in final form and no similar paper has been or is being submitted elsewhere.

A Low Degree Non–Conforming Approximation of the Steady Stokes Problem with an Eddy Viscosity

F. Boyer, F. Dardalhon, C. Lapuerta, and J.-C. Latché

Abstract In the context of Large Eddy Simulation, the use of a turbulence model brings the question of the implementation of the eddy–viscosity. In this communication, we propose to assess the discretization of the diffusive term based on a low–order non–conforming finite element. For this, we build a manufactured solution of the incompressible steady Stokes problem, for which the turbulent viscosity is given either by the Smagorinsky or WALE models. Numerical tests are performed for both models with the finite element approximation and the MAC scheme.

Keywords Large Eddy Simulation, WALE and Smagorinsky models, incompressible steady Stokes equations, low-order finite element approximation
MSC2010: 65N30, 76D05

1 Introduction

In the context of turbulence modelling, there is an increasing interest in the Large Eddy Simulation approach (LES), resulting from the augmentation of the computer resources. LES modelling solves large turbulent structures, while small–scale effects are modelled (see [2, 13]). The approach consists in averaging the Navier–Stokes equations in space (by convolution), and then commuting this filtering operation (denoted with the overbar symbol) with space and time derivatives. This

---

F. Dardalhon, C. Lapuerta, and J.-C. Latché
Institut de Radioprotection et de Sûreté Nucléaire (IRSN), BP3 - 13115 Saint Paul–lez–Durance CEDEX, France, e-mail: [fanny.dardalhon,celine.lapuerta,jean--claude.latche]@irsn.fr

F. Boyer and F. Dardalhon
Université Paul Cézanne, LATP, FST Saint–Jérôme, Case Cour A, Avenue Escadrille Normandie–Niemen, 13397 Marseille Cedex 20, France, e-mail: fboyer@cmi.univ-mrs.fr

yields balance equations, which keep the same form as the original ones, for the resolved (large scales) velocity $\overline{\mathbf{u}}$ and pressure $\overline{p}$. Due to the presence of the nonlinear convection term, the unclosed quantity $-\mathrm{div}(\overline{\mathsf{T}})$, with $\overline{\mathsf{T}} = \overline{\mathbf{u}\mathbf{u}} - \overline{\mathbf{u}}\,\overline{\mathbf{u}}$ appears at the right–hand side, and must be modelled, *i.e.* recast as a function of the unknowns $\overline{\mathbf{u}}$ and $\overline{p}$.

The key issue in the LES approach is thus to find a suitable expression for the subgrid–scale tensor $\overline{\mathsf{T}}$. A common assumption is to suppose a proportional relation between $\overline{\mathsf{T}}$ and the deformation tensor $\overline{\mathsf{S}}$:

$$\overline{\mathsf{T}} = -2\,\nu_t\,\overline{\mathsf{S}}, \quad \text{with} \quad \overline{\mathsf{S}}_{i,j} = \frac{1}{2}(\partial_j \overline{\mathbf{u}}_i + \partial_i \overline{\mathbf{u}}_j), \quad \forall i,j \in \{1,\cdots,d\},$$

the scalar ν_t being referred to as the turbulent viscosity. We propose to study here two subgrid–scale models often encountered in the literature, namely the Smagorinsky [14] and WALE [10] models.

The Smagorinsky model is the most frequently used because of its quite simple form and reads:

$$\nu_t = (C_s\,\overline{\Delta})^2\,\sqrt{2\,\mathrm{Trace}(\overline{\mathsf{S}}\,\overline{\mathsf{S}}^T)} = (C_s\,\overline{\Delta})^2\,\left(2\sum_{i,j \in \{1,\cdots,d\}} \overline{\mathsf{S}}_{i,j}\,\overline{\mathsf{S}}_{i,j}\right)^{\frac{1}{2}}, \quad (1)$$

where C_s is a constant adjusted as a function of the flow and $\overline{\Delta}$ is the cut–off length scale, usually identified to a characteristic size of the cell. However, the viscosity obtained in this way behaves like $\mathcal{O}(1)$ near a wall, contrary to the scaling $\nu_t = \mathcal{O}(y^3)$, where y stands for the distance to the wall, which may be inferred by asymptotic analysis [13]. So this model dissipates the large scales too much near a wall.

The WALE model (Wall Adaptating Local Eddy–viscosity) aims at solving this problem and reads:

$$\nu_t = (C_w\overline{\Delta})^2\,\frac{\left(\sum_{i,j} \overline{\varsigma}_{i,j}\,\overline{\varsigma}_{i,j}\right)^{3/2}}{\left(\sum_{i,j} \overline{\mathsf{S}}_{i,j}\,\overline{\mathsf{S}}_{i,j}\right)^{5/2} + \left(\sum_{i,j} \overline{\varsigma}_{i,j}\,\overline{\varsigma}_{i,j}\right)^{5/4}}, \quad (2)$$

C_w being a real constant adjusted as a function of the flow and

$$\overline{\varsigma} = \frac{1}{2}\left(\nabla\overline{\mathbf{u}}^2 + (\nabla\overline{\mathbf{u}}^2)^T\right) - \frac{1}{d}\,\mathrm{Trace}(\nabla\overline{\mathbf{u}}^2)\,I_d,$$

I_d being the $d \times d$ identity matrix. Asymptotic analysis of Eq. (2) shows that the proper behaviour $\mathcal{O}(y^3)$ of the viscosity is recovered, without any near–wall modification, which makes this model particularly attractive to deal with complex geometries.

As a first step toward the construction of a scheme for LES equations, we propose in this paper to study the discretization of the nonlinear (due to the presence of the subgrid model) diffusion term of the momentum balance equation. To this purpose, we address a simplified problem, namely the steady incompressible Stokes problem obtained by dropping the time derivative and convection terms in the original Navier-Stokes equations:

$$\begin{cases} -\mathrm{div}(2\nu S(\bar{\mathbf{u}})) + \nabla \bar{p} = \bar{\mathbf{f}} & \text{in } \Omega, \\ \mathrm{div}\,\bar{\mathbf{u}} = 0 & \text{in } \Omega, \\ \bar{\mathbf{u}} = 0 & \text{on } \partial\Omega, \end{cases} \quad (3)$$

where $S(\bar{\mathbf{u}}) = \frac{1}{2}(\nabla\bar{\mathbf{u}} + \nabla\bar{\mathbf{u}}^T)$ is the symmetric part of the gradient of $\bar{\mathbf{u}}$ and $\bar{\mathbf{f}}$ is a known forcing term. This problem is posed in Ω, an open, connected, bounded domain of $\mathbb{R}^d$ ($d = 2, 3$), supposed to be polygonal for the sake of simplicity. The effective viscosity ν is equal to the sum of the laminar and turbulent viscosities denoted by ν_l and ν_t, respectively, the latter one being given as a function of the velocity by Eqs. (1) or (2) with a coefficient $\bar{\Delta}$ supposed here to be fixed (*i.e.* independent of the mesh). Since the velocity is prescribed to zero on the whole boundary, the pressure must be supposed to be mean-valued to obtain a well-posed problem.

Two approaches are considered for the spatial discretization: low order finite element (Rannacher–Turek element) and MAC scheme. We focus the paper on the finite element version, the description of the MAC scheme used for comparison in the numerical experiments being given in [6–8]. The obtained schemes are assessed by numerical experiments, using a manufactured solution technique.

The outline of the article is as follows. After the introduction of the Rannacher–Turek finite element (Sect. 2), we describe the resulting discretization of Problem (3), *i.e.*, essentially, the proposed discrete expression for the Smagorinsky or WALE subgrid viscosity (Sect. 3). Numerical tests are reported in Sect. 4.

To alleviate the notations, we drop in the remainder of this paper the overbar symbol to denote the averaged fields.

2 Mesh and discrete spaces

Let $\mathcal{M}$ be a decomposition of the domain Ω into quadrangles, supposed to be regular in the usual sense of the finite element literature [4]. We denote by $\mathcal{E}$ the set of all faces σ of the mesh; by $\mathcal{E}_{ext}$ the set of faces included in the boundary of Ω, by $\mathcal{E}_{int}$ the set of internal faces (*i.e.* $\mathcal{E} \setminus \mathcal{E}_{ext}$) and by $\mathcal{E}(K)$ the faces of a particular cell $K \in \mathcal{M}$. By $|K|$ and $|\sigma|$ we denote the measure, respectively, of the control volume K and of the face σ.

The space discretization relies on the Rannacher–Turek mixed finite element. The degrees of freedom for the velocity are located at the mass center of the faces of the mesh, and we use the version of the element where they represent the average of the velocity over the face. The set of degrees of freedom thus reads:

$$\{\mathbf{u}_{\sigma,i},\ \sigma \in \mathcal{E},\ i = 1, \cdots, d\}.$$

The discrete functional space over a cell K is obtained through the usual Q_1 mapping from the space $span\ \{1, (x_i)_{i=1,\ldots,d}, (x_i^2 - x_{i+1}^2)_{i=1,\ldots,d-1}\}$ over the reference element. The approximation for the velocity is non–conforming: the space X_h is composed of discrete functions which are discontinuous through an edge, but the jump of their integral is imposed to be zero. We denote by $\varphi_\sigma^{(i)}$ the vector shape function associated to $\mathbf{u}_{\sigma,i}$, which, by definition, reads $\varphi_\sigma^{(i)} = \varphi_\sigma\,\mathbf{e}^{(i)}$, where φ_σ is the Rannacher–Turek scalar shape function and $\mathbf{e}^{(i)}$ is the ith vector of the canonical basis of $\mathbb{R}^d$, and we define $\mathbf{u}_\sigma$ by $\mathbf{u}_\sigma = \sum_i \mathbf{u}_{\sigma,i}\,\mathbf{e}^{(i)}$. With these definitions, we have:

$$\mathbf{u} = \sum_{\sigma \in \mathcal{E}} \sum_{i=1,\cdots,d} \mathbf{u}_{\sigma,i}\,\varphi_\sigma^{(i)}(\mathbf{x}) = \sum_{\sigma \in \mathcal{E}} \mathbf{u}_\sigma\,\varphi_\sigma(\mathbf{x}), \quad \text{for a.e. } \mathbf{x} \in \Omega.$$

Dirichlet boundary conditions are built in the definition of the discrete velocity space X_h by fixing $\mathbf{u}_{\sigma,i} = 0$ for all faces in $\mathcal{E}_{\text{ext}}$ and any component i.

The pressure is piecewise constant, and its degrees of freedom are denoted by p_K for any cell $K \in \mathcal{M}$. We denote by M_h the discrete pressure space.

3 The scheme

In this section, we begin with describing the approximation of the turbulent viscosity, which is chosen piecewise constant by cell, and we then present the discretization of Problem (3).

Expression of the cell viscosity ν_K for the Smagorinsky model – We propose to study two discretizations of the term S in Eq. (1) of the turbulent viscosity. The first one consists in approximating the expression $\text{Trace}(\mathbf{S}\,\mathbf{S}^T)$ by its mean value over a cell K:

$$\overline{\mathbf{S}^2}^K = \frac{1}{|K|} \int_K \mathbf{S}(\mathbf{u}) : \mathbf{S}(\mathbf{u})\,d\mathbf{x}.$$

The second approach is to compute the mean value of the velocity gradient over K and then to use it in the definition of S:

$$\overline{\mathbf{S}_{ij}}^K = \frac{1}{2}\left(\overline{\partial_j \mathbf{u}_i}^K + \overline{\partial_i \mathbf{u}_j}^K\right) = \frac{1}{2}\left(\frac{1}{|K|}\int_K \partial_j \mathbf{u}_i\,d\mathbf{x} + \frac{1}{|K|}\int_K \partial_i \mathbf{u}_j\,d\mathbf{x}\right). \quad (4)$$

Finally, the expression of the effective viscosity for both approximations is:
- for the method 1:
$$\nu_K = \nu_l + (C_s \overline{\Delta})^2 \left(2 \overline{S^2}^K\right)^{\frac{1}{2}}, \qquad (5)$$
- for the method 2:
$$\nu_K = \nu_l + (C_s \overline{\Delta})^2 \left(2 \sum_{i,j} \overline{S_{ij}}^K \overline{S_{ij}}^K\right)^{\frac{1}{2}}. \qquad (6)$$

These discretizations are different since the discrete velocity field is not piecewise affine. Hovever, as reported hereafter in Sect. 4, they give similar results, so only Method 2 is retained for the discretization of the WALE model, to avoid the computation of integrals needing high order quadrature formulas.

Expression of the cell viscosity ν_K for the WALE model – The discretization of the tensor ς in a cell $K \in \mathcal{M}$ is:

$$\overline{\varsigma_{ij}}^K = \frac{1}{2} \sum_{\ell \in \{1,\cdots,d\}} \left(\overline{\partial_\ell \mathbf{u}_i}^K \overline{\partial_j \mathbf{u}_\ell}^K + \overline{\partial_i \mathbf{u}_\ell}^K \overline{\partial_\ell \mathbf{u}_j}^K\right)$$

$$- \left(\frac{1}{d} \sum_{m,n \in \{1,\cdots,d\}} \overline{\partial_m \mathbf{u}_n}^K \overline{\partial_n \mathbf{u}_m}^K\right) \delta_{i,j}, \quad \forall i,j \in \{1,\cdots,d\}, \quad (7)$$

where δ is the Kronecker symbol. Using the approximations of S and ς given respectively by (4) and (7), the effective viscosity on K reads:

$$\nu_K = \nu_l + (C_w \overline{\Delta})^2 \frac{\left(\sum_{i,j} \overline{\varsigma_{ij}}^K \overline{\varsigma_{ij}}^K\right)^{3/2}}{\left(\sum_{i,j} \overline{S_{ij}}^K \overline{S_{ij}}^K\right)^{5/2} + \left(\sum_{i,j} \overline{\varsigma_{ij}}^K \overline{\varsigma_{ij}}^K\right)^{5/4}}.$$

Discretization of Problem (3) – The scheme for the solution of Problem (3) consists in searching for $\mathbf{u} \in X_h$ and $p \in M_h$ such that the mean value of p over Ω is zero and:

for $1 \leq i \leq d$ and for any $\sigma \in \mathcal{E}_{\text{int}}$,

$$\sum_{K \in \mathcal{M}} 2\nu_K \int_K \mathbf{S}(\mathbf{u}) : \mathbf{S}(\varphi_\sigma^{(i)}) \, dx - \sum_{K \in \mathcal{M}} \int_K p \, \text{div}(\varphi_\sigma^{(i)}) \, dx = \int_\Omega \mathbf{f} \cdot \varphi_\sigma^{(i)} dx,$$

$$\forall K \in \mathcal{M}, \quad \int_K \text{div}(\mathbf{u}) \, dx = 0. \tag{8}$$

4 Numerical tests

In this section, we build a manufactured solution to Problem (3) in 2D, and compare the results obtained with the considered discretizations to the analytical solution and to the discrete solutions obtained with the MAC scheme. The simulations are performed with the ISIS software based on the platform PELICANS, both developed at IRSN [9, 11].

Description of the numerical test – The computational domain Ω is the unit square $(0, 1)^2$ and we calculate the forcing term $\mathbf{f}$ such that the exact velocity and pressure fields, $\mathbf{u}_{exact}$ and p_{exact}, are given by:

$$\mathbf{u}_{exact} = \mathbf{curl}(\sin(\pi x) \sin(\pi y)), \quad p_{exact} = \cos(\pi x) \sin(\pi y).$$

Note that $\mathbf{u}_{exact}$ indeed satisfies homogeneous Dirichlet boundary conditions on $\partial \Omega$, and the mean value over Ω of p_{exact} is zero.

We take $\nu_l = 10^{-3}$ and the coefficient $C_w \overline{\Delta}$ in the expression of ν_t (Eqs. (1) and (2)) is set to $C_s \overline{\Delta} = 0.007$ for the Smagorinsky model and $C_w \overline{\Delta} = 0.009$ for the WALE model, which yields a turbulent and a laminar viscosity of the same range. The Smagorinsky and WALE viscosities obtained for $\mathbf{u}_{exact}$ are plotted on Fig. 1. The profiles are quite different, and one remarks that, as expected, the turbulent viscosity vanishes near the wall with the WALE model while it does not decrease with the Smagorinsky model.

The nonlinear problem (8) is solved using an iterative process analog to a time marching algorithm of pressure correction type [5], computing at each step the value of the turbulent viscosity from the beginning-of-step velocity. The steady state is supposed to be reached when velocity and pressure increments are small enough.

The discrete L^2–norm defined by

$$\|\mathbf{u}\|_0^2 = \sum_K \frac{|K|}{4} \sum_\sigma |\mathbf{u}_\sigma|^2$$

is used to measure the spatial error for $n \times n$ structured uniform meshes, with $n = 10, 20, 40$ and 80.

A Low Degree Non–Conforming Approximation of the Steady Stokes Problem

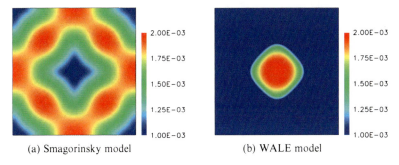

(a) Smagorinsky model   (b) WALE model

Fig. 1 Repartition of the effective viscosity

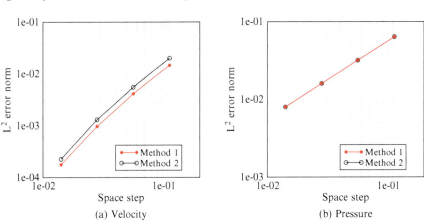

(a) Velocity   (b) Pressure

Fig. 2 L^2 error norm for the velocity and the pressure as a function of the space step for both discretizations of the Smagorinsky viscosity: Method 1 corresponds to Eq. (5), Method 2 to Eq. (6)

Comparison of both implementations for the Smagorinsky model – On Fig. 2, the spatial error in L^2–norm is plotted for both methods for the computation of the Smagorinsky viscosity. Both implementations give about the same accuracy. Consequently, Method 2 is chosen for further numerical experiments, because its implementation is simpler.

Comparison of the finite element approach and the MAC scheme for both models – On Fig. 3 and Fig. 4, 'FE' and 'FV' represent the discretization chosen, namely the Rannacher–Turek Finite Element and the MAC scheme (Finite Volume) respectively. Both discretizations seem to lead to the same order of convergence in space, that is 2 for the velocity and 1 for the pressure, for the Smagorinsky model. The FE discretization is more accurate than the MAC scheme but, for a given mesh, the number of degrees of freedom for the velocity for the FE discretization is twice (for $d = 2$) greater than for the MAC approximation (the number of degrees of freedom for the pressure being the same in both cases). For the WALE model, results look similar, with a more irregular convergence for the velocity.

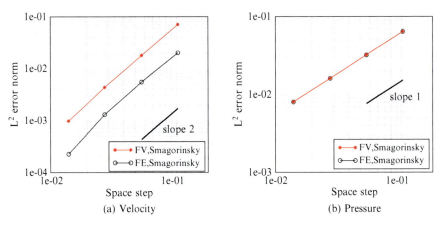

Fig. 3 L^2 error norm for the velocity and the pressure as a function of the space step for the Smagorinsky model

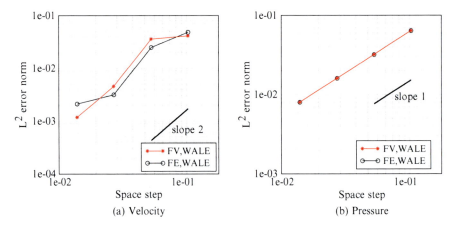

Fig. 4 L^2 error norm for the velocity and the pressure as a function of the space step for the WALE model

5 Conclusion

As a conclusion, the space discretizations retained for the Smagorinsky model and the WALE model give satisfactory results for the considered steady nonlinear Stokes problem, both for the finite element method and for the MAC scheme. Next steps will be to extend the scheme to the complete Navier–Stokes equations with the same subgrid models (see [1, 3] for a kinetic energy preserving discretization of the convection term) and assess it on the academic test of the plane channel, before turning to more complex industrial applications.

References

1. G. Ansanay-Alex, F. Babik, J.-C. Latché, D. Vola: An L2-stable Approximation of the Navier-Stokes Convection Operator for Low-Order Non-Conforming Finite Elements. International Journal for Numerical Methods in Fluids, online (2010).
2. L.C. Berselli, T. Illiescu, W.J. Layton: Mathematics of Large Eddy Simulation of Turbulent Flows. Springer (2006).
3. F. Boyer, F. Dardalhon, C. Lapuerta, J.-C. Latché: Stability of a Crank-Nicolson Pressure Correction Scheme based on Staggered Discretizations, in preparation (2011).
4. P.G. Ciarlet: Handbook of Numerical Analysis Volume II: Finite Elements Methods – Basic Error Estimates for Elliptic Problems. North-Holland (1991).
5. F. Dardalhon, J.-C. Latché, S. Minjeaud: Analysis of a Projection Method for Low-Order Non-Conforming Finite Elements. Submitted (2011).
6. F.H. Harlow, J.E. Welsh: Numerical Calculation of Time-Dependent Viscous Incompressible Flow of Fluid with Free Surface. Physics of Fluids, **8**, 2182–2189 (1965).
7. R. Herbin, J.-C. Latché: A Kinetic Energy Control in the MAC Discretization of Compressible Navier-Stokes Equations. International Journal of Finite Volumes, **7**(2) (2010).
8. R. Herbin, W. Kheriji, J.-C. Latché: Discretization of Dissipation Terms with the MAC Scheme. Finite Volumes for Complex Appplications VI, Prague (2011).
9. ISIS: A CFD Computer Code for the Simulation of Reactive Turbulent Flows. https://gforge.irsn.fr/gf/project/isis.
10. F. Nicoud, F. Ducros: Subgrid-Scale Stress Modelling Based on the Square of the Velocity Gradient Tensor. Flow, Turbulence And Combustion, **62**, 183–200 (1999).
11. PELICANS: Collaborative Development Environment. https://gforge.irsn.fr/gf/project/pelicans.
12. R. Rannacher, S. Turek: Simple Nonconforming Quadrilateral Stokes Element. Numerical Methods for Partial Differential Equations, **8**, 97–111 (1992).
13. P. Sagaut: Large Eddy Simulation for Incompressible Flows. Scientific Computation, Springer-Verlag Berlin Heidelberg (2006).
14. J. Smagorinsky: General Circulation Experiments with the Primitive Equations. Monthly Weather Review, **91**(3), 99-164 (1963).

The paper is in final form and no similar paper has been or is being submitted elsewhere.

Some Abstract Error Estimates of a Finite Volume Scheme for the Wave Equation on General Nonconforming Multidimensional Spatial Meshes

Abdallah Bradji

Abstract A general class of nonconforming meshes has been recently studied for sationary anisotropic heterogeneous diffusion problems, see [2]. The aim of this contribution is to deal with error estimates, using this new class of meshes, for the wave equation. We present an implicit time scheme to approximate the wave equation. We prove that, when the discrete flux is calculated using a stabilized discrete gradient, the convergence order is $h_\mathscr{D} + k$, where $h_\mathscr{D}$ (resp. k) is the mesh size of the spatial (resp. time) discretization. This estimate is valid for discrete norms $\mathbb{L}^\infty(0, T; H_0^1(\Omega))$ and $\mathscr{W}^{1,\infty}(0, T; L^2(\Omega))$ under the regularity assumption $u \in \mathscr{C}^3([0, T]; \mathscr{C}^2(\overline{\Omega}))$ for the exact solution u. These error estimates are useful because they allow to obtain approximations to the exact solution and its first derivatives of order $h_\mathscr{D} + k$.

Keywords second order hyperbolic equation, wave equation, non–conforming grid, SUSHI scheme, implicit scheme, discrete gradient
MSC2010: 65M08, 65M15

1 Motivation and aim of this paper

We consider the wave equation, as a model for second order hyperbolic equations:

$$u_{tt}(x,t) - \Delta u(x,t) = f(x,t), \quad (x,t) \in \Omega \times (0,T), \tag{1}$$

where Ω is an open polygonal bounded subset in $\mathbb{R}^d$, $T > 0$, and f is a given function.
An initial condition is given by: for given functions u^0 and u^1 defined on Ω

Abdallah Bradji
Department of Mathematics, university of Annaba–Algeria, e-mail: bradji@cmi.univ-mrs.fr

J. Fořt et al. (eds.), *Finite Volumes for Complex Applications VI – Problems & Perspectives*, Springer Proceedings in Mathematics 4,
DOI 10.1007/978-3-642-20671-9_19, © Springer-Verlag Berlin Heidelberg 2011

$$u(x,0) = u^0(x) \text{ and } u_t(x,0) = u^1(x) \ x \in \Omega, \tag{2}$$

Homogeneous Dirichlet boundary conditions are given by

$$u(x,t) = 0, \ (x,t) \in \partial\Omega \times (0,T). \tag{3}$$

2 Definition of the scheme

The discretization of Ω is performed using the mesh $\mathscr{D} = (\mathscr{M}, \mathscr{E}, \mathscr{P})$ described in [2, Definition 2.1] which we recall here for the sake of completeness.

Definition 1. (Definition of the spatial mesh, cf. [2, Definition 2.1, Page 1012]) Let Ω be a polyhedral open bounded subset of $\mathbb{R}^d$, where $d \in \mathbb{N} \setminus \{0\}$, and $\partial\Omega = \overline{\Omega} \setminus \Omega$ its boundary. A discretisation of Ω, denoted by $\mathscr{D}$, is defined as the triplet $\mathscr{D} = (\mathscr{M}, \mathscr{E}, \mathscr{P})$, where:

1. $\mathscr{M}$ is a finite family of non empty connected open disjoint subsets of Ω (the "control volumes") such that $\overline{\Omega} = \cup_{K \in \mathscr{M}} \overline{K}$. For any $K \in \mathscr{M}$, let $\partial K = \overline{K} \setminus K$ be the boundary of K; let $m(K) > 0$ denote the measure of K and h_K denote the diameter of K.
2. $\mathscr{E}$ is a finite family of disjoint subsets of $\overline{\Omega}$ (the "edges" of the mesh), such that, for all $\sigma \in \mathscr{E}$, σ is a non empty open subset of a hyperplane of $\mathbb{R}^d$, whose $(d-1)$–dimensional measure is strictly positive. We also assume that, for all $K \in \mathscr{M}$, there exists a subset $\mathscr{E}_K$ of $\mathscr{E}$ such that $\partial K = \cup_{\sigma \in \mathscr{E}_K} \overline{\sigma}$. For any $\sigma \in \mathscr{E}$, we denote by $\mathscr{M}_\sigma = \{K; \sigma \in \mathscr{E}_K\}$. We then assume that, for any $\sigma \in \mathscr{E}$, either $\mathscr{M}_\sigma$ has exactly one element and then $\sigma \subset \partial\Omega$ (the set of these interfaces, called boundary interfaces, denoted by $\mathscr{E}_{\text{ext}}$) or $\mathscr{M}_\sigma$ has exactly two elements (the set of these interfaces, called interior interfaces, denoted by $\mathscr{E}_{\text{int}}$). For all $\sigma \in \mathscr{E}$, we denote by x_σ the barycentre of σ. For all $K \in \mathscr{M}$ and $\sigma \in \mathscr{E}$, we denote by $\mathbf{n}_{K,\sigma}$ the unit vector normal to σ outward to K.
3. $\mathscr{P}$ is a family of points of Ω indexed by $\mathscr{M}$, denoted by $\mathscr{P} = (x_K)_{K \in \mathscr{M}}$, such that for all $K \in \mathscr{M}$, $x_K \in K$ and K is assumed to be x_K–star-shaped, which means that for all $x \in K$, the property $[x_K, x] \subset K$ holds. Denoting by $d_{K,\sigma}$ the Euclidean distance between x_K and the hyperplane including σ, one assumes that $d_{K,\sigma} > 0$. We then denote by $\mathscr{D}_{K,\sigma}$ the cone with vertex x_K and basis σ.

The time discretization is performed with a constant time step $k = \frac{T}{N+1}$, where $N \in \mathbb{N}^*$, and we shall denote $t_n = nk$, for $n \in [\![0, N+1]\!]$. Throughout this paper, the letter C stands for a positive constant independent of the parameters of the space and time discretizations and its values may be different in different appearance.

We define the space $\mathscr{X}_\mathscr{D}$ as the set of all $((v_K)_{K \in \mathscr{M}}, (v_\sigma)_{\sigma \in \mathscr{E}})$, and $\mathscr{X}_{\mathscr{D},0} \subset \mathscr{X}_\mathscr{D}$ is the set of all $v \in \mathscr{X}_\mathscr{D}$ such that $v_\sigma = 0$ for all $\sigma \in \mathscr{E}_{\text{ext}}$. Let $H_\mathscr{M}(\Omega) \subset \mathbb{L}^2(\Omega)$ be the space of piecewise constant functions on the control volumes of the mesh $\mathscr{M}$. For all $v \in \mathscr{X}_\mathscr{D}$, we denote by $\Pi_\mathscr{M} v \in H_\mathscr{M}(\Omega)$ the function defined by $\Pi_\mathscr{M} v(x) = v_K$, for a.e. $x \in K$, for all $K \in \mathscr{M}$.

For all $\varphi \in \mathscr{C}(\Omega)$, we define $\mathscr{P}_\mathscr{D}\varphi = ((\varphi(x_K))_{K\in\mathscr{M}}, (\varphi(x_\sigma))_{\sigma\in\mathscr{E}}) \in \mathscr{X}_\mathscr{D}$. We denote by $\mathscr{P}_\mathscr{M}\varphi \in H_\mathscr{M}(\Omega)$ the function defined by $\mathscr{P}_\mathscr{M}\varphi(x) = \varphi(x_K)$, for a.e. $x \in K$, for all $K \in \mathscr{M}$.

In order to analyze the convergence, we need to consider the size of the discretization $\mathscr{D}$ defined by $h_\mathscr{D} = \sup\{\operatorname{diam}(K), K \in \mathscr{M}\}$ and the regularity of the mesh given by $\theta_\mathscr{D} = \max\left(\max_{\sigma\in\mathscr{E}_{\mathrm{int}},K,L\in\mathscr{M}}\frac{d_{K,\sigma}}{d_{L,\sigma}}, \max_{K\in\mathscr{M},\sigma\in\mathscr{E}_K}\frac{h_K}{d_{K,\sigma}}\right)$. The scheme we want to consider in this note (A general framework will be detailed in a future paper.) is based on the use of the discrete gradient given in [2]. For $u \in \mathscr{X}_\mathscr{D}$, we define, for all $K \in \mathscr{M}$

$$\nabla_\mathscr{D} u(x) = \nabla_{K,\sigma} u, \quad \text{a. e. } x \in \mathscr{D}_{K,\sigma}, \tag{4}$$

where $\mathscr{D}_{K,\sigma}$ is the cone with vertex x_K and basis σ and

$$\nabla_{K,\sigma} u = \nabla_K u + \left(\frac{\sqrt{d}}{d_{K,\sigma}}(u_\sigma - u_K - \nabla_K u \cdot (x_\sigma - x_K))\right)\mathbf{n}_{K,\sigma}, \tag{5}$$

where $\nabla_K u = \dfrac{1}{m(K)} \sum_{\sigma\in\mathscr{E}_K} m(\sigma)(u_\sigma - u_K)\mathbf{n}_{K,\sigma}$ and d is the space dimension.

We define the finite volume approximation for (1)–(3) as $(u_\mathscr{D}^n)_{n=0}^{N+1} \in \mathscr{X}_{\mathscr{D},0}^{N+2}$ with $u_\mathscr{D}^n = ((u_K^n)_{K\in\mathscr{M}}, (u_\sigma^n)_{\sigma\in\mathscr{E}})$, for all $n \in [\![0, N+1]\!]$ and

1. discretization of the initial conditions (2):

$$\langle u_\mathscr{D}^0, v\rangle_F = -\left(\Delta u^0, \Pi_\mathscr{M} v\right)_{\mathbb{L}^2(\Omega)}, \quad \forall v \in \mathscr{X}_{\mathscr{D},0}, \tag{6}$$

and

$$\langle \frac{u_\mathscr{D}^1 - u_\mathscr{D}^0}{k}, v\rangle_F = -\left(\Delta u^1, \Pi_\mathscr{M} v\right)_{\mathbb{L}^2(\Omega)}, \quad \forall v \in \mathscr{X}_{\mathscr{D},0}, \tag{7}$$

2. discretization of equation (1): for any $n \in [\![1, N]\!]$, $v \in \mathscr{X}_{\mathscr{D},0}$

$$\left(\Pi_\mathscr{M} \partial^2 u_\mathscr{D}^{n+1}, \Pi_\mathscr{M} v\right)_{\mathbb{L}^2(\Omega)} + \langle u_\mathscr{D}^{n+1}, v\rangle_F = \sum_{K\in\mathscr{M}} m(K) f_K^n v_K, \tag{8}$$

where

$$\langle u, v\rangle_F = \int_\Omega \nabla_\mathscr{D} u(x) \cdot \nabla_\mathscr{D} v(x)\,dx, \quad \forall u, v \in \mathscr{X}_\mathscr{D}, \tag{9}$$

$$\partial^2 v^{n+1} = \frac{v^{n+1} - 2v^n + v^{n-1}}{k^2}, \quad \forall n \in [\![1, N]\!], \tag{10}$$

$$f_K^n = \frac{1}{k m(K)} \int_{t_n}^{t_{n+1}} \int_K f(x,t)\,dx\,dt, \tag{11}$$

and $(\cdot,\cdot)_{\mathbb{L}^2(\Omega)}$ denotes the $\mathbb{L}^2$ inner product.

The main result of the present contribution is the following theorem.

Theorem 1. *(Error estimates for the finite volume scheme (6)–(11)) Let Ω be a polyhedral open bounded subset of $\mathbb{R}^d$, where $d \in \mathbb{N} \setminus \{0\}$, and $\partial\Omega = \overline{\Omega} \setminus \Omega$ its boundary. Assume that the solution (weak) of (1)–(3) satisfies $u \in \mathscr{C}^3([0,T]; \mathscr{C}^2(\overline{\Omega}))$. Let $k = \frac{T}{N+1}$, with $N \in \mathbb{N}^*$, and denote by $t_n = nk$, for $n \in [\![0, N+1]\!]$. Let $\mathscr{D} = (\mathscr{M}, \mathscr{E}, \mathscr{P})$ be a discretization in the sense of [2, Definition 2.1]. Assume that $\theta_{\mathscr{D}}$ satisfies $\theta \geq \theta_{\mathscr{D}}$. Then there exists a unique solution $(u_{\mathscr{D}}^n)_{n=0}^{N+1} \in \mathscr{X}_{\mathscr{D}, \mathscr{B}}^{N+2}$ for problem (6)–(11). For each $n \in [\![0, N+1]\!]$, let us define the error $e_{\mathscr{M}}^n \in H_{\mathscr{M}}(\Omega)$ by:*

$$e_{\mathscr{M}}^n = \mathscr{P}_{\mathscr{M}} u(\cdot, t_n) - \Pi_{\mathscr{M}} u_{\mathscr{D}}^n. \tag{12}$$

Then, the following error estimates hold

- *discrete $\mathbb{L}^\infty(0, T; H_0^1(\Omega))$–estimate: for all $n \in [\![0, N+1]\!]$*

$$\|e_{\mathscr{M}}^n\|_{1,2,\mathscr{M}} \leq C(k + h_{\mathscr{D}}) \|u\|_{\mathscr{C}^3([0,T]; \mathscr{C}^2(\overline{\Omega}))}. \tag{13}$$

- *discrete $\mathscr{W}^{1,\infty}(0, T; \mathbb{L}^2(\Omega))$–estimate: for all $n \in [\![1, N+1]\!]$*

$$\|\partial^1 e_{\mathscr{M}}^n\|_{\mathbb{L}^2(\Omega)} \leq C(k + h_{\mathscr{D}}) \|u\|_{\mathscr{C}^3([0,T]; \mathscr{C}^2(\overline{\Omega}))}, \tag{14}$$

where $\partial^1 v^n = \frac{1}{k}(v^n - v^{n-1})$.

- *error estimate in the gradient approximation: for all $n \in [\![0, N+1]\!]$*

$$\|\nabla_{\mathscr{D}} u_{\mathscr{D}}^n - \nabla u(\cdot, t_n)\|_{\mathbb{L}^2(\Omega)} \leq C(k + h_{\mathscr{D}}) \|u\|_{\mathscr{C}^3([0,T]; \mathscr{C}^2(\overline{\Omega}))}. \tag{15}$$

The following lemma will help us to prove Theorem 1

Lemma 1. *Let Ω be a polyhedral open bounded subset of $\mathbb{R}^d$, where $d \in \mathbb{N} \setminus \{0\}$, and $\partial\Omega = \overline{\Omega} \setminus \Omega$ its boundary. Let $k = \frac{T}{N+1}$, with $N \in \mathbb{N}^*$, and denote by $t_n = nk$, for $n \in [\![0, N+1]\!]$. Let $\mathscr{D} = (\mathscr{M}, \mathscr{E}, \mathscr{P})$ be a discretization in the sense of [2, Definition 2.1]. Assume that $\theta_{\mathscr{D}}$ satisfies $\theta \geq \theta_{\mathscr{D}}$. Assume in addition that there exists $(\eta_{\mathscr{D}}^n)_{n=0}^{N+1} \in \mathscr{X}_{\mathscr{D}}^{N+2}$ such that for any $n \in [\![1, N]\!]$, for all $v \in \mathscr{X}_{\mathscr{D}}$*

$$\left(\Pi_{\mathscr{M}} \partial^2 \eta_{\mathscr{D}}^{n+1}, \Pi_{\mathscr{M}} v\right)_{\mathbb{L}^2(\Omega)} + \langle \eta_{\mathscr{D}}^{n+1}, v \rangle_F = \sum_{K \in \mathscr{M}} \mathrm{m}(K) \mathscr{S}_K^n v_K, \tag{16}$$

where $\mathscr{S}_K^n \in \mathbb{R}$, for all $n \in [\![1, N]\!]$ and for all $K \in \mathscr{M}$. Then the following estimate holds, for all $j \in [\![1, N]\!]$.

$$\|\Pi_{\mathscr{M}} \partial^1 \eta_{\mathscr{D}}^{j+1}\|_{\mathbb{L}^2(\Omega)}^2 + C |\eta_{\mathscr{D}}^{j+1}|_{\mathscr{X}}^2$$
$$\leq C \left(\|\Pi_{\mathscr{M}} \partial^1 \eta_{\mathscr{D}}^1\|_{\mathbb{L}^2(\Omega)}^2 + |\eta_{\mathscr{D}}^1|_{\mathscr{X}}^2 + (\mathscr{S})^2 \right), \tag{17}$$

where

$$\mathscr{S} = \max\left\{\left(\sum_{K \in \mathcal{M}} m(K)\left(\mathscr{S}_K^n\right)^2\right)^{\frac{1}{2}}, \; n \in [\![1, N]\!]\right\}. \tag{18}$$

Proof. Taking $v = \partial^1 \eta_{\mathcal{D}}^{n+1}$ in (16) and summing the result over $n \in [\![1, j]\!]$, where $j \in [\![1, N]\!]$, we get

$$\sum_{n=1}^{j}\left(\Pi_{\mathcal{M}} \partial^2 \eta_{\mathcal{D}}^{n+1}, \Pi_{\mathcal{M}} \partial^1 \eta_{\mathcal{D}}^{n+1}\right)_{L^2(\Omega)} + \sum_{n=1}^{j}\langle \eta_{\mathcal{D}}^{n+1}, \partial^1 \eta_{\mathcal{D}}^{n+1}\rangle_F$$

$$= \sum_{n=1}^{j} \sum_{K \in \mathcal{M}} m(K) \mathscr{S}_K^n \partial^1 \eta_K^{n+1}. \tag{19}$$

We need the following two rules

$$\left(\Pi_{\mathcal{M}} \partial^2 \eta_{\mathcal{D}}^{n+1}, \Pi_{\mathcal{M}} \partial^1 \eta_{\mathcal{D}}^{n+1}\right)_{L^2(\Omega)} = \frac{1}{2k}\|\alpha_{\mathcal{D}}^{n+1} - \alpha_{\mathcal{D}}^n\|_{L^2(\Omega)}^2$$

$$+ \frac{1}{2k}\left(\|\alpha_{\mathcal{D}}^{n+1}\|_{L^2(\Omega)}^2 - \|\alpha_{\mathcal{D}}^n\|_{L^2(\Omega)}^2\right), \tag{20}$$

where $\alpha_{\mathcal{D}}^n = \Pi_{\mathcal{M}} \partial^1 \eta_{\mathcal{D}}^n$ and

$$\langle \eta_{\mathcal{D}}^{n+1}, \partial^1 \eta_{\mathcal{D}}^{n+1}\rangle_F = \frac{1}{2k}\langle \eta_{\mathcal{D}}^{n+1} - \eta_{\mathcal{D}}^n, \eta_{\mathcal{D}}^{n+1} - \eta_{\mathcal{D}}^n\rangle_F$$

$$+ \frac{1}{2k}\left\{\langle \eta_{\mathcal{D}}^{n+1}, \eta_{\mathcal{D}}^{n+1}\rangle_F - \langle \eta_{\mathcal{D}}^n, \eta_{\mathcal{D}}^n\rangle_F\right\}. \tag{21}$$

Identities (20)–(21) yield

$$\sum_{n=1}^{j}\left(\Pi_{\mathcal{M}} \partial^2 \eta_{\mathcal{D}}^{n+1}, \Pi_{\mathcal{M}} \partial^1 \eta_{\mathcal{D}}^{n+1}\right)_{L^2(\Omega)} + \sum_{n=1}^{j}\langle \eta_{\mathcal{D}}^{n+1}, \partial^1 \eta_{\mathcal{D}}^{n+1}\rangle_F$$

$$\geq \frac{1}{2k}\left(\|\alpha_{\mathcal{D}}^{j+1}\|_{L^2(\Omega)}^2 + \langle \eta_{\mathcal{D}}^{j+1}, \eta_{\mathcal{D}}^{j+1}\rangle_F\right) - \frac{1}{2k}\left(\|\alpha_{\mathcal{D}}^1\|_{L^2(\Omega)}^2 + \langle \eta_{\mathcal{D}}^1, \eta_{\mathcal{D}}^1\rangle_F\right).$$

This with (19) and [2, Lemma 4.2] implies

$$\frac{1}{2k}\left(\|\alpha^{j+1}\|_{L^2(\Omega)}^2 + C|\eta_{\mathcal{D}}^{j+1}|_{\mathcal{X}}^2\right) \leq \sum_{n=1}^{j} \sum_{K \in \mathcal{M}} m(K) \mathscr{S}_K^n \partial^1 \eta_K^{n+1}$$

$$+ \frac{1}{2k}\left(\|\alpha_{\mathcal{D}}^1\|_{L^2(\Omega)}^2 + C|\eta_{\mathcal{D}}^1|_{\mathcal{X}}^2\right). \tag{22}$$

Multiplying both sides of the previous inequality by $2k$ and using the Cauchy Schwarz inequality, we get

$$\|\Pi_{\mathcal{M}} \partial^1 \eta_{\mathcal{D}}^{j+1}\|_{L^2(\Omega)}^2 + C|\eta_{\mathcal{D}}^{j+1}|_{\mathcal{X}}^2 \leq 2k\mathcal{S} \sum_{n=1}^{j} \|\Pi_{\mathcal{M}} \partial^1 \eta_{\mathcal{D}}^{n+1}\|_{L^2(\Omega)}$$

$$+ \|\Pi_{\mathcal{M}} \partial^1 \eta_{\mathcal{D}}^1\|_{L^2(\Omega)}^2 + C|\eta_{\mathcal{D}}^1|_{\mathcal{X}}^2, \quad (23)$$

where $\mathcal{S}$ is given by (18).
This with the inequality $ab \leq \frac{T}{k}a^2 + \frac{k}{T}b^2$, (23) implies, for all $j \in [\![1, N]\!]$

$$\|\Pi_{\mathcal{M}} \partial^1 \eta_{\mathcal{D}}^{j+1}\|_{L^2(\Omega)}^2 + C|\eta_{\mathcal{D}}^{j+1}|_{\mathcal{X}}^2 \leq \frac{2k}{T} \sum_{n=2}^{j} \left(\|\Pi_{\mathcal{M}} \partial^1 \eta_{\mathcal{D}}^n\|_{L^2(\Omega)}^2 + C|\eta_{\mathcal{D}}^n|_{\mathcal{X}}^2 \right)$$

$$+ 2\|\Pi_{\mathcal{M}} \partial^1 \eta_{\mathcal{D}}^1\|_{L^2(\Omega)}^2 + C|\eta_{\mathcal{D}}^1|_{\mathcal{X}}^2 + 8T^2(\mathcal{S})^2. \quad (24)$$

Using the discrete version of the Gronwall's Lemma, (24) implies estimate (17).

Sketch of the proof of Theorem 1: The uniqueness of $\left(u_{\mathcal{D}}^n\right)_{n \in [\![0, N+1]\!]}$ satisfying (6)–(11) can be deduced from the [2, Lemma 4.2]. As usual, we can use this uniqueness to prove the existence. To prove (13)–(15), we compare the solution $\left(u_{\mathcal{D}}^n\right)_{n \in [\![0, N+1]\!]}$ satisfying (6)–(11) with the solution (it exists and it is unique thanks to [2, Lemma 4.2]): for any $n \in [\![0, N+1]\!]$, find $\bar{u}_{\mathcal{D}}^n \in \mathcal{X}_{\mathcal{D},0}$ such that, see (9)

$$\langle \bar{u}_{\mathcal{D}}^n, v \rangle_F = - \sum_{K \in \mathcal{M}} v_K \int_K \Delta u(x, t_n) dx, \quad \forall v \in \mathcal{X}_{\mathcal{D},0}. \quad (25)$$

Taking $n = 0$ in (25), using the fact that $u(\cdot, 0) = u^0(\cdot)$, and comparing this with (6), we get the following property which will be used below

$$\bar{u}_{\mathcal{D}}^0 = u_{\mathcal{D}}^0. \quad (26)$$

One remarks that the solution of (25) is the same one of [1, (12)], one can use error estimates [1, (13), (15), and (16)] as error estimates for the solution of (25). Writing (25) in the step $n + 1$ and substracting the result from (8) to get

$$\left(\Pi_{\mathcal{M}} \partial^2 \eta_{\mathcal{D}}^{n+1}, \Pi_{\mathcal{M}} v\right)_{L^2(\Omega)} + \langle \eta_{\mathcal{D}}^{n+1}, v \rangle_F = \sum_{K \in \mathcal{M}} m(K) \mathcal{S}_K^n v_K, \quad (27)$$

where $\eta_{\mathcal{D}}^n = u_{\mathcal{D}}^n - \bar{u}_{\mathcal{D}}^n$, for all $n \in [\![0, N+1]\!]$ and

$$\mathcal{S}_K^n = \frac{1}{k m(K)} \int_{t_n}^{t_{n+1}} \int_K f(x, t) dx \, dt + \frac{1}{m(K)} \int_K \Delta u(x, t_{n+1}) dx - \partial^2 \bar{u}_K^{n+1}. \quad (28)$$

Equation (27) with Lemma 1 implies that, for all $n \in [\![1, N]\!]$

$$\|\Pi_{\mathcal{M}} \partial^1 \eta_{\mathcal{D}}^{n+1}\|_{L^2(\Omega)}^2 + C|\eta_{\mathcal{D}}^{n+1}|_{\mathcal{X}}^2$$
$$\leq C \left(\|\Pi_{\mathcal{M}} \partial^1 \eta_{\mathcal{D}}^1\|_{L^2(\Omega)}^2 + |\eta_{\mathcal{D}}^1|_{\mathcal{X}}^2 + (\mathcal{S})^2 \right). \tag{29}$$

To estimate the terms on the right hand side of the previous inequality, we consider

$$\xi_{\mathcal{D}}^n = \bar{u}_{\mathcal{D}}^n - \mathcal{P}_{\mathcal{D}} u(\cdot, t_n), \quad \forall n \in [\![0, N+1]\!]. \tag{30}$$

It is useful to remark that (recall that $\eta_{\mathcal{D}}^n = u_{\mathcal{D}}^n - \bar{u}_{\mathcal{D}}^n$)

$$u_{\mathcal{D}}^n - \mathcal{P}_{\mathcal{D}} u(\cdot, t_n) = \eta_{\mathcal{D}}^n + \xi_{\mathcal{D}}^n. \tag{31}$$

1. *Estimate of* $\|\Pi_{\mathcal{M}} \partial^1 \eta_{\mathcal{D}}^1\|_{L^2(\Omega)}$: using (31), we get (recall that $u_t(\cdot, 0) = u^1(\cdot)$)

$$\|\Pi_{\mathcal{M}} \partial^1 \eta_{\mathcal{D}}^1\|_{L^2(\Omega)} \leq \sum_{i=1}^{4} \mathbb{T}_i, \tag{32}$$

where

$$\mathbb{T}_1 = \|\Pi_{\mathcal{M}} \partial^1 \xi_{\mathcal{D}}^1\|_{L^2(\Omega)}, \quad \mathbb{T}_2 = \|\Pi_{\mathcal{M}} \partial^1 u_{\mathcal{D}}^1 - u^1\|_{L^2(\Omega)},$$
$$\mathbb{T}_3 = \|u_t(\cdot, 0) - \partial^1 u(\cdot, t_1)\|_{L^2(\Omega)}, \text{ and } \mathbb{T}_4 = \|\partial^1 u(\cdot, t_1) - \mathcal{P}_{\mathcal{M}} \partial^1 u(\cdot, t_1)\|_{L^2(\Omega)}.$$

Estimate [1, (15)], when $j = 1$, with (30) leads to

$$\mathbb{T}_1 \leq C h_{\mathcal{D}} \|u\|_{\mathcal{C}^1([0,T]; \mathcal{C}^2(\overline{\Omega}))}. \tag{33}$$

Equation (7) can be written as

$$\langle \partial^1 u_{\mathcal{D}}^1, v \rangle_F = -\left(\Delta u^1, \Pi_{\mathcal{M}} v \right)_{L^2(\Omega)}, \quad \forall v \in \mathcal{X}_{\mathcal{D},0}. \tag{34}$$

This with [2, (4.25)] and the triangle inequality implies that

$$\mathbb{T}_i \leq C (k + h_{\mathcal{D}}) \|u\|_{\mathcal{C}^1([0,T]; \mathcal{C}^2(\overline{\Omega}))}, \quad \forall i \in [\![2, 4]\!]. \tag{35}$$

Thanks to (32), (33), and (35), we have

$$\|\Pi_{\mathcal{M}} \partial^1 \eta_{\mathcal{D}}^1\|_{L^2(\Omega)} \leq C (k + h_{\mathcal{D}}) \|u\|_{\mathcal{C}^1([0,T]; \mathcal{C}^2(\overline{\Omega}))}. \tag{36}$$

2. *Estimate of* $|\eta_{\mathcal{D}}^1|_{\mathcal{X}}$: let us first remark that thanks to (6) and (7), we have

$$\langle u_{\mathcal{D}}^1, v \rangle_F = -\left(\Delta (u^0 + k u^1), \Pi_{\mathcal{M}} v \right)_{L^2(\Omega)}, \quad \forall v \in \mathcal{X}_{\mathcal{D},0}. \tag{37}$$

In order to bound $|\eta_\mathcal{D}^1|_\mathscr{X} = |u_\mathcal{D}^1 - \bar{u}_\mathcal{D}^1|_\mathscr{X}$, we use the triangle inequality to get

$$|\eta_\mathcal{D}^1|_\mathscr{X} \leq |u_\mathcal{D}^1 - \mathscr{P}_\mathcal{D}(u^0 + ku^1)|_\mathscr{X} + |\mathscr{P}_\mathcal{D}(u^0 + ku^1) - \mathscr{P}_\mathcal{D} u(\cdot, t_1)|_\mathscr{X}$$
$$+ |\mathscr{P}_\mathcal{D} u(\cdot, t_1) - \bar{u}_\mathcal{D}^1|_\mathscr{X}. \tag{38}$$

This with the proof of [2, (4.29)] and suitable Taylor expansions, we get

$$|\eta_\mathcal{D}^1|_\mathscr{X} \leq C(k + h_\mathcal{D}) \|u\|_{\mathscr{C}^2([0,T]; \mathscr{C}^2(\overline{\Omega}))}. \tag{39}$$

3. *Estimate of* $\mathscr{S}$: substituting f by $u_{tt} - \Delta u$, see (1), in the expansion of $\mathscr{S}_K^n$, we get

$$\mathscr{S}_K^n = \frac{1}{km(K)} \int_{t_n}^{t_{n+1}} \int_K u_{tt}(x,t) dx\, dt - \frac{1}{km(K)} \int_{t_n}^{t_{n+1}} \int_K \Delta(x,t) dx\, dt$$
$$+ \frac{1}{m(K)} \int_K \Delta u(x, t_{n+1}) dx - \partial^2 \bar{u}_K^{n+1}. \tag{40}$$

Thanks to the Taylor expansion and [1, (15)], when $j = 2$, we have

$$\mathscr{S} \leq C(k + h_\mathcal{D}) \|u\|_{\mathscr{C}^3([0,T]; \mathscr{C}^2(\overline{\Omega}))}. \tag{41}$$

Gathering now (29), (36), (39), and (41) yields, for all $n \in [\![2, N+1]\!]$

$$\|\Pi_\mathcal{M} \partial^1 \eta_\mathcal{D}^n\|_{L^2(\Omega)} \leq C(k + h_\mathcal{D}) \|u\|_{\mathscr{C}^3([0,T]; \mathscr{C}^2(\overline{\Omega}))}, \tag{42}$$

and

$$|\eta_\mathcal{D}^n|_\mathscr{X} \leq C(k + h_\mathcal{D}) \|u\|_{\mathscr{C}^3([0,T]; \mathscr{C}^2(\overline{\Omega}))}. \tag{43}$$

We now combine (42)–(43) with [1, (13), (15), and (16)] to prove the required estimates (13)–(15).

– *Proof of estimate* (13): estimate (43) with [2, (4.6)] implies

$$\|\Pi_\mathcal{M} \eta_\mathcal{D}^n\|_{1,2,\mathcal{M}} \leq C(k + h_\mathcal{D}) \|u\|_{\mathscr{C}^3([0,T]; \mathscr{C}^2(\overline{\Omega}))}, \quad \forall n \in [\![2, N+1]\!]. \tag{44}$$

This with (31), the fact that $\Pi_\mathcal{M} \xi_\mathcal{D}^n = \Pi_\mathcal{M} \bar{u}_\mathcal{D}^n - \mathscr{P}_\mathcal{M} u(\cdot, t_n)$, estimate [1, (13)], and the triangle inequality implies estimate (13) for all $n \in [\![2, N+1]\!]$. The case when $n = 1$ in (13) can be proved by gathering (39), [2, (4.6)], and the case $n = 1$ of [1, (13)]. Property (26) with the case $n = 0$ of [1, (13)] yields the case $n = 0$ of (13).

– *Proof of estimate* (14): the case when $n \in [\![2, N+1]\!]$ of (14) can be proved by gathering (42), the case when $j = 1$ in [1, (15)], and the triangle inequality. The case $n = 1$ of (14) can be proved by gathering (36), the case when $n = 1$ and $j = 1$ in [1, (15)], and the triangle inequality.

– *Proof of estimate* (15): gathering (39) and (43), and [2, Lemma 4.2] leads to

$$\|\nabla_{\mathscr{D}}\eta_{\mathscr{D}}^n\|_{L^2(\Omega)} \leq C\,(k+h_{\mathscr{D}})\|u\|_{\mathscr{C}^3([0,T];\,\mathscr{C}^2(\overline{\Omega}))},\ \forall\,n\in [\![1,N+1]\!]. \quad (45)$$

Combining (45), [1, (16)], and the triangle inequality yields (15) for all $n \in [\![1, N+1]\!]$. The case $n = 0$ of (15) can be deduced directly from the case $n = 0$ of [1, (16)] by using (26). □

References

1. Bradji, A., Fuhrmann, J.: Error estimates of the discretization of linear parabolic equations on general nonconforming spatial grids. C. R. Math. Acad. Sci. Paris **348**/19-20, 1119–1122 (2010).
2. Eymard, R., Gallouët, T., Herbin, R.: Discretization of heterogeneous and anisotropic diffusion problems on general nonconforming meshes SUSHI: a scheme using stabilization and hybrid interfaces. IMA J. Numer. Anal. **30**/4, 1009–1043 (2010).

The paper is in final form and no similar paper has been or is being submitted elsewhere.

A Convergent Finite Volume Scheme for Two-Phase Flows in Porous Media with Discontinuous Capillary Pressure Field

K. Brenner, C. Cancès, and D. Hilhorst

Abstract We consider an immiscible incompressible two-phase flow in a porous medium composed of two different rocks. The flows of oil and water are governed by the Darcy–Muskat law and a capillary pressure law, where the capillary pressure field may be discontinuous at the interface between the rocks. Using the concept of multi-valued phase pressures, we introduce a notion of weak solution for the flow, and prove the convergence of a finite volume approximation towards a weak solution.

Keywords Finite volume, two phase flow, discontinuous capillary pressures
MSC2010: 65M112, 76M12, 35K65, 76S05

1 The Continuous Problem

1.1 Multivalued Phase Pressures

Consider a heterogeneous porous medium, represented by a polygonal domain $\Omega \subset \mathbb{R}^d$, built of two homogeneous and isotropic subdomains, represented by polygonal domains $\Omega_1, \Omega_2 \subset \mathbb{R}^d$. We assume that $\overline{\Omega_1 \cup \Omega_2} = \overline{\Omega}$ and $\Omega_1 \cap \Omega_2 = \emptyset$, and we denote by Γ the interface between the two rocks, i.e. $\overline{\Gamma} = \partial \Omega_1 \cap \partial \Omega_2$. We consider two immiscible incompressible phases (e.g. water and oil), whose flows within Ω_i are described by the conservation of mass equations together with the

---

Konstantin Brenner and Danielle Hilhorst
CNRS and Université Paris-Sud 11, 91405 Orsay Cedex, e-mail: konstantin.brenner@math.u-psud.fr, danielle.hilhorst@math.u-psud.fr

Clément Cancès
LJLL, UPMC, 75252 Paris cedex 05, e-mail: cances@ann.jussieu.fr

Darcy–Muskat law:

$$\phi_i \partial_t s - \nabla \cdot (\eta_{o,i}(s)(\nabla p_o - \rho_o \mathbf{g})) = 0, \tag{1}$$

$$-\phi_i \partial_t s - \nabla \cdot (\eta_{w,i}(s)(\nabla p_w - \rho_w \mathbf{g})) = 0, \tag{2}$$

where s denotes the oil saturation of the fluid, $\phi_i > 0$ the porosity of Ω_i, the oil mobility $\eta_{o,i}$ is a Lipschitz continuous increasing function on $[0, 1]$ satisfying $\eta_{o,i}(0) = 0$, while the water mobility $\eta_{w,i}$ is Lipschitz continuous, decreasing on $[0, 1]$ and such that $\eta_{w,i}(1) = 0$. The density of the phase α ($\alpha \in \{o, w\}$) is denoted by ρ_α, and $\mathbf{g}$ is the gravity vector. Assume first that both phases coexist, i.e. $s \in (0, 1)$, then each phase has its own pressure denoted by p_α. Classically, they are supposed to be linked by the capillary pressure relation

$$p_o - p_w = \pi_i(s), \tag{3}$$

where the capillary pressure function π_i is supposed to be increasing and to belong to $\mathscr{C}^1((0, 1); \mathbb{R}) \cap L^1(0, 1)$. Since the equation (1) degenerates, there is no control on the oil pressure p_o on $\{s = 0\} \cap \Omega_i$, excepted that, because of (3), one has $p_o \leq p_w + \pi_i(0)$. Similarly, on $\{s = 1\} \cap \Omega_i$, $p_w \leq p_o - \pi_i(1)$. In these cases, the pressure has to be considered as multivalued, i.e.

$$s = 0 \Leftrightarrow p_o = [-\infty, p_w + \pi_i(0)], \qquad s = 1 \Leftrightarrow p_w = [-\infty, p_o - \pi_i(1)]. \tag{4}$$

We deduce from (4) that the capillary pressure function π_i has to be extended into the monotone graph $\tilde{\pi}_i$, already introduced in [3,5], defined by

$$\tilde{\pi}_i(s) = \begin{cases} [-\infty, \pi_i(0)] & \text{if } s = 0, \\ \pi_i(s) & \text{if } s \in (0, 1), \\ [\pi_i(1), +\infty] & \text{if } s = 1. \end{cases} \tag{5}$$

Note that there exists a continuous non-decreasing reciprocal function on $\mathbb{R}$, which we denote by $\tilde{\pi}_i^{-1}$.

At the interface Γ, we prescribe the continuity of the multivalued phase pressures

$$p_{\alpha,1} \cap p_{\alpha,2} \neq \emptyset, \qquad (\alpha \in \{o, w\}) \tag{6}$$

where $p_{\alpha,i}$ denote the trace of the pressure of the phase α on Γ from Ω_i. It is worth noticing that the condition (6) is equivalent to the continuity of the mobile phases prescribed in [8]. The volume conservation of each phase yields

$$\sum_{i=1,2} \eta_{\alpha,i}(s_i)(\nabla p_{\alpha,i} - \rho_\alpha \mathbf{g}) \cdot \mathbf{n}_i = 0, \tag{7}$$

where $\mathbf{n}_i$ denote the outward normal to $\partial \Omega_i$ w.r.t. Ω_i. In order to close the problem, we prescribe the initial condition

$$s_0 \in L^\infty(\Omega), \quad 0 \le s_0 \le 1 \text{ a.e. in } \Omega, \tag{8}$$

and the null-flux boundary condition on $\partial\Omega_i \cap \partial\Omega$:

$$\eta_{\alpha,i}(s_i)(\nabla p_{\alpha,i} - \rho_\alpha \mathbf{g}) \cdot \mathbf{n}_i = 0. \tag{9}$$

1.2 Reformulation of the Problem

We define the fractional flow function $f_i(s) = \frac{\eta_{o,i}(s)}{\eta_{o,i}(s) + \eta_{w,i}(s)}$. We introduce the Kirchhoff transform $\varphi_i(s)$ and the global pressure P defined by

$$\varphi_i(s) = \int_0^s f_i(a) \eta_{w,i}(a) \pi_i'(a) da, \tag{10}$$

$$P = p_w + \lambda_{w,i}(\pi) = p_o + \lambda_{o,i}(\pi) \quad \text{for some } \pi \in \tilde{\pi}_i(s), \tag{11}$$

where $\lambda_{w,i}(\pi) = \int_0^\pi f_i \circ \tilde{\pi}_i^{-1}(p) dp$ and $\lambda_{o,i}(\pi) = \lambda_{w,i}(\pi) - \pi$. Classical computations (see e.g. [7]) allow the rewrite the equations (1) as

$$\phi_i \partial_t s - \nabla \cdot (\eta_{o,i}(s)(\nabla P - \rho_o \mathbf{g}) + \nabla \varphi_i(s)) = 0, \tag{12}$$

while the sum of the equations (1) and (2) yields

$$-\nabla \cdot (M_i(s)\nabla P - \zeta_i(s)\mathbf{g}) = 0, \tag{13}$$

where $M_i(s) = \eta_{o,i}(s) + \eta_{w,i}(s) \ge \alpha_M > 0$ and $\zeta_i(s) = \eta_{o,i}(s)\rho_o + \eta_{w,i}(s)\rho_w$. At the interface, the relations (6) have to be replaced (see [6]) by

$$\exists \pi \in \tilde{\pi}_1(s_1) \cap \tilde{\pi}_2(s_2) \text{ s.t. } P_1 - \lambda_{w,1}(\pi) = P_2 - \lambda_{w,2}(\pi). \tag{14}$$

We solve the problem on the domain $Q = \Omega \times (0, T)$ for some $T > 0$, and we define $Q_i = \Omega_i \times (0, T)$.

Definition 1 (weak solution). A pair (s, P) is said to be a weak solution of the problem if

1. $s \in L^\infty(Q)$ with $0 \le s \le 1$ a.e. in Q, $\varphi_i(s)$ and P belong to $L^2((0,T); H^1(\Omega_i))$;
2. there exists a measurable function π mapping $\Gamma \times (0, T)$ to $\overline{\mathbb{R}}$ such that

$$\pi \in \tilde{\pi}_1(s_1) \cap \tilde{\pi}_2(s_2) \text{ and } P_1 - \lambda_{w,1}(\pi) = P_2 - \lambda_{w,2}(\pi);$$

3. for all $\psi \in C_c^\infty(\overline{\Omega} \times [0, T))$,

$$\iint_Q \phi s \partial_t \psi \, dx \, dt + \int_\Omega \phi s_0 \psi(\cdot, 0) \, dx \, dt$$

$$- \sum_{i=1,2} \iint_{Q_i} (\eta_{o,i}(s)(\nabla P - \rho_o \mathbf{g}) + \nabla \varphi_i(s)) \cdot \nabla \psi \, dx \, dt = 0, \quad (15)$$

$$\iint_Q (M_i(s) \nabla P - \zeta_i(s) \mathbf{g}) \cdot \nabla \psi \, dx \, dt = 0. \quad (16)$$

Because of the choice of the boundary conditions, the global pressure P is only defined up to a constant. In order to eliminate this degree of freedom, we prescribe that

$$\int_\Omega P(x, t) \, dx = 0, \quad \forall t > 0. \quad (17)$$

The equation (13) can be reformulated as

$$\nabla \cdot \mathbf{q} = 0, \text{ with } \mathbf{q} = -M_i(s) \nabla P + \zeta_i(s) \mathbf{g}, \quad (18)$$

while (12) can be rewritten under the form

$$\phi_i \partial_t s + \nabla \cdot (\mathbf{q} f_i(s) + \gamma_i(s) \mathbf{g} - \nabla \varphi_i(s)) = 0, \quad (19)$$

with $\gamma_i(s) = (\rho_o - \rho_w) \eta_{w,i}(s) f_i(s)$.

2 The Finite Volume Scheme

Since nonlinear test functions are necessary for proving the convergence of the scheme, we must restrict our study to spatial discretizations satisfying an orthogonality condition, as developed in [9].

Definition 2 (admissible discretization of Q).

1. An admissible discretization of Ω is given by $(\mathcal{T}, \mathcal{E}, (x_K)_{K \in \mathcal{T}})$ where for all $K \in \mathcal{T}$, K is an open polygonal subset of Ω such that $K \subset \Omega_i$ for some i. We define $\mathcal{T}_i = \{K \subset \Omega_i\}$, and we assume that $\overline{\Omega}_i = \bigcup_{K \in \mathcal{T}_i} \overline{K}$. For $K, L \in \mathcal{T}$ with $K \neq L$, then either the $(d-1)$-Lebesgue measure of $\overline{K} \cap \overline{L}$ is 0, or there exists $\sigma \in \mathcal{E}_K \cap \mathcal{E}_L$ (denoted by $\sigma = K|L$) such that $\overline{\sigma} = \overline{K} \cap \overline{L}$. For all $K \in \mathcal{T}$, there exists $\mathcal{E}_K \subset \mathcal{E}$ such that $\partial K = \bigcup_{\sigma \in \mathcal{E}_K} \overline{\sigma}$. Moreover, $\mathcal{E} = \bigcup_{K \in \mathcal{T}} \mathcal{E}_K$. We define $\mathcal{E}_\Gamma = \{\sigma \in \mathcal{E} : \sigma \subset \Gamma\}$, $\mathcal{E}_i = \{\sigma \in \mathcal{E} : \sigma \subset \Omega_i\}$ and $\mathcal{E}_{\text{ext}} = \{\sigma \in \mathcal{E} : \sigma \subset \partial \Omega\}$, and set $\mathcal{E}_{K,\Gamma} = \mathcal{E}_K \cap \mathcal{E}_\Gamma$, $\mathcal{E}_{K,i} = \mathcal{E}_K \cap \mathcal{E}_i$. The family of points $(x_K)_{K \in \mathcal{T}}$ is such that $x_K \in K$ and if $\sigma = K|L$, the straight line $(x_K x_L)$ is orthogonal to σ. We denote by $d_{K,L}$ the distance between x_K and x_L, and by $d_{K,\sigma}$ the distance between x_K and $\sigma \in \mathcal{E}_K$. For all $K \in \mathcal{T}$ and $\sigma \in \mathcal{E}$ we denote by $m(K)$ and $m(\sigma)$ the corresponding Lebesgue measures.

2. Let N be a positive integer, and $\delta t = T/N$; then a uniform discretization of $(0, T)$ is given by the family $(t^n)_{n \in \{0,...,N\}}$, where $t^n = n\delta t$.
3. A discretization $\mathscr{D} = \left(\mathscr{T}, \mathscr{E}, (x_K)_{K \in \mathscr{T}}, (t^n)_{n \in \{0,...,N\}}\right)$ of Q is said to be admissible if $(\mathscr{T}, \mathscr{E}, (x_K)_{K \in \mathscr{T}})$ is an admissible discretization of Ω and $(t^n)_n$ is a uniform discretization of $(0, T)$.

For a given admissible discretization $\mathscr{D} = (\mathscr{T}, \mathscr{E}, (x_K)_{K \in \mathscr{T}}, (t^n)_{n \in \{0,...,N\}})$ of Q, we define the quantities

$$\text{size}(\mathscr{T}) = \max_{K \in \mathscr{T}} \text{diam}(K), \quad \text{reg}(\mathscr{T}) = \max_{i=1,2} \max_{K \in \mathscr{T}} \left(\sum_{\sigma = K|L \in \mathscr{E}_{K,i}} \frac{m(\sigma)d_{K,L}}{m(K)} \right),$$

and

$$\text{size}(\mathscr{D}) = \max(\text{size}(\mathscr{T}), \delta t), \quad \text{reg}(\mathscr{D}) = \text{reg}(\mathscr{T}).$$

Remark 1. The choice of uniform time steps is not necessary, and all the results presented here can be adapted to the case of nonuniform time steps.

For $K \in \mathscr{T}_i$, we define $g_K(s) = g_i(s)$ for all functions g whose definition depends on the subdomain Ω_i, as for example $\phi_i, \varphi_i, M_i, f_i, \ldots$.

We propose a fully implicit cell-centered finite volume scheme for the problem, whose unknowns at each time step are $(s_K^n, P_K^n)_{K \in \mathscr{T}}$ and an interface unknown $(\pi_\sigma^n)_{\sigma \in \mathscr{E}_\Gamma}$. For all $\sigma \in \mathscr{E}_{K,\Gamma}$, we define $s_{K,\sigma}^n = \tilde{\pi}_K^{-1}(\pi_\sigma^n)$, so that, if $\sigma = K|L$, one directly has that

$$\pi_\sigma^n \in \tilde{\pi}_K(s_{K,\sigma}^n) \cap \tilde{\pi}_L(s_{L,\sigma}^n).$$

The total flux balance equation (18) is discretized by

$$\sum_{\sigma \in \mathscr{E}_K} m(\sigma) Q_{K,\sigma}^n = 0, \quad \forall n \in \{1, \ldots, N\}, \forall K \in \mathscr{T}, \tag{20}$$

with

$$Q_{K,\sigma}^n = \begin{cases} \frac{M_{K,L}(s_K^n, s_L^n)}{d_{K,L}} \left(P_K^n - P_L^n\right) + \mathscr{R}\left(Z_{K,\sigma}; s_K^n, s_L^n\right) & \text{if } \sigma = K|L \in \mathscr{E}_{K,i}, \\ \frac{M_K(s_K^n)}{d_{K,\sigma}} \left(P_K^n - P_{K,\sigma}^n\right) + \mathscr{R}\left(Z_{K,\sigma}; s_K^n, s_{K,\sigma}^n\right) & \text{if } \sigma \in \mathscr{E}_{K,\Gamma}, \\ 0 & \text{if } \sigma \in \mathscr{E}_{K,\text{ext}}, \end{cases} \tag{21}$$

where $M_{K,L}(s_K^n, s_L^n) = M_{L,K}(s_L^n, s_K^n)$ is an average of $M_K(s_K^n)$ and $M_L(s_L^n)$. For example, we can suppose, as in [10] that it is given by the harmonic mean

$$M_{K,L}(s_K^n, s_L^n) = \frac{M_K(s_K^n) M_K(s_L^n) d_{K,L}}{d_{L,\sigma} M_K(s_K^n) + d_{K,\sigma} M_K(s_L^n)}.$$

The function $Z_{K,\sigma}$ is defined by $Z_{K,\sigma}(s) = \zeta_K(s) \mathbf{g} \cdot \mathbf{n}_{K,\sigma}$, where $\mathbf{n}_{K,\sigma}$ denotes the outward normal to σ with respect to K. For a function f, we denote by $\mathscr{R}(f; a, b)$

the Riemann solver

$$\mathscr{R}(f;a,b) = \begin{cases} \min_{c\in[a,b]} f(c) & \text{if } a \leq b, \\ \max_{c\in[b,a]} f(c) & \text{if } b \leq a. \end{cases}$$

The oil-flux balance equation (19) is discretized in the form

$$\phi_K \frac{s_K^n - s_K^{n-1}}{\delta t} m(K) + \sum_{\sigma \in \mathscr{E}_K} m(\sigma) F_{K,\sigma}^n = 0, \quad \forall n \in \{1,\ldots,N\}, \forall K \in \mathscr{T}, \tag{22}$$

with

$$F_{K,\sigma}^n = \begin{cases} Q_{K,\sigma}^n \, f_K(\overline{s}_{K,\sigma}^n) + \mathscr{R}(G_{K,\sigma}; s_K^n, s_L^n) + \frac{\varphi_K(s_K^n) - \varphi_K(s_L^n)}{d_{K,L}} & \text{if } \sigma = K|L \in \mathscr{E}_{K,i}, \\ Q_{K,\sigma}^n \, f_K(\overline{s}_{K,\sigma}^n) + \mathscr{R}(G_{K,\sigma}; s_K^n, s_{K,\sigma}^n) + \frac{\varphi_K(s_K^n) - \varphi_K(s_{K,\sigma}^n)}{d_{K,\sigma}} & \text{if } \sigma \in \mathscr{E}_{K,\Gamma}, \\ 0 & \text{if } \sigma \in \mathscr{E}_{K,\text{ext}}, \end{cases} \tag{23}$$

where $G_{K,\sigma}(s) = \gamma_K(s)\mathbf{g} \cdot \mathbf{n}_{K,\sigma}$ and $\overline{s}_{K,\sigma}^n$ is the upstream value defined by

$$\overline{s}_{K,\sigma}^n = \begin{cases} s_K^n & \text{if } Q_{K,\sigma}^n \geq 0, \\ s_L^n & \text{if } Q_{K,\sigma}^n < 0 \text{ and } \sigma = K|L \in \mathscr{E}_{K,i}, \\ s_{K,\sigma}^n & \text{if } Q_{K,\sigma}^n < 0 \text{ and } \sigma \in \mathscr{E}_{K,\Gamma}. \end{cases} \tag{24}$$

The interface values $(\pi_\sigma^n, P_{K,\sigma}^n, P_{L,\sigma}^n)$ for $\sigma = K|L \in \mathscr{E}_\Gamma$ are defined by the following nonlinear system:

$$P_{K,\sigma}^n - \lambda_{w,K}(\pi_\sigma^n) = P_{L,\sigma}^n - \lambda_{w,L}(\pi_\sigma^n). \tag{25}$$

$$Q_{K,\sigma}^n + Q_{L,\sigma}^n = 0, \tag{26}$$

$$F_{K,\sigma}^n + F_{L,\sigma}^n = 0. \tag{27}$$

Note that since the equations (25) and (26) are linear with respect to $P_{K,\sigma}^n$ and $P_{L,\sigma}^n$, one can eliminate these interface values, only keeping π_σ^n. We impose the discrete counterpart of (17), that is

$$\sum_{K \in \mathscr{T}} m(K) P_K^n = 0, \quad \forall n \in \{1,\ldots,N\}. \tag{28}$$

The discrete initial data is given by:

$$s_K^0 = \frac{1}{m(K)} \int_K s_0(x) dx, \quad \forall K \in \mathscr{T},$$

so that $0 \leq s_K^0 \leq 1$.

Proposition 1 (existence of a discrete solution). *For all $n \in \{1, \ldots, N\}$, there exists $\left((s_K^n)_{K \in \mathcal{T}}, (P_K^n)_{K \in \mathcal{T}}, (\pi_\sigma^n)_{\sigma \in \mathcal{E}_\Gamma}\right)$ satisfying the relations (20)–(28). Moreover,*

$$0 \leq s_K^n \leq 1, \quad \forall K \in \mathcal{T}. \tag{29}$$

The proof of Proposition 1 will be given in the forthcoming paper [2].

For an admissible discretization $\mathcal{D}$ of Q, we denote by $s_\mathcal{D}$ and $P_\mathcal{D}$ the piecewise constant functions defined almost everywhere by

$$s_\mathcal{D}(x,t) = s_K^n, \quad P_\mathcal{D}(x,t) = P_K^n \quad \text{if } (x,t) \in K \times (t^{n-1}, t^n].$$

We consider now a sequence $(\mathcal{D}_m)_{m \geq 0}$ of admissible discretizations of Q in the sense of Definition 2 such that $\text{size}(\mathcal{D}_m)$ tends to 0 and $\text{reg}(\mathcal{D}_m)$ remains uniformly bounded as m tends to ∞. We denote by $(s_{\mathcal{D}_m}, P_{\mathcal{D}_m})_m$ a corresponding sequence of discrete solutions, whose existence has been stated in Proposition 1.

Theorem 1 (main result). *There exists a weak solution (s, P) in the sense of Definition 1 such that, up to a subsequence,*

$$s_{\mathcal{D}_m} \to s \text{ and a.e. in } Q \text{ as } m \to \infty,$$

$$P_{\mathcal{D}_m} \to P \text{ weakly in } L^2(Q) \text{ as } m \to \infty.$$

The proof of Theorem 1 that we will present in the forthcoming paper [2] is based on compactness arguments, using the material developed in [9, 10]. The proof adapts the steps that are given in [6] for the continuous frame.

3 Numerical Results

We consider a model porous medium $\Omega = (0,1)^2$ composed of two layers $\Omega_1 = \{(x,y) \in \Omega \mid y < \Gamma(x)\}$ and $\Omega_2 = \{(x,y) \in \Omega \mid y > \Gamma(x)\}$, which have different capillary pressure laws. The fluid densities are given by $\rho_o = 0.81$, $\rho_w = 1$, and $\mathbf{g} = -9.81 \mathbf{e}_y$. We suppose that the porosity is such that $\phi_i = 1, i \in \{1, 2\}$, and we define the oil and water mobilities by

$$\eta_{o,i}(s) = 0.5 s^2 \quad \text{and} \quad \eta_{w,i} = (1-s)^2, \ i \in \{1, 2\}.$$

Moreover we suppose that the capillary pressure curves have the form

$$\pi_1(s) = s \quad \text{and} \quad \pi_2(s) = 0.5 + s.$$

and that the initial saturation is given by

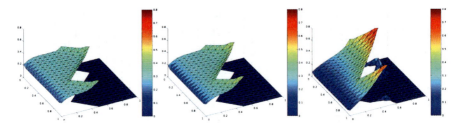

Fig. 1 Saturation for $t = 0.06$, $t = 0.11$ and $t = 0.6$

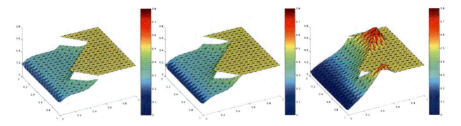

Fig. 2 Capillary pressure for $t = 0.06$, $t = 0.11$ and $t = 0.6$

$$s_0(x) = \begin{cases} 0.3 & \text{if } x \in \Omega_1, \\ 0 & \text{otherwise.} \end{cases}$$

The flow is driven by buoyancy, making the oil move along $\mathbf{e}_y$ until it reaches the interface Γ. For $t \leq 0.11$, oil can not access the domain Ω_2, since the capillary pressure $\pi_1(s_1)$ is lower than the threshold value $\pi_2(0) = 0.5$, which is called *the entry pressure*, see the Fig. 2. Hence the saturation (see the Fig. 1) below the interface s_1 increases, as well as the capillary pressure $\pi_1(s_1)$. As soon as the capillary pressure $\pi_1(s_1)$ reaches the entry pressure $\pi_2(0)$, oil starts to penetrate in the domain Ω_2. Nevertheless, as pointed out in [1,4], a finite quantity of oil remains trapped under the rock discontinuity. This phenomenon is called *oil trapping*.

Acknowledgements This work was supported by the GNR MoMaS (PACEN/CNRS, ANDRA, BRGM, CEA, EdF, IRSN), France.

References

1. M. Bertsch, R. Dal Passo, and C. J. van Duijn. Analysis of oil trapping in porous media flow. *SIAM J. Math. Anal.*, 35:245–267, 2003.
2. K. Brenner, C. Cancès, and D. Hilhorst. Convergence of a finite volume approximation of an immiscible two-phase flow in porous media with discontinuous capillary pressure field. In preparation.

3. F. Buzzi, M. Lenzinger, and B. Schweizer. Interface conditions for degenerate two-phase flow equations in one space dimension. *Analysis*, 29:299–316, 2009.
4. C. Cancès. Finite volume scheme for two-phase flow in heterogeneous porous media involving capillary pressure discontinuities. *M2AN*, 43:973–1001, 2009.
5. C. Cancès, T. Gallouët, and A. Porretta. Two-phase flows involving capillary barriers in heterogeneous porous media. *Interfaces Free Bound.*, 11(2):239–258, 2009.
6. C. Cancès and M. Pierre. An existence result for multidimensional immiscible two-phase flows with discontinuous capillary pressure fields. HAL : hal-00518219, 2010.
7. G. Chavent and J. Jaffré. *Mathematical Models and Finite Elements for Reservoir Simulation*, volume 17. North-Holland, Amsterdam, stud. math. appl. edition, 1986.
8. G. Enchéry, R. Eymard, and A. Michel. Numerical approximation of a two-phase flow in a porous medium with discontinuous capillary forces. *SIAM J. Numer. Anal.*, 43(6):2402–2422, 2006.
9. R. Eymard, T. Gallouët, and R. Herbin. Finite volume methods. Ciarlet, P. G. (ed.) et al., in Handbook of numerical analysis. North-Holland, Amsterdam, pp. 713–1020, 2000.
10. A. Michel. A finite volume scheme for two-phase immiscible flow in porous media. *SIAM J. Numer. Anal.*, 41(4):1301–1317 (electronic), 2003.

The paper is in final form and no similar paper has been or is being submitted elsewhere.

Uncertainty Quantification for a Clarifier–Thickener Model with Random Feed

Raimund Bürger, Ilja Kröker, and Christian Rohde

Abstract The continuous sedimentation process in a clarifier–thickener can be described by a scalar nonlinear conservation law for the solid volume fraction. The flux is discontinuous with respect to space due to the feed mechanism. Typically the feed flux cannot be given in an exact manner. To quantify uncertainty the unknown solid concentration and the feed bulk flow are expressed by polynomial chaos. A deterministic hyperbolic system for a finite number of stochastic moments is constructed. For the resulting high-dimensional system a simple finite volume scheme is presented. Numerical experiments cover one- and two-dimensional situations.

Keywords Clarifier–Thickener model, Polynomial chaos, Uncertainty quantification, Galerkin projection, Finite–Volume method
MSC2010: 65M08, 68U20, 35R60

1 Introduction

Modelling uncertainty is important in many technical applications. Straightforward Monte-Carlo computations are easy but computationally inefficient or even impossible. The quantification of randomness by stochastic Galerkin or collocation methods seems to be more promising in many situations as this leads to deterministic models for at least a finite number of stochastic moments (cf. [MK05] for an overview).

Raimund Bürger
CI^{2}MA and Departamento de Ingeniería Matemática, Universidad de Concepción, Casilla 160-C Concepcion, Chile, e-mail: rburger@ing-mat.udec.cl

Ilja Kröker and Christian Rohde
IANS, Universität Stuttgart, Pfaffenwaldring 57, D-70569 Stuttgart, Germany,
e-mail: ikroeker|crohde@mathematik.uni-stuttgart.de

Roughly speaking, there is by now a well-understood theory for models that can be described by linear partial differential equations. What concerns nonlinear problems –we are interested in hyperbolic conservation laws– first steps have been done just recently [Abg07, PDL09, TLMNE10].

As a prototype model in this field we consider a clarifier–thickener (CT) model for the continuous fluid-solid separation of suspensions under gravity. The CT model provides an idealized description of secondary settling tanks in waste water treatment or of thickeners in mineral processing [BCBT99]. Typically, many input parameters can not be described with deterministic accuracy but behave noisily. We take into account two stochastic dimensions: the uncertainty of the rate of inflow of feed suspension and that of the fraction of solid material. This uncertainty produces a hyperbolic equation with a doubly random flux function. To be precise, consider the longitudinal-infinite vessel $D := \mathbb{R} \times S \subset \mathbb{R}^d$ with the cross-sectional domain $S \subset \mathbb{R}^{d-1}$ and coordinates $\mathbf{x} = (x_1, x_2, \ldots, x_d)^T$. The longitudinal direction is aligned with gravity. For a final time $T > 0$ we search then as the unknown the solid volume fraction $u : D_T := D \times (0, T) \to [0, 1]$. According to [BKRT04, BWC00] the sedimentation process can be modelled by the initial value problem

$$u_t(\mathbf{x},t,\omega) + \mathrm{div}\big(\mathbf{h}(\mathbf{x},t,u(\mathbf{x},t,\omega))\big) = \delta(x_1) Q_F(t,\omega_1) u_F(t,\omega_2) \text{ in } D_T \times \Omega, \quad (1)$$
$$u(.,0) = 0 \quad \text{in } D.$$

The nonlinear flux is given by

$$\mathbf{h}(\mathbf{x},t,u) = \mathbf{q}(\mathbf{x},t)u + (\chi_{(-1,1) \times S}(\mathbf{x}) b(u), 0, \cdots, 0)^T,$$

where b is the given nonlinear batch flux density function. The vector field $\mathbf{q} = \mathbf{q}(\mathbf{x},t) \in \mathbb{R}^d$ is the volume average flow velocity which satisfies a coupled Navier–Stokes-like system [BWC00]. For simplicity, we assume $\mathbf{q}$ to be a given deterministic quantity whose transversal components vanish on $\mathbb{R} \times \partial D$. Furthermore, $\chi_{(-1,1) \times S}$ is the characteristic function for the set $(-1, 1) \times S$. This choice describes the upper overflow boundary and the lower discharge boundary of the vessel. The right-hand side in (1) models the stochastic feed process. For probability measures P_1, P_2 let $\Omega = ((\Omega_1, P_1), (\Omega_2, P_2))$ be the vector-valued probability space. By $Q_F = Q_F(t, \omega_1) > 0$, $\omega_1 \in \Omega_1$, we denote the random feed rate and by $u_F = u_F(t, \omega_2) \in [0, 1]$, $\omega_2 \in \Omega_2$, the feed solid volume fraction. For the idealized vessel we assume that the feed source is distributed over the whole cross section $\{0\} \times S$, i.e. δ denotes the Dirac function in (1). As we will show below, the complete feed term in (1) can be rewritten as part of the flux such that (1) gets the form of a nonlinear conservation law with discontinuous flux. To our knowledge such a situation has not yet been treated in the framework of uncertainty quantification.

In Sect. 2 we detail the model and introduce an approximation for the stochastic process u by a polynomial chaos (PC-) ansatz. A numerical scheme for the PC-system on the base of the Lax–Friedrichs approach is presented. Note that the

Engquist–Osher flux, which is usually applied for problems with discontinuous flux, cannot be used for the higher-dimensional PC-system. Finally, in Sect. 3 numerical experiments are displayed.

2 A Polynomial Chaos Approach for Discontinuous Fluxes

2.1 Formulation of the Model

For notational simplicity we choose $d = 1$ (i.e. $S = \emptyset$) in (1) and use $x = x_1$ for the remaining vertical coordinate. The source term is formally rewritten as

$$\delta(x) Q_F(t, \omega_1) u_F(t, \omega_2) = (H(x) Q_F(t, \omega_1) u_F(t, \omega_2))_x, \tag{2}$$

where H denotes the Heaviside function. Following [BKRT04] we obtain then the flux formulation form

$$u_t(x, t, \omega) + \big(g(x, t, u, \omega)\big)_x = 0 \qquad \text{in } \mathbb{R} \times (0, T) \times \Omega. \tag{3}$$

The flux function g is determined for $t \in (0, T)$ and $\omega \in \Omega$ by the flux in (1) (see assumptions below) minus the flux in (2). This leads to

$$g(x, t, u, \omega) := \begin{cases} q_L(t, \omega_1)(u - u_F(t, \omega_2)) & \text{for } x < -1, \\ q_L(t, \omega_1)(u - u_F(t, \omega_2)) + b(u) & \text{for } -1 < x < 0, \\ q_R(t, \omega_1)(u - u_F(t, \omega_2)) + b(u) & \text{for } 0 < x < 1, \\ q_R(u - u_F(t, \omega_2)) & \text{for } x > 1. \end{cases} \tag{4}$$

To obtain this representation, firstly we have made for $q = q(x, t)$ the ansatz

$$q(x, t) = \begin{cases} q_L(t, \omega_1) & \text{for } x < 0, \\ q_R & \text{for } x > 0, \end{cases} \quad q_L(\cdot, \omega_1) \in C^1([0, T]), q_L(\cdot, \omega_1) < 0, q_R > 0.$$

Stochasticity is solely attached to q_L. Secondly, to ensure global conservativity, we have chosen $Q_F(t, \omega_1) = q_R - q_L(t, \omega_1)$.

The flux (4) has discontinuities for $x \in \{-1, 0, 1\}$. We will not directly work with (3) but expand the equation to a system. For $x \in \mathbb{R}, t \in [0, T), \omega_1 \in \Omega_1$ we define

$$\gamma^1(x, t, \omega_1) := \begin{cases} q_L(t, \omega_1) & \text{for } x < 0, \\ q_R & \text{for } x > 0, \end{cases} \qquad \gamma^2(x, t) := \begin{cases} 1 & \text{for } x \in (-1, 1), \\ 0 & \text{for } x \notin (-1, 1). \end{cases} \tag{5}$$

With the flux $f(t,u,\gamma^1,\gamma^2,\omega_2) := \gamma^1(\cdot,\omega_1)(u - u_F(t,\omega_2)) + \gamma^2 b(u)$ we can understand (3), (4) as a (only seemingly trivial) system of balance laws

$$u(x,t,\omega)_t + \bigl(f(t,u,\gamma^1,\gamma^2,\omega_2)\bigr)_x = 0,$$
$$\gamma_t^1(x,t,\omega_1) = H(-x)q_{L,t}(t,\omega_1), \qquad \gamma_t^2(x,t) = 0 \tag{6}$$

for the unknown vector $(u,\gamma^1,\gamma^2)^T \in [0,1] \times \mathbb{R}^2$.

2.2 Polynomial Chaos Representation

Let $\theta = \theta(\omega) = (\theta_1(\omega_1), \theta_2(\omega_2))^T \in \mathbb{R}^2$ be a vector of i.i.d. (independent identically distributed) random variables. Define

$$\psi_{jk}(\theta) = \phi_j(\theta_1)\phi_k(\theta_2) \quad (j,k \in \mathbb{N}_0),$$

where ϕ_k is the k-th Legendre polynomial. Then $\{\psi_{jk}(\theta)\}_{j,k\in\mathbb{N}_0}$ is a family of $L^2(\Omega_1 \times \Omega_2)$-orthonormal polynomials in the sense

$$\langle \psi_{jk}(\theta), \psi_{lm}(\theta)\rangle_{L^2(\Omega)} := \int_{\Omega_1}\int_{\Omega_2} \psi_{jk}(\theta)\psi_{lm}(\theta)\,dP_1(\omega_1)dP_2(\omega_2) = \delta_{jk}\delta_{lm}. \tag{7}$$

We recall that for some second order random field $w = w(x,t,\omega)$ the polynomial chaos (PC-) representation

$$w(x,t,\omega) = \sum_{j,k\in\mathbb{N}_0} w^{jk}(x,t)\psi_{jk}(\theta(\omega)), \quad w^{jk} := \int_{\Omega_1}\int_{\Omega_2} w\psi_{jk}\,dP_1(\omega_1)dP_2(\omega_2) \tag{8}$$

holds [GS91]. For the sake of a more handsome notation let $w^0, \ldots, w^P$ for $P = P(M) = (M+1)(M+2)/2 - 1$ be an arbitrary but fixed re-indexing of the set $\{w^{jk} \mid j,k \in \mathbb{N}_0, j+k \leq M\}$. The M-th order approximation of $w(x,t,\omega)$ in (8) is given by

$$(\Pi^P w)(x,t,\omega) := \sum_{p=0}^{P} w^p(x,t)\psi_p(\theta(\omega)).$$

The standard stochastic Galerkin approach (for the first equation in (6)) reads as follows. For $M \in \mathbb{N}_0$ find $u^0, \ldots, u^P : D \times (0,T) \to \mathbb{R}$ such that

$$\int_{\Omega_1}\int_{\Omega_2}\left(\Pi^P u + \bigl(\Pi_2^M \gamma^1 \bigl(\Pi^P u - \Pi_1^M u_F\bigr) + \gamma^2 b\bigl(\Pi^P u\bigr)\bigr)_x\right)\Psi_q\,dP_1(\omega_1)dP_2(\omega_2) = 0 \tag{9}$$

holds for $q = 0,\ldots,P$. We used for the given, stochastically one-dimensional approximation of u_F the notation $\Pi_1^M u_F$. An analogous formulation holds for the unknown (stochastically one-dimensional) approximation $\Pi_2^M \gamma^1$ of γ^1.
Using now the orthogonality from (7) the equations (9) can be written in the form

$$u_t^p + \left(\sum_{m=0}^M \sum_{q=0}^P \gamma^{1^m} u^q c_{mqp} - \sum_{m,l=0}^M \gamma^{1^m} u_F^l(t) d_{mlp} + \gamma^2 \mathbb{E}\left[b\left(\Pi^P u\right) \psi_p\right] \right)_x = 0, \tag{10}$$

with $p = 0,\ldots,P$. Here $\mathbb{E}$ denotes the expectation value and

$$\begin{aligned} c_{mqp} &= \int_{\Omega_1} \int_{\Omega_2} \phi_m(\omega_2) \psi_q(\omega) \psi_p(\omega) \, dP(\omega_1) \, dP(\omega_2), \\ d_{mlp} &= \int_{\Omega_1} \int_{\Omega_2} \phi_m(\omega_2) \phi_l(\omega_1) \psi_p(\omega) \, dP(\omega_1) \, dP(\omega_2). \end{aligned} \tag{11}$$

Below we choose b to be a polynomial such that the expectation in (10) can be computed exactly.

We obtain finally from (10) and equations for $\gamma^{1^0},\ldots,\gamma^{1^M}$ the $(P+M+3)$-dimensional PC-system. Using the definition of the coefficients in (11) and the (weak) hyperbolicity of (6) it can be shown that the PC-system (10) is weakly hyperbolic.

2.3 1D Finite–Volume Method

The PC-system (10) is quite general and it appears hard to construct e.g. a Godunov-type solver. Therefore, at least in this paper, we use the simple Lax–Friedrichs method on a uniform mesh with cells $[x_{i-1/2}, x_{i+1/2})$, $i \in \mathbb{Z}$ and $\Delta x = x_{i+1/2} - x_{i-1/2}$. Restricting to the u-components $u^0,\ldots,u^P$ we have for time step $\Delta t^n > 0$ the scheme

$$u_i^{p,n+1} = u_i^{p,n} - \frac{\Delta t^n}{\Delta x} \left(F_{i+1/2}^{p,n} - F_{i-1/2}^{p,n} \right) \quad (i \in \mathbb{Z}, n \in \mathbb{N}, p = 0,\ldots,P),$$

$$F_{i+1/2}^{p,n} := \frac{1}{2} \left(f^p(t^n, u_i^{0,n}, \ldots, u_i^{P,n}, \gamma_i^{1^{0,n}}, \ldots, \gamma_i^{1^{M,n}}, \gamma_i^{2,n}) \right.$$
$$\left. + f^p(t^n, u_{i+1}^{0,n}, \ldots, u_{i+1}^{P,n}, \gamma_{i+1}^{1^{0,n}}, \ldots, \gamma_{i+1}^{1^{M,n}}, \gamma_{i+1}^{2,n}) \right) + \frac{\Delta x}{2 \Delta t^n} (u_{i+1}^{p,n} - u_i^{p,n}).$$

The function f^p is defined by

$$f^P(t, u^0, \ldots, u^P, \gamma^{1^0}, \ldots, \gamma^{1^M}, \gamma^2) =$$

$$\sum_{m=0}^{M}\sum_{q=0}^{P} \gamma^{1^m} u^q c_{mqp} - \sum_{m,l=0}^{M} \gamma^{1^m} u_F^l(t) d_{mlp} + \gamma^2 \mathbb{E}\left[b\left(\Pi^P u\right) \psi_p \right].$$

Initial values are obtained from $u_i^0 = \ldots = u_i^{P,0} = \gamma_i^{1^{1,0}} = \ldots = \gamma_i^{1^{M,0}} = 0$ (cf. (1)) and averaging γ^1, γ^2 from (5) for $\gamma_i^{1^{0,0}}, \gamma_i^{2,0}$.

3 Numerical Experiments

Example 1: [1D Computation with one random dimension]
We consider the problem (3) with the batch flux function $b(w) := \frac{27}{4} w((1-w)^2)$ [BKRT04] and $u_0 = 0$. The solid volume feed fraction u_F satisfies

$$u_F(t, \omega_1) := 0.6 + 0.2\theta(\omega_2),$$

such that θ is uniformly distributed on $[0, 1]$. Consequently the random variable u_F has the expectation 0.7. No further uncertainty is assumed. We choose $q_L = -1$, $q_R = 0.6$. Figure 1 shows (total view and blow-up close to inflow) the numerical solution with $P = 5$ together with the numerical solution of the deterministic problem using $u_F \equiv 0.7$ and the numerical Monte-Carlo approach with 5000 samples computed with $\Delta x = 0.01$. We use Lax–Friedrichs method for our computation. Almost no differences can be detected.

This is confirmed by the subsequent table which displays the $L^1(\mathbb{R})$-difference between the Monte-Carlo sample solution and the PC-approach for $P = 1, \ldots, 6$.

P	1	2	3	4	5	6
L^1-Error	1.1372e-02	1.5566e-02	3.2322e-03	1.4975e-03	8.5714e-04	5.0671e-04

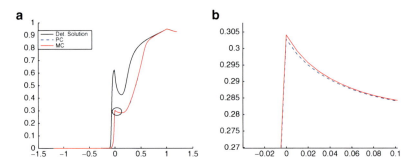

Fig. 1 Solid volume fraction for the deterministic case using the expectation value of the feeding rate, PC-solution, and Monte-Carlo samples. Blow-up in the right figure

We here observe a clear convergence for a reasonable number of stochastic modes.

Example 2: [1D Computation with two random dimensions]
We choose the same setting as in Example 1 but introduce the second random dimension in the suspension feed rate via

$$q_L(t, \omega_1) = -1.2 + 0.4\theta(\omega_1).$$

Again let θ be uniformly distributed on the interval $[0, 1]$. Figure 2 shows the numerical solution with $M = 3$ and $P = 9$. This is compared with the numerical solution of the deterministic problem using the expectation values $q_L = -1$ and $u_F \equiv 0.7$, and the numerical Monte-Carlo approach with 50000 samples at time $T = 1$.

Already for this low random (and spatial) dimension we immediately attain the limits of available computing power. The table below shows the computing time of the PC-approach.

M (P)	1 (2)	2 (5)	3 (9)	4 (14)	5 (20)
cpu-time [s]	1.3721e+03	3.9463e+03	1.2037e+04	3.5001e+04	6.6399e+04

Example 3: [2D Computation with one random dimension]
Let us consider the CT problem (1) for $d = 2$ and $S = (-1.2, 1.2)$, with flux components $h_1(\mathbf{x}, t, u, \omega) = g(x_1, t, u, \omega)$ defined in (4) and $h_2(\mathbf{x}, t, u, \omega) = 0.02*\cos(\frac{\pi x_2}{0.6})u$ This corresponds not to a realistic velocity field $\mathbf{q}$ but we understand this example as a test case for the uncertainty quantification. The batch flux function b, solid volume feed fraction $u_F(t, \omega_1)$, and q_L, q_R are as in Example 1. For the numerical approximation we use an adaptive finite-volume method based on

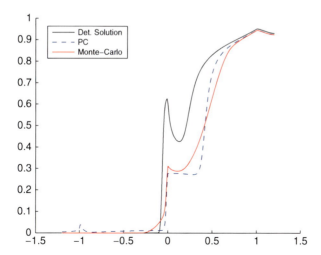

Fig. 2 Solid volume fraction for the deterministic case using the expectation values for solid fraction feeding rate and suspension feeding rate, PC-solution, and Monte-Carlo samples

unstructured triangular meshes with the Lax–Friedrichs flux (cf. [Krö08]). Initially 4608 triangles are used.

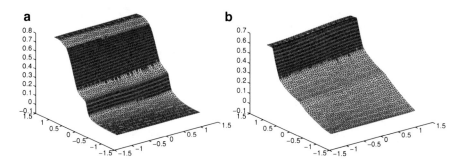

Fig. 3 Solid volume fraction for the deterministic case using the expectation values for solid fraction feeding rate and suspension feeding rate (a), and PC-solution (b) at time $T = 1$

Figure 3(a) shows a deterministic computation with $u_F = 0.7$ and the PC-solution with $P = 7$ (Fig. 3(b)). As in the 1D computations the PC-solution is much smoother and does not develop a peak close to the inlet. As a consequence the adaptive algorithm uses a coarser grid for the PC-solutions. To be specific, at $T = 1$ we had 11826 triangles for the deterministic computation, 8280 for $P = 7$, and 4608 for $P = 1$ (no refinement). Because of the long computation time of each deterministic solution, the computational effort of the Monte-Carlo simulation with a considerable number of samples significantly is higher then the computational effort of the PC approach.

Acknowledgements R. B. acknowledges support by Fondecyt project 1090456, BASAL project CMM, Universidad de Chile and Centro de Investigación en Ingeniería Matemática (CI^{2}MA), Universidad de Concepción. I. K. and C. R. would like to thank the German Research Foundation (DFG) for financial support of the project within the Cluster of Excellence in Simulation Technology (EXC 310/1) at the University of Stuttgart.

References

[Abg07]   R. Abgrall. A simple, flexible and generic deterministic appoarch to uncertainty quantifications in non linear problems: application to fluid flow problems. 2007.

[BCBT99]   M.C. Bustos, F. Concha, R. Bürger, and E. M. Tory. *Sedimentation and thickening*, volume 8 of *Mathematical Modelling: Theory and Applications*. Kluwer Academic Publishers, Dordrecht, 1999. Phenomenological foundation and mathematical theory.

[BKRT04]   R. Bürger, K. H. Karlsen, N. H. Risebro, and J. D. Towers. Well-posedness in BV_t and convergence of a difference scheme for continuous sedimentation in ideal clarifier-thickener units. *Numer. Math.*, 97(1):25–65, 2004.

[BWC00]   R. Bürger, W. L. Wendland, and F. Concha. Model equations for gravitational sedimentation-consolidation processes. *ZAMM Z. Angew. Math. Mech.*, 80(2): 79–92, 2000.

[GS91]   R. G. Ghanem and P. D. Spanos. *Stochastic finite elements: a spectral approach.* Springer-Verlag, New York, 1991.

[Krö08]   I. Kröker. Finite volume methods for conservation laws with noise. In *Finite volumes for complex applications V*, pages 527–534. ISTE, London, 2008.

[MK05]   H. G. Matthies and A. Keese. Galerkin methods for linear and nonlinear elliptic stochastic partial differential equations. *Comput. Methods Appl. Mech. Engrg.*, 194(12-16):1295–1331, 2005.

[PDL09]   G. Poëtte, B. Després, and D. Lucor. Uncertainty quantification for systems of conservation laws. *J. Comput. Phys.*, 228(7):2443–2467, 2009.

[TLMNE10] J. Tryoen, O. Le Maître, M. Ndjinga, and A. Ern. Intrusive Galerkin methods with upwinding for uncertain nonlinear hyperbolic systems. *J. Comput. Phys.*, 229(18):6485–6511, 2010.

The paper is in final form and no similar paper has been or is being submitted elsewhere.

Asymptotic preserving schemes in the quasi-neutral limit for the drift-diffusion system

Chainais-Hillairet Claire and Vignal Marie-Hélène

Abstract We are interested in the drift-diffusion system near quasi-neutrality. For this system, classical explicit schemes are decoupled but subject to severe numerical constraints in the quasi-neutral regime. By constrast, the implicit discretizations are unconditionally stable but non linearly coupled. Then, an iterative method must be used yielding a large numerical cost. Here, we propose a new decoupled asymptotic preserving scheme. We perform one and two dimensional numerical experiments which show its good behavior.

Keywords drift-diffusion, asymptotic preserving schemes, quasi-neutral regime
MSC2010: 65M08, 65M12

1 Presentation of the problem

Let $\Omega \subset \mathbb{R}^d$ ($d \geq 1$) be an open bounded domain describing the geometry of a semiconductor device. The unknowns of the linear drift-diffusion system are the density of electrons and holes, N and P, and the electrostatic potential Ψ. It writes:

$$\partial_t N + \mathrm{div}(-\nabla N + N \nabla \Psi) = 0 \text{ on } \Omega \times [0, T], \tag{1a}$$

$$\partial_t P + \mathrm{div}(-\nabla P - P \nabla \Psi) = 0 \text{ on } \Omega \times [0, T], \tag{1b}$$

$$-\lambda^2 \Delta \Psi = P - N + C \text{ on } \Omega \times [0, T], \tag{1c}$$

Chainais-Hillairet Claire
Laboratoire P. Painlevé, UMR CINRS 8524, Université Lille 1, 59655 Villeneuve d'Ascq Cédex, e-mail: Claire.Chainais@math.univ-lille1.fr

Vignal Marie-Hélène
Institut de Mathématiques de Toulouse, UMR 5219, Université Paul Sabatier, Toulouse 3, 118 route de Narbonne, 31062 Toulouse Cedex 9, e-mail: mhvignal@math.univ-toulouse.fr

where C is the given doping profile non depending on t. The parameter λ comes from the scaling of the physical model. It is called the rescaled Debye length and is given by the ratio of the Debye length to the size of the domain. The Debye length measures the typical scale of electric interactions in the semiconductor.

The system (1) is supplemented with initial conditions N_0, P_0 and with mixed boundary conditions: Dirichlet boundary conditions on Γ^D (N^D, P^D and Ψ^D) and homogeneous Neumann boundary conditions on Γ^N (with $\partial\Omega = \Gamma^D \cup \Gamma^N$).

We are interested in the so-called quasi-neutral regime. This regime occurs when the parameter λ tends to zero. There has been an intense literature on the rigorous quasi-neutral limit of the drift-diffusion model; we can refer for instance to [9] for a zero doping profile C and to [10] for a regular doping profile.

Many different numerical methods have been already developed for the approximation of (1); see for instance [1] and [12, 13] in the non linear case. The convergence of some finite volume schemes has been proved in [2, 3]. But, up to our knowledge, all the schemes are studied in the case $\lambda = 1$. In this paper, we focus on the behavior of schemes in the quasi-neutral limit, that means when λ tends to zero. In this regime, the local electric charge vanishes everywhere. However, simultaneously, very high frequency oscillations, of order $1/\lambda^2$, are triggered. When a standard explicit scheme is used, the scale of these very high frequency oscillations must be resolved by the time step. Hence, the time step must be smaller than λ^2 otherwise a numerical instability appears. The satisfaction of this constraint requires huge computational resources which makes the explicit methods unusable.

Here, the purpose is to define numerical schemes free of such constraints. For a given time step, we look for schemes which may be used as well as for values of λ of order 1 and for values of λ as small as possible. Furthermore, these schemes must preserve the behavior of the continuous problem in the quasi-neutral limit ($\lambda \to 0$). Such schemes are called asymptotic preserving schemes, this name has been introduced in [11] for relaxation limits of kinetic systems. Asymptotic preserving schemes in the quasi-neutral limit have been developed in [5] for the Euler-Poisson problem and in [6,7] for the Vlasov-Poisson system. For the drift-diffusion model, implicit strategies have been proposed in [15].

This paper is organized as follows. In Section 2, we present the formal quasi-neutral limit of the drift-diffusion system. Then, in Section 3, we recall two classical schemes and discuss their stability. Section 4 is devoted to the presentation of a new scheme for the drift-diffusion model. Finally, in Section 5, we conclude with numerical simulations.

2 The formal quasi-neutral limit

Formally, passing to the limit $\lambda \to 0$ in system (1) gives the quasi-neutral drift-diffusion system. It is constituted of the mass equations (1a), (1b) and of the quasi-neutrality constraint $P - N + C = 0$. The Poisson equation is lost, and the electrostatic potential becomes the Lagrange multiplier of this constraint. In order to obtain an explicit equation for the potential we subtract the mass equations (1a), (1b) and we remark that thanks to the quasi-neutrality constraint $P - N = -C$. This yields an elliptic equation for the potential: $-\mathrm{div}((P + N)\nabla\Psi) = -\Delta C$.

Let us perform the same transformations on the original drift-diffusion system. We begin by subtracting the mass equations. Then, remarking that, thanks to Poisson equation, $\partial_t(P - N) = \partial_t(P - N + C) = \partial_t(-\lambda^2 \Delta\Psi)$, we obtain

$$-\lambda^2 \partial_t \Delta\Psi - \mathrm{div}((P + N)\nabla\Psi) = \Delta(P - N). \tag{2}$$

Following [5], we call this equation the reformulated Poisson equation. If P and N are constant, this equation is an order one differential equation on the quantity $-\Delta\Psi$. And, we can note that solutions oscillate in time at the period λ^2.

Thus, an explicit discretization of the electric force terms in (1) will give an explicit discretization of equation (2) and so a stability non uniform in λ. By contrast, an implicit discretization of these terms will give an implicit discretization of (2) and so a stability uniform in λ. This remark will be used in Section 4 for the construction of our decoupled asymptotic preserving scheme.

3 "Classical" schemes

In this section, we present the classical schemes used for the discretization of the drift-diffusion system. The mesh is given by $\mathcal{T}$, a family of control volumes, $\mathcal{E}$, a family of edges and $\mathcal{P} = (x_K)_{K \in \mathcal{T}}$ a family of points. We assume that the mesh is admissible in the sense of [8]. The set of edges will be split into $\mathcal{E} = \mathcal{E}_{int} \cup \mathcal{E}_{ext}$ and for the exterior edges, we distinguish the edges included in Γ^D from the edges included in Γ^N: $\mathcal{E}_{ext} = \mathcal{E}_{ext}^D \cup \mathcal{E}_{ext}^N$. For a given control volume $K \in \mathcal{T}$, we define $\mathcal{E}_K$ the set of its edges, which is also split into $\mathcal{E}_K = \mathcal{E}_{K,int} \cup \mathcal{E}_{K,ext}^D \cup \mathcal{E}_{K,ext}^N$.

For all edge $\sigma \in \mathcal{E}$, we define $d_\sigma = d(x_K, x_L)$ if $\sigma = K|L \in \mathcal{E}_{int}$ and $d_\sigma = d(x_K, \sigma)$ if $\sigma \in \mathcal{E}_{K,int}$. Then, the transmissibility coefficient is defined by $\tau_\sigma = m(\sigma)/d_\sigma$, for all $\sigma \in \mathcal{E}$.

Let Δt be the time step. A finite volume scheme for (1) writes:

$$m(K)\frac{N_K^{n+1} - N_K^n}{\Delta t} + \sum_{\sigma \in \mathcal{E}_K} \mathcal{F}_{K,\sigma}^{n+1} = 0, \forall K \in \mathcal{T}, \forall n \geq 0,$$

$$m(K)\frac{P_K^{n+1} - P_K^n}{\Delta t} + \sum_{\sigma \in \mathcal{E}_K} \mathcal{G}_{K,\sigma}^{n+1} = 0, \forall K \in \mathcal{T}, \forall n \geq 0,$$

$$-\lambda^2 \sum_{\sigma \in \mathcal{E}_K} \tau_\sigma D\Psi_{K,\sigma}^n = m(K)(P_K^n - N_K^n + C_K), \forall K \in \mathcal{T}, \forall n \geq 0.$$

It remains to define the numerical fluxes $D\Psi_{K,\sigma}^n, \mathcal{F}_{K,\sigma}^{n+1}$ and $\mathcal{G}_{K,\sigma}^{n+1}$. As usually, we set $D\Psi_{K,\sigma}^n = \Psi_L^n - \Psi_K^n$ if $\sigma = K|L$, $D\Psi_{K,\sigma}^n = \Psi_\sigma^D - \Psi_K^n$ if $\sigma \in \mathcal{E}_{K,ext}^D$ and $D\Psi_{K,\sigma}^n = 0$

elsewhere. The numerical approximations of the convection-diffusion fluxes in (1a) and (1b), $\mathscr{F}_{K,\sigma}^{n+1}$ and $\mathscr{G}_{K,\sigma}^{n+1}$, are written with the following compact form:

$$\mathscr{F}_{K,\sigma}^{n+1} = \tau_\sigma \left(B(-D\Psi_{K,\sigma}^m) N_K^{n+1} - B(D\Psi_{K,\sigma}^m) N_L^{n+1} \right), \forall \sigma \in \mathscr{E}_{int}, \sigma = K|L \quad (3a)$$

$$\mathscr{G}_{K,\sigma}^{n+1} = \tau_\sigma \left(B(D\Psi_{K,\sigma}^m) P_K^{n+1} - B(-D\Psi_{K,\sigma}^m) P_L^{n+1} \right), \forall \sigma \in \mathscr{E}_{int}, \sigma = K|L. \quad (3b)$$

If $\sigma \in \mathscr{E}_{K,ext}^D$, we replace N_L^{n+1} by N_σ^D in (3a) and P_L^{n+1} by P_σ^D in (3b). If $\sigma \in \mathscr{E}_{K,ext}^N$, we set $\mathscr{F}_{K,\sigma}^{n+1} = \mathscr{G}_{K,\sigma}^{n+1} = 0$.

The case $m = n$ corresponds to a semi-implicit and decoupled scheme: at each time step $(N_K^{n+1})_{K \in \mathscr{T}}$, $(P_K^{n+1})_{K \in \mathscr{T}}$, and $(\Psi_K^{n+1})_{K \in \mathscr{T}}$, are obtained by solving three linear systems. With $m = n + 1$, we write a fully implicit scheme. For the function B, we may choose either $B(x) = 1 - \min(x, 0)$ or $B(x) = x/(\exp(x) - 1)$ with $B(0) = 1$. The first choice corresponds to a classical two-points discretization of the diffusion with an upwinding for the convection. With the Bernoulli function, we get the Scharfetter-Gummel scheme. One main advantage of this last choice, well-known in semiconductor device simulation, is that the scheme is order 2 in space (see [14]). Moreover, as shown in [4], the Scharfetter-Gummel scheme satisfies some crucial properties like energy and energy dissipation decrease.

The decoupled scheme ($m = n$) has been studied in [2] for $B(x) = 1 - \min(x, 0)$ and the convergence has been established (for the nonlinear drift-diffusion system). The proof can be extended to the Scharfetter-Gummel scheme (in the linear case). However, in [2], the convergence proof has been done for $\lambda^2 = 1$ and in fact all the a priori estimates (leading to stability, compactness and convergence) depend on λ^2. More precisely, when there is no doping profile or when the doping profile is constant in space, there exists uniform in time L^∞ estimates on the densities N and P (see [10]). In this case, the L^∞ estimates holds at the discrete level, but only under a condition of the form: $\Delta t \leq D\lambda^2$ with $D \in \mathbb{R}$. It means that such a scheme might not be used for small values of λ.

Let us now consider the fully implicit scheme ($m = n + 1$). In this case, existence of a solution to the scheme can be proved via a fixed point theorem. Moreover, when the doping profile is constant in space, we can prove that the scheme is unconditionally stable. However, the implementation of the scheme needs the resolution of a nonlinear system of equations at each iteration. This might be done using a Newton's method. It has a numerical cost and the solution is computed up to a precision criterion.

In the next section, we propose a new scheme with the same numerical cost as the decoupled scheme, but remaining stable and consistent when λ tends to 0.

4 Construction of an asymptotic preserving scheme

Following the remark given in Section 2, let us first consider the following semi-discretization of (1) in which the electric force terms are discretized implicitly.

$$\frac{N^{n+1} - N^n}{\Delta t} + \operatorname{div}(-\nabla N^n + N^n \nabla \Psi^{n+1}) = 0 \text{ on } \Omega \times [0, T], \qquad (4a)$$

$$\frac{P^{n+1} - P^n}{\Delta t} + \operatorname{div}(-\nabla P^n - P^n \nabla \Psi^{n+1}) = 0 \text{ on } \Omega \times [0, T], \qquad (4b)$$

$$-\lambda^2 \Delta \Psi^{n+1} = P^{n+1} - N^{n+1} + C \text{ on } \Omega \times [0, T]. \qquad (4c)$$

We eliminate P^{n+1} and N^{n+1} in (4c) using their expression respectively given in (4b) and (4a). It yields:

$$-\lambda^2 \Delta \Psi^{n+1} - \Delta t \operatorname{div}((P^n + N^n)\nabla \Psi^{n+1}) = P^n - N^n + C + \Delta t \Delta(P^n - N^n). \quad (5)$$

The semi-discretization given by (4a), (4b) and (5) is uniformally stable in λ but not unconditionaly stable. Then, in order to construct an unconditionally stable semi-discretization we just have to change the discretizations (4a), (4b) into the implicit semi-discretizations of the mass equations.

This corresponds to the following fully discrete scheme:

$$m(K) \frac{N_K^{n+1} - N_K^n}{\Delta t} + \sum_{\sigma \in \mathscr{E}_K} \mathscr{F}_{K,\sigma}^{n+1} = 0, \forall K \in \mathscr{T}, \forall n \geq 0, \qquad (6a)$$

$$m(K) \frac{P_K^{n+1} - P_K^n}{\Delta t} + \sum_{\sigma \in \mathscr{E}_K} \mathscr{G}_{K,\sigma}^{n+1} = 0, \forall K \in \mathscr{T}, \forall n \geq 0, \qquad (6b)$$

$$-\sum_{\sigma \in \mathscr{E}_K} \tau_\sigma (\lambda^2 + \Delta t (P_\sigma^n + N_\sigma^n)) D\Psi_{K,\sigma}^{n+1} = m(K)(P_K^n - N_K^n + C_K)$$

$$+ \Delta t \sum_{\sigma \in \mathscr{E}_K} \tau_\sigma (DP_{K,\sigma}^n - DN_{K,\sigma}^n) \forall K \in \mathscr{T}, \forall n \geq 0, \quad (6c)$$

with the values (3a), (3b) and $m = n + 1$ for the numerical fluxes $\mathscr{F}_{K,\sigma}^{n+1}$, $\mathscr{G}_{K,\sigma}^{n+1}$. The interface values, P_σ^n and N_σ^n are defined by taking the mean value between the values of N^n and P^n at two neighboring control volumes. Let us also note that we keep an implicit discretization on N and P in (6a) and (6b) in order to avoid any CFL condition on the time step.

We stress that our scheme is decoupled. It means that, at each time step, if the values $(N_K^n)_{K \in \mathscr{T}}$, $(P_K^n)_{K \in \mathscr{T}}$ are known,

- we first compute $(\Psi_K^{n+1})_{K \in \mathscr{T}}$ by solving the linear system (6c), whose matrix and right-hand-side depend on N^n and P^n,
- then we compute $(N_K^{n+1})_{K \in \mathscr{T}}$ and $(P_K^{n+1})_{K \in \mathscr{T}}$ solutions of the linear systems (6a) and (6b), whose matrices depend on Ψ^{n+1}.

The matrices from (6a) and (6b) are identical to that obtained in the classical decoupled scheme. They are M-matrices, which ensure the positivity at N^n and P^n for all n (starting with positive initial and boundary conditions). However, the numerical analysis of the scheme (6) is not straightforward and is in progress. In the next section, we present the results of numerical simulations in which we compare our new decoupled scheme to the fully implicit scheme. We will focus on the behavior when the rescaled Debye length tends to 0.

5 Numerical experiments

Test case 1. The first test case is a one-dimensional test case ($\Omega =]0, 1[$). The doping profile is a continuous function satisfying $C(x) = -1$ for $0 \leq x \leq 0.4$, $C(x) = +1$ for $0.6 \leq x \leq 1$ and $C(x)$ affine on $[0.4, 0.6]$. The initial and the boundary conditions satisfy the quasi-neutrality condition $P + C - N = 0$, in order to avoid any boundary or initial time layers:

$$N^D = 0, P^D = 1, \Psi^D = 0 \text{ in } x = 0, \quad N^D = 1, P^D = 0, \Psi^D = 4 \text{ in } x = 1, \tag{7a}$$

$$N_0(x) = \max(C(x), 0) \quad P_0(x) = -\min(C(x), 0). \tag{7b}$$

With a time step $\Delta t = 10^{-3}$, we run computations with the fully implicit scheme and with the new one for different values of λ^2 on a mesh made of 100 cells. The solution is computed at the final time $T = 1$. For the Newton's method used in the fully implicit scheme the precision criterion is set to 10^{-10} and the maximal number of iterations to 60. In Table 1, we present the CPU times needed by both schemes and also the relative error between the two solutions in a discrete L^2-norm.

We note that the CPU time needed by the new scheme is almost independent of λ. For the fully implicit scheme, we see that for $\lambda^2 \leq 10^{-6}$ the CPU time has a ratio 3 with those of the new scheme. For smaller values of λ^2, it appears some default of convergence of the Newton's method with the given time step for the fully implicit scheme. However, the new scheme still works and we show on Fig. 1(a) the density profiles obtained for $\lambda^2 = 10^{-14}$.

Table 1 Comparison of the fully implicit scheme with the new scheme for the Test Case 1

λ^2	CPU time fully implicit	CPU time new scheme	ratio	relative error on N	relative error on P	relative error on Ψ
1	1.92	0.64	3.00	1.32e-08	1.32e-08	5.94e-09
1e-2	1.82	0.59	3.08	5.73e-06	5.73e-06	2.98e-06
1e-4	2.07	0.59	3.51	2.77e-04	2.77e-04	1.99e-04
1e-6	1.67	0.60	2.78	5.15e-04	5.15e-04	5.70e-04
1e-8	51.46	0.60	85.77	5.24e-04	5.24e-04	5.88e-04

Test Case 2. We change the doping profile for a discontinuous doping profile: $C(x) = -1$ for $x \leq 0.5$ and $C(x) = +1$ for $x \geq 0.5$. We keep (7) as initial

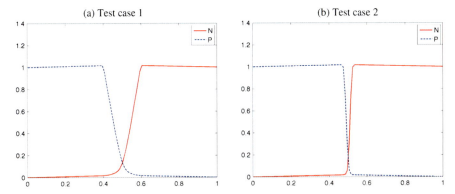

Fig. 1 Density profiles computed by the new scheme for $\lambda^2 = 10^{-14}$ on a mesh made of 100 cells, with $\Delta t = 10^{-3}$

and boundary conditions. The numerical results, presented in Table 2, are similar to those of Test Case 1. We just observe that the relative errors are bigger. This is due to the discontinuity appearing in the density profiles (due to the discontinuity in C): the two schemes do not capture the discontinuity similarly. However, we still note that the new scheme has the same efficiency up to very small values of λ. On Fig. 1(b), we present the density profiles obtained for $\lambda^2 = 10^{-14}$.

Table 2 Comparison of the fully implicit scheme with the new scheme for the Test Case 2

λ^2	CPU time fully implicit	CPU time new scheme	ratio	relative error on N	relative error on P	relative error on Ψ
1	2.09	0.67	3.12	1.31e-08	1.31e-08	5.89e-09
1e-2	1.88	0.60	3.13	7.50e-06	7.50e-06	4.22e-06
1e-4	2.15	0.61	3.52	1.36e-02	1.36e-02	9.51e-03
1e-6	1.73	0.61	2.84	1.03e-01	1.03e-01	6.07e-02
1e-8	51.51	0.60	85.85	1.08e-01	1.08e-01	6.23e-02

Test Case 3. We consider now the simulation of a two-dimensional forward PN diode. The device is made of two different regions: a P-region with a doping profile equal to -1 and an N-region with a doping profile equal to 1 (see [3]). We use a triangular mesh made of 896 triangles and we set the time step $\Delta t = 5 \cdot 10^{-4}$.

Table 3 shows the efficiency of the new scheme. It really runs faster than the fully implicit scheme. Moreover, the fully implicit scheme did not give results for values of λ^2 less that 10^{-3}, while the new scheme still works. We show on Fig. 2, the density profiles obtained with the new scheme for $\lambda^2 = 10^{-10}$.

As a conclusion, we recall that we have proposed in this paper a new scheme for the drift-diffusion system, whose efficiency is independent of the value of the rescaled Debye length. This scheme can be used at the quasi-neutral limit. Numerical analysis of the scheme is in progress.

Table 3 Comparison of the fully implicit scheme with the new scheme for the Test Case 3

λ^2	CPU time fully implicit	CPU time new scheme	ratio	relative error on N	relative error on P	relative error on Ψ
1	203.28	14.68	13.85	1.13e-01	2.78e-01	2.54e-03
1e-1	219.85	14.52	15.14	8.54e-02	2.19e-01	3.01e-02
1e-2	310.72	14.52	21.40	3.21e-02	1.00e-01	4.50e-02
1e-3	718.09	14.68	48.92	4.84e-02	8.30e-02	7.49e-02

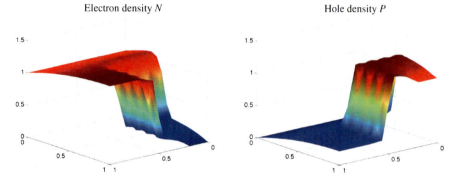

Fig. 2 Test case 3. Density profiles computed by the new scheme for $\lambda^2 = 10^{-10}$ on a mesh made of 896 triangles, with $\Delta t = 5 \cdot 10^{-4}$

References

1. Brezzi F., Marini L.D., Pietra P.: Two-dimensional exponential fitting and applications to drift-diffusion models. SIAM J. Numer. Anal. **26**, 1342–1355 (1989).
2. Chainais-Hillairet, C., Liu, J.-G., Peng, Y.-J.: Finite volume scheme for multi-dimensional drift-diffusion equations and convergence analysis. M2AN **37(2)**,319–338 (2003).
3. Chainais-Hillairet, C., Peng, Y.-J.: Finite volume approximation for degenerate drift-diffusion system in several space dimensions. M3AS **14(3)**, 461–481 (2004).
4. Chatard, M.: Asymptotic behavior of the Scharfetter-Gummel scheme for the drift-diffusion model. Submitted to this conference.
5. Crispel P., Degond P., Vignal M.-H.: An asymptotic preserving scheme for the two-fluid EulerPoisson model in the quasineutral limit, J. Comput. Phys. **223**, 208–234 (2007).
6. Degond P., Deluzet F., Navoret L.: An asymptotically stable particle-in-cell (PIC) scheme for collisionless plasma simulations near quasineutrality. C.R.Acad. Sci. Paris Ser. I **343**, 613–618 (2006).
7. Degond P., Deluzet F., Navoret L., Sun A-B, Vignal M.-H.: Asymptotic-Preserving Particle-In-Cell method for the Vlasov-Poisson system near quasineutrality. Journal of Computational Physics, **229(16)**, 5630–5652 (2010).
8. Eymard, R., Gallouët, T., Herbin, R.: Finite volume methods. In: Handbook of numerical analysis **VII**, pp. 713–1020. North-Holland, Amsterdam (2000).
9. Gasser I.: The initial time layer problem and the quasineutral limit in a nonlinear drift diffusion model for semiconductors, NoDEA, **8 (3)**, 237–249 (2001).
10. Gasser, I., Levermore, C.D., Markowich, P.A., Schmeiser, C.: The initial time layer problem and the quasineutral limit in the semi-conductor drift-diffusion model. Euro. Jnl of Applied Mathematics **12**, 497–512 (2001).

11. Jin S.: Efficient Asymptotic-Preserving (AP) Schemes for Some Multiscale Kinetic Equations. SIAM J. Sci. Comp. **21(441)** (1999).
12. Jüngel A.: Numerical approximation of a drift-diffusion model for semiconductors with nonlinear diffusion. ZAMM Z. Angew. Math. Mech. **75**, 783–799 (1995).
13. Jüngel A., Pietra P.: A discretization scheme for a quasi-hydrodynamic semiconductor model. Math. Models Methods Appl. Sci. **7**, 935–955 (1997).
14. Lazarov, R.D., Mishev, I.D., Vassilevski, P.S.: Finite volume methods for convection-diffusion problems. SIAM J. Numer. Anal. **33-1**, 31–55 (1996).
15. Ventzek P.L., Hoekstra R., Kushner M.: Two-dimensional modeling of high plasma density inductively coupled sources for materials processing. J. Vac. Sci. Tech. B **12**, 461–477 (1994).

The paper is in final form and no similar paper has been or is being submitted elsewhere.

A Posteriori Error Estimates for Unsteady Convection–Diffusion–Reaction Problems and the Finite Volume Method

Nancy Chalhoub, Alexandre Ern, Tony Sayah, and Martin Vohralík

Abstract We derive a posteriori error estimates for the discretization of the unsteady linear convection–diffusion–reaction equation approximated with the cell-centered finite volume method in space and the backward Euler scheme in time. The estimates are based on a locally postprocessed approximate solution preserving the conservative fluxes and are established in the energy norm. We propose an adaptive algorithm which ensures the control of the total error with respect to a user-defined relative precision and refines the meshes adaptively while equilibrating the time and space contributions to the error. Numerical experiments illustrate the theory.

Keywords a posteriori estimate, unsteady convection–diffusion–reaction, cell-centered finite volumes, mesh adaptation
MSC2010: 65N15, 76M12, 76S05

1 Introduction

We consider the time-dependent linear convection–diffusion–reaction equation

---

Nancy Chalhoub and Alexandre Ern
Université Paris-Est, CERMICS, Ecole des Ponts, 77455 Marne-la-Vallée, France,
e-mail: nancy.chalhoub@gmail.com, ern@cermics.enpc.fr

Tony Sayah
Faculté des Sciences, Université Saint-Joseph, B.P. 11-514 Riad El Solh, Beirut 1107 2050, Lebanon, e-mail: tsayah@fs.usj.edu.lb

Martin Vohralík
UPMC Univ. Paris 06, UMR 7598, Laboratoire Jacques-Louis Lions, 75005, Paris, France &
CNRS, UMR 7598, Laboratoire Jacques-Louis Lions, 75005, Paris, France,
e-mail: vohralik@ann.jussieu.fr

$$\partial_t u - \nabla \cdot (S \nabla u) + \nabla \cdot (\boldsymbol{\beta} u) + ru = f \quad \text{a.e. in } Q_T := \Omega \times (0, T), \tag{1a}$$

$$u(\cdot, 0) = u_0 \quad \text{a.e. in } \Omega, \tag{1b}$$

$$u = 0 \quad \text{a.e. on } \partial\Omega \times (0, T). \tag{1c}$$

Here S is the diffusion–dispersion tensor, $\boldsymbol{\beta}$ is the velocity field, r is the reaction function, f is the source term, $\Omega \subset \mathbb{R}^d$, $d \geq 2$, is the space domain which we suppose polyhedral, and $(0, T)$ is the time interval. We suppose that $S = (S_{i,j})$ with $S_{i,j} \in L^\infty(Q_T)$, $1 \leq i, j \leq d$, is a symmetric, bounded, and uniformly positive definite tensor (we suppose that $S_{i,j}$ are piecewise constant on space-time meshes defined below), $\boldsymbol{\beta} \in C^0([0, T]; [W^{1,\infty}(\Omega)]^d)$, $r \in L^\infty(Q_T)$, $f \in L^2(Q_T)$, and $u_0 \in L^2(\Omega)$.

Several works have studied a posteriori error estimates for the cell-centered finite volume method. Ohlberger derives in [7] estimates in the L^1-norm. Nicaise [6] establishes a posteriori energy-norm estimates using Morley-type interpolants of the original piecewise constant finite volume approximation. Guaranteed flux-based estimates were established in [8] and extended in [3] to the parabolic case. Estimates for vertex-centered unsteady convection–diffusion–reaction problems were derived in [1] and [5].

The purpose of this work is to derive guaranteed a posteriori error estimates for the discretization of (1a)–(1c) by the cell-centered finite volume method in space and the backward Euler scheme in time. We allow for time-varying meshes.

2 Notation and Continuous Problem

2.1 Notation

We consider a strictly increasing sequence of discrete times $\{t^n\}_{0 \leq n \leq N}$ such that $t^0 = 0$ and $t^N = T$. For all $1 \leq n \leq N$, we define $\tau^n := t^n - t^{n-1}$ and $I^n := (t^{n-1}, t^n]$. On each time interval I^n, we consider partition $\mathscr{T}^n$ of Ω such that $\overline{\Omega} = \bigcup_{K \in \mathscr{T}^n} K$. For simplicity, we assume that the meshes are simplicial and matching (in the sense that they do not contain hanging nodes). For $1 \leq n \leq N$, $\mathscr{T}^{n-1,n}$ is a common refinement of $\mathscr{T}^{n-1}$ and $\mathscr{T}^n$. For all $0 \leq n \leq N$ and all $K \in \mathscr{T}^n$, h_K denotes the diameter of K. We denote by $c_{S,K}^n$ the smallest eigenvalue of S on K and by $c_{\beta,r,K}^n$ the essential minimum of $\frac{1}{2}\nabla \cdot \boldsymbol{\beta} + r$ on $K \times I^n$. We denote by $\mathscr{E}_K$ the set of the sides of $K \in \mathscr{T}^n$, and we fix $\mathbf{n}_{K,\sigma}$ as the unit normal vector to a side σ outward to K.

We denote by $(\cdot, \cdot)_S$ the $L^2(S)$ inner product, by $\|\cdot\|_S$ the associated norm (when $S = \Omega$, the index is dropped), and by $|S|$ the Lebesgue measure of S. Next, we set $\mathbf{H}(\text{div}, S) := \{\mathbf{v} \in L^2(S); \nabla \cdot \mathbf{v} \in L^2(S)\}$. Moreover, we use the "broken Sobolev space" $H^1(\mathscr{T}^n) := \{\varphi \in L^2(\Omega); \varphi|_K \in H^1(K) \; \forall K \in \mathscr{T}^n\}$. Finally, we use the Raviart–Thomas–Nédélec space $\mathbf{RTN}^0(\mathscr{T}^n) := \{\mathbf{v}_h \in \mathbf{H}(\text{div}, \Omega); \mathbf{v}_h|_K \in$

RTN$^0(K)$ $\forall K \in \mathscr{T}^n\}$ where **RTN**$^0(K) := [\mathbb{P}_0(K)]^d + \mathbf{x}\mathbb{P}_0(K)$. For W, a vector space of functions defined on Ω, we define $\mathscr{P}^1_\tau(W)$ (respectively $\mathscr{P}^0_\tau(W)$) as the vector space of functions v defined on Q_T such that $v(\cdot, t)$ takes values in W and is continuous and piecewise affine (respectively constant) in time.

Because of the nonconformity of the cell-centered finite volume method, we introduce, for all $0 \leq n \leq N$, the broken gradient operator ∇^n such that for a function $v \in H^1(\mathscr{T}^n)$, $\nabla^n v \in [L^2(\Omega)]^d$ is defined as $(\nabla^n v)|_K := \nabla(v|_K)$ for all $K \in \mathscr{T}^n$. The broken gradient operator $\nabla^{n-1,n}$ on the mesh $\mathscr{T}^{n-1,n}$ is defined similarly.

2.2 Continuous Problem

Let $X := L^2(0, T; H^1_0(\Omega))$, $X' = L^2(0, T; H^{-1}(\Omega))$, and $Y := \{v \in X; \partial_t v \in X'\}$. The weak solution u of the problem (1a)–(1c) is such that $u \in Y$ with $u(\cdot, 0) = u_0$. For a.e. $t \in (0, T)$ and for all $\varphi \in H^1_0(\Omega)$, there holds

$$\langle \partial_t u, \varphi \rangle(t) + (S\nabla u, \nabla \varphi)(t) + (\nabla \cdot (\boldsymbol{\beta} u), \varphi)(t) + (ru, \varphi)(t) = (f, \varphi)(t), \quad (2)$$

where $\langle \cdot, \cdot \rangle$ stands for the duality pairing between $H^{-1}(\Omega)$ and $H^1_0(\Omega)$.

For $y \in X$, we introduce the space-time energy norm $\|y\|^2_X := \int_0^T \|\|y\|\|^2(t) dt$, where $\|\|y\|\|^2 := \|S^{\frac{1}{2}} \nabla y\|^2 + \|(\frac{1}{2}\nabla \cdot \boldsymbol{\beta} + r)^{\frac{1}{2}} y\|^2$. We extend the energy norm to discrete functions using the broken gradient.

3 The Cell-centered Finite Volume Schemes and Postprocessing

A general cell-centered finite volume scheme for the problem (1a)–(1c) can be written in the following form: for all $1 \leq n \leq N$, find $\overline{u}^n_h := (u^n_K)_{K \in \mathscr{T}^n}$, such that

$$\frac{1}{\tau^n}(\overline{u}^n_h - u^{n-1}_h, 1)_K + \sum_{\sigma \in \mathscr{E}_K} S^n_{K,\sigma} + \sum_{\sigma \in \mathscr{E}_K} W^n_{K,\sigma} + r^n_K(\overline{u}^n_h, 1)_K = f^n_K |K| \quad \forall K \in \mathscr{T}^n, \quad (3)$$

where $f^n_K = \frac{1}{\tau^n} \int_{I^n} (f(\cdot,t), 1)_K / |K| dt$, $r^n_K = \frac{1}{\tau^n} \int_{I^n} (r(\cdot,t), 1)_K / |K| dt$, $S^n_{K,\sigma}$ and $W^n_{K,\sigma}$ are, respectively, the diffusive and convective fluxes through a side σ of an element K, and u^{n-1}_h is the postprocessed solution that we define below.

For $1 \leq n \leq N$, we reconstruct a conforming convective flux $\boldsymbol{\psi}^n$ and a conforming diffusive flux $\boldsymbol{\theta}^n$ such that $\boldsymbol{\psi}^n, \boldsymbol{\theta}^n \in \mathbf{RTN}^0(\mathscr{T}^n)$ and verifying

$$\langle \boldsymbol{\psi}^n \cdot \mathbf{n}_{K,\sigma}, 1 \rangle_\sigma = W_{K,\sigma}^n \quad \forall K \in \mathcal{T}^n,\ \forall \sigma \in \mathcal{E}_K, \tag{4}$$

$$\langle \boldsymbol{\theta}^n \cdot \mathbf{n}_{K,\sigma}, 1 \rangle_\sigma = S_{K,\sigma}^n \quad \forall K \in \mathcal{T}^n,\ \forall \sigma \in \mathcal{E}_K. \tag{5}$$

We refer to [4, 8] for more details on such construction. We define $\boldsymbol{\theta}$ and $\boldsymbol{\psi}$ in $\mathcal{P}_\tau^0(\mathbf{H}(\mathrm{div}, \Omega))$ by $\boldsymbol{\theta}|_{I^n} := \boldsymbol{\theta}^n$ and $\boldsymbol{\psi}|_{I^n} := \boldsymbol{\psi}^n$.

Following [8], we introduce a piecewise quadratic approximation u_h^n for all $1 \leq n \leq N$ verifying for all $K \in \mathcal{T}^n$,

$$-\mathbf{S}\nabla u_h^n|_K = \boldsymbol{\theta}^n|_K, \tag{6}$$

$$(u_h^n, 1)_K = |K| u_K^n. \tag{7}$$

When $\mathbf{S} = \nu I d$, u_h^n lies in the space $\mathbf{P}_{1,2}(\mathcal{T}^n)$ which is $\mathbf{P}_1(\mathcal{T}^n)$ enriched elementwise with $\sum_{i=1}^d x_i^2$. Finally, we set u_h^0 the L^2-projection of u_0 onto $\mathbf{P}_{1,2}(\mathcal{T}^n)$.

Because of the nonconformity of u_h^n, i.e., of the fact that $u_h^n \in H^1(\mathcal{T}^n)$, $u_h^n \notin H_0^1(\Omega)$, we define an averaging interpolate $s^n = I_{\mathrm{av}}(u_h^n) \in H_0^1(\Omega)$ of u_h^n that verifies

$$(s^n, 1)_K = (u_h^n, 1)_K \quad \forall K \in \mathcal{T}^{n,n+1},\ \forall 0 \leq n \leq N, \tag{8}$$

with the convention $\mathcal{T}^{N,N+1} := \mathcal{T}^N$. We refer to [3] for the details on such construction. Finally, we consider $u_{h,\tau} \in P_\tau^1(H^1(\mathcal{T}^n))$ and $s \in P_\tau^1(H_0^1(\Omega))$. They are defined by the values u_h^n and s^n for all $0 \leq n \leq N$. We set $\partial_t^n v = \partial_t v|_{I^n}$. An important consequence of this construction is the following, cf. [3],

$$(\partial_t^n s, 1)_K = (\partial_t^n u_{h,\tau}, 1)_K \quad \forall K \in \mathcal{T}^n. \tag{9}$$

4 A Posteriori Error Estimate

Our a posteriori estimate bounds the energy error between the weak solution u and the approximate solution $u_{h,\tau}$. We use the postprocessed solution instead of the original piecewise constant solution since the latter has a zero broken gradient and therefore is not suitable for energy norm estimates.

Let $1 \leq n \leq N$ and $K \in \mathcal{T}^n$. We define the *residual estimator* as

$$\eta_{\mathrm{R},K}^n := m_K^n \|\widetilde{f}^n - \partial_t^n s - \nabla \cdot \boldsymbol{\theta}^n - \nabla \cdot \boldsymbol{\psi}^n - r_K^n s^n\|_K. \tag{10}$$

Here $\widetilde{f}^n = \frac{1}{\tau^n}\int_{I^n} f(\cdot, t)dt$ and $m_K^n := \min\{C_{\mathrm{P},K} h_K (c_{\mathrm{S},K}^n)^{-\frac{1}{2}}, (c_{\beta,r,K}^n)^{-\frac{1}{2}}\}$ is the constant from the inequality

$$\|\varphi - \varphi_K\|_K \leq m_K^n \|\|\varphi\|\|_K \quad \forall K \in \mathcal{T}^n,\ \forall \varphi \in H^1(K), \tag{11}$$

shown in [8]. Here, $\varphi_K := (\varphi, 1)_K / |K|$ and $C_{P,K} := 1/\pi$ is the constant from the Poincaré inequality (recall that K are convex). We define the *flux estimator* as

$$\eta_{F,K}^n(t) := \|S^{\frac{1}{2}}\nabla s + S^{-\frac{1}{2}}\theta^n - S^{-\frac{1}{2}}\beta s + S^{-\frac{1}{2}}\psi^n\|_K. \tag{12}$$

Furthermore, we define the following *nonconformity estimator*

$$\eta_{NC,K}^n(t) := \||u_{h,\tau} - s\||_K. \tag{13}$$

Let $\overline{m}^n := \min\{C_{F,\Omega} h_\Omega (c_{S,\Omega}^n)^{-\frac{1}{2}}, (c_{\beta,r,\Omega}^n)^{-\frac{1}{2}}\}$, where $C_{F,\Omega}$ is the Friedrichs inequality constant detailed in [5]. The *quadrature estimator* is given by

$$\eta_{Q,K}^n(t) := \overline{m}^n \|f - \widetilde{f}^n - rs + r_K^n s^n\|_K. \tag{14}$$

Finally, we define the *initial condition estimator* as

$$\eta_{IC} := 2^{-\frac{1}{2}} \|s^0 - u^0\|. \tag{15}$$

We now state and prove our main result concerning the error upper bound.

Theorem 1 (Energy norm a posteriori estimate). *Let $\eta_{R,K}^n$, $\eta_{F,K}^n$, $\eta_{NC,K}^n$, $\eta_{Q,K}^n$, and η_{IC} be defined by (10) and (12)–(15). Then,*

$$\|u - u_{h,\tau}\|_X \leq \eta := \left\{ \sum_{n=1}^N \int_{I^n} \sum_{K \in \mathcal{T}^n} (\eta_{R,K}^n + \eta_{F,K}^n(t))^2 dt \right\}^{\frac{1}{2}} + \eta_{IC}$$

$$+ \left\{ \sum_{n=1}^N \int_{I^n} \sum_{K \in \mathcal{T}^n} (\eta_{Q,K}^n(t))^2 dt \right\}^{\frac{1}{2}} + \left\{ \sum_{n=1}^N \int_{I^n} \sum_{K \in \mathcal{T}^n} (\eta_{NC,K}^n(t))^2 dt \right\}^{\frac{1}{2}}.$$

Proof. For $s \in Y$, we define $\mathcal{R}(s)$ in X' by $\langle \mathcal{R}(s), \varphi \rangle := \int_0^T \{(f - \partial_t s - \nabla \cdot (\beta s) - rs, \varphi) - (S\nabla s, \nabla \varphi)\}(t) dt$, for all $\varphi \in X$. We obtain

$$\frac{1}{2}\|u - s\|^2(T) = \frac{1}{2}\|u^0 - s^0\|^2 + \int_0^T \langle \partial_t(u-s), u-s \rangle(t) dt,$$

which yields

$$\|u - s\|_X^2 \leq \frac{1}{2}\|u^0 - s^0\|^2 + \langle \mathcal{R}(s), u-s \rangle.$$

Using the definition of the dual norm yields $\|u-s\|_X^2 \leq \|\mathcal{R}(s)\|_{X'} \|u-s\|_X + \frac{1}{2}\|u^0 - s^0\|^2$. Since $x^2 \leq ax + b^2$ implies $x \leq a + b$, $(a,b > 0)$, we infer

$$\|u - s\|_X \leq \|\mathcal{R}(s)\|_{X'} + 2^{-\frac{1}{2}}\|u^0 - s^0\|. \tag{16}$$

For $1 \leq n \leq N$, set $\langle \mathscr{R}^n(s), \varphi \rangle := T_R^n(\varphi) + T_F^n(\varphi) + T_Q^n(\varphi)$ with

$$T_R^n(\varphi) := \sum_{K \in \mathscr{T}^n} (\widetilde{f}^n - \partial_t^n s - \nabla \cdot \boldsymbol{\theta}^n - \nabla \cdot \boldsymbol{\psi}^n - r_K^n s^n, \varphi)_K,$$

$$T_F^n(\varphi) := -(S\nabla s + \boldsymbol{\theta}^n + \boldsymbol{\psi}^n - \boldsymbol{\beta} s, \nabla \varphi),$$

$$T_Q^n(\varphi) := \sum_{K \in \mathscr{T}^n} (f - \widetilde{f}^n - rs + r_K^n s^n, \varphi)_K.$$

First, we have $T_R^n(\varphi) = T_R^n(\varphi - \Pi_0 \varphi)$, where $\Pi_0 \varphi|_K := \varphi_K$ for all K, using $(\widetilde{f}^n - \partial_t^n s - \nabla \cdot \boldsymbol{\theta}^n - \nabla \cdot \boldsymbol{\psi}^n - r_K^n s^n, 1)_K = 0$ from (3), (4), (5), and (7)–(9). Hence, $T_R^n(\varphi) \leq \sum_{K \in \mathscr{T}^n} \eta_{R,K}^n |||\varphi|||_K$ using the Cauchy–Schwarz inequality and (11). Moreover, $T_F^n(\varphi)$ is bounded by $\sum_{K \in \mathscr{T}^n} \eta_{F,K}^n |||\varphi|||_K$ using the Cauchy–Schwarz inequality, and $T_Q^n(\varphi)$ is bounded by $\left\{\sum_{K \in \mathscr{T}^n} (\eta_{Q,K}^n)^2\right\}^{1/2} |||\varphi|||$ as in [5]. Using (16), the definition of $\mathscr{R}(s)$, and the Cauchy–Schwarz and triangle inequalities concludes the proof.

In order to make the calculation efficient, it is important to distinguish the space and time errors. To this purpose, the flux estimator $\eta_{F,K}^n(t)$ is split into two contributions using the triangle inequality. We define, for all $1 \leq n \leq N$,

$$(\eta_{sp}^n)^2 := 4 \sum_{K \in \mathscr{T}^n} \left\{ \tau^n (\eta_{R,K}^n + \eta_{F,1,K}^n)^2 + \int_{I^n} (\eta_{NC,K}^n)^2(t) dt \right\},$$

$$(\eta_{tm}^n)^2 := 4 \sum_{K \in \mathscr{T}^n} \left\{ \int_{I^n} \|S^{\frac{1}{2}} \nabla(s - s^n) - S^{-\frac{1}{2}}(\boldsymbol{\beta} s - \boldsymbol{\beta}^n s^n)\|_K^2(t) dt + \int_{I^n} (\eta_{Q,K}^n(t))^2 dt \right\},$$

where $\boldsymbol{\beta}^n := \frac{1}{\tau^n} \int_{I^n} \boldsymbol{\beta}(\cdot, t) dt$ and $\eta_{F,1,K}^n := \|S^{\frac{1}{2}} \nabla s^n + S^{-\frac{1}{2}} \boldsymbol{\theta}^n - S^{-\frac{1}{2}} \boldsymbol{\beta}^n s^n + S^{-\frac{1}{2}} \boldsymbol{\psi}^n\|_K$.

Proceeding as in [3], we obtain

Theorem 2 (A posteriori estimate distinguishing the space and time errors). *There holds*

$$\|u - u_{h,\tau}\|_X \leq \left\{ \sum_{n=1}^N \{(\eta_{sp}^n)^2 + (\eta_{tm}^n)^2\} \right\}^{1/2} + \eta_{IC}.$$

5 A Space-time Adaptive Time-marching Algorithm

We present here an adaptive algorithm based on our a posteriori error estimates which ensures that the relative energy error between the exact and the approximate solutions is below a prescribed tolerance ε. At the same time, it intends to equilibrate the space and time estimators η_{sp}^n and η_{tm}^n. Recalling Theorem 2 and neglecting η_{IC}

we aim at achieving

$$\frac{\sum_{n=1}^{N}\{(\eta_{\text{sp}}^n)^2 + (\eta_{\text{tm}}^n)^2\}}{\sum_{n=1}^{N} \|u_{h,\tau}\|_{X(t^{n-1},t^n)}^2} \leq \varepsilon^2. \tag{17}$$

On a given time level t^{n-1}, we set $\textbf{Crit} := \varepsilon \frac{\|u_{h,\tau}\|_{X(t^{n-1},t^n)}}{\sqrt{2}}$ and we choose the space mesh $\mathcal{T}^n$ and the time step τ^n such that $\eta_{\text{sp}}^n \leq \textbf{Crit}$ and $\eta_{\text{tm}}^n \leq \textbf{Crit}$. For practical implementation purposes and because of computer limitations, we introduce maximal refinement level parameters N_{sp} and N_{tm}. The actual algorithm is as follows:

```
Choose an initial mesh 𝒯⁰, an initial time step τ⁰, and set t⁰ = 0
Set n = 1 and t¹ = t⁰ + τ⁰
Loop in time: While tⁿ≤T
    Set 𝒯ⁿ* := 𝒯ⁿ⁻¹
    Do
        Solve uₕⁿ* = Sol(uₕⁿ⁻¹,τⁿ⁻¹,𝒯ⁿ*)
        Estimate η_sp^n and η_tm^n
        Refine the elements K ∈ 𝒯ⁿ* where η_sp,K^n ≥ Ref η_sp^n and such
            that their level of refinement is less than N_sp
    While {η_sp^n ≥ Crit or η_sp^n is much larger than η_tm^n}
    If {η_tm^n ≥ Crit or η_tm^n is much larger than η_sp^n and when
        the level of time refinement is less than N_tm}
        Set tⁿ = tⁿ - τⁿ⁻¹ and τⁿ⁻¹ = τⁿ⁻¹/2
    Else
        Save the approximate solution uₕⁿ := uₕⁿ*, the mesh 𝒯ⁿ := 𝒯ⁿ*,
            and the time step τⁿ, and set n = n + 1
```

In this version we are only refining the elements and time steps where the estimated error is large. In a later version, we will also coarsen elements and time steps where the estimated error is small.

6 Numerical Experiments

We consider (1a)–(1c) on $\Omega = (0,3) \times (0,3)$ with $S = \nu Id$, $\boldsymbol{\beta} = (\beta_1, \beta_2)$, $r = 0$, and $f = 0$, where $\nu > 0$ determines the amount of diffusion. The initial condition u_0, as well as the Dirichlet boundary condition, are given by the exact solution

$$u(x,y,t) = \frac{1}{200\nu t + 1} e^{-50 \frac{(x-x_0-\beta_1 t)^2 + (y-y_0-\beta_2 t)^2}{200\nu t + 1}}.$$

Here $x_0 = 0.33$, $y_0 = 1.125$, $\beta_1 = 0.8$, and $\beta_2 = 0.4$. We set $T = 0.6$. We use the DDFV method detailed in [2]. We neglect the additional error from the inhomogeneous Dirichlet boundary condition. We consider two cases $\nu = 0.1$ and $\nu = 0.001$. We start from an initial time step $\tau = 0.05$ and an initial mesh of 336 triangles and we refine uniformly by dividing the time step by 2 and each triangle into 4 subelements. Tables 1 and 2 show the actual and estimated energy error where η is the upper bound from Theorem 1, as well as the contribution of each estimator to the upper bound. Specifically, we define the global-in-time and global-in-space version of the estimators, $(\eta_R)^2 := \sum_{n=1}^{N} \tau^n \sum_{K \in \mathcal{T}^n} (\eta_{R,K}^n)^2$, $(\eta_{NC})^2 := \sum_{n=1}^{N} \int_{I^n} \sum_{K \in \mathcal{T}^n} (\eta_{NC,K}^n(t))^2 dt$ and $(\eta_F)^2 := \sum_{n=1}^{N} \int_{I^n} \sum_{K \in \mathcal{T}^n} (\eta_{F,K}^n(t))^2 dt$.

Table 1 Convergence results with uniform refinement in the case $\nu = 0.1$

$\|u - u_{h,\tau}\|_X$	η	η_R	η_F	η_{NC}	$\frac{\eta}{\|u - u_{h,\tau}\|_X}$
0.0625	0.2070	0.0420	0.0910	0.0600	3.3102
0.0366	0.1299	0.0242	0.0613	0.0327	3.5464
0.0199	0.0662	0.0065	0.0328	0.0179	3.3182
0.0104	0.0335	0.0017	0.0167	0.0095	3.2104

Table 2 Convergence results with uniform refinement in the case $\nu = 0.001$

$\|u - u_{h,\tau}\|_X$	η	η_R	η_F	η_{NC}	$\frac{\eta}{\|u - u_{h,\tau}\|_X}$
0.0342	1.6490	0.3894	1.0875	0.0101	48.2496
0.0286	1.2341	0.2175	0.8354	0.0091	43.2175
0.0221	0.7992	0.0701	0.5541	0.0083	36.1332
0.0158	0.4773	0.0226	0.3312	0.0076	30.2736

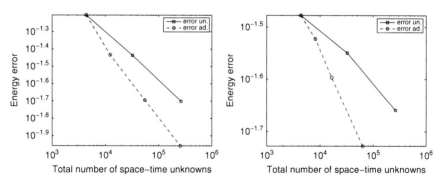

Fig. 1 Energy error in adaptive and uniform refinement for $\nu = 0.1$ (left) and $\nu = 0.001$ (right)

We next compare the uniform and adaptive refinement strategies. We note that the refinement maintains the conformity of the mesh. Figure 1 shows that we obtain a better precision in the adaptive strategy for much fewer space–time unknowns. Figure 2 depicts the approximate solution at the final time for $\nu = 0.001$ obtained

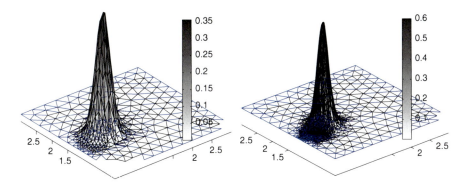

Fig. 2 Approximate solution with adaptive refinement: $N_{sp} = N_{tm} = 2$ (left), $N_{sp} = N_{tm} = 4$ (right)

with adaptive refinement for $N_{sp} = N_{tm} = 2$, and $N_{sp} = N_{tm} = 4$. We can see that in the second case the approximate solution better approximates the exact solution.

Acknowledgements Nancy Chalhoub was supported by a joint fellowship from Ecole des Ponts ParisTech and CNRS Lebanon.

References

1. Amaziane, B. and Bergam, A. and El Ossmani, M. and Mghazli, Z.: A posteriori estimators for vertex centred finite volume discretization of a convection-diffusion-reaction equation arising in flow in porous media. Internat. J. Numer. Methods Fluids **59**, 259–284, (2009)
2. Domelevo, K. and Omnes, P.: A finite volume method for the Laplace equation on almost arbitrary two-dimensional grids. M2AN Math. Model. Numer. Anal. **39**, 1203–1249 (2005)
3. Ern, A. and Vohralík, M.: A posteriori error estimation based on potential and flux reconstruction for the heat equation. SIAM J. Numer. Anal. **48**, 198–223 (2010)
4. Eymard, R. and Gallouët, T. and Herbin, R.: Finite volume approximation of elliptic problems and convergence of an approximate gradient. Appl. Numer. Math. **37**, 31–53 (2001)
5. Hilhorst, D. and Vohralík, M.: A posteriori error estimates for combined finite volume–finite element discretizations of reactive transport equations on nonmatching grids. Comput. Methods Appl. Mech. Engrg. **200**, 597–613 (2011)
6. Nicaise, S.: A posteriori error estimations of some cell centered finite volume methods for diffusion-convection-reaction problems. SIAM J. Numer. Anal. **44**, 949–978 (2006)
7. Ohlberger, M.: A posteriori error estimate for finite volume approximations to singularly perturbed nonlinear convection–diffusion equations. Numer. Math. **87**, 737–761 (2001)
8. Vohralík, M.: Residual flux-based a posteriori error estimates for finite volume and related locally conservative methods. Numer. Math. **11**, 121–158 (2008)

The paper is in final form and no similar paper has been or is being submitted elsewhere.

Large Time-Step Numerical Scheme for the Seven-Equation Model of Compressible Two-Phase Flows

Christophe Chalons, Frédéric Coquel, Samuel Kokh, and Nicole Spillane

Abstract We consider the seven-equation model for compressible two-phase flows and propose a large time-step numerical scheme based on a time implicit-explicit Lagrange-Projection strategy introduced in Coquel *et al.* [6] for Euler equations. The main objective is to get a Courant-Friedrichs-Lewy (CFL) condition driven by (slow) contact waves instead of (fast) acoustic waves.

Keywords Compressible two-phase flows, Baer-Nunziato model, seven-equation model, large time-step numerical scheme
MSC2010: 76T99, 74S10

1 Introduction

We are interested in the computation of compressible two-phase flows with the so-called *two-fluid two-pressure* or *seven-equation* model. It was first proposed in Baer & Nunziato [4] and has since aroused more and more interest, see for instance Embid & Baer [7], Stewart & Wendroff [13], Abgrall & Saurel [11], Gallouët, Hérard & Seguin [8], Andrianov & Warnecke [3], Karni *et al.* [9] Schwendeman, Wahle & Kapila [12], Munkejord [10], Tokareva & Toro [14], Ambroso, Chalons,

---

Christophe Chalons and Samuel Kokh
CEA-Saclay, 91191 Gif-sur-Yvette, France, e-mail: christophe.chalons@cea.fr, samuel.kokh@cea.fr

Frédéric Coquel
CNRS & CMAP, U.M.R. 7641, Ecole Polytechnique, Route de Saclay, 91128 Palaiseau Cedex, France, e-mail: frederic.coquel@cmap.polytechnique.fr

Nicole Spillane
Laboratoire J.-L. Lions, UPMC Univ Paris 06, BC 187, 75252 Paris cedex 05, France,
e-mail: spillane@ann.jussieu.fr

Coquel & Galié [1], Ambroso, Chalons & Raviart [2], and the references therein. One of the main features of this model is that it is hyperbolic, at least in the context of subsonic flows. In particular, an interesting property is that the seven-equation model possesses seven *real* eigenvalues given by $\lambda_k^\pm(\mathbf{u}) = u_k \pm c_k$, $\lambda_k^0(\mathbf{u}) = u_k$ and $\lambda_I(\mathbf{u}) = u_I$, where u_k denote the velocities of both phases $k = 1, 2$, c_k the sound speeds, u_I an interfacial velocity and $\mathbf{u}$ the vector of unknowns.

However from a numerical point of view, the seven-equation model raises some issues. The first difficulty is related to the large size of the model and as a consequence to the Riemann problem that is difficult to solve, even approximately. The second difficulty comes from the presence of nonconservative products and more precisely the fact that the model cannot be equivalently recast in full conservative form. However, the nonconservative products naturally vanish when the void fractions α_k are locally constant in space, and the model coincides in that case with two (decoupled) gas dynamics systems. This property will be used in the numerical strategy.

Numerous papers are devoted to the numerical study of two-fluid two-pressure models, see again for instance [8], [3], [9], [12], [10], [14], [1], [2] and the references therein. Many of the proposed methods are based on time-explicit, exact or approximate, Godunov-type methods (Roe or Roe-like scheme, HLL or HLLC scheme...). For stability reasons, the time steps Δt involved in such methods are subject to a usual Courant-Friedrichs-Lewy (CFL) condition that reads

$$\max_{k,\mathbf{u}}(|\lambda_k^\pm(\mathbf{u})|, |\lambda_k^0(\mathbf{u})|, |\lambda_I(\mathbf{u})|) \Delta t \leq 0.5 \Delta x,$$

where Δx represents the space step. It is then clear that the definition of Δt is driven by the fastest eigenvalues $\lambda_k^\pm(\mathbf{u})$, associated with the so-called acoustic waves.

In many applications, like for instance in two-phase flows involved in nuclear reactors, the acoustic waves are not predominant physical phenomena. A CFL condition based on the most influent waves, the so-called contact waves associated with eigenvalues $\lambda_k^0(\mathbf{u})$ and $\lambda_I(\mathbf{u})$ would be more adapted. The idea is then to propose a time-implicit treatment of the (fast) acoustic waves, in order to get rid of a too restrictive CFL condition, together with an explicit treatment of the (slow) contact waves in order to preserve accuracy. This was recently proposed in Coquel et al. [6] in the context of Euler equations, using a Lagrange-Projection approach. This approach is well-adapted as it naturally decouples the fast and slow waves in the Lagrange and Projection steps respectively.

In this paper, we propose a first attempt to extend this approach to the seven-equation model. The idea is to operate a relevant operator splitting between the conservative and nonconservative parts of the original model, in order to make Euler systems for each phase appear. The latter parts are treated as in [6]. The nonconservative products are then discretized so as to maintain conservativity properties of the model on each partial mass, on the total momentum and total energy. Numerical results are proposed. We underline that this work is still in progress.

2 Governing equations

The model under consideration in this contribution reads as follows:

$$\begin{cases} \partial_t \alpha_k + u_I \partial_x \alpha_k = 0, & t > 0, \quad x \in \mathbb{R}, \\ \partial_t \alpha_k \rho_k + \partial_x \alpha_k \rho_k u_k = 0, \\ \partial_t \alpha_k \rho_k u_k + \partial_x \alpha_k (\rho_k u_k^2 + p_k) - p_I \partial_x \alpha_k = 0, \\ \partial_t \alpha_k \rho_k e_k + \partial_x \alpha_k (\rho_k e_k + p_k) u_k - p_I u_I \partial_x \alpha_k = 0, \end{cases} \quad (1)$$

with $k = 1, 2$. Here, α_k, ρ_k, u_k, e_k and p_k denote the volume fraction, density, velocity, specific total energy and pressure of the phase $k = 1, 2$. The two phases are assumed to be non-miscible that is $\alpha_1 + \alpha_2 = 1$. The structure of (1) is the one of two gas dynamics systems for each phase, coupled with a transport equation on the void fraction α_k at speed u_I. We note that nonconservative products involving the interfacial pressure p_I and velocity u_I (to be precised later on) and the space derivative of the void fractions α_k are present in the momentum and energy equations. These terms act as coupling terms in the evolution of the two phases. Source terms like external forces, pressure and velocity relaxations, dissipation, heat conduction, phase changes and heat exchanges between the two phases are not taken into account.

Each phase is provided with an equation of state $p_k = p_k(\rho_k, \varepsilon_k)$, where $\varepsilon_k = e_k - u_k^2/2$ is the specific internal energy. So far as the definitions of u_I and p_I are concerned, we follow [8] and first observe that the characteristic speeds of (1) are always real and given by u_I, u_k, $u_k \pm c_k$, $k = 1, 2$, where c_k denotes the speed of sound in phase k. System (1) turns out to be only weakly hyperbolic since there are not enough eigenvectors to span the entire space when $u_I = u_k \pm c_k$ for some index k (resonance occurs). When (1) is hyperbolic, one can easily check that similarly to the classical gas dynamics equations, the characteristic fields associated with the eigenvalues $u_k \pm c_k$ are nonlinear while the one associated with u_k is linearly degenerate. Regarding the characteristic field associated with u_I, it is generally required to be linearly degenerate in practice. This property holds as soon as

$$u_I = \beta u_1 + (1 - \beta) u_2, \quad \beta = \frac{\chi \alpha_1 \rho_1}{\chi \alpha_1 \rho_1 + (1 - \chi) \alpha_2 \rho_2} \quad (2)$$

where $\chi \in [0, 1]$ is a constant (we refer to [8] for the details), which gives a natural definition for the interfacial velocity u_I. The usual choices for χ are $0, 1/2$ and 1. Regarding the interfacial pressure p_I, we set $p_I = \mu p_1 + (1 - \mu) p_2$, $\mu \in [0, 1]$. The choice of the coefficient μ is not detailed here (see again [8]) but is related to the ability to provide the system with an entropy balance equation. Indeed, it can be proved that for a specific choice of μ, smooth solutions of (1) verify the conservation law $\partial_t \eta + \partial_x q = 0$, where (η, q) plays the role of a mathematical entropy pair.

3 A natural operator splitting

The starting point is to propose an equivalent form of (1) where the space derivatives of $\alpha_k p_k$ and $\alpha_k p_k u_k$ are decomposed using a chain rule:

$$\begin{cases} \partial_t \alpha_k + u_I \partial_x \alpha_k = 0, \\ \partial_t \alpha_k \rho_k + \partial_x \alpha_k \rho_k u_k = 0, \\ \partial_t \alpha_k \rho_k u_k + \partial_x \alpha_k \rho_k u_k^2 + \alpha_k \partial_x p_k + (p_k - p_I) \partial_x \alpha_k = 0, \\ \partial_t \alpha_k \rho_k e_k + \partial_x \alpha_k \rho_k e_k u_k + \alpha_k \partial_x p_k u_k + (p_k u_k - p_I u_I) \partial_x \alpha_k = 0. \end{cases} \quad (3)$$

We then suggest to split (3) into two independent and *quasi-classical* gas dynamics equations (their Lagrangian forms will be seen to be *classical*), namely

$$\begin{cases} \partial_t \alpha_k = 0, \\ \partial_t \alpha_k \rho_k + \partial_x \alpha_k \rho_k u_k = 0, \\ \partial_t \alpha_k \rho_k u_k + \partial_x \alpha_k \rho_k u_k^2 + \alpha_k \partial_x p_k = 0, \\ \partial_t \alpha_k \rho_k e_k + \partial_x \alpha_k \rho_k e_k u_k + \alpha_k \partial_x p_k u_k = 0, \end{cases} \quad (4)$$

and into the following genuinely nonconservative system:

$$\begin{cases} \partial_t \alpha_k + u_I \partial_x \alpha_k = 0, \\ \partial_t \alpha_k \rho_k = 0, \\ \partial_t \alpha_k \rho_k u_k + (p_k - p_I) \partial_x \alpha_k = 0, \\ \partial_t \alpha_k \rho_k e_k + (p_k u_k - p_I u_I) \partial_x \alpha_k = 0. \end{cases} \quad (5)$$

This transformation aims at proposing in the next section an implicit-explicit Lagrange-Projection scheme similar to [6], and at treating separately the nonconservative products. Note from now on that the overall algorithm will be conservative on the partial mass $\alpha_k \rho_k$, total momentum $\alpha_1 \rho_1 u_1 + \alpha_2 \rho_2 u_2$ and on the total energy $\alpha_1 \rho_1 e_1 + \alpha_2 \rho_2 e_2$, as it is expected from the original form (1) of the model.

4 Numerical approximation

This section is devoted to the discretization of (1), using (4) and (5). Let us introduce a time step $\Delta t > 0$ and a space step $\Delta x > 0$ that we assume to be constant for simplicity. We set $\lambda = \Delta t / \Delta x$ and define the mesh interfaces $x_{j+1/2} = j\Delta x$ for $j \in \mathbb{Z}$, and the intermediate times $t^n = n\Delta t$ for $n \in \mathbb{N}$. In the sequel, $\mathbf{u}_j^n = (\alpha_1, \mathbf{u}_1, \mathbf{u}_2)_j^n$ where $(\mathbf{u}_k)_j^n = (\alpha_k \rho_k, \alpha_k \rho_k u_k, \alpha_k \rho_k e_k)_j^n$ denotes the approximate value of the unknowns at time t^n and on the cell $\mathscr{C}_j =]x_{j-1/2}, x_{j+1/2}[$.

Implicit-explicit discretization of (4). We first recall that (4) is made of two independent quasi-classical gas dynamics systems, whose eigenvalues are given by

$u_k \pm c_k$, u_k and 0. As already stated, our aim is to propose an implicit treatment of the fast waves $u_k \pm c_k$, and an explicit treatment of u_k. With this in mind, we follow [6] and adopt a Lagrange-Projection scheme, coupled with a pressure relaxation strategy that is well adapted to this purpose. A Lagrange-Projection splitting strategy applied to (4) amounts to introducing the Lagrangian system

$$\begin{cases} \partial_t \alpha_k = 0, \\ \partial_t \alpha_k \rho_k + \alpha_k \rho_k \partial_x u_k = 0, \\ \partial_t \alpha_k \rho_k u_k + \alpha_k \rho_k u_k \partial_x u_k + \alpha_k \partial_x p_k = 0, \\ \partial_t \alpha_k \rho_k e_k + \alpha_k \rho_k e_k \partial_x u_k + \alpha_k \partial_x p_k u_k = 0, \end{cases} \text{ or equivalently } \begin{cases} \partial_t \alpha_k = 0, \\ \partial_t \tau_k - \tau_k \partial_x u_k = 0, \\ \partial_t u_k + \tau_k \partial_x p_k = 0, \\ \partial_t e_k + \tau_k \partial_x p_k u_k = 0, \end{cases} \quad (6)$$

with $\tau_k = 1/\rho_k$, and the transport (or projection) system

$$\begin{cases} \partial_t \alpha_k = 0, \\ \partial_t \alpha_k \rho_k + u_k \partial_x \alpha_k \rho_k = 0, \\ \partial_t \alpha_k \rho_k u_k + u_k \partial_x \alpha_k \rho_k u_k = 0, \\ \partial_t \alpha_k \rho_k e_k + u_k \partial_x \alpha_k \rho_k e_k = 0. \end{cases} \quad (7)$$

We note that (6) coincides with two classical gas dynamics systems written in Lagrangian coordinates, the eigenvalues of which are given by $\pm c_k$ and 0. This system is treated using a pressure relaxation approach that consists in introducing a linearized pressure π_k (see for instance [5] and especially the references therein), such that $(\pi_k)_j^n = (p_k)_j^n$, and in solving the partial differential system

$$\begin{cases} \partial_t \alpha_k = 0, \\ \partial_t \tau_k - \tau_k \partial_x u_k = 0, \\ \partial_t u_k + \tau_k \partial_x \pi_k = 0, \\ \partial_t \pi_k + a_k^2 \tau_k \partial_x u_k = 0, \\ \partial_t e_k + \tau_k \partial_x \pi_k u_k = 0, \end{cases} \text{ or equivalently } \begin{cases} \partial_t \alpha_k = 0, \\ \partial_t I_k = 0, \\ \partial_t w_k^+ + a_k \tau_k \partial_x w_k^+ = 0, \\ \partial_t w_k^- - a_k \tau_k \partial_x w_k^- = 0, \\ \partial_t e_k + \tau_k \partial_x \pi_k u_k = 0, \end{cases} \quad (8)$$

where $w_k^\pm = \pi_k \pm a_k u_k$, $I_k = \pi_k + a_k^2 \tau_k$, and a_k is a constant satisfying the subcharacteristic condition $a_k > \rho_k c_k$. A natural time-implicit discretization of (8) is

$$\begin{cases} (\alpha_k)_j^{n+1} = (\alpha_k)_j^n, \\ (I_k)_j^{n+1} = (I_k)_j^n, \\ (w_k^+)_j^{n+1} = (w_k^+)_j^n - \lambda(\tau_k)_j^n a_k\left((w_k^+)_j^{n+1} - (w_k^+)_{j-1}^{n+1}\right), \\ (w_k^-)_j^{n+1} = (w_k^-)_j^n + \lambda(\tau_k)_j^n a_k\left((w_k^-)_{j+1}^{n+1} - (w_k^-)_j^{n+1}\right), \\ (e_k)_j^{n+1} = (e_k)_j^n - \lambda(\tau_k)_j^n\left((\pi_k u_k)_{j+1/2}^{n+1} - (\pi_k u_k)_{j-1/2}^{n+1}\right), \end{cases} \quad (9)$$

with $(\pi_k u_k)_{j+1/2}^{n+1} = (\pi_k)_{j+1/2}^{n+1}(u_k)_{j+1/2}^{n+1}$ and

$$(\pi_k)_{j+1/2}^{n+1} = \frac{1}{2}\left((w_k^+)_j^{n+1} + (w_k^-)_j^{n+1}\right), \quad (u_k)_{j+1/2}^{n+1} = \frac{1}{2a_k}\left((w_k^+)_j^{n+1} - (w_k^-)_j^{n+1}\right).$$

The updated values of u_k, τ_k and ρ_k are recovered from the formulas $u_k = (w_k^+ - w_k^-)/2a_k$, $\pi_k = (w_k^+ + w_k^-)/2$, $\tau_k = (I_k - \pi_k)/a_k^2$ and $\rho_k = 1/\tau_k$. The computation of $(w_k^\pm)_j^{n+1=}$ is cheap and amounts to solving a tridiagonal system of linear equations, while the time-implicit definition of $(e_k)_j^{n+1=}$ explicitly follows. Then, the transport equations involved in (7) are associated with the following classical time-explicit update formula

$$(\mathbf{u}_k)_j^{n+1-} = (\mathbf{u}_k)_j^{n+1=} + \lambda \Big(
\max((u_k)_{j-1/2}^{n+1=}, 0)(\mathbf{u}_k)_{j-1}^{n+1=} - \min((u_k)_{j+1/2}^{n+1=}, 0)(\mathbf{u}_k)_{j+1}^{n+1=}
+ [\min((u_k)_{j+1/2}^{n+1=}, 0) - \max((u_k)_{j-1/2}^{n+1=}, 0)](\mathbf{u}_k)_j^{n+1=}
\Big),$$
(10)

and of course $(\alpha_k)_j^{n+1-} = (\alpha_k)_j^{n+1=}$.

Discretization of (5). Our objective is to propose a consistent approximation of (5) such that the overall algorithm is conservative for each partial mass, for the total momentum and for the total energy, as already motivated. First of all and similarly to (10), the transport equation associated with α_k is treated as follows:

$$(\alpha_k)_j^{n+1} = (\alpha_k)_j^{n+1=} + \lambda \Big(
\max((u_I)_{j-1/2}^{n+1=}, 0)(\alpha_k)_{j-1}^{n+1=} - \min((u_I)_{j+1/2}^{n+1=}, 0)(\alpha_k)_{j+1}^{n+1=}
+ [\min((u_I)_{j+1/2}^{n+1=}, 0) - \max((u_I)_{j-1/2}^{n+1=}, 0)](\alpha_k)_j^{n+1=}
\Big)$$

where $(u_I)_{j+1/2}^{n+1=} = \beta_{j+1/2}^{n+1=}(u_1)_{j+1/2}^{n+1=} + (1 - \beta_{j+1/2}^{n+1=})(u_2)_{j+1/2}^{n+1=}$ and for instance $\beta_{j+1/2}^{n+1=} = \frac{1}{2}(\beta_j^{n+1=} + \beta_{j+1}^{n+1=})$. We set $(\alpha_k \rho_k)_j^{n+1} = (\alpha_k \rho_k)_j^{n+1-}$ for the partial mass, so that only the treatments of the momentum and total energy of each phase are now left. We propose

$$\frac{(\alpha_k \rho_k u_k)_j^{n+1} - (\alpha_k \rho_k u_k)_j^{n+1-}}{\Delta t} + ((\overline{p_k})_j - (\overline{p_I})_j)\frac{(\alpha_k)_{j+1/2}^n - (\alpha_k)_{j-1/2}^n}{\Delta x} = 0,$$

$$\frac{(\alpha_k \rho_k e_k)_j^{n+1} - (\alpha_k \rho_k e_k)_j^{n+1-}}{\Delta t} + ((\overline{p_k u_k})_j - (\overline{p_I u_I})_j)\frac{(\alpha_k)_{j+1/2}^n - (\alpha_k)_{j-1/2}^n}{\Delta x} = 0.$$

In order to get the expected overall conservativity properties, we pay a particular attention to the definitions of $(\overline{p_k})_j$, $(\overline{p_I})_j$, $(\overline{p_k u_k})_j$ and $(\overline{p_I u_I})_j$. For any consistent definition of the flux $(\alpha_k)_{j+1/2}^n$, we set with $\kappa_j \in [0, 1]$

$$\begin{cases} (\alpha_k)_j^n = \kappa_j (\alpha_k)_{j+1/2}^n + (1-\kappa_j)(\alpha_k)_{j-1/2}^n, \\ (\overline{p_k})_j = (1-\kappa_j)(\pi_k)_{j+1/2}^{n+1=} + \kappa_j (\pi_k)_{j-1/2}^{n+1=}, \\ (\overline{p_k u_k})_j = (1-\kappa_j)(\pi_k u_k)_{j+1/2}^{n+1=} + \kappa_j (\pi_k u_k)_{j-1/2}^{n+1=}, \end{cases}$$

and

$$\begin{cases} (\overline{p_I})_j = \mu_{j+1/2}^{n+1=}(\overline{p_1})_j + (1-\mu_{j+1/2}^{n+1=})(\overline{p_2})_j, \text{ with } \mu_{j+1/2}^{n+1=} = \tfrac{1}{2}\left(\mu_j^{n+1=} + \mu_{j+1}^{n+1=}\right) \\ (\overline{u_I})_j = \beta_{j+1/2}^{n+1=}(\overline{u_1})_j + (1-\beta_{j+1/2}^{n+1=})(\overline{u_2})_j, \text{ with } (\overline{u_k})_j = (\overline{p_k u_k})_j/(\overline{p_k})_j, \\ (\overline{p_I u_I})_j = (\overline{p_I})_j (\overline{u_I})_j. \end{cases}$$

We choosed in practice $(\alpha_k)_{j+1/2}^n = (\alpha_k)_j^n$ or equivalently $\kappa_j = 1$.

With such definitions, it can be proved that under a suitable CFL condition based on the velocities u_k and u_I only, and not on the acoustic waves $u_k \pm c_k$, the void fractions $(\alpha_k)_j^{n+1}$ belong to $(0, 1)$ if $(\alpha_k)_j^n$ do. We can also prove that under the same restriction on the time step $(\rho_k)_j^{n+1}$ is positive, as well as $(\varepsilon_k)_j^{n+1-}$ and $(p_k)_j^{n+1-}$. Unfortunately, the positivity of $(\varepsilon_k)_j^{n+1}$ and $(p_k)_j^{n+1}$ is not proved at the moment.

5 Numerical experiments

For the sake of illustration, we present in this section the results given by our algorithm on three Riemann problems, see the Fig. 1. They are all taken from [2] and are fully described therein. Space and time orders of accuracy are one. The first one (top left) corresponds to an isolated contact discontinuity propagating with a positive velocity, while the second one (top right) and the third one (bottom) involve several distinct waves. The scheme we propose here is denoted LP implicit and is compared with its explicit version (which amounts to replacing (9) by its time-explicit version) and the well-known Rusanov scheme (see [8]). We observe that our approach is clearly less diffusive around the contact discontinuities since the CFL condition is well-adapted to the corresponding speed of propagation, but more diffusive around the acoustic waves since it is implicit. Table 1 gives for each test case the number of iterations needed to perform the computations. As expected, the gain is important when using the proposed implicit-explicit algorithm and the corresponding CFL restriction based on the material waves (instead of the acoustic waves as for the explicit scheme). A careful evaluation of the CPU cost necessitates an additional programming effort that has not been implemented yet.

Table 1 Number of time-iterations for each test case

	Test 1	Test 2	Test 3
Rusanov	4231	550	2630
LP explicit	4297	551	2631
LP implicit	63	41	151

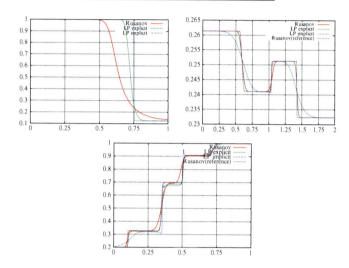

Fig. 1 Comparison of several schemes with a reference solution (density profile)

Acknowledgements Part of this work has been achieved within the framework of the NEPTUNE project, supported by CEA, EDF, IRSN, AREVA-NP.

References

1. A. Ambroso, C. Chalons, F. Coquel, T. Galié. Relaxation and numerical approximation of a two fluid two pressure diphasic model. *M2AN*, vol. 43, pp. 1063-1097, (2009).
2. A. Ambroso, C. Chalons and P.-A. Raviart. A Godunov-type method for the seven-equation model of compressible two-phase flow. *LJLL report number R10020*, http://www.ljll.math.upmc.fr/publications/2010/R10020.php, (2010).
3. N. Andrianov and G. Warnecke. The Riemann problem for the Baer-Nunziato two-phase flow model. *Journal of Computational Physics*, vol. 195, pp. 434-464, (2004).
4. M.R. Baer and J.W. Nunziato, A two phase mixture theory for the deflagration to detonation transition in reactive granular materials. *Int. J. Mult. Flows*, vol. 12(6), pp. 861-889, (1986).
5. C. Chalons and J.-F. Coulombel, Relaxation approximation of the Euler equations. *J. Math. Anal. Appl.*, vol. 348(2), pp. 872-893, (2008).
6. F. Coquel, Q.-L. Nguyen, M. Postel and Q.-H. Tran, Entropy-satisfying relaxation method with large time-steps for Euler IBVPs. *Math. Comp*, vol. 79, pp. 1493-1533, (2010).
7. P. Embid and M. Baer, Mathematical analysis of a two-phase continuum mixture theory, *Contin. Mech. Thermodyn.* vol. 4(4), pp. 279-312, (1992).
8. T. Gallouët, J.-M. Hérard and N. Seguin. Numerical modeling of two-phase flows using the two-fluid two-pressure approach. *M3AS*, vol. 14(5), pp. 663-700, (2004).

9. S. Karni, E. Kirr, A. Kurganov and G. Petrova, Compressible two-phase flows by central and upwind schemes, *M2AN*, vol. 38(3), pp. 477-493, (2004).
10. S.T. Munkejord, Comparison of Roe-type methods for solving the two-fluid model with and without pressure relaxation, *Computers and Fluids*, vol. 36, pp. 1061-1080, (2007).
11. R. Saurel and R. Abgrall, A multiphase Godunov method for compressible multifluid and multiphase flows, *J. Comput. Phys.*, vol. 150, pp. 425-467, (1999).
12. D.-W. Schwendeman, C.-W. Wahle and A.-K. Kapila. The Riemann problem and a high-resolution Godunov method for a model of compressible two-phase flow. *Journal of Computational Physics*, vol. 212, pp. 490-526, (2006).
13. B. Stewart and B. Wendroff, Two-phase flow : models and methods, *J. Comput. Phys.*, vol. 56, pp. 363-409, (1984).
14. S.-A. Tokareva and E.-F. Toro, HLLC-type Riemann solver for the Baer-Nunziato equations of compressible two-phase flow, *J. Comput. Phys.*, to appear, (2010).

The paper is in final form and no similar paper has been or is being submitted elsewhere.

Asymptotic Behavior of the Scharfetter–Gummel Scheme for the Drift-Diffusion Model

Marianne CHATARD

Abstract The aim of this work is to study the large-time behavior of the Scharfetter–Gummel scheme for the drift-diffusion model for semiconductors. We prove the convergence of the numerical solutions to an approximation of the thermal equilibrium. We also present numerical experiments which underline the preservation of long-time behavior.

Keywords Drift-diffusion system, finite volume scheme, thermal equilibrium.
MSC2010: 65M08, 76X05, 82D37.

1 Introduction

In the modeling of semiconductor devices, the drift-diffusion system is widely used as it simplifies computations while giving an accurate description of the device physics.

Let $\Omega \subset \mathbb{R}^d$ ($d \geq 1$) be an open and bounded domain describing the geometry of the semiconductor device. The isothermal drift-diffusion system consists of two continuity equations for the electron density $N(x,t)$ and the hole density $P(x,t)$, and a Poisson equation for the electrostatic potential $V(x,t)$:

$$\begin{cases} \partial_t N - \mathrm{div}(\nabla N - N\nabla V) = 0 & \text{on } \Omega \times (0,T), \\ \partial_t P - \mathrm{div}(\nabla P + P\nabla V) = 0 & \text{on } \Omega \times (0,T), \\ \lambda^2 \Delta V = N - P - C & \text{on } \Omega \times (0,T), \end{cases} \quad (1)$$

Marianne CHATARD
Université Blaise Pascal - Laboratoire de Mathématiques UMR 6620 - CNRS - Campus des Cézeaux, B.P. 80026 63177 Aubière cedex, e-mail: Marianne.Chatard@math.univ-bpclermont.fr

where $C(x)$ is the doping profile, which is assumed to be a given datum, and λ is the Debye length arising from the scaling of the physical model. We supplement these equations with initial conditions $N_0(x)$ and $P_0(x)$ and physically motivated boundary conditions: Dirichlet boundary conditions $\overline{N}$, $\overline{P}$ and $\overline{V}$ on ohmic contacts Γ^D and homogeneous Neumann boundary conditions on insulating boundary segments Γ^N.

There is an extensive literature on numerical schemes for the drift-diffusion equations: finite difference methods, finite elements methods, mixed exponential fitting finite elements methods, finite volume methods (see [1]). The Scharfetter–Gummel scheme is widely used to approximate the drift-diffusion equations in the linear case. It has been proposed and studied in [7] and [10]. It preserves steady-state, and is second order accurate in space (see [9]).

The purpose of this paper is to study the large time behavior of the numerical solution given by the Scharfetter–Gummel scheme for the transient linear drift-diffusion model (1). Indeed, it has been proved by H. Gajewski and K. Gärtner in [5] that the solution to the transient system (1) converges to the thermal equilibrium state as $t \to \infty$ if the boundary conditions are in thermal equilibrium. A. Jüngel extends this result to a degenerate model with nonlinear diffusivities in [8].

The thermal equilibrium is a particular steady-state for which electron and hole currents, namely $\nabla N - N \nabla V$ and $\nabla P + P \nabla V$, vanish.

If the Dirichlet boundary conditions satisfy $\overline{N}, \overline{P} > 0$ and

$$\log(\overline{N}) - \overline{V} = \alpha_N \text{ and } \log(\overline{P}) + \overline{V} = \alpha_P \text{ on } \Gamma^D, \tag{2}$$

the thermal equilibrium is defined by

$$\begin{cases} \Delta V^{eq} = \exp(\alpha_N + V^{eq}) - \exp(\alpha_P - V^{eq}) - C & \text{on } \Omega, \\ N^{eq} = \exp(\alpha_N + V^{eq}), \ P^{eq} = \exp(\alpha_P - V^{eq}) & \text{on } \Omega, \end{cases} \tag{3}$$

with the same boundary conditions as (1).

Our aim is to prove that the solution of the Scharfetter–Gummel scheme converges to an approximation of the thermal equilibrium as $t \to +\infty$. Long-time behavior of solutions to discretized drift-diffusion systems have been studied in [5], [2] and [6], using estimates of the energy.

In the sequel, we will suppose that the following hypotheses are fulfilled:

(H1) $\overline{N}, \overline{P}$ are traces on $\Gamma^D \times (0, T)$ of functions, also denoted $\overline{N}$ and $\overline{P}$, such that $\overline{N}, \overline{P} \in H^1(\Omega \times (0, T)) \cap L^\infty(\Omega \times (0, T))$ and $\overline{N}, \overline{P} \geq 0$ a.e.,
(H2) $N_0, P_0 \in L^\infty(\Omega)$ and $N_0, P_0 \geq 0$ a.e.,
(H3) there exist $0 < m \leq M$ such that: $m \leq \overline{N}, N_0, \overline{P}, P_0 \leq M$,
(H4) $\overline{N}, \overline{P}$ and $\overline{V}$ satisfy the compatibility condition (2).

2 Numerical schemes

In this section, we present the finite volume schemes for the time evolution drift-diffusion system (1) and for the thermal equilibrium (3).
An admissible mesh of Ω is given by a family $\mathcal{T}$ of control volumes (open and convex polygons in 2-D, polyhedra in 3-D), a family $\mathcal{E}$ of edges in 2-D (faces in 3-D) and a family of points $(x_K)_{K \in \mathcal{T}}$ which satisfy Definition 5.1 in [4]. It implies that the straight line between two neighboring centers of cells (x_K, x_L) is orthogonal to the edge $\sigma = K|L$.
In the set of edges $\mathcal{E}$, we distinguish the interior edges $\sigma \in \mathcal{E}_{int}$ and the boundary edges $\sigma \in \mathcal{E}_{ext}$. We split $\mathcal{E}_{ext}$ into $\mathcal{E}_{ext} = \mathcal{E}_{ext}^D \cup \mathcal{E}_{ext}^N$ where $\mathcal{E}_{ext}^D$ is the set of Dirichlet boundary edges and $\mathcal{E}_{ext}^N$ is the set of Neumann boundary edges. For a control volume $K \in \mathcal{T}$, we denote by $\mathcal{E}_K$ the set of its edges, $\mathcal{E}_{int,K}$ the set of its interior edges, $\mathcal{E}_{ext,K}^D$ the set of edges of K included in Γ^D and $\mathcal{E}_{ext,K}^N$ the set of edges of K included in Γ^N.
The size of the mesh is defined by $\Delta x = \max_{K \in \mathcal{T}} (\text{diam}(K))$.
We denote by d the distance in $\mathbb{R}^d$ and m the measure in $\mathbb{R}^d$ or $\mathbb{R}^{d-1}$.
We also need some assumption on the mesh:

$$\exists \xi > 0 \text{ s. t. } \text{d}(x_K, \sigma) \geq \xi \text{d}(x_K, x_L) \text{ for } K \in \mathcal{T}, \text{ for } \sigma = K|L \in \mathcal{E}_{int,K}.$$

For all $\sigma \in \mathcal{E}$, we define the transmissibility coefficient $\tau_\sigma = \dfrac{\text{m}(\sigma)}{d_\sigma}$, where $d_\sigma = \text{d}(x_K, x_L)$ for $\sigma = K|L \in \mathcal{E}_{int}$ and $d_\sigma = \text{d}(x_K, \sigma)$ for $\sigma \in \mathcal{E}_{ext}$.
Let $(\mathcal{T}, \mathcal{E}, (x_K)_{K \in \mathcal{T}})$ be an admissible discretization of Ω and let us define the time step Δt, $N_T = E(T/\Delta t)$ and the increasing sequence $(t^n)_{0 \leq n \leq N_T}$, where $t^n = n\Delta t$, in order to get a space-time discretization $\mathcal{D}$ of $\Omega \times (0, T)$. The size of the space-time discretization $\mathcal{D}$ is defined by $\delta = \max(\Delta x, \Delta t)$.
First of all, the initial conditions and the doping profile are approximated by $(N_K^0, P_K^0, C_K)_{K \in \mathcal{T}}$ by taking the mean values of N_0, P_0 and C on each cell K. The numerical boundary conditions $(N_\sigma^{n+1}, P_\sigma^{n+1}, V_\sigma^{n+1})_{n \geq 0, \sigma \in \mathcal{E}_{ext}^D}$ are also given by the mean values of $(\overline{N}, \overline{P}, \overline{V})$ on $\sigma \times [t^n, t^{n+1}[$.

2.1 The scheme for the thermal equilibrium

We compute an approximation $(N_K^{eq}, P_K^{eq}, V_K^{eq})_{K \in \mathcal{T}}$ of the thermal equilibrium (N^{eq}, P^{eq}, V^{eq}) defined by (3) with the finite volume scheme proposed by C. Chainais-Hillairet and F. Filbet in [2]:

$$\begin{cases} \lambda^2 \sum_{\sigma \in \mathcal{E}_K} \tau_\sigma DV_{K,\sigma}^{eq} = m(K) \left(\exp(\alpha_N + V_K^{eq}) - \exp(\alpha_P - V_K^{eq}) - C_K \right) & \forall K \in \mathcal{T}, \\ N_K^{eq} = \exp(\alpha_N + V_K^{eq}), \quad P_K^{eq} = \exp(\alpha_P - V_K^{eq}) & \forall K \in \mathcal{T}, \end{cases} \quad (4)$$

where for a given function f and $(U_K)_{K \in \mathcal{T}}$, $Df(U)_{K,\sigma}$ is defined by:

$$Df(U)_{K,\sigma} = \begin{cases} f(U_L) - f(U_K) & \text{if } \sigma = K|L \in \mathcal{E}_{int,K}, \\ f(U_\sigma) - f(U_K) & \text{if } \sigma \in \mathcal{E}_{ext,K}^D, \\ 0 & \text{if } \sigma \in \mathcal{E}_{ext,K}^N. \end{cases}$$

Assuming that the boundary conditions satisfy hypotheses (H1)–(H4), the scheme (4) admits a unique solution (see [2]).

2.2 The scheme for the transient model

The Scharfetter–Gummel scheme for the system (1) is defined by:

$$\begin{cases} m(K) \dfrac{N_K^{n+1} - N_K^n}{\Delta t} + \sum_{\sigma \in \mathcal{E}_K} \mathcal{F}_{K,\sigma}^{n+1} = 0, & \forall K \in \mathcal{T}, \forall n \geq 0, \\ m(K) \dfrac{P_K^{n+1} - P_K^n}{\Delta t} + \sum_{\sigma \in \mathcal{E}_K} \mathcal{G}_{K,\sigma}^{n+1} = 0, & \forall K \in \mathcal{T}, \forall n \geq 0, \quad (5) \\ \lambda^2 \sum_{\sigma \in \mathcal{E}_K} \tau_\sigma DV_{K,\sigma}^n = m(K) \left(N_K^n - P_K^n - C_K \right), & \forall K \in \mathcal{T}, \forall n \geq 0, \end{cases}$$

with for all $\sigma \in \mathcal{E}_K$

$$\mathcal{F}_{K,\sigma}^{n+1} = \tau_\sigma \left(B\left(-DV_{K,\sigma}^{n+1}\right) N_K^{n+1} - B\left(DV_{K,\sigma}^{n+1}\right) N_\sigma^{n+1} \right), \quad (6)$$

$$\mathcal{G}_{K,\sigma}^{n+1} = \tau_\sigma \left(B\left(DV_{K,\sigma}^{n+1}\right) P_K^{n+1} - B\left(-DV_{K,\sigma}^{n+1}\right) P_\sigma^{n+1} \right), \quad (7)$$

where B is the Bernoulli function defined by:

$$B(x) = \frac{x}{e^x - 1} \quad \text{for } x \neq 0, \quad B(0) = 1. \quad (8)$$

We consider a fully implicit discretization in time to avoid the restrictive stability condition $\Delta t \leq \lambda^2 / M$.
Using a fixed point theorem, we can prove the following result:

Theorem 1. *Let us assume (H1)–(H4) and $C = 0$. Then there exists a solution $\{(N_K^n, P_K^n, V_K^n), K \in \mathcal{T}, 0 \leq n \leq N_T + 1\}$ to the scheme (5)–(6)–(7), and moreover we have*

$$0 < m \leq N_K^n, \quad P_K^n \leq M, \quad \forall K \in \mathcal{T}, \quad \forall n \geq 0. \quad (9)$$

3 Asymptotic behavior of the Scharfetter–Gummel scheme

We may now state our main result.

Theorem 2. *Let us assume (H1)–(H4) and $C = 0$. Then solution $(N_\delta, P_\delta, V_\delta)$ given by the scheme (5)–(6)–(7) satisfies for each $K \in \mathcal{T}$*

$$\left(N_K^n, P_K^n, V_K^n\right) \longrightarrow \left(N_K^{eq}, P_K^{eq}, V_K^{eq}\right) \text{ as } n \to +\infty,$$

where $\left(N_K^{eq}, P_K^{eq}, V_K^{eq}\right)_{K \in \mathcal{T}}$ is an approximation to the solution of the steady-state equation (3) given by (4).

The proof is based, as in the continuous case (see [5] and [8]), on an energy estimate and a control of its dissipation, given in Proposition 1 which is valid even if $C \neq 0$. Nevertheless to prove rigorously the convergence to equilibrium, we need the uniform lower bound (9) on N and P which holds under the restrictive assumption $C = 0$.

In the last section, we perform some numerical experiments and observe a convergence to steady-state even when this condition is not satisfied.

3.1 Notations and definitions

For $U = (U_K)_{K \in \mathcal{T}}$, we define the H^1-seminorm as follows:

$$|U|_{1,\Omega}^2 = \sum_{\substack{\sigma \in \mathcal{E}_{int} \\ \sigma = K|L}} \tau_\sigma |DU_{K,\sigma}|^2 + \sum_{K \in \mathcal{T}} \sum_{\sigma \in \mathcal{E}_{ext,K}} \tau_\sigma |DU_{K,\sigma}|^2$$

Since the study of the large time behavior of the scheme (5)–(6)–(7) is based on an energy estimate with the control of its dissipation, let us introduce the discrete version of the deviation of the total energy from the thermal equilibrium:

$$\mathcal{E}^n = \sum_{K \in \mathcal{T}} m(K) \left(H(N_K^n) - H(N_K^{eq}) - \log(N_K^{eq})\left(N_K^n - N_K^{eq}\right)\right)$$

$$+ \sum_{K \in \mathcal{T}} m(K) \left(H(P_K^n) - H(P_K^{eq}) - \log(P_K^{eq})(P_K^n - P_K^{eq})\right)$$

$$+ \frac{\lambda^2}{2} |V^n - V^{eq}|_{1,\Omega}^2.$$

Since $s \mapsto H(s) = \int_1^s \log(\tau) d\tau$ is defined and convex on $\mathbb{R}_+$, we have $\mathcal{E}^n \geq 0$ for all $n \geq 0$. We also introduce the discrete version of the energy dissipation:

$$\mathscr{I}^n = \sum_{\substack{\sigma \in \mathscr{E}_{int} \\ \sigma = K|L}} \tau_\sigma \min\left(N_K^n, N_L^n\right) \left[D\left(\log\left(N^n\right) - V^n\right)_{K,\sigma}\right]^2$$

$$+ \sum_{K \in \mathscr{T}} \sum_{\sigma \in \mathscr{E}_{ext,K}} \tau_\sigma \min\left(N_K^n, N_\sigma^n\right) \left[D\left(\log\left(N^n\right) - V^n\right)_{K,\sigma}\right]^2$$

$$+ \sum_{\substack{\sigma \in \mathscr{E}_{int} \\ \sigma = K|L}} \tau_\sigma \min\left(P_K^n, P_L^n\right) \left[D\left(\log\left(P^n\right) + V^n\right)_{K,\sigma}\right]^2$$

$$+ \sum_{K \in \mathscr{T}} \sum_{\sigma \in \mathscr{E}_{ext,K}} \tau_\sigma \min\left(P_K^n, P_\sigma^n\right) \left[D\left(\log\left(P^n\right) + V^n\right)_{K,\sigma}\right]^2.$$

3.2 Energy estimate

The following Proposition gives the control of energy and dissipation. With this result, Theorem 2 can be proved in the same way as Theorem 2.2 in [2].

Proposition 1. *Under hypotheses (H1)–(H4), we have for all $n \geq 0$:*

$$0 \leq \mathscr{E}^{n+1} + \Delta t \mathscr{I}^{n+1} \leq \mathscr{E}^n.$$

Proof. Firstly, using the convexity of H and (4), we get

$$\mathscr{E}^{n+1} - \mathscr{E}^n \leq \sum_{K \in \mathscr{T}} m(K) \left(\log\left(N_K^{n+1}\right) - \alpha_N - V_K^{eq}\right)\left(N_K^{n+1} - N_K^n\right)$$

$$+ \sum_{K \in \mathscr{T}} m(K) \left(\log\left(P_K^{n+1}\right) - \alpha_P + V_K^{eq}\right)\left(P_K^{n+1} - P_K^n\right)$$

$$+ \frac{\lambda^2}{2} |V^{n+1} - V^{eq}|_{1,\Omega}^2 - \frac{\lambda^2}{2} |V^n - V^{eq}|_{1,\Omega}^2,$$

and then, by adding $V_K^{n+1} - V_K^{n+1}$ in the two first sums, we have

$$\mathscr{E}^{n+1} - \mathscr{E}^n \leq T_1 + T_2 + T_3,$$

where

$$T_1 = \sum_{K \in \mathscr{T}} m(K) \left(\log\left(N_K^{n+1}\right) - \alpha_N - V_K^{n+1}\right)\left(N_K^{n+1} - N_K^n\right),$$

$$T_2 = \sum_{K \in \mathscr{T}} m(K) \left(\log\left(P_K^{n+1}\right) - \alpha_P + V_K^{n+1}\right)\left(P_K^{n+1} - P_K^n\right),$$

$$T_3 = \sum_{K \in \mathcal{T}} m(K) \left(V_K^{n+1} - V_K^{eq}\right) \left(N_K^{n+1} - N_K^n - P_K^{n+1} + P_K^n\right)$$

$$+ \frac{\lambda^2}{2} |V^{n+1} - V^{eq}|_{1,\Omega}^2 - \frac{\lambda^2}{2} |V^n - V^{eq}|_{1,\Omega}^2.$$

Using the scheme (5) and an integration by parts, we get that $T_3 \leq 0$ and

$$T_1 = \Delta t \sum_{\substack{\sigma \in \mathcal{E}_{int} \\ \sigma = K|L}} \tau_\sigma \mathcal{R}_{K,\sigma}^{n+1} + \Delta t \sum_{K \in \mathcal{T}} \sum_{\sigma \in \mathcal{E}_{ext,K}^D} \tau_\sigma \mathcal{R}_{K,\sigma}^{n+1},$$

where for $\sigma = K|L$,

$$\mathcal{R}_{K,\sigma}^{n+1} = \left(D \log\left(N^{n+1}\right)_{K,\sigma} - DV_{K,\sigma}^{n+1}\right)\left(B\left(-DV_{K,\sigma}^{n+1}\right)N_K^{n+1} - B\left(DV_{K,\sigma}^{n+1}\right)N_L^{n+1}\right).$$

We now prove that

$$\mathcal{R}_{K,\sigma}^{n+1} \leq \mathcal{S}_{K,\sigma}^{n+1} := -\min\left(N_K^{n+1}, N_L^{n+1}\right)\left(D\log\left(N^{n+1}\right)_{K,\sigma} - DV_{K,\sigma}^{n+1}\right)^2.$$

Indeed, applying the property $B(-x) - B(x) = x$, we obtain

$$\mathcal{R}_{K,\sigma}^{n+1} - \mathcal{S}_{K,\sigma}^{n+1} = \left(D\log\left(N^{n+1}\right)_{K,\sigma} - DV_{K,\sigma}^{n+1}\right) \times$$

$$\left[\left(B\left(-DV_{K,\sigma}^{n+1}\right) - B\left(-D\log\left(N^{n+1}\right)_{K,\sigma}\right)\right)\left(N_K^{n+1} - \min\left(N_K^{n+1}, N_L^{n+1}\right)\right)\right.$$

$$- \left(B\left(DV_{K,\sigma}^{n+1}\right) - B\left(D\log\left(N^{n+1}\right)_{K,\sigma}\right)\right)\left(N_L^{n+1} - \min\left(N_K^{n+1}, N_L^{n+1}\right)\right)$$

$$\left. + B\left(-D\log\left(N^{n+1}\right)_{K,\sigma}\right)N_K^{n+1} - B\left(D\log\left(N^{n+1}\right)_{K,\sigma}\right)N_L^{n+1}\right].$$

Now, since B is non-increasing on $\mathbb{R}$, the two first terms are non positive, and by using the definition (8) of B, the third term is equal to zero. Then we can conclude that

$$T_1 \leq \Delta t \sum_{\substack{\sigma \in \mathcal{E}_{int} \\ \sigma = K|L}} \tau_\sigma \mathcal{S}_{K,\sigma}^{n+1} + \Delta t \sum_{K \in \mathcal{T}} \sum_{\sigma \in \mathcal{E}_{ext,K}^D} \tau_\sigma \mathcal{S}_{K,\sigma}^{n+1},$$

and we obtain in the same way a similar estimate for T_2. To sum up, we have

$$\mathcal{E}^{n+1} - \mathcal{E}^n \leq T_1 + T_2 \leq -\Delta t \, \mathcal{I}^{n+1},$$

which completes the proof. $\square$

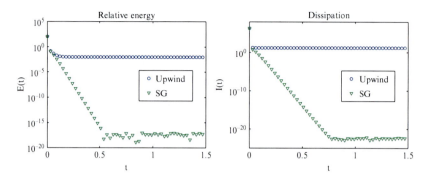

Fig. 1 Evolution of the relative energy $\mathscr{E}^n$ and its dissipation $\mathscr{I}^n$ in log-scale

4 Numerical experiments

We present here a test case for a geometry corresponding to a PN-junction in 1D. The doping profile is piecewise constant, equal to $+1$ in the N-region $]0.5, 1[$ and -1 in the P-region $]0, 0.5[$. The Debye length is $\lambda = 10^{-2}$.
In Fig. 1 we compare the relative energy $\mathscr{E}^n$ and its dissipation $\mathscr{I}^n$ obtained with the the Scharfetter–Gummel scheme (5) and with the scheme studied by C. Chainais-Hillairet, J. G. Liu and Y. J. Peng in [3], where the diffusion terms are discretized classically and the convection terms are discretized with upwind fluxes. With the Scharfetter–Gummel scheme, we observe that $\mathscr{E}^n$ and $\mathscr{I}^n$ converge to zero when $n \to \infty$, which is in keeping with Theorem 2. On the contrary, the upwind scheme, which does not preserve thermal equilibrium, is not very satisfying to reflect the long time behavior of the solution.

Acknowledgements The author is partially supported by the European Research Council ERC Starting Grant 2009, project 239983-NuSiKiMo.

References

1. F. Brezzi, L.D. Marini, S. Micheletti, P. Pietra, R. Sacco, and S. Wang. Discretization of semiconductor device problems. I. In *Handbook of numerical analysis. Vol. XIII*, pages 317–441. North-Holland, Amsterdam, 2005.
2. C. Chainais-Hillairet and F. Filbet. Asymptotic behavior of a finite volume scheme for the transient drift-diffusion model. *IMA J. Numer. Anal.*, 27(4):689–716, 2007.
3. C. Chainais-Hillairet, J.G. Liu, and Y.J. Peng. Finite volume scheme for multi-dimensional drift-diffusion equations and convergence analysis. *M2AN Math. Model. Numer. Anal.*, 37(2):319–338, 2003.
4. R. Eymard, T. Gallouët, and R. Herbin. Finite volume methods. In *Handbook of numerical analysis*, volume VII, pages 713–1020. North-Holland, Amsterdam, 2000.
5. H. Gajewki and K. Gärtner. On the discretization of Van Roosbroeck's equations with magnetic field. *Z. Angew. Math. Mech.*, 76(5):247–264, 1996.

6. A. Glitzky. Exponential decay of the free energy for discretized electro-reaction-diffusion systems. *Nonlinearity*, 21(9):1989–2009, 2008.
7. A.M. Il'in. A difference scheme for a differential equation with a small parameter multiplying the highest derivative. *Math. Zametki*, 6:237–248, 1969.
8. A. Jüngel. Qualitative behavior of solutions of a degenerate nonlinear drift-diffusion model for semiconductors. *Math. Models Methods Appl. Sci.*, 5(5):497–518, 1995.
9. R. D. Lazarov, Ilya D. Mishev, and P. S. Vassilevski. Finite volume methods for convection-diffusion problems. *SIAM J. Numer. Anal.*, 33(1):31–55, 1996.
10. D.L. Scharfetter and H.K. Gummel. Large signal analysis of a silicon Read diode. *IEEE Trans. Elec. Dev.*, 16:64–77, 1969.

The paper is in final form and no similar paper has been or is being submitted elsewhere.

A Finite Volume Solver for Radiation Hydrodynamics in the Non Equilibrium Diffusion Limit

D. Chauveheid, J.-M.Ghidaglia, and M. Peybernes

Abstract We derive an Implicit Explicit finite volume scheme for the computation of radiation hydrodynamics. The convective part is handled through a classical upwind method while the reactive and diffusive parts are discretized thanks to a centered scheme. These results are compared to semi-analytic solutions obtained by Lowrie and Edwards [10].

Keywords Cell centered finite volume scheme, Radiation hydrodynamics, Implicit/Explicit schemes
MSC2010: 65M08, 76M12, 76N99, 80A99

1 Introduction

Radiation hydrodynamics models are of interest for many applications *e.g.* astrophysics, inertial confinement fusion (ICF) and other flows with very high temperatures. One of the major difficulties for these multi-physics problems is the presence of multiple time scales. From the numerical point of view, this leads to build implicit-explicit schemes with respect to time. The implicit part is here to handle small time scales while the explicit one takes care of larger time scales. In our context, the small time scales result from the radiation transport part (diffusion) while larger time scales come from purely hydrodynamical phenomena (entropy and pressure waves). Our strategy consists in relying on classical cell centered Finite

---

Daniel Chauveheid and Mathieu Peybernes
CEA, DAM, DIF, F-91297 Arpajon, France, e-mail: daniel.chauveheid@cmla.ens-cachan.fr, mathieu.peybernes@cea.fr

Jean-Michel Ghidaglia
CMLA, ENS Cachan and CNRS UMR 8635, 61 avenue du Président Wilson, F-94235 CACHAN CEDEX, e-mail: jmg@cmla.ens-cachan.fr

Volume schemes based on approximate Riemann solver (namely Flux Schemes see Ghidaglia [5]) for the hydrodynamics part and on an implicit Finite Volume scheme for the radiative one.

This article is a first step towards the derivation of a multi-material solver, that is studying flows with two or more different materials. For example in the ICF applications, we have at least two materials in presence, a metal (Gold) and a highly compressed gas (a mixture of Deuterium and Tritium). The multi material version of the scheme (Chauveheid [3]), relies on a generalization of the method of Braeunig *et al.* [1]. The latter method computes sharp interfaces between non miscible materials whose computation uses directional splitting. Hence in this paper, although we solely address the case of one material, we shall use cartesian meshes.

The governing equations, in non dimensional form (Lowrie and Edwards [10], Lowrie and Morel [9]), read in $3D$ as:

$$\frac{\partial \rho}{\partial t} + \nabla \cdot (\rho \mathbf{u}) = 0, \qquad (1)$$

$$\frac{\partial (\rho \mathbf{u})}{\partial t} + \nabla \cdot \left(\rho \mathbf{u} \otimes \mathbf{u} + \left(p + \mathscr{P}_0 \frac{\mathscr{E}_r}{3} \right) Id \right) = 0, \qquad (2)$$

$$\frac{\partial (\rho E)}{\partial t} + \nabla \cdot ((\rho E + p) \mathbf{u}) = -\mathscr{P}_0 \left(\sigma(T^4 - \mathscr{E}_r) + \mathbf{u} \cdot \nabla \frac{\mathscr{E}_r}{3} \right), \qquad (3)$$

$$\frac{\partial \mathscr{E}_r}{\partial t} + \nabla \cdot (\mathscr{E}_r \mathbf{u}) + \frac{\mathscr{E}_r}{3} \nabla \cdot \mathbf{u} = \nabla \cdot (\kappa \nabla \mathscr{E}_r) + \sigma(T^4 - \mathscr{E}_r), \qquad (4)$$

where we denote by ρ the density, $\mathbf{u}$ the velocity field, p the hydrodynamic pressure, related to the density ρ and the internal energy e by an equation of state: $EOS(p, \rho, e) = 0$. The hydrodynamic specific energy $E = e + \frac{1}{2} \|\mathbf{u}\|^2$ is the sum of the specific internal energy e and the kinetic energy, T is the material temperature. The radiative energy is denoted by $\mathscr{E}_r$ and we define the radiation temperature by $T_r^4 = \mathscr{E}_r$. Finally, $\mathscr{P}_0$ is a non dimensional number ([9, 10]). This system is non conservative but adding (3) and $\mathscr{P}_0$ (4) we readily obtain the total energy conservation law:

$$\frac{\partial (\rho E + \mathscr{P}_0 \mathscr{E}_r)}{\partial t} + \nabla \cdot \left(\left(\rho E + p + 4 \mathscr{P}_0 \frac{\mathscr{E}_r}{3} \right) \mathbf{u} \right) = \mathscr{P}_0 \nabla \cdot (\kappa \nabla \mathscr{E}_r). \qquad (5)$$

Then, introducing the radiative entropy (as done in [2]) $S_r \equiv T_r^3$, we can rewrite (4) as

$$\frac{\partial S_r}{\partial t} + \nabla \cdot (S_r \mathbf{u}) = \frac{3}{4 T_r} \left[\nabla \cdot (\kappa \nabla \mathscr{E}_r) + \sigma(T^4 - \mathscr{E}_r) \right]. \qquad (6)$$

The system (1), (2), (5) and (6) is conservative as far as convection terms are concerned. Equation (6) is a non linear heat equation for the radiative temperature T_r. This variable is therefore diffused and the non conservative product appearing in the right hand side of this equation should not induce non uniqueness of solutions.

A Finite Volume Solver for Radiation Hydrodynamics

2 Numerical scheme

We use an operator splitting which consists in solving first the left-hand side of (1), (2), (5) and (6) by means of an upwind explicit finite volume scheme. Then, the diffusion-reaction part is discretized thanks to a centered implicit finite volume scheme. This kind of technique is often referred to as IMEX method (for Implicit/Explicit), see for example [7, 8].

We consider a regular cartesian grid and split also the space differential operators, that is to say we solve successively the x-derivative terms, the y-derivative terms and the z-derivative term.

Therefore, and without loss of generality, we deal only with $1D$ schemes, corresponding to what is done direction by direction. From now on, we call x the generic direction that we are looking at.

2.1 Cell centered upwind Finite Volume scheme for the convection operator

We denote by $v = (\rho, \rho\mathbf{u}, \rho E + \mathscr{P}_0 \mathscr{E}_r, S_r)$ the conservative variables for the convective part of the system (1), (2), (5) and (6), and $F(v)$ the flux matrix such that:

$$F(v) \cdot \mathbf{n} \equiv (\rho(\mathbf{u} \cdot \mathbf{n}), \rho\mathbf{u}(\mathbf{u} \cdot \mathbf{n}) + (p + \mathscr{P}_0 \mathscr{E}_r/3)\mathbf{n}, S_r(\mathbf{u} \cdot \mathbf{n})), \qquad (7)$$

is the normal flux in the direction $\mathbf{n} \in \mathbb{S}^{d-1}$, d being the physical space dimension.
With these notations, the left-hand side of equations (1)-(2)-(5)-(6) reads:

$$\frac{\partial v}{\partial t} + \nabla \cdot F(v) = 0. \qquad (8)$$

The integration of (8) over a control volume $K_{i,j,k} = [x_i, x_{i+1}] \times [y_j, y_{j+1}] \times [z_k, z_{k+1}]$, keeping only the terms corresponding to the derivation in the generic x-direction, leads to a system of ordinary differential equations:

$$\frac{dV_{K_{i,j,k}}}{dt} + \frac{1}{|K_{i,j,k}|} \left(A_{i+1/2,j,k} \phi(v_{i+1,j,k}, v_{i,j,k}) - A_{i-1/2,j,k} \phi(v_{i,j,k}, v_{i-1,j,k})\right) = 0, \qquad (9)$$

where $\phi(v_{i+1,j,k}, v_{i,j,k})$ denotes the numerical flux at the interface between volumes $K_{i,j,k}$ and $K_{i+1,j,k}$. $A_{i+1/2,j,k}$ is the measure of the edge located at $x_{i+1/2} \equiv \frac{x_i + x_{i+1}}{2}$.

The Characteristic Flux Finite Volume (CFFV) scheme. The CFFV scheme [4] consists in choosing, for the numerical flux in (9), the following value:

$$\phi(v, w, \mathbf{n}) = \frac{F(v) + F(w)}{2} \cdot \mathbf{n} - \mathscr{U}(u, v, \mathbf{n}) \frac{F(w) - F(v)}{2} \cdot \mathbf{n}. \qquad (10)$$

Here, $\mathbf{n} = e_x$, for the generic x-direction. $\mathcal{U}(u, v, \mathbf{n})$ is the sign matrix of the jacobian $\frac{\partial F(v) \cdot \mathbf{n}}{\partial v}$, in the sense that it has the same eigenvectors as, and its eigenvalues are the signs of those of $\frac{\partial F(v) \cdot \mathbf{n}}{\partial v}$. Namely, when $\frac{\partial F(v) \cdot \mathbf{n}}{\partial v}$ reads $L(diag(\lambda_i))R$ (which is the case for hyperbolic problems), with λ_i the eigenvalues, R right eigenvectors, and L left eigenvectors such that $LR = Id$, we have $\mathcal{U}(u, v, \mathbf{n}) = L(diag(sign(\lambda_i)))R$.

The boundary conditions use the normal flux method, we refer to [6].

Eigenelements. The jacobian matrix $\frac{\partial F(v) \cdot \mathbf{n}}{\partial v}$ of the normal flux (7) is found to be equal to:

$$\begin{pmatrix} 0 & \mathbf{n} & 0 & 0 \\ K\mathbf{n} - \mathbf{u}(\mathbf{u} \cdot \mathbf{n}) & \mathbf{u} \otimes \mathbf{n} - k\mathbf{n} \otimes \mathbf{u} + (\mathbf{u} \cdot \mathbf{n})Id & k\mathbf{n} & \frac{4}{9}\mathcal{P}_0 T_r (1 - 3k)\mathbf{n} \\ (K - (H + \frac{4\mathcal{P}_0 \mathcal{E}_r}{3\rho}))\mathbf{u} \cdot \mathbf{n} & (H + \frac{4\mathcal{P}_0 \mathcal{E}_r}{3\rho})\mathbf{n} - k(\mathbf{u} \cdot \mathbf{n})\mathbf{u} & \mathbf{u} \cdot \mathbf{n}(k+1) & \frac{4}{9}\mathcal{P}_0 T_r (1 - 3k)\mathbf{u} \cdot \mathbf{n} \\ -\frac{T_r^3}{\rho}\mathbf{u} \cdot \mathbf{n} & \frac{T_r^3}{\rho}\mathbf{n} & 0 & \mathbf{u} \cdot \mathbf{n} \end{pmatrix}$$

Its eigenvalues are as follows:

$$\begin{cases} \lambda_1(v, \mathbf{n}) = \mathbf{u} \cdot \mathbf{n} - c_s, \\ \lambda_2(v, \mathbf{n}) = \cdots = \lambda_{d+2}(v, \mathbf{n}) = \mathbf{u} \cdot \mathbf{n}, \\ \lambda_{d+3}(v, \mathbf{n}) = \mathbf{u} \cdot \mathbf{n} + c_s. \end{cases} \quad (11)$$

with $k = \frac{1}{\rho T}\left(\frac{\partial p}{\partial s}\right)_\rho$, $c^2 = \left(\frac{\partial p}{\partial \rho}\right)_s$, s being the material entropy, $H = E + \frac{p}{\rho}$, $K = c^2 + k(\|\mathbf{u}\|^2 - H)$ and $c_s^2 = c^2 + \mathcal{P}_0 \frac{4\mathcal{E}_r}{9\rho}$.

The right eigenvectors associated to these eigenvalues can be taken equal to:

$$\begin{cases} r_1(v, \mathbf{n}) = (1, \mathbf{u} - c_s\mathbf{n}, H + \frac{4\mathcal{P}_0 \mathcal{E}_r}{3\rho} - c_s\mathbf{u} \cdot \mathbf{n}, \frac{T_r^3}{\rho}), \\ r_{d+1}(v, \mathbf{n}) = (1, \mathbf{u}, H - \frac{c^2}{k}, 0), \\ r_{d+2}(v, \mathbf{n}) = (\mathcal{P}_0 T_r, \mathcal{P}_0 T_r \mathbf{u}, \mathcal{P}_0 T_r (H - 3c^2), -\frac{9}{4}c^2), \\ r_{d+3}(v, \mathbf{n}) = (1, \mathbf{u} + c_s\mathbf{n}, H + \frac{4\mathcal{P}_0 \mathcal{E}_r}{3\rho} + c_s\mathbf{u} \cdot \mathbf{n}, \frac{T_r^3}{\rho}), \\ r_2(v, \mathbf{n}) = (0, \mathbf{n}_2^\perp, \mathbf{u} \cdot \mathbf{n}_2^\perp), \cdots, r_d(v, \mathbf{n}) = (0, \mathbf{n}_d^\perp, \mathbf{u} \cdot \mathbf{n}_d^\perp). \end{cases} \quad (12)$$

where $\mathbf{n}_2^\perp \cdots \mathbf{n}_d^\perp$ is an orthonormal basis of the hyperplane orthogonal to $\mathbf{n}$.

The dual basis is then:

$$\begin{cases} \ell_1(v, \mathbf{n}) = \frac{1}{2c_s^2}(K + c_s\mathbf{u} \cdot \mathbf{n}, -k\mathbf{u} - c_s\mathbf{n}, k, \frac{4}{9}\mathcal{P}_0 T_r (1 - 3k)), \\ \ell_{d+1}(v, \mathbf{n}) = \frac{k}{c^2}(H - \|\mathbf{u}\|^2, \mathbf{u}, -1, \frac{4}{9}\mathcal{P}_0 T_r)), \\ \ell_{d+2}(v, \mathbf{n}) = \frac{4}{9\rho c^2 c_s^2}(T_r^3 K, -kT_r^3 \mathbf{u}, kT_r^3, -\rho c^2 - \frac{4}{9}\mathcal{P}_0 k\mathcal{E}_r), \\ \ell_{d+3}(v, \mathbf{n}) = \frac{1}{2c_s^2}(K - c_s\mathbf{u} \cdot \mathbf{n}, -k\mathbf{u} + c_s\mathbf{n}, k, \frac{4}{9}\mathcal{P}_0 T_r (1 - 3k)), \\ \ell_2(v, \mathbf{n}) = (-\mathbf{u} \cdot \mathbf{n}_2^\perp, \mathbf{n}_2^\perp, 0, 0), \cdots, \ell_d(v, \mathbf{n}) = (-\mathbf{u} \cdot \mathbf{n}_d^\perp, \mathbf{n}_2^\perp, 0, 0). \end{cases} \quad (13)$$

A Finite Volume Solver for Radiation Hydrodynamics

Time discretization and stability condition. We use the explicit Euler's scheme to discretize the time derivative in (9) and then the Courant condition for the linearized scheme reads:

$$\max_{i,j,k} |\lambda_{i,j,k}^n| \frac{A_{i,j,k}}{|K_{i,j,k}|} \Delta t^n \leq CFL \leq 1, \tag{14}$$

where $A_{i,j,k}$ is either $\Delta x_i \Delta y_j$, $\Delta y_j \Delta z_k$ or $\Delta z_k \Delta x_i$ depending on the direction we solve.

2.2 Implicit centered finite volume scheme for the diffusion equation

The diffusion part consists in the following system:

$$\frac{\partial \rho}{\partial t} = 0, \tag{15}$$

$$\frac{\partial (\rho \mathbf{u})}{\partial t} = 0, \tag{16}$$

$$\frac{\partial (\rho E)}{\partial t} = -\mathscr{P}_0 \sigma (T^4 - \mathscr{E}_r), \tag{17}$$

$$\frac{\partial \mathscr{E}_r}{\partial t} = \nabla \cdot (\kappa \nabla \mathscr{E}_r) + \sigma (T^4 - \mathscr{E}_r). \tag{18}$$

Since (18) is a heat equation, if we want to use reasonable time step (governed by the Courant Friedrichs Lewy condition (14)), we have to make use of an implicit time discretization.

Writing $E = C_v T + \|\mathbf{u}\|^2/2$, and using (15) and (16), we can show that (17) reduces to an ODE for the temperature T:

$$\rho C_v \frac{\partial T}{\partial t} = -\mathscr{P}_0 \sigma (T^4 - \mathscr{E}_r). \tag{19}$$

The scheme then reads:

$$\rho_i^n C_v \frac{T_i^{n+1} - T_i^n}{\Delta t^n} = -\mathscr{P}_0 \sigma_i^n ((T_i^{n+1})^4 - \mathscr{E}_{r,i}^{n+1}), \tag{20}$$

$$\frac{\mathscr{E}_{r,i}^{n+1} - \mathscr{E}_{r,i}^n}{\Delta t^n} - 2 \frac{\kappa_{i+1/2}^n \frac{\mathscr{E}_{r,i+1}^{n+1} - \mathscr{E}_{r,i}^{n+1}}{\Delta x_{i+1}} - \kappa_{i-1/2}^n \frac{\mathscr{E}_{r,i}^{n+1} - \mathscr{E}_{r,i-1}^{n+1}}{\Delta x_i}}{(\Delta x_i + \Delta x_{i+1})} = \sigma_i^n ((T_i^{n+1})^4 - \mathscr{E}_{r,i}^{n+1}), \tag{21}$$

$$\frac{2}{\kappa_{i+1/2}^n} = \frac{1}{\kappa_i^n} + \frac{1}{\kappa_{i+1}^n}.$$

It can be shown by a motonicity argument that this system has a unique solution. It is then solved by the Newton method, mainly because of the nonlinear terms, and the GMRES algorithm ([11]) at each Newton iteration to solve the linear system.

3 Numerical results

In this section, we present numerical simulations of radiative shock solutions. These results are compared to semi-analytic solutions obtained following the method described in [10].

We initialize a Riemann problem setting the left-state (subscript 0) to $\rho_0 = 1, Tr_0 = 1, T_0 = 1, u_0 = \mathcal{M}$, for a given $\mathcal{M}$ (some different values are chosen for the tests), and the right-state (subscript 1) is obtained by solving the so-called "overall jump conditions" ([10]), and taking material and radiative temperatures equal to each other:

$$\rho_0 u_0 = \rho_1 u_1 \tag{22}$$

$$\rho_0 u_0^2 + p_0 + \mathscr{P}_0 \frac{T_{r,0}^4}{3} = \rho_1 u_1^2 + p_1 + \mathscr{P}_0 \frac{T_{r,1}^4}{3} \tag{23}$$

$$u_0(\rho_0 E_0 + p_0 + \frac{4}{3}\mathscr{P}_0 T_{r,0}^4) = u_1(\rho_1 E_1 + p_1 + \frac{4}{3}\mathscr{P}_0 T_{r,1}^4) \tag{24}$$

Here, $\kappa = 1, \sigma = 10^6$ and $\mathscr{P}_0 = 10^{-4}$.

We take perfect gas equation of state $p = \frac{\rho T}{\gamma}$, with $\gamma = 5/3$.

Figure 1 shows a continuous solution computed over 128 cells. Solutions of Figs. 2 to 4 undergo discontinuities. For these simulations, a finer mesh is used to capture the solutions.

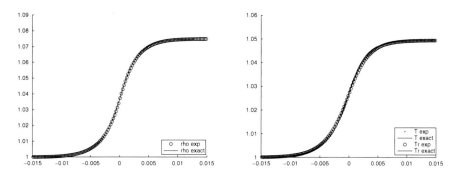

Fig. 1 Solution for density, temperature and radiative temperature for $\mathcal{M} = 1.05$. Comparison with semi-analytic solutions. Number of cells: 128

A Finite Volume Solver for Radiation Hydrodynamics

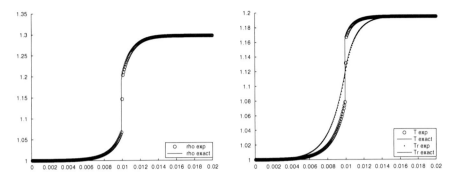

Fig. 2 Solution for density, temperature and radiative temperature for $\mathcal{M} = 1.2$. Comparison with semi-analytic solutions. Number of cells: 256

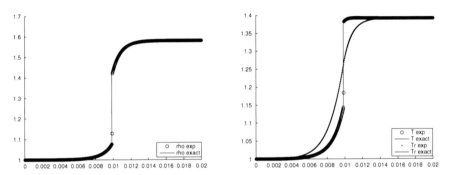

Fig. 3 Solution for density, temperature and radiative temperature for $\mathcal{M} = 1.4$. Comparison with semi-analytic solutions. Number of cells: 512

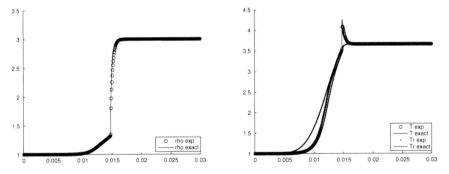

Fig. 4 Solution for density, temperature and radiative temperature for $\mathcal{M} = 3$. Comparison with semi-analytic solutions. Number of cells: 512

Numerical and theoretical results are in good agreement. The conservative formulation chosen in (6) seems to be relevant with regard to these particular physical solutions.

4 Conclusion

As said in the introduction, this work is a first step towards the derivation of a multi material $3D$ solver for multi material radiative hydrodynamics. In this paper we have presented our method for the single material case and shown that on physically relevant non trivial solutions, our solver behaves well. The extension for multi material flows is in progress (Chauveheid [3]). The method presented here was designed in order to make this extension as simple as possible. In fact it only remains to extend the so-called condensate techniques of Braeunig *et. al.* [1] to radiative flows.

References

1. Braeunig J.-P., Desjardins B., Ghidaglia J.-M., A totally Eulerian Finite Volume solver for multi-material fluid flows, Eur. J. Mech. B/Fluids, Vol. 28, pp. 475-485, 2009.
2. Buet, C., Despres B.: Asymptotic preserving and positive schemes for radiation hydrodynamic. J. Comput. Phys. **215**, 717–740, 2006.
3. Chauveheid D., Thesis, École normale supérieure, in preparation.
4. Ghidaglia J.-M., Kumbaro A., Le Coq G.: On the numerical solution to two fluid models *via* a cell centered finite volume method, Eur. J. Mech. B/Fluids, 20, 841-867, 2001.
5. Ghidaglia J.-M., Flux schemes for solving monlinear systems of conservation laws, *in* Innovative Methods for Numerical Solution of Partial Differential Equations, Chattot J.J. and Hafez M. Eds, pp 232-242, WORLD SCIENTIFIC, Singapore, 2001.
6. Ghidaglia J.-M., Pascal F.: The normal flux method at the boundary for multidimensional finite volume approximations in CFD. Eur. J. Mech. B/Fluids, Vol. 24(1), pp. 1-17, 2005.
7. Kadioglu, S.Y. , Knoll, D.A., Lowrie, R.B., Rauenzahn, R.M.: A second order self-consistent IMEX method for radiation hydrodynamics. J. Comput. Phys. **229**, 8313–8332, 2010.
8. Kadioglu, S.Y. , Knoll: A fully second order implicit/explicit time integration technique for hydrodynamics plus nonlinear heat conduction problems. J. Comput. Phys. **229**, 3237–3249 (2010).
9. Lowrie, R.B., Morel, J.E.: The coupling of radiation and hydrodynamics. Astrophys. J. **521**, 423–450, 1999.
10. Lowrie, R.B. , Edwards J.D.: Radiative shock solutions with grey nonequilibrium diffusion. Shock Waves **18**, 129–143, 2008.
11. Saad, Y.: Iterative methods for sparse linear systems 2nd edition, SIAM, 2003.

The paper is in final form and no similar paper has been or is being submitted elsewhere.

An Extension of the MAC Scheme to some Unstructured Meshes

Eric Chénier, Robert Eymard, and Raphaèle Herbin

Abstract We give a variational formulation of the standard MAC scheme for the approximation of the Navier-Stokes problem. This allows an extension of the MAC scheme to locally refined Cartesian meshes. A numerical example is presented, which shows an efficient computation of the solution of the Navier-Stokes problem for a general 2D or 3D domain, using locally refined meshes.

Keywords MAC scheme, incompressible Navier-Stokes, non conforming grid
MSC2010: 65N08,76D05

1 Introduction

Our aim is the approximation on an unstructured mesh, of the weak solution to the steady-state Navier-Stokes equations, defined by

$$\begin{cases} u \in E(\Omega), \ p \in L^2(\Omega) \text{ with } \int_\Omega p(x) \mathrm{d}x = 0, \\ \int_\Omega \nabla u(x) : \nabla v(x) \mathrm{d}x + \mathrm{R} \int_\Omega (u(x) \cdot \nabla) u(x) \cdot v(x) \mathrm{d}x \\ \qquad - \int_\Omega p(x) \mathrm{div} v(x) \mathrm{d}x = \int_\Omega f(x) \cdot v(x) \mathrm{d}x, \ \forall v \in H_0^1(\Omega)^d, \end{cases} \quad (1)$$

where

E. Chénier and R. Eymard
Université Paris-Est, e-mail: eric.chenier@univ-mlv.fr, robert.eymard@univ-mlv.fr

R. Herbin
Université Aix-Marseille, e-mail: Raphaele.Herbin@latp.univ-mrs.fr

J. Fořt et al. (eds.), *Finite Volumes for Complex Applications VI – Problems & Perspectives*, Springer Proceedings in Mathematics 4,
DOI 10.1007/978-3-642-20671-9_27, © Springer-Verlag Berlin Heidelberg 2011

$d \in \{2, 3\}$ denotes the space dimension,
Ω is an open polygonal bounded and connected subset of $\mathbb{R}^d$,
with Lipschitz-continuous boundary $\partial \Omega$,

$$R \in [0, +\infty), \; f \in L^2(\Omega)^d,$$
$$E(\Omega) := \{v = (v^{(i)})_{i=1,\dots,d} \in H_0^1(\Omega)^d, \text{div} v = 0 \text{ a.e. in } \Omega\},$$

and, for all $u, v \in H_0^1(\Omega)^d$ and for a.e. $x \in \Omega$, $\nabla u(x) : \nabla v(x) = \sum_{i=1}^{d} \nabla u^{(i)}(x) \cdot \nabla v^{(i)}(x)$. The approximation of Problem (1) may be performed with several schemes among which the MAC scheme: see e.g. [7] for a presentation of its implementation and [3–6] for its mathematical analysis; the MAC scheme is very popular, in particular because it is simple and needs no stabilisation procedure. Its main drawback is that it only holds on domains which can be gridded by rectangular conforming meshes, in the sense that no hanging node is permitted. This paper is devoted to the presentation of a simple way to extend this scheme to any geometry and to possibly refined meshes, while keeping simplicity and convergence properties. In Sect. 2, we first write a discrete variational formulation of the standard MAC scheme on the Stokes problem, which is (1) with $R = 0$. Thanks to this formulation, we are able in Sect. 3 to extend this scheme to more complex geometries and to the Navier-Stokes equation (1). Section 3 proposes a numerical example on a non-rectangular domain, using local refinement along the boundary of the domain.

2 The standard MAC scheme for the Stokes equations

Let us consider in this section the standard MAC scheme for the approximation of the Stokes problem, that is (1) with $R = 0$. We then consider the following case and notations, as depicted in Fig. 1. Let us consider the unit square: $\Omega =]0, 1[\times]0, 1[$, let N and M be two positive integers. With the notations of Fig.1, we denote by $\mathcal{M}$ the set of pressure grid cells:

$$\mathcal{M} = \left\{]x_{i-\frac{1}{2}}, x_{i+\frac{1}{2}}[\times]y_{j-\frac{1}{2}}, y_{j+\frac{1}{2}}[, \; 1 \leq i \leq N, \; 1 \leq j \leq M \right\},$$

and by $\mathcal{E} = \mathcal{E}^{(1)} \cup \mathcal{E}^{(2)}$ the set of the edges of the mesh, where $\mathcal{E}^{(1)}$ (resp. $\mathcal{E}^{(2)}$) is the set of vertical (resp. horizontal) edges, associated to the x (resp. y) component of the velocity. In order to define the normal velocity flux from one cell to a neighbouring one, we introduce, for any pair $\sigma, \sigma' \in \mathcal{E}^{(k)}$, $k = 1$ or 2, the transmissivity $\tau_{\sigma,\sigma'}$ between cell $K_\sigma^{(k)}$ and cell $K_{\sigma'}^{(k)}$:

An Extension of the MAC Scheme to some Unstructured Meshes

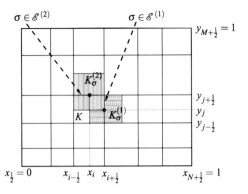

Fig. 1 Notations for the standard MAC scheme

$$\tau_{\sigma,\sigma'} = \frac{|\partial K_\sigma^{(k)} \cap \partial K_{\sigma'}^{(k)}|}{d(x_\sigma, x_{\sigma'})}, \qquad (2)$$

For instance, for a vertical edge $\sigma = \{x_{i-\frac{1}{2}}\} \times]y_{j-\frac{1}{2}}, y_{j+\frac{1}{2}}[\in \mathcal{E}^{(1)}$, one has:

$$\tau_{\sigma,\sigma'}^{(1)} = \begin{cases} \dfrac{y_{j+\frac{1}{2}} - y_{j-\frac{1}{2}}}{x_{i+\frac{1}{2}} - x_{i-\frac{1}{2}}} & \text{if } \sigma' = \{x_{i+\frac{1}{2}}\} \times]y_{j-\frac{1}{2}}, y_{j+\frac{1}{2}}[, \\ \dfrac{x_i - x_{i-1}}{y_{j+1} - y_j} & \text{if } \sigma' = \{x_{i-\frac{1}{2}}\} \times]y_{j+\frac{1}{2}}, y_{j+\frac{3}{2}}[. \end{cases} \qquad (3)$$

Denoting by $(e^{(k)})_{k=1,\dots,d}$ the canonical orthonormal basis of $\mathbb{R}^d$ and, for $K \in \mathcal{M}$, $n_{K,\sigma}$ the unit normal vector to σ outward to K, the MAC scheme then reads:

Find $(u_\sigma)_{\sigma \in \mathcal{E}} \subset \mathbb{R}$, $(p_K)_{K \in \mathcal{M}} \subset \mathbb{R}$; $\sum_{K \in \mathcal{M}} |K| p_K = 0$,

$$\sum_{k=1}^{2} \sum_{\sigma \in \mathcal{E}_K^{(k)}} |\sigma| u_\sigma e^{(k)} \cdot n_{K,\sigma} = 0, \ \forall K \in \mathcal{M}, \qquad (4a)$$

$$-\sum_{\sigma' \in \mathcal{E}^{(k)}} \tau_{\sigma,\sigma'}^{(k)}(u_{\sigma'} - u_\sigma) + |\sigma|(p_{L_\sigma} - p_{M_\sigma}) = \int_{K_\sigma^{(k)}} f^{(k)}(x) dx, \ \forall \sigma \in \mathcal{E}^{(k)}, k = 1, 2, \qquad (4b)$$

where L_σ and $M_\sigma \in \mathcal{M}$ are the two cells which share $\sigma \in \mathcal{E}^{(k)}$ as an edge, and such that $e^{(k)}$ is oriented from L_σ and M_σ, and where the value of u_σ is set to 0 on all exterior edges.

In order to extend the MAC scheme, the idea is to rewrite (4a) and (4b) under a variational formulation. We first define $H_{\mathcal{M}}(\Omega)$ as the set of piecewise functions constant in $K \in \mathcal{M}$, and $H_{\mathcal{E}}^{(k)}(\Omega)$ as the set of piecewise functions which are constant in K_σ, for $\sigma \in \mathcal{E}^{(k)}$, and which are meant to approximate the kth component of the velocity. We finally denote by $H_{\mathscr{E}}(\Omega)$ the set of all $v = (v^{(k)})_{k=1,\ldots,d}$ with $v^{(k)} \in H_{\mathscr{E}}^{(k)}(\Omega)$. We then define the discrete divergence by:

$$\operatorname{div}_K v = \frac{1}{|K|} \sum_{\sigma \in \mathscr{E}_K} |\sigma| v_{K,\sigma}, \quad \forall K \in \mathscr{M}, \ \forall v \in H_{\mathscr{E}}(\Omega), \tag{5}$$

where, denoting by $\mathscr{E}_{\text{int}}$ (resp. $\mathscr{E}_{\text{ext}}$) the set of internal (resp. boundary) edges,

$$v_{K,\sigma} = \begin{cases} v_\sigma n_\sigma \cdot n_{K,\sigma} & \forall \sigma \in \mathscr{E}_K \cap \mathscr{E}_{\text{int}}, \\ 0 & \forall \sigma \in \mathscr{E}_K \cap \mathscr{E}_{\text{ext}}, \end{cases} \quad \forall K \in \mathscr{M}, \ \forall v \in H_{\mathscr{E}}(\Omega), \tag{6}$$

where n_σ denotes the basis vector e to which σ is orthogonal. Using (5), we may define the following operator:

$$\operatorname{div}_{\mathscr{D}} v(x) = \operatorname{div}_K v, \quad \text{for a.e. } x \in K, \ \forall K \in \mathscr{M}, \ \forall v \in H_{\mathscr{E}}(\Omega), \tag{7}$$

and remark that (4a) can be written

$$\operatorname{div}_{\mathscr{D}} u(x) = 0, \quad \text{for a.e. } x \in \Omega. \tag{8}$$

Next, for $k = 1, \ldots, d$, we define an inner product on the space $H_{\mathscr{E}}^{(k)}$:

$$\langle u, v \rangle_k = \sum_{\{\sigma,\sigma'\} \subset \mathscr{E}^{(k)}} \tau_{\sigma,\sigma'}^{(k)} (u_\sigma - u_{\sigma'})(v_\sigma - v_{\sigma'}), \quad \forall u, v \in H_{\mathscr{E}}^{(k)}(\Omega); \tag{9}$$

this allows the definition of the following inner product on $H_{\mathscr{E}}(\Omega)$ which is expected to approximate $\int_\Omega \nabla u(x) : \nabla v(x) dx$:

$$\langle u, v \rangle_{\mathscr{E}} = \sum_{k=1}^{d} \langle u^{(k)}, v^{(k)} \rangle_k, \quad \forall u, v \in H_{\mathscr{E}}(\Omega). \tag{10}$$

We then obtain, multiplying (4b) by v_σ and summing on $k = 1, 2$ and $\sigma \in \mathscr{E}^{(k)}$,

$$\langle u, v \rangle_{\mathscr{E}} - \int_\Omega p(x) \operatorname{div}_{\mathscr{D}} v(x) dx = \int_\Omega f(x) \cdot v(x) dx, \quad \forall v \in H_{\mathscr{E}}(\Omega), \tag{11}$$

A discrete variational formulation of the MAC scheme (4) is therefore:

Find $u \in H_{\mathscr{E}}(\Omega)$ and $p \in H_{\mathscr{M}}(\Omega)$ s. t. $\sum_{K \in \mathscr{M}} |K| p_K = 0$ and (8) and (11) hold.

$$\tag{12}$$

3 The extended MAC scheme for the Navier-Stokes equations

We extend the standard MAC scheme to cases where all internal edges (2D) or faces (3D) whose normal is parallel to a basis vector $e^{(k)}$, such as the pressure grid depicted in Fig. 2 (left). Because of possibly hanging nodes, we may no longer define the velocity meshes by dual rectangles, but use instead the Voronoi cells associated with the barycentres of the edges $(x_\sigma)_{\sigma \in \mathscr{E}}$; they are defined as follows:

$$K_\sigma^{(k)} = \{x \in \Omega, d(x, x_\sigma) < d(x, x_{\sigma'}), \sigma' \in \mathscr{E}^{(k)} \setminus \{\sigma\}\}, \forall \sigma \in \mathscr{E}^{(k)},$$

Note that in the case of a uniform rectangular mesh, the Voronoi cells thus defined are equal to the velocity cells defined in the previous section. However, this is no longer true if a non uniform mesh is used, even in the conforming case; indeed, in this latter case, the Voronoi cells $K_\sigma^{(k)}$ are again rectangles, but they are not equal to the rectangular cells $K_\sigma^{(k)}$ defined previously. In the case of hanging nodes, they are no longer rectangular, as can be seen in Fig. 2, where we depict the pressure mesh, the horizontal and vertical velocity grids.

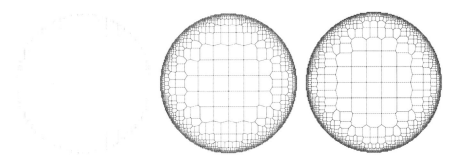

Fig. 2 The pressure and velocity grids

The diffusion term is again approximated by the discrete inner product defined by (9)-(10)-(2), but the expression of $\tau_{\sigma,\sigma'}$ given by (2) can no longer be written as in (3) for non rectangular cells. For Voronoï cells $K_\sigma^{(k)}$ and $K_{\sigma'}^{(k)}$ separated by a (dual) edge ε, such as those depicted in Fig. 3, one has

$$\tau_{\sigma,\sigma'} = \frac{|\varepsilon|}{d_\varepsilon} \quad (13)$$

where $|\varepsilon|$ denotes the length of the edge ε shared by $K_\sigma^{(k)}$ and $K_{\sigma'}^{(k)}$, and $d_\varepsilon = d(x_\sigma, x_{\sigma'})$ the distance between the two cell centres x_σ and $x_{\sigma'}$, which are also the barycentres of the edges σ and σ'. We can again define $H_{\mathscr{M}}(\Omega)$

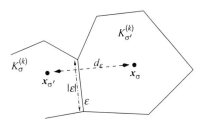

Fig. 3 Notations for a velocity cell

as the set of piecewise functions constant on the pressure cells $K \in \mathcal{M}$, the set $H_{\mathcal{E}}^{(k)}(\Omega)$ of piecewise constant functions on the grid cells K_σ, for $\sigma \in \mathcal{E}_{\text{int}}^{(k)} \cup \mathcal{E}_{\text{ext}}$ which vanish on any grid cell K_σ for a boundary edge $\sigma \subset \partial\Omega$; this discrete set is the space of functions meant to approximate the k-th component of the velocity. We finally denote by $H_{\mathcal{E}}(\Omega)$ the set of all $\mathbf{v} = (v^{(k)})_{k=1,\ldots,d}$ with $v^{(k)} \in H_{\mathcal{E}}^k(\Omega)$. The extended MAC scheme for the Stokes equations ($R = 0$) is again (5)-(12), with the new definition (13) for $\tau_{\sigma,\sigma'}$.

In order to write this generalized scheme for the Navier-Stokes equations, we need to add the discretization of the nonlinear term $\int_\Omega (\mathbf{u}(\mathbf{x}) \cdot \nabla)\mathbf{u}(\mathbf{x}) \cdot \mathbf{v}(\mathbf{x})d\mathbf{x}$. For $\mathbf{u}, \mathbf{v}, \mathbf{w} \in H_{\mathcal{E}}(\Omega)$, we define the discrete nonlinear convection term

$$b_{\mathcal{E}}(\mathbf{u}, \mathbf{v}, \mathbf{w}) = \sum_{K \in \mathcal{M}} \sum_{\substack{\sigma \in \mathcal{E}_K \\ \mathcal{M}_\sigma = \{K, L\}}} |\sigma| u_{K,\sigma} \frac{\Pi_K \mathbf{v} + \Pi_L \mathbf{v}}{2} \cdot \Pi_K \mathbf{w},$$

where $u_{K,\sigma}$ is defined by (6) $\Pi_K \mathbf{v}$ is a reconstruction of the full velocity on each pressure cell K defined by its components $(\Pi_K \mathbf{v})_k, k = 1, \ldots, d$:

$$(\Pi_K \mathbf{v})_k = \frac{1}{\sum_{\sigma \in \mathcal{E}_K \cap \mathcal{E}^{(k)}} |K_\sigma^{(k)}|} \sum_{\sigma \in \mathcal{E}_K \cap \mathcal{E}^{(k)}} |K_\sigma^{(k)}| v_\sigma.$$

The extended MAC scheme for the Navier-Stokes equation then reads:

$$\begin{cases} \text{Find } \mathbf{u} \in H_{\mathcal{E}}(\Omega) \text{ and } p \in H_{\mathcal{M}}(\Omega) \text{ s. t. } \sum_{K \in \mathcal{M}} |K| p_K = 0, \\ \text{div}_{\mathcal{D}} \mathbf{u}(\mathbf{x}) = 0, \text{ for a.e. } \mathbf{x} \in \Omega. \\ \langle \mathbf{u}, \mathbf{v} \rangle_{\mathcal{E}} - \int_\Omega p(\mathbf{x}) \text{div}_{\mathcal{D}} \mathbf{v}(\mathbf{x}) d\mathbf{x} + R \, b_{\mathcal{E}}(\mathbf{u}, \mathbf{u}, \mathbf{v}) = \int_\Omega \mathbf{f}(\mathbf{x}) \cdot \mathbf{v}(\mathbf{x}) d\mathbf{x}, \forall \mathbf{v} \in H_{\mathcal{E}}(\Omega). \end{cases}$$

With this scheme, a control over the discrete kinetic energy can be obtained, which allows to prove some discrete H^1 estimates on the velocity. Then an L^2 estimate is proved for the discrete pressure, using the standard Necas lifting, which is particularly easy thanks to the staggered grids. The proof of convergence is then completed, considering the interpolation of regular test functions. Details may be found in [2].

4 Numerical example

We consider a problem where the continuous solution of the Navier–Stokes equations (1) with $R = 1$ is given by:

$$\bar{u}_1(x_1, x_2) = 2\pi \sin^2(\pi x_1) \cos(\pi x_2) \sin(\pi x_2)$$
$$\bar{u}_2(x_1, x_2) = -2\pi \cos(\pi x_1) \sin(\pi x_1) \sin^2(\pi x_2)$$
$$\bar{p}(x_1, x_2) = \sin^2(\pi x_1) \sin^2(\pi x_2)$$

in a circle with centre $(0, 0)$ and radius 0.45. We consider four meshes for the mass conservation $\mathcal{M}_j$, $j = 0, \ldots, 3$, defined in the following way:

1. a structured square 10×10 is given on the square $[0, 1] \times [0, 1]$,
2. for $i = 0, \ldots, 3$, let us split in 4 control volumes each grid block whose centre (x_1, x_2) satisfies

$$\sqrt{(x_1 - 0.5)^2 + (x_2 - 0.5)^2} \geq 0.45 - 0.25/2^i,$$

3. for $i = 0, \ldots, j$, let us split in 4 control volumes each grid block K,
4. get rid of all the control volumes K with centre (x_1, x_2) such that

$$\sqrt{(x_1 - 0.5)^2 + (x_2 - 0.5)^2} > 0.45.$$

Let us denote card$(\mathcal{M}_j)$ the number of control volumes of the mesh $\mathcal{M}_j$. We get that card$(\mathcal{M}_0) = 1604$, card$(\mathcal{M}_1) = 6416$, card$(\mathcal{M}_2) = 25592$ and card$(\mathcal{M}_3) = 102324$. The L^2 errors of unknowns u_1, u_2, p, respectively denoted by $e_2(u_1), e_2(u_2), e_2(p)$, are respectively computed in the Voronoi grids associated to the velocity components and in $\mathcal{M}_j$.

Left part of Fig. (4) shows the errors $\log 10(e_2(u_1))$ and $\log 10(e_2(p))$ with respect to $\log 10(1/\sqrt{\text{card}(\mathcal{M}_j)})$ for $j = 0, \ldots, 3$. On right part of Fig. (4) are

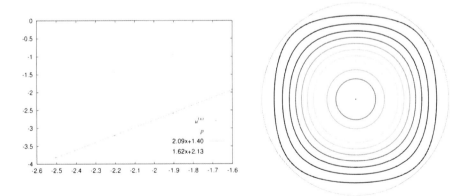

Fig. 4 Left: The L^2 error with respect to the number of control volumes. Right: Stream lines

plotted the stream lines for the finest mesh. The velocity components and the pressure are respectively shown in Figs. (5), (6) and (7). Although the velocity

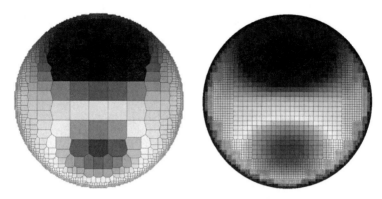

Fig. 5 Horizontal component of the velocity for $j = 0$ and $j = 2$

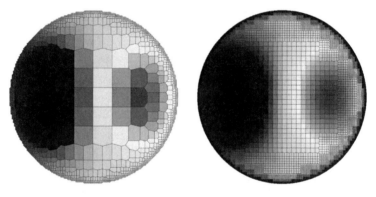

Fig. 6 Vertical component of the velocity for $j = 0$ for $j = 2$

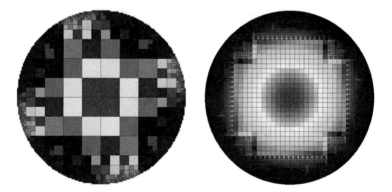

Fig. 7 Pressure for $j = 0$ and $j = 2$

fields are accurately computed on the coarsest mesh, the pressure fields show oscillations where neighbouring control volumes have contrasted sizes. However, these oscillations disappear while refining the mesh.

5 Conclusion

The generalised MAC scheme seems very efficient on meshes which are parallel to the axes. In particular, the scheme keeps a five-point stencil on all non-refined regions. It can also be extended to more general non-structured grids. However for these latter grids, the stencil may become large, which can be a problem when solving the linear systems in the Newton iteration.

References

1. P. Blanc. Convergence of a finite volume scheme on a MAC mesh for the Stokes problem with right-hand side in H^{-1}. In *Finite volumes for complex applications IV*, pages 133–142. ISTE, London, 2005.
2. E. Chénier, R. Eymard and R. Herbin. The MAC scheme on general meshes. in preparation, 2011.
3. V. Girault and J. Lopez. Finite-element error estimates for the MAC scheme., *IMA J. Numer. Anal., 16, 3, 247–379, 1996.*
4. R. Nicolaïdes. Analysis and convergence of the mac scheme i: The linear problem. *SIAM J. Numer. Anal.*, 29:1579–1591, 1992.
5. R. Nicolaïdes and X. Wu. Analysis and convergence of the mac scheme ii, Navier-Stokes equations. *Math. Comp.*, 65:29–44, 1996.
6. D. Shin and J.C. Strikwerda Inf-sup conditions for finite-difference approximations of the Stokes equations. *J. Austral. Math. Soc. Ser. B, 39, 1 121–134, 1997.*
7. S.V. Patankar. *Numerical heat transfer and fluid flow. Series in Computational Methods in Mechanics and Thermal Sciences*, volume XIII. Washington - New York - London: Hemisphere Publishing Corporation; New York. McGraw-Hill Book Company, 1980.

The paper is in final form and no similar paper has been or is being submitted elsewhere.

Multi-dimensional Optimal Order Detection (MOOD) — a Very High-Order Finite Volume Scheme for Conservation Laws on Unstructured Meshes

S. Clain, S. Diot, and R. Loubère

Abstract The Multi-dimensional Optimal Order Detection (MOOD) method is an original Very High-Order Finite Volume (FV) method for conservation laws on unstructured meshes. The method is based on an *a posteriori* degree reduction of local polynomial reconstructions on cells where prescribed stability conditions are not fulfilled. Numerical experiments on advection and Euler equations problems are drawn to prove the efficiency and competitiveness of the MOOD method.

Keywords MOOD, high-order, finite volume, unstructured meshes, conservation laws
MSC2010: 65M08, 65Z05, 76M12

1 Introduction

The Multi-dimensional Optimal Order Detection has been introduced in [6] as an original High-Order Finite Volume method for conservation laws on unstructured meshes. As multi-dimensional MUSCL [2–4, 8] or ENO/WENO methods [1, 7, 10], the MOOD method is based on a high-order space discretization with local polynomial reconstructions coupled with a high-order TVD Runge–Kutta method for time discretization.

S. Clain
Departamento de Matemática e Aplicações, Campus de Gualtar - 4710-057 Braga Campus de Azurm - 4800-058 Guimares, Portugal, e-mail: clain@math.uminho.pt

S. Diot, and R. Loubère
Institut de Mathématiques de Toulouse, Université de Toulouse, France,
e-mail: steven.diot@math.univ-toulouse.fr, raphael.loubere@math.univ-toulouse.fr

The main difference between classical high-order methods and the MOOD one is that the limitation procedure is done *a posteriori*. Inside a time step, a first solution is computed with numerical fluxes evaluated from unlimited high-order polynomial reconstructions. Then polynomial degrees are reduced on cells where prescribed stability conditions are not fulfilled and the solution is re-evaluated. That iterative procedure provides a solution which respects the stability constraints.

The present article is devoted to an extension of the MOOD method to a sixth-order space discretization on triangular meshes. Numerical tests for the advection problem and Euler equations with gravity are given in last section.

2 Framework

We consider the scalar hyperbolic equation defined on a bounded polygonal domain $\Omega \subset \mathbb{R}^2$ written in its conservative form

$$\partial_t u + \nabla \cdot F(u) = 0, \qquad (1)$$

$$u(\cdot, 0) = u_0,$$

where $u = u(\mathbf{x}, t)$ is the unknown function with $t > 0$, $\mathbf{x} \in \Omega$, F is the physical flux and u_0 stands for the initial condition. We consider a triangular tessellation of Ω where K_i is a generic triangle with centroid $\mathbf{c}_i$. Moreover $\mathbf{n}_{ij}$ is the unit normal vector of edge e_{ij} from K_i to K_j and q_{ij}^r, $r = 1, 2, 3$, are the Gaussian quadrature points of e_{ij}. Finally $\underline{v}(i)$ (resp. $\overline{v}(i)$) is the index set of cells which share an edge (resp. an edge or a node with K_i). This notation is summarized in Fig. 1.

We recall the generic first-order Finite Volume discretization of (1)

$$u_i^{n+1} = u_i^n - \Delta t \sum_{j \in \underline{v}(i)} \frac{|e_{ij}|}{|K_i|} G(u_i^n, u_j^n, \mathbf{n}_{ij}), \qquad (2)$$

where u_i^n is an approximation of the mean value of u on cell K_i at time t^n and $|e_{ij}|$, $|K_i|$ stand for the edge length and the cell surface respectively. We assume that the numerical flux $G(u_i^n, u_j^n, \mathbf{n}_{ij})$ satisfies the consistency and monotonicity properties such that, under an adequate CFL condition, the following Discrete Maximum Principle (DMP) is fulfilled

$$\min_{j \in \overline{v}(i)} (u_i^n, u_j^n) \leq u_i^{n+1} \leq \max_{j \in \overline{v}(i)} (u_i^n, u_j^n). \qquad (3)$$

Only few modifications of (2) are needed to get the following High-Order Finite Volume scheme

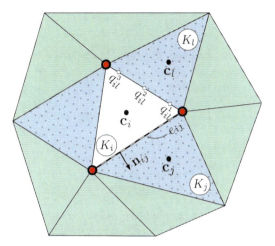

Fig. 1 Mesh notation. Index set $\underline{v}(i)$ corresponds to blue cells with dots and $\overline{v}(i)$ corresponds to every non-white cells

$$u_i^{n+1} = u_i^n - \Delta t \sum_{j \in \underline{v}(i)} \frac{|e_{ij}|}{|K_i|} \sum_{r=1}^{3} \xi_r G(u_{ij,r}^n, u_{ji,r}^n, \mathbf{n}_{ij}), \qquad (4)$$

namely the use of a sixth-order Gaussian quadrature rule with weights ξ_r ($r = 1, 2, 3$) and the replacement of u_i^n (resp. u_j^n) by $u_{ij,r}^n$ (resp. $u_{ji,r}^n$) which is an approximation of $u(q_{ij}^r, t^n)$ from the high-order polynomial reconstruction on K_i (resp. K_j). Notice that the high-order scheme (4) corresponds to a convex combination of the first-order one (2), that is important from a practical point of view for an easy and effective implementation.

It is well known that methods based on high-order reconstructions without limiting procedure produce spurious oscillations in the vicinity of discontinuities. In order to prevent such oscillations, the today's effective high-order methods (MUSCL, WENO...) use *a priori* limitation procedures.
The Multi-dimensional Optimal Order Detection (MOOD) method breaks away from this approach through an original effective iterative procedure based on an *a posteriori* detection of such unphysical oscillations (see Fig. 2). The details of MOOD method are recalled in next section

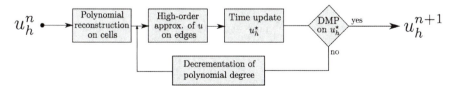

Fig. 2 A simplistic view of the Multi-dimensional Optimal Order Detection concept

3 MOOD method

For the sake of clarity, we only consider a forward Euler method and one quadrature point per edge. Consequently we denote by u_{ij} (resp. u_{ji}) the high-order approximation of u on edge e_{ij} from cell K_i (resp. K_j).

3.1 Basics

Polynomial reconstruction.

High-order approximations of the solution at quadrature points are mandatory. To this end, multi-dimensional polynomial reconstructions from mean values are carried out. There exist several techniques [1, 5] to obtain such reconstructions, but we choose to use the one from [7] where a over-determined linear system is solved using a QR decomposition. The reconstructed polynomial of arbitrary high-order $d_{max} + 1$ has the form

$$\widetilde{u}(x,y) = \bar{u} + \sum_{1 \leq \alpha+\beta \leq d_{max}} \mathcal{R}_{\alpha\beta}\left((x-c_x)^\alpha(y-c_y)^\beta - \frac{1}{|K|}\int_K (x-c_x)^\alpha(y-c_y)^\beta \, dxdy\right),$$

where (c_x, c_y) is the centroid of a generic cell K and $\mathcal{R}_{\alpha\beta}$ are the unknowns polynomial coefficients. In this way mean value on K is conserved and the truncation of all terms of degree $\alpha + \beta > \bar{d}$ produces a relevant approximation of u as a polynomial of degree $\bar{d} \leq d_{max}$.

At least $\mathcal{N}(d) = (d+1)(d+2)/2 - 1$ neighbors are needed to perform reconstructions. However for the sake of robustness at least $1.5 \times \mathcal{N}(d)$ elements are involved. We first take the neighbors by nodes of K and then the neighbors by faces of already picked elements. Lastly, since the condition number of the generated system is dependent of spatial characteristic length, we use the technique proposed in [5] to overcome this problem.

CellPD and EdgePD.

We recall the fundamental notions introduced in [6].

- d_i is the Cell Polynomial Degree (**CellPD**) which represents the degree of the polynomial reconstruction on cell K_i.
- d_{ij} and d_{ji} are the Edge Polynomial Degrees (**EdgePD**) which correspond to the effective degrees used to respectively build u_{ij} and u_{ji} on both sides of edge e_{ij}.

We now detail the MOOD method using both notions in the case of the scalar problem (1).

3.2 Algorithm for the scalar case.

The MOOD method consists of the following iterative procedure which details the concept depicted in Fig. 2.

1. **CellPD initialization.** Each CellPD is initialized with d_{max}.
2. **EdgePD evaluation.** Each EdgePD is set up as the minimum of the two neighboring CellPD.
3. **Quadrature points evaluation.** Each u_{ij} is evaluated with the polynomial reconstruction of degree d_{ij}.
4. **Mean values update.** The updated values $u_h^\star$ are computed using the finite volume scheme (4).
5. **DMP test.** The DMP criterion is checked on each cell K_i

$$\min_{j \in \bar{v}(i)} (u_i^n, u_j^n) \le u_i^\star \le \max_{j \in \bar{v}(i)} (u_i^n, u_j^n). \tag{5}$$

If $u_i^\star$ does not satisfy (5) the CellPD is decremented, $d_i := \max(0, d_i - 1)$.
6. **Stopping criterion.** If all cells satisfy the DMP property, the iterative procedure stops with $u_h^{n+1} = u_h^\star$ else go to Step 2.

Since only problematic cells and their neighbors in the compact stencil $\underline{v}(i)$ have to be checked and re-updated during the iterative MOOD procedure, the computational cost is dramatically reduced.

3.3 Algorithm for the Euler equations case.

We now extend the MOOD method to the Euler system, namely

$$\partial_t \begin{pmatrix} \rho \\ \rho u \\ \rho v \\ E \end{pmatrix} + \partial_x \begin{pmatrix} \rho u \\ \rho u^2 + p \\ \rho u v \\ u(E + p) \end{pmatrix} + \partial_y \begin{pmatrix} \rho v \\ \rho u v \\ \rho v^2 + p \\ v(E + p) \end{pmatrix} = 0, \tag{6}$$

where ρ, $\mathbf{V} = (u, v)$ and p are the density, velocity and pressure respectively while the total energy per unit volume E is given by

$$E = \rho \left(\frac{1}{2} \mathbf{V}^2 + e \right), \quad \mathbf{V}^2 = u_1^2 + u_2^2, \quad e = \frac{p}{\rho(\gamma - 1)},$$

where e is the specific internal energy and γ the ratio of specific heats.

The reconstruction is classically done on the primitive variables ρ, u, v, p while $U = (\rho, \rho u, \rho v, E)$ and we use the same CellPD and EdgePD for all variables in a cell. In other words, the two notions are linked to cells and edges and not affected by the number of variables. Furthermore steps 5 and 6 of the previous MOOD algorithm are substituted with the following stages.

5. **Density DMP test.** The DMP criterion is checked on the density

$$\min_{j \in \overline{v}(i)} (\rho_i^n, \rho_j^n) \leq \rho_i^\star \leq \max_{j \in \overline{v}(i)} (\rho_i^n, \rho_j^n). \tag{7}$$

If $\rho_i^\star$ does not satisfy (7) the CellPD is decremented, $d_i := \max(0, d_i - 1)$.

6. **Pressure positivity test.** The pressure positivity is checked and if $p_i^\star \leq 0$ and d_i has not been altered by step 5 then the CellPD is decremented, $d_i := \max(0, d_i - 1)$.

7. **Stopping criterion.** If for all $i \in \mathcal{E}_{el}$, d_i has not been altered by steps 5 and 6 then the iterative procedure stops with $U_h^{n+1} = U_h^\star$ else go to step 2.

4 Numerical results

The reader should refer to [6] for a study on the effective convergence rate and for more hydrodynamics test cases. In this paper, we restrict the presentation to two representative tests.

Scalar case

We first deal with the classical Solid Body Rotation (see [6] for details) test case for the advection problem. We plot in Fig. 3 isolines top views of the solution obtained with the MOOD method applied to different polynomial degrees and meshes. Method name, triangles number and computational times are embedded in each figure. Time is given in relative time units (r.t.u) where MOOD-P1 is taken as reference with 100 r.t.u.

First solutions obtained on the 5190 cells mesh (3 top) clearly show that the MOOD method is able to handle high-order polynomials with a great improvement of solutions while enforcing a strict DMP. Then for the sake of comparison, results with lower degrees on finer meshes are given in the bottom line of Fig. 3. Finally notice that the computational cost increase is mainly due to the reconstruction step. However since profiles are not smooth the DMP is often violated and the iterative procedure cost more than in a smooth case. For example a sixth-order unlimited version of the scheme costs 586 r.t.u., thus the iterative procedure costs about a third of the total time of the MOOD-P5 computation.

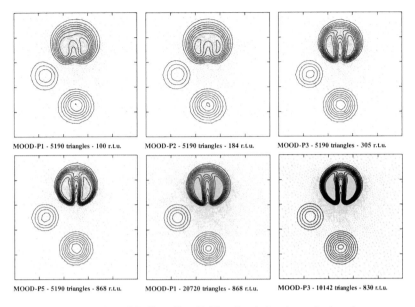

Fig. 3 Solid Body Rotation. 10 isolines (0 to 1). Time in relative time units (r.t.u.)

Euler equations case

For the system case, a Rayleigh–Taylor Instability for the Euler equations with gravity is considered. The reader should refer to [9] for complete description of the test case. A zoom on the pattern of the unstructured symmetric triangular mesh of 28800 cells and the density solutions for MOOD-P1, MOOD-P3 and MOOD-P5 are plotted in Fig. 4.

As for the scalar case the MOOD method is plainly able to improve the solution through the use of high-order polynomial reconstructions. From a computational cost point of view, computational times given in Fig. 4 prove that the MOOD iterative procedure is effective since the time raise from a degree to a bigger one is mainly due to the reconstruction cost itself.

Decrementation procedure

In Table 1, we give the mean percentage over all the calculation of polynomial degrees actually used to compute the solution, *i.e.* the CellPD at the end of the iterative procedure. Three test cases are taken as examples (see [6] for details), the Solid Body Rotation of Fig. 3 with MOOD-P3, the classical Double Mach Reflection on a 57600 cells uniform mesh with MOOD-P2 and the Mach 3 Wind Tunnel on a 4978 cells Delaunay mesh with MOOD-P3. Results show that only few cells are affected by the *a posteriori* limitation.

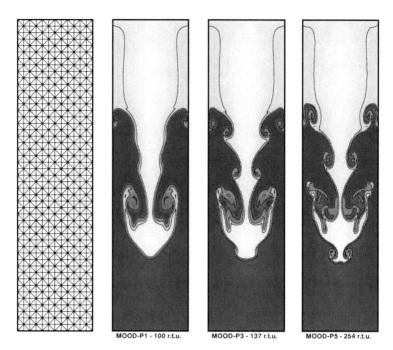

Fig. 4 Rayleigh–Taylor Instability. Density. 5 isolines from 0.8 (dark) to 2.3 (light)

Test case	P0	P1	P2	P3
Solid Body Rotation	7.16%	0.78%	0.64%	91.42%
Double Mach Reflection	5.69%	0.72%	93.69%	—
Mach 3 Wind Tunnel	3.02%	0.36%	0.16%	96.46%

Fig. 5 Mean percentage of polynomial degrees actually used with MOOD method

References

1. R. Abgrall, On Essentially Non-oscillatory Schemes on Unstructured Meshes: Analysis and Implementation, J. Comput. Phys. **114** 45–58 (1994)
2. T. J. Barth, Numerical methods for conservation laws on structured and unstructured meshes, VKI March 2003 Lectures Series
3. T. J. Barth, D. C. Jespersen, The design and application of upwind schemes on unstructured meshes, AIAA Report 89-0366 (1989)
4. T. Buffard, S. Clain, Monoslope and Multislope MUSCL Methods for unstructured meshes, J. Comput. Phys. **229** 3745-3776 (2010)
5. O. Friedrich, Weighted Essentially Non-Oscillatory Schemes for the Interpolation of Mean Values on Unstructured Grids, J. Comput. Phys. 144 (1998) 194–212.
6. S. Clain, S. Diot, R. Loubère A high-order finite volume method for systems of conservation laws — Multi-dimensional Optimal Order Detection (MOOD), accepted in J. Comput. Phys. (2011)

7. C. F. Ollivier-Gooch, Quasi-ENO Schemes for Unstructured Meshes Based on Unlimited Data-Dependent Least-Squares Reconstruction, J. Comput. Phys. **133** 6–17 (1997)
8. J. S. Park, S.-H. Yoon, C. Kim, Multi-dimensional limiting process for hyperbolic conservation laws on unstructured grids, J. Comput. Phys. **229** 788–812 (2010)
9. J. Shi, Y-T Zhang, C-W Shu, Resolution of high order WENO schemes for complicated flow structures, J. Comput. Phys. **186** 690–696 (2003)
10. W. R. Wolf , J. L. F. Azevedo, High-order ENO and WENO schemes for unstructured grids, International Journal for Numerical Methods in Fluids, **55** Issue 10 917—943 (2007)

The paper is in final form and no similar paper has been or is being submitted elsewhere.

A Relaxation Approach for Simulating Fluid Flows in a Nozzle

Frédéric Coquel, Khaled Saleh, and Nicolas Seguin

Abstract We present here a Godunov-type scheme to simulate one-dimensional flows in a nozzle with variable cross-section. The method relies on the construction of a relaxation Riemann solver designed to handle all types of flow regimes, from subsonic to supersonic flows, as well as resonant transonic flows. Some computational results are also provided, in which this relaxation method is compared with the classical Rusanov scheme and a modified Rusanov scheme.

Keywords Relaxation scheme, Godunov-type scheme, resonant transonic flows.
MSC2010: 76M12, 76H05, 76S05, 65M12

1 Introduction

In this paper, we are interested in the numerical approximation of the solutions of a model describing one-dimensional barotropic flows in a nozzle. In this model, ρ and w are respectively the density and the velocity of the fluid while α stands for the cross-section of the nozzle, which is assumed to be constant in time. Under the classical assumption that α is small with respect to a characteristic length in the

---

Frédéric Coquel
CMAP, UMR 7641, Ecole Polytechnique, route de Saclay, F-91128 Palaiseau,
e-mail: frederic.coquel@cmap.polytechnique.fr

Khaled Saleh
EDF R&D, MFEE, 6 Quai Watier, F-78400 Chatou
and
UPMC & CNRS, UMR 7598, LJLL, F-75005 Paris, e-mail: khaled.saleh@edf.fr, saleh@ann.jussieu.fr

Nicolas Seguin
UPMC & CNRS, UMR 7598, LJLL, F-75005 Paris, e-mail: seguin@ann.jussieu.fr

mainstream direction, the flow can be supposed to be one-dimensional and described by the following set of partial differential equations:

$$\begin{aligned} &\partial_t \alpha = 0, \\ &\partial_t (\alpha \rho) + \partial_x (\alpha \rho w) = 0, \\ &\partial_t (\alpha \rho w) + \partial_x (\alpha \rho w^2 + \alpha p(\tau)) - p(\tau) \partial_x \alpha = 0, \end{aligned} \qquad t > 0, \ x \in \mathbb{R}, \qquad (1)$$

where $\tau = \rho^{-1}$ is the specific volume and $\tau \mapsto p(\tau)$ is a barotropic pressure law (satisfying $p'(\tau) < 0$ and $p''(\tau) > 0$). System (1) takes the condensed form:

$$\partial_t \mathbb{U} + \partial_x \mathbf{f}(\mathbb{U}) + \mathbf{c}(\mathbb{U}) \partial_x \mathbb{U} = 0, \qquad (2)$$

where the state vector is $\mathbb{U} = (\alpha, \alpha\rho, \alpha\rho w)^T$. The solutions are sought in the phase space of positive solutions defined as

$$\Omega = \{\mathbb{U} = (\alpha, \alpha\rho, \alpha\rho w)^T \in \mathbb{R}^3, \alpha > 0, \alpha\rho > 0\}. \qquad (3)$$

We recall the properties of this model:

- **Property 1.1 (Hyperbolicity)** *System (1) admits, for $\mathbb{U}$ in Ω, the following eigenvalues*

$$\lambda_0(\mathbb{U}) = 0, \qquad \lambda_1(\mathbb{U}) = w - c(\tau), \qquad \lambda_2(\mathbb{U}) = w + c(\tau), \qquad (4)$$

where $c(\tau) = \tau\sqrt{-p'(\tau)}$. The system is hyperbolic (i.e. the corresponding eigenvectors span $\mathbb{R}^3$) if and only if $|w| \neq c(\tau)$. Besides, the fields associated with the λ_1 and λ_2 eigenvalues are genuinely non-linear while the field associated with λ_0 is linearly degenerate.

- **Property 1.2 (Entropy)** *The entropy solutions of system (1) satisfy the following inequality in the weak sense*

$$\partial_t (\alpha\rho\mathscr{E}) + \partial_x (\alpha\rho\mathscr{E} w + \alpha p(\tau) w) \leq 0 \qquad (5)$$

where $\mathscr{E} = \frac{w^2}{2} + e(\tau)$ is the total energy and where the function $\tau \mapsto e(\tau)$ is given by $e'(\tau) = -p(\tau)$.

The Godunov scheme for this model is difficult to implement because the Riemann problem for system (1) is hard to solve due to the non linearities of the pressure law (giving rise to the genuinely non-linear acoustic fields), to the absence of a satisfactory definition of the non-conservative product $p(\tau)\partial_x \alpha$ and to the resonance phenomenon that appears for transonic flows causing the model to lose hyperbolicity [5]. For these reasons, we rather follow the classical approach of [7] and design an approximate Riemann solver, relying on a relaxation method. With this end in view, the solutions of system (1) are approximated by the solutions of the following enlarged relaxation system in the limit of a vanishing positive parameter ε:

$$\begin{aligned}
&\partial_t \alpha^\varepsilon = 0, \\
&\partial_t (\alpha\rho)^\varepsilon + \partial_x (\alpha\rho w)^\varepsilon = 0, \\
&\partial_t (\alpha\rho w)^\varepsilon + \partial_x (\alpha\rho w^2 + \alpha\pi(\tau,\mathcal{T}))^\varepsilon - \pi(\tau,\mathcal{T})^\varepsilon \partial_x \alpha^\varepsilon = 0, \\
&\partial_t (\alpha\rho\mathcal{T})^\varepsilon + \partial_x (\alpha\rho\mathcal{T}w)^\varepsilon = \frac{1}{\varepsilon}(\alpha\rho)^\varepsilon(\tau-\mathcal{T})^\varepsilon,
\end{aligned} \qquad t > 0,\ x \in \mathbb{R}, \quad (6)$$

with a linearization of the pressure law given by $\pi(\tau,\mathcal{T}) = p(\mathcal{T}) + a^2(\mathcal{T}-\tau)$. The variable $\mathcal{T}$ is an additionnal unknown relaxing towards the specific volume τ in the limit $\varepsilon \searrow 0$, and the constant a is a numerical parameter that must be taken large enough so as to guarantee the non-linear stability of the numerical approximation. The state vector for the relaxation system is $\mathbb{W} = (\alpha, \alpha\rho, \alpha\rho w, \alpha\rho\mathcal{T})^T$ and the solutions are sought in the phase space

$$\Omega^r = \{\mathbb{W} = (\alpha, \alpha\rho, \alpha\rho w, \alpha\rho\mathcal{T})^T \in \mathbb{R}^4, \alpha > 0, \alpha\rho > 0, \alpha\rho\mathcal{T} > 0\}. \quad (7)$$

The following property motivates the introduction of this relaxation system

Property 1.3 (Hyperbolicity) *The convective part of (6) admits, for $\mathbb{W}$ in Ω^r, the following eigenvalues*

$$\sigma_0(\mathbb{W}) = 0, \quad \sigma_1(\mathbb{W}) = w - a\tau, \quad \sigma_2(\mathbb{W}) = w, \quad \sigma_3(\mathbb{W}) = w + a\tau. \quad (8)$$

The system is hyperbolic (i.e. the corresponding eigenvectors span $\mathbb{R}^4$) if and only if $|w| \neq a\tau$, and all the fields are linearly degenerate.

2 The Riemann problem for the relaxation system

In this section, we give the main ideas leading to the construction of solutions to the Riemann problem for the convective part of the relaxation system (6). Being given $\mathbb{W}_L$ and $\mathbb{W}_R$ two states in Ω^r, we look for solutions of

$$\begin{cases} \partial_t \mathbb{W} + \partial_x \mathbf{g}(\mathbb{W}) + \mathbf{d}(\mathbb{W})\partial_x \mathbb{W} = 0, \\ \mathbb{W}(x,0) = \mathbb{W}_L \quad \text{if} \quad x < 0 \quad \text{and} \quad \mathbb{W}_R \quad \text{if} \quad x > 0. \end{cases} \quad (9)$$

As all the characteristic fields are linearly degenerate, the solution turns out to be simpler to construct than a solution of the Riemann problem for the equilibrium system (1). Indeed, the solution is sought in the form of a self-similar function consisting in constant intermediate states separated by contact discontinuities. The linear degeneracy of the fields provides natural jump relations across each discontinuity and yields a set of equations eventually leading to the expressions of the wave speeds and intermediate states. However, some issues related to the resonance phenomenon still need to be handled with care (see [2] for details).

We show that the solutions can be expressed in terms of the physical data $\mathbb{V}_L = (\rho_L, w_L, \mathcal{T}_L)$ and $\mathbb{V}_R = (\rho_R, w_R, \mathcal{T}_R)$ (i.e. all the initial data excluding the cross-section α) and of the ratio of left and right initial sections $\nu := \frac{\alpha_L}{\alpha_R}$. More precisely, we introduce the following quantities depending only on $(\mathbb{V}_L, \mathbb{V}_R)$

$$w^\sharp := \frac{1}{2}(w_L + w_R) - \frac{1}{2a}(\pi_R - \pi_L), \tag{10}$$

$$\tau_L^\sharp := \tau_L + \frac{1}{a}(w^\sharp - w_L) = \tau_L + \frac{1}{2a}(w_R - w_L) - \frac{1}{2a^2}(\pi_R - \pi_L), \tag{11}$$

$$\tau_R^\sharp := \tau_R - \frac{1}{a}(w^\sharp - w_R) = \tau_R + \frac{1}{2a}(w_R - w_L) + \frac{1}{2a^2}(\pi_R - \pi_L), \tag{12}$$

where $w^\sharp$ has the dimension of a speed and $\tau_L^\sharp$, $\tau_R^\sharp$ the dimension of specific volumes. These quantities appear in the explicit expressions of the solutions and it can be proved that these specific volumes need to be positive in order to guarantee the positivity of the solutions. In the numerical applications however, a will be chosen large for stability matters (see Sect. 4) and it will always be possible to impose the positivity of $\tau_L^\sharp$ and $\tau_R^\sharp$ by taking a large enough.

The main result of this section is the existence theorem for the Riemann problem.

Theorem 2.1 *Let $\mathbb{W}_L$ and $\mathbb{W}_R$ be two positive states in Ω^r. Assume that a is such that $\tau_L^\sharp > 0$ and $\tau_R^\sharp > 0$. Then the Riemann problem (9) admits a positive self-similar solution whatever the ratio $\nu = \frac{\alpha_L}{\alpha_R}$ is.*

Sketch of the proof *(see [2] for details).* The proof consists in the effective construction of a solution. For the relaxation system, the eigenvalues are not naturally ordered because of the existence of a standing wave, and a resonance phenomenon does appear for transonic flows. Therefore, in order to construct solutions, we investigate all admissible wave configurations (including sonic and supersonic ones) and for each admissible ordering of the eigenvalues, we determine sufficient conditions on the initial states $\mathbb{W}_L$ and $\mathbb{W}_R$ for the solution to have this particular ordering. Eventually, we check *a posteriori* that the determined conditions totally cover the whole space of initial conditions $\Omega^r \times \Omega^r$. □

Figure 1 represents the map of the admissible solutions given by Theorem 2.1 with respect to the initial states $\mathbb{W}_L$ and $\mathbb{W}_R$. The right part of the chart corresponds to the solutions with positive material speed, while the left part depicts the symmetric configurations with negative material speed.

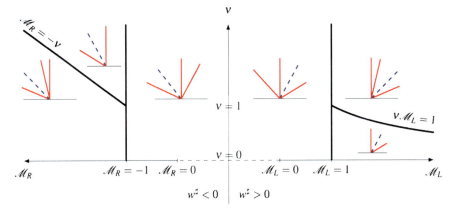

Fig. 1 Wave configuration of the solution of the Riemann problem (9) with respect to $\mathbb{W}_L$ and $\mathbb{W}_R$. $\mathcal{M}_L = \frac{w_L}{a\tau_L}$ and $\mathcal{M}_R = \frac{w_R}{a\tau_R}$ are the Mach numbers of the initial left and right states $\mathbb{W}_L$ and $\mathbb{W}_R$. The material wave is represented by a dashed line

3 Numerical approximation

In this section, we derive a numerical scheme from the relaxation approximation introduced in Sect. 1, the aim being to approximate the weak solutions of a Cauchy problem associated with system (1):

$$\begin{cases} \partial_t \mathbb{U} + \partial_x \mathbf{f}(\mathbb{U}) + \mathbf{c}(\mathbb{U})\partial_x \mathbb{U} = 0, \\ \mathbb{U}(x,0) = \mathbb{U}_0(x). \end{cases} \quad (13)$$

Let Δx be a space step and Δt a time step. The space is partitioned into cells $\mathbb{R} = \bigcup_{j \in \mathbb{Z}} C_j$ with $C_j = [x_{j-\frac{1}{2}}, x_{j+\frac{1}{2}}[$, where $x_{j+\frac{1}{2}} = (j+\frac{1}{2})\Delta x$ are the cell interfaces. At the discrete times $t^n = n\Delta t$, the solution of (13) is approximated on each cell C_j by a constant value denoted by $\mathbb{U}_j^n = \left(\alpha_j^n, (\alpha\rho)_j^n, (\alpha\rho w)_j^n\right)^T$. We now describe the two-step splitting method associated with the relaxation system (6) in order to calculate the values of the approximate solution at time t^{n+1} $(\mathbb{U}_j^{n+1})_{j \in \mathbb{Z}}$ from those at time t^n.

Step 1: Time evolution $(t^n \to t^{n+1,-})$

We first introduce the piecewise constant approximate solution of the relaxation system at time t^n: $x \mapsto \mathbb{W}(x, t^n) = \mathbb{W}_j^n$ in C_j with $\mathbb{W}_j^n = \left(\alpha_j^n, (\alpha\rho)_j^n, (\alpha\rho w)_j^n, (\alpha\rho\mathcal{T})_j^n\right)$, where $\mathcal{T}_j^n := \tau_j^n$, i.e. $\mathbb{W}_j^n$ is at equilibrium. Then, the following Cauchy problem is **exactly solved** for $t \in [0, \Delta t]$ with Δt small enough (see condition (15) below)

$$\begin{cases} \partial_t \widetilde{\mathbb{W}} + \partial_x \mathbf{g}(\widetilde{\mathbb{W}}) + \mathbf{d}(\widetilde{\mathbb{W}}) \partial_x \widetilde{\mathbb{W}} = 0, \\ \widetilde{\mathbb{W}}(x, 0) = \mathbb{W}(x, t^n). \end{cases} \quad (14)$$

Since the initial condition $x \mapsto \mathbb{W}(x, t^n)$ is piecewise constant, the exact solution of (14) is obtained by gluing together the solutions of the Riemann problems set at each cell interface $x_{j+\frac{1}{2}}$, provided that these solutions do not interact during the period Δt, i.e. provided the following classical CFL condition

$$\frac{\Delta t}{\Delta x} \max_{\mathbb{W}} |\sigma_i(\mathbb{W})| < \frac{1}{2}, \quad i \in \{0, \ldots, 3\}, \quad (15)$$

for all $\mathbb{W}$ under consideration. More precisely, if (x, t) is in $[x_j, x_{j+1}] \times [0, \Delta t]$, then

$$\widetilde{\mathbb{W}}(x, t) = \mathbb{W}_r \left(\frac{x - x_{j+1/2}}{t}; a_{j+1/2}, \mathbb{W}_j^n, \mathbb{W}_{j+1}^n \right), \quad (16)$$

where $(x, t) \mapsto \mathbb{W}_r \left(\frac{x}{t}; a, \mathbb{W}_L, \mathbb{W}_R \right)$ is the self-similar solution of the Riemann problem constructed in Sect. 1, which clearly depends on the local choice of the parameter a. Then, in order to define a piecewise constant approximate solution at time $t^{n+1,-}$, the solution $\widetilde{\mathbb{W}}(x, t)$ is averaged on each cell C_j at time Δt:

$$\mathbb{W}(x, t^{n+1,-}) = \mathbb{W}_j^{n+1,-} := \frac{1}{\Delta x} \int_{x_{j-\frac{1}{2}}}^{x_{j+\frac{1}{2}}} \widetilde{\mathbb{W}}(x, \Delta t) dx, \quad \forall x \in C_j, \quad \forall j \in \mathbb{Z}. \quad (17)$$

Step 2: Instantaneous relaxation ($t^{n+1,-} \to t^{n+1}$)

The second step consists in sending ε to zero instantaneously in the piecewise constant function $\mathbb{W}(x, t^{n+1,-})$ obtained at the end of the first step. This amounts to imposing $\mathcal{T}_j^{n+1} := \tau_j^{n+1}$, thus we have

$$\mathbb{W}_j^{n+1} = \left(\alpha_j^{n+1,-}, (\alpha \rho)_j^{n+1,-}, (\alpha \rho w)_j^{n+1,-}, \alpha_j^{n+1,-} \right)^T. \quad (18)$$

Finally, the new cell value at time t^{n+1} of the approximate solution reads

$$\mathbb{U}_j^{n+1} = \left(\alpha_j^{n+1,-}, (\alpha \rho)_j^{n+1,-}, (\alpha \rho w)_j^{n+1,-} \right)^T. \quad (19)$$

We can prove that this two-step relaxation method can be equivalently rewritten in the form of a Godunov-type finite volume scheme [7].

4 Non-linear stability of the scheme

Non-linear stability issues are usually dealt with through a so-called *discrete entropy inequality*, which is the discrete counterpart of the entropy inequality (5) satisfied by the weak solutions of the model. We have the following definition:

Definition 4.1 *We say that a numerical scheme satisfies a discrete entropy inequality if there exists a numerical entropy flux $G(\mathbb{U}_L, \mathbb{U}_R)$ which is consistent with the exact entropy flux $\mathscr{G} = \alpha \rho \mathscr{E} w + \alpha p(\tau) w$ (in the sense that $G(\mathbb{U}, \mathbb{U}) = \mathscr{G}(\mathbb{U})$ for all $\mathbb{U}$) such that, under some CFL condition, the discrete values $(\mathbb{U}_j^n)_{j \in \mathbb{Z}, n \in \mathbb{N}}$ computed by the scheme automatically satisfy*

$$(\alpha \rho \mathscr{E})(\mathbb{U}_j^{n+1}) - (\alpha \rho \mathscr{E})(\mathbb{U}_j^n) + \frac{\Delta t}{\Delta x}(G(\mathbb{U}_j^n, \mathbb{U}_{j+1}^n) - G(\mathbb{U}_{j-1}^n, \mathbb{U}_j^n)) \leq 0. \quad (20)$$

As seen in Sect. 3, under the CFL condition (15), the different Riemann problems at each interface do not interact and the parameter $a = a_{j+\frac{1}{2}}$ can be chosen locally interface by interface. Usually, if $a_{j+\frac{1}{2}}$ is large enough, so as to satisfy a so-called Whitham condition (see [1]), then a discrete entropy inequality (20) is guaranteed. In order to define $a_{j+\frac{1}{2}}$, we propose a weak Whitham-like condition that handles the resonance phenomenon and still guarantees a discrete entropy inequality under the CFL condition (15) (see [2] for details).

5 Numerical tests

In this section, we run the relaxation scheme described in Sect. 3 on a Riemann problem that contains the standing wave associated with the constant cross-section α, a left-going λ_1-rarefaction wave, a sonic right-going λ_1-rarefaction wave and a right-going λ_2-shock. The chosen pressure law is an ideal gas barotropic pressure law $p(\tau) = \tau^{-\gamma}$, with $\gamma = 3$. The left and right initial conditions are given by $\alpha_L = 3.0$, $\rho_L = 1.0$, $w_L = 0$, $\alpha_R = 1.0$, $\rho_R = 0.1$, and $w_R = 0$. The outcome of the relaxation method is compared with two other numerical schemes. The first one is the classical Rusanov scheme where the cross-section α is preserved throughout time:

$$\alpha_j^{n+1} := \alpha_j^n. \quad (21)$$

The second one is a modification of the Rusanov scheme that consists in applying the scheme to the whole state vector $\mathbb{U}$ (including the cross-section α) causing α to be dissipated:

$$\alpha_j^{n+1} := \alpha_j^n - \frac{\Delta t}{\Delta x}\left(q_{j+\frac{1}{2}}^n - q_{j-\frac{1}{2}}^n\right), \quad (22)$$

with $q_{j+\frac{1}{2}}^n = -r(\mathbb{U}_j^n)(\alpha_{j+1}^n - \alpha_j^n)$ where the scalar $r(\mathbb{U}_j^n)$ is the maximal value of the spectral radius of the Jacobian matrices $(\nabla \mathbf{f} + \mathbf{c})(\mathbb{U}_k^n)$ for $k = j, j+1$.

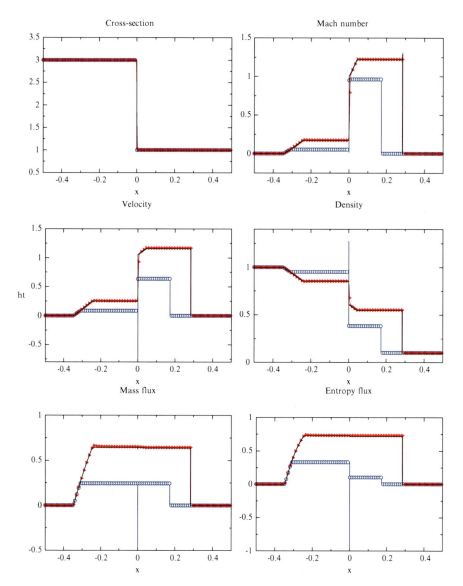

Fig. 2 Solution of the Riemann problem at time $T = 0.2$. Space step $\Delta x = 10^{-5}$. Straight line: relaxation scheme, circles: classical Rusanov scheme, triangles: Rusanov scheme with dissipation of the cross-section

In Fig. 2, we can see that, due to a smoothing effect, the dissipation of the cross-section α provides a notable improvement for the Rusanov scheme (see [4] and [8] for different approaches to improve the Rusanov scheme). The L^1-norm of the error on α, at the final time T, vanishes as the space step Δx goes to zero (with $\Delta t / \Delta x$ constant) with the order $\mathcal{O}(\Delta x^{1/2})$.

Acknowledgements The second author receives a financial support by ANRT through an EDF-CIFRE contract 529/2009. Computational facilities were provided by EDF. The third author is partially supported by the LRC Manon (Modélisation et Approximation Numérique Orientées pour l'énergie Nucléaire – CEA/DM2S-LJLL).

References

1. F. Bouchut. *Nonlinear Stability of Finite Volume Methods for Hyperbolic Conservation Laws.* Birkhauser. Frontiers in Mathematics. 2004.
2. F. Coquel, K. Saleh, N. Seguin. Relaxation and numerical approximation for fluid flows in a nozzle. *Preprint to be published.*
3. C.M. Dafermos. *Hyperbolic Conservation Laws in Continuum Physics.* Springer-Verlag. Grundlehren der mathematischen Wissenschaften. **Vol 325**. 2000.
4. L. Girault, J-M. Hérard. A two-fluid hyperbolic model in a porous medium. *M2AN*, **Vol 44(6)**, pp 1319-1348, 2010.
5. P. Goatin, P.G. LeFloch. The Riemann problem for a class of resonant hyperbolic systems of balance laws. *Ann. Inst. H. Poincaré Anal. Non Linéaire 21*, **no 6**, pp 881-902, 2004.
6. E. Godlewski, P-A. Raviart. *Numerical Approximation of Hyperbolic Systems ofConservation Laws.* Springer-Verlag. Applied Mathematical Sciences. **Vol 118**. 1996.
7. A. Harten, P.D. Lax & B. Van Leer. On upstream differencing and Godunov-type schemes for hyperbolic conservation laws. *Comm. Math. Sci.* **Vol 1**. pp 763-796. 2003.
8. D. Kröner, M.D. Thanh. Numerical solution to compressible flows in a nozzle with variable cross-section. *SIAM J. Numer. Anal.*, **Vol 43(2)**, pp 796-824, 2006.
9. P.G. LeFloch, M.D. Thanh. The Riemann problem for fluid flows in a nozzle with discontinuous cross-section. *Comm. Math. Sci.* **Vol 1**, pp 763-796, 2003.

The paper is in final form and no similar paper has been or is being submitted elsewhere.

A CeVeFE DDFV scheme for discontinuous anisotropic permeability tensors

Yves Coudière, Florence Hubert, and Gianmarco Manzini

Abstract In this work we derive a formulation for discontinuous diffusion tensor for the Discrete Duality Finite Volume (DDFV) framework that is exact for affine solutions. In fact, DDFV methods can naturally handle anisotropic or non-linear problems on general distorted meshes. Nonetheless, a special treatment is required when the diffusion tensor is discontinuous across an internal interfaces shared by two control volumes of the mesh. In such a case, two different gradients are considered in the two subdiamonds centered at that interface and the flux conservation is imposed through an auxiliary variable at the interface.

Keywords Finite volume schemes, Darcy flow
MSC2010: 65N08, 76S05

1 Introduction

In this proceeding we propose a Discrete Duality Finite Volume (DDFV) method that can handle *discontinuous* permeability coefficients. This method is a variant of the DDFV formulation proposed by Y. Coudière and F. Hubert in [6] to extend to three-dimensional (3D) problems the original two-dimensional finite volume schemes by F. Hermeline [11] and K. Domelevo and P. Omnès [9]. In the DDFV approach the diffusive flux is approximated using a piecewise constant

---

Yves Coudière
Laboratoire Jean Leray, Nantes, FRANCE, e-mail: Yves.Coudiere@univ-nantes.fr

Florence Hubert
LATP, Université de Provence, Marseille, FRANCE, e-mail: fhubert@cmi.univ-mrs.fr

Gianmarco Manzini
IMATI and CESNA-IUSS, Pavia, ITALY, e-mail: gm.manzini@gmail.com

approximation of the solution gradient over a set of edge-based cells called *diamond cells*. In the two dimensional formulation, the gradient is approximated by a formula that requires the vertex values of the scalar solution. Following the DDFV approach, such vertex values are the solution of another finite volume method whose control volumes are built around the vertices. Therefore, the resulting scheme combines two distinct finite volume methods for the cell unknowns and the vertex unknowns on two overlapping meshes. Effectiveness and efficiency of such coupled finite volume formulation are documented in [5, 10].

Several generalizations of the two-dimensional DDFV formulation have been proposed in the literature; it is worth mentioning the works by F. Hermeline in [12], C. Pierre in [8, 13], and B. Andreianov and collaborators in [1–4]. Here, we consider the alternative construction proposed in [6], which uses two families of additional unknowns. In the first family, the unknowns are located at the vertices of the mesh and are the solution of a finite volume method whose control volumes are built around the vertices. In the second family, the unknowns are located at the centers of mesh edges and faces and are the solution of a finite volume method whose control volumes are built around such geometric objects. Therefore, the resulting scheme couples three distinct finite volume methods through a 3D gradient formula that generalizes the 2D one on a set of special cells, the so called *diamond cells*, built around edges and faces as will be discussed in the next sub-section.

The outline of the paper is as follows. In Sect. 2 we present a short review of the DDFV method. In Sect. 3 we present the numerical treatment that we propose for the case of discontinuous permeabilities. In Sect. 6 we offer final remarks and conclusions.

2 The Discrete Duality Finite volume formulation

Meshes

Given a general finite volume mesh $\mathcal{M}$ of the computational domain Ω, composed of polyhedra, three additional polyhedral partitions of Ω are built, denoted by $\mathcal{N}$, $\mathcal{FE}$ and $\mathcal{D}$, hereafter described.

We denote the control volumes of the initial mesh $\mathcal{M}$ by K or L. The set $\partial \mathcal{M}$ gathers the boundary faces, which we consider as degenerated control volumes, and we complete the initial mesh as $\overline{\mathcal{M}} = \mathcal{M} \cup \partial \mathcal{M}$. We associate a set of points $x_K \in K$ with the control volumes in $\overline{\mathcal{M}}$; specifically, in the current applications we use the arithmetic average of the vertex position vectors for each polyhedral cell. We denote the vertices, the edges, and the faces of mesh $\mathcal{M}$ by x_A, E and F, respectively, and we define some additional points: the center of gravity x_F of each face F and the midpoint x_E of each edge E. These points are ordered following the relation

$$x_A \prec x_E \prec x_F \prec x_K \quad \text{which means that} \quad x_A \subset \partial E, \quad E \subset \partial F, \quad F \subset \partial K.$$

A CeVeFE DDFV scheme for discontinuous anisotropic permeability tensors

The 3D gradient formula that we will introduce in the next subsection provides a piecewise constant approximation of the solution gradient on the mesh $\mathscr{D}$, which is the set of *diamond cells* D. To each one of the pairs "(edge, face)" (E, F) related by $x_E \prec x_F$ there corresponds a different diamond cell D that we define as follows. Cell D is the convex polyhedra with vertices $x_A, x_B, x_E, x_F, x_K, x_L$, where x_A and x_B denote the endpoints of E, K and L the two cells sharing the common face F. Specifically, it holds that $D = \text{hull}(x_A, x_F, x_B, x_K) \cup \text{hull}(x_A, x_F, x_B, x_L)$. We associate with each diamond cell D the point $x_D = \frac{1}{2}(x_E + x_F) \in D$.

We partition each diamond cell into eight tetrahedra sharing x_D as common vertex and having the remaining three vertices chosen within the pairs (x_A, x_B), (x_E, x_F) and (x_K, x_L), respectively. Formally, we denote the eight possible combinations by

$$D = \text{hull}\left(x_D, \begin{pmatrix} x_A \\ x_B \end{pmatrix}, \begin{pmatrix} x_E \\ x_F \end{pmatrix}, \begin{pmatrix} x_K \\ x_L \end{pmatrix}\right), \quad \text{with} \quad \begin{pmatrix} x_A \\ x_B \end{pmatrix} \prec x_E \prec x_F \prec \begin{pmatrix} x_K \\ x_L \end{pmatrix}.$$

We assume the six vertices x_K, x_L, x_A, x_B and x_E, x_F of the diamond cell D(E, F) to be ordered in such a way that $\Delta_{EF} := \det(x_B - x_A, x_F - x_E, x_L - x_K) > 0$. Thus, the measure of D is $|D| = \frac{1}{6}\Delta_{EF}$.

We denote the control volume associated with a vertex x_A of the mesh by A. This control volume is built by gathering the contributions (i.e., sub-tetraedra) of the diamond cells that share vertex x_A as:

$$A = \bigcup_{D \in D_A} \text{hull}\left(x_D, x_A, \begin{pmatrix} x_E \\ x_F \end{pmatrix}, \begin{pmatrix} x_K \\ x_L \end{pmatrix}\right),$$

where $D_A = \{D \in \mathscr{D}, \text{ such that } x_A \prec x_E \prec x_F\}$ for x_A fixed. The resulting finite volume partition of Ω, denoted by $\mathscr{N}$, forms the *vertex mesh*. The vertex mesh is split into interior and boundary controls volumes, respectively denoted by $\mathscr{N}$ and $\partial\mathscr{N}$; formally, it holds that $\overline{\mathscr{N}} = \mathscr{N} \cup \partial\mathscr{N}$.

Similarly, we associate a control volume denoted either by F or by E, with the point x_F (face center) or the point x_E (edge midpoint) in accordance with the following formula:

$$E = \bigcup_{D \in D_E} \text{hull}\left(x_D, \begin{pmatrix} x_A \\ x_B \end{pmatrix}, x_E, \begin{pmatrix} x_K \\ x_L \end{pmatrix}\right), \quad F = \bigcup_{D \in D_F} \text{hull}\left(x_D, \begin{pmatrix} x_A \\ x_B \end{pmatrix}, x_F, \begin{pmatrix} x_K \\ x_L \end{pmatrix}\right),$$

where $D_E = \{D \in \mathscr{D}, \text{ with } x_E \prec x_F\}$ with x_E fixed and $D_F = \{D \in \mathscr{D}, \text{ with } x_E \prec x_F\}$ with x_F fixed. The resulting finite volume partition of Ω, denoted by $\overline{\mathscr{FE}}$, is the *face-edge mesh*. This partition contains both control volumes associated with the faces and the edges of the initial mesh and is split into the interior and boundary controls volumes, respectively denoted by $\mathscr{FE}$ and $\partial\mathscr{FE}$; formally, it holds that $\overline{\mathscr{FE}} = \mathscr{FE} \cup \partial\mathscr{FE}$.

The 3D "Cell-Vertex-Face/Edge" DDFV Scheme

We say that $u^{\mathcal{T}} = (u^{\mathcal{M}}, u^{\mathcal{N}}, u^{\mathcal{FE}})$ is a *discrete function* on Ω whenever its three components are piecewise constant functions on the meshes $\mathcal{M}$, $\mathcal{N}$ and $\mathcal{FE}$, respectively, and take the form

$$u^{\mathcal{M}} = \sum_{K \in \mathcal{M}} u_K \chi_K, \quad u^{\mathcal{N}} = \sum_{A \in \mathcal{N}} u_A \chi_A, \quad u^{\mathcal{FE}} = \sum_{F \in \mathcal{F}} u_F \chi_F + \sum_{E \in \mathcal{E}} u_E \chi_E.$$

Let X denote the set of the degrees of freedom of the form

$$u^{\mathcal{T}} = \left((u_K)_{K \in \mathcal{M}}, (u_A)_{A \in \mathcal{N}}, (u_E)_{E \in \mathcal{E}}, (u_F)_{F \in \mathcal{F}} \right).$$

In order to take into account the Dirichlet boundary conditions, this set is supplemented by the boundary data

$$\delta u^{\mathcal{T}} = \left((u_K)_{x_K \in \partial \mathcal{M}}, (u_A)_{x_A \in \partial \mathcal{N}}, (u_E)_{x_E \in \partial \mathcal{FE}}, (u_F)_{x_F \in \partial \mathcal{FE}} \right),$$

which form the set ∂X. We will search the numerical approximation to the scalar solution field u in the product set $(u^{\mathcal{T}}, \delta u^{\mathcal{T}}) \in X \times \partial X$. Note that X, ∂X, and $X \times \partial X$ can be given the algebraic structure of a linear space after introducing (in the obvious way) the addition of two elements of the set and the multiplication of an element of the set by a real number.

The gradient of the discrete unknown $u^{\mathcal{T}}$, denoted by $\nabla^{\mathcal{T}} u^{\mathcal{T}}$, is a constant vector field on each diamond cell and is identified with a piecewise constant vector field on mesh $\mathcal{D}$. It depends on the boundary data $\delta u^{\mathcal{T}}$ and can be written as $\nabla^{\mathcal{T}}_{\delta u} u^{\mathcal{T}} = \sum_{D \in \mathcal{D}} \nabla^D_{\delta u} u^{\mathcal{T}} \chi_D$ where

$$\nabla^D_{\delta u} u^{\mathcal{T}} = \frac{1}{3|D|} \left((u_L - u_K) N_{KL} + (u_B - u_A) N_{AB} + (u_F - u_E) N_{EF} \right). \tag{1}$$

for any $D \in \mathcal{D}$ and with the vectors $N_{KL} = \frac{1}{2}(x_B - x_A) \times (x_F - x_E)$, $N_{AB} = \frac{1}{2}(x_F - x_E) \times (x_L - x_K)$ and $N_{EF} = \frac{1}{2}(x_L - x_K) \times (x_B - x_A)$. This procedure defines a gradient operator, denoted by $\nabla^{\mathcal{T}}_{\delta u}$, mapping the discrete space X onto the space of the discrete vector fields $\nabla^{\mathcal{T}} u^{\mathcal{T}}$, which we conveniently denote by $\mathbf{Q}$.

Using the gradient formula we define the flux through each interface of the control volumes of the three meshes $\mathcal{M}$, $\mathcal{N}$ and $\mathcal{FE}$. The three finite volume schemes are written by using a discrete divergence operator that maps each vector field in $\mathbf{Q}$ to a triple of scalar functions in X. Formally, we introduce the operator

$$\text{div}^{\mathcal{T}} : \xi = (\xi_D)_{D \in \mathcal{D}} \in \mathbf{Q} \mapsto (\text{div}^{\mathcal{M}} \xi, \text{div}^{\mathcal{N}} \xi, \text{div}^{\mathcal{FE}} \xi) \in X$$

where $\text{div}^{\mathcal{M}} \xi = (\text{div}_K \xi)_K$, $\text{div}^{\mathcal{N}} \xi = (\text{div}_A \xi)_A$ and $\text{div}^{\mathcal{FE}} \xi = \{(\text{div}_E \xi)_E, (\text{div}_F \xi)_F\}$ are given by

$$|K|\text{div}_K\xi = \sum_{D\in D_K} \xi_D \cdot N_{KL}, \quad |A|\text{div}_A\xi = \sum_{D\in D_A} \xi_D \cdot N_{AB}, \tag{2}$$

$$|E|\text{div}_E\xi = \sum_{D\in D_E} \xi_D \cdot N_{EF}, \quad |F|\text{div}_F\xi = \sum_{D\in D_F} \xi_D \cdot (-N_{EF}). \tag{3}$$

In the previous statements, the symbols D_K, D_A, D_E, D_F refer to the diamond cells which overlap the cells labeled by the corresponding subscripted indices K, A, E, and L.

Since each of the $\text{div}_C\xi$ approximates $\frac{1}{|C|}\int_C \text{div}\xi$ (for $C = K, A, E, F$), the right hand side of the discrete problem is given by the piecewise constant projection of the function f onto the space X, $\pi^{\mathcal{T}} f = \{(f_K)_{K\in\mathcal{M}}, (f_A)_{A\in\mathcal{N}}, (f_E, f_F)_{E\in\mathcal{E},F\in\mathcal{F}}\}$ with $f_C = \frac{1}{|C|}\int_C f(x)dx$ for any cell $C = K \in \mathcal{M}$ or $A \in \mathcal{N}$ or F or $E \in \mathcal{F}\mathcal{E}$.

Finally, the DDFV scheme reads as

$$-\text{div}^{\mathcal{T}}(\mathbf{K}_D \nabla^D_{\delta u} u^{\mathcal{T}}) = \pi^{\mathcal{T}} f \tag{4}$$

where $\mathbf{K}_D = \frac{1}{|D|}\int_D \mathbf{K}(x)dx$ is defined piecewise on the diamond cells. The scheme in (4) originates a symmetric and positive-definite linear system of equations (see [6] for a thourough discussion of the other properties). Assembling the matrix of the system amounts to gathering the local contributions of the discrete gradient associated to each diamond cell. These contributions are explicitly taken into account by the local Gram matrix

$$\mathbb{K}_D = \begin{pmatrix} \mathbf{K}_D N_{KL} \cdot N_{KL} & \mathbf{K}_D N_{KL} \cdot N_{AB} & \mathbf{K}_D N_{KL} \cdot N_{EF} \\ \mathbf{K}_D N_{AB} \cdot N_{KL} & \mathbf{K}_D N_{AB} \cdot N_{AB} & \mathbf{K}_D N_{AB} \cdot N_{EF} \\ \mathbf{K}_D N_{EF} \cdot N_{KL} & \mathbf{K}_D N_{EF} \cdot N_{AB} & \mathbf{K}_D N_{EF} \cdot N_{EF} \end{pmatrix}$$

The right hand side in (4) is split similarly in elementary contributions on the eight tetrahedra that compose the diamond cells D.

3 Treatment of discontinuous permeability tensors

The case of a discontinuous permeability tensor in the DDFV framework deserves a special treatment that we discuss in this subsection. Let us suppose that the permeability tensor is discontinuous across the interfaces of the control volumes of mesh $\mathcal{M}$. We decompose each diamond cell into two sub-diamonds D_K and D_L, i.e., $D = D_K \cup D_L$, where D_K is the union of the four tetrahedra with vertices x_D, x_K, the third vertex being x_A or x_B, and the fourth vertex being x_E or x_F.

Then, we introduce an additional degree of freedom at x_D, the center of the diamond cell, and we write a gradient formula that is exact for affine functions on the two sub-diamonds. We obtain the two following formulas

$$\nabla_K^{\mathcal{T}} u^{\mathcal{T}} = \frac{1}{3|D_K|} \left((u_D - u_K) N_{KL} + (u_B - u_A) N_{AB}^K + (u_F - u_E) N_{EF}^K \right)$$

$$\nabla_L^{\mathcal{T}} u^{\mathcal{T}} = \frac{1}{3|D_L|} \left((u_L - u_D) N_{KL} + (u_B - u_A) N_{AB}^L + (u_F - u_E) N_{EF}^L \right)$$

using the geometric vectors $N_{AB}^K = \frac{1}{2}(x_F - x_E) \times (x_D - x_K)$, $N_{AB}^L = \frac{1}{2}(x_F - x_E) \times (x_L - x_D)$, $N_{EF}^K = \frac{1}{2}(x_D - x_K) \times (x_B - x_A)$, $N_{EF}^L = \frac{1}{2}(x_L - x_D) \times (x_B - x_A)$, and introducing the two volume factors $|D_K| = \frac{1}{6} \det(x_B - x_A, x_F - x_E, x_D - x_K)$ and $|D_L| = \frac{1}{6} \det(x_B - x_A, x_F - x_E, x_L - x_D)$. Also, we remark that $|D| = |D_K| + |D_L|$, $N_{AB} = N_{AB}^K + N_{AB}^L$, $N_{EF} = N_{EF}^K + N_{EF}^L$ and it holds that $|D_K|N_{AB}^L - |D_L|N_{AB}^K = |D_K|N_{AB} - |D|N_{AB}^L$.

Let $\mathbf{K}_{D_K} = \frac{1}{|D_K|} \int_{D_K} \mathbf{K}(x) dx$ and $\mathbf{K}_{D_L} = \frac{1}{|D_L|} \int_{D_L} \mathbf{K}(x) dx$ be the constant approximation of the diffusion tensor on the two sub-diamonds D_K and D_L. We determine the additional unknown u_D in terms of the other local degrees of freedom u_K, u_L, u_A, u_B, u_E and u_F by imposing that

$$\mathbf{K}_{D_K} \nabla_K^{\mathcal{T}} u^{\mathcal{T}} \cdot N_{KL} = \mathbf{K}_{D_L} \nabla_L^{\mathcal{T}} u^{\mathcal{T}} \cdot N_{KL},$$

which is the flux conservation through the common face $D_K | D_L$. Moreover, let us introduce the following geometric factors that also depend on the permeability coefficients:

$$\beta_{KL} = +|D_L|\mathbf{K}_{D_K} N_{KL} \cdot N_{KL} + |D_K|\mathbf{K}_{D_L} N_{KL} \cdot N_{KL}$$

$$\beta_{AB} = -|D_L|\mathbf{K}_{D_K} N_{AB}^K \cdot N_{KL} + |D_K|\mathbf{K}_{D_L} N_{AB}^L \cdot N_{KL}$$

$$\beta_{EF} = -|D_L|\mathbf{K}_{D_K} N_{EF}^K \cdot N_{KL} + |D_K|\mathbf{K}_{D_L} N_{EF}^L \cdot N_{KL}$$

A straightforward calculation yields the formula for u_D

$$u_D = \frac{|D_L|\mathbf{K}_{D_K} N_{KL} \cdot N_{KL}}{\beta_{KL}} u_K + \frac{|D_K|\mathbf{K}_{D_L} N_{KL} \cdot N_{KL}}{\beta_{KL}} u_L + \frac{\beta_{AB}}{\beta_{KL}} (u_B - u_A) + \frac{\beta_{EF}}{\beta_{KL}} (u_F - u_E),$$

and the formulas for the numerical gradients:

$$3\nabla_K^{\mathcal{T}} u^{\mathcal{T}} = \frac{\mathbf{K}_{D_L} N_{KL} \cdot N_{KL}}{\beta_{KL}} N_{KL} (u_L - u_K) + \left(\frac{\beta_{AB}}{|D_K|\beta_{KL}} N_{KL} + \frac{1}{|D_K|} N_{AB}^K \right) (u_B - u_A)$$
$$+ \left(\frac{\beta_{EF}}{|D_K|\beta_{KL}} N_{KL} + \frac{1}{|D_K|} N_{EF}^K \right) (u_F - u_E),$$

$$3\nabla_L^{\mathcal{T}} u^{\mathcal{T}} = \frac{\mathbf{K}_{D_K} N_{KL} \cdot N_{KL}}{\beta_{KL}} N_{KL} (u_L - u_K) + \left(\frac{\beta_{AB}}{|D_L|\beta_{KL}} N_{KL} + \frac{1}{|D_L|} N_{AB}^L \right) (u_B - u_A)$$
$$+ \left(\frac{\beta_{EF}}{|D_L|\beta_{KL}} N_{KL} + \frac{1}{|D_L|} N_{EF}^L \right) (u_F - u_E).$$

Finally, we define the divergence operator for a discrete vector field which is piecewise constant on $D_K \cap D_L$ and may be discontinuous across $D_K|D_L$ as

$$|K|\mathrm{div}_K \xi^{\mathcal{D}} = \sum_{D|K} \xi_{D_K} \cdot N_{KL} = \sum_{D|K} \xi_{D_L} \cdot N_{KL} = \sum_{D|K} \left(\frac{|D_K|}{|D|} \xi_{D_K} + \frac{|D_L|}{|D|} \xi_{D_L} \right) \cdot N_{KL}, \quad (5)$$

$$|A|\mathrm{div}_A \xi^{\mathcal{D}} = \sum_{D|A} (\xi_{D_K} \cdot N_{AB}^K + \xi_{D_L} \cdot N_{AB}^L), \quad (6)$$

$$|E|\mathrm{div}_E \xi^{\mathcal{D}} = \sum_{D|E} (\xi_{D_K} \cdot N_{EF}^K + \xi_{D_L} \cdot N_{EF}^L), \quad (7)$$

$$|F|\mathrm{div}_F \xi^{\mathcal{D}} = \sum_{D|D_F} (\xi_{D_K} \cdot (-N_{EF}^K) - \xi_{D_L} \cdot (-N_{EF}^L)). \quad (8)$$

The DDFV method for the discontinuous case follows by using (5)-(8) with the approximate permeability tensors $\mathbf{K}_{D_K}$ and $\mathbf{K}_{D_L}$ instead of (2)-(3) in the scheme formulation (4). Let $\xi^{\mathcal{D}} = \mathbf{K}^{\mathcal{D}} \nabla^{\mathcal{T}} u^{\mathcal{T}}$ and evaluate the quantities:

$$\left(\frac{|D_K|}{|D|} \xi_{D_K} + \frac{|D_L|}{|D|} \xi_{D_L} \right) \cdot N_{KL} = \alpha_{KL \cdot KL}(u_L - u_K) + \alpha_{KL \cdot AB}(u_B - u_A) + \alpha_{KL \cdot EF}(u_F - u_E)$$

$$\xi_{D_K} \cdot N_{AB}^K + \xi_{D_L} \cdot N_{AB}^L = \alpha_{AB \cdot KL}(u_L - u_K) + \alpha_{AB \cdot AB}(u_B - u_A) + \alpha_{AB \cdot EF}(u_F - u_E)$$

$$\xi_{D_K} \cdot N_{EF}^K + \xi_{D_L} \cdot N_{EF}^L = \alpha_{EF \cdot KL}(u_L - u_K) + \alpha_{EF \cdot AB}(u_B - u_A) + \alpha_{EF \cdot EF}(u_F - u_E)$$

using the entries of the coefficient matrix

$$\mathbf{K}_D^{new} = \begin{pmatrix} \alpha_{KL \cdot KL} & \alpha_{KL \cdot AB} & \alpha_{KL \cdot EF} \\ \alpha_{AB \cdot KL} & \alpha_{AB \cdot AB} & \alpha_{AB \cdot EF} \\ \alpha_{EF \cdot KL} & \alpha_{EF \cdot AB} & \alpha_{EF \cdot EF} \end{pmatrix}.$$

Since $\mathbf{K}_D^{new}$ is a 3×3 symmetric elements we have only six independent entries, which after a straightforward calculations are given by:

$$\alpha_{KL \cdot KL} = \frac{1}{3} \frac{\mathbf{K}_{D_K} N_{KL} \cdot N_{KL} \, \mathbf{K}_{D_L} N_{KL} \cdot N_{KL}}{\beta_{KL}},$$

$$\alpha_{KL \cdot AB} = \frac{\mathbf{K}_{D_K} N_{KL} \cdot N_{AB}^K \, \mathbf{K}_{D_L} N_{KL} \cdot N_{KL} + \mathbf{K}_{D_L} N_{KL} \cdot N_{AB}^L \, \mathbf{K}_{D_K} N_{KL} \cdot N_{KL}}{\beta_{KL}},$$

$$\alpha_{KL \cdot EF} = \frac{1}{3} \frac{\mathbf{K}_{D_K} N_{KL} \cdot N_{EF}^K \, \mathbf{K}_{D_L} N_{KL} \cdot N_{KL} + \mathbf{K}_{D_L} N_{KL} \cdot N_{EF}^L \, \mathbf{K}_{D_K} N_{KL} \cdot N_{KL}}{\beta_{KL}},$$

$$\alpha_{AB \cdot AB} = \frac{1}{3} \left(-\frac{\beta_{AB}^2}{|D_K||D_L|\beta_{KL}} + \frac{1}{|D_K|} \mathbf{K}_{D_K} N_{AB}^K \cdot N_{AB}^K + \frac{1}{|D_L|} \mathbf{K}_{D_L} N_{AB}^L \cdot N_{AB}^L \right),$$

$$\alpha_{\text{AB·EF}} = \frac{1}{3}\left(-\frac{\beta_{AB}\beta_{EF}}{|D_K||D_L|\beta_{KL}} + \frac{1}{|D_K|}\mathbf{K}_{D_K} N^K_{\text{EF}} \cdot N^K_{\text{AB}} + \frac{1}{|D_L|}\mathbf{K}_{D_L} N^L_{\text{EF}} \cdot N^L_{\text{AB}}\right),$$

$$\alpha_{\text{EF·EF}} = \frac{1}{3}\left(-\frac{\beta^2_{EF}}{|D_K||D_L|\beta_{KL}} + \frac{1}{|D_K|}\mathbf{K}_{D_K} N^K_{\text{EF}} \cdot N^K_{\text{EF}} + \frac{1}{|D_L|}\mathbf{K}_{D_L} N^L_{\text{EF}} \cdot N^L_{\text{EF}}\right).$$

and the remaining coefficients are determined by symmetry, i.e., $\alpha_{\text{AB·KL}} = \alpha_{\text{KL·AB}}$, $\alpha_{\text{EF·KL}} = \alpha_{\text{KL·EF}}$, and $\alpha_{\text{EF·AB}} = \alpha_{\text{AB·EF}}$.

4 Conclusions

In this work, we discussed how a discontinuous permeability can be treated in the numerical framework offered by the DDFV method. Whenever the discontinuity is across an internal interfaces shared by two control volumes of the primal mesh, two different gradients are considered on the two subdiamonds centered at that interface. Introducing an auxiliary variable at the interface and imposing flux conservation makes it possible to derive a formula for both gradients that is exact for affine functions. Then, a DDFV method can be formulated using a discrete divergence operator to express the flux balance on the overlapping meshes for primal control volumes, vertex control volumes and face-edge control volumes. The numerical experiments in [7] show the effectiveness of the method.

Acknowledgements The work of the first two authors was supported by Groupement de Recherche MOMAS. The work of the third author was partially supported by the Italian MIUR through the program PRIN2008.

References

1. Andreianov, B., Bendahmane, M., Hubert, F.: On 3D DDFV discretization of gradient and divergence operators. Part II. (2011). HAL, http://hal.archives-ouvertes.fr/hal-00567342
2. Andreianov, B., Bendahmane, M., Hubert, F., Krell, S.: On 3D DDFV discretization of gradient and divergence operators. Part I. (2011) HAL, http://hal.archives-ouvertes.fr/hal-00355212.
3. Andreianov, B., Bendahmane, M., Karlsen, K.: A gradient reconstruction formula for finite-volume schemes and discrete duality. FVCA5, Wiley, (2008).
4. Andreianov, B., Hubert, F., Krell, S.: Benchmark 3D: a version of the DDFV scheme with cell/vertex unknowns on general meshes, this volume (2011).
5. Boyer, F., Hubert, F.: Benchmark on anisotropic problems, the DDFV discrete duality finite volumes and m-DDFV schemes. In: R. Eymard, J.M. Hérard (eds.) FVCA5, Wiley, (2008).
6. Coudière, Y., Hubert, F.: A 3D discrete duality finite volume method for nonlinear elliptic equation (2010) HAL, URL: http://hal.archives-ouvertes.fr/hal-00456837/fr.
7. Coudière, Y., Hubert, F., Manzini, G.: Benchmark 3D: CeVeFE-DDFV, a discrete duality scheme with cell/vertex/face+edge unknowns. (this volume) (2011).
8. Coudière, Y., Pierre, C., Rousseau, O., Turpault, R.: A 2D/3D discrete duality finite volume scheme. Application to ECG simulation. International Journal on Finite Volumes (2009) **6**(1).

9. Domelevo, K., Omnès, P.: A finite volume method for the laplace equation on almost arbitrary two-dimensional grids. M2AN, Math. Model. Numer. Anal. (2005) **39**(6), 1203–1249.
10. Herbin, R., Hubert, F.: Benchmark on discretization schemes for anisotropic diffusion problems on general grids. FVCA5, Wiley, (2008).
11. Hermeline, F.: Approximation of diffusion operators with discontinuous tensor coefficients on distorted meshes. Comp. Meth. Appl. Mech. Eng. (2003) **192**(16), 1939–1959.
12. Hermeline, F.: A finite volume method for approximating 3D diffusion operators on general meshes. Journal of computational Physics (2009) **228**(16), 5763–5786.
13. Pierre, C.: Modélisation et simulation de l'activité électrique du coeur dans le thorax, analyse numérique et méthodes de volumes finis. Ph.D. thesis, Université de Nantes (2005).

The paper is in final form and no similar paper has been or is being submitted elsewhere.

Multi-Water-Bag Model And Method Of Moments For The Vlasov Equation

Anaïs Crestetto and Philippe Helluy

Abstract The kinetic Vlasov-Poisson model is very expensive to solve numerically. It can be approximated by a multi-water-bag model in order to reduce the complexity. This model amounts to solve a set of Burgers equations, which can be done easily by finite volume methods. However, the solution is naturally multivalued (filamentation). The multivalued solution can be computed by a moment method. We present here several numerical experiments.

Keywords Vlasov-Poisson, water-bag approximation, Burgers equation, multivalued solution
MSC2010: 35Q83, 44A60, 65M08

1 Introduction

Kinetic equations are used in several domains, such as plasma physics or bubble flows in gases or liquids. The distribution function depends on space and time but also on an additional velocity variable. Computations are thus very expensive. It is of great interest to reduce the complexity of the resolution by using fluid methods.

We consider a plasma containing ions of positive charge and electrons of negative charge. Ions are much heavier than electrons so that we can neglect their displacement and assume that their density n_0 is constant. The electrons move following the system of Vlasov-Poisson in a periodic domain in **x**:

Anaïs Crestetto
IRMA, University of Strasbourg & INRIA Nancy-Grand Est, e-mail: crestetto@math.unistra.fr

Philippe Helluy
IRMA, University of Strasbourg, e-mail: helluy@math.unistra.fr

$$\partial_t f(\mathbf{x}, \mathbf{v}, t) + \mathbf{v} \cdot \nabla_\mathbf{x} f(\mathbf{x}, \mathbf{v}, t) + \frac{q}{m} \mathbf{E}(\mathbf{x}, t) \cdot \nabla_\mathbf{v} f(\mathbf{x}, \mathbf{v}, t) = 0, \qquad (1)$$

$$\operatorname{div} \mathbf{E}(\mathbf{x}, t) = \rho(\mathbf{x}, t) = \frac{q}{m} \left(\int f(\mathbf{x}, \mathbf{v}, t) \, d\mathbf{v} - n_0 \right), \qquad (2)$$

$$\int \mathbf{E}(\mathbf{x}, t) \, d\mathbf{x} = 0, \qquad (3)$$

$$f(\mathbf{x}, \mathbf{v}, 0) = f_0(\mathbf{x}, \mathbf{v}) \approx n_0. \qquad (4)$$

The unknowns are the distribution function of electrons f, and the electric field $\mathbf{E}$. The electric field depends on space and time $\mathbf{x}, t$, while the distribution function depends also on an additional velocity variable $\mathbf{v}$. The charge and the mass of one electron are noted $q < 0$ and $m > 0$ respectively. Without loss of generality, we can take $\frac{q}{m} = -1$ and $n_0 = 1$. In one dimension of space, this system becomes:

$$\partial_t f(x, v, t) + v \partial_x f(x, v, t) - E(x, t) \partial_v f(x, v, t) = 0, \qquad (5)$$

$$\partial_x E(x, t) = \rho(x, t) = 1 - \int f(x, v, t) \, dv, \qquad (6)$$

$$\int E(x, t) \, dx = 0, \qquad (7)$$

$$f(x, v, 0) = f_0(x, v) \approx 1. \qquad (8)$$

The distribution function f is initially a perturbation of the equilibrium n_0. After simple calculations, Equation (6) can also be written

$$\partial_t E(x, t) = \int v f(x, v, t) \, dv - \int v f(0, v, t) \, dv. \qquad (9)$$

In higher dimensions, Equation (6) would be replaced by a Poisson equation, assuming that $E = -\nabla \Phi$, where Φ is the electric potential.

The solution can be stable (for example in the case of Landau damping). But since there is no dissipation, the solution can become unstable and filamentation can appear.

In order to reduce the complexity of our model, we propose to approximate the kinetic equation by fluid models. We consider two possibilities:

- the multi-water-bag model,
- the method of moments.

Then we propose numerical approximations of these models by simple finite volume schemes. The numerical results will be compared to those obtained with a full resolution of the kinetic model by the popular Particle-In-Cell (PIC) method (described for example in [3]).

2 Multi-water-bag model

The multi-water-bag model, detailed in [2], generalizes the water-bag model (see for example [1]). It consists of replacing f by a piecewise constant approximation in the velocity variable. Each piece is called a "water-bag". It is possible to compute only the boundaries of the water-bags.

2.1 Presentation of the model

Let N be an integer, $v_j^+(x,t)$ and $v_j^-(x,t)$ velocities, A_j constants, $j = 1, \ldots, N$, such that we can write:

$$f(x,v,t) = \sum_{j=1}^{N} A_j \left(H\left(v_j^+(x,t) - v\right) - H\left(v_j^-(x,t) - v\right) \right), \quad (10)$$

where H is the Heaviside function: $H(u) = \begin{cases} 0 \text{ if } u < 0, \\ 1 \text{ if } u > 0. \end{cases}$

Injecting this expression in the Vlasov equation, and assuming that there is no two $v_j^\pm(x,t)$ equal, we obtain the following system:

$$\partial_t v_j^\pm + v_j^\pm \partial_x v_j^\pm + E = 0, \quad \forall j = 1, \ldots, N, \quad (11)$$

$$\partial_t E = -\sum_{j=1}^{N} A_j \left(\frac{v_j^{+^2}(x,0) - v_j^{-^2}(x,0)}{2} - \frac{v_j^{+^2}(x,t) - v_j^{-^2}(x,t)}{2} \right). \quad (12)$$

The velocities follow a Burgers equation with a source term. Instead of evolving the distribution function f, we evolve these velocities. The natural solution can become multivalued (filamentation). The weak entropy solution is only an approximation in this context, when shocks appear.

2.2 Numerical scheme

We discretize our domain: $x_i = x_0 + i\Delta x$, with Δx being the spacial step, and consider a time step Δt such that $t^n = n\Delta t$. We evolve each velocity independently by using, for example, the Godunov scheme:

$$v_{j,i}^{\pm,n+1} = v_{j,i}^{\pm,n} - \frac{\Delta t}{\Delta x} \left(\frac{\left(v_{j,i+\frac{1}{2}}^{\pm,\star}\right)^2}{2} - \frac{\left(v_{j,i-\frac{1}{2}}^{\pm,\star}\right)^2}{2} \right) - \Delta t E_i^n, \quad (13)$$

where $v_{j,i}^{\pm,n} \simeq v_j^{\pm}(x_i, t^n)$ and $E_i^n \simeq E(x_i, t^n)$. We compute the $v_j^{\pm,*}$ with an exact Riemann solver.

We compute the electric field with the following scheme:

$$E_i^{n+1} = E_i^n - \Delta t \sum_{j=1}^{N} A_j \left(\frac{\left(v_{j,i}^{+,0}\right)^2 - \left(v_{j,i}^{-,0}\right)^2}{2} - \frac{\left(v_{j,i}^{+,n}\right)^2 - \left(v_{j,i}^{-,n}\right)^2}{2} \right). \quad (14)$$

In higher dimension, this step should be replaced by the numerical resolution of a Poisson equation.

2.3 Remarks

Before the shock, the weak solution and the multivalued solution coincide. After the shock, the natural solution is multivalued, filaments or branches appear, and the weak solution becomes discontinuous. In Sect. 4, we compare numerically the two kinds of solutions. It is also possible to compute several branches by the method of moments [4].

3 Method of moments

The method of moments, presented in [5,6], consists in taking the first moments of the equation that we have to solve in order to reduce the number of variables. The system is closed by assuming that the distribution function is made of water-bags.

3.1 Presentation of the method

Definition 1. The moment M_k of order k of f is defined by:

$$M_k(x,t) = \int_{-\infty}^{+\infty} v^k f(x,v,t) \, dv. \quad (15)$$

Taking the $2N$ first moments of the Vlasov equation:

$$\int v^k \partial_t f(x,v,t) \, dv + \int v^{k+1} \partial_x f(x,v,t) \, dv - E(x,t) \int v^k \partial_v f(x,v,t) \, dv = 0, \quad (16)$$

we obtain the following system of moments, coupled with the equation for the electric field:

$$\partial_t M_0 + \partial_x M_1 = 0 \tag{17}$$

$$\partial_t M_k + \partial_x M_{k+1} + kEM_{k-1} = 0, \text{ for } k = 1, \ldots, 2N-1, \tag{18}$$

$$\partial_t E(x,t) = M_1(x,t) - M_1(0,t). \tag{19}$$

We now have a system of $2N+1$ equations, in which the velocity no longer appears. There are $2N+2$ unknowns: the moments M_k for $k = 0, \ldots, 2N$ and E. We have to close this system by finding an expression for M_{2N}.

3.2 Closure relation

We represent f by water-bags for closing the system:

$$f(x,v,t) = \sum_{j=1}^{N} A_j \left(H\left(v_j^+(x,t) - v\right) - H\left(v_j^-(x,t) - v\right) \right). \tag{20}$$

We obtain:

$$M_k(x,t) = \sum_{j=1}^{N} A_j \frac{v_j^{+^{k+1}}(x,t) - v_j^{-^{k+1}}(x,t)}{k+1}, \quad \forall k = 0, \ldots, 2N, \tag{21}$$

and thus have an expression of M_{2N}, assuming that we know the $v_j^\pm$.

3.3 Numerical scheme

At time t^n, we know the $2N$ first moments of f and the electric field. We solve the system:

$$M_k(x,t) = \sum_{j=1}^{N} A_j \frac{v_j^{+^{k+1}}(x,t) - v_j^{-^{k+1}}(x,t)}{k+1}, \quad \forall k = 0, \ldots, 2N-1, \tag{22}$$

to obtain $v_j^\pm$, at time t^n. In general, the system may have several solutions. Uniqueness can be recovered through an entropy argument [4].

To solve this system, we can use the Newton method. But we have no rigorous result of convergence.

When the problem is such that f can be written:

$$f(x,v,t) = \sum_{j=1}^{2N} (-1)^{j-1} H\left(a_j^+(x,t) - v\right), \quad (23)$$

where $-\infty < a_{2N} \leq \cdots \leq a_1 < +\infty$, we can use the algorithm described in [4, 7], which solves rigorously such a system. It is then possible to catch numerically the multivalued solutions.

For approximating the system of moments, we use a natural kinetic scheme. Formally, we write:

$$\partial_t f(x,v,t) + v^+ \partial_x f(x,v,t) + v^- \partial_x f(x,v,t) - E(x,t) \partial_v f(x,v,t) = 0, \quad (24)$$

where $v^+ = \max(0, v)$ and $v^- = \min(0, v)$. Denoting Δx as the space step, Δt as the time step, and f_i^n as the approximation of $f(x_i, v, t^n)$, we use the following upwind discretization for the Vlasov equation:

$$\frac{f_i^{n+1} - f_i^n}{\Delta t} + \frac{1}{\Delta x}\left(v^+\left(f_i^n - f_{i-1}^n\right) + v^-\left(f_{i+1}^n - f_i^n\right)\right) - E_i^n \partial_v f_i^n = 0. \quad (25)$$

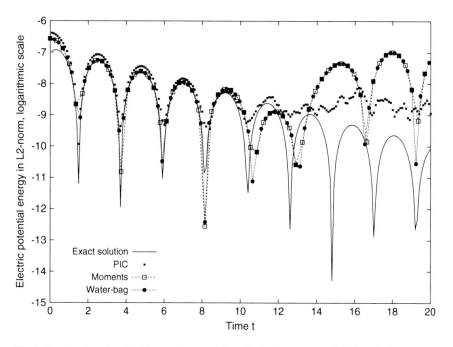

Fig. 1 Landau damping. Multi-water-bag model, method of moments and PIC method compared to the exact solution

We multiply it by v^k and integrate it in v:

$$\frac{1}{\Delta t} \int v^k \left(f_i^{n+1} - f_i^n\right) dv$$
$$+ \frac{1}{\Delta x} \int \left(v^{+k+1} \left(f_i^n - f_{i-1}^n\right) + v^{-k+1} \left(f_{i+1}^n - f_i^n\right)\right) dv \qquad (26)$$
$$- E_i^n \int v^k \partial_v f_i^n \, dv = 0.$$

We obtain a finite volume scheme:

$$\frac{M_{k,i}^{n+1} - M_{k,i}^n}{\Delta t} + \frac{1}{\Delta x} \left(F\left(M_{k,i}^n, M_{k,i+1}^n\right) - F\left(M_{k,i-1}^n, M_{k,i}^n\right)\right) - k E_i^n M_{k-1,i}^n = 0. \qquad (27)$$

The scheme for the electric field is:

$$\frac{E_i^{n+1} - E_i^n}{\Delta t} = M_{1,i}^n - M_{1,0}^n. \qquad (28)$$

4 Numerical results

We validate our models on classical test cases: Landau damping and two stream instability, and obtain good decrease rates for the electric potential energy, when N is big enough. An example for a Landau damping test case is given in Fig. 1, with $N = 5$.

We are now interested in solutions that can initially exactly be depicted by the multi-water-bag model, and that are unstable. We compare the three methods in Fig. 2, for $N = 1$:

- the method of moments with an approximation by water-bags,
- the multi-water-bag model with the scheme of Godunov,
- the Particle-In-Cell (PIC) method, considered as the reference.

Before the shock, the two fluid methods describe precisely the solution. After the shock, they describe the main part of the solution, but cannot catch the filaments. More test cases will be presented at the conference, with higher N.

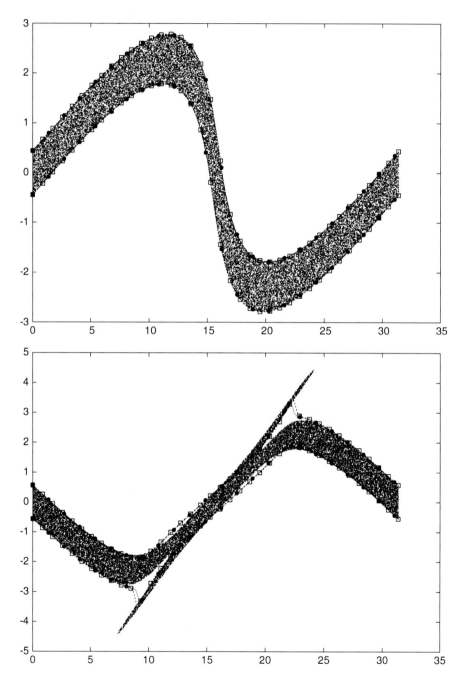

Fig. 2 Test case for $N = 1$ in the phase space at times $T = 2$ and $T = 5$. Multi-water-bag model (black circles), method of moments (empty squares) and PIC method (dots)

References

1. Bertrand, P., Feix, M. R.: Non linear electron plasma oscillation: the water bag model. Phys. Lett. **28A**, 68–69 (1968)
2. Besse, N., Berthelin, F., Brenier, Y., Bertrand, P.: The multi-water-bag equations for collisionless kinetic modeling. Kinetic and Related Models **2**, 39–80 (2009)
3. Birdsall, C. K. and Langdon, A. B.: Plasma Physics via Computer Simulation. Institute of Physics (1991)
4. Brenier, Y. and Corrias, L.: A kinetic formulation for multi-branch entropy solutions of scalar conservation laws. Annales de l'Institut Henri Poincaré. Analyse Non Linéaire **15**, 169–190 (1998)
5. Desjardins, O., Fox, R. O., Villedieu, P.: A quadrature-based moment method for dilute fluid-particle flows. Journal of Computational Physics **227**, 2514–2539 (2008)
6. Fox, R. O., Laurent, F., Massot, M.: Numerical simulation of spray coalescence in an Eulerian framework: direct quadrature method of moments and multi-fluid method. Journal of Computational Physics **227**, 3058–3088 (2008)
7. Gosse, L. and Runborg, O.: Resolution of the finite Markov moment problem. Comptes Rendus Mathématique. Académie des Sciences. Paris **12**, 775–780 (2005)

The paper is in final form and no similar paper has been or is being submitted elsewhere.

Comparison of Upwind and Centered Schemes for Low Mach Number Flows

Thu–Huyen DAO, Michael NDJINGA, and Frédéric MAGOULES

Abstract In this paper, fully implicit schemes are used for the numerical simulation of compressible flows at low Mach number. The compressible Navier–Stokes equations are discretized classically using the finite volume framework and a Roe type scheme for the convection flux. Though explicit Godunov type schemes are inaccurate for low Mach number flows on Cartesian meshes, we claim that their implicit counterpart can be more precise for that type of flow. Numerical evidence from the lid driven cavity benchmark shows that the centered implicit scheme can capture low Mach vortices, unlike the upwind scheme. We also propose a Scaling strategy based on the convection spectrum to reduce the computational cost and accelerate the convergence of both linear system and Newton scheme iterations.

Keywords Low Mach number, centered scheme, upwind scheme, compressible flows, scaling preconditioner
MSC2010: 35L65, 65F35, 76E19, 76M12

1 Introduction

Accurate numerical simulation of compressible flows at low Mach number is of great practical importance in the design and safety analysis of nuclear reactors and core thermal-hydraulic studies (see [6] and [7]). The numerical solutions of the

---

Thu–Huyen DAO
CEA–Saclay, DEN, DM2S, SFME, LGLS, F–91191 Gif–sur–Yvette, France and MAS, Ecole Centrale Paris, 92295 Châtenay–Malabry, France, e-mail: thu-huyen.dao@cea.fr

Michael NDJINGA
CEA–Saclay, DEN, DM2S, SFME, LGLS, F–91191 Gif–sur–Yvette, France, e-mail: michael.ndjinga@cea.fr

Frédéric MAGOULES
MAS, Ecole Centrale Paris, 92295 Châtenay–Malabry, France,
e-mail: frederic.magoules@hotmail.fr

corresponding two-phase flow models are based on Riemann approximate solvers which are robust and can efficiently capture shock wave solutions using an upwind strategy. However, when the flow is at low Mach number, especially on Cartesian meshes, these schemes are inaccurate, and corrections have to be made to capture the correct dynamics (see for example [8]). In [5], a detailed analysis of the behavior of Godunov type schemes applied to the compressible Euler system at low Mach number is proposed. The upwind part of the Roe scheme is identified as bringing excessive numerical diffusion and several corrections are proposed. These corrections aim at reducing the numerical diffusion of the explicit schemes, as well as maintaining their stability.

In this paper we present a more general strategy that could be easily applied to simulate various multiphase models at low Mach number. Such a strategy is inspired by single phase analysis and is first tested on the compressible Navier–Stokes equations in the present paper. In order to reduce the numerical diffusion, we consider a scheme that is order two in space such as the implicit centered scheme, already studied for example in [9].

In Sect. 2, we briefly recall the mathematical model and the considered numerical schemes. In Sect. 4 we give numerical evidence that the centered implicit scheme is much less diffusive than the upwind scheme (whether explicit or implicit) and can capture low Mach vortices. In order to reduce the computational cost involved by the resolution of many linear systems, Sect. 3 presents preconditioning strategy based on the scaling of the linear system matrix. This strategy is based on the underlying hyperbolic operator and could be applied to other set of equations.

2 Mathematical model and Numerical method

2.1 Mathematical model

The model consists of the following three balance laws for the mass, the momentum and the energy:

$$\begin{cases} \frac{\partial \rho}{\partial t} + \nabla . \mathbf{q} = 0 \\ \frac{\partial \mathbf{q}}{\partial t} + \nabla . \left(\mathbf{q} \otimes \frac{\mathbf{q}}{\rho} + p \mathbb{I}_d \right) - \nu \Delta(\frac{\mathbf{q}}{\rho}) = 0 \\ \frac{\partial (\rho E)}{\partial t} + \nabla . \left[(\rho E + p) \frac{\mathbf{q}}{\rho} \right] - \lambda \Delta T = 0 \end{cases} \quad (1)$$

where ρ is the density, $\mathbf{v}$ the velocity, $\mathbf{q} = \rho \mathbf{v}$ the momentum, p the pressure, ρe the internal energy, $\rho E = \rho e + \frac{\|\mathbf{q}\|^2}{2\rho}$ the total energy, T the absolute temperature, ν the viscosity and λ the thermal conductivity. We close the system (1) by the ideal gas law $p = (\gamma - 1)\rho e$. For the sake of simplicity, we consider constant viscosity and conductivity, and neglect the contribution of viscous forces in the energy equation. By denoting $U = (\rho, \mathbf{q}, \rho E)^t$ the vector of conserved variables, the Navier–Stokes system (1) can be written as a nonlinear system of conservation laws:

$$\frac{\partial U}{\partial t} + \nabla \cdot \left(\mathscr{F}^{conv}(U)\right) + \nabla \cdot \left(\mathscr{F}^{diff}(U)\right) = 0, \quad (2)$$

where $\mathscr{F}^{conv}(U) = \begin{pmatrix} \mathbf{q} \\ \mathbf{q} \otimes \frac{\mathbf{q}}{\rho} + p\mathbb{I}_d \\ (\rho E + p)\frac{\mathbf{q}}{\rho} \end{pmatrix}$, $\mathscr{F}^{diff}(U) = \begin{pmatrix} 0 \\ -\nu \nabla(\frac{\mathbf{q}}{\rho}) \\ -\lambda \nabla T \end{pmatrix}$.

2.2 Numerical method

The conservation form (2) enables to define the concept of weak solutions, which can be discontinuous ones. Discontinuous solutions such as shock waves are of great importance in transient calculations. In order to correctly capture shock waves, one needs a robust, low diffusive conservative scheme. The finite volume framework is the best appropriate setup to build such schemes as it enables to write discrete equations that express the conservation laws at each cell (see for example [1]).

We decompose the computational domain into N disjoint cells C_i with volume v_i. Two neighboring cells C_i and C_j have a common boundary ∂C_{ij} with area s_{ij}. We denote $N(i)$ the set of neighbors of a given cell C_i and $\mathbf{n}_{ij}$ the exterior unit normal vector of ∂C_{ij}. Integrating the system (2) over C_i and setting $U_i(t) = \frac{1}{v_i} \int_{C_i} U(x,t)dx$ and $U_i^n = U_i(n\Delta t)$, the discretized equations can be written:

$$\frac{U_i^{n+1} - U_i^n}{\Delta t} + \sum_{j \in N(i)} \frac{s_{ij}}{v_i} \left(\vec{\Phi}_{ij}^{conv} + \vec{\Phi}_{ij}^{diff}\right) = 0. \quad (3)$$

with: $\vec{\Phi}_{ij}^{conv} = \frac{1}{s_{ij}} \int_{\partial C_{ij}} \mathscr{F}^{conv}(U).\mathbf{n}_{ij} ds$, $\vec{\Phi}_{ij}^{diff} = \frac{1}{s_{ij}} \int_{\partial C_{ij}} \mathscr{F}^{diff}(U).\mathbf{n}_{ij} ds$.

To approximate the convection numerical flux $\vec{\Phi}_{ij}^{conv}$ we solve an approximate Riemann problem at the interface ∂C_{ij}. Using the Roe local linearisation of the fluxes [2], we obtain the following formula:

$$\vec{\Phi}_{ij}^{conv} = \frac{\mathscr{F}^{conv}(U_i) + \mathscr{F}^{conv}(U_j)}{2}.\mathbf{n}_{ij} - \mathscr{D}(U_i, U_j)\frac{U_j - U_i}{2} \quad (4)$$

$$= \mathscr{F}^{conv}(U_i)\mathbf{n}_{ij} + A^-(U_i, U_j)(U_j - U_i), \quad (5)$$

where $\mathscr{D}$ is an upwinding matrix, $A(U_i, U_j)$ the Roe matrix and $A^- = \frac{A-\mathscr{D}}{2}$. The choice $\mathscr{D} = 0$ gives the centered scheme, whereas $\mathscr{D} = |A|$ gives the upwind scheme. For the Euler equations, we can build $A(U_i, U_j)$ explicitly using the Roe averaged state (see [1]).

The diffusion numerical flux $\vec{\Phi}_{ij}^{diff}$ is approximated on structured meshes using the formula:

$$\vec{\Phi}_{ij}^{diff} = D(\frac{U_i + U_j}{2})(U_j - U_i) \tag{6}$$

with the matrix $D(U) = \begin{pmatrix} 0 & 0 & 0 & 0 \\ \frac{v\mathbf{q}}{\rho^2} & \frac{-v}{\rho}\mathbb{I}_d & 0 \\ \frac{\lambda}{c_v}(\frac{c_v T}{\rho} - \frac{||\mathbf{q}||^2}{2\rho^3}) & \frac{\mathbf{q}^t \lambda}{\rho^2 c_v} & -\frac{\lambda}{c_v \rho} \end{pmatrix}$, where c_v is the heat capacity at constant volume.

2.3 Newton scheme

Finally, since $\sum_{j \in N(i)} \mathscr{F}^{conv}(U_i).\mathbf{n}_{ij} = 0$, using (5) and (6) the equation (3) of the numerical scheme becomes:

$$\frac{U_i^{n+1} - U_i^n}{\Delta t} + \sum_{j \in N(i)} \frac{S_{ij}}{v_i}\{(A^- + D)(U_i^{n+1}, U_j^{n+1})\}(U_j^{n+1} - U_i^{n+1}) = 0. \tag{7}$$

The system (7) is nonlinear. We use the following Newton iterative method to obtain the required solutions:

$$\frac{\delta U_i^{k+1}}{\Delta t} + \sum_{j \in N(i)} \frac{S_{ij}}{v_i} \left[(A^- + D)(U_i^k, U_j^k)\right] \left(\delta U_j^{k+1} - \delta U_i^{k+1}\right)$$

$$= -\frac{U_i^k - U_i^n}{\Delta t} - \sum_{j \in N(i)} \frac{S_{ij}}{v_i} \left[(A^- + D)(U_i^k, U_j^k)\right] (U_j^k - U_i^k),$$

where $\delta U_i^{k+1} = U_i^{k+1} - U_i^k$ is the variation of the k-th iterate that approximate the solution at time $n + 1$. Defining the unknown vector $\mathscr{U} = (U_1, \ldots, U_N)^t$, each Newton iteration for the computation of $\mathscr{U}$ at time step $n + 1$ requires the numerical solution of the following linear system:

$$\mathscr{A}(\mathscr{U}^k)\delta \mathscr{U}^{k+1} = b(\mathscr{U}^n, \mathscr{U}^k). \tag{8}$$

2.4 The low Mach problem

When the flow is smooth and the Mach number $\frac{||\mathbf{v}||}{c}$ (where $c = \sqrt{\frac{\gamma p}{\rho}}$ is the sound speed) is small, the solutions of the system (1) should behave as those of an incompressible Navier–Stokes model (see [10]). However, in general, Godunov type schemes do not preserve the asymptotic behavior and generate spurious solutions when applied to low Mach number flows (see [5]). The analysis presented in [5] suggests that the inaccuracies originate from the anisotropy of the upwind matrix $\mathscr{D}$, and various " Low Mach Schemes " are proposed in the explicit context. In order to avoid the stability issue, we propose to use implicit schemes and to consider the simpler case $\mathscr{D} = 0$ (no upwinding). The resulting centered scheme can be applied to any system of conservation law, and we present in Sect. 4 our first numerical experiments.

3 Description of the Scaling strategy

The larger the time step, the worse the condition number of the matrix $\mathscr{A}$ in (8). As a consequence, it is important to apply a preconditioner before solving the linear system. The most popular choice is the Incomplete LU factorisation (later named ILU, see [3] for more details). The error made by the approximate factorisation using an ILU preconditioner depends on the size of the off diagonal coefficients of the matrix. For a better performance of the preconditioner, it is desirable that off diagonal entries of the matrix have small magnitudes.

As we are interested in convection dominated flows, the main contributions to the matrix $\mathscr{A}$ come from the convective part discretisation of the equations through the matrix A^-. Unfortunately, the coefficients of the Roe matrix have very different magnitudes for low Mach number flows. Consequently, A^- and hence $\mathscr{A}$ have coefficients with very different magnitudes.

We are now going to detail a procedure that scales the matrix coefficients so that they have the same magnitude. The matrix A^- can be expressed using a complete eigenstructure decomposition of the Roe matrix: $A = \sum_k \lambda_k L^k \otimes R^k$. The three eigenvalues of the Roe matrix are $v_n + c$, v_n (multiplicity d), and $v_n - c$. As we are interested in flows at low Mach number, we can assume $\mathbf{v} \approx 0$ and in that case the eigenvalues of A become $\lambda^- = -c$, $\lambda_v = 0$, and $\lambda^+ = +c$. The right and left eigenvectors $R^\pm$ and $L^\pm$ associated to the sound waves are:

$$R^\pm = (1, \pm c\mathbf{n}, \frac{c^2}{\gamma - 1})^t, \qquad L^\pm = \frac{1}{2}(0, \pm\frac{1}{c}\mathbf{n}, \frac{\gamma - 1}{c^2})^t. \qquad (9)$$

We have:

$$A^- = -cL^- \otimes R^- \qquad \text{for the upwind scheme,}$$

$$A^- = \frac{1}{2}(cL^+ \otimes R^+ - cL^- \otimes R^-) \text{ for the centered scheme.}$$

One sees from (9) that the disequilibrium in A^- coefficients comes from the difference in the magnitude of the components of the left and right eigenvectors of A.

If we multiply A^- to the left (respectively to the right) by a diagonal matrix with the coefficients $d_{sca} = diag(1, c\mathbf{n}, \frac{c^2}{\gamma-1})$, respectively $d_{sca}^{-1} = diag(1, \frac{1}{c}\mathbf{n}, \frac{\gamma-1}{c^2})$ ($\mathbf{n}$ is the unit normal vector), we obtain vectors and matrices with better balanced coefficients:

$$d_{sca}^{-1} R^{\pm} = (1, \pm\mathbf{n}, 1)^t, \qquad d_{sca} L^{\pm} = (0, \pm\mathbf{n}, 1)^t,$$

$$L^{\pm} \otimes R^{\pm} = \frac{1}{2}\begin{pmatrix} 0 & 0 & 0 \\ \pm\frac{1}{c}\mathbf{n} & \mathbf{n}\otimes\mathbf{n} & \pm\frac{c}{\gamma-1}\mathbf{n} \\ \frac{\gamma-1}{c^2} & \pm\frac{\gamma-1}{c}\mathbf{n}^t & 1 \end{pmatrix}, \quad d_{sca} L^{\pm} \otimes R^{\pm} d_{sca}^{-1} = \frac{1}{2}\begin{pmatrix} 0 & 0 & 0 \\ \pm\mathbf{n} & \mathbf{n}\otimes\mathbf{n} & \pm\mathbf{n} \\ 1 & \pm\mathbf{n}^t & 1 \end{pmatrix}$$

Any mesh can be associated with two diagonal matrices D_{sca} and D_{sca}^{-1} having the size of the mesh and containing the successive coefficients of the local matrices d_{sca} and d_{sca}^{-1}. Instead of solving system (8), one can rather solve:

$$\tilde{\mathscr{A}}\mathscr{V} = \tilde{b}, \qquad (10)$$

where $\tilde{\mathscr{A}} = D_{sca}\mathscr{A}D_{sca}^{-1}$, $\mathscr{V} = D_{sca}\mathscr{U}$ and $\tilde{b} = D_{sca}b$. System (10) can be resolved more easily using an ILU preconditioner. Once the solution $\mathscr{V}$ is obtained we compute $D_{sca}^{-1}\mathscr{V}$ to obtain the original unknown vector $\mathscr{U}$.

4 Numerical results

4.1 Upwind scheme vs Centered one

Figures 1 and 2 present the streamlines of the steady state result obtained using either the upwind or the centered schemes to discretize the convective part of the Navier–Stokes equations with the fully implicit scheme presented in Sect. 2.2. Our test case is a lid driven cavity flow at Reynolds number 400 solved on a cartesian 50×50 cell mesh. This case is described in [4], with the correct solution given by an incompressible solver. The lid speed is $1\,m/s$, the maximum Mach number of the flow is 0.008. The Roe approximate Riemann solver [2] employed for the convection fluxes is known to have problem solving such low Mach number flows when the scheme is explicit, especially on multidimensional cartesian meshes (see [5]). It can be seen on Fig. 1 that the upwind scheme does not capture the correct streamlines. However, on Fig. 2, it can be seen that the implicit centered scheme is much less diffusive and captures the correct solution with its expected three vortices.

4.2 Assessment of the Scaling strategy

We now study the performance of our numerical methods on the same lid driven cavity test case presented in Sect. 4.1. In this section, we vary the time step

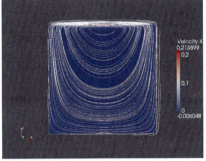

Fig. 1 Steady state, upwind scheme

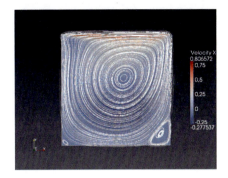

Fig. 2 Steady state, centered scheme

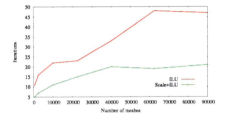

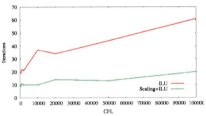

Fig. 3 Number of GMRES iterations for the upwind scheme, CFL 1000

Fig. 4 Number of GMRES iterations for the upwind scheme, mesh 100 × 100

(CFL number) and the mesh size. We also compare the direct solver with the iterative one and the effect of different preconditioners on the resolution of the linear systems.

Considering first the upwind scheme, we remark that the ILU preconditioner with no level of fill-in performs well. Figs. 3 and 4, show the average number of GMRES iterations at each Newton iteration. We observe that the use of our Scaling strategy presented in Sect. 3 reduces more than twice the iteration number.

When we use the centered scheme, the system matrix has a poor diagonal, and ILU preconditioner with no fill-in is not efficient in preconditioning the linear system. One needs to use an incomplete factorisation with two levels of fill-in to solve linear system up to the CFL 100, and the Scaling strategy enables to save a considerable number of iterations (Fig. 5). Beyond that value, only a direct solver is able to solve the system. However, one can remark that the Scaling strategy enables a reduction of the number of Newton iterations using a direct solver (Fig. 6). We also stress that the steady state solution obtained with very large CFL numbers is still accurate and displays the expected vortices.

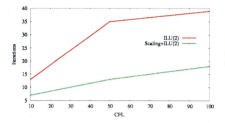 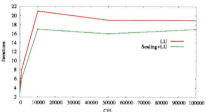

Fig. 5 Number of GMRES iterations for the centered scheme, mesh 50 × 50

Fig. 6 Number of Newton iterations for the centered scheme, mesh 50 × 50s

5 Conclusion and Perspectives

In this paper, two simple and general fully implicit schemes have been presented for the simulation of compressible Navier–Stokes equations at low Mach number. We have shown that the centered scheme is able to capture low Mach vortices unlike the upwind scheme. However, ILU preconditioning performs better with the upwind scheme than with the centered scheme. Thanks to the particular features of Roe matrix for compressible Navier–Stokes equations, we have proposed a preconditioning strategy Scaling+ILU that considerably reduces the computation time. The centered scheme and the scaling strategy can be applied to other sets of equations than Navier–Stokes. Study of these techniques applied to two-phase flow models will follow.

References

1. E. Godlewski, P.A. Raviart, *Numerical Approximation of Hyperbolic Systems of Conservation Laws* Springer Verlag, 1996.
2. P.L Roe, Approximate Riemann solvers, parameter vectors and difference schemes *J. Comput. Phys.*, 43 (1981), 537-372.
3. Michele Benzi, Preconditioning Techniques for Large Linear Systems: A Survey *J. Comput. Phys.*, 182 (2002), 418-477.
4. U. Ghia, K.N. Ghia, C.T. Shin, High-Re Solutions for Incompressible Flow Using the Navier-Stokes Equations and a Multigrid Method *J. Comput. Phys.*, 48 (1982), p 387-411.
5. S. Dellacherie, Analysis of Godunov type schemes applied to the compressible Euler system at low Mach number *J. Comput. Phys.*, 229(2010), 701-727.
6. I. Toumi, A. Bergeron, D. Gallo, and D. Caruge, FLICA-4: a three-dimensional two-phase flow computer code with advanced numerical methods for nuclear applications *Nucl. Eng. Design*, 200 (2000), p 139-155.
7. P. Fillion, A. Chanoine, S. Dellacherie, A. Kumbaro, FLICA-OVAP: a New Platform for Core Thermal-hydraulic Studies *NURETH-13* Japon, Sep 27-Oct 2, (2009).
8. H. Guillard, C. Viozat, On the behavior of upwind schemes in the low Mach number limit *Comput. Fluids* 28 (1999).

9. J.-A. Désidéri, P. W. Hemker, Convergence Analysis of the Defect-Correction Iteration for Hyperbolic Problems *SIAM J. Sci. Comput.*, Vol. 16 (1995), No.1, pp. 88-118.
10. S. Schochet, Fast Singular Limits of Hyperbolic PDEs *Journal of Differential Equations* 114 (1994).

The paper is in final form and no similar paper has been or is being submitted elsewhere.

On the Godunov Scheme Applied to the Variable Cross-Section Linear Wave Equation

Stéphane Dellacherie and Pascal Omnes

Abstract We investigate the accuracy of the Godunov scheme applied to the variable cross-section acoustic equations. Contrarily to the constant cross-section case, the accuracy issue of this scheme in the low Mach number regime appears even in the one-dimensional case; on the other hand, we show that it is possible to construct another Godunov type scheme which is accurate in the low Mach number regime.

Keywords Variable cross-section, wave equation, low Mach, Godunov scheme
MSC2010: 35L05, 35L45, 65M08

1 Introduction

It is well-known that Godunov type schemes suffer from an accuracy problem at low Mach number. The analysis of this scheme applied to the linear wave equation has shown that this problem already occurs for such a simple submodel, except in the one-dimensional case [1]. However, it has also been proved that in higher dimensions, simplicial meshes perform much better than rectangular meshes [2]. These results are obtained by the analysis of the invariant space of the discrete wave operator: when this invariant space is rich enough to approach well the invariant space of the continuous wave operator (that is to say the incompressible fields), then the Godunov scheme is accurate at low Mach number. With the same type of analysis, we show in the present work that accuracy problems may already occur in the one-dimensional case for the variable cross-section linear wave equation, if one

---

Stéphane Dellacherie
CEA, DEN, DM2S, SFME F-91191 Gif-sur-Yvette, France, e-mail: stephane.dellacherie@cea.fr

Pascal Omnes
CEA, DEN, DM2S, SFME F-91191 Gif-sur-Yvette, France and LAGA, Université Paris 13, 99 Av. J.-B. Clément, F-93430 Villetaneuse, France, e-mail: pascal.omnes@cea.fr

is not careful about the expression of the diffusion terms inherent to the Godunov scheme. This equation may be seen as a simple model for diphasic flows in which the volumic fraction plays the role of the variable cross-section.

2 The variable cross-section wave equation

For regular solutions, the dimensionless barotropic Euler system with variable cross-section may be written as

$$\partial_t(A\rho) + \nabla \cdot (A\rho u) = 0 \quad \text{and} \quad \rho(\partial_t u + u \cdot \nabla u) + \frac{\nabla p}{M^2} = 0, \quad (1)$$

where the Mach number M is supposed to be small and where $p = p(\rho)$ with $p'(\rho) > 0$. Denoting by a_* a reference sound velocity, and setting

$$\rho(t,x) := \rho_* \left[1 + \frac{M}{Aa_*} s(t,x) \right], \quad (2)$$

system (1) may be written, after some simplifications

$$\partial_t q + \mathcal{H}(q) + \frac{\mathcal{L}_{A,M}}{M}(q) = 0 \quad (3)$$

with

$$\begin{cases} q = (s, u)^T, & \text{(a)} \\ \mathcal{H}(q) = \left(\nabla \cdot (us), (u \cdot \nabla)u \right)^T, & \text{(b)} \\ \mathcal{L}_{A,M}(q) = \left(a_* \nabla \cdot (Au), \frac{p'[\rho_*(1 + \frac{M}{Aa_*}s)]}{a_* + \frac{M}{A}s} \nabla \left(\frac{s}{A} \right) \right)^T. & \text{(c)} \end{cases} \quad (4)$$

2.1 The linear wave equation with variable cross-section

When A is bounded by below and by above independently of M, we formally have that $\frac{M}{Aa_*} s(t,x) \ll 1$ in (2) and $\mathcal{O}(\|\mathcal{L}_{A,M}(q)\|) = 1$ in (3) when $\mathcal{O}(\|q\|) = 1$. In that case, (3) contains a transport contribution whose characteristic time scale is a $\mathcal{O}(1)$ and a non-linear acoustic contribution whose characteristic time scale is a $\mathcal{O}(M)$, like in the usual barotropic low Mach number Euler system. In that case, at least in a first approach, we may drop the transport contribution and study the linearized cross-section acoustic equation which reads

$$\partial_t q + \frac{L_A}{M} q = 0 \quad \text{with} \quad L_A q = a_* \left(\nabla \cdot (Au), \nabla \left(\frac{s}{A} \right) \right)^T. \tag{5}$$

3 Basic properties of the variable cross-section linear wave equation

3.1 General properties

In this section, we are interested in basic properties of (5) solved on a periodic torus $\mathbb{T}^{d \in \{1,2,3\}}$. For this, we define the energy space

$$(L_A^2(\mathbb{T}^d))^{1+d} := \left\{ q := (s, u)^T \text{ such that } \int_{\mathbb{T}^d} s^2 \frac{dx}{A(x)} + \int_{\mathbb{T}^d} |u|^2 A(x) dx < +\infty \right\}$$

endowed with the scalar product

$$\langle q_1, q_2 \rangle_A = \int_{\mathbb{T}^d} s_1 s_2 \frac{dx}{A(x)} + \int_{\mathbb{T}^d} u_1 \cdot u_2 \, A(x) dx. \tag{6}$$

On the other hand, we set

$$\begin{cases} \mathcal{E}_A = \left\{ q := (s, u)^T \in (L_A^2(\mathbb{T}^d))^{1+d} \text{ such that } s = aA, \, a \in \mathbb{R} \text{ and } \nabla \cdot (Au) = 0 \right\}, \\ \mathcal{E}^\perp = \left\{ q := (s, u)^T \in (L_A^2(\mathbb{T}^d))^{1+d} \right. \\ \qquad \qquad \text{such that } \int_{\mathbb{T}^d} s \, dx = 0 \text{ and } \exists \phi \in H^1(\mathbb{T}^d), u = \nabla \phi \right\}. \end{cases}$$

We remark that $\mathcal{E}_A \perp \mathcal{E}^\perp$ for the scalar product (6). We shall admit the following extension of the Hodge decomposition $(L_A^2(\mathbb{T}^d))^{1+d} = \mathcal{E}_A \oplus \mathcal{E}^\perp$. Moreover, we have

$$\mathcal{E}_A = Ker L_A. \tag{7}$$

Finally, for all $q \in (L_A^2(\mathbb{T}^d))^{1+d}$, we define the energy $E_A := \langle q, q \rangle_A$. The following lemma is an easy extension of the energy conservation property to the variable cross-section case:

Lemma 1. *Let $q(t, x)$ be the solution of (5) on $\mathbb{T}^{d \in \{1,2,3\}}$. Then:*

$$E_A(t \geq 0) = E_A(t = 0).$$

We also have the following result:

Lemma 2. *Let $q(t, x)$ be the solution of (5) on $\mathbb{T}^{d \in \{1,2,3\}}$ with initial condition q^0. Then:*
1) $\forall q^0 \in \mathcal{E}_A : q(t \geq 0) \in \mathcal{E}_A$;
2) $\forall q^0 \in \mathcal{E}^\perp : q(t \geq 0) \in \mathcal{E}^\perp$.

Proof of Lemma 2: The first point is a direct consequence of (7). The second point is inferred from the first item, from Lemma 1, and from the following Lemma, a proof of which may be found in the appendix A of [1].

Lemma 3. *Let $\mathcal{A}$ be a linear isometry in a Hilbert space $\mathbb{H}$ and let $\mathcal{E}$ be a linear subspace of $\mathbb{H}$. Then:* $\mathcal{A}\mathcal{E} = \mathcal{E} \implies \mathcal{A}\mathcal{E}^\perp \subset \mathcal{E}^\perp$.

3.2 The one-dimensional case

In the particular case of the one-dimensional geometry, equation (5) is now set in $\mathbb{T}^{d=1}$ and writes

$$\partial_t q + \frac{L_A}{M} q = 0 \tag{8}$$

with

$$L_A q = a_* \left(\partial_x (Au) , \partial_x \left(\frac{s}{A} \right) \right)^T. \tag{9}$$

The subspaces $\mathcal{E}_A$ and $\mathcal{E}^\perp$ are now characterized by

$$\begin{cases} \mathcal{E}_A = \left\{ q := (s, u)^T \in (L_A^2(\mathbb{T}^1))^2 \text{ such that } s = aA \text{ and } u = \frac{b}{A}, (a,b) \in \mathbb{R}^2 \right\}, \\ \mathcal{E}^\perp = \left\{ q := (s, u)^T \in (L_A^2(\mathbb{T}^1))^2 \text{ such that } \int_{\mathbb{T}^d} s \, dx = \int_{\mathbb{T}^d} u \, dx = 0 \right\}. \end{cases}$$

In the one-dimensional case, we remark that, as soon as $A'(x) \neq 0$, the variables s and u do not play the same role, while when $A = 1$, they do play symmetrical roles.

4 Numerical approximation in the one-dimensional geometry

We now consider the numerical approximation of (8)–(9) on a mesh with N cells $[x_{i-1/2}, x_{i+1/2}]$ of constant size Δx. We denote by x_i the midpoints of the cells and by $u_i(t)$ and $s_i(t)$ the numerical approximation of u and s in the cell $[x_{i-1/2}, x_{i+1/2}]$.

4.1 A first numerical scheme

Integrating (8) over the cell $[x_{i-1/2}, x_{i+1/2}]$, we obtain

$$\begin{cases} \dfrac{d}{dt}s_i + \dfrac{a_*}{M} \cdot \dfrac{A_{i+1/2}u_{i+1/2}(t) - A_{i-1/2}u_{i-1/2}(t)}{\Delta x} = 0, \\[2mm] \dfrac{d}{dt}u_i + \dfrac{a_*}{M} \cdot \dfrac{\dfrac{s_{i+1/2}(t)}{A_{i+1/2}} - \dfrac{s_{i-1/2}(t)}{A_{i-1/2}}}{\Delta x} = 0, \end{cases} \quad (10)$$

where $A_{i+1/2} := A(x_{i+1/2})$ and where the interface values $(s_{i+1/2}(t), u_{i+1/2}(t))$ are determined by the solution of a Riemann problem (R.P.) based on the equation

$$\partial_t q + \dfrac{a_*}{M}\left(A_{i+1/2}\partial_x u, \dfrac{1}{A_{i+1/2}}\partial_x s\right)^T = 0,$$

which amounts to locally neglect the variations of A in (8). The left and right initial states of the R. P. being $(s_i(t), u_i(t))$ and $(s_{i+1}(t), u_{i+1}(t))$ respectively, its solution is

$$\begin{cases} s_{i+1/2} = \tfrac{1}{2}(s_i + s_{i+1}) + \dfrac{A_{i+1/2}}{2}(u_i - u_{i+1}), \\[2mm] u_{i+1/2} = \dfrac{1}{2A_{i+1/2}}(s_i - s_{i+1}) + \tfrac{1}{2}(u_i + u_{i+1}). \end{cases} \quad (11)$$

Plugging (11) into (10), we obtain the following scheme

$$\begin{cases} \dfrac{d}{dt}s_i + \dfrac{a_*}{M} \cdot \dfrac{A_{i+1/2}(u_i + u_{i+1}) - A_{i-1/2}(u_{i-1} + u_i)}{2\Delta x} = \dfrac{a_*}{2M\Delta x}(s_{i+1} - 2s_i + s_{i-1}), \\[2mm] \dfrac{d}{dt}u_i + \dfrac{a_*}{M} \cdot \dfrac{\dfrac{(s_i + s_{i+1})}{A_{i+1/2}} - \dfrac{(s_{i-1} + s_i)}{A_{i-1/2}}}{2\Delta x} = \dfrac{a_*}{2M\Delta x}(u_{i+1} - 2u_i + u_{i-1}), \end{cases} \quad (12)$$

whose first-order modified equation is given by

$$\partial_t q + \dfrac{L_A}{M} q = \left(\nu_s \partial_{xx}^2 s, \nu_u \partial_{xx}^2 u\right)^T \quad (13)$$

with $(\nu_s, \nu_u) = \dfrac{a_* \Delta x}{2M}(1, 1)$. This shows that for all non trivial $(\nu_s, \nu_u) \in \mathbb{R}^2$, the space $\mathscr{E}_A$ is no more invariant as soon as $A' \neq 0$. In particular, even when $\nu_u = 0$, the space $\mathscr{E}_A$ is not invariant as soon as $A' \neq 0$: this property stresses the fact that the Godunov scheme, as well as its low Mach modification obtained by simply removing the dissipative term in the right-hand side of the second equation of (12) like in [1, 2], may not be accurate at low Mach number, including in the 1D case, contrarily to the case $A' = 0$.

4.2 Study of a second numerical scheme

In order to propose a numerical scheme which will be accurate at low Mach number, we proceed like in [1]. That is to say:

- First, we try to modify the diffusion term in (13) such that the new equation preserves the invariance of $\mathscr{E}_A$.
- Then, we identify a numerical scheme whose modified equation corresponds to the equation with the new diffusion term.

To these two points, we add something new with respect to what is done in [1]: we shall show that it is possible to define a Godunov type scheme which corresponds to the numerical scheme proposed in the second point above. This stresses the fact that it is possible to build a particular Godunov scheme that is accurate at low Mach number for the linear wave equation with variable cross-section, if one discretizes equation (8) in a adequate set of variables. Another interest of this scheme is that it doesn't suffer from any checkerboard mode (see [3] when $A = 1$).

4.2.1 Modification of the diffusion term

Let us replace the diffusion term

$$\left(v_s \partial_{xx}^2 s , \ v_u \partial_{xx}^2 u \right)^T \qquad (14)$$

in equation (13) by the diffusion term

$$\left(v_s \partial_x \left[A \partial_x \left(\frac{s}{A} \right) \right] , \ v_u \partial_x \left[\frac{1}{A} \partial_x (Au) \right] \right)^T \qquad (15)$$

with $(v_s, v_u) \in \mathbb{R}^2$. Then, by construction, the space $\mathscr{E}_A$ is invariant for equation

$$\partial_t q + \frac{L_A}{M} q = \left(v_s \partial_x \left[A \partial_x \left(\frac{s}{A} \right) \right] , \ v_u \partial_x \left[\frac{1}{A} \partial_x (Au) \right] \right)^T . \qquad (16)$$

Moreover, we have the following result:

Lemma 4. *Let $q(t, x)$ be solution of (16) over $\mathbb{T}^1$. Then:*

$$E_A(t \geq 0) \leq E_A(t = 0).$$

A numerical scheme associated to (16) is then likely to be stable.

Proof of Lemma 4: Let $q(t, x)$ be solution of (16). There holds

$$\frac{1}{2}\frac{d}{dt}E_A = v_s \int_{\mathbb{T}^d} s\partial_x \left[A\partial_x\left(\frac{s}{A}\right)\right]\frac{dx}{A(x)} + v_u \int_{\mathbb{T}^d} u\partial_x\left[\frac{1}{A}\partial_x(Au)\right]A(x)dx$$

$$= -v_s \int_{\mathbb{T}^d}\left[\partial_x\left(\frac{s}{A}\right)\right]^2 A(x)dx - v_u\int_{\mathbb{T}^d}[\partial_x(Au)]^2 \frac{dx}{A(x)} \leq 0.$$

This proves that $E_A(t \geq 0) \leq E_A(t = 0)$. □

4.2.2 Identifying the numerical scheme

A numerical scheme whose modified equation corresponds to (16) is given by

$$\begin{cases} \dfrac{d}{dt}s_i + \dfrac{a_*}{M} \cdot \dfrac{A_{i+1}u_{i+1} - A_{i-1}u_{i-1}}{2\Delta x} = \\[6pt] \qquad \dfrac{a_*}{2M\Delta x}\left[\dfrac{A_{i+1/2}}{A_{i+1}}s_{i+1} - \left(\dfrac{A_{i+1/2} + A_{i-1/2}}{A_i}\right)s_i + \dfrac{A_{i-1/2}}{A_{i-1}}s_{i-1}\right] \quad (a) \\[14pt] \dfrac{d}{dt}u_i + \dfrac{a_*}{M} \cdot \dfrac{\dfrac{s_{i+1}}{A_{i+1}} - \dfrac{s_{i-1}}{A_{i-1}}}{2\Delta x} = \\[6pt] \qquad \dfrac{a_*}{2M\Delta x}\left[\dfrac{A_{i+1}}{A_{i+1/2}}u_{i+1} - A_i\left(\dfrac{1}{A_{i+1/2}} + \dfrac{1}{A_{i-1/2}}\right)u_i + \dfrac{A_{i-1}}{A_{i-1/2}}u_{i-1}\right] \quad (b) \end{cases} \quad (17)$$

where $A_i := A(x_i)$. A discrete analogue of Lemma 4 may be proved through "discrete integration by parts" and shows that the scheme is stable and that the discrete invariant space is the set

$$\mathcal{E}_A^h = \left\{q := (s, u)^T \in (\mathbb{R}^N)^2 \text{ such that } s_i = aA_i \text{ and } u_i = \frac{b}{A_i}, \ (a,b) \in \mathbb{R}^2\right\},$$

which admits the orthogonal set

$$(\mathcal{E}^h)^\perp = \left\{q := (s, u)^T \in (\mathbb{R}^N)^2 \text{ such that } \sum_{i=1}^N s_i = \sum_{i=1}^N u_i = 0\right\}$$

for the discrete scalar product $\langle q_1, q_2\rangle_A^h := \sum_{i=1}^N \Delta x\left(\dfrac{(s_1)_i (s_2)_i}{A_i} + (u_1)_i(u_2)_i A_i\right).$

4.2.3 The associated Godunov scheme

It is possible to obtain scheme (17) from (10) by the following process: we set

$$\widetilde{q} := (r, j)^T, \quad r := \frac{s}{A}, \quad j := Au$$

and solve the Riemann Problem based on the equation

$$\partial_t \widetilde{q} + \frac{a_*}{M} \left(\partial_x \left(\frac{j}{A_{i+1/2}} \right), \partial_x (A_{i+1/2} r) \right)^T = 0$$

with initial left and right states given by $\left(\frac{s_i}{A_i}, A_i u_i \right)^T$ and $\left(\frac{s_{i+1}}{A_{i+1}}, A_{i+1} u_{i+1} \right)^T$ respectively. This provides an expression for $(r_{i+1/2}, j_{i+1/2})^T$ which is plugged into (10) for the evaluation of $\frac{s_{i+1/2}}{A_{i+1/2}}$ and $A_{i+1/2} u_{i+1/2}$, and we obtain (17).

5 Numerical results in 1D

In this section, we chose $A(x) = \frac{1}{4} \sin(2\pi x) + \frac{1}{2}$. As an initial condition, we choose $s^0(x) = A(x)$ and $u^0(x) = 1/A(x)$. At the discrete level, we choose $s_i^0 = A(x_i)$ and $u_i^0 = 1/A(x_i)$, so that the initial condition belongs to $\mathscr{E}_A^h$. Then, with (17), this initial condition is left unchanged for all times, as is the case with the continuous solution. On the other hand, with (12), the solution $(s_i(t), u_i(t))_{i \in [1,N]}^T$ has a non zero component in the space $(\mathscr{E}^h)^\perp$ as soon as $t > 0$, which may be computed by an orthogonal projection. Figure 1 shows the discrete weighted L^2

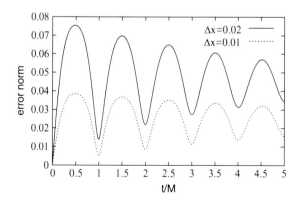

Fig. 1 norm of the spurious component for $M = 10^{-4}$ as a function of t/M

norm of this spurious component in $(\mathscr{E}^h)^\perp$ as a function of time scaled by M, with $M = 10^{-4}$ and for two different values of Δx. The size of the spurious wave grows up from 0 at $t = 0$ to $\mathscr{O}(\Delta x)$ at $t = \mathscr{O}(M)$, which is much greater than $\mathscr{O}(M)$, since $M \ll \Delta x$.

References

1. Dellacherie, S.: Analysis of Godunov type schemes applied to the compressible Euler system at low Mach number. J. Comp. Phys. **229**(4), 978–1016 (2010)
2. Dellacherie, S., Omnes, P., Rieper, F.: The influence of cell geometry on the Godunov scheme applied to the linear wave equation. J. Comp. Phys. **229**(14), 5315–5338 (2010)
3. Dellacherie, S.: Checkerboard modes and wave equation. In: Proc. of the Algoritmy 2009 Conference on Scientic Computing (March 15-20, 2009, Vysoke Tatry, Podbanske, Slovakia), pp. 71-80 (2009)

The paper is in final form and no similar paper has been or is being submitted elsewhere.

Towards stabilization of cell-centered Lagrangian methods for compressible gas dynamics

Bruno Després and Emmanuel Labourasse

Abstract We propose a sub-cell procedure for the stabilization of cell-centered Lagrangian numerical schemes for the computation of compressible gas dynamics. This procedure is intended to stabilize the mesh, indeed cell-centered schemes are already stable for shocks since they are based on a Riemann solver technology. In this work we focus on the basic principles and on the compatibility with the entropy. We show that a sub-cell decomposition into four triangles is always mesh-stable provided the scheme is entropy increasing. Numerical examples serve as illustration. We also discuss the consistency issue.

Keywords Finite Volume, Lagrangian scheme, stabilization
MSC2010: 65M08, 65M12, 65Z99, 76M12

1 Introduction

Cell centered Finite Volume numerical methods for the calculation of the equations of Lagrangian gas dynamics [1]

$$\begin{cases} \partial_t \rho + \nabla \cdot (\rho \mathbf{u}) = 0, \\ \partial_t (\rho \mathbf{u}) + \nabla \cdot (\rho \mathbf{u} \otimes \mathbf{u}) + \nabla p = 0, \\ \partial_t (\rho e) + \nabla \cdot (\rho \mathbf{u} e + p \mathbf{u}) = 0, \\ \partial_t (\rho S) + \nabla \cdot (\rho \mathbf{u} S) \geq 0. \end{cases} \quad (1)$$

Bruno Després
Lab. LJLL-UPMC, France, e-mail: despres@ann.jussieu.fr

Emmanuel Labourasse
CEA, DAM, DIF, 91297 Arpajon, France

receive increasing interest nowadays due to three facts: 1) these are cell centered methods, like any Finite Volume scheme this is easy to handle in a multidimensional code [4, 9]; 2) in the context of compressible gas dynamics new corner based cell centered Riemann solvers have been developed which make these methods appropriate for shock calculations; 3) remeshing and projection techniques are easy to developed for such algorithms (but not that easy to optimize). However these methods suffer for the clue of all numerical methods on moving grids [2, 8, 10]. The mesh may become pathological (tangling) due to physical features of the flow (vortex, shear). Spurious modes, like checkerboard modes, may indeed be responsible of local negative volumes. This problem does not show up in dimension one but is the rule in dimension two and three.

In this paper, we focus on the underlying subcell finite volume structure in the context of finite Volume discretization and on possible ideas which can be used to develop an unconditionally stable lagrangian algorithm for compressible gas dynamics. Notice that other stabilization processes have been developed for Lagrangian hydrodynamics discretization [2], using subcell modeling [3] and also for ALE techniques which are another way to stabilize Lagrangian calculations [7].

2 An example

In order to establish a guideline for further developments, we want first to consider the example of a Sod shock tube problem computed with two different meshes and with the GLACE scheme [4]. We display a zoom in the shocked zone is in Fig. 1.

One sees a difference between triangles and quadrangles. Meshes made with triangles are stable in the sense that the volume of all cells always remain positive. Numerical observation with quadrangles show that the total volume is also positive, but the local volumes may become negative. This phenomenon is generated by the numerical scheme used to move the mesh. In our case the scheme is the cell-centered Glace scheme. But this behavior is common to any Lagrangian scheme.

In what follows we propose to introduce some aspects of the computation with triangles into the computation with quadrangles in order to improve the robustness.

3 Subcell modeling

To overcome the difficulties encountered with quadrangles we propose to consider a subcell modeling. The idea is to consider one cell at time step n. The volume of the cell is referred to as V^n. The total mass in the cell is

$$M = V^n \rho^n.$$

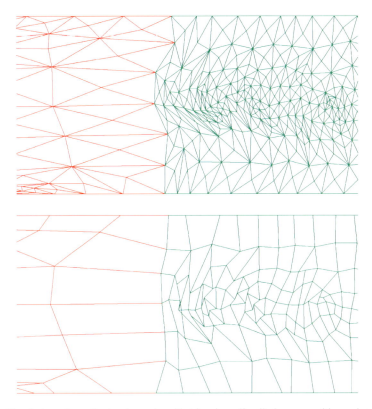

Fig. 1 The first mesh on the top is made with triangles: all cells have a positive volume. The second mesh on the bottom is made with quadrangles: pathological cells are visible to rows on the right of the interface between the coarse mesh and the fine mesh. The calculation stops

Notice that the total mass does not depend upon n since the scheme is Lagrangian. The momentum in the cell is the vector

$$\mathbf{I}^n = V^n \rho^n \mathbf{u}^n$$

and the total energy is

$$E^n = V^n \rho^n e^n.$$

The internal energy is

$$\varepsilon^n = E^n - \frac{1}{2} |\mathbf{I}^n|^2.$$

Next we split the cell into triangles. For example the quadrangle of Fig. 2 is split into four triangles $\overline{V} = \overline{T_1 \cup T_2 \cup T_3 \cup T_4}$ where the center O is simply the average of the corners

$$0 = \frac{1}{4}(A + B + C + D).$$

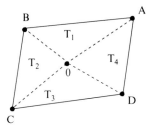

Fig. 2 Decomposition of a quad into four triangles

Then we decide arbitrarily that the total mass is split into four equal parts and affected in each triangle

$$m_i = \frac{1}{4}M$$

and the same for the internal energy

$$\varepsilon_i^n = \frac{1}{4}\varepsilon^n.$$

In our mind it is absolutely essential to characterize this very simple operation at the thermodynamical level, that is for the entropy variable. We present here a somewhat naive procedure for doing this.

Assume for example a perfect gas pressure law $p = (\gamma - 1)\rho\varepsilon$ and $S = \log(\varepsilon\rho^{-\gamma})$. The entropy in T_i is

$$S_i = \log\left(\varepsilon_i \rho_i^{-\gamma}\right).$$

The density in triangle T_i is

$$\rho_i = \frac{m_i}{|T_i|} = \frac{1}{4}\frac{m}{|T_i|} = \frac{1}{4}f_i\rho$$

where $f_i = \frac{|T_i|}{V}$ is the volume fraction. So the entropy in T_i is

$$S_i = \log(\varepsilon\rho^{-\gamma}) - \gamma \log f_i + (\gamma - 1)\log 4.$$

The new total entropy in the cell is

$$\overline{S} = \frac{S_1 + S_2 + S_3 + S_4}{4} = S - \frac{\gamma}{4}\sum_i \log f_i + (\gamma - 1)\log 4.$$

This very basic example shows that subcell modeling is somehow equivalent to modifying the entropy S into a new entropy variable

$$\overline{S} = S + \varphi(f_1, \ldots, f_4). \tag{2}$$

Definition 1. For any pressure law, we say that the subcell model (2) is entropy consistent if the function φ satisfies two conditions: it is a concave function and

$$f_1 f_2 f_3 f_4 = 0 \text{ implies that } \varphi = -\infty.$$

We say that φ is a subzonal entropy.

In order to stabilize a given Lagrangian computation, the general principle is to introduce a subzonal entropy in the numerical method which can be either staggered [2] or cell-centered as in our case [4]. The next proposition explains a fundamental advantage of subzonal entropies.

Proposition 1. *Consider a Lagrangian scheme which has the property to be entropy consistent, that the entropy in the cell increases from time step n to time step $n+1$ $S^{n+1} \geq S^n$.*

Assume that we are able to modify this scheme in order to introduce a subzonal entropy in a way such that

$$\overline{S}^{n+1} \geq \overline{S}^n. \tag{3}$$

Then the mesh is never pathological.

Indeed by continuity a pathological mesh is such that one volume fraction tends to zero, becomes zero and after become negative. In this case $\varphi^{n+1} \approx -\infty$ and it is in contradiction with the assumption (3).

Once the general framework of a thermodynamically consistent subcell model has been identified, it remains to use these new subzonal entropies in the scheme. In our case we focus on the GLACE scheme [4] which is cell-centered and of Finite Volume type. It is quite technical and there are many options, this is why we prefer to skip this issue in this presentation. However we managed to stabilize some computations which were unstable before. We present a numerical result that has been computed with the function $\varphi = \sum_i \log f_i$ in order to illustrate the numerical performances of the method proposed in this work. One sees on Fig. 3 that the robustness of the simulation is achieved with a result which is still physically correct.

Another major point is the convergence (as the mesh size tends to zero) of the new scheme with the new entropy. Indeed such a modification could be a source of major errors and of some fundamental inconsistency with the real problem (1). However since the proposed procedure is compatible with the idea of a geometric sub-cell modeling which is a kind of interpolation between a computation with quadrangles and a computation between triangles, it is reasonable to think the numerical solution is indeed consistent. At least the numerical test displayed in Fig. 3 shows the correctness of the result. In some situations it is also possible to show that this procedure is weakly consistent as proposed in [5]: the key property is to show consistency in the mimetic sense of the scheme [6].

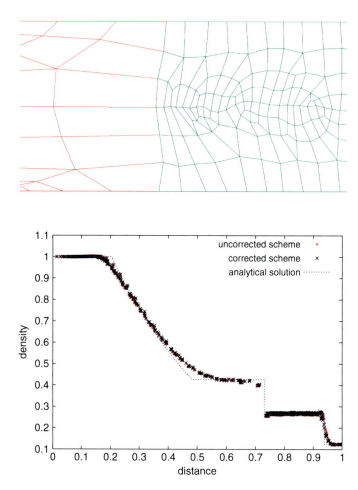

Fig. 3 Top: the quad mesh of Fig. 1 with the subzonal entropy; the mesh is no more pathological. Bottom: we plot the reference density of the Sod shock tube problem, the density calculated with the classical scheme and the density calculated with the subzonal entropy: there all agree which means that the subzonal entropy does not perturb the accuracy for this particular problem

Acknowledgements he first author kindly acknowledges the support of CEA.

References

1. D.J. Benson. Computational methods in Lagrangian and Eulerian hydrocodes. *Comp. Meth. Appl. Mech. Eng.*, 99:235–394, 1992.
2. E.J. Caramana, D.E. Burton, M.J. Shashkov, and P.P. Whalen. The construction of compatible hydrodynamics algorithms utilizing conservation of total energy. *J. Comput. Phys.*, 146:227–262, 1998.

3. E.J. Caramana and M.J. Shashkov. Elimination of Artificial Grid Distortion and Hourglass-Type Motions by Means of Lagrangian Subzonal Masses and Pressures. *J. Comput. Phys.*, 142:521–561, 1998.
4. G. Carré, S. Delpino, B. Després, and E. Labourasse. A cell-centered Lagrangian hydrodynamics scheme on general unstructured meshes in arbitrary dimension. *J. Comput. Phys.*, 228:5160–5183, 2009.
5. B. Després. Weak consistency of the cell-centered lagrangian glace scheme on general meshes in any dimension. *CMAME*, 199:2669–2679, 2010.
6. B. Després and E. Labourasse. Subzonal entropy stabilization of cell-centered lagrangian methods. in preparation.
7. Liska, Shashkov, Vchal, and Wendroff. Optimization-based synchronized flux-corrected conservative interpolation (remapping) of mass and momentum for arbitrary lagrangian-eulerian methods. *J. Comput. Phys.*, 229(5):1467–1497, 2010.
8. R. Loubere, J. Ovadia, and R. Abgrall. A Lagrangian discontinuous Galerkin type method on unstructured meshes to solve hydrodynamics problems. *Int. J. Num. Meth. in Fluids*, 2000.
9. P.H. Maire, R. Abgrall, J. Breil, and J. Ovadia. A cell-centered lagrangian scheme for 2D compressible flow problems. *Siam J. Sci. Comp.*, 29, 2007.
10. G. Scovazzi, E. Love, and M.J. Shashkov. Multi-scale Lagrangian shock hydrodynamics on Q1/P0 finite elements: Theoretical framework and two-dimensional computations. *Comp. Meth. in Applied Mech. and Eng.*, 197:1056–1079, 2008.

The paper is in final form and no similar paper has been or is being submitted elsewhere.

Hybrid Finite Volume Discretization of Linear Elasticity Models on General Meshes

Daniele A. Di Pietro, Robert Eymard, Simon Lemaire, and Roland Masson

Abstract This paper presents a new discretization scheme for linear elasticity models using only one degree of freedom per face corresponding to the normal component of the displacement. The scheme is based on a piecewise constant gradient construction and a discrete variational formulation for the displacement field. The tangential components of the displacement field are eliminated using a second order linear interpolation. Our main motivation is the coupling of geomechanical models and porous media flows arising in reservoir or CO2 storage simulations. Our scheme guarantees by construction the compatibility condition between the displacement discretization and the usual cell centered finite volume discretization of the Darcy flow model. In addition it applies on general meshes possibly non conforming such as Corner Point Geometries commonly used in reservoir and CO2 storage simulations.

Keywords Hybrid finite volumes, linear elasticity, general meshes
MSC2010: 74S10, 74B05

1 Introduction

The oil production in unconsolidated, highly compacting porous media (such as Ekofisk or Bachaquero) induces a deformation of the pore volume which (i) modifies significantly the production, and (ii) may have severe consequences such as surface subsidence or damage of well equipments. This explains the growing interest in reservoir modeling for simulations coupling the reservoir Darcy multiphase flow with the geomechanical deformation of the porous media [3]. Similarly,

---

D.A. Di Pietro, S. Lemaire, and R. Masson
IFP Énergies nouvelles, FRANCE, e-mail: dipietrd, simon.lemaire, roland.masson@ifpen.fr

R. Eymard
Université Paris-Est Marne-la-Vallée, FRANCE, e-mail: robert.eymard@univ-mlv.fr

poromechanical models are also used in CO2 storage simulations to predict the over pressure induced by the injection of CO2 in order to assess the mechanical integrity of the storage in the injection phase.

The most commonly used geometry in reservoir and CO2 storage models is the so called Corner Point Geometry or CPG [7]. Although the CPG discretization is initially a structured hexahedral grid, vertical edges of the cells may typically collapse to account for the erosion of the geological layers and vertices may be dedoubled and slide along the coordlines (*i.e.* straight lines orthogonal to the geological layers) to model faults. In addition non conforming local grid refinement is used in near well regions. The resulting mesh is unstructured, non conforming, it includes all the degenerate cells obtained from hexahedra by collapsing edges, and hence it is not adapted to conforming finite element discretizations.

The objective of this paper is to introduce a new discretization scheme for linear elasticity equations which should

- apply on general meshes possibly non conforming;
- guarantee the stability of the coupling with Darcy flow models using cell centered finite volume discretization for the Darcy equation [1].

Our discretization is based on the family of Hybrid Finite Volume schemes introduced for diffusion problems on general meshes in [5] and also closely related to Mimetic Finite Difference schemes [6] as shown in [4]. The degrees of freedom are defined by the displacement vector $\mathbf{u}_\sigma$ at the center of gravity of each face σ of the mesh as well as the displacement vector $\mathbf{u}_K$ at a given point $\mathbf{x}_K$ of each cell K of the mesh. Following [5], a piecewise constant gradient is built and can be readily used to define the discrete variational formulation which mimics the continuous variational formulation for the displacement vector field. In order to stabilize this formulation and to reduce the degrees of freedom, the tangential components of the displacement are interpolated in terms of the neighbouring normal components. Numerical experiments on two dimensional and three dimensional meshes show that the resulting discretization is stable and convergent. In addition, this discretization satisfies by construction the compatibility condition or LBB (see [2], [1]) condition for poroelastic models when coupled with a cell centered finite volume scheme for the Darcy flow equation.

2 Discretization of Linear Elasticity Models

Let Ω be a d-dimensional polygonal or polyhedral domain ($d = 2$ or 3) and let us consider the following linear elasticity problem in infinitesimal strain theory:

$$\begin{cases} \mathbf{div}\,(\sigma(\mathbf{u})) + \mathbf{f} = \mathbf{0} & \text{on } \Omega, \\ \mathbf{u} = \mathbf{u}^D & \text{on } \partial\Omega^D, \\ \sigma(\mathbf{u}) \cdot \mathbf{n} = \mathbf{g} & \text{on } \partial\Omega^N, \end{cases} \quad (1)$$

where $\mathbf{u} \in \mathbb{R}^d$ is the unknown displacement field and D and N the two exponents standing respectively for Dirichlet and Neumann boundary conditions. $\sigma(\mathbf{u})$ is the Cauchy stress tensor and is given by Hooke's law $\sigma(\mathbf{u}) = 2\mu\varepsilon(\mathbf{u}) + \lambda \text{tr}(\varepsilon(\mathbf{u}))\,\mathbf{1}$, where μ, λ are the Lamé parameters and $\varepsilon(\mathbf{u}) = \frac{1}{2}\left(\nabla\mathbf{u} + \nabla\mathbf{u}^T\right)$ is the infinitesimal strain tensor.

2.1 Hybrid Finite Volume Discretization

The simulation domain Ω is discretized by a set of polygonal or polyhedral control volumes $K \in \mathcal{K}$, such that $\overline{\Omega} = \bigcup_{K \in \mathcal{K}} K$. The set of faces of the mesh is denoted by $\mathcal{E}$ and splits into boundary interfaces $\mathcal{E}_{ext}$ and inner interfaces $\mathcal{E}_{int}$. Among the boundary interfaces, we denote by $\mathcal{E}_{ext}^D$ and $\mathcal{E}_{ext}^N$ the subsets of boundary faces verifying Dirichlet or Neumann conditions. The center of gravity of the face σ is denoted by $\mathbf{x}_\sigma$ and its $d-1$ dimensional measure by $|\sigma|$. A point $\mathbf{x}_K$ is defined inside each cell K of the mesh. The set of faces of each cell K is denoted by $\mathcal{E}_K$, and the distance between $\mathbf{x}_K$ and σ by $d_{K,\sigma}$. The cone of base $\sigma \in \mathcal{E}_K$ and top $\mathbf{x}_K$ is denoted by K_σ.

Brief reminder of the hybrid finite volume discretization of a scalar diffusion problem (see [5]):

We first define the discrete hybrid spaces $V = \{(v_K \in \mathbb{R})_{K \in \mathcal{K}}, (v_\sigma \in \mathbb{R})_{\sigma \in \mathcal{E}}\}$ and $V^0 = \{v \in V \mid v_\sigma = 0 \ \forall \sigma \in \mathcal{E}_{ext}^D\}$. V^0 is endowed with the discrete $H_{0,D}^1(\Omega)$ norm

$$\|v\|_{V^0} = \left(\sum_{K \in \mathcal{K}} \sum_{\sigma \in \mathcal{E}_K} \frac{|\sigma|}{d_{K,\sigma}} |v_\sigma - v_K|^2 \right)^{\frac{1}{2}}. \tag{2}$$

Then, following [5], a discrete gradient is defined on each cone K_σ which only depends on v_K and $v_{\sigma'}$ for $\sigma' \in \mathcal{E}_K$. This gradient is exact on linear functions and satisfies a weak convergence property. According to [5], it can be written

$$\nabla_{K_\sigma} v = \sum_{\sigma' \in \mathcal{E}_K} (v_{\sigma'} - v_K) \mathbf{y}_K^{\sigma\sigma'} \quad \forall v \in V, \tag{3}$$

where $\mathbf{y}_K^{\sigma\sigma'} \in \mathbb{R}^d$ only depends on the geometry.

Hybrid finite volume discretization of the linear elasticity model:

As we did above, we introduce $\mathbf{V} = \{(\mathbf{v}_K \in \mathbb{R}^d)_{K \in \mathcal{K}}, (\mathbf{v}_\sigma \in \mathbb{R}^d)_{\sigma \in \mathcal{E}}\}$ as the discrete hybrid space. With an equivalent definition for $\mathbf{V}^0$, the discrete norm is now defined as $\|\mathbf{v}\|_{V^0}^2 = \sum_{i=1}^d \|v_i\|_{V^0}^2$. Let us also introduce the space $\mathbf{W} = \left\{ (\mathbf{w}_K \in \mathbb{R}^d)_{K \in \mathcal{K}}, (\mathbf{w}_\sigma \in \mathbb{R}^d)_{\sigma \in \mathcal{E}_{ext}^D}, (w_\sigma^n \in \mathbb{R})_{\sigma \in \mathcal{E}_{int} \cup \mathcal{E}_{ext}^N} \right\}$ and the following projection operator $P_\mathbf{W} : \mathbf{V} \to \mathbf{W}$, $\mathbf{v} \mapsto \left((\mathbf{v}_K)_{K \in \mathcal{K}}, (\mathbf{v}_\sigma)_{\sigma \in \mathcal{E}_{ext}^D}, (\mathbf{v}_\sigma \cdot \mathbf{n}_\sigma)_{\sigma \in \mathcal{E}_{int} \cup \mathcal{E}_{ext}^N} \right)$, where $\mathbf{n}_\sigma$ is a unit vector normal to σ which orientation is fixed once and for all. Let us finally define the space $\mathbf{W}^0 = P_\mathbf{W}(\mathbf{V}^0)$ endowed with the discrete norm $\|\mathbf{w}\|_{\mathbf{W}^0} = \inf_{\mathbf{v} \in \mathbf{V}^0 \mid P_\mathbf{W}(\mathbf{v}) = \mathbf{w}} \|\mathbf{v}\|_{\mathbf{V}^0}$.

The main novelty of our discretization lies in the definition of a linear interpolation operator $\mathbf{I} : \mathbf{W} \to \mathbf{V}$. This linear interpolation operator must be second order accurate to preserve the order of approximation of the scheme and interpolant in the sense that $P_\mathbf{W}(\mathbf{I}(\mathbf{w})) = \mathbf{w}$ for all $\mathbf{w} \in \mathbf{W}$. It must also be local in the sense that it computes the displacement vector $\mathbf{v}_\sigma$ at a given face $\sigma \in \mathcal{E}_{int} \cup \mathcal{E}_{ext}^N$ in terms of a given number of normal components $\mathbf{v}_{\sigma'} \cdot \mathbf{n}_{\sigma'}$ taken on a stencil $\mathcal{S}_\sigma \subset \mathcal{E}$ of neighbouring faces σ' of σ (with $\sigma \in \mathcal{S}_\sigma$). An example of construction of such an interpolator is given in subsection 2.3.

The use of the interpolation operator $\mathbf{I}$ will bring two crucial improvements to the discretization: first a drastic reduction of the degrees of freedom and secondly a stabilization of the discretization.

Finally, generalizing the scalar framework to the vectorial case of the linear elasticity model, we introduce a piecewise constant discrete gradient for each cone K_σ:

$$\nabla_{K_\sigma} \mathbf{v} = \sum_{\sigma' \in \mathcal{E}_K} (\mathbf{v}_{\sigma'} - \mathbf{v}_K) \otimes \mathbf{y}_K^{\sigma\sigma'} \quad \forall \mathbf{v} \in \mathbf{V}. \tag{4}$$

2.2 Discrete Variational Formulation

Starting from (1), we deduce a discrete weak formulation of the problem in $\mathbf{W}^0$.

Setting $\varepsilon_{K_\sigma}(\mathbf{v}) = \frac{1}{2} \left(\nabla_{K_\sigma} \mathbf{v} + \nabla_{K_\sigma} \mathbf{v}^T \right)$ and $\sigma_{K_\sigma}(\mathbf{v}) = 2\mu \varepsilon_{K_\sigma}(\mathbf{v}) + \lambda \text{tr}(\varepsilon_{K_\sigma}(\mathbf{v})) \mathbf{1}$ for all $\mathbf{v} \in \mathbf{V}$, we introduce the discrete bilinear form on $\mathbf{W} \times \mathbf{W}$

$$a_\mathscr{D}(\mathbf{u}, \mathbf{v}) = \sum_{K \in \mathcal{K}} \sum_{\sigma \in \mathcal{E}_K} |K_\sigma| \sigma_{K_\sigma}(\mathbf{I}(\mathbf{u})) : \varepsilon_{K_\sigma}(\mathbf{I}(\mathbf{v})). \tag{5}$$

Then, the discrete variational formulation reads: find $\mathbf{u} \in \mathbf{W}$ such that $\mathbf{u}_\sigma = \mathbf{u}_\sigma^D$ for all $\sigma \in \mathcal{E}_{ext}^D$ and such that, for all $\mathbf{v} \in \mathbf{W}^0$,

$$a_\mathscr{D}(\mathbf{u}, \mathbf{v}) = \sum_{K \in \mathcal{K}} |K| \mathbf{f}_K \cdot \mathbf{v}_K + \sum_{\sigma \in \mathcal{E}_{ext}^N} |\sigma| \mathbf{g}_\sigma \cdot \mathbf{I}(\mathbf{v})_\sigma, \tag{6}$$

where $\mathbf{u}_\sigma^D = \frac{1}{|\sigma|} \int_\sigma \mathbf{u}^D d\sigma$, $\mathbf{f}_K = \frac{1}{|K|} \int_K \mathbf{f} dx$ and $\mathbf{g}_\sigma = \frac{1}{|\sigma|} \int_\sigma \mathbf{g} d\sigma$ are average values.

It is important to keep in mind that numerical experiments show that without interpolation, the bilinear form $a_\mathscr{D}$ on the space $\mathbf{V} \times \mathbf{V}$ leads to an unstable scheme

Hybrid Finite Volume Discretization of Linear Elasticity Models

with vanishing eigenvalues on triangular or tetrahedral meshes with mixed Neumann Dirichlet boundary conditions.

Note also that for the solution of the linear system (6), the unknowns $\mathbf{u}_K$ can easily be eliminated without any fill in, reducing the degrees of freedom to the face normal components of the displacement.

2.3 Interpolation of the tangential components of the displacement

Given a face $\sigma \in \mathcal{E}_{int} \cup \mathcal{E}_{ext}^N$, for each component $i \in [\![1, d]\!]$ of the displacement field $\mathbf{u}_\sigma$, we look for a linear interpolation of the form

$$\bar{u}_\sigma^i(\mathbf{x}) = \sum_{j=1}^{d} \alpha_\sigma^{ij} x_j + \beta_\sigma^i. \tag{7}$$

In order to determine the $d(d+1)$ coefficients $(\alpha_\sigma^{ij})_{i,j \in [\![1,d]\!]}$, $(\beta_\sigma^i)_{i \in [\![1,d]\!]}$ as linear combinations of normal components $\mathbf{u}_{\sigma'} \cdot \mathbf{n}_{\sigma'}$, $\sigma' \in \mathcal{S}_\sigma$, we look for a set $\mathcal{S}_\sigma$ of $d(d+1)$ neighbouring faces σ' of the face σ, with $\sigma \in \mathcal{S}_\sigma$ and such that the system of equations $\bar{\mathbf{u}}_\sigma(\mathbf{x}_{\sigma'}) \cdot \mathbf{n}_{\sigma'} = \mathbf{u}_{\sigma'} \cdot \mathbf{n}_{\sigma'}$ is non singular. The set $\mathcal{S}_\sigma$ is built using the following greedy algorithm:

1. Initialization: for a given number $k > d(d+1)$, we select the k closest neighbouring faces of the face σ which are sorted from the closest to the farthest using the distance between the face center and $\mathbf{x}_\sigma$: $\sigma_0 = \sigma, \sigma_1, \cdots, \sigma_{k-1}$. We set $\mathcal{S}_\sigma = \{\sigma\}$ and $q = 1, l = 0$;
2. while $q < d(d+1)$ and $l < k - 1$:
   - $l = l + 1$;
   - if the equation $\bar{\mathbf{u}}_\sigma(\mathbf{x}_{\sigma_l}) \cdot \mathbf{n}_{\sigma_l} = \mathbf{u}_{\sigma_l} \cdot \mathbf{n}_{\sigma_l}$ is linearly independent with the set of equations $\bar{\mathbf{u}}_\sigma(\mathbf{x}_{\sigma'}) \cdot \mathbf{n}_{\sigma'} = \mathbf{u}_{\sigma'} \cdot \mathbf{n}_{\sigma'}$ for all $\sigma' \in \mathcal{S}_\sigma$, then $\mathcal{S}_\sigma = \{\sigma_l\} \cup \mathcal{S}_\sigma$, $q = q + 1$;
3. if $q < d(d+1)$, the algorithm is rerun with a larger value of k.

Note that since $\sigma \in \mathcal{S}_\sigma$, the interpolation operator satisfies as required the property $P_\mathbf{W}(\mathbf{I}(\mathbf{u})) = \mathbf{u}$ for all $\mathbf{u} \in \mathbf{W}$.

2.4 Compatibility condition with cellwise constant pressure for poroelastic models

Let $L_0^2(\Omega)$ be the subspace of functions of $L^2(\Omega)$ with vanishing mean values. For the sake of simplicity but without any loss of generality, we consider here

$\partial\Omega^D = \partial\Omega$ and $\mathbf{u}^D = \mathbf{0}$. It is well known (see [1]) that the well-posedness of linear poroelasticity models relies on the well-posedness of the following saddle point problem: find $(\mathbf{u}, p) \in H_0^1(\Omega)^d \times L_0^2(\Omega)$ such that

$$\begin{cases} a(\mathbf{u}, \mathbf{v}) + b(\mathbf{v}, p) = (\mathbf{f}, \mathbf{v})_{L^2(\Omega)^d} & \text{for all } \mathbf{v} \in H_0^1(\Omega)^d, \\ b(\mathbf{u}, q) = (h, q)_{L^2(\Omega)} & \text{for all } q \in L_0^2(\Omega), \end{cases} \quad (8)$$

where a is the bilinear form of the linear elasticity model and $b(\mathbf{v}, p) = -(\text{div }\mathbf{v}, p)_{L^2(\Omega)}$. The stability of this saddle point problem results from the coercivity of a and the following LBB condition (see [2]) which guarantees the existence and uniqueness of the solution: $\inf_{p \in L_0^2(\Omega)} \sup_{\mathbf{v} \in H_0^1(\Omega)^d} \frac{b(\mathbf{v}, p)}{\|\mathbf{v}\|_{H_0^1(\Omega)^d} \|p\|_{L^2(\Omega)}} \geq \gamma > 0$.

The following theorem states that the LBB condition holds on the discrete spaces $\mathbf{W}^0 \times M_0$, where M_0 is the space of cellwise constant functions on the mesh $\mathcal{K}$ with vanishing mean values endowed with the $L^2(\Omega)$ norm, and for the discrete divergence operator defined by:

$$b_{\mathcal{D}}(\mathbf{w}, p) = -(\text{div}_{\mathcal{D}}\mathbf{w}, p)_{L^2(\Omega)} = -\sum_{K \in \mathcal{K}} p_K \sum_{\sigma \in \mathcal{E}_K \cap \mathcal{E}_{int}} |\sigma| w_\sigma^n (\mathbf{n}_\sigma \cdot \mathbf{n}_{K,\sigma}), \quad (9)$$

for all $(\mathbf{w}, p) \in \mathbf{W}^0 \times M_0$, and where $\mathbf{n}_{K,\sigma}$ is the normal to the face σ outward K. It implies that, assuming the coercivity of $a_{\mathcal{D}}$, the coupling of our discretization for the elasticity model with a cell centered finite volume scheme for the Darcy pressure equation will lead to a convergent and stable scheme for the poroelastic model.

Theorem 1. *The bilinear form $b_{\mathcal{D}}$ defined on $\mathbf{W}^0 \times M_0$ satisfies the LBB condition*

$$\inf_{p \in M_0} \sup_{\mathbf{w} \in \mathbf{W}^0} \frac{b_{\mathcal{D}}(\mathbf{w}, p)}{\|\mathbf{w}\|_{\mathbf{W}^0} \|p\|_{L^2(\Omega)}} \geq \gamma_{\mathcal{D}} > 0, \quad (10)$$

with a constant $\gamma_{\mathcal{D}}$ depending only on usual regularity parameters of the mesh.

Proof. From the continuous LBB condition, for all $p \in M_0$, there exists a displacement field $\mathbf{v} \in H_0^1(\Omega)^d$ such that $b(\mathbf{v}, p) \geq \gamma \|\mathbf{v}\|_{H_0^1(\Omega)^d} \|p\|_{L^2(\Omega)}$. Let $\mathbf{u}$ be the element of $\mathbf{V}^0$ such that $\mathbf{u}_K = \frac{1}{|K|} \int_K \mathbf{v}\, d\mathbf{x}$ for $K \in \mathcal{K}$ and $\mathbf{u}_\sigma = \frac{1}{|\sigma|} \int_\sigma \mathbf{v}\, d\sigma$ for $\sigma \in \mathcal{E}$. Then, $\mathbf{w} = P_{\mathbf{W}}(\mathbf{u}) \in \mathbf{W}^0$ satisfies $b_{\mathcal{D}}(\mathbf{w}, p) = b(\mathbf{v}, p)$.

Since it is shown in [9] that $\|\mathbf{u}\|_{\mathbf{V}^0} \leq \kappa \|\mathbf{v}\|_{H_0^1(\Omega)^d}$, with a constant κ depending on usual regularity parameters of the mesh, and since we have by definition the inequality $\|\mathbf{w}\|_{\mathbf{W}^0} \leq \|\mathbf{u}\|_{\mathbf{V}^0}$, the discrete LBB condition holds with $\gamma_{\mathcal{D}} = \frac{\gamma}{\kappa}$. □

3 Numerical experiments

In this section, the convergence of the scheme is tested on the linear elasticity model with exact solution

$$u_i = e^{\cos\left(\sum_{j=1}^d \chi^{ij} x_j\right)}, \qquad i = 1, \cdots, d. \tag{11}$$

The right hand side and the Dirichlet boundary conditions are defined by the exact solution. The Lamé parameters λ and μ are set to 1.

The tests have been held using an object oriented C++ implementation which original approach is described in [8]. The relative l^2 error on the displacement and on the gradient of the displacement are plotted function of the number of inner faces. In the computation of these errors, the cellwise constant discrete solution and discrete gradient are used. We first consider the triangular meshes (mesh family 1), the Cartesian grids (mesh family 2), the local grid refinement (mesh family 3) and the Kershaw meshes (mesh family 4) from the FVCA5 2D benchmark. The exact solution is defined by

$$\chi = \begin{pmatrix} 1 & 1 \\ 2 & -1 \end{pmatrix}.$$

The results presented on Fig. 1 show the good convergence behaviour of the scheme. The expected order of convergence is reached for all the meshes (for the solution and its gradient) and is even exceeded for the gradient on Cartesian grids. Next, let us consider the Cartesian grids (mesh family A), the randomly distorted grids (mesh family AA), and the tetrahedral meshes (mesh family B) from the FVCA6 3D benchmark. The exact solution is defined by

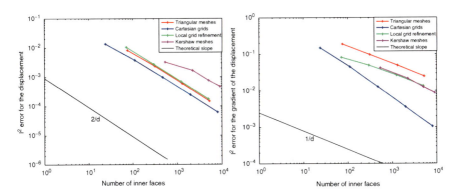

Fig. 1 l^2 error for the displacement and for the gradient of the displacement function of the number of inner faces for the triangular meshes, the Cartesian grids, the local grid refinement and the Kershaw meshes

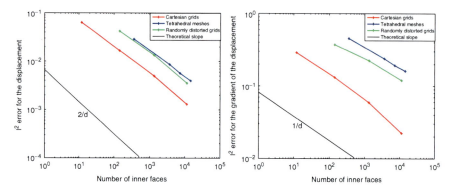

Fig. 2 l^2 error for the displacement and for the gradient of the displacement function of the number of inner faces for the Cartesian grids, the randomly distorted grids and the tetrahedral meshes

$$\chi = \begin{pmatrix} 1 & 1 & 1 \\ 2 & 1 & -1 \\ -1 & 1 & 2 \end{pmatrix}.$$

The results exhibited on Fig. 2 show again the good convergence behaviour of the scheme with the expected order on the three meshes for both the discrete solution and its gradient.

4 Conclusion

In this paper, a new discretization method has been introduced for linear elasticity using only one degree of freedom per face. It applies to general polygonal and polyhedral meshes possibly non conforming. In addition this discretization satisfies the compatibility condition when coupled with cell centered finite volume schemes for the Darcy equation in poroelastic models.

First numerical experiments in 2D and 3D exhibit the stability and convergence of the scheme. In the near future, further testings will be performed on CPG grids with erosions, local grid refinement and faults, to assess the potential of this scheme for reservoir and CO2 storage simulations.

References

1. M.A. Murad and A.F.D. Loula. On Stability and Convergence of Finite Element Approximations of Biot's Consolidation Problem. *Int. Jour. Numer. Eng.*, 37:645–667, 1994.
2. F. Brezzi and M. Fortin. Mixed and Hybrid Finite Element Methods. *Springer-Verlag, New-York*, 1991.

3. A. Settari and F.M. Mourits. Coupling of Geomechanics and Reservoir simulation models. *Comp. Meth. Adv. Geomech., Siriwardane and Zaman and Balkema, Rotterdam*, 2151–2158, 1994.
4. J. Droniou, R. Eymard, T. Gallouët and R. Herbin. A unified approach to mimetic finite difference, hybrid finite volume and mixed finite volume methods. *Math. Models Methods Appl. Sci.*, 20(2):265–295, 2010.
5. R. Eymard, T. Gallouët and R. Herbin. Discretisation of heterogeneous and anisotropic diffusion problems on general non-conforming meshes, SUSHI: a scheme using stabilisation and hybrid interfaces. *IMA J. Numer. Anal.*, 30(4):1009–1043, 2010. See also http://hal.archives-ouvertes.fr/.
6. F. Brezzi, K. Lipnikov and V. Simoncini. A family of mimetic finite difference methods on polygonal and polyhedral meshes. *Math. Models Methods Appl. Sci.*, 15:1533–1553, 2005.
7. D.K. Ponting. Corner Point Geometry in reservoir simulation. *In Clarendon Press, editor*, Proc. ECMOR I, 45–65, Cambridge, 1989.
8. D.A. Di Pietro and J.M. Gratien. Lowest order methods for diffusive problems on general meshes: A unified approach to definition and implementation. *These proceedings*, 2011.
9. R. Eymard, T. Gallouët and R. Herbin. Finite Volume Methods. *Handbook of Numerical Analysis*, 7:713–1020, 2000.

The paper is in final form and no similar paper has been or is being submitted elsewhere.

An A Posteriori Error Estimator for a Finite Volume Discretization of the Two-phase Flow

Daniele A. Di Pietro, Martin Vohralík, and Carole Widmer

Abstract We derive a posteriori error estimates for a multi-point finite volume discretization of the two-phase Darcy problem. The proposed estimators yield a fully computable upper bound for the selected error measure. The estimate also allows to distinguish, estimate separately, and compare the linearization and algebraic errors and the time and space discretization errors. This enables, in particular, to design a discretization algorithm so that all the sources of error are properly balanced. Namely, the linear and nonlinear solvers can be stopped as soon as the algebraic and linearization errors drop to the level at which they do not affect to the overall error. This can lead to significant computational savings, since performing an excessive number of unnecessary iterations can be avoided. Similarly, the errors in space and in time can be equilibrated by time step and local mesh adaptivity.

Keywords Finite volumes, a posteriori error estimates, darcy model, fully computable upperbound, twophase flow.
MSC2010: 65M08, 65M50, 76S05.

1 The two-phase flow model

Let $\Omega \subset \mathbb{R}^d$, $d \geq 1$, denote a bounded connected polygonal domain and let $t_F > 0$. Let w denote the wetting phase (e.g., water) and o the non-wetting phase (e.g., oil), and let there be given sources $f_o, f_w \in L^2((0, t_F); L^2(\Omega))$ and a (constant) porosity

---

Daniele A. Di Pietro and Carole Widmer
IFP Energies nouvelles, 1&4, avenue du Bois-Préau, Rueil-Malmaison, France,
e-mail: dipietrd@ifpen.fr, carole.widmer@ifpen.fr

Martin Vohralík
UPMC Univ. Paris 06, UMR 7598, Laboratoire J.-L. Lions, 75005, Paris, France & CNRS, UMR 7598, Laboratoire J.-L. Lions, 75005, Paris, France, e-mail: vohralik@ann.jussieu.fr

$\phi \in (0, 1]$. We consider the two-phase flow (see, e.g., [3]): Find $\mathbf{U} := \{P, S_o, S_w\}$, with P the pressure and S_p, $p \in \{o, w\}$, the saturations, such that

$$\partial_t(\phi S_o) + \nabla \cdot (\nu_o(P, S_o)\mathbf{u}_o(P, S_o)) = f_o \quad \text{in } \Omega \times (0, t_F),$$
$$\partial_t(\phi S_w) + \nabla \cdot (\nu_w(P, S_w)\mathbf{u}_w(P, S_w)) = f_w \quad \text{in } \Omega \times (0, t_F), \quad (1)$$
$$S_o + S_w = 1 \quad \text{in } \Omega \times (0, t_F).$$

For $p \in \{o, w\}$, ν_p denotes here the mobility of the phase p defined as the ratio of the relative permeability to the viscosity. In (1), $\mathbf{u}_o$ and $\mathbf{u}_w$ are such that

$$\mathbf{u}_p(P, S_p) := -K\nabla\left(P + P_{c,p}(S_p)\right), \quad \text{for } p \in \{o, w\}, \text{ in } \Omega \times (0, t_F), \quad (2)$$

where $P_{c,p}(S_p)$ is the capillary pressure and K denotes a piecewise constant, uniformly elliptic tensor-valued field corresponding to the absolute permeability. To find some example of the physics laws (capillarity pressure, phase mobility) or of the absolute permeability see [7].

Problem (1) is complemented by the initial conditions:

$$S_o(\cdot, 0) = S_o^0 \text{ and } P(\cdot, 0) = P^0, \quad \text{in } \Omega, \quad (3)$$

as well as by no-flow boundary conditions:

$$\mathbf{u}_p(P, S_p) \cdot \mathbf{n}_\Omega = 0, \quad \text{in } \partial\Omega \times (0, t_F). \quad (4)$$

The purpose of this paper is to propose fully computable a posteriori error estimates for the discretization of (1)–(4) by cell-centered finite volume methods in space and the backward Euler scheme in time. In particular, we consider the multi-point finite volume method proposed in [1]. Using a dual error norm is motivated by, e.g., [8]. Developing the ideas of [4–6, 9], we in particular separate the estimate into contributions representing the *space discretization error, time discretization error, linearization error*, and *algebraic error*. Then, at each time step, the linearization algorithm and the iterative algebraic solver can be stopped as soon as the corresponding errors no longer affect the total error, and space and the time errors can be equilibrated.

2 Discretization by the finite volume method

2.1 Notations

Let $\mathcal{T} = \{T\}$ denotes a partition of Ω into simplices or rectangular parallelepipeds (the extension to general polygonal meshes is possible via the introduction of simplicial submeshes). For rectangular parallelepipeds, we further assume that K

is diagonal to perform $H(\text{div}; \Omega)$-conforming reconstructions. For every element $T \in \mathcal{T}$, we denote by $|T|$ its measure and by h_T its diameter. Let $\mathcal{F} = \{\sigma\}$ be the set of faces of the mesh and, for all $T \in \mathcal{T}$, set $\mathcal{F}_T := \{\sigma \in \mathcal{F} \mid \sigma \subset \partial T\}$. The time discretization is defined by a strictly increasing sequence of discrete times $\{t^n\}_{0 \le n \le N}$ such that $t^0 = 0$ and $t^N = t_F$. For $1 \le n \le N$, we define the time interval $I_n := (t^{n-1}, t^n]$ and the time step $\tau^n := t^n - t^{n-1}$.

2.2 The finite volume scheme

The discrete problem reads: For all $1 \le n \le N$, all $T \in \mathcal{T}$, and all $p \in \{o, w\}$, find $\mathbf{U}_T^n := \{P_T^n, S_{o,T}^n, S_{w,T}^n\}$ such that

$$\phi \frac{|T|}{\tau^n}\left(S_{p,T}^n - S_{p,T}^{n-1}\right) + \sum_{\sigma \in \mathcal{F}_T} \nu_p(P_{T_p^*(\sigma)}^{n-1}, S_{p,T_p^*(\sigma)}^{n-1}) F_{p,T,\sigma}^n - f_{p,T}^n = 0, \quad (5)$$

where $f_{p,T}^n = (f_p^n, 1)_T$ and $f_p^n = \frac{1}{\tau^n}\int_{t^{n-1}}^{t^n} f_p(t)\,dt$. We set $P_T^0 := (P^0, 1)_T / |T|$, $S_{o,T}^0 := (S_o^0, 1)_T / |T|$, and impose $S_{o,T}^n + S_{w,T}^n = 1$ for all $0 \le n \le N$. Furthermore, $F_{p,T,\sigma}^n = F_{p,T,\sigma}(\{\mathbf{U}_{T'}^n\}_{\mathcal{S}_\sigma})$ is a multi-point approximation of the flux of the phase p leaving $T \in \mathcal{T}$ through the face $\sigma \in \mathcal{F}_T$ that depends on the unknowns associated to the elements of the face stencil $\mathcal{S}_\sigma \subset \mathcal{T}$. The numerical flux is assumed to be conservative, i.e., for all internal faces $\sigma \subset \partial T_1 \cap \partial T_2$, there holds $F_{p,T_1,\sigma}^n = -F_{p,T_2,\sigma}^n$. The upwind cell $T_p^*(\sigma)$ is equal to T_1 if $F_{p,T_1,\sigma}^n \ge 0$, to T_2 otherwise. For boundary faces $\sigma \subset \partial T \cap \partial \Omega$, $F_{p,T,\sigma}^n = 0$ to honor the no-flow boundary condition (4), and we can leave $T_p^*(\sigma)$ undefined.

For all $0 \le n \le N$ and $T \in \mathcal{T}$, the unknown $S_{w,T}^n$ is eliminated using the local volume conservation equation $S_{o,T}^n + S_{w,T}^n = 1$. We introduce the reduced set of unknowns $\overline{\mathbf{U}}^n := \{\mathbf{P}^n, \mathbf{S}_o^n\}$, where $\mathbf{P}^n = \{P_T^n\}_{T \in \mathcal{T}}$ and $\mathbf{S}_o^n = \{S_{o,T}^n\}_{T \in \mathcal{T}}$. With a little abuse of notation, for a function $\xi(S_w)$, we write $\xi(S_o)$ to mean $\xi(1 - S_o)$. As a consequence, $\nu_w(S_o)$ and $P_{c,w}(S_o)$ are equal to $\nu_w(1 - S_o)$, $P_{c,w}(1 - S_o)$ and $\mathbf{u}_w(P, 1 - S_o)$ respectively. Equation (5) becomes, for all $1 \le n \le N$, all $T \in \mathcal{T}$, and all $p \in \{o, w\}$

$$D_{p,T}^n(\overline{\mathbf{U}}^n) = 0, \text{ with,} \quad (6)$$

$$D_{p,T}^n(\overline{\mathbf{U}}^n) := \phi \frac{|T|}{\tau^n}(-1)^j (S_{o,T}^n - S_{o,T}^{n-1}) + \sum_{\sigma \in \mathcal{F}_T} \nu_p(P_{T_p^*(\sigma)}^{n-1}, S_{o,T_p^*(\sigma)}^{n-1}) F_{p,T,\sigma}^n - f_{p,T}^n, \quad (7)$$

where $j = 1$ if $p = w$ and 0 otherwise.

2.3 Linearization

Problem (6) is a system of nonlinear algebraic equations that can be solved using the Newton algorithm. For a fixed $1 \leq n \leq N$, let $\overline{\mathbf{U}}^{n,0}$ be given (typically, $\overline{\mathbf{U}}^{n,0} = \overline{\mathbf{U}}^{n-1}$). For $1 \leq k$, a new estimate $\overline{\mathbf{U}}^{n,k}$ is computed from the previous $\overline{\mathbf{U}}^{n,k-1}$ by solving the following system of linear algebraic equations: For all $T \in \mathscr{T}$ and all $p \in \{\mathrm{o}, \mathrm{w}\}$,

$$\sum_{T' \in \mathscr{T}} \frac{\partial \mathbf{D}_{p,T}^n}{\partial \overline{\mathbf{U}}_{T'}} \left(\overline{\mathbf{U}}_{T'}^{n,k} - \overline{\mathbf{U}}_{T'}^{n,k-1} \right) = -\mathbf{D}_{p,T}^n(\overline{\mathbf{U}}^{n,k-1}), \tag{8}$$

where $\overline{\mathbf{U}}_T^{n,k} = \{P_T^{n,k}, S_{\mathrm{o},T}^{n,k}\}$ denotes the approximate solutions in T at the n-th time step and k-th Newton iteration. We suppose that (8) is solved using an iterative linear solver. For a fixed $1 \leq n \leq N$ and $k \geq 1$, let $\overline{\mathbf{U}}^{n,k,0}$ be given (typically, $\overline{\mathbf{U}}^{n,k,0} = \overline{\mathbf{U}}^{n,k-1}$). Then, at a given step $i \geq 1$, we have, for all $T \in \mathscr{T}$ and $p \in \{\mathrm{o}, \mathrm{w}\}$,

$$\sum_{T' \in \mathscr{T}} \frac{\partial \mathbf{D}_{p,T}^n}{\partial \overline{\mathbf{U}}_{T'}} \left(\overline{\mathbf{U}}_{T'}^{n,k,i} - \overline{\mathbf{U}}_{T'}^{n,k-1} \right) + \mathbf{D}_{p,T}^n(\overline{\mathbf{U}}^{n,k-1}) = \mathbf{R}_{p,T}^{n,k,i}, \tag{9}$$

where $\mathbf{R}_{p,T}^{n,k,i}$ is the algebraic residual, while $\overline{\mathbf{U}}_T^{n,k,i} = \{P_T^{n,k,i}, S_{\mathrm{o},T}^{n,k,i}\}$ denotes the approximate solution at the n-th time step, k-th Newton iteration, and i-th linear solver iteration.

3 A posteriori error estimate

3.1 Space-time approximate solutions

Let, for $0 \leq n \leq N$ and $p \in \{\mathrm{o}, \mathrm{w}\}$, $S_{p,h}^n$ be the piecewise constant function such that $S_{p,h|T} = S_{p,T}$ for all $T \in \mathscr{T}$. We introduce the space-time function $S_{p,h\tau}$ continuous and piecewise affine in time, and such that $S_{p,h\tau}(t^n) = S_{p,h}^n$ for $0 \leq n \leq N$. In order to give a meaning to the gradient operator appearing in (2), we need to postprocess the approximate cell pressures $\{P_T^n\}_{T \in \mathscr{T}}$ and capillary pressures $\{P_{\mathrm{c},p,T}^n\}_{T \in \mathscr{T}}$, $P_{\mathrm{c},p,T}^n := P_{\mathrm{c},p}(S_{p,T}^n)$, $p \in \{\mathrm{o}, \mathrm{w}\}$. As in [5, 6, 9], we introduce an elementwise postprocessing of $\{P_T^n\}_{T \in \mathscr{T}}$ and $\{P_{\mathrm{c},p,T}^n\}_{T \in \mathscr{T}}$, $1 \leq n \leq N$, yielding piecewise quadratic functions $\tilde{P}_h^n$ and $\tilde{P}_{\mathrm{c},p,h}^n$ ($\tilde{P}_h^0$ is given by a projection of the initial pressure P^0). As for the saturations, $\tilde{P}_{h\tau}$ and $\tilde{P}_{\mathrm{c},p,h\tau}$ are the space-time functions, continuous and piecewise affine in time, and such that $\tilde{P}_{p,h\tau}(t^n) := \tilde{P}_h^n$ and $\tilde{P}_{\mathrm{c},p,h\tau}(t^n) := \tilde{P}_{\mathrm{c},p,h}^n$, respectively.

3.2 Error measure

Set $X := L^2((0, t_F); H^1(\Omega))$. For $\varphi \in X$, let $\|\varphi\|_X^2 := \int_0^{t_F} \|\nabla\varphi\|^2 dt$ and $\|\cdot\|$ denotes the L^2-norm on Ω. We suppose that the solution (P, S_o, S_w) of the problem (1)–(4) has the necessary regularity to permit the following weak formulation characterization: For all $\varphi \in X$, and all $p \in \{o, w\}$,

$$\int_0^{t_F} \{\langle \partial_t(\phi S_p), \varphi \rangle - \left(\nu_p(P, S_p)\mathbf{u}_p(P, S_p), \nabla\varphi\right)_\Omega - (f_p, \varphi)_\Omega\} \, dt = 0. \quad (10)$$

The aim of the following measure is to evaluate the residual of the approximate solution and the nonconformity of the approximate pressure (i.e., the facts that $(\tilde{P}_{h\tau}, S_{o,h\tau}, S_{w,h\tau})$ do not satisfy (10) and that $\tilde{P}_{h\tau} \notin X$ in general). Note that if $S_{p,h\tau}$ coincide with S_p, $p \in \{o, w\}$, and $\tilde{P}_{h\tau}$ with P, the error measure equals zero:

$$\||(S_p - S_{p,h\tau}, P - \tilde{P}_{h\tau})\||$$

$$:= \sup_{\varphi \in X, \|\varphi\|_X = 1} \int_0^{t_F} \Big\{ \langle \partial_t(\phi S_p) - \partial_t(\phi S_{p,h\tau}), \varphi \rangle$$
$$- \left(\nu_p(P, S_p)\mathbf{u}_p(P, S_p) - \nu_p(\tilde{P}_{h\tau}, S_{p,h\tau})\mathbf{u}_p(\tilde{P}_{h\tau}, S_{p,h\tau}), \nabla\varphi\right) \Big\} dt \quad (11)$$

$$+ \inf_{\delta \in X} \left\{ \int_0^{t_F} \left\|\nu_p(\tilde{P}_{h\tau}, S_{p,h\tau})\mathbf{u}_p(\tilde{P}_{h\tau}, S_{p,h\tau}) - \nu_p(\delta, S_{p,h\tau})\mathbf{u}_p(\delta, S_{p,h\tau})\right\|^2 dt \right\}^{\frac{1}{2}}.$$

3.3 A posteriori error estimate

We let $\mathbf{RTN}(T) := [\mathbb{P}_0(T)]^d + \mathbb{P}_0(T)\mathbf{x}$ and $\mathbf{RTN}(T) := [\mathbb{P}_0(T)]^d + [\mathbb{P}_0(T)]^d\mathbf{x}$, on simplices and on rectangular parallelepipeds respectively, and we introduce the Raviart–Thomas–Nédélec space

$$\mathbf{RTN}(\mathcal{T}) := \{\mathbf{v}_h \in \mathbf{H}(\text{div}, \Omega) \mid \mathbf{v}_{h|T} \in \mathbf{RTN}(T), \forall T \in \mathcal{T}\}.$$

Following [2, 4, 5, 9], in order to obtain an estimate on (11), we introduce for $1 \le n \le N$ and $p \in \{o, w\}$ the flux reconstructions $\boldsymbol{\theta}_{p,h}^n \in \mathbf{RTN}(\mathcal{T})$ such that for $1 \le n \le N$, $T \in \mathcal{T}$, $T' \in \mathcal{T}_T$, $(T \cap T' = \sigma_{T,T'})$, and $p \in \{o, w\}$,

$$\langle \boldsymbol{\theta}_{p,h}^n \cdot \mathbf{n}_T \mid_{\sigma_{T,T'}}, 1 \rangle_{\sigma_{T,T'}} := \nu_p(P_{T_p^*(\sigma)}^{n-1}, S_{o,T_p^*(\sigma)}^{n-1}) F_{p,T,\sigma}^n. \quad (12)$$

The following local conservation property is obtained by the Green theorem from (6) and (12):

$$(f_p^n - \partial_t(\phi S_{p,h\tau}) - \nabla \cdot \boldsymbol{\theta}_{p,h}^n, 1)_T = 0. \quad (13)$$

Let us now define the *residual estimators* $\eta^n_{R,T,p}$, the *diffusive flux estimators* $\eta^n_{DF,T,p}$, and the *nonconformity estimators* $\eta^n_{NC,T,p}$ as

$$\eta^n_{R,T,p} := \frac{h_T}{\pi} \| f_p - \partial_t(\phi S_{p,h\tau}) - \nabla \cdot \boldsymbol{\theta}^n_{p,h} \|_T,$$

$$\eta^n_{DF,T,p}(t) := \left\| \boldsymbol{\theta}^n_{p,h} - \nu_p(\tilde{P}_{h\tau}, S_{p,h\tau}) \mathbf{u}_p(\tilde{P}_{h\tau}, S_{p,h\tau})(t) \right\|_T, \quad (14)$$

$$\eta^n_{NC,T,p}(t) := \left\| \nu_p(\tilde{P}_{h\tau}, S_{p,h\tau}) \mathbf{u}_p(\tilde{P}_{h\tau}, S_{p,h\tau})(t) - \nu_p(\delta_{h\tau}, S_{p,h\tau}) \mathbf{u}_p(\delta_{h\tau}, S_{p,h\tau})(t) \right\|_T.$$

Here $\delta_{h\tau} \in X$ is continuous and piecewise affine in time and such that $\delta_{h\tau}(t^n) = \delta^n_h$, with $\delta^n_h := \mathcal{I}_{av}(\tilde{P}^n_h)$ for all $0 \le n \le N$; $\mathcal{I}_{av}$ is an averaging operator as in [5, 6, 9].

Theorem 1 (Guaranteed a posteriori error estimate). *Let $p \in \{o, w\}$. Then*

$$|||(S_p - S_{p,h\tau}, P - \tilde{P}_{h\tau})||| \le \left\{ \sum_{n=1}^N \int_{I_n} \sum_{T \in \mathcal{T}} (\eta^n_{R,T,p} + \eta^n_{DF,T,p}(t))^2 \, dt \right\}^{\frac{1}{2}} + \left\{ \sum_{n=1}^N \int_{I_n} \sum_{T \in \mathcal{T}} (\eta^n_{NC,T,p}(t))^2 \, dt \right\}^{\frac{1}{2}}. \quad (15)$$

Proof. The proof is straightforward using the definition of the error measure (11) and following the techniques of [5]. The second term in (15) clearly issues from the second term in the right hand-side of (11). We thus only have to prove that the first term is an upper bound on the first term in the right hand-side of (11). Let $\varphi \in X$, $\|\varphi\|_X = 1$, and $p \in \{o, w\}$. Set $\mathbf{w}_p := \nu_p(P, S_p) \mathbf{u}_p(P, S_p)$ and $\mathbf{w}_{p,h\tau} := \nu(\tilde{P}_{h\tau}, S_{p,h\tau}) \mathbf{u}_p(\tilde{P}_{h\tau}, S_{p,h\tau})$. Then using the characterization of the weak solution (10),

$$\int_0^{t_F} \{\langle \partial_t(\phi S_p) - \partial_t(\phi S_{p,h\tau}), \varphi \rangle - (\mathbf{w}_p - \mathbf{w}_{p,h\tau}, \nabla\varphi)\} dt$$
$$= \int_0^{t_F} \{(f_p - \partial_t(\phi S_{p,h\tau}), \varphi) + (\mathbf{w}_{p,h\tau}, \nabla\varphi)\} dt.$$

Let now $1 \le n \le N$ be given. Adding and subtracting $(\boldsymbol{\theta}^n_{p,h}, \nabla\varphi)$, using the Green theorem, the local conservativity property (13), the Poincaré inequality, and the Cauchy–Schwarz inequality, we obtain

$$(f_p, \varphi) - (\partial_t(\phi S_{p,h\tau}), \varphi) + (\mathbf{w}_{p,h\tau}, \nabla\varphi)$$
$$= (f_p - \partial_t(\phi S_{p,h\tau}) - \nabla \cdot \boldsymbol{\theta}^n_{p,h}, \varphi) + (\mathbf{w}_{p,h\tau} - \boldsymbol{\theta}^n_{p,h}, \nabla\varphi)$$
$$= (f_p - \partial_t(\phi S_{p,h\tau}) - \nabla \cdot \boldsymbol{\theta}^n_{p,h}, \varphi - \Pi_0\varphi) + (\mathbf{w}_{p,h\tau} - \boldsymbol{\theta}^n_{p,h}, \nabla\varphi)$$
$$\le \sum_{T \in \mathcal{T}} (\eta^n_{R,T,p} + \eta^n_{DF,T,p}(t)) \|\nabla\varphi\|_T,$$

where Π_0 denotes the L^2-orthogonal projection onto piecewise constants on $\mathcal{T}$. The assertion follows by the Cauchy–Schwarz inequality and by $\|\varphi\|_X = 1$. □

3.4 Identification of different components of the error

Let $1 \leq n \leq N$, $T \in \mathcal{T}$, and $p \in \{\text{o}, \text{w}\}$. In Section 2.2, we define the nonlinear system (6) and we solve it in Section 2.3 using an iterative solver for the Newton algorithm. Let assume we are at the n-th time step, k-th Newton step and i-th linearization step. We introduce the following notations:

$$A_{p,T}^{n,k,i} := \phi \frac{|T|}{\tau^n}\left[(S_{p,T}^{n,k,i} - S_{p,T}^{n,k-1}) - S_{p,T}^{n-1}\right], \quad B_{p,T,\sigma}^{n,k,i} := \nu_p(P_{T_p^*(\sigma)}^{n,k-1}, S_{p,T_p^*(\sigma)}^{n,k-1})F_{p,T,\sigma}^{n,k,i}.$$

The linear system (9) is then equivalent to the following sum of diagonal terms and face fluxes:

$$\frac{\partial A_{p,T}^{n,k,i}}{\partial \overline{\mathbf{U}}_T} + \sum_{\sigma \in \mathcal{F}_T} \sum_{T' \in \mathcal{S}_\sigma} \frac{\partial B_{p,T,\sigma}^{n,k,i}}{\partial \overline{\mathbf{U}}_{T'}} + \mathbf{D}_{p,T}^n(\overline{\mathbf{U}}^{n,k-1}) = \mathbf{R}_{p,T}^{n,k,i}. \tag{16}$$

Let us now define a linearization flux $\overline{\boldsymbol{\theta}}_{p,h}^{n,k,i} \in \mathbf{RTN}(\mathcal{T})$ and algebraic solver flux $\mathbf{r}_{p,h}^{n,k,i} \in \mathbf{RTN}(\mathcal{T})$ such that $\boldsymbol{\theta}_{p,h}^{n,k,i} := \overline{\boldsymbol{\theta}}_{p,h}^{n,k,i} + \mathbf{r}_{p,h}^{n,k,i}$ and such that

$$\langle \overline{\boldsymbol{\theta}}_{p,h}^{n,k,i} \cdot \mathbf{n}_T \mid_{\sigma_{T,T'}}, 1\rangle_{\sigma_{T,T'}} := \sum_{T' \in \mathcal{F}_T} \frac{\partial B_{p,T,\sigma}^{n,k,i}}{\partial \overline{\mathbf{U}}_{T'}} \quad \text{and} \quad (\nabla \cdot \mathbf{r}_{p,h}^{n,k,i}, 1)_T = -\mathbf{R}_{p,T}^{n,k,i}. \tag{17}$$

Note that $\overline{\boldsymbol{\theta}}_{p,h}^{n,k,i}$ is fully specified; $\mathbf{r}_{p,h}^{n,k,i}$ can be constructed as in [6]. This gives

$$(f_p^n - \partial_t(\phi S_{p,h\tau}^{k,i}) - \nabla \cdot \overline{\boldsymbol{\theta}}_{p,h}^{n,k,i}, 1)_T = (\nabla \cdot \mathbf{r}_{p,h}^{n,k,i}, 1)_T, \quad p \in \{\text{o}, \text{w}\}. \tag{18}$$

We can now define the same estimators as in (14) and we have:

$$\eta_{\text{R},T,p}^{n,k,i} + \eta_{\text{DF},T,p}^{n,k,i}(t) + \eta_{\text{NC},T,p}^{n,k,i}(t) \leq \eta_{\text{tm},T,p}^{n,k,i}(t) + \eta_{\text{sp},T,p}^{n,k,i}(t) + \eta_{\text{lin},T,p}^{n,k,i}(t) + \eta_{\text{alg},T,p}^{n,k,i},$$

with

$$\eta_{\text{tm},T,p}^{n,k,i}(t) := \left\|\nu_p(\tilde{P}_{h\tau}^{k,i}, S_{p,h\tau}^{k,i})\mathbf{u}_p(\tilde{P}_{h\tau}^{k,i}, S_{p,h\tau}^{k,i})(t) - \nu_p(\tilde{P}_h^{n,k,i}, S_{p,h}^{n,k,i})\mathbf{u}_p(\tilde{P}_h^{n,k,i}, S_{p,h}^{n,k,i})\right\|_T,$$

$$\eta_{\text{sp},T,p}^{n,k,i}(t) := \eta_{\text{R},T,p}^{n,k,i} + \eta_{\text{NC},T,p}^{n,k,i}(t), \tag{19}$$

$$\eta_{\text{lin},T,p}^{n,k,i}(t) := \left\|\nu_p(\tilde{P}_h^{n,k,i}, S_{p,h}^{n,k,i})\mathbf{u}_p(\tilde{P}_h^{n,k,i}, S_{p,h}^{n,k,i}) - \overline{\boldsymbol{\theta}}_{p,h}^{n,k,i}\right\|_T,$$

$$\eta_{\text{alg},T,p}^{n,k,i} := \|\mathbf{r}_p^{n,k,i}\|_T.$$

3.5 Adaptive algorithm

To solve the nonlinear system (6), let us introduce the following algorithm, for $1 \leq n \leq N$.

1. Choose initial saturations $\mathbf{S}_o^{n,0}$ and pressures $\mathbf{P}^{n,0}$ according to (3). Typically, we put $\mathbf{S}_o^{n,0} = \mathbf{S}_o^{n-1}$ and $\mathbf{P}^{n,0} = P^{n-1}$. Set $k = 1$.
2. Set up the linear system (8).

    a. Choose some initial saturation $\mathbf{S}_o^{n,k,0}$ and pressure $\mathbf{P}^{n,k,0}$. Typically, we let $\mathbf{S}_o^{n,k,0} = \mathbf{S}_o^{n,k-1}$ and $\mathbf{P}^{n,k,0} = \mathbf{P}^{n,k-1}$. Set $i = 1$.
    b. Perform a step of a chosen iterative method for the solution of (8), starting from $\mathbf{S}_o^{n,k,i-1}$ and $\mathbf{P}^{n,k,i-1}$. This gives approximations $\mathbf{S}_o^{n,k,i}$ and $\mathbf{P}^{n,k,i}$.
    c. Postprocess locally the pressures $\mathbf{P}^{n,k,i}$.
    d. Construct the fluxes $\overline{\boldsymbol{\theta}}_{p,h}^{n,k,i} \in \mathrm{RTN}(\mathcal{T})$, $p \in \{\mathrm{o},\mathrm{w}\}$, according to Section 3.4.
    e. For $p \in \{\mathrm{o},\mathrm{w}\}$, from the algebraic residual vectors $\mathbf{R}_p^{n,k,i}$ construct the fluxes $\mathbf{r}_{p,h}^{n,k,i} \in \mathrm{RTN}(\mathcal{T})$, as described in Section 3.4.
    f. We evaluate all the indicators (19) and define their global versions by their Hilbertian sums. The convergence criterion for the linear solver is:

    $$\eta_{\mathrm{alg},p}^{n,k,i} \leq \gamma_{\mathrm{alg}}(\eta_{\mathrm{sp},p}^{n,k,i} + \eta_{\mathrm{tm},p}^{n,k,i} + \eta_{\mathrm{lin},p}^{n,k,i}), \qquad p \in \{\mathrm{o},\mathrm{w}\}. \tag{20}$$

    Here, $0 < \gamma_{\mathrm{alg}} \leq 1$ is a user-given weight, typically close to 1. Criterion (20) expresses that there is no need to continue with the algebraic solver iterations if the overall error is dominated by the other components. If (20) is reached, set $\mathbf{S}_o^{n,k} := \mathbf{S}_o^{n,k,i}$ and $\mathbf{P}^{n,k} := \mathbf{P}^{n,k,i}$. If not, $i := i + 1$ and go back to step 2(b).

3. The convergence criterion for the nonlinear solver is:

$$\eta_{\mathrm{lin},p}^{n,k,i} \leq \gamma_{\mathrm{lin}}(\eta_{\mathrm{sp},p}^{n,k,i} + \eta_{\mathrm{tm},p}^{n,k,i}), \qquad p \in \{\mathrm{o},\mathrm{w}\}. \tag{21}$$

Here $0 < \gamma_{\mathrm{lin}} \leq 1$ is a user-given weight, typically close to 1. Criterion (21) expresses that there is no need to continue with the linearization iterations if the overall error is dominated by the other components. If criterion (21) is reached, finish. If not, $k := k + 1$ and go back to step 1.

Additionally, for all $1 \leq n \leq N$, the space and time estimators $\eta_{\mathrm{sp},p}^n$ and $\eta_{\mathrm{tm},p}^n$ should be made of similar size.

References

1. L. AGÉLAS, D.A. DI PIETRO, AND R. MASSON, *A symmetric and coercive finite volume scheme for multiphase porous media flow with applications in the oil industry*, (2008), pp. 35–52.
2. C. CANCÈS AND M. VOHRALÍK, *A posteriori error estimtate for immiscible incompressible two-phase flows*. In preparation, 2011.
3. Z. CHEN, G. HUAN, AND Y. MA, *Computational methods for multiphase flows in porous media*, Computational Science & Engineering, Society for Industrial and Applied Mathematics (SIAM), Philadelphia, PA, 2006.
4. L. EL ALAOUI, A. ERN, AND M. VOHRALÍK, *Guaranteed and robust a posteriori error estimates and balancing discretization and linearization errors for monotone nonlinear problems*, Comput. Methods Appl. Mech. Engrg., (2010). DOI 10.1016/j.cma.2010.03.024.
5. A. ERN AND M. VOHRALÍK, *A posteriori error estimation based on potential and flux reconstruction for the heat equation*, SIAM J. Numer. Anal., 48 (2010), pp. 198–223.
6. P. JIRÁNEK, Z. STRAKOŠ, AND M. VOHRALÍK, *A posteriori error estimates including algebraic error and stopping criteria for iterative solvers*, SIAM J. Sci. Comput., 32 (2010), pp. 1567–1590.
7. C. MARLES, *Cours de production, Tome 4*, Technip, 1984.
8. R. VERFÜRTH, *Robust a posteriori error estimates for nonstationary convection-diffusion equations*, SIAM J. Numer. Anal., 43 (2005), pp. 1783–1802.
9. M. VOHRALÍK, *A posteriori error estimates, stopping criteria, and adaptivity for two-phase flows*. In preparation, 2011.

The paper is in final form and no similar paper has been or is being submitted elsewhere.

A Two-Dimensional Relaxation Scheme for the Hybrid Modelling of Two-Phase Flows

Kateryna Dorogan, Jean-Marc Hérard, and Jean-Pierre Minier

Abstract Recently, a new relaxation scheme for hybrid modelling of two-phase flows has been proposed. This one allows to obtain stable unsteady approximations for a system of partial differential equations containing non-smooth data. This paper is concerned with a two-dimensional extension of the present method, in which two alternative relaxation schemes are compared. A short stability analysis is given.

Keywords Finite Volumes, Relaxation schemes, Hybrid methods
MSC2010: 76M12, 65M12

1 Introduction

This paper deals with the modelling and the numerical simulation of polydispersed turbulent two-phase flows, where one phase is a turbulent fluid (considered to be a continuum) and the other appears as separate inclusions carried by the fluid (solid particles, droplets or bubbles). Such a kind of flows can be encountered in many industrial situations (combustion, water sprays, smokes) and in some environmental problems. Despite the need of their accurate prediction, the physical complexity of these processes is so broad that existing methods are either too expensive (in calculation cost) or not sufficiently accurate. A hybrid approach recently proposed in [2] enables to reach an acceptable compromise between the

---

Kateryna Dorogan
EDF R&D, MFEE, 6 quai Watier, F-78400 Chatou
and
LATP, CMI, 39 rue Joliot Curie, F-13453 Marseille, e-mail: kateryna.dorogan@edf.fr

Jean-Marc Hérard and Jean-Pierre Minier
EDF R&D, MFEE, 6 quai Watier, F-78400 Chatou, e-mail: jean-marc.herard@edf.fr, jean-pierre.minier@edf.fr

physical realism and a cheap numerical treatment. For two-phase flows, it consists in coupling two classic approaches (Eulerian and Lagrangian) in the particle phase description. This method allows to gather the advantages of classic approaches: high level of physical description, lower calculation costs, correct treatment of non-linearities and polydispersity, expected values free from statistical error. From now on, "L" and "E" superscripts respectively refer to all quantities calculated with the Lagrangian and Eulerian descriptions, and subscript "p" is used for the particle phase. The Lagrangian part of the particle phase description is given by the stochastic differential equations:

$$dZ_i(t) = A_i(t, Z, f(t;z), Y)dt + \sum_j B_{ij}(t, Z, f(t;z), Y)dW_j(t), \quad (1)$$

where $f(t;z)$ stands for the probability density function (pdf) of the particle state vector $Z = (x_p, U_p, U_s)$ with $x_p(t)$ the particle position, $U_p(x_p(t), t)$ the particle velocity, $U_s(x_p(t), t)$ the fluid velocity seen at the particle position and the local relative velocity $U_r = U_s - U_p$, whereas external mean fields, i.e. the fluid mean fields defined at particle locations [9, 10] are denoted by Y. A_i and B_{ij} represent the drift vector and the diffusion matrix, and $W(t)$ the vector of independent Wiener processes. Here we assume that the particles are only influenced by the drag and the gravity forces. Then, using corresponding Fokker-Planck equation we deduce from (1) a system of equations for the mean particle concentration α_p^E and flow rate $\alpha_p^E \langle U_{p,i}^E \rangle$, which represents an Eulerian description of the particle phase:

$$\partial_t \alpha_p^E + \partial_{x_i} \left(\alpha_p^E \langle U_{p,i}^E \rangle \right) = 0$$

$$\partial_t \left(\alpha_p^E \langle U_{p,i}^E \rangle \right) + \partial_{x_j} \left(\alpha_p^E \left(\langle U_{p,i}^E \rangle \langle U_{p,j}^E \rangle + \langle u_{p,i} u_{p,j} \rangle^L \right) \right) = \alpha_p^E \left(g_i + \left\langle \frac{U_{r,i}^L}{\tau_p^L} \right\rangle \right)$$
(2)

Usually, only one among the two systems (1), (2) is solved. However, in this case we are faced with shortcomings of the standard methods. In fact, system (1) contains a bias-error and thus needs calculations with a larger number of particles, whereas the Reynolds stress term $\langle u_{p,i} u_{p,j} \rangle^L$ in system (2) is not closed. The new hybrid approach consists in solving both of these systems at the same time. Thus, the terms with superscript "L", calculated with a better accuracy in the Lagrangian part of the model, are provided to the Eulerian part (2). The latter, in turn, gives the values of $\langle U_{p,i}^E \rangle$ free from statistical error, that enable computations with a smaller number of particles in (1). Hence, for the same accuracy, the total calculation cost is reduced with reference to the pure Lagrangian approach. However, such a coupling introduces noisy quantities (computed by the stochastic equations) in the Eulerian part of the model, which presents an important convective part and thus requires a stabilization. A specific relaxation approach was proposed in [3,4] in order to tackle this problem in a one-dimensional case. It relies both on upwinding techniques and relaxation tools [8] and it allows to obtain stable unsteady approximations of solutions of system (2), even with noisy data $\langle u_{p,i} u_{p,j}\rangle^L$. Actually, two slightly distinct relaxation systems were examined and compared in references [3,4]. The present paper is concerned with a two-dimensional extension of these relaxation approaches. In section 2, we propose two forms of the relaxation system that are very similar and give the motivation for such a choice. Some stability results are presented in section 3 and we briefly describe the numerical treatment and results in section 4. We recall that the density of particles is constant.

2 Relaxation approach in a two-dimensional framework

From now on, we omit the superscripts "E" and subscripts "p", and introduce the -constant- density of particles ρ_p. Thus we denote by $\rho = \alpha_p^E \rho_p$ the mean density distribution of the particles in the domain, by $U_i = \langle U_{p,i}^E \rangle, i = 1, 2$ the mean particle velocity. Hence, for given non-smooth values of the Lagrangian Reynolds stress tensor $\underline{\underline{R}}_{ij}^L = \langle u_{p,i} u_{p,j}\rangle^L$, we want to compute stable approximations of solutions of:

$$\partial_t \rho + \partial_{x_j}(\rho U_j) = 0$$
$$\partial_t(\rho U_i) + \partial_{x_j}(\rho U_i U_j) + \partial_{x_j}(\rho R_{ij}^L) = \rho g_i + \rho \langle U_{r,i}/\tau_p\rangle^L \quad (3)$$

By construction, $\underline{\underline{R}}_{ij}^L$ complies with the *realisability condition*: $\underline{x}^t \underline{\underline{R}}^L \underline{x} \geq 0$ for all $\underline{x} \in \mathcal{R}^2$. Since non-smooth external data $\underline{\underline{R}}_{ij}^L$ are introduced in the system (3), we are formally interested in finding discontinuous solutions. In order to overcome this difficulty, a relaxation technique was proposed in [5,6], which is in fact grounded on ideas developed in [1]. It consists in introducing new variables $\underline{\underline{R}}_{ij}$ (that are expected to relax towards $\underline{\underline{R}}_{ij}^L$ when a given relaxation time scale τ_p^R tends to 0), and supplementary partial differential equations that govern the time evolution of the Reynolds stresses $\underline{\underline{R}}_{ij}$, in such a way that the new relaxation system is hyperbolic and preserves the realizability of solutions ($\underline{x}^t \underline{\underline{R}} \underline{x} \geq 0$ for $\underline{x} \in \mathcal{R}^2$). On the basis of [1,7], the following relaxation system naturally arises:

$$\partial_t \rho + \partial_{x_j}(\rho U_j) = 0$$
$$\partial_t(\rho U_i) + \partial_{x_j}(\rho U_i U_j) + \partial_{x_j}(\rho R_{ij}) = 0 \quad (4)$$
$$\partial_t(\rho R_{ij}) + \partial_{x_k}(\rho U_k R_{ij}) + \rho(R_{ik}\partial_{x_k} U_j + R_{jk}\partial_{x_k} U_i) = \rho(R_{ij}^L - R_{ij})/\tau_p^R$$

Since this system is invariant under frame rotation, we consider the reference frame $(\underline{n}, \underline{\tau})$: $\underline{n} = (n_x, n_y)$, $\underline{\tau} = (-n_y, n_x)$, such that $n_x^2 + n_y^2 = 1$, for a given interface whose normal is $\underline{n}$. We also introduce: $U_n = \underline{U}.\underline{n}$, $U_\tau = \underline{U}.\underline{\tau}$, $R_{nn} = \underline{n}^t.\underline{\underline{R}}.\underline{n}$, $R_{n\tau} = \underline{n}^t.\underline{\underline{R}}.\underline{\tau} = \underline{\tau}^t.\underline{\underline{R}}.\underline{n} = R_{\tau n}$, $R_{\tau\tau} = \underline{\tau}^t.\underline{\underline{R}}.\underline{\tau}$. When neglecting transverse variations (i.e. $\forall \phi : \partial \phi/\partial \tau = 0$), the relaxation system corresponding to system

(4) written in terms of variable $Z^t = (\rho, U_n, U_\tau, \rho R_{nn}, \rho R_{n\tau}, S)$ takes the following form for smooth solutions:

$$\partial_t Z + A_n(Z)\partial_n Z = \mathscr{S}(Z), \qquad (5)$$

with: $Z = Z(t, x_n)$, $S = \left((\rho R_{nn})(\rho R_{\tau\tau}) - (\rho R_{n\tau})^2\right)/\rho^4$ and noting $\vartheta(x,t) = 1/\rho(x,t)$:

$$A_n(Z) = \begin{pmatrix} U_n & \rho & 0 & 0 & 0 & 0 \\ 0 & U_n & 0 & \vartheta & 0 & 0 \\ 0 & 0 & U_n & 0 & \vartheta & 0 \\ 0 & \Psi_{nn} & 0 & U_n & 0 & 0 \\ 0 & 2\rho R_{n\tau} & \Phi_{n\tau} & 0 & U_n & 0 \\ 0 & 0 & 0 & 0 & 0 & U_n \end{pmatrix}, \quad \mathscr{S}(Z) = \begin{pmatrix} 0 \\ 0 \\ 0 \\ \rho(R_{nn}^L - R_{nn})/\tau_p^R \\ \rho(R_{n\tau}^L - R_{n\tau})/\tau_p^R \\ (S^L - S)/\tau_p^R \end{pmatrix}$$

where: $\quad \Psi_{nn} = 3\rho R_{nn}, \quad \Phi_{n\tau} = \rho R_{nn}. \qquad (6)$

Eigenvalues of the homogeneous part of system (5) are:

$$\lambda_{1,6} = U_n \pm c_1, \quad \lambda_{2,5} = U_n \pm c_2, \quad \lambda_3 = \lambda_4 = U_n, \qquad (7)$$

with $c_1^2 = \Psi_{nn}/\rho = 3R_{nn}$ and $c_2^2 = \Phi_{n\tau}/\rho = c_1^2/3$. Thus, system (5) is hyperbolic (unless vacuum occurs in the solution) if $\Psi_{nn} > 0$ and $\Phi_{n\tau} > 0$, thus if $R_{nn} > 0$. This first method associated with the choice (6), and refered to as (**A1**), takes advantage of the hyperbolic structure of the set of PDE that governs Eulerian Reynolds stress components, while assuming classical closure laws [1]. Actually, we note that system (5) is characterized by four linearly-degenerate (LD) fields associated with $\lambda_{2,3,4,5}$ and by two genuinely non-linear (GNL) fields associated with $\lambda_{1,6}$. Details can be found in [6, 7]. A nice feature is that the whole set of partial differential equations in the evolution step preserves the realisability of the Reynolds stress tensor R_{ij}, both at the continuous and the discrete levels. This is in fact mandatory since eigenvalues remain real if and only if the quadratic form $n_i R_{ij} n_j$ remains positive (see the form of c_1, c_2 above). *However, a drawback in this approach is due to the true non-conservative form of the governing equations for the Reynolds stress components in (5).* Thus, non-conservative products that are active in genuinely non-linear fields are not uniquely defined.

This has motivated the introduction of a second form for $(\Psi_{nn}, \Phi_{n\tau})$ - corresponding to (**A2**). The aim is to comply with specifications (i,ii): (i) the relaxation system should be hyperbolic, (ii) jump conditions in the relaxation system should be uniquely defined, field by field. The idea is to introduce functions which are close to (6), but such that non-conservative products are only effective through linearly degenerate fields. Introducing $(R_{nn})_0 > 0$ and choosing functions $(\Psi_{nn}, \Phi_{n\tau})$ as:

$$\Psi_{nn} = 3\rho_0^2(R_{nn})_0\vartheta, \quad \Phi_{n\tau} = \Psi_{nn}/3 = \rho_0^2(R_{nn})_0\vartheta, \qquad (8)$$

we note that the relaxation system corresponding to *system (5) with the choice (8)* is hyperbolic and it is characterized by 6 LD fields; thus the jump relations are uniquely defined. Eigenvalues of the homogeneous part of the modified system (5) are now:

$$\lambda'_{1,6} = U_n \pm c'_1, \quad \lambda'_{2,5} = U_n \pm c'_2, \quad \lambda'_3 = \lambda'_4 = U_n, \tag{9}$$

with $(c'_1)^2 = 3(c'_2)^2 = 3\rho_0^2(R_{nn})_0 \vartheta^2$. This method associated with the choice (8) will be refered to as **(A2)**. We provide below some properties of approaches **(A1, A2)**, assuming that the initial conditions are physically relevant:

$$\rho > 0, \quad \underline{x}^t.\underline{\underline{R}}.\underline{x} > 0. \tag{10}$$

Property 1 (Existence and Uniqueness of the solution of the Riemann problem for A1). *The Riemann problem associated with (5), (6), approximate jump relations given in [6, 7], and initial conditions for left and right states Z_L, Z_R in agreement with condition (10), admits a unique solution if:*

$$(U_n)_R - (U_n)_L < \sqrt{3}\left(\sqrt{(R_{nn})_L} + \sqrt{(R_{nn})_R}\right). \tag{11}$$

The solution is composed of six constant states $Z_L, Z_1, Z_2, Z_3, Z_4, Z_R$ separated by 2 GNL waves associated with $\lambda_{1,6}$ and 4 LD waves associated with $\lambda_{2,3,4,5}$.

Property 2 (Existence and Uniqueness of the solution of the Riemann problem for A2). *Assume that $\rho_0^2(R_{nn})_0 \geq 0$ is such that it satisfies the Wave Ordering Condition (WOC): $\lambda'_1 < \lambda'_2 < \lambda'_3 = \lambda'_4 < \lambda'_5 < \lambda'_6$. Then the Riemann problem associated with (5), (8) and initial conditions Z_L, Z_R satisfying (10), admits a unique solution composed of six constant states $Z_L, Z'_1, Z'_2, Z'_3, Z'_4, Z_R$ separated by 6 LD waves. The WOC is the same as in the pure one-dimensional framework (see [3, 4]).*

Property 3 (Positivity of interface values of the density).
• The realisability in approach **(A1)** is ensured by condition (11) and density intermediate states are such that: $\rho_1 = \rho_2 > 0$, $\rho_3 = \rho_4 > 0$;
• For **(A2)**, the latter condition (11) is replaced by the WOC, that guarantees the positvity of the densities in intermediate states: $\rho'_1 = \rho'_2 > 0$, $\rho'_3 = \rho'_4 > 0$.

Remark 1 (Positivity of interface values of Reynolds stresses). *In approach* **(A1)**, *the realisability of the Reynolds stress tensor is required to ensure the hyperbolicity property for the corresponding relaxation system and, at the same time, is preserved by the very construction of this system. In approach* **(A2)** *the realisability of Reynolds stresses in the intermediate states is not preserved for any initial condition; however, the hyperbolicity of the relaxation system in* **(A2)** *holds since $\vartheta^2 > 0$ and the realisability is recovered through the instantaneous relaxation step.*

3 Stability properties of approaches A1, A2

We focus now on the evolution step in the relaxation procedure, thus on the homogeneous system corresponding to the left hand side of (4). In order to give an estimation of the mean kinetic energy, which characterises the initial system of equations (3), we focus only on smooth solutions (we assume that: $\rho(\underline{x},t), U_i(\underline{x},t), R_{ij}(\underline{x},t) \in \mathscr{C}^1, i,j = 1,2$), and we study the evolution of the "total" energy in the relaxation system (4). Let us denote by

$$\mathscr{E}_1(t) = \frac{1}{2}\int_\Omega \rho U_i^2(\underline{x},t)d\Omega \quad \text{and} \quad \mathscr{E}_2(t) = \frac{1}{2}\int_\Omega \rho\, tr(\underline{\underline{R}})(\underline{x},t)d\Omega, \quad i=1,2. \tag{12}$$

the kinetic energy of the drift (the mean motion) and the energy of the fluctuating particle motion. The total particle energy is given by $\mathscr{E}(t) = \mathscr{E}_1(t) + \mathscr{E}_2(t)$. We also assume that: $\forall \underline{x} \in \partial\Omega \quad \underline{U}.\underline{n} = 0$.

Property 4 (Energy estimation for A1). *We define:* $\delta = R_{11}R_{22} - R_{12}^2$ *and we assume that* $\delta(\underline{x} \in \partial\Omega, t > t_0) > 0, \delta(\underline{x} \in \Omega, t_0) > 0$. *Then smooth solutions of the homogeneous relaxation system corresponding to approach (A1) (left-hand side of system (4)) satisfy the following energy estimate:*

$$0 \leq \mathscr{E}_1(t) = \mathscr{E}(t_0) - \mathscr{E}_2(t) \leq \mathscr{E}(t_0), \quad since \quad \mathscr{E}_2(t) \geq 0. \tag{13}$$

An important ingredient in the proof is linked with the fact that the governing equation of $X = \delta/\rho^2$ reads:

$$\partial_t X + (\underline{U} \cdot \nabla)X = 0. \tag{14}$$

However we can only give a partial estimation for approach (**A2**). Actually, for the system corresponding to (5), (8), we must introduce a modified definition of the total energy in a *pure 1D framework* in order to get some estimation (see [4]):

$$\tilde{\mathscr{E}} = \mathscr{E}_1(t) + \mathscr{E}_2(t) + \int_\Omega \frac{\rho(a_0^2\vartheta - 3\rho R_{nn})^2}{16a_0^2}d\Omega, \quad \text{with } a_0^2 = 3\rho_0^2(R_{nn})_0, \tag{15}$$

Remark 2 (Energy estimation for A2). *In a one-dimensional framework, smooth solutions of the homogeneous relaxation system corresponding to (5), (8) satisfy:*

$$0 \leq \mathscr{E}_1(t) \quad \text{and} \quad \mathscr{E}_1(t) + \mathscr{E}_2(t) \leq \tilde{\mathscr{E}}(t_0). \tag{16}$$

4 Numerical algorithm and results

In order to compute the approximations of solutions of system (3) at each time step, the Finite Volume method relies on a classical fractional step method, which proceeds in three distinct steps (Evolution/Instantaneous Relaxation/Sources):

- **Step 1 (Evolution):** Starting from $\rho^n, (\rho U_i)^n, (\rho R_{ij})^n$, compute approximate solutions $\rho^{n+1,-}, (\rho U_i)^{n+1,-}, (\rho R_{ij})^{n+1,-}$ of the homogeneous system corresponding to the left hand side of (4) at time t^{n+1}, using an approximate Godunov solver for (**A1**) [7] or an exact Godunov solver for (**A2**) (using property 2, see [3,4]).

- **Step 2 (Relaxation):** restore local values of the Reynolds stresses $R_{ij} = R_{ij}^L$:

$$\rho^{n+1} = \rho^{n+1,-}, \quad (\rho U_i)^{n+1} = (\rho U_i)^{n+1,-}, \quad (\rho R_{ij})^{n+1} = \rho^{n+1}(R_{ij}^L)^{n+1}.$$

- **Step 3 (Sources):** account for physical source terms (right hand side of (3)).

An extensive validation of both methods (**A1, A2**) has been achieved in [3, 4] in the one-dimensional framework, by computing the L^1 norm of the error for analytical solutions of Riemann problems associated with the homogeneous part of (3), assuming specific forms for $R_{ij}^L = r_{ij}(\rho, U)$. We only show here a few computations and we put emphasis on the main conclusions.

Analytical test cases: In order to validate the two approaches (**A1, A2**), we consider some test cases where analytical solutions are known and we focus especially on the most difficult configurations. Assuming the following closure relation:

$$R_{ij}^L = S_0 \rho^{\gamma-1} \delta_{ij}$$

with $S_0 = 10^5$ and $\gamma = 3$ (this value of γ corresponds to the isentropic case arising in [1,7]), we focus on two 1D Riemann problems. The computational domain is a square $[-1, 1]^2$, the time step is in agreement with the CFL condition (CFL = 0.49), and the regular meshes contain from 2×10^2 up to 2×10^5 cells. The figures below (Fig. 1) represent the L^1-norm of the errors w.r.t. the mesh size. *On the whole, both methods (A1, A2) guarantee the correct convergence of approximations, even when the solution contains strong shocks.* This is very encouraging and not obvious since (**A1**) involves non conservative products in GNL fields, which means that we might expect to retrieve convergence of approximations towards *wrong* shock solutions. Though we have no proof at all for that, the fact that the scheme preserves the conservative form of the first two equations of (4) might explain this good behaviour. Moreover, we note (see Figs. 1) that *we retrieve the classical h^1 convergence* since no LD wave is involved here. Whereas (**A1**) and (**A2**) schemes exhibit almost the same accuracy, *(A2) seems to be a little bit more stable than (A1). Eventually, both schemes can handle vacuum occurence and strong shock waves.*

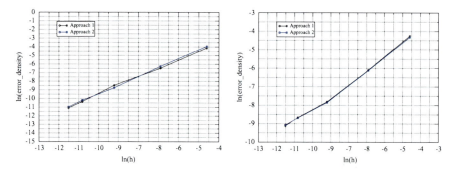

Fig. 1 L1 convergence curves for symmetric double shock (left) and symmetric double rarefaction waves with vacuum occurence (right). Coarser mesh: 200 cells; finer mesh: 200000 cells

Numerical results with noisy Reynolds stresses: We choose the initial conditions of a subsonic shock tube problem and we plug noisy Reynolds stresses in the system of equations (3) at each time step in the cells that belong to the region $(x, y) \in [-0.25, 0.25] \times [-1, 1]$ with: $R_{ij}^L = S_0 \rho^{\gamma-1}(1 + rms(0.5 - rand(0, 1)))\delta_{ij}$, where rms stands for the noise intensity and $rand$ allow to manage the noise amplitude. The noisy region is not developping in time (Fig. 2). The same remark holds for other values of the noise intensity. Other test cases with noisy data [3] show that the noise is independent of the mesh refinement. Eventually, the difference between noisy approximations and those without a noise is increasing with rms in a linear manner. Both methods are stable.

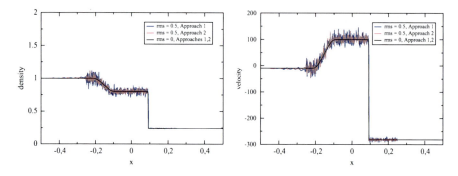

Fig. 2 Approximations of the density (left) and the velocity (right) with rms = 0.5 and rms = 0 in time. Mesh size: 1000 cells in the x-direction

Acknowledgements The first author receives a financial support through an EDF-CIFRE contract 203/2009. Computational facilities were provided by EDF.

References

1. Berthon, C., Coquel, F., Hérard, J.M., Uhlmann, M.: An approximate solution of the Riemann problem for a realisable second-moment turbulent closure. Shock Waves, **11**, 245–269 (2002)
2. Chibbaro, S., Hérard, J.M., Minier, J.P.: A novel Hybrid Moments/Moments-PDF method for turbulent two-phase flows. Final Technical Report Activity Marie Curie Project. TOK project LANGE Contract MTKD-CT-2004 509849 (2006)
3. Dorogan, K., Hérard, J.M., Minier, J.P.: Development of a new scheme for hybrid modelling of gas-particle two-phase flows. EDF report H-I81-2010-2352-EN, unpublished, 1–50 (2010)
4. Dorogan, K., Hérard, J.M., Minier, J.P.: A relaxation scheme for hybrid modelling of gas-particle flows. Submitted (2011)
5. Hérard, J.M.: A relaxation tool to compute hybrid Euler-Lagrange compressible models. AIAA paper 2006-2872 (2006) http://aiaa.org
6. Hérard, J.M., Minier, J.P., Chibbaro, S.: A Finite Volume scheme for hybrid turbulent two-phase flow models. AIAA paper 2007-4587, http://aiaa.org
7. Hérard, J.M., Uhlmann, M., Van der Velden, D.: Numerical techniques for solving hybrid Eulerian Lagrangian models for particulate flows. EDF report H-I81-2009-3961-EN (2009)
8. Jin, S., Xin, Z.: The relaxation schemes for systems of conservation laws in arbitrary space dimensions. Comm. Pure Appl. Math., **48**, 235–276 (1995)
9. Minier, J.P., Peirano, E.: The pdf approach to polydispersed turbulent two-phase flows. Physics reports, **352**, 1–214 (2001)
10. Peirano, E., Chibbaro, S., Pozorski, J., Minier, J.P.: Mean-field/PDF numerical approach for polydispersed turbulent two-phase flows. Prog. Ene. Comb. Sci., **32**, 315–371 (2006)

The paper is in final form and no similar paper has been submitted elsewhere.

Finite Volume Method for Well-Driven Groundwater Flow

Milan Dotlić, Dragan Vidović, Milan Dimkić, Milenko Pušić, and Jovana Radanović

Abstract Finite volume method for well-driven porous media flow which uses a computational mesh tailored for finite elements is presented. It replaces one-dimensional elements used to model well drains in the original mesh with one-dimensional cells. It does not modify the original mesh by adding or moving nodes. It can handle the discontinuous anisotropic hydraulic conductivity. Special discretization of the flux between the porous medium and the drain is proposed. Numerical results are compared to an analytical solution.

Keywords Finite volume method, well-driven flow, porous media, privileged routes, 1d elements
MSC2010: 76S05, 65N08

1 Introduction

A significant number of finite element codes for well-driven groundwater flow simulation is available (FEFLOW [1], HydroGeoSphere [6], PAKP–Lizza [2], etc). Well drains are represented in these codes as arrays of one-dimensional elements, i.e. mesh edges. Triangulators, such as Triangle [5], allow the user to specify the exact location of these drains prior to triangulation, and place the mesh nodes and the edges at the specified locations.

Milan Dotlić, Dragan Vidović, Milan Dimkić, and Jovana Radanović
Jaroslav Černi Institute, Jaroslava Černog 80, 11226 Pinosava, Belgrade, Serbia,
e-mail: milandotlic@gmail.com, draganvid@gmail.com jdjcerni@jcerni.co.rs, jovanaradanovic@gmail.com

Milenko Pušić
University of Belgrade, Faculty of Mining and Geology, Đušina 7, 11000 Belgrade, Serbia,
e-mail: mpusic@ptt.rs

This arrangement is appropriate for finite element method because it associates the discrete variables with the mesh nodes. Cell-centered finite volume methods associate the discrete variables with mesh cells. Thus, matching the cell center with the exact well location and representing a drain as an array of cells would be more appropriate.

Since numerous tools exist to construct finite element meshes, our goal is to find a suitable way to use these meshes with finite volumes.

One possibility to obtain a suitable finite volume mesh is to construct a dual of a finite element mesh. However, the conductivity, which is associated with the finite element mesh cells and may be discontinuous between the cells, will now be associated with the nodes of the dual mesh. In finite volumes the conductivity is also associated with cells. Therefore, some kind of interpolation must be performed in order to compute the dual cell conductivity, which may introduce significant error because the conductivity may vary by several orders of magnitude between geological layers.

Another possibility is to associate a fictive one-dimensional cell with each 1d element. These new cells are connected with the surrounding three-dimensional cells by one-dimensional faces, and with each other by zero-dimensional faces. In order to compute non-zero fluxes and finite hydraulic heads, physical surfaces and volumes of the real drain portions must be associated with these new entities.

In this paper we present details of such discretization. Groundwater flow equation is given in Section 2, together with boundary conditions and a well clogging model. Interpretation of the mesh is explained in Section 3, and the flux and the boundary conditions discretization is specified. Obtained numerical results are compared to an analytical solution in Section 4, and a correction to the flux discretization between the porous medium and a drain is proposed.

2 Problem formulation

Correlation between the hydraulic head gradient ∇h and the flux density $\mathbf{q}$ is known as Darcy's law

$$\mathbf{q} = -K \nabla h, \tag{1}$$

where the hydraulic conductivity K is in general a symmetric positive piecewise-continuous anisotropic tensor.

Mass conservation is expressed trough the groundwater flow equation

$$S \frac{\partial h}{\partial t} = -\nabla \cdot \mathbf{q}, \tag{2}$$

where S is the specific storage. Substituting (1) into (2) results in a form suitable for solving

$$S \frac{\partial h}{\partial t} = \nabla \cdot (K \nabla h). \tag{3}$$

Domain boundary is divided in two parts $\partial\Omega = \Gamma_D \cup \Gamma_N$, $\Gamma_D \cap \Gamma_N = \emptyset$, and Dirichlet and Neumann boundary conditions are specified:

$$h = g_D \quad \text{on } \Gamma_D, \tag{4}$$

$$(-K\nabla h)\mathbf{n} = g_N \quad \text{on } \Gamma_N, \tag{5}$$

where $\mathbf{n}$ is the outward unit normal to $\partial\Omega$. In addition to these common boundary conditions, either the hydraulic head or the total flux per unit of time Q is specified in each well.

Initial condition is

$$h|_{t=0} = h_0, \tag{6}$$

and the final time is $t = T$.

Drain clogging, which happens due to complex mechanical, chemical, and biological processes [3, 4], results in a colmated layer along the drain wall, which causes an additional hydraulic resistance. This can be expressed as

$$\mathbf{q} \cdot \mathbf{n} = \Psi(h_f - h_w), \tag{7}$$

where h_w is the hydraulic head inside the well, h_f is the hydraulic head just outside the colmated layer, $\mathbf{n}$ is the unit normal to the drain wall pointing inside, $\Psi = K_c/d_c$ is the transfer coefficient, K_c is the unknown conductivity of the colmated layer, and d_c is it's unknown thickness.

3 Discretization

Integrating (3) over polyhedral control volume T, and using the divergence theorem and implicit Euler time integration results in

$$|T|S\frac{h^{n+1} - h^n}{\Delta t} = \sum_{f \in \partial T} \chi_{T,f} Q_f^{n+1}, \quad Q_f = \int_f \mathbf{q} \cdot \mathbf{n}_f ds, \tag{8}$$

where Q_f is the flux through face f, $\mathbf{n}_f$ is a unit vector normal to face f fixed once and for all, and $\chi_{T,f} = 1$ if $\mathbf{n}_f$ points outside of T, or -1 otherwise. At boundary faces, fixed normal vectors point outside.

3.1 Drains

It is assumed that well drains coincide with the mesh edges (see Fig. 1). Each well may have one or more connected drains.

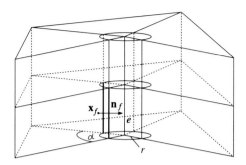

Fig. 1 Drain discretization

A cylindrical cell called 1d cell is associated with each edge e belonging to a well drain. This cell is logically plugged into the grid by defining interfaces between it and the surrounding cells, but nodes are not added or moved. A volume $r^2\pi|e|$, where r is the drain radius and $|e|$ is the edge length, is associated with this cell. Each volume T sharing the edge e is reduced by $r^2\frac{\alpha}{2}|e|$, where α is the angle between the faces of the cell T sharing the edge e.

Interface f between the 3d cell T and a 1d cell is called 1d face. This is a portion of the cylinder with surface $\alpha r|e|$. Unit normal $\mathbf{n}_f$ belongs to the bisector plain of the angle α and it points into the cylinder. Center $\mathbf{x}_f$ of the face f is obtained by shifting the centroid of the 1d cell by vector $-r\mathbf{n}_f$.

Interfaces between the 1d cells are so-called 0d faces. These are circles with radius r associated with nodes where the drain edges meet.

If more drains meet in a node, a 0d cell is introduced. This cell has zero volume and it is connected with each of the drains by a 0d face.

Hydraulic head or the total well flux is specified at a boundary 0d face, which may be at a drain end, or it may be an extra face introduced in a 0d cell in order to impose a boundary condition.

Hagen–Poiseuille law is used for a flow through a pipe, which means that within the drains we take $k = r^2/8$, $K = k\rho g/\mu$, where $\rho = 1000\text{kg/m}^3$ is the water density, $\mu = 0.001307\text{kg/(ms)}$ is the dynamic viscosity of water, and $g = 9.81\text{m/s}^2$.

3.2 Flux

It is assumed that K is continuous within cells. Possible discontinuities happen along faces. At the interface between a 3d cell and a drain, K is discontinuous.

If f is an internal face and if K is continuous in f, then

$$Q_f = -|f|\|K\mathbf{n}\|\frac{h_{out} - h_{in}}{\|\mathbf{x}_{out} - \mathbf{x}_{in}\|}, \tag{9}$$

where $|f|$ is the face f area, and $\mathbf{x}_{in}$ and $\mathbf{x}_{out}$ are the centroids of the cells sharing the face f such that $\mathbf{n}$ points from cell in to cell out. This is the most basic finite volume flux discretization used here for simplicity, and it is not very accurate. A more accurate non-linear flux discretization [7] is planned.

If K is discontinuous in f, then two one-sided flux approximations

$$Q_f = -|f| \|K_{out}\mathbf{n}\| \frac{h_{out} - h_f}{\|\mathbf{x}_{out} - \mathbf{x}_f\|}, \qquad Q_f = -|f| \|K_{in}\mathbf{n}\| \frac{h_f - h_{in}}{\|\mathbf{x}_f - \mathbf{x}_{in}\|} \qquad (10)$$

are combined to compute the face hydraulic head and eliminate it from (10):

$$Q_f = -|f| \frac{\Psi_{in} \Psi_{out}}{\Psi_{in} + \Psi_{out}} (h_{out} - h_{in}), \qquad (11)$$

$$\Psi_{in} = \frac{\|K_{in}\mathbf{n}\|}{\|\mathbf{x}_{in} - \mathbf{x}_f\|}, \qquad \Psi_{out} = \frac{\|K_{out}\mathbf{n}\|}{\|\mathbf{x}_f - \mathbf{x}_{out}\|}. \qquad (12)$$

At the interface between a drain and a 3d cell, transfer coefficient Ψ defined in (7) is substituted in (11) instead of Ψ_{out}.

Second formula in (10) is inadequate when the drain radius is much smaller than the mesh size. This is demonstrated in Section 4, and a correction is proposed.

3.3 Boundary conditions

If f is a boundary face, one-sided flux approximation is used:

$$Q_f = -|f| \|K\mathbf{n}\| \frac{h_f - h_{in}}{\|\mathbf{x}_f - \mathbf{x}_{in}\|}. \qquad (13)$$

Flux Q_f or hydraulic head h_f imposed at f is used trough this relation. If f is a 0d face, then $K = K_{drain}$.

4 Flux correction

Example 1. If K is constant, then

$$h(\rho) = A + B \ln \rho \qquad (14)$$

is a stationary solution of (3), where ρ is the distance from the well central axis. If domain Ω is a cylinder with radius R and a well of radius r at the center (see Fig. 2), then A and B can be found from the requirement that $h(r) = h_r$ and $h(R) = h_R$,

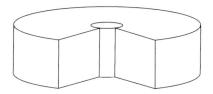

Fig. 2 Example domain

for some specified h_r and h_R. The resulting solution is

$$h(\rho) = \frac{h_r \ln \frac{R}{\rho} + h_R \ln \frac{\rho}{r}}{\ln \frac{R}{r}}. \tag{15}$$

Total well flux is

$$Q = 2\pi K H \frac{h_R - h_r}{\ln \frac{R}{r}}, \tag{16}$$

where H is the cylinder height. Axial well resistance has been neglected here.

In order to incorporate the resistance of the colmated layer, we compute Ψ from (7) using the exact flux obtained in (16), so that we obtain a desired hydraulic head decrease in the well $h_r - h_w$ due to colmation.

We choose $K = 10^{-4}$, $R = 20$, $H = 10$, $h_R = 100$, $h_r = 60$, $h_w = 55$, and compute fluxes for two different well radii, $r = 0.5$ and $= 0.01$, to test the scheme in cases when the well radius is close to the mesh size, and when it is much smaller than the mesh size. The computational mesh with the maximal base triangle area of 0.5 is shown in Fig. 3.

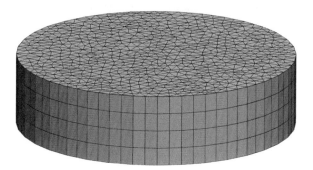

Fig. 3 Computational grid

Exact hydraulic head is specified at the inner and the outer cylinder. Zero flux is specified at the flat boundaries.

The exact and the numerical fluxes are given in Table 1. If the well radius is much smaller than the mesh size, the flux is about ten times smaller than what it should

Table 1 Exact and numerical fluxes for $r = 0.01$ and $r = 0.5$, with and without the proposed correction

	Exact Q	Numerical Q	Q with correction
$r = 0.5$	0.06813	0.06634	0.06894
$r = 0.01$	0.03306	0.00322	0.03312

be. The reason is that $h'(\rho)$ is very sharp at $\rho = r$, and it is not well approximated with a finite difference.

To allow the computation on coarse meshes, we replace R in (15) with the distance between $\mathbf{x}_{in}$ and the central well axis $\rho(\mathbf{x}_{in})$, and use the derivative of this formula to derive a replacement for the second formula in (10) for the case of a 3d-1d cell interface:

$$Q_f = -|f| \|K_{in}\mathbf{n}\| \frac{h_f - h_{in}}{r \ln \frac{\rho(\mathbf{x}_{in})}{r}}. \qquad (17)$$

Results presented in the last column of Table 1 show that the total well flux computed with this correction is much more accurate.

Example 2. Another stationary analytical solution of (3) can be obtained by superposing a linear solution $h = cy$ to (15), where c is an arbitrary constant:

$$h = \frac{h_r \ln \frac{R}{\rho} + h_R \ln \frac{\rho}{r}}{\ln \frac{R}{r}} + cy. \qquad (18)$$

The fluxes in the well which are due to the linear term cancel out, and the total well flux is the same as in Example 1. This solution is not constant at $\rho = r$ and we cannot set such a boundary condition in our method. However, presuming that r is small, h varies little around the well, thus specifying a constant h_r should give a close approximation. We take $c = 1$ and specify the exact hydraulic head at $\rho = R$. The obtained total well flux given in Table 2 shows that correction (17) improves the accuracy even if formula (15) on which the correction is based is not the exact solution, because the influence of the well is still dominant near the well.

Table 2 Numerical fluxes in Example 2 for $r = 0.01$ with and without the correction

Q without correction	Q with correction
0.00322	0.03309

5 Conclusion

We have presented a finite volume method for well-driven groundwater flow that uses a computational grid tailored for a finite element method, in which well drains are represented by one-dimensional elements. In its interpretation of the grid, our method adds faces and cells that correspond to well drains with geometry that is not fully resolved in the original grid. However, the original grid is not modified in the sense that nodes are not added or moved.

We compared the obtained numerical fluxes with the analytical solution in cases when the well radius is close to the grid resolution, and also when it is much smaller than the grid resolution, and we found that the match is poor for the small radius case. We proposed a correction to the discretization of the flux between the 3d porous medium and a drain. Total well flux obtained with this correction is very accurate in all cases, bearing in mind that the inaccurate linear two-point flux discretization was used.

References

1. H.-J.G. Diersch. *FEFLOW 5.3 user's manual*. WASY GmbH, Berlin, Germany, 2006.
2. M. Dimkić, M. Pušić, D. Vidović, N. Filipović, V. Isailović, and B. Majkić. Numerical model assessment of radial-well ageing. *ASCE's Journal of computing in civil engineering*, 25(1):43–49, 2010.
3. M. Dimkić and M. Pušić. Preporuke za projektovanje bunara uzevši u obzir kolmiranje gvoždjem na osnovu iskustva beogradskog izvorišta. *Gradjevinski kalendar*, 40:430–496, 2008.
4. M. Dimkić, M. Pušić, V. Obradović, and D. Djurić. Several natural indicators of radial well ageing at the belgrade groundwater source, part 2. *Water Science and Technology*, 2011. submitted WST-S-10-02140[1].
5. J. R. Shewchuk. Triangle: Engineering a 2d quality mesh generator and delaunay triangulator. *Computational Geometry: Theory and Applications*, 22(1):21–74, 2002.
6. R. Therrien, R.G. McLaren, E.A. Sudicky, and S.M. Panday. *HydroGeoSphere - A three-dimensional Numerical Model Describing Fully-integrated Subsurface and Surface Flow and Solute Transport*. Groundwater Simulations Group, 2010. Available on Internet:www.hydrogeosphere.org/hydrosphere.pdf.
7. D. Vidović, M. Dimkić, and M. Pušić. Accelerated non-linear finite volume method for diffusion. *Journal of Computational Physics*, 2011. doi: 10.1016/j.jcp.2011.01.016.

The paper is in final form and no similar paper has been or is being submitted elsewhere.

Adaptive Reduced Basis Methods for Nonlinear Convection–Diffusion Equations

Martin Drohmann, Bernard Haasdonk, and Mario Ohlberger

Abstract Many applications from science and engineering are based on parametrized evolution equations and depend on time-consuming parameter studies or need to ensure critical constraints on the simulation time. For both settings, model order reduction by the reduced basis methods is a suitable means to reduce computational time. In this proceedings, we show the applicability of the reduced basis framework to a finite volume scheme of a parametrized and highly nonlinear convection–diffusion problem with discontinuous solutions. The complexity of the problem setting requires the use of several new techniques like parametrized empirical operator interpolation, efficient a posteriori error estimation and adaptive generation of reduced data. The latter is usually realized by an adaptive search for base functions in the parameter space. Common methods and effects are shortly revised in this presentation and supplemented by the analysis of a new strategy to adaptively search in the time domain for empirical interpolation data.

Keywords Finite volume methods, model reduction, reduced basis methods, empirical interpolation
MSC2010: 65M08, 65J15, 65Y20

1 Introduction

Reduced basis (RB) methods are popular methods for model order reduction of problems with parametrized partial differential equations that need to be solved for many parameters. Such scenarios might occur in parameter studies, optimization,

---

M. Drohmann and M. Ohlberger
Institute of Computational and Applied Mathematics, University of Muenster, Einsteinstr. 62, 48149 Muenster, e-mail: mdrohmann,ohlberger@uni-muenster.de

B. Haasdonk
Institute of Applied Analysis and Numerical Simulation, University of Stuttgart, 70569 Stuttgart, e-mail: haasdonk@mathematik.uni-stuttgart.de

control, inverse problems or statistical analysis for a given parametrized problem. Such problems deal with different solutions $u_h(\mu) \in \mathscr{W}_h$ from a high dimensional discrete function space $\mathscr{W}_h \subset L^2(\Omega)$ which are characterized by a parameter $\mu \in \mathscr{P} \subset \mathbb{R}^p$. For evolution problems, a discrete solution forms a series of what we call "snapshot solutions" $u_h^k(\mu)$ indexed by a time-step number $k = 0, \ldots, K$.

By applying the reduced basis method, these solution trajectories need to be computed for a few parameters only and can then be used to span a problem-specific subspace $\mathscr{W}_{\text{red}} \subset \mathscr{W}_h$. If this subspace captures a broad solution variety, a numerical scheme based on this reduced basis space $\mathscr{W}_{\text{red}}$ can produce reduced solutions $u_{\text{red}}(\mu) \in \mathscr{W}_{\text{red}}$ very inexpensively for every parameter $\mu \in \mathscr{P}$. In case of nonlinear discretizations or complex dependencies of the equations on the parameter, the reduced scheme requires an empirical interpolation method [1] to efficiently interpolate operator evaluations in a low-dimensional discrete function space.

The applicability of the reduced scheme has been successfully demonstrated for stationary, instationary, linear and nonlinear problems mainly based on finite element schemes (cf. [7] and the references therein). In this presentation, we focus on a scalar, but highly nonlinear convection–diffusion problem: For a given parameter $\mu \in \mathscr{P}$ determine solutions $u = u(x, t; \mu)$ fulfilling

$$\partial_t u + \nabla \cdot (\mathbf{v}(u; \mu)u) - \nabla \cdot (d(u; \mu)\nabla u) = 0 \qquad \text{in } \Omega \times [0, T_{\max}] \qquad (1)$$

$$u(0; \mu) = u_0(\mu) \qquad \text{in } \Omega \times \{0\} \qquad (2)$$

plus Dirichlet boundary conditions $u(\mu) = u_{\text{dir}}(\mu)$ on $\Gamma_{\text{dir}} \times [0, T_{\max}]$, Neumann boundary conditions $(\mathbf{v}(u; \mu)u - d(u; \mu)\nabla u) \cdot \mathbf{n} = u_{\text{neu}}(\mu)$ on $\Gamma_{\text{neu}} \times [0, T_{\max}]$ with suitable parametrized functions $\mathbf{v}(\cdot; \mu) \in C(\mathbb{R}, \mathbb{R}^d)$ and $d(\cdot; \mu) \in C(\mathbb{R}, \mathbb{R}^+)$.

For complex data functions, solutions of this problem can depend on the parameter in a highly nonlinear way, and the convection term can lead to a variety of solution snapshots which is difficult to capture by a linear subspace $\mathscr{W}_{\text{red}}$. This makes the construction of the reduced basis space $\mathscr{W}_{\text{red}}$ difficult and therefore requires sophisticated construction algorithms for the reduced data. After elaborating on the empirical operator interpolation and the reduced basis scheme for problem (1)-(2) in Section 2, we provide an overview of such algorithms in Section 3 with a focus on the time-adaptive construction of interpolation for the empirical interpolation. In Section 4, we numerically discuss the effects and costs of the introduced algorithms based on a finite volume discretization of a Buckley–Leverett type problem.

2 Reduced basis method

In this section, we present a reduced basis method for general operator based discretizations of equations (1), (2). We show that the reduced scheme depends both in memory and computational complexity on the low dimensions of suitable reduced spaces only and can therefore be efficiently evaluated. We first introduce the basic approach, and discuss the main ingredients to efficiently compute the reduced solutions at the end of this section. For a more detailed presentation, we refer to [2].

As a starting point for the reduced basis scheme, we assume a high dimensional discretization scheme producing for each parameter $\mu \in \mathscr{P}$ a sequence of solution snapshots $u_h^k(\mu)$ stemming from an H-dimensional discrete function space $\mathscr{W}_h$. The sequence indices $k = 0, \ldots, K$ correspond to strictly increasing time steps $t^k := k \Delta t$ from the interval $[0, T_{\max}]$, where $\Delta t > 0$ is a global time step size. For the high-dimensional scheme, first, the initial data is projected on the discrete function space $\mathscr{W}_h$ yielding a discrete solution $u_h^0(\mu) = \mathscr{P}_h [u_0(\mu)]$, where $\mathscr{P}_h : L^2(\Omega) \to \mathscr{W}_h$ is a projection operator. Subsequently, equations of the form

$$(\mathrm{Id} + \Delta t \mathscr{L}_I(\mu)) \left[u_h^{k+1}(\mu) \right] - (\mathrm{Id} + \Delta t \mathscr{L}_E(\mu)) \left[u_h^k(\mu) \right] = 0 \qquad (3)$$

are solved with the Newton–Raphson method. The operators $\mathscr{L}_I(\mu), \mathscr{L}_E(\mu) : \mathscr{W}_h \to \mathscr{W}_h$ describe the explicit and implicit discretization terms of a first order Runge–Kutta scheme. For our numerical experiments presented in Section 4, the operators implement finite volume fluxes for the diffusive respectively convective terms.

For the reduced basis scheme, we first assume a given reduced basis space $\mathscr{W}_{\mathrm{red}} \subset \mathscr{W}_h$ of dimension $N \ll H$. This space is spanned by selected solution snapshots and its construction implies a computationally expensive preprocessing step. This allows to solve for reduced solutions $u_{\mathrm{red}}^k(\mu) \in \mathscr{W}_{\mathrm{red}}$. These are computed by projection of the initial data on the reduced basis space and with the same evolution scheme as in (3), but with the operators $\mathscr{L}_I(\mu), \mathscr{L}_E(\mu)$ substituted by reduced counterparts

$$\mathscr{L}_{\mathrm{red},I}^{k+1}(\mu) := \mathscr{P}_{\mathrm{red}} \circ \mathscr{I}_{M^{k+1}}^{k+1} \circ \mathscr{L}_I(\mu) \quad \text{and} \quad \mathscr{L}_{\mathrm{red},E}^k(\mu) := \mathscr{P}_{\mathrm{red}} \circ \mathscr{I}_{M^k}^k \circ \mathscr{L}_E(\mu) \qquad (4)$$

at each time instance $k = 0, \ldots, K-1$. Here, $\mathscr{P}_{\mathrm{red}} : \mathscr{W}_h \to \mathscr{W}_{\mathrm{red}}$ is a further projection operator and the actual operator evaluations are substituted by approximations in a further low dimensional function space $\mathscr{W}_M \subset \mathscr{W}_h$. This approximation, the so-called *empirical operator interpolation*, is denoted by $\mathscr{I}_M \circ \mathscr{L}$ and shortly summarized in the next subsection. Note that in this scheme the empirical operator interpolation and therefore also the reduced function spaces can vary over time.

Empirical operator interpolation and offline/online splitting The idea of empirical interpolation was first introduced in [1]. The empirical operator interpolation presented here is extracted from [2].

The principal idea is to interpolate functions $v_h \in \mathscr{W}_h$ in a *collateral reduced basis space* $\mathscr{W}_M$ spanned by basis functions $q_m, m = 1, \ldots, M$ with exact evaluations at interpolation points $x_m \in T_M$, i.e.

$$\mathscr{I}_M [v_h](x_m) = \sum_{m=1}^{M} \sigma_m q_m(x_m) = v_h(x_m), \qquad (5)$$

where the coefficients can be determined easily because the construction process for the basis functions ensures for each $m = 0, \ldots, M$ that the condition $q_m(x_{m'}) = 0$ is fulfilled for all $m' < m$. By optimizing the collateral reduced basis space such that it well approximates operator evaluations $\mathscr{L}_h(\boldsymbol{\mu})[v_h] \in \mathscr{W}_h$ of a parameterized discrete operator $\mathscr{L}_h(\boldsymbol{\mu})$ on solution snapshots v_h, we obtain an approximation $\mathscr{I}_M[\mathscr{L}_h(\boldsymbol{\mu})[v_h]] \approx \mathscr{L}_h(\boldsymbol{\mu})[v_h]$ which can be computed by evaluating the operator locally at M given interpolation points. If such an evaluation depends only on a few degrees of freedom of the argument function (H independent Dof-dependence) and $M \ll H$, the interpolation can be computed very efficiently. The interpolant is therefore suitable for the reduced basis method. Furthermore, it can be verified that the same argumentation also applies to Fréchet derivatives of discrete operators fulfilling the H independent Dof-dependence. This result is needed for the efficient implementation of the Newton–Raphson method. In Section 3.2, we summarize how the discrete function space $\mathscr{W}_M$ can be constructed by a greedy search algorithms in a finite set of operator evaluations.

In order to evaluate the reduced numerical scheme efficiently, the high dimensional data needs to be precomputed in an expensive offline phase and to be reduced to low-dimensional matrices and vectors. Afterwards, every Newton step of a reduced simulation can be computed with complexity $\mathscr{O}(NM^2 + N^3)$ including the costs of the linear equation solver. In [2], the computations leading to these results are presented in detail. The same article also introduces an efficiently computable a posteriori error estimator $\eta(\boldsymbol{\mu})$ estimating the error

$$\max_{k=0,\ldots,K} \left\| u_h^k(\boldsymbol{\mu}) - u_{\text{red}}^k(\boldsymbol{\mu}) \right\| \leq \eta(\boldsymbol{\mu}) \tag{6}$$

for a suitable problem-specific norm $\|\cdot\|$.

3 Adaptive basis generation strategies

In this section, we give an introduction on how reduced basis functions and empirical interpolation data are constructed by algorithms that greedily search in a finite subset of the parameter space for new basis functions. For complex parameter sets or complex dependencies of the solution on the parameter, these algorithms, however, can result in very large reduced basis spaces and therefore make the speed advantages of the reduced simulations obsolete. For this reason, we also discuss variations of the algorithms adapting the parameter search set during the basis construction which lead to smaller and better basis spaces.

3.1 POD-greedy algorithm

The "POD-greedy" algorithm introduced in [4] is used to generate the reduced basis space $\mathscr{W}_{\text{red}}$. Its purpose is to minimize the error $\|u_h(\mu) - u_{\text{red}}(\mu)\|$ for all $\mu \in \mathscr{P}$ in a suitable problem-specific norm. We assume the existence of an estimator $\eta(\mu)$ as introduced in (6), a finite training set $M_{\text{train}} \subset \mathscr{P}$ and an initial choice for the reduced basis $\Phi_{N_0} := \{\varphi_n\}_{n=1}^{N_0}$. For evolution problems, the span of this initial reduced basis usually comprises all initial data functions. Then, the reduced basis can be iteratively extended by searching for the parameter $\mu_{\max} := \arg\max_{\mu \in M_{\text{train}}} \eta(\mu)$ of the worst approximated trajectory, and adding the first and most significant mode gained from a proper orthogonal decomposition of this trajectory's projection errors $\{u_h^k(\mu_{\max}) - \mathscr{P}_{\text{red}}[u_h^k(\mu_{\max})]\}_{k=0}^{K}$ as a new basis function. This algorithm is repeated, until $\eta(\mu_{\max})$ falls beneath a given tolerance.

Adaptation techniques: The basic algorithm described above depends on a fixed initial choice for the training subset M_{train}. In case of complex dependencies of the solution trajectories on the parameter, the reduced basis approximation can therefore turn out to be very bad for parameters not in the training set. In [6] this problem is addressed by adaptively refining the parameter space if indicated by bad approximations from a further validation training set.

Other variations of the POD-Greedy algorithm adaptively partition the parameter space and construct different reduced bases for each of these partitions [3,5] leading to faster reduced simulations at the cost of a more expensive offline phase.

3.2 Time-adaptive empirical operator interpolation

The construction of the collateral reduced basis space and corresponding interpolation points follows a similar idea like the "POD-greedy" algorithms. For the empirical interpolation of an operator $\mathscr{I}_M \circ \mathscr{L}_h$, the interpolation error $\left\| v_h - \sum_{j=1}^{M} \mathscr{I}_M[v_h] \right\|$ is minimized over all $v_h \in \mathbf{L} := \{\mathscr{L}_h(\mu)[u_h^k(\mu)] \mid \mu \in \mathscr{P}, k = 0, \cdots, K - 1\}$. Analogously to the reduced basis generation, we define a finite subset $L_{\text{train}} \subset \mathbf{L}$ and pick one of this set's snapshots as an initial collateral reduced basis function. The extension step for the empirical interpolation then looks as follows:

1. Find the approximation with the worst error $v_M \leftarrow \arg\sup_{v_h \in L_{\text{train}}} \|u_h - \mathscr{I}_M[v_h]\|$.
2. Compute the residual between v_M and its interpolant $r_M \leftarrow v_M - \mathscr{I}_M[v_h]$.
3. Find the interpolation point maximizing the residual $x_M \leftarrow \arg\sup_{x \in X_h} |r_M(x)|$.
4. Normalize to construct new reduced basis space function $q_M \leftarrow \frac{r_M}{r_M(x_M)}$.

These steps are repeated until the maximum interpolation error falls beneath a given tolerance. We call this algorithm EIDETAILED in the sequel.

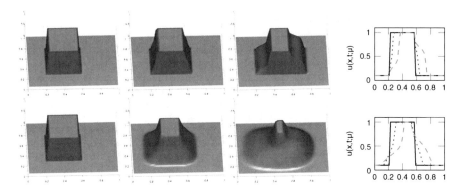

Fig. 1 Detailed simulation solution snapshots at time instants $t = 0.0$ (first column), $t = 0.1$ (second column), $t = 0.3$ (third column) and for different parameters $\mu = (0, 0.1, 0.4)$ (first row) and $\mu = (2, 0.1, 0.4)$ (second row). The last column shows the reduced solution on cross-sections at $y = 0.5$ for the time instants $t = 0$ (solid line), $t = 0.1$ (dotted line), $t = 0.3$ (dashed line)

Adaptation techniques: The adaptation techniques mentioned in Section 3.1 can also be applied to the empirical interpolation algorithm EIDETAILED, but so far no actual implementation for this is known to us. Supplementary to the adaptive search in the parameter space, we propose to build different collateral reduced basis spaces for different time instant sets $\mathscr{K} \subset \{0, \ldots, K-1\}$. As this time-adaptation strategy is the main focus of this article, we want to give a detailed description of the algorithm:

procedure TIMESLICEDEI($\mathscr{W}_{\text{init}}, \mathscr{K}, L_{\text{train}}^{\mathscr{K}}$)
  $\mathscr{W}_M \leftarrow$ EIDETAILED($\mathscr{W}_{\text{init}}, L_{\text{train}}^{\mathscr{K}}, M_{\max}, \varepsilon_{tol}$)
  if ε_{tol} reached **then**
    $M^k \leftarrow M$ and $\mathscr{W}_{M^k}^k \leftarrow \mathscr{W}_M$ for all $k \in \mathscr{K}$.
  else if card($\mathscr{K}$) $\leq 2c_{\min}$ **then**
    $\mathscr{W}_{M^k}^k \leftarrow$ EIDETAILED($\mathscr{W}_M, L_{\text{train}}^{\mathscr{K}}, \infty, \varepsilon_{tol}$) for all $k \in \mathscr{K}$.
  else % *maximum number of extensions $M_{\max}$ reached*
    $\mathscr{K}_1, \mathscr{K}_2 \leftarrow$ SPLITTIMEINTERVAL($\mathscr{K}, \mathscr{W}_M$)
    TIMESLICEDEI($\mathscr{W}_M^{\mathscr{K}_1}, L_{\text{train}}^{\mathscr{K}_1}$)
    TIMESLICEDEI($\mathscr{W}_M^{\mathscr{K}_2}, L_{\text{train}}^{\mathscr{K}_2}$)
  end if
end procedure

The training sets $L_{\text{train}}^{\mathscr{K}}$ are restrictions of the full training set L_{train} to operator evaluations on solutions snapshots at time steps t^k for $k \in \mathscr{K}$. Likewise $\mathscr{W}_M^{\mathscr{K}}$ is a restriction of the discrete space $\mathscr{W}_M$ build only out of solution snapshots with time indices stemming from $\mathscr{K}$. This strategy reduces the computation time, as no computed reduced basis function needs to be thrown away. The method SPLITTIMEINTERVAL splits the interval $\mathscr{K}$ such that afterwards the spaces $\mathscr{W}_M^{\mathscr{K}_1}$ and $\mathscr{W}_M^{\mathscr{K}_2}$ are of equal dimension. The threshold $c_{\min}$ asserts a lower bound on the size of the time intervals.

Table 1 Comparison of the number of bases, the reduced basis sizes averaged over sub-intervals, offline time, averaged online reduced simulation times and maximum errors for non-adaptive and adaptive runs with threshold $c_{min} = 5$, and $= 1$. The average online run-times and maximum errors are obtained from 20 simulations with randomly selected parameters μ

adaptation	no. of bases	ø-dim(CRB)	offline time[h]	ø-runtime[s]	max. error
no	1	350	1.47	6.79	$5.88 \cdot 10^{-4}$
yes, $c_{min} = 5$	11	223.09	2.08	4.06	$5.80 \cdot 10^{-4}$
yes, $c_{min} = 1$	26	198.42	8.40	3.38	$5.75 \cdot 10^{-4}$

4 Example: Buckley–Leverett equation

We consider a Buckley–Leverett type problem in two space dimensions fulfilling the equations (1)-(2) on a rectangular domain $\Omega := [0, 1]^2$ with initial data function $u_0(\mu) = c_{low} + (1 - c_{low})\chi_{[0.2,0.6] \times [0.25,0.75]}$, velocity vector $\mathbf{v}(u; \mu) = (0, 1)^t f(u; \mu)$ and diffusion $d(u; \mu) = KD(s; \mu)$. Here $f(u; \mu) = \frac{u^3}{\mu_1} \cdot \left(\frac{u^3}{\mu_1} + \frac{(1-u)^3}{\mu_2} \right)^{-1}$ denotes the fractional flow rate, $D(u; \mu) = \frac{(1-u)^3}{\mu_2} f(u; \mu) p'_c(u; \mu)$ the capillary diffusion for a capillary pressure $p_c(u; \mu) = u^{-\lambda}$. The variable parameters are chosen as $\mu := (K, c_{low}, \lambda)$ and the parameter space is given by $\mathscr{P} := [1, 2] \times [0, 0.1] \times [0.1, 0.4]$. The scalar viscosities are fixed at $\mu_1 = \mu_2 = 5$. At the boundary of the domain a Dirichlet condition applies with $u_{\mathcal{N}_{dir}}(\mu) = c_{low}$.

Discretization The problem is discretized with a standard finite volume scheme comprising an explicitly computed Engquist–Osher flux for the convective terms and an implicit discretization of the diffusive terms. The underlying grid has a dimension of $H = 25 \times 25$ grid cells and the time interval $[0, T_{max}]$ is discretized by 60 uniformly distributed time steps. Fig. 1 illustrates solution snapshots for two different parameters with different diffusion levels $K = 0$ respectively $K = 2$. The cross-section plots in the last column show the expected behaviour of combinations of rarefaction waves and smoothed shocks.

Offline phase In order to assess the effects of the adaptation algorithms, the reduced basis algorithms are run three times, once without the time adaptive empirical operator interpolation and two times with adaptation, but different thresholds c_{min} to bound the time interval size from below. The results concerning reduced basis sizes, offline and reduced simulation time, are summarized in Table 1.

In order to assure that the generated reduced basis leads to equally small reduction errors for all parameters of the parameter space, the parameter training set for the "POD-greedy" algorithm has been adapted with a validation set of randomly chosen parameters μ in both runs. In the test runs, after three refinement steps the training parameter set comprises 255 elements, and the chosen validation ratio of 1.4 is assured after the maximum error for the training parameters has fallen beneath the targeted level of $5 \cdot 10^{-4}$. The target interpolation error for the empirical interpolation was set to 10^{-6} in all runs. This error is reached with an average number of 198

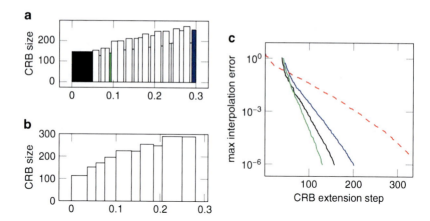

Fig. 2 Illustration of basis sizes on time intervals after adaptation with (a) $c_{min} = 1$ and (b) $c_{min} = 5$. Plot (c) illustrates the error decrease during generation of bases on three intervals marked with the same color in plot (a). The dashed line graph shows the slower decrease for a single basis without adaptation

respectively 223 basis functions in the adaptive cases, and 350 basis functions without adaptation. In the adaptive runs, the time interval has been decomposed into 11 respectively 26 sub-intervals (cf. Fig 2(a)&(b)). Fig. 2(c) illustrates the error decrease during the generation of the reduced spaces for selected time intervals (dashed lines) for the run with $c{min} = 1$. It can be observed that the slopes for the error graphs are much steeper than in the non-adaptive case illustrated with a dashed line. Because of the larger variation of the solutions for larger time steps, however, the basis on the last interval [0.29, 0.30] still shows the slowest error decrease. Fig. 2(a+b) show that for both adaptive runs the bases dimensions for all intervals stay significantly below the non-adaptive basis size of 350.

Conclusion We observed that the adaptive search in the time domain can lead to faster reduced simulations. However, the costs of 26 generated basis spaces for an average dimension reduction by a factor of approximately 0.56 turned out to be very expensive. We therefore advice to combine the time domain search with a parameter domain search to obtain a further improvement of the method.

Acknowledgements This work was supported by the German Science Foundation (DFG) under the contract number OH 98/2-1. The second author was supported by the Baden Württemberg Stiftung gGmbH.

References

1. Barrault, M., Maday, Y., Nguyen, N., Patera, A.: An 'empirical interpolation' method: application to efficient reduced-basis discretization of partial differential equations. C. R. Math. Acad. Sci. Paris Series I **339**, 667–672 (2004)

2. Drohmann, M., Haasdonk, B., Ohlberger, M.: Reduced Basis Approximation for Nonlinear Parametrized Evolution Equations based on Empirical Operator Interpolation. Tech. rep., FB10, University of Münster (2010)
3. Eftang, J.L., Patera, A.T., Rønquist, E.M.: An hp Certified Reduced Basis Method for Parametrized Parabolic Partial Differential Equations. Technical report, MIT, Cambridge, 2009. Submitted to SISC
4. Haasdonk, B. and Ohlberger, M.: Reduced basis method for finite volume approximations of parametrized evolution equations. M2AN Math. Model. Numer. Anal., **42**(2):277-302 (2008)
5. Haasdonk, B., Dihlmann, M., Ohlberger, M.: A training set and multiple bases generation approach for parametrized model reduction based on adaptive grids in parameter space. Tech. rep., University of Stuttgart (submitted) (2010)
6. Haasdonk, B., Ohlberger, M.: Adaptive basis enrichment for the reduced basis method applied to finite volume schemes. In: Proc. 5th International Symposium on Finite Volumes for Complex Applications, pp. 471–478 (2008)
7. Patera, A., Rozza, G.: Reduced Basis Approximation and a Posteriori Error Estimation for Parametrized Partial Differential Equations. MIT (2007). http://augustine.mit.edu/methodology/methodology_bookPartI.htm. Version 1.0, Copyright MIT 2006-2007, to appear in (tentative rubric) MIT Pappalardo Graduate Monographs in Mechanical Engineering

The paper is in final form and no similar paper has been or is being submitted elsewhere.

Adaptive Time-Space Algorithms for the Simulation of Multi-scale Reaction Waves

Max Duarte, Marc Massot, Stéphane Descombes, and Thierry Dumont

Abstract We present a new resolution strategy for multi-scale reaction waves based on adaptive time operator splitting and space adaptive multiresolution, in the context of localized and stiff reaction fronts. The main goal is to perform computationally efficient simulations of the dynamics of multi-scale phenomena under study, considering large simulation domains with conventional computing resources. We aim at time-space accuracy control of the solution and splitting time steps purely dictated by the physics of the phenomenon and not by stability constraints associated with mesh size or source time scales. Numerical illustrations are provided for 2D and 3D combustion applications modeled by reaction-convection-diffusion equations.

Keywords time adaptive integration, space adaptive multiresolution, combustion
MSC2010: 65M08, 65M50, 65Z05, 65G20

1 Introduction

Numerical simulations of multi-scale phenomena are commonly used for modeling purposes in many applications such as combustion, chemical vapor deposition, or air pollution modeling. In general, all these models raise several difficulties created by the high number of unknowns, the wide range of temporal scales due to large and

---

M. Duarte and M. Massot
Laboratoire EM2C - UPR CNRS 288, Ecole Centrale Paris, Grande Voie des Vignes, 92295 Chatenay-Malabry Cedex, France, email: {max.duarte,marc.massot}@em2c.ecp.fr

S. Descombes
Laboratoire J. A. Dieudonné - UMR CNRS 6621, Université de Nice - Sophia Antipolis, Parc Valrose, 06108 Nice Cedex 02, France, e-mail: sdescomb@unice.fr

T. Dumont
Institut Camille Jordan - UMR CNRS 5208, Université de Lyon, 43 Boulevard du 11 novembre 1918, 69622 Villeurbanne Cedex, France, e-mail: tdumont@math.univ-lyon1.fr

detailed chemical kinetic mechanisms, as well as steep spatial gradients associated with localized fronts of high chemical activity. In this context, faced with the induced stiffness of these time dependent problems, a natural stumbling block to perform 3D simulations with all scales resolution is either the unreasonably small time step due to stability requirements or the unreasonable memory and computing time required by implicit methods. Furthermore, an accurate description of such spatial multi-scale phenomena would also lead to large and sometimes unfeasible computation domains, if no adaptive meshing technique is used.

To overcome these difficulties, we present a new numerical strategy with a time operator splitting that considers dedicated high order time integration methods for reaction, diffusion and convection problems, in order to build a time operator splitting scheme that exploits efficiently the special features of each problem. Based on recent theoretical studies of numerical analysis, such a strategy leads to a splitting time step which is not restricted neither by the fastest scales in the source term nor by restrictive stability limits of diffusive or convective steps, but only by the physics of the phenomenon. Moreover, this splitting time step is dynamically adapted taking into account local error estimates [4]. The time integration is performed over a dynamic adapted grid obtained by multiresolution techniques in a finite volumes framework [2, 9, 11], which on the one hand, yield important savings in computing resources and on the other hand, allow to somehow control the spatial accuracy of the compressed representation based on a solid mathematical background.

Even though, the strategy was developed for the resolution of general multi-scale phenomena in various domains as biomedical applications [7] or nonlinear chemical dynamics [6], we will focus here on multidimensional combustion problems at large Reynolds numbers in order to assess the capability of the method. The paper is organized as follows: section 2 describes briefly the numerical strategy, based on spatial adaptive multiresolution and second order adaptive time integration. Physical configuration and modeling equations are presented in section 3 for laminar premixed flames interacting with vortices, along with 2D and 3D numerical illustrations. We end in the last part with some concluding remarks.

2 Construction of the Numerical Strategy

We detail briefly the developed operator splitting strategy with splitting time step adaptation, and some fundamental aspects of the adaptive multiresolution method.

2.1 Adaptive Time Operator Splitting

Given a general convection-reaction-diffusion system of equations

$$\partial_t \mathbf{u} - \partial_\mathbf{x} \left(\mathbf{F}(\mathbf{u}) + \mathbf{D}(\mathbf{u}) \partial_\mathbf{x} \mathbf{u} \right) = \mathbf{f}(\mathbf{u}), \quad \mathbf{x} \in \mathbb{R}^d, \, t > 0, \qquad (1)$$

with $\mathbf{u}(0, \mathbf{x}) = \mathbf{u}_0(\mathbf{x})$, where $\mathbf{F}, \mathbf{f} : \mathbb{R}^m \to \mathbb{R}^m$ and $\mathbf{u} : \mathbb{R} \times \mathbb{R}^d \to \mathbb{R}^m$, with diffusion matrix $\mathbf{D}(\mathbf{u})$: a tensor of order $d \times d \times m$; an operator splitting procedure allows to consider dedicated solvers for the reaction part which is decoupled from the other physical phenomena like convection, diffusion or both, for which there also exist dedicated numerical methods. These dedicated methods chosen for each subsystem are then responsible for dealing with the fast scales associated with each one of them, in a separate manner, while the reconstruction of the global solution by the splitting scheme should guarantee an accurate description with error control of the global physical coupling, without being related to the stability constraints of the numerical resolution of each subsystem.

A second order Strang scheme is then implemented [12]

$$\mathscr{S}^{\Delta t}(\mathbf{u}_0) = \mathscr{R}^{\Delta t/2} \mathscr{D}^{\Delta t/2} \mathscr{C}^{\Delta t} \mathscr{D}^{\Delta t/2} \mathscr{R}^{\Delta t/2}(\mathbf{u}_0), \qquad (2)$$

where operators $\mathscr{R}$, $\mathscr{D}$, $\mathscr{C}$ indicate respectively the independent resolution of the reaction, diffusion and convection problems with Δt defined as the splitting time step. Usually, for propagating reaction waves where for instance, the speed of propagation is much slower than some of the chemical scales, the fastest scales are not directly related to the global physics of the phenomenon, and thus, larger splitting time steps might be considered. Nevertheless, order reductions may then appear due to short-life transients associated to fast variables and in these cases, it has been proved in [5] that better performances are expected while ending the splitting scheme by operator $\mathscr{R}$ or in a more general case, the part involving the fastest time scales of the phenomenon.

An adaptive splitting time step strategy, based on a local error estimate at the end of each Δt, is implemented in order to control the accuracy of computations. A second, embedded and lower order Strang splitting method $\widetilde{\mathscr{S}}^{\Delta t}$ was developed [4] in order to dynamically calculate a local error estimate that should verify

$$\left\| \mathscr{S}^{\Delta t}(\mathbf{u}_0) - \widetilde{\mathscr{S}}^{\Delta t}(\mathbf{u}_0) \right\| \approx \mathscr{O}(\Delta t^2) < \eta_{\text{split}}, \qquad (3)$$

in order to accept current computation with Δt, and thus, the new splitting time step is given by

$$\Delta t_{\text{new}} = \Delta t \sqrt{\frac{\eta_{\text{split}}}{\left\| \mathscr{S}^{\Delta t}(\mathbf{u}_0) - \widetilde{\mathscr{S}}^{\Delta t}(\mathbf{u}_0) \right\|}}. \qquad (4)$$

The choice of suitable time integration methods to approximate numerically $\mathscr{R}$, $\mathscr{D}$ and $\mathscr{C}$ during each Δt is mandatory not only to guarantee the theoretical framework of the numerical analysis but also to take advantage of the particular features of each independent subproblem. A new operator splitting for reaction-diffusion systems was recently introduced [6], which considers a high fifth order, A-stable, L-stable method like Radau5 [8], based on implicit Runge-Kutta schemes for stiff ODEs, that solves with a local cell by cell approach the reaction term: a system of stiff ODEs without spatial coupling. On the other hand, a high fourth order method was chosen, like ROCK4 [1], based on explicit stabilized Runge-Kutta schemes which features extended stability domains along the negative real axis, very appropriate for diffusion problems because of the usual predominance of negative real eigenvalues. Both methods incorporate adaptive time integration tools, similar to (4), in order to control accuracy for given η_{Radau5} and η_{ROCK4}.

An explicit high order in time and space one step monotonicity preserving scheme OSMP [3] is used as convective scheme. It combines monotonicity preserving constraints for non-monotone data to avoid extrema clipping, with TVD features to prevent spurious oscillations around discontinuities or sharp spatial gradients. Classical CFL stability restrictions are though imposed during each splitting time step Δt. Notice that the overall combination of explicit treatment of spatial phenomena as convection and diffusion, with local implicit integration of stiff reaction implies important savings in computing time and memory resources. For the reaction, local treatment plus adaptive time stepping allow to discriminate cells of high reactive activity in the neighborhood of the localized wavefront, saving as a consequence a large quantity of integration time.

2.2 Mesh Refinement Technique

We are concerned with the propagation of reacting wavefronts, hence important reactive activity as well as steep spatial gradients are localized phenomena. This implies that if we consider the resolution of reactive problem, a considerable amount of computing time is spent on nodes that are practically at (partial) equilibrium. Moreover, there is no need to represent these quasi-stationary regions with the same spatial discretization needed to describe the reaction front, so that convection and diffusion problems might also be solved over a smaller number of nodes. An adapted mesh obtained by a multiresolution process which discriminates the various space scales of the phenomenon, turns out to be a very convenient solution to overcome these difficulties [6, 7].

In practice, if one considers a set of nested spatial grids from the coarsest to the finest one, a multiresolution transformation allows to represent a discretized function as values on the coarsest grid plus a series of local estimates at all other levels of such nested grids. These estimates correspond to the wavelet coefficients of a wavelet decomposition obtained by inter-level transformations, and retain the information on local regularity when going from a coarse to a finer grid. Hence, the main idea is to use the decay of the wavelet coefficients to obtain information on local regularity of the solution: lower wavelet coefficients are associated to local regular spatial configurations and vice-versa. This representation yields to a thresholding process that builds dynamically the corresponding adapted grid on which the solutions are represented; then the error committed by the multiresolution transformation is proportional to η_{MR}, where η_{MR} is a threshold parameter [2, 9].

3 Numerical Illustration

In these illustrating examples, we are concerned with the numerical simulation of premixed flames interacting with vortex structures: a pair of counter rotating vortices in a 2D configuration and a 3D toroidal vortex. This is usually a difficult problem to solve because of the localized and stiff reactive fronts, even more with large Reynolds numbers. Nevertheless, in order to properly evaluate the proposed strategy we consider only time evolution problems for which the hydrodynamics is not solved but a large Reynolds number velocity field is imposed. Based on a model presented in [10], we consider that the chemistry may be modeled by a global, single step, irreversible reaction characterized by an Arrhenius law; and a thermodiffusive approach of laminar flame theory is adopted in order to decouple velocity field computation from determination of species mass fractions and temperature. Known solutions of incompressible Navier-Stokes equations may then be imposed, and the problem is reduced to solving the standard species and energy balance equations.

Following [10], a progress variable $c(x, y, t)$ is introduced:

$$c = \frac{T - T_o}{T_b - T_o}, \tag{5}$$

where subscripts $(\ )_o$ and $(\ )_b$ indicate respectively, fresh mixture zone and burnt product zone; and we finally obtain for a 2D configuration

$$\frac{\partial c}{\partial t_\star} + u_\star \frac{\partial c}{\partial x_\star} + v_\star \frac{\partial c}{\partial y_\star} - \left(\frac{\partial^2 c}{\partial x_\star^2} + \frac{\partial^2 c}{\partial y_\star^2}\right) = \mathrm{Da}(1 - c)\exp\left(-\frac{T_a}{T_o(1 + \tau c)}\right), \tag{6}$$

where Da is a Damköhler number, T_a the activation energy, $\tau = T_b/T_o - 1$, and $(\ )_\star$ indicates dimensionless variables. The velocity field $(u_\star(t), v_\star(t))$ is deduced analytically and imposed into (6), considering a 2D viscous core vortex with a dimensionless azimuthal velocity of the form:

$$v_{\theta\star}(r_\star, t_\star) = \frac{\mathrm{Re}\,\mathrm{Sc}}{r_\star}\left(1 - \exp\left(-\frac{r_\star^2}{4\,\mathrm{Sc}\,t_\star}\right)\right), \tag{7}$$

with $r_\star(x_\star, y_\star)$, the distance to the vortex center, Reynolds and Schmidt numbers.

Figure 1 shows the interaction of the premixed flame with two counter rotating vortices modeled each one of them by (7), centered at $(-0.25, -0.5)$ and $(0.25, -0.5)$ for a 2D spatial domain of $[-1, 1]^2$. The upper (red) and lower (blue) regions correspond respectively to burnt product ($c = 1$) and fresh mixture ($c = 0$) zones. The corresponding adapted mesh tightens around the stiff regions and propagates along the wavefronts.

The following modeling values were considered into (6) and (7): $\mathrm{Da} = 2.5 \times 10^9$, $T_a = 20000$ K, $T_o = 300$ K, $T_b = 2315.4$ K, $\tau \approx 6.72$, $\mathrm{Sc} = 1$ and $\mathrm{Re} = 1000$. The initial condition corresponds to a planar premixed flame at $y = -0.5$ and the

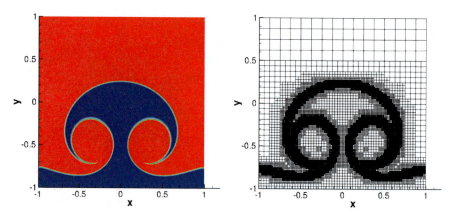

Fig. 1 2D premixed flame interacting with two counter rotating vortices. Solution of variable c at $t_\star = 4 \times 10^{-4}$ (left) and corresponding adapted mesh (right). Finest grid: 1024^2

phenomenon is studied over a time domain of $[0, 4 \times 10^{-3}]$. The MR procedure considers a set of 10 grids, equivalent to $1024^2 = 1048576$ cells on the finest grid. MR and adaptive splitting time step tolerances were set to $\eta_{MR} = 10^{-2}$ and $\eta_{split} = 10^{-3}$, with $\eta_{Radau5} = \eta_{ROCK4} = 10^{-5}$.

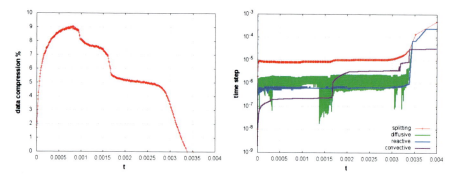

Fig. 2 2D premixed flame interacting with two counter rotating vortices. Time evolution of data compression in the solution representation (left) and splitting, diffusive, reactive and convective time steps (right). Finest grid: 1024^2

Figure 2 shows data compression obtained by MR representation of the solution, measured as the percentage of active cells with respect to the finest grid representation; in this case, lower than 9% of 1024^2. On the other hand, splitting time step starts from an initial value set to 10^{-8} in order to handle correctly the initial sudden apparition of the vortices, that evolves rapidly to a final quasi stable value of 10^{-5}, which indicates the decoupling degree achieved within the accuracy prescribed to describe the global propagating phenomenon. The corresponding convective time step with CFL = 1 illustrates the time scale decoupling obtained by a splitting technique and highlights the eventual inconveniences of solving (6) considering all phenomena at once. The same conclusion is valid concerning reactive and diffusive time steps. By the way, larger convective time steps are used thanks to the adapted grid representation which allows to discriminate locally large velocity values (in this case $|u_\star|, |v_\star| \approx 40000$) from the refined regions around the wavefront, as we can see in the "jumps" of convective time steps in Fig. 2. Reactive time steps correspond to cells at the wavefront (for furthest cells, reactive time steps are equal to splitting ones), while lower diffusive time steps are needed in order to fulfill each splitting time step, which explains the "oscillations". Diffusive time steps might take values beyond classical stability constraints (of the order of 10^{-6} for explicit RK4 [8] and eigenvalues of -2.2×10^6), and it is finally set by the accuracy criterion.

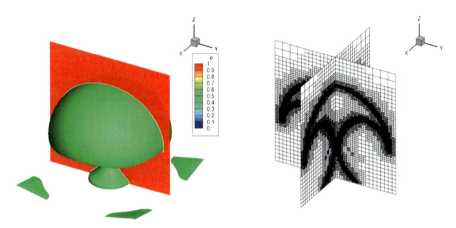

Fig. 3 3D premixed flame interacting with a toroidal vortex. Solution of variable c at $t_\star = 1.1 \times 10^{-3}$ showing isosurface $c = 0.5$ (left) and corresponding adapted mesh (right). Finest grid: 256^3

This resolution technique has a straightforward extension to 3D configurations. Figure 3 shows the interaction of the premixed flame with a toroidal vortex modeled by (7) centered at $\sqrt{x_\star^2 + y_\star^2} = 0.25, z_\star = -0.5$ for a 3D spatial domain of $[-1, 1]^3$. The modeling and tolerance parameters are taken equal to the 2D case and the MR procedure considers a set of 8 grids, equivalent to $256^3 = 16777216$ cells on the finest grid. The splitting time step shows the same behavior as for the previous case with same order of values, while the data compression is lower than 17%, taking into account that a lower scale discrimination is available with 8 different grids. All the computations have been performed on a AMD Shanghai processor of 2.7 GHz with memory capacity of 4 GB. Computing times for the 2D and 3D configurations were about of 0h57m and 14h40m, respectively.

4 Concluding Remarks

The present work proposes a new numerical approach which is shown to be computationally efficient. It couples adaptive multiresolution techniques with a new operator splitting strategy with high order time integration methods to properly solve the entire spectrum of scales of each phenomenon. The splitting time step is chosen on the sole basis of the structure of the continuous system and its decoupling capabilities, but not related to stability requirements of the numerical methods involved in order to integrate each subsystem, even if stiffness is present. The global accuracy of the simulation is controlled and dynamically evaluated based on theoretical and numerical results. As a consequence, the resulting highly compressed data representations as well as the accurate and feasible resolution of these stiff phenomena prove that large computational domains previously out of reach can be successfully simulated with conventional computing resources. At this stage of development, the same numerical strategy can be coupled to a hydrodynamics solver, considering though that an important amount of work is still in progress concerning programming features such as data structures and parallelization strategies.

Acknowledgements This research was supported by fundamental project grants from ANR *Séchelles* (project leader S. Descombes), and DIGITEO *MUSE* (project leader M. Massot). M. Duarte has a Ph.D. grant from Mathematics (INSMI) and Engineering (INSIS) Institutes of CNRS and supported by INCA project. Authors express special thanks to Christian Tenaud (LIMSI-CNRS) for providing the basis of the multiresolution kernel of MR CHORUS (DI 03760-01).

References

1. Abdulle, A.: Fourth order Chebyshev methods with recurrence relation. SIAM J. Sci. Comput. **23**, 2041–2054 (2002)
2. Cohen, A., Kaber, S., Müller, S., Postel, M.: Fully adaptive multiresolution finite volume schemes for conservation laws. Math. of Comp. **72**, 183–225 (2003)
3. Daru, V., Tenaud, C.: High order one-step monotonicity-preserving schemes for unsteady compressible flow calculations. Journal of Computational Physics **193**(2), 563–594 (2004)
4. Descombes, S., Duarte, M., Dumont, T., Louvet, V., Massot, M.: Adaptive time splitting method for multi-scale evolutionary PDEs. Confluentes Mathematici (to app.) (2011)
5. Descombes, S., Massot, M.: Operator splitting for nonlinear reaction-diffusion systems with an entropic structure: Singular perturbation and order reduction. Numer. Math. **97**(4), 667–698 (2004)
6. Duarte, M., Massot, M., Descombes, S., Tenaud, C., Dumont, T., Louvet, V., Laurent, F.: New resolution strategy for multi-scale reaction waves using time operator splitting, space adaptive multiresolution and dedicated high order implicit/explicit time integrators. Submitted to SIAM J. Sci. Comput., available on HAL (http://hal.archives-ouvertes.fr/hal-00457731) (2010)
7. Duarte, M., Massot, M., Descombes, S., Tenaud, C., Dumont, T., Louvet, V., Laurent, F.: New resolution strategy for multi-scale reaction waves using time operator splitting and space adaptive multiresolution: Application to human ischemic stroke. ESAIM Proc. (to app.) (2011)

8. Hairer, E., Wanner, G.: Solving ordinary differential equations II, second edn. Springer-Verlag, Berlin (1996). Stiff and differential-algebraic problems
9. Harten, A.: Multiresolution algorithms for the numerical solution of hyperbolic conservation laws. Comm. Pure and Applied Math. **48**, 1305–1342 (1995)
10. Laverdant, A., Candel, S.: Computation of diffusion and premixed flames rolled up in vortex structures. Journal of Propulsion and Power **5**, 134–143 (1989)
11. Müller, S.: Adaptive multiscale schemes for conservation laws, vol. 27. Springer-Verlag, Heidelberg (2003)
12. Strang, G.: On the construction and comparison of difference schemes. SIAM J. Numer. Anal. **5**, 506–517 (1968)

The paper is in final form and no similar paper has been or is being submitted elsewhere.

Dispersive wave runup on non-uniform shores

Denys Dutykh, Theodoros Katsaounis, and Dimitrios Mitsotakis

Abstract Historically the finite volume methods have been developed for the numerical integration of conservation laws. In this study we present some recent results on the application of such schemes to dispersive PDEs. Namely, we solve numerically a representative of Boussinesq type equations in view of important applications to the coastal hydrodynamics. Numerical results of the runup of a moderate wave onto a non-uniform beach are presented along with great lines of the employed numerical method (see D. Dutykh *et al.* (2011) [6] for more details).

Keywords dispersive wave, runup, Boussinesq equations, shallow water
MSC2010: 65M08, 76B15

1 Introduction

The simulation of water waves in realistic and complex environments is a very challenging problem. Most of the applications arise from the areas of coastal and naval engineering, but also from natural hazards assessment. These applications may require the computation of the wave generation [5, 12], propagation [17], interaction with solid bodies, the computation of long wave runup [16, 18] and even the extraction of the wave energy [15]. Issues like wave breaking, robustness

---

Denys Dutykh
LAMA, UMR 5127 CNRS, Université de Savoie, Campus Scientifique, 73376 Le Bourget-du-Lac Cedex, France, e-mail: Denys.Dutykh@univ-savoie.fr

Theodoros Katsaounis
Department of Applied Mathematics, University of Crete, Heraklion, 71409 Greece Inst. of App. and Comp. Math. (IACM), FORTH, Heraklion, 71110, Greece, e-mail: thodoros@tem.uoc.gr

Dimitrios Mitsotakis
IMA, University of Minnesota, Minneapolis MN 55455, USA, e-mail: dmitsot@gmail.com

of the numerical algorithm in wet-dry processes along with the validity of the mathematical models in the near-shore zone are some basic problems in this direction [11]. During past several decades the classical Nonlinear Shallow Water Equations (NSWE) have been essentially employed to face these problems [7]. Mathematically, these equations represent a system of conservation laws describing the propagation of infinitely long waves with a hydrostatic pressure assumption. The wave breaking phenomenon is commonly assimilated to the formation of shock waves (or hydraulic jumps) which is a common feature of hyperbolic PDEs. Consequently, the finite volume (FV) method has become the method of choice for these problems due to its excellent intrinsic conservative and shock-capturing properties [3, 7].

In the present article we report on recent results concerning the extension of the finite volume method to dispersive wave equations steming essentially from water wave modeling [4, 6, 14].

2 Mathematical model and numerical methods

Consider a cartesian coordinate system in two space dimensions (x, z) to simplify notations. The z-axis is taken vertically upwards and the x-axis is horizontal and coincides traditionally with the still water level. The fluid domain is bounded below by the bottom $z = -h(x)$ and above by the free surface $z = \eta(x, t)$. Below we will also need the total water depth $H(x, t) := h(x) + \eta(x, t)$. The flow is supposed to be incompressible and the fluid is inviscid. An additional assumption of the flow irrotationality is made as well.

In the pioneering work of D.H. Peregrine (1967) [14] the following system of Boussinesq type equations has been derived:

$$\eta_t + \big((h + \eta)u\big)_x = 0, \tag{1}$$

$$u_t + uu_x + g\eta_x - \frac{h}{2}(hu)_{xxt} + \frac{h^2}{6}u_{xxt} = 0, \tag{2}$$

where $u(x, t)$ is the depth averaged fluid velocity, g is the gravity acceleration and underscripts (u_x, η_t) denote partial derivatives.

In our recent study [6] we proposed an improved version of this system which contains higher order nonlinear terms which should be neglected from asymptotic point of view and can be written in conservative variables $(H, Q) = (H, Hu)$ as:

$$H_t + Q_x = 0, \tag{3}$$

$$\Big(\big(1 + \frac{1}{3}H_x^2 - \frac{1}{6}HH_{xx}\big)Q_t - \frac{1}{3}H^2 Q_{xxt} - \frac{1}{3}HH_x Q_{xt}\Big) + \Big(\frac{Q^2}{H} + \frac{g}{2}H^2\Big)_x = gHh_x. \tag{4}$$

Obviously the linear characteristics of both systems (1), (2) and (3), (4) coincide since they differ only by nonlinear terms.

However, this modification has several important implications onto structural properties of the obtained system. First of all, the magnitude of the dispersive terms tends to zero when we approach the shoreline $H \to 0$. This property corresponds to our physical representation of the wave shoaling and runup process. On the other hand, the resulting system becomes invariant under vertical translations (subgroup G_5 in Theorem 4.2, T. Benjamin & P. Olver (1982) [2]):

$$z \leftarrow z + d, \quad \eta \leftarrow \eta - d, \quad h \leftarrow h + d, \quad u \leftarrow u, \tag{5}$$

where d is some constant. This property is straightforward to check since we use only the total water depth variable $H = h + \eta$ which remains invariant under transformation (5).

Remark 1. In this paper we will consider the initial-boundary value problem posed in a bounded domain $I = [b_1, b_2]$ with reflective boundary conditions. In this case one needs to impose boundary conditions only in one of the two dependent variables, cf. [8]. In the case of reflective boundary conditions it is sufficient to take $u(b_1, t) = u(b_2, t) = 0$.

2.1 Finite volume discretization

Let $\mathscr{T} = \{x_i\}$, $i \in \mathbb{Z}$ denotes a partition of $\mathbb{R}$ into cells $C_i = (x_{i-\frac{1}{2}}, x_{i+\frac{1}{2}})$ where $x_i = (x_{i+\frac{1}{2}} + x_{i-\frac{1}{2}})/2$ denotes the midpoint of C_i. Let $\Delta x_i = x_{i+\frac{1}{2}} - x_{i-\frac{1}{2}}$ be the length of the cell C_i, $\Delta x_{i+\frac{1}{2}} = x_{i+1} - x_i$. (Here, we consider only uniform grids with $\Delta x_i = \Delta x_{i+\frac{1}{2}} = \Delta x$.)

The governing equations (3), (4) can be recast in the following vector form:

$$[\mathbf{D}(\mathbf{v}_t)] + [\mathbf{F}(\mathbf{v})]_x = \mathbf{S}(\mathbf{v}),$$

where

$$\mathbf{D}(\mathbf{v}_t) = \begin{pmatrix} H_t \\ (1 + \tfrac{1}{3}H_x^2 - \tfrac{1}{6}HH_{xx})Q_t - \tfrac{1}{3}H^2 Q_{xxt} - \tfrac{1}{3}HH_x Q_{xt} \end{pmatrix}, \tag{6}$$

$$\mathbf{F}(\mathbf{v}) = \begin{pmatrix} Q \\ \tfrac{Q^2}{H} + \tfrac{g}{2}H^2 \end{pmatrix}, \quad \mathbf{S}(\mathbf{v}) = \begin{pmatrix} 0 \\ gHh_x \end{pmatrix}. \tag{7}$$

We denote by H_i and U_i the corresponding cell averages. To discretize the dispersive terms in (6) we consider the following approximations:

$$\frac{1}{\Delta x}\int_{x_{i-\frac{1}{2}}}^{x_{i+\frac{1}{2}}}\left[1+\frac{1}{3}(H_x)^2-\frac{1}{6}HH_{xx}\right]Q\,dx \approx$$

$$\left(1+\frac{1}{3}\left(\frac{H_{i+1}-H_{i-1}}{2\Delta x}\right)^2-\frac{1}{6}H_i\frac{H_{i+1}-2H_i+H_{i-1}}{\Delta x^2}\right)Q_i,$$

$$\frac{1}{\Delta x}\int_{x_{i-\frac{1}{2}}}^{x_{i+\frac{1}{2}}}\frac{1}{3}HH_xQ_x\,dx \approx \frac{1}{3}H_i\frac{H_{i+1}-H_{i-1}}{2\Delta x}\frac{Q_{i+1}-Q_{i-1}}{2\Delta x},$$

$$\frac{1}{\Delta x}\int_{x_{i-\frac{1}{2}}}^{x_{i+\frac{1}{2}}}\frac{1}{3}H^2Q_{xx}\,dx \approx \frac{1}{3}H_i^2\frac{Q_{i+1}-2Q_i+Q_{i-1}}{\Delta x^2}.$$

We note that we approximate the reflective boundary conditions by taking the cell averages of u on the first and the last cell to be $u_0 = u_{N+1} = 0$. We do not impose explicitly boundary conditions on H. The reconstructed values on the first and the last cell are computed using neighboring ghost cells and taking odd and even extrapolation for u and H respectively. These specific boundary conditions appeared to reflect incident waves on the boundaries while conserving the mass.

This discretization leads to a linear system with tridiagonal matrix denoted by **L** that can be inverted efficiently by a variation of Gauss elimination for tridiagonal systems with computational complexity $O(n)$, n-being the dimension of the system. We note that on the dry cells the matrix becomes diagonal since H_i is zero on dry cells. For the time integration the explicit third-order TVD-RK method is used. In the numerical experiments we observed that the fully discrete scheme is stable and preserves the positivity of H during the runup under a mild restriction on the time step Δt.

Therefore, the semidiscrete problem of (6) - (7) is written as a system of ODEs in the form:

$$\mathbf{L}_i\mathbf{V}_{it} + \frac{1}{\Delta x}(\mathscr{F}_{i+\frac{1}{2}} - \mathscr{F}_{i-\frac{1}{2}}) = \frac{1}{\Delta x}\mathbf{S}_i,$$

where $\mathbf{L}_i$ is the i-th row of matrix **L** and $\mathscr{F}_{i+\frac{1}{2}}$ can be chosen as one of the numerical flux functions [6] (in computations presented below we choose the FVCF flux [9]). In the sequel we will use the KT and the CF numerical fluxes. In this case the Jacobian of **F** is given by the matrix

$$A = \begin{pmatrix} 0 & 1 \\ gH - (Q/H)^2 & 2Q/H \end{pmatrix},$$

and the eigenvalues are $\lambda_{1,2} = Q/H \pm \sqrt{gH}$. Therefore, the characteristic numerical flux [9] takes the form

$$\mathscr{F}_{i+\frac{1}{2}} = \frac{\mathbf{F}(\mathbf{V}^L_{i+\frac{1}{2}}) + \mathbf{F}(\mathbf{V}^R_{i+\frac{1}{2}})}{2} - \mathbf{U}(\mu)\frac{\mathbf{F}(\mathbf{V}^R_{i+\frac{1}{2}}) - \mathbf{F}(\mathbf{V}^L_{i+\frac{1}{2}})}{2},$$

where $\boldsymbol{\mu} = (\mu_1, \mu_2)^T$ are the Roe average values,

$$\mu_1 = \frac{H^L_{i+\frac{1}{2}} + H^R_{i+\frac{1}{2}}}{2}, \quad \mu_2 = \frac{\sqrt{H^L_{i+\frac{1}{2}}}U^L_{i+\frac{1}{2}} + \sqrt{H^R_{i+\frac{1}{2}}}U^R_{i+\frac{1}{2}}}{\sqrt{H^L_{i+\frac{1}{2}}} + \sqrt{H^R_{i+\frac{1}{2}}}}$$

and

$$\mathbf{U}(\boldsymbol{\mu}) = \begin{pmatrix} \frac{s_2(\mu_2+c)-s_1(\mu_2-c)}{2c} & \frac{s_1-s_2}{2c} \\ \frac{(s_2-s_1)(\mu_2^2-c^2)}{2c} & \frac{s_1(\mu_2+c)-s_2(\mu_2-c)}{2c} \end{pmatrix}, \quad c = \sqrt{g\mu_1}, \quad s_i = \text{sign}(\lambda_i).$$

For more details on the discretization and reconstruction procedures, (that are based on the hydrostatic reconstruction, [1]), we refer to our complete work on this subject [6].

3 Numerical results

In the present section we show a numerical simulation of a solitary wave runup onto a non-uniform sloping beach. More precisely, we add a small pond along the slope. As our results indicate, this small complication is already sufficient to develop some instabilities which remain controlled in our simulations.

As an initial condition we used an approximate solitary wave solution of the following form:

$$\eta_0(x) = A_s \text{sech}^2\big(\lambda(x-x_0)\big), \quad u_0(x) = -c_s \frac{\eta_0(x)}{1+\eta_0(x)},$$

where A_s is the amplitude relative to the constant water depth taken to be unity in our study. The solitary wave speed c_s along with the wavelength λ are given here:

$$\lambda = \sqrt{\frac{3A_s}{4(1+A_s)}}, \quad c_s = \sqrt{g}\frac{\sqrt{6(1+A_s)}}{\sqrt{3+2A_s}} \cdot \frac{\sqrt{(1+A_s)\log(1+A_s)-A_s}}{A_s}.$$

The solitary wave is centered initially at $x_0 = 10.62$ and has amplitude $A_s = 0.08$. The constant slope β is equal to $2.88°$. The sketch of the computational domain can be found in [6].

In numerical simulations presented below we used a uniform space discretization with $\Delta x = 0.025$ and very fine time step $\Delta t = \Delta x/100$ to guarantee the accuracy and stability during the whole simulation.

Snapshots of numerical results are presented on Figs. 1 – 6. We present simultaneously three different computational results:

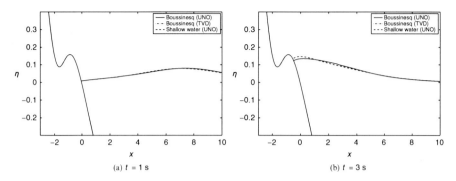

Fig. 1 Solitary wave aproaching a sloping beach with a pond

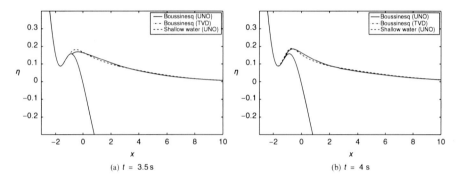

Fig. 2 Beginning of the pond inundation

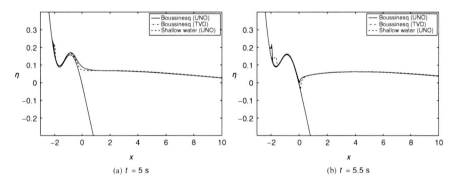

Fig. 3 A part of the wave mass is trapped in the pond volume

Dispersive wave runup on non-uniform shores

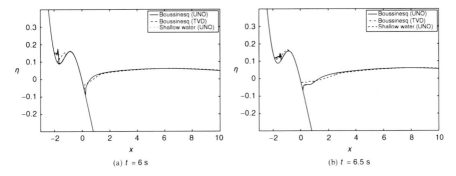

Fig. 4 Wave oscillations in the pond

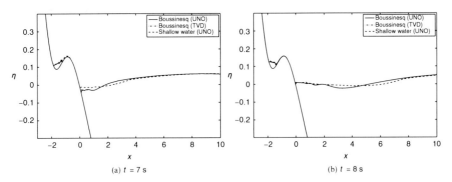

Fig. 5 Stabilization of wave oscillations

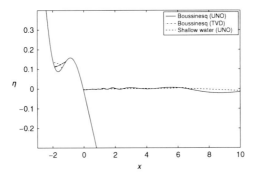

Fig. 6 The whole system is tending to the rest position ($t = 10$ s)

- Modified Peregrine system solved with UNO2 reconstruction [10]
- The same system with classical MUSCL TVD2 scheme [13]
- Nonlinear Shallow Water Equations (NSWE) with UNO2 scheme [10]

Surprisingly good agreement was obtained among all three numerical models. Presumably, the complex runup process under consideration is governed essentially by nonlinearity. However, on Figs. 1(b) and 2(a) the amplitude predicted by NSWE is slightly overestimated.

On Figs. 3(b) – 4(b) some oscillations (due to the small-dispersion effect characterizing dispersive wave breaking procedures) can be observed. However, their amplitude remains small for all times and does not produce any blow up phenomena. Later these oscillations decay tending gradually to the "lake at the rest" state (see Figs. 5, 6).

In the specific experiment a friction term could be beneficial to reduce the amplitude of oscillations (or damp them out completely). However, we prefer to present the computational results of our model without adding any ad-hoc term to show its original performance.

4 Conclusions

In this study we presented an improved version of the Peregrine system which is particularly suited for the simulation of dispersive waves runup. This system allows for the description of higher amplitude waves due to improved nonlinear characteristics. Better numerical stability properties have been obtained since most of the dispersive terms tend to zero when we approach the shoreline. Consequently, our model naturally degenerates to classical Nonlinear Shallow Water Equations (NSWE) for which the runup simulation technology is completely mastered nowadays. However we underline that there is no artificial parameter to turn off dispersive terms. Their importance is naturally governed by the underlying physical process.

Acknowledgements D. Dutykh acknowledges the support from French Agence Nationale de la Recherche, project MathOcean (Grant ANR-08-BLAN-0301-01) and Ulysses Program of the French Ministry of Foreign Affairs under the project 23725ZA. The work of Th. Katsaounis was partially supported by European Union FP7 program Capacities(Regpot 2009-1), through ACMAC (http://acmac.tem.uoc.gr).

References

1. Audusse, E., Bouchut, F., Bristeau, O., Klein, R., Perthame, B.: A fast and stable well-balanced scheme with hydrostatic reconstruction for shallow water flows. SIAM J. of Sc. Comp. **25**, 2050–2065 (2004)
2. Benjamin, T., Olver, P.: Hamiltonian structure, symmetries and conservation laws for water waves. J. Fluid Mech **125**, 137–185 (1982)

3. Delis, A.I., Katsaounis, T.: Relaxation schemes for the shallow water equations. Int. J. Numer. Meth. Fluids **41**, 695–719 (2003)
4. Dutykh, D., Dias, F.: Dissipative Boussinesq equations. C. R. Mecanique **335**, 559–583 (2007)
5. Dutykh, D., Dias, F.: Water waves generated by a moving bottom. In: A. Kundu (ed.) Tsunami and Nonlinear waves. Springer Verlag (Geo Sc.) (2007)
6. Dutykh, D., Katsaounis, T., Mitsotakis, D.: Finite volume schemes for dispersive wave propagation and runup. Accepted to Journal of Computational Physics **http://hal.archives-ouvertes.fr/hal-00472431/** (2011)
7. Dutykh, D., Poncet, R., Dias, F.: Complete numerical modelling of tsunami waves: generation, propagation and inundation. Submitted **http://arxiv.org/abs/1002.4553** (2010)
8. Fokas, A.S., Pelloni, B.: Boundary value problems for Boussinesq type systems. Math. Phys. Anal. Geom. **8**, 59–96 (2005)
9. Ghidaglia, J.M., Kumbaro, A., Coq, G.L.: On the numerical solution to two fluid models via cell centered finite volume method. Eur. J. Mech. B/Fluids **20**, 841–867 (2001)
10. Harten, A., Osher, S.: Uniformly high-order accurate nonscillatory schemes, I. SIAM J. Numer. Anal. **24**, 279–309 (1987)
11. Hibberd, S., Peregrine, D.: Surf and run-up on a beach: a uniform bore. J. Fluid Mech. **95**, 323–345 (1979)
12. Kervella, Y., Dutykh, D., Dias, F.: Comparison between three-dimensional linear and nonlinear tsunami generation models. Theor. Comput. Fluid Dyn. **21**, 245–269 (2007)
13. van Leer, B.: Towards the ultimate conservative difference scheme V: a second order sequel to Godunov' method. J. Comput. Phys. **32**, 101–136 (1979)
14. Peregrine, D.H.: Long waves on a beach. J. Fluid Mech. **27**, 815–827 (1967)
15. Simon, M.: Wave-energy extraction by a submerged cylindrical resonant duct. Journal of Fluid Mechanics **104**, 159–187 (1981)
16. Tadepalli, S., Synolakis, C.E.: The run-up of N-waves on sloping beaches. Proc. R. Soc. Lond. A **445**, 99–112 (1994)
17. Titov, V., González, F.: Implementation and testing of the method of splitting tsunami (MOST) model. Tech. Rep. ERL PMEL-112, Pacific Marine Environmental Laboratory, NOAA (1997)
18. Titov, V.V., Synolakis, C.E.: Numerical modeling of tidal wave runup. J. Waterway, Port, Coastal, and Ocean Engineering **124**, 157–171 (1998)

The paper is in final form and no similar paper has been or is being submitted elsewhere.

MAC Schemes on Triangular Meshes

Robert Eymard, Jürgen Fuhrmann, and Alexander Linke

Abstract We present numerical results for two generalized MAC schemes on triangular meshes, which are based on staggered meshes using the Delaunay–Voronoi duality. In the first one, the pressures are defined at the vertices of the mesh, and the discrete velocities are tangential to the edges of the triangles. In the second one, the pressures are defined in the triangles, and the discrete velocities are normal to the edges of the triangles. In both cases, convergence results are obtained.

Keywords Navier–Stokes, MAC scheme, Delaunay mesh
MSC2010: 76D07, 65N08

1 Introduction

We consider in this paper two different generalizations of the classical MAC scheme [1] for the incompressible Stokes problem

$$
\begin{aligned}
-\Delta \boldsymbol{u} + \nabla p &= \boldsymbol{f} & \boldsymbol{x} \in \Omega, \\
\nabla \cdot \boldsymbol{u} &= 0 & \boldsymbol{x} \in \Omega, \\
\int_\Omega p \, d\boldsymbol{x} &= 0 \\
\boldsymbol{u} &= 0 & \boldsymbol{x} \in \partial\Omega.
\end{aligned} \quad (1)
$$

R. Eymard
Université Paris–Est, Paris, France, e-mail: robert.eymard@univ-mlv.fr

J. Fuhrmann and A. Linke
Weierstrass Institute, Berlin, Germany, e-mail: juergen.fuhrmann@wias-berlin.de, alexander.linke@wias-berlin.de

We assume that $f \in L^2(\Omega)^2$ holds, where Ω is an open polygonal bounded and connected subset of $\mathbb{R}^2$ without holes and with boundary $\partial\Omega$.

The MAC scheme [1] is based on a staggered approach on structured grids, where the velocity and the pressure control volumes are dual to each other and have square or rectangular shape. Since the scheme is staggered, the pressure is not prone to instabilities. In this situation, convergence proofs for the linear Stokes and the nonlinear Navier–Stokes problems (with small data assumption) have been presented by Nicolaides [2, 3]. But in spite of its success, this scheme has the main drawback that complex geometries cannot be well approximated by structured grids. Therefore, several attempts have been made to generalize it for unstructured grids, see e.g., [4], where the unstructured simplex grid possesses the Delaunay property. Then the dual Voronoi grid can be defined in a sensible way, and two different staggered approaches are possible, where the pressure is discretized either in the triangles or at the vertices of the mesh:

1. in the first scheme, in the sequel called *tangential velocity scheme*, the velocity is approximated by its tangential values along the edges of the triangles, whereas the pressures are approximated at the vertices of the triangles;
2. in the second scheme, in the sequel called *normal velocity scheme*, the velocity is approximated by its normal values to the edges of the triangles, whereas the pressures are approximated at the center of the triangles.

For these generalized MAC schemes on unstructured grids, no convergence proofs have been found up to now. Therefore, we will present in this paper an appropriate discrete weak formulation of the problem, which allows to give a convergence proof [5]. Moreover, we show experimental orders of convergence in appropriate norms for a test problem with known analytical solution. It is worth noticing that for both schemes, the discrete rotation operator is consistent, but not the discrete divergence operator. In order to obtain a consistent discrete rotation operator for the tangential velocity scheme, the locations of the discrete velocity degrees of freedom are imagined as the midpoints of the triangle edges. On the other hand, in order to obtain a consistent discrete rotation operator for the normal velocity scheme, the locations of the discrete velocity degrees of freedom are imagined as the midpoints of the Voronoi faces. We note that for the tangential velocity scheme, the proposed discretization of $\nabla \cdot \boldsymbol{u} = 0$ exactly coincides with the discrete solenoidal condition allowing to prove a discrete maximum principle for the Voronoi finite volume method for convective transport of a dissolved species in the velocity field $\boldsymbol{u}$ [6].

The structure of the paper is as follows: In the second section, the notions of a Delaunay mesh and its dual, the Voronoi mesh, are introduced, and related quantities are defined. With these notions, discrete divergence and rotation operators for the tangential and the normal scheme are introduced, and both discretization schemes for the incompressible Stokes equations are presented. In the third section, a numerical example exhibits the convergence properties of both schemes on structured and unstructured grids. Experimental convergence rates for the tangential and normal scheme are given for the corresponding discrete L^2 norms for the velocities, the corresponding discrete L^2 norms for the pressure, and the corresponding discrete norms for the discrete rotation of the velocities.

2 Definition of the schemes

We define primal and dual meshes of the domain Ω as follows:

1. The set $\mathscr{T}$ is the finite set of disjoint triangles (considered as open subsets of $\mathbb{R}^2$) such that $\bigcup_{T \in \mathscr{T}} \overline{T} = \overline{\Omega}$. It is considered as the primal mesh. We denote by h_{mesh} the greatest diameter of all triangles. For all $T \in \mathscr{T}$, the point $\boldsymbol{x}_T$, defined as the center of the circumcircle of T, is such that $\boldsymbol{x}_T \in T$.
2. The set $\mathscr{V}$ contains the vertices of all the triangles (and therefore of the edges of the triangles). For all $\boldsymbol{y} \in \mathscr{V}$, we denote by V_y the Voronoi box around the vertex $\boldsymbol{y} \in \mathscr{V}$, defined as $V_y = \{\boldsymbol{x} \in \Omega, |\boldsymbol{x} - \boldsymbol{y}| < |\boldsymbol{x} - \boldsymbol{y}'| \text{ for all } \boldsymbol{y}' \in \mathscr{V}, \boldsymbol{y}' \neq \boldsymbol{y}\}$. The set of Voronoi boxes is considered as the dual mesh.
3. The set $\mathscr{E}$ contains all the edges of the triangles, and is such that, for all $\sigma \in \mathscr{E}$, either σ is located on the boundary of Ω (we denote by $\mathscr{E}_{\text{bnd}}$ the set of these boundary edges), either σ is common to two neighboring triangles (we denote by $\mathscr{E}_{\text{int}}$ the set of these interior edges). We then denote by $\boldsymbol{x}_\sigma$ the middle of σ and by θ_{mesh} the infimum of all quantities $|\boldsymbol{x}_\sigma - \boldsymbol{x}_T|/h_T$, for all triangles T, and $h_T/h_{T'}$, for any pair of neighboring triangles T and T'.

We note that the circumcenter condition $\boldsymbol{x}_T \in T \; \forall T \in \mathscr{T}$ implies that the triangulation is acute, and, therefore, Delaunay. In agreement with the numerical results, we believe that it is possible to weaken the conditions on the triangulation to boundary conforming Delaunay meshes, i.e. Delaunay meshes with the additional property that $\boldsymbol{x}_T \in \Omega \; \forall T \in \mathscr{T}$ [7].

For every edge σ, we define a fixed orientation, which is given by a unit vector $\boldsymbol{t}_\sigma$ parallel to σ, and we define $\boldsymbol{n}_\sigma$ the normal vector to σ, obtained from $\boldsymbol{t}_\sigma$ by a rotation with angle $\pi/2$ in the counterclockwise sense (this rotation operator will be denoted as $\rho_{\frac{\pi}{2}}$, see Fig. 1). We further assume that the edges $\sigma \in \mathscr{E}_{\text{bnd}}$ at the border of Ω build a counterclockwise path around Ω. Then, for any edge $\sigma \in \mathscr{E}_{\text{bnd}}$ the exterior of Ω is located to the right of σ. For every $T \in \mathscr{T}$ we denote by $\mathscr{E}_T$ the set of edges of the triangle T, and we denote, for any $\sigma \in \mathscr{E}_T$, by $\boldsymbol{t}_{T,\sigma}$ the unit vector parallel to σ oriented in the counterclockwise sense around T, by $\boldsymbol{n}_{T,\sigma}$ the unit vector normal to σ and outward to T, and by $D_{T,\sigma}$ the cone with basis σ and vertex $\boldsymbol{x}_T$. For any $\sigma \in \mathscr{E}_{\text{int}}$, let T and T' be the two neighboring triangles such that σ is an edge of T and T'. We denote by $\sigma^\perp$ the segment $[\boldsymbol{x}_T, \boldsymbol{x}_{T'}]$ and by $D_\sigma = D_{T,\sigma} \cup D_{T',\sigma}$. For any $\sigma \in \mathscr{E}_{\text{bnd}}$, let T be the triangle such that σ is an edge of T. We then denote by $\sigma^\perp$ the segment $[\boldsymbol{x}_T, \boldsymbol{x}_\sigma]$ and by $D_\sigma = D_{T,\sigma}$.

For any $\boldsymbol{y} \in \mathscr{V}$, we denote by $\mathscr{E}_y$ the set of all the edges where $\boldsymbol{y}$ is a vertex of, and we denote, for any $\sigma \in \mathscr{E}_y$, by $\boldsymbol{t}_{y,\sigma}$ the unit vector parallel to σ oriented

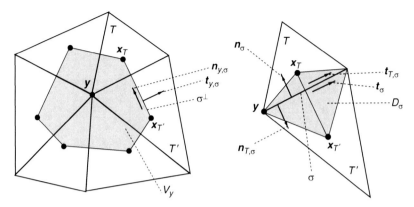

Fig. 1 Notations for the mesh: Left: the Voronoi box associated to a vertex. Right: Zoom on a diamond

from y to the other vertex of σ and by $\boldsymbol{n}_{y,\sigma}$ the unit vector normal to σ and in the counterclockwise sense around y.

The space of degrees of freedom at edges, vertices and triangles are respectively defined by $X_{\mathscr{E}} = \mathbb{R}^{\mathscr{E}}$, $X_{\mathscr{V}} = \mathbb{R}^{\mathscr{V}}$ and $X_{\mathscr{T}} = \mathbb{R}^{\mathscr{T}}$.

For the tangential velocity scheme, the degrees of freedom for the velocity represent the tangential velocity components $v \cdot \boldsymbol{t}_\sigma$ at the midpoint of the edges $\sigma \in \mathscr{E}$, which are oriented in the direction $\boldsymbol{t}_\sigma$. The degrees of freedom for the pressure represent the pressure at the vertices of the triangulation. The space

$$\dot{X}_{\mathscr{E}} = \{v \in X_{\mathscr{E}}, v_\sigma = 0, \forall \sigma \in \mathscr{E}_{\text{ext}}\} \qquad (2)$$

represents the degrees of freedom for the velocity, when homogeneous Dirichlet boundary conditions are prescribed at the boundary edges. We introduce the following discrete differential operators:

$$\mathrm{rot}_T v = \frac{1}{|T|} \sum_{\sigma \in \mathscr{E}_T} |\sigma| v_\sigma \boldsymbol{t}_\sigma \cdot \boldsymbol{t}_{T,\sigma} \qquad \forall v \in X_{\mathscr{E}}, \forall T \in \mathscr{T},$$

$$\mathrm{div}_y v = \frac{1}{|V_y|} \sum_{\sigma \in \mathscr{E}_y} |\sigma^\perp| v_\sigma \boldsymbol{t}_\sigma \cdot \boldsymbol{t}_{y,\sigma} \qquad \forall v \in X_{\mathscr{E}}, \forall y \in \mathscr{V}.$$

Then the tangential velocity scheme reads:

find $(v, p) \in \dot{X}_{\mathcal{E}} \times X_{\mathcal{V}}$ such that

$$\sum_{T \in \mathcal{T}} |T| \mathrm{rot}_T v \mathrm{rot}_T w - \sum_{y \in \mathcal{V}} |V_y| p_y \mathrm{div}_y w = \sum_{\sigma \in \mathcal{E}} 2w_\sigma \int_{D_\sigma} f \cdot t_\sigma \mathrm{d}x, \quad \forall w \in \dot{X}_{\mathcal{E}}$$

$$\sum_{y \in \mathcal{V}} |V_y| p_y = 0,$$

$$\mathrm{Div}_y v = 0, \quad \forall y \in \mathcal{V}.$$

For the normal velocity scheme, the degrees of freedom for the velocity represent the normal velocity components $v \cdot n_\sigma$ at the midpoints of the Voronoi faces $\sigma^\perp$ for all $\sigma \in \mathcal{E}$, which are oriented in the direction n_σ, and the degrees of freedom for the pressure represent the pressure at the center of the triangles. Using the discrete differential operators

$$\mathrm{rot}_y v = \frac{1}{|V_y|} \sum_{\sigma \in \mathcal{E}_y} |\sigma^\perp| v_\sigma n_\sigma \cdot n_{y\sigma} \qquad \forall v \in X_{\mathcal{E}}, \forall y \in \mathcal{V},$$

$$\mathrm{div}_T v = \frac{1}{|T|} \sum_{\sigma \in \mathcal{E}_T} |\sigma| v_\sigma n_\sigma \cdot n_{T\sigma} \qquad \forall v \in X_{\mathcal{E}}, \forall T \in \mathcal{T},$$

the normal velocity scheme writes:

find $(v, p) \in \dot{X}_{\mathcal{E}} \times X_{\mathcal{T}}$ such that

$$\sum_{y \in \mathcal{V}} |V_y| \mathrm{rot}_y v \mathrm{rot}_y w - \sum_{T \in \mathcal{T}} |T| p_T \mathrm{div}_T w = \sum_{\sigma \in \mathcal{E}} 2w_\sigma \int_{D_\sigma} f \cdot n_\sigma \mathrm{d}x, \quad \forall w \in \dot{X}_{\mathcal{E}}$$

$$\sum_{T \in \mathcal{T}} |T| p_T = 0,$$

$$\mathrm{div}_T v = 0, \quad \forall T \in \mathcal{T}.$$

3 Numerical results

In order to investigate numerically the convergence rate that can be achieved with the extended MAC schemes introduced above, we define an academic Stokes problem on two sequences of meshes. We remark that we achieved the same experimental convergence rates for the full nonlinear Navier–Stokes equations [5], where the nonlinear term was discretized in rotational form. The problem is posed on $\Omega = [0, 1]^2$, has homogeneous Dirichlet boundary conditions and reads

$$v = \begin{pmatrix} 2(x-1)^2 x^2 (y-1) y (2y-1) \\ -2(2x-1)(x-1) x (y-1)^2 y^2 \end{pmatrix},$$
$$p = x^3 + y^3 - 0.5.$$

The vector f is computed such that v and p fulfill the Stokes equations (1).

In the first sequence of meshes, every mesh is build up from small squares, where the side length of such a square defines the mesh size. Every square in the mesh is split into two triangles. This mesh is not admissible in the strict sense of the above definition, since the circumcenters of these two triangles coincide. But this does not pose any problem, since in this degenerated case, the discrete method is equivalent to a method where the squares take over the role of triangles, and the diagonals of the squares can be removed from the above considerations, as the measure of their corresponding Voronoi faces are zero. At the same time, on these meshes, triangle edge midpoints and Voronoi face midpoints coincide. This fact will result in superior convergence behavior on these meshes in comparison to "purely" triangular meshes.

In Table 1 we show some information about the degrees of freedom in these square meshes. The last two columns of this Table show some quite interesting information. The penultimate column reveals that the tangential velocity scheme is quite efficient in terms of degrees of freedom, since the ratio between the number of degrees of freedom corresponding to discretely divergence-free velocities and the total number of degrees of freedom is about 0.5. For the normal velocity scheme, the corresponding ratio is only 0.20.

Table 1 Number of edges, vertices and triangles in different square meshes. The penultimate column shows the ratio between discretely divergence-free degrees of freedom and the total number of degrees of freedom for the tangential velocity scheme. The last column shows the ratio between discretely divergence-free degrees of freedom and the total number of degrees of freedom for the normal velocity scheme

mesh size	$|E|$	$|V|$	$|T|$	$\frac{|E|-|V|}{|E|+|V|}$	$\frac{|E|-|T|}{|E|+|T|}$
$\frac{1}{32}$	2945	1024	1922	0.484	0.210
$\frac{1}{64}$	12033	4096	7938	0.492	0.205
$\frac{1}{128}$	48641	16384	32258	0.496	0.203
$\frac{1}{256}$	195585	65536	130050	0.498	0.201
$\frac{1}{512}$	784385	262144	522242	0.499	0.201
$\frac{1}{1024}$	3141633	1048576	2093058	0.500	0.200

The second sequence of meshes are made up of isotropic, unstructured boundary conforming Delaunay meshes. They have been generated by the mesh generator TRIANGLE [8]. We remark, that this approach does not guarantee that the triangulation is acute. In Table 2 we show some information about the degrees of freedom in these triangle meshes. An approximate mesh size was defined according to the largest triangle area that the mesh generator was allowed to generate within a mesh.

From Tables 1 and 2 we recognize that the degrees of freedom of corresponding meshes in the two mesh families are quite similar, such that the definition of the mesh size for unstructured meshes seems to be reasonable. The two schemes are

Table 2 Number of edges, vertices and triangles in different Delaunay meshes generated by the mesh generator TRIANGLE[8]. The penultimate column shows the ratio between discretely divergence-free degrees of freedom and the total number of degrees of freedom for the tangential velocity scheme. The last column shows the ratio between discretely divergence-free degrees of freedom and the total number of degrees of freedom for the normal velocity scheme.

mesh size	$|E|$	$|V|$	$|T|$	$\frac{|E|-|V|}{|E|+|V|}$	$\frac{|E|-|T|}{|E|+|T|}$
$\frac{1}{32}$	3121	1084	2038	0.484	0.210
$\frac{1}{64}$	12326	4195	8132	0.492	0.205
$\frac{1}{128}$	48664	16393	32272	0.496	0.203
$\frac{1}{256}$	194879	65302	129578	0.498	0.201
$\frac{1}{512}$	779506	260519	518988	0.499	0.201
$\frac{1}{1024}$	3114404	1039501	2074904	0.500	0.200

implemented within the framework of the software package PDELIB[9]. All the discrete linear systems are solved with the direct solver PARDISO[10, 11].

In Figs. 2 and 3, for both schemes and series of meshes, we plot various measures of the error between the discrete solution and a projection of the exact solution onto the grid. We used two different projections for both schemes. For the tangential velocity scheme, we evaluate the tangential velocities at the edge midpoints and assign them to the corresponding velocity degrees of freedom. For the normal velocity scheme, we evaluate the normal velocities at the Voronoi face midpoints and assign them to the corresponding velocity degrees of freedom, likewise.

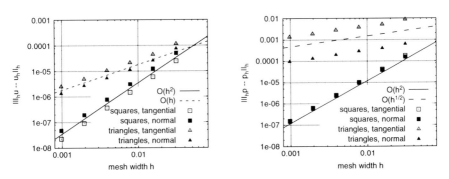

Fig. 2 Discrete L^2-norm of the error between the projected exact solution and the discrete solution. Left: velocity, right: pressure

We start the discussion with the approximation of the velocity, see Fig. 2, left. We observe similar behavior for the two discretization schemes proposed. On triangular

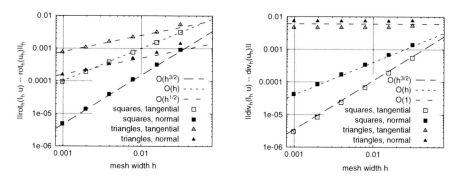

Fig. 3 Discrete L^2-norm of the discrete vector calculus operators applied to the difference between the projected exact velocity and the velocity component of the discrete solution. Left: rotational, right: divergence

meshes the convergence order is approximately $O(h)$. On square meshes, we gain an order of magnitude in the convergence rate in comparison to the triangular meshes.

Also, concerning the approximation orders of the pressure, both schemes behave in a similar way, including second order convergence on square meshes, see Fig. 2, right. We observe that on triangular meshes, the convergence order drops to $O(h^{\frac{1}{2}})$. At the same time, the accuracy of the normal velocity scheme on triangular meshes is better by a factor of ≈ 10 in comparison to the tangential velocity scheme.

As shown by the mathematical analysis [5], the discrete rotation is convergent for both schemes. This is confirmed by Fig. 3 (left), where we observe the convergence of the difference between the discrete rotation of the discrete solution and the discrete rotation of the projected exact solution. On square meshes, for the normal velocity scheme, the L^2 norm of this difference exhibits $O(h^{\frac{3}{2}})$-convergence, while the convergence order of the tangential velocity scheme is only $O(h)$. On triangular meshes, both schemes exhibit $O(h^{\frac{1}{2}})$ convergence with an advantage for the normal velocity scheme concerning the constants.

By construction, for both schemes, the discrete divergence of the velocity component of the discrete solution is zero. Therefore, the error shown in Fig. 3 (right) coincides with the discrete divergence of the projected exact velocity. On square meshes, for both schemes, the discrete divergence operator is consistent, since mid points of an edge coincide with mid points of the orthogonal Voronoi faces. Therefore, the discrete divergence converges on square meshes to zero with order $O(h^{1.5})$ for the tangential velocity scheme and $O(h)$ for the normal velocity scheme. On the triangular meshes, edge mid points and Voronoi face mid points do not coincide and the discrete divergence operator is not consistent resulting in no convergence at all if it is applied to the projection of the velocity component of the exact solution.

We note that the convergence behavior on the boundary conforming Delaunay meshes, which are not acute, is consistent with the theoretical considerations which for technical reasons had been constrained to acute triangulations.

References

1. F. H. Harlow and J. E. Welch. Numerical calculation of time-dependent viscous incompressible flow of fluid with free surface. *Physics of fluids*, 8(12):2182–2189, 1965.
2. R. A. Nicolaides. Analysis and convergence of the MAC scheme. I. The linear problem. *SIAM J. Numer. Anal.*, 29(6):1579–1591, 1992.
3. R. A. Nicolaides and X. Wu. Analysis and convergence of the MAC scheme. II. Navier-Stokes equations. *Math. Comp.*, 65(213):29–44, 1996.
4. J. Nicolaides, T. A. Porsching, and C. A. Hall. Covolume methods in computational fluid dynamics. In M. Hafez and K. Oshma, editors, *Computation Fluid Dynamics Review*, pages 279–299. John Wiley and Sons, New York, 1995.
5. J. Fuhrmann R. Eymard and A. Linke. Extended MAC schemes on Delaunay meshes for the incompressible Navier-Stokes equations, 2011. In preparation.
6. J. Fuhrmann, A. Linke, and H. Langmach. Mass conservative coupling between fluid flow and solute transport. In *Finite Volumes for Complex Application VI*. Springer, 2011.
7. H. Si, K. Gärtner, and J. Fuhrmann. Boundary conforming Delaunay mesh generation. *Comput. Math. Math. Phys.*, 50:38–53, 2010.
8. J. Shewchuk. Triangle: A two-dimensional quality mesh generator and Delaunay triangulator. http://www.cs.cmu.edu/ quake/triangle.html, University of California at Berkeley.
9. J. Fuhrmann et al. Pdelib. www.wias-berlin.de/software/pdelib/.
10. O. Schenk, K. Gärtner, and W. Fichtner. Efficient sparse LU factorization with left-right looking strategy on shared memory multiprocessors. *BIT*, 40(1):158–176, 1999.
11. O. Schenk, K. Gärtner, G. Karypis, S. Röllin, and M. Hagemann. PARDISO Solver Project. URL: http://www.pardiso-project.org, 2010. Retrieved 2010-02-15.

The paper is in final form and no similar paper has been or is being submitted elsewhere.

Multiphase Flow in Porous Media Using the VAG Scheme

Robert Eymard, Cindy Guichard, Raphaèle Herbin, and Roland Masson

Abstract We present the use of the Vertex Approximate Gradient scheme for the simulation of multiphase flow in porous media. The porous volume is distributed to the natural grid blocks and to the vertices, hence leading to a new finite volume mesh. Then the unknowns in the control volumes may be eliminated, and a 27-point scheme results on the vertices unknowns for a hexahedral structured mesh. Numerical results show the efficiency of the scheme in various situations, including miscible gas injection.

Keywords two-phase flow in porous media, vertex approximate gradient scheme, reservoir simulation.
MSC2010: 65M08,76S05

1 Introduction

Simulation of multiphase flow in porous media is a complex task, which has been the object of several works over a long period of time, see the reference books [12] and [3]. Several types of numerical schemes have been proposed in the past decades. Those which are implemented in industrial codes are mainly built upon cell centred approximations and discrete fluxes, in a framework which is also that of the method

---

R. Eymard
Université Paris-Est, France, e-mail: robert.eymard@univ-mlv.fr

C. Guichard
Université Paris-Est and IFP Energies nouvelles, France, e-mail: cindy.guichard@ifpen.fr

R. Herbin
Université Aix-Marseille, France, e-mail: raphaele.herbin@latp.univ-mrs.fr

R. Masson
IFP Energies nouvelles, France, e-mail: roland.masson@ifpen.fr

we propose here. Let us briefly sketch this framework. The 3D simulation domain Ω is meshed by control volumes $X \in \mathcal{M}$. Let us denote by Λ the diffusion matrix (which is a possibly full matrix depending on the point of the domain).

For each control volume $X \in \mathcal{M}$, the set of neighbours $Y \in \mathcal{N}_X$ is the set of all control volumes involved in the mass balance in X, which means that the following approximation formula is used: $-\int_X \nabla \cdot \Lambda \, grad \, p \, dx \simeq \sum_{Y \in \mathcal{N}_X} F_{X,Y}(P)$, where $P = (p_Z)_{Z \in \mathcal{M}}$ is the family of all pressure unknowns in the control volumes, and where the flux $F_{X,Y}(P)$, between control volumes X and Y, is a linear function of the components of P which ensures the following conservativity property:

$$F_{X,Y}(P) = -F_{Y,X}(P). \tag{1}$$

Such a linear function, which is expected to vanish on constant families, may be defined by

$$F_{X,Y}(P) = \sum_{Z \in \mathcal{M}_{X,Y}} a_{X,Y}^Z p_Z, \tag{2}$$

where the family $(a_{X,Y}^Z)_{Z \in \mathcal{M}_{X,Y}}$ and $\mathcal{M}_{X,Y} \subset \mathcal{M}$ are such that $\sum_{Z \in \mathcal{M}_{X,Y}} a_{X,Y}^Z = 0$.

Assuming N_c constituents and N_α phases, the discrete balance laws then read

$$\frac{\Phi_X}{\delta t}(A_{X,i}^{(n+1)} - A_{X,i}^{(n)}) + \sum_{\alpha=1}^{N_\alpha} \sum_{Y \in \mathcal{N}_X} M_{X,Y,i}^{(n+1),\alpha} F_{X,Y}^{(n+1),\alpha} = 0, \ \forall i = 1, \ldots, N_c, \tag{3}$$

$$F_{X,Y}^{(n+1),\alpha} = F_{X,Y}(P^{(n+1),\alpha}) - \rho_{X,Y}^{(n+1),\alpha} g \cdot (x_Y - x_X), \ \forall \alpha = 1, \ldots, N_\alpha,$$

where n is the time index, δt is the time step, Φ_X is the porous volume of the control volume $X \in \mathcal{M}$, $A_{X,i}$ represents the accumulation of constituent i in the control volume X per unit pore volume (assumed to take into account the dependence of the porosity with respect to the pressure), $M_{X,Y,i}^\alpha$ is the amount of constituent i transported by phase α from the control volume X to the control volume Y (generally computed by taking the upstream value with respect to the sign of $F_{X,Y}$), P^α is the family of the pressure unknowns of phase α, g is the gravity acceleration, $\rho_{X,Y}^\alpha$ is the bulk density of phase α between control volumes X and Y and x_X is the centre of control volume X. In addition to these relations, the differences between the phase pressures are ruled by capillary pressure laws. Thermodynamical equilibrium and standard closure relations are used.

When applying scheme (3), one should be very wary of the use of conformal finite elements in the case of highly heterogeneous media. Indeed, assuming that the control volumes are vertex centred with vertices located at the interfaces between different media, then the porous volume concerned by the flow of very permeable medium includes that of non permeable medium. This may lead to surprisingly wrong results on the component velocities. A possible interpretation of these poor results is that, when seen as a set of discrete balance laws, the finite element method provides the same amount of impermeable and permeable porous volume for the accumulation term for a node located at a heterogeneous interface.

We present in this paper the use of a new scheme, called Vertex Approximate Gradient (VAG) scheme [8, 9], which can be implemented in (3) so that the components velocities are correctly approximated, thanks to a special choice of the control volumes and of the discrete fluxes, which respect to the form (2). The purpose of respecting the form (3)-(2) is to be able to plug it easily into an existing reservoir code, say Multi-Point Flux Approximation (MPFA), by simply redefining the control volumes and the coefficients $a_{X,Y}^Z$ of the discrete flux.

Although part of this scheme is vertex centred, we show that the solution obtained on a very heterogeneous medium with a coarse mesh remains accurate. This is a great advantage of this scheme, which is also always coercive, symmetric, and leads to a 27-stencil on hexahedral structured meshes. In addition the VAG scheme is very efficient on meshes with tetrahedra since the scheme can then be written with the nodal unknowns only, thus inducing a reduction of the number of degrees of freedom by a factor 5 compared with cell centred finite volume schemes such as MPFA schemes [1, 2, 4, 5].

2 Presentation of the scheme

The VAG scheme is described in [8,9], and its gradient scheme properties are related to those presented in [7]; therefore we focus here on the use of this scheme for a multiphase flow simulation of the form (3). Let $\mathcal{M}$ be a general mesh of Ω, defined by a set $\mathcal{G}$ of grid blocks and the set $\mathcal{V}$ of their vertices; this is a mesh of control volumes in the sense of the preceding section: a control volume is either a grid block $K \in \mathcal{G}$ or a vertex $v \in \mathcal{V}$. In particular, a porous volume must be associated to each control volume, i.e. to each grid block and to each vertex. Finally a flux $F_{X,Y}$ from the control volume X to the control volume Y must be specified.

Any given grid block $K \in \mathcal{G}$ has, say, N_K vertices; let us denote by $\mathcal{V}_K \subset \mathcal{V}$ the set of these vertices. We wish to define a flux between neighbouring control volumes $X = K$ and $Y = v \in \mathcal{V}_K$, and between neighbouring control volumes $X = v \in \mathcal{V}_K$ and $Y = K \in \mathcal{G}_v = \{Y = K \in \mathcal{G} \text{ such that } v \in \mathcal{V}_K\}$; for this purpose, we introduce a local discrete gradient $\nabla_{K,v}(P_K) \in \mathbb{R}^3$ (see [8,9] for the precise definitions), which only depends on the values $P_K = (P_{K,v})_{v \in \mathcal{V}_K} = (p_v - p_K)_{v \in \mathcal{V}_K}$. We then introduce the matrices $(A_K^{v,v'})_{v,v' \in \mathcal{V}_K}$, which are defined by the following relation

$$\frac{|K|}{N_K} \sum_{v \in \mathcal{V}_K} \Lambda_K \nabla_{K,v} P_K \cdot \nabla_{K,v} Q_K = \sum_{v \in \mathcal{V}_K} \sum_{v' \in \mathcal{V}_K} A_K^{v,v'} P_{K,v'} Q_{K,v}, \quad \forall P_K, Q_K \in \mathbb{R}^{\mathcal{V}_K}.$$

The flux from control volume $X = K$ to control volume $Y = v$ is then given by

$$F_{X,Y}(P) = F_{K,v}(P) = - \sum_{v' \in \mathcal{V}_K} A_K^{v,v'} (p_{v'} - p_K),$$

which is of the same form as (2); using (1), we get $F_{Y,X}(P) = -F_{X,Y}(P)$. Let us now turn to the definition of porous volumes for all $X \in \mathcal{M}$. The question is to associate to each vertex a porous volume in such a way that the component velocities are well approximated. Let us denote by $\Phi_K = \int_K \Phi(x)\,\mathrm{d}x$ the total porous volume of each grid block $K \in \mathcal{G}$. We shall then take out a little bit of this porous volume of each grid block to associate it with the control volumes of the vertices. In order to obtain a systematic way to redistribute the porous volume between the grid blocks and the vertices, we define a first indicator of the transmissivity between K and v by $B_{K,v} = \sum_{v' \in \mathcal{V}_K} A_K^{v,v'} > 0$, $\forall K \in \mathcal{G}_v$, and then, for a global small value $\mu \in]0,1[$ (for example, $\mu = 0.05$), a weighted relative transmissivity (which is larger for permeable regions than for impermeable ones):

$$\widetilde{B}_{K,v} = \mu \frac{B_{K,v}}{\sum_{L \in \mathcal{G}_v} B_{L,v}}, \quad \forall v \in \mathcal{V}, \forall K \in \mathcal{G}_v, \tag{4}$$

Note that it might be expected that a too small value for μ lead to some numerical problems; nevertheless, such consequences have not been observed within the range $\mu \in [0.01, 0.05]$. The total porous volume can then be redistributed between all control volumes $X \in \mathcal{M}$, that is between the grid blocks and the vertices, by the following relations:

$$\Phi_X = \begin{cases} \sum_{K \in \mathcal{G}_v} \widetilde{B}_{K,v} \widetilde{\Phi}_K & \text{if } X = v \in \mathcal{V}, \\ \widetilde{\Phi}_K (1 - \sum_{v \in \mathcal{V}_K} \widetilde{B}_{K,v}) & \text{if } X = K \in \mathcal{G}. \end{cases}$$

Hence, we distribute a small amount of the porous volume of K to its vertices, in a conservative way; indeed, by construction, we get that

$$\sum_{X \in \mathcal{M}} \Phi_X = \sum_{K \in \mathcal{G}} \widetilde{\Phi}_K,$$

with all $\Phi_X > 0$, provided that the value μ be chosen sufficiently small. We can remark that:

1. the porous volume of a vertex $v \in \mathcal{V}$ located at the interface between high and low permeability regions is mainly extracted from the higher permeability region,
2. the part of the lower permeability region distributed to the vertices is reduced by the factor μ.

We recall that we keep the property ensured in the monophasic case on the full system: indeed, the linear systems issued from Newton's method may be solved by first eliminating all unknowns $K \in \mathcal{G}$, and then solve a 27-point system on $v \in \mathcal{V}$.

3 Numerical applications

3.1 Heterogeneous case

The first example is the injection of CO_2, considered as immiscible with the liquid phase, at the middle point of an isotropic and heterogeneous reservoir, with size $[-100, 100] \times [0, 50] \times [0, 45]$ m^3. The reservoir includes three 15 m-thick layers. The top and bottom layers are assumed to be weakly permeable ($|\Lambda| = 10^{-16}$ m^2) and the medium layer is much more permeable ($|\Lambda| = 10^{-12}$ m^2). A regular coarse $100 \times 10 \times 15$ mesh is used for the simulation (depicted in Fig. 1). The values

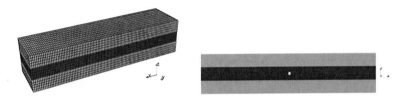

Fig. 1 First example. Left: mesh and layers. Right: the well is depicted at the centre of the section $y = 25$ m, illustrated by the white block

$\mu = 0.01$ and $\mu = 0.05$ have been tested in (4), without significant influence on the results both in terms of accuracy and CPU time. The results of the VAG scheme are compared to those obtained using the two-point flux approximation (TPFA) scheme, which is available on such a regular mesh. We observe in Fig. 2 that the numerical diffusion along the axes of the mesh leads, after a short injection time, to a distorted profile of the gas saturation in the case of the TPFA scheme, known as Grid Orientation Effect (GOE), see also [10]. This phenomenon is clearer in

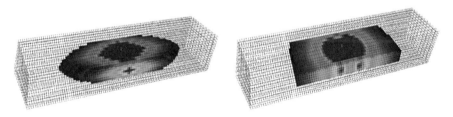

Fig. 2 View of the gas saturation in the reservoir, after a short injection time. Farthest to the well: $S = 0.001$. Closest to the well: $S = 0.042$. Left: TPFA scheme. Right: VAG scheme

the profile of the saturation at the end of the gas injection. We see in Fig. 3 the important GOE due to the TPFA scheme, whereas this effect is nearly invisible in the results obtained with the VAG scheme. Moreover, this distortion, also visible in the vertical section (Fig. 4), is again corrected using the VAG scheme. These results can be explained by the construction of the fluxes. In fact, after elimination of the cell centred unknowns, the resulting scheme on the vertex unknowns has a 27-point stencil, whereas it remains a 7-point scheme on the control volumes unknowns using the TPFA scheme.

Fig. 3 Gas saturation at the end of the gas injection. Section $z = 22.5\ m$. Farthest to the well : $S = 0$. Closest to the well : $S = 1$. Left: TPFA scheme. Right: VAG scheme

Fig. 4 Gas saturation at the end of the gas injection. Section $y = 25\ m$. Farthest to the well : $S = 0$. Closest to the well : $S = 1$. Left: TPFA scheme. Right: VAG scheme

3.2 Near-Well case

In the second example, we consider the numerical simulation of the injection of CO_2 in near-well regions for a deviated well. A hexahedral radial part is connected to the outside boundary either by a hexahedral mesh or a hybrid mesh (using both pyramids and tetrahedra) as illustrated in Fig. 5. The number of cells is roughly the same for both types of grids. This family of meshes is also used in the 3D benchmark on monophasic diffusion [11]. The medium is homogeneous, but anisotropic. We consider that the CO_2 can be dissolved in the aqueous phase.

We consider in Figs. 7 and 6 the mass outflow rate of CO_2 in the two phases at the outer boundary using the VAG scheme and the MPFA O-scheme on both types of grids. The values 0.01 and 0.05 have been tested for the parameter μ used in (4) and the results are almost the same. In order to keep the output clearer, the curves are only plotted for $\mu = 0.05$. We observe that the VAG scheme produces results which are not very sensitive to the type of the grid. On the contrary, the MPFA O-scheme shows a significant sensitivity to the type the grid, since the production of CO_2 is slowed down by the use of the tetrahedral mesh.

We finally remark that there are 74 679 cell unknowns and 74 800 nodal unknowns for the hexahedral mesh, to be compared with 77 599 cell unknowns (including 28 704 tetrahedra) and only 37 883 nodal unknowns for the hybrid mesh.

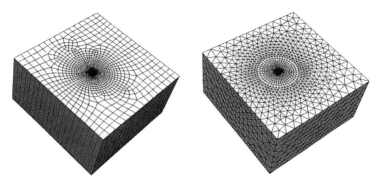

Fig. 5 Near-well grid : the hexahedral mesh (left) and the hybrid mesh (right)

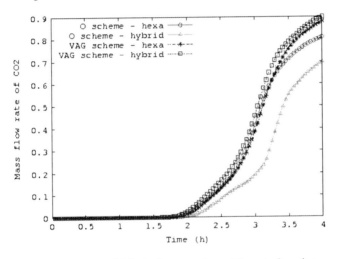

Fig. 6 Outflowing mass flow rate of CO_2 in the water phase at the outer boundary

As stated in the introduction, we see on this example that computing costs of the VAG scheme may be reduced in the case of meshes with tetrahedra.

4 Conclusion

The above numerical results show that the VAG scheme seems to be an efficient scheme for multiphase flow simulation of a heterogeneous anisotropic reservoir; it features the following properties:

1. it may be implemented, without any additional cost, into an MPFA industrial code;

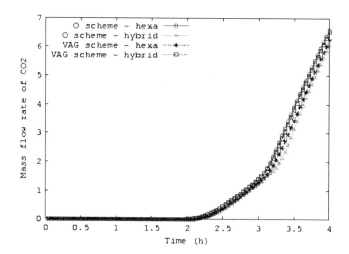

Fig. 7 Outflowing mass flow rate of CO_2 in the gas phase at the outer boundary

2. it leads to a 27-point compact stencil and a symmetric and coercive operator for the treatment of the diffusion terms, even in the case of distorted meshes and heterogeneous and anisotropic diffusion,
3. its cost is considerably reduced in the case of meshes with tetrahedra compared with cell centred MPFA schemes;
4. it remains accurate on coarse meshes thanks to a well-chosen distribution of the porous volume between the centre of the control volumes and the vertices;
5. since a pore volume is assigned to the Neumann boundary nodes, the Neumann conditions are obtained by writing the conservation and closure equations as in the inner control volumes.

Full scale reservoir simulations will be performed in order to confirm the efficiency of the method.

Acknowledgements Work sponsored by GNR MOMAS and ANR VFSitCom.

References

1. I. Aavatsmark, T. Barkve, O. Bøe and T. Mannseth. Discretization on non-orthogonal, quadrilateral grids for inhomogeneous, anisotropic media. *J. Comput. Phys.*, 127 (1):2-14, 1996.
2. I. Aavatsmark, GT. Eigestad, BT. Mallison and JM. Nordbotten. A compact multipoint flux approximation method with improved robustness. *Numer. Methods Partial Diff. Eqns.*, 27 (5):1329-1360, 2008.
3. K. Aziz and A. Settari. Petroleum reservoir simulation. *Chapman & Hall*, 1979.
4. MG. Edwards. Unstructured, control-volume distributed, full-tensor finite-volume schemes with flow based grids. *Computational Geosciences*, 6 (3):433-452, 2002.

5. MG. Edwards and CF. Rogers. Finite volume discretization with imposed flux continuity for the general tensor pressure equation. *Computational Geosciences*, 2 (4):259-290, 1998.
6. R. Eymard, T. Gallouët, and R. Herbin. Discretisation of heterogeneous and anisotropic diffusion problems on general non-conforming meshes, sushi: a scheme using stabilisation and hybrid interfaces. *IMA J. Numer. Anal.*, 30(4):1009–1043, 2010. see also http://hal.archives-ouvertes.fr/.
7. R. Eymard and R. Herbin. Gradient schemes for diffusion problem. *in FVCA VI proc.*, Prague, June 6-10, 2011.
8. R. Eymard, C. Guichard, and R. Herbin. Small-stencil 3D schemes for diffusive flows in porous media. *submitted*, 2010. see also http://hal.archives-ouvertes.fr/.
9. R. Eymard, C. Guichard and R. Herbin. Benchmark 3D: the VAG scheme. *in FVCA VI proc.*, Prague, June 6-10, 2011
10. R. Eymard, C. Guichard and R. Masson. Grid orientation effect and multipoint flux approximation. *in FVCA VI proc.*, Prague, June 6-10, 2011.
11. R. Herbin and F. Hubert. Benchmark 3D on discretization schemes for anisotropic diffusion problem on general grids. *in FVCA VI proc.*, Prague, June 6-10, 2011.
12. D. Peaceman. Fundamentals of numerical reservoir simulation. *Elsevier*, 1977.

The paper is in final form and no similar paper has been or is being submitted elsewhere.

Grid Orientation Effect and MultiPoint Flux Approximation

Robert Eymard, Cindy Guichard, and Roland Masson

Abstract Some cases of nonlinear coupling between a diffusion equation, related to the computation of a pressure field within a porous medium, and a convection equation, related to the conservation of a species, lead to the apparition of the so-called grid orientation effect. We propose in this paper a new procedure to eliminate this Grid Orientation Effect, only based on the modification of the stencil of the discrete version of the convection equation. Numerical results show the efficiency and the accuracy of the method.

Keywords Grid Orientation Effect, Two Phase Porous Media Flow, Finite Volume Methods
MSC2010: 76S05, 65M08, 76E06

1 Introduction

In the 1980's, numerous papers have been concerned with the so-called grid orientation effect, in the framework of oil reservoir simulation. This effect is due to the anisotropy of the numerical diffusion induced by the upstream weighting scheme, and the computation of a pressure field, solution to an elliptic equation in which the diffusion coefficient depends on the value of the convected unknown. This problem has been partly solved in the framework of industrial codes, in which the meshes are structured and regular (mainly based on squares and cubes). The literature on

R. Eymard
Université Paris-Est, France, e-mail: robert.eymard@univ-mlv.fr

C. Guichard (work supported by ANR VFSitCom)
Université Paris-Est and IFP Energies nouvelles, France, e-mail: cindy.guichard@ifpen.fr

R. Masson
IFP Energies nouvelles, France, e-mail: roland.masson@ifpen.fr

this problem is huge, and is impossible to exhaustively quote; let us only cite [3, 4, 6, 10, 11] and references therein. In the 2000's, a series of new schemes have been introduced in order to compute these coupled problems on general grids [1, 2, 5, 8]. But, in most of the cases, the non regular meshes conserve structured directions, although the shape of the control volumes is no longer that of a regular cube. This is the case for the Corner Point Geometries [9] widely used in industrial reservoir simulations. The control volumes which are commonly used in 3D reservoir simulations are generalised "hexahedra", in the sense that each of them is neighboured by 6 other control volumes. In this case, the stencil for the pressure resolution may have a 27-point stencil (using for instance a MPFA scheme). Nevertheless, selecting a 27-point stencil instead of a 7-point stencil for the pressure resolution has no influence on the Grid Orientation Effect, which results from the stencil used in upstream weighted mass exchanges coupled with the pressure resolution.

In order to overcome this problem, we study here a generalisation of methods consisting in increasing the stencil of the convection equation, without modifying the pressure equation. The method will be presented on a simplified problem, modelling immiscible two-phase flow within a porous medium. Let $\Omega \subset \mathbb{R}^d$ (with $d = 2$ or 3) be the considered space domain. We consider the following two-phase flow problem in Ω:

$$\begin{cases} u_t - \operatorname{div}(k_1(u) \Lambda \nabla p) = 0 \\ (1-u)_t - \operatorname{div}(k_2(u) \Lambda \nabla p) = 0, \end{cases} \quad (1)$$

where $u(x,t) \in [0, 1]$ is the saturation of phase 1 (for example water), and therefore $1 - u(x,t)$ is the saturation of phase 2, k_1 is the mobility of phase 1 (increasing function such that $k_1(0) = 0$), k_2 is the mobility of phase 2 (decreasing function such that $k_2(1) = 0$), and p is the common pressure of both phases (the capillary pressure is assumed to be negligible in front of the pressure gradients due to injection and production wells) and we consider a horizontal medium with permeability tensor Λ. It is therefore possible to see System (1) as the coupling of an elliptic problem with unknown p and a nonlinear scalar hyperbolic problem with unknown u:

$$\begin{cases} m(u) = k_1(u) + k_2(u), \ f(u) = \dfrac{k_1(u)}{m(u)} \\ \operatorname{div} F = 0 \text{ with } F = -m(u) \Lambda \nabla p \\ u_t + \operatorname{div}(f(u) F) = 0 \end{cases} \quad (2)$$

We then consider a MultiPoint Flux Approximation finite volume scheme for the approximation of Problem (1), coupled with an upstream weighting scheme for the mass exchanges. Such a scheme may be written:

$$F_{K,L}^{(n)} = m_{KL}^{(n)} \sum_{M \in \mathcal{M}} a_{KL}^M p_M^{(n+1)} \text{ with } \sum_{M \in \mathcal{M}} a_{KL}^M = 0 \quad (3)$$

$$\sum_{L \in \mathcal{N}_K} F_{K,L}^{(n)} = 0 \quad (4)$$

$$F_{K,L}^{(n)} + F_{L,K}^{(n)} = 0 \quad (5)$$

$$|K|\left(u_K^{(n+1)} - u_K^{(n)}\right) + \delta t^n \sum_{L \in \mathcal{N}_K} \left(f(u_K^{(n)})(F_{K,L}^{(n)})^+ - f(u_L^{(n)})(F_{K,L}^{(n)})^-\right) = 0. \quad (6)$$

In the above system, we denote by $\mathcal{M}$ the finite volume mesh of Ω, K, L are control volumes, $\mathcal{N}_K$ is the set of the neighbours of K (i.e. control volumes exchanging fluid mass with K), n is the time index and δt^n is the time step ($\delta t^n = t^{(n+1)} - t^{(n)}$), $p_M^{(n)}$ and $u_M^{(n)}$ are respectively the pressure and the saturation in control volume M at time $t^{(n)}$. The coefficients a_{KL}^M are computed with respect to the geometry of the mesh and to Λ. The value $m_{KL}^{(n)}$ is any average value (arithmetic or harmonic) of the values $m(u_K^{(n)})$ and $m(u_L^{(n)})$. Then $F_{K,L}^{(n)}$ is the approximation of $F \cdot n$ at the interface $K|L$ between control volumes K and L at time step n, and, for all real a, the values a^+ and a^- are respectively defined by $\max(a,0)$ and $\max(-a,0)$.

The set $\mathcal{N}_K$ of the neighbours of K is classically defined as all the control volumes which have a common face with K. But, as we show in this paper, this notion may be relaxed. Defining the notion of "stencil" $S \subset \mathcal{M}^2$ by $S = \{(K, L) \in \mathcal{M}^2, L \in \mathcal{N}_K\}$, this stencil is then equal to the set of all $(K, L) \in \mathcal{M}^2$ such that $F_{K,L}^{(n)}$ may be different from 0. In view of (5), S must verify the symmetry property

$$S \subset \mathcal{M}^2 \text{ and } \forall (K, L) \in S, (L, K) \in S. \quad (7)$$

As we stated in the introduction, the drawback of the use of this stencil for practical problems, where $F_{K,L}^{(n)}$ is computed from the resolution of a pressure equation, is that it leads to the Grid Orientation Effect. Therefore, we want to replace (6) by

$$|K|\left(u_K^{(n+1)} - u_K^{(n)}\right) + \delta t^n \sum_{L \in \widehat{\mathcal{N}}_K} \left(f(u_K^{(n)})(\widehat{F}_{K,L}^{(n)})^+ - f(u_L^{(n)})(\widehat{F}_{K,L}^{(n)})^-\right) = 0, \quad (8)$$

where the new stencil $\widehat{S}$, defined by $\widehat{S} = \{(K, L) \in \mathcal{M}^2, L \in \widehat{\mathcal{N}}_K\}$, is such that the Grid Orientation Effect is suppressed. In (8), the values of the fluxes $(\widehat{F}_{K,L}^{(n)})_{(K,L) \in \widehat{S}}$ will be set such that the two following properties hold: the flux continuity holds

$$\widehat{F}_{K,L}^{(n)} + \widehat{F}_{L,K}^{(n)} = 0, \ \forall (K, L) \in \widehat{S}, \quad (9)$$

and the balance in the control volumes is the same as that satisfied by the fluxes $(F_{K,L}^{(n)})_{(K,L) \in S}$:

$$\sum_{L, (K,L) \in \widehat{S}} \widehat{F}_{K,L}^{(n)} = \sum_{L, (K,L) \in S} F_{K,L}^{(n)}, \ \forall K \in \mathcal{M}. \quad (10)$$

In view of (15), we again prescribe the symmetry property

$$\widehat{S} \subset \mathcal{M}^2 \text{ and } \forall (K,L) \in \widehat{S}, \ (L,K) \in \widehat{S}. \tag{11}$$

The section 2 of this paper is devoted to the description of a method for constructing $\widehat{F}_{K,L}^{(n)}$ for a given stencil $\widehat{S}$, which ensures properties (9) and (10) (corresponding, for a given n, to (15) and (16) below). The application of this method to the case of an initial five-point pattern stencil S and of a nine-point stencil $\widehat{S}$ is detailed in Section 3. Then numerical tests show the efficiency of the method to fight the Grid Orientation Effect (section 4).

2 Construction of $\widehat{F}_{K,L}$ in the new stencil $\widehat{S}$

The method presented in this section concerns the reconstruction of the fluxes, which has to be applied to each time step. Hence, for the simplicity of notation, we drop the index n in this section. For a stencil $\widehat{S} \subset \mathcal{M}^2$ such that (11) holds and for given $(K,L) \in \mathcal{M}^2$, the set $\widehat{\mathcal{P}}_{K,L}$ of the paths from K to L following $\widehat{S}$ is defined by

$$\widehat{\mathcal{P}}_{K,L} := \left\{ P = \begin{cases} (K_i, K_{i+1}), i = 1, \ldots, N-1 \text{ with } K_1 = K, \ K_N = L \\ \text{and } K_i \neq K_j \text{ for } i \neq j = 1, \cdots, N \end{cases} \subset \widehat{S} \right\}. \tag{12}$$

We denote by $\sharp \widehat{\mathcal{P}}_{K,L}$ the cardinality of $\widehat{\mathcal{P}}_{K,L}$, i.e. the number of paths P from K to L following $\widehat{S}$. For any $P = \{(K_i, K_{i+1}), i = 1, \ldots, N-1\} \in \widehat{\mathcal{P}}_{K,L}$, we denote by $P^{\leftarrow}$ the inverse path from L to K following $\widehat{S}$, defined by $P^{\leftarrow} = \{(K_{i+1}, K_i), i = 1, \ldots, N-1\}$.

We may now state the following result.

Lemma 1 (New stencil and fluxes). *Let $\mathcal{M}$ be a finite set, let $S \subset \mathcal{M}^2$ be given such that (7) holds. Let $(F_{K,L})_{(K,L) \in S}$ be a family such that the property*
$$F_{K,L} + F_{L,K} = 0, \ \forall (K,L) \in \mathcal{M}^2$$
holds. Let $\widehat{S} \subset \mathcal{M}^2$ be given such that (11) holds and such that
$$\forall (K,L) \in S, \ \sharp \widehat{\mathcal{P}}_{K,L} > 0.$$
For all $(K,L) \in S$, let $(F_{K,L}^P)_{P \in \widehat{\mathcal{P}}_{K,L}}$ be a family such that
$$\forall (K,L) \in S, \ \sum_{P \in \widehat{\mathcal{P}}_{K,L}} F_{K,L}^P = F_{K,L},$$
satisfying the property

$$\forall (K,L) \in S, \ \forall P \in \widehat{\mathcal{P}}_{K,L}, \ F_{K,L}^P + F_{L,K}^{P^{\leftarrow}} = 0. \tag{13}$$

Then the family $(\widehat{F}_{K,L})_{(K,L) \in \widehat{S}}$, defined by

$$\forall (I,J) \in \widehat{S}, \ \widehat{F}_{I,J} = \sum_{(K,L) \in S} \sum_{P \in \widehat{\mathcal{P}}_{K,L}} \xi_{I,J,P} F_{K,L}^P, \tag{14}$$

where $\xi_{I,J,P}$ is such that $\xi_{I,J,P} = 1$ if $(I, J) \in P$ and $\xi_{I,J,P} = 0$ otherwise, satisfies

$$\widehat{F}_{K,L} + \widehat{F}_{L,K} = 0, \ \forall (K, L) \in \widehat{S}, \tag{15}$$

and

$$\sum_{L, (K,L) \in \widehat{S}} \widehat{F}_{K,L} = \sum_{L, (K,L) \in S} F_{K,L}, \ \forall K \in \mathcal{M}. \tag{16}$$

Proof. Firstly, using definitions, for a given $(I, J) \in \widehat{S}$, we have $(J, I) \in \widehat{S}$ and $\widehat{F}_{J,I} = \sum_{(L,K) \in S} \sum_{P \in \widehat{\mathcal{P}}_{L,K}} \xi_{J,I,P} F_{L,K}^{P}$. Then, thanks to the following equivalences

$$\begin{cases} (L, K) \in S \iff (K, L) \in S \\ P \in \widehat{\mathcal{P}}_{L,K} \iff P^{\leftarrow} \in \widehat{\mathcal{P}}_{K,L} \\ (J, I) \in P \iff (I, J) \in P^{\leftarrow}, \end{cases}$$

and using (13), we can rewrite $\widehat{F}_{J,I}$ as follows

$$\widehat{F}_{J,I} = -\sum_{(K,L) \in S} \sum_{P \in \widehat{\mathcal{P}}_{K,L}} \xi_{I,J,P} F_{K,L}^{P} = -\widehat{F}_{I,J},$$

which proves (15).

Secondly, for a given $I \in \mathcal{M}$, by reordering the sums, we can write that

$$\sum_{J, (I,J) \in \widehat{S}} \widehat{F}_{I,J} = \sum_{J, (I,J) \in \widehat{S}} \sum_{(K,L) \in S} \sum_{P \in \widehat{\mathcal{P}}_{K,L}} \xi_{I,J,P} F_{K,L}^{P} = \sum_{(K,L) \in S} \sum_{P \in \widehat{\mathcal{P}}_{K,L}} \chi_{I,P} F_{K,L}^{P}$$

where $\chi_{I,P} = \sum_{J, (I,J) \in \widehat{S}} \xi_{I,J,P}$ is equal to 1 if there exists $J \in \mathcal{M}$ such that $(I, J) \in P$ (therefore $I \neq L$), and to 0 otherwise. Note that, for $(K, L) \in S$ with $K \neq I$ and for $P \in \widehat{\mathcal{P}}_{K,L}$ with $\chi_{I,P} = 1$, we have $I \neq L$, $(L, K) \in S$, $P^{\leftarrow} \in \widehat{\mathcal{P}}_{L,K}$ and $\chi_{I,P^{\leftarrow}} = 1$. So, using (13), we obtain

$$\sum_{(K,L) \in S \text{ s.t. } K \neq I} \sum_{P \in \widehat{\mathcal{P}}_{K,L}} \chi_{I,P} F_{K,L}^{P} = 0.$$

Therefore we can write

$$\sum_{J, (I,J) \in \widehat{S}} \widehat{F}_{I,J} = \sum_{L, (I,L) \in S} \sum_{P \in \widehat{\mathcal{P}}_{I,L}} \chi_{I,P} F_{I,L}^{P} = \sum_{L, (I,L) \in S} \sum_{P \in \widehat{\mathcal{P}}_{I,L}} F_{I,L}^{P} = \sum_{L, (I,L) \in S} F_{I,L},$$

which proves (16).

3 Application to an initial five-point stencil on a structured quadrilateral mesh

Let us assume, taking the example of a 2D situation, that the initial stencil S is a five-point stencil, defined on a regular quadrilateral mesh

$$S = \{(K, L) \in \mathcal{M}^2, \overline{K} \text{ and } \overline{L} \text{ have a common edge}\}, \tag{17}$$

and that the new stencil $\widehat{S}$ is the nine-point stencil (see the figure below), defined by

$$\widehat{S} = S \cup \{(K, L) \in \mathcal{M}^2, \overline{K} \text{ and } \overline{L} \text{ have a common point}\}. \tag{18}$$

Then we define $(F_{K,L}^P)_{P \in \widehat{\mathscr{P}}_{K,L}}$, for all $P \in \widehat{\mathscr{P}}_{K,L}$ and all $(K, L) \in S$ (remark that in this case, $S \subset \widehat{S}$):

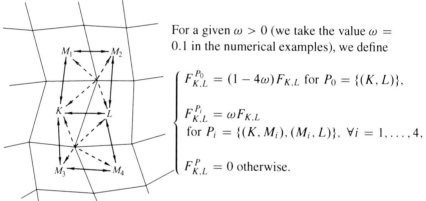

For a given $\omega > 0$ (we take the value $\omega = 0.1$ in the numerical examples), we define

$$\begin{cases} F_{K,L}^{P_0} = (1 - 4\omega) F_{K,L} \text{ for } P_0 = \{(K, L)\}, \\ F_{K,L}^{P_i} = \omega F_{K,L} \\ \quad \text{for } P_i = \{(K, M_i), (M_i, L)\}, \forall i = 1, \ldots, 4, \\ F_{K,L}^P = 0 \text{ otherwise.} \end{cases}$$

Assuming that this procedure has been applied to all initial five-point connection, let us give the resulting values of $\widehat{F}_{K,L}$ deduced from (14) in two cases:

$$\begin{cases} \widehat{F}_{K,L} = (1 - 4\omega) F_{K,L} \\ \widehat{F}_{K,M_2} = \omega (F_{K,L} + F_{L,M_2} + F_{K,M_1} + F_{M_1,M_2}). \end{cases}$$

4 Numerical results

The numerical tests presented here are inspired by [7]. The domain is defined by $\Omega = [-0.5, 0.5] \times [-0.5, 0.5] \times [-0.15, 0.15]$. The permeability $\Lambda(x), x \in \Omega$ is equal to 1 if the distance from x to the vertical axis $0z$ is lower than 0.48, and to 10^{-3} otherwise (see Fig. 1), which ensures the confinement of the flow in the cylinder with axis $0z$ and radius 0.48. We use two Cartesian grids, the second one deduced from the first one by a rotation of angle $\theta = \frac{\pi}{6}$ with axis Oz. The number of cells in each

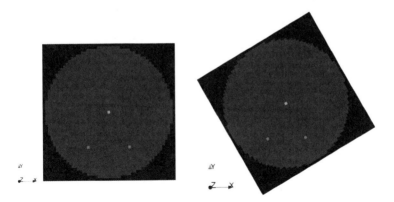

Fig. 1 The two meshes used. In grey scale, the highest permeability zone, in black the lower permeability zone. Squares indicate wells

direction (x, y, z) are $N_x = N_y = 51$ and $N_z = 3$. At the initial state, the reservoir is assumed to be saturated by the oil phase. Water is injected at the origin by an injection well. Two production wells, denoted by P_1 and P_2, are respectively located at the points $(-0.3\cos\frac{\pi}{3}, -0.3\sin\frac{\pi}{3}, 0)$ and $(0.3\cos\frac{\pi}{3}, -0.3\sin\frac{\pi}{3}, 0)$ (that means that the three wells are numerically taken into account as source terms in the middle layer of the mesh). The oil and water properties are respectively denoted by the index o and w. The viscosity ratio between the two phases is given by $\mu_o/\mu_w = 100$ and, the density ratio is given by $\rho_o/\rho_w = 0.8$. We use Corey-type relative permeability, $k_{r_w} = S_w^4$ and $k_{r_o} = S_o^2$. We use the method described in Sections 2 and 3, with $\omega = 0.1$ for all grid blocks which are inscribed in the cylinder (this value, also used in [6], provides the less sensitive numerical results with respect to the grid orientation). The same value for the time step is used for all the computations, which are stopped once a given quantity of water has been injected. Note that, in the mesh depicted on the right part of Fig. 1, the line (P_2, O) is the axis $0y$ of the mesh. We then see on Fig. 2 the resulting contours of the saturation. We observe that the results obtained using the method described in Sections 2 and 3 look very similar in the two grids, whereas the ones obtained using the five-point stencil are strongly distorted by the Grid Orientation Effect.

5 Conclusion

The method presented in this paper is a natural extension of the nine-point schemes defined some decades ago on regular grids. Its advantage is that it applies on the structured but not regular grids used in reservoir simulation, in association with MultiPoint Flux Approximation finite volume schemes. It demands no further modification to the standard industrial codes, since the modification are only the definition of new coefficients a_{KL}^M used in (3).

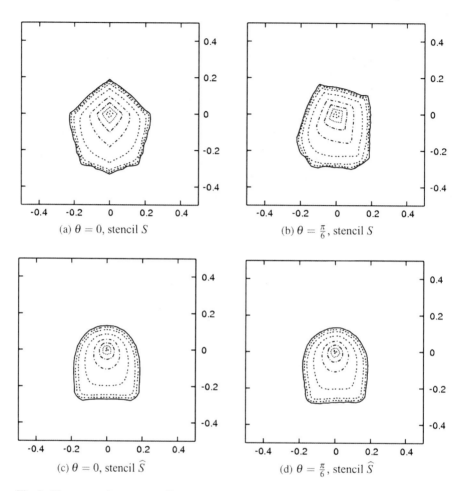

Fig. 2 Water saturation contours $S_w = 0.1, 0.2, \ldots, 1$ at the same time

References

1. Aavatsmark, I., Eigestad, G.T.: Numerical convergence of the MPFA O-method and U-method for general quadrilateral grids. Int. J. Numer. Meth. Fluids **51**, 939–961 (2006)
2. Agelas, L., Masson, R.: Convergence of the finite volume MPFA O scheme for heterogeneous anisotropic diffusion problems on general meshes. C. R. Math. **346**, 1007–1012 (2008)
3. Aziz, K., Ramesh, A.B., Woo, P.T.: Fourth SPE comparative solution project: comparison of steam injection simulators. J. Pet. Tech. **39**, 1576–1584 (1987)
4. Corre, B., Eymard, R., Quettier, L.: Applications of a thermal simulator to field cases, SPE ATCE (1984)
5. Dawson, C., Sun, S., Wheeler, M.F.: Compatible algorithms for coupled flow and transport. Comput. Meth. Appl. Mech. Eng. **193**, 2565–2580 (2004)
6. Eymard, R., Sonier, F.: Mathematical and Numerical Properties of Control-Volumel Finite-Element Scheme for Reservoir Simulation. SPE Reservoir Eng. **9**, 283–289 (1994)

7. Keilegavlen, E., Kozdon, J., Mallison, B.T.: Monotone Multi-dimensional Upstream Weighting on General Grids. Proceeding of ECMOR XII (2010)
8. Lipnikov, K., Moulton, J.D., Svyatskiy, D.: A multilevel multiscale mimetic (M3) method for two-phase flows in porous media. J. Comput. Phys. **14**, 6727–6753 (2008)
9. D.K. Ponting. Corner Point Geometry in reservoir simulation. *In Clarendon Press, editor*, Proc. ECMOR I, 45–65, Cambridge, 1989
10. Vinsome, P., Au, A.: One approach to the grid orientation problem in reservoir simulation. Old SPE J. **21**, 160–161 (1981)
11. Yanosik, J.L., McCracken, T.A.: A nine-point, finite-difference reservoir simulator for realistic prediction of adverse mobility ratio displacements. Old SPE J. **19**, 253–262 (1979)

The paper is in final form and no similar paper has been or is being submitted elsewhere.

Gradient Schemes for Image Processing

Robert Eymard, Angela Handlovičová, Raphaèle Herbin, Karol Mikula, and Olga Stašová

Abstract We present a gradient scheme (which happens to be similar to the MPFA finite volume O-scheme) for the approximation to the solution of the Perona-Malik model regularized by a time delay and to the solution of the nonlinear tensor anisotropic diffusion equation. Numerical examples showing properties of the method and applications in image filtering are discussed.

Keywords advection equation, semi-implicit scheme, finite volume method
MSC2010: 35L04, 65M08, 65M12

1 Introduction

A series of methods for image processing are based on the use of approximate solutions to equations of the type

$$u_t - \text{div} \ (G(u,x,t)\nabla u) = r(x,t), \text{ for a.e. } (x,t) \in \Omega \times]0, T[\qquad (1)$$

with the initial condition

$$u(x,0) = u_{\text{ini}}(x), \text{ for a.e. } x \in \Omega, \qquad (2)$$

Robert Eymard
Université Paris-Est, 5 boulevard Descartes Champs-sur-Marne F-77454 Marne la Vallée, France, e-mail: robert.eymard@univ-mlv.fr

Raphaèle Herbin
Centre de Mathmatiques et Informatique, Université de Provence, 39 rue Joliot Curie, 13453 Marseille 13, France, e-mail: raphaele.herbin@cmi.univ-mrs.fr

Angela Handlovičová, Karol Mikula, and Olga Stašová
Department of Mathematics, Slovak University of Technology, Radlinského 11, 81368 Bratislava, Slovakia, e-mail: angela@math.sk, mikula@math.sk, stasova@math.sk

and the homogeneous Neumann boundary condition

$$G(u,x,t)\nabla u(x,t) \cdot \mathbf{n}_{\partial\Omega}(x) = 0, \text{ for a.e. } (x,t) \in \partial\Omega \times \mathbb{R}_+, \quad (3)$$

where Ω is an open bounded polyhedron in $\mathbb{R}^d$, $d \in \mathbb{N}^*$, with boundary $\partial\Omega$, $T > 0$, $u_{\text{ini}} \in L^2(\Omega)$, $r \in L^2(\Omega \times]0,T[)$, and G is such that, for all $v \in L^2(\Omega)$ and a.e. $(x,t) \in \Omega \times]0,T[$, $G(v,x,t)$ is a self-adjoint linear operator with eigenvalues in $(\underline{\lambda}, \overline{\lambda})$ with $0 < \underline{\lambda} \leq \overline{\lambda}$, and $G(v,x,t)$ is continuous with respect to v and measurable with respect to x,t. In image processing applications, u_{ini} represents an original noisy image, the solution $u(x,t)$ represents its filtering which depends on scale parameter t and $d = 2$ for $2D$ image filtering, $d = 3$ for $3D$ image or $2D$+time movie filtering and $d = 4$ for $3D$+time filtering of spatio-temporal image sequences.

The image processing methods based on approximations of equation (1) differ by definition of the function G. The first such model was proposed by Perona-Malik in 1987 [9], and nowadays, its regularization (by spatial convolution) due to Catte, Lions, Morel and Coll [2] is usually used. The regularized equation has the following form

$$\partial_t u - \nabla.(g(|\nabla G_\sigma * u|)\nabla u) = 0 \quad (4)$$

where $g(s)$ is a Lipschitz continuous decreasing function, $g(0) = 1$, $0 < g(s) \to 0$ for $s \to \infty$, $G_\sigma \in C^\infty(\mathbb{R}^d)$ is a smoothing kernel, e.g. the Gauss function or mollifier with a compact support, for which $\int_{\mathbb{R}^d} G_\sigma(x)dx = 1$. Thanks to convolution, the nonlinearity in difusion term depends on the unknown function u, opposite to the original Perona-Malik equation (without convolution) where it depends on the gradient of solution. For the regularized model, the finite volume scheme were suggested and convergence and error estimates were proved in [3, 8].

Next interesting image processing model with the structure of equation (1) is the so-called nonlinear tensor anisotropic diffusion introduced by Weickert [11]. In that case, the matrix $G(u,x,t)$ represents the so-called diffusion tensor depending on the eigenvalues and eigenvectors of the (regularized) structure tensor

$$J_\rho(\nabla u_{\bar{\imath}}) = G_\rho * (\nabla u_{\bar{\imath}} \nabla u_{\bar{\imath}}^T), \quad (5)$$

where

$$u_{\bar{\imath}}(x,t) = (G_{\bar{\imath}} * u(\cdot,t))(x) \quad (6)$$

and $G_{\bar{\imath}}$ and G_ρ are Gaussian kernels. In computer vision, the matrix $J_\rho = \begin{pmatrix} a & b \\ b & c \end{pmatrix}$, which is symmetric and positive semidefinite, is also known as the interest operator or second moment matrix. If we denote $x = (x_2,x_2)$ we can write $a = G_\rho * \left(\frac{\partial G_{\bar{\imath}}}{\partial x_1} * u\right)^2$, $b = G_\rho * \left(\left(\frac{\partial G_{\bar{\imath}}}{\partial x_1} * u\right)\left(\frac{\partial G_{\bar{\imath}}}{\partial x_2} * u\right)\right)$ and $c = G_\rho * \left(\frac{\partial G_{\bar{\imath}}}{\partial x_2} * u\right)^2$. The orthogonal set of eigenvectors (v,w) of J_ρ corresponding to its eigenvalues

(μ_1, μ_2), $\mu_1 \geq \mu_2$, is such that the orientation of the eigenvector w, which corresponds to the smaller eigenvalue μ_2, gives the so-called coherence orientation. This orientation has the lowest fluctuations in image intensity. The diffusion tensor G in equation (1) is then designed to steer a smoothing process such that the filtering is strong along the coherence direction w and increasing with the coherence defined by difference of eigenvalues $(\mu_1 - \mu_2)^2$. To that goal, G must possess the same eigenvectors $v = (v_1, v_2)$ and $w = (-v_2, v_1)$ as the structure tensor $J_\rho(\nabla u_{\bar{t}})$ and the eigenvalues of G can be chosen as follows

$$\kappa_1 = \alpha, \quad \alpha \in (0, 1), \; \alpha \ll 1, \tag{7}$$

$$\kappa_2 = \begin{cases} \alpha, & \text{if } \mu_1 = \mu_2, \\ \alpha + (1-\alpha)\exp\left(\frac{-C}{(\mu_1-\mu_2)^2}\right), & C > 0 \quad \text{else}. \end{cases}$$

So, the matrix G is finally defined by

$$G = ABA^{-1}, \quad \text{where} \quad A = \begin{pmatrix} v_1 & -v_2 \\ v_2 & v_1 \end{pmatrix} \quad \text{and} \quad B = \begin{pmatrix} \kappa_1 & 0 \\ 0 & \kappa_2 \end{pmatrix}. \tag{8}$$

By the construction, again thanks to convolutions, we see that diffusion matrix depends nonlinearly on the solution u and it satisfies smoothness, symmetry and uniform positive definitness properties. The so-called diamond-cell finite volume schemes for the nonlinear tensor anisotropic diffusion were suggested and analyzed in [6, 7].

In this paper, we use a new class of finite volume schemes, the so-called gradient schemes [5], for solving image processing models based on equation (1). Moreover, we suggest and study numerically new type of regularization of the classical Perona-Malik approach by considering the gradient information from delayed time $t - \bar{t}$. We called this model time-delayed Perona-Malik equation, and consider (1) with $u_{\text{ini}} \in H^1(\Omega)$, and we define $u(x,t) = u_{\text{ini}}(x)$ for $x \in \Omega$ and $t < 0$ and function G is defined by

$$G(u, x, t) = \max\left(\frac{1}{1 + |\nabla u(x, t - \bar{t})|^2}, \alpha\right) \tag{9}$$

where $\bar{t}$ is a time delay and $\alpha > 0$ is a parameter. It turns out that for any $k \in \mathbb{N}$ in the time interval $]k\bar{t}, (k+1)\bar{t}[$, G is a given function of (x, t) only, which leads to a construction of efficient linear numerical scheme for this type of problems.

2 Gradient scheme approximation

In order to describe the scheme, we now introduce some notations for the space discretisation, see the Fig. 1.

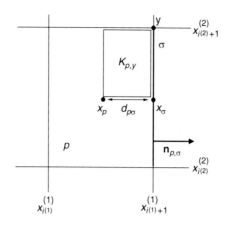

Fig. 1 Notations for the meshes

1. A rectangular discretisation of Ω is defined by the increasing sequences $a_i = x_0^{(i)} < x_1^{(i)} < \ldots < x_{n^{(i)}}^{(i)} = b_i$, $i = 1, \ldots, d$.
2. We denote by

$$\mathcal{M} = \left\{]x_{i^{(1)}}^{(1)}, x_{i^{(1)}}^{(1)}+1[\times \ldots \times]x_{i^{(d)}}^{(d)}, x_{i^{(d)}}^{(d)}+1[,\ 0 \le i^{(1)} < n^{(1)},\ \ldots,\ 0 \le i^{(d)} < n^{(d)} \right\}$$

the set of the control volumes. The elements of $\mathcal{M}$ are denoted $p, q, \ldots$. We denote by $\boldsymbol{x}_p$ the centre of p. For any $p \in \mathcal{M}$, let $\partial p = \overline{p} \setminus p$ be the boundary of p; let $|p| > 0$ denote the measure of p and let h_p denote the diameter of p and $h_{\mathcal{D}}$ denote the maximum value of $(h_p)_{p \in \mathcal{M}}$.

3. We denote by $\mathcal{E}_p$ the set of all the faces of $p \in \mathcal{M}$, by $\mathcal{E}$ the union of all $\mathcal{E}_p$, and for all $\sigma \in \mathcal{E}$, we denote by $|\sigma|$ its $(d-1)$-dimensional measure. For any $\sigma \in \mathcal{E}$, we define the set $\mathcal{M}_\sigma = \{p \in \mathcal{M}, \sigma \in \mathcal{E}_p\}$ (which has therefore one or two elements), we denote by $\mathcal{E}_p$ the set of the faces of $p \in \mathcal{M}$ (it has $2d$ elements) and by $\boldsymbol{x}_\sigma$ the centre of σ. We then denote by $d_{p\sigma} = |\boldsymbol{x}_\sigma - \boldsymbol{x}_p|$ the orthogonal distance between $\boldsymbol{x}_p$ and $\sigma \in \mathcal{E}_p$ and by $\mathbf{n}_{p,\sigma}$ the normal vector to σ, outward to p.
4. We denote by $\mathcal{V}_p$ the set of all the vertices of $p \in \mathcal{M}$ (it has 2^d elements), by $\mathcal{V}$ the union of all $\mathcal{V}_p$, $p \in \mathcal{M}$. For $y \in \mathcal{V}_p$, we denote by $K_{p,y}$ the rectangle whose faces are parallel to those of p, and whose the set of vertices contains $\boldsymbol{x}_p$ and y. We denote by $\mathcal{V}_\sigma$ the set of all vertices of $\sigma \in \mathcal{E}$ (it has 2^{d-1} elements), and by $\mathcal{E}_{p,y}$ the set of all $\sigma \in \mathcal{E}_p$ such that $y \in \mathcal{V}_\sigma$ (it has d elements).
5. We define the set $X_{\mathcal{D}}$ of all $u = ((u_p)_{p \in \mathcal{M}}, (u_{\sigma,y})_{\sigma \in \mathcal{E}, y \in \mathcal{V}_\sigma})$, where all u_p and $u_{\sigma,y}$ are real numbers.
6. We denote, for all $u \in X_{\mathcal{D}}$, by $\Pi_{\mathcal{D}} u \in L^2(\Omega)$ the function defined by the constant value u_p a.e. in $p \in \mathcal{M}$.
7. For $u \in X_{\mathcal{D}}$, $p \in \mathcal{M}$ and $y \in \mathcal{V}_p$, we denote by

$$\nabla_{p,y} u = \frac{2}{|p|} \sum_{\sigma \in \mathcal{E}_{p,y}} |\sigma|(u_{\sigma,y} - u_p)\mathbf{n}_{p,\sigma} = \sum_{\sigma \in \mathcal{E}_{p,y}} \frac{u_{\sigma,y} - u_p}{d_{p\sigma}} \mathbf{n}_{p,\sigma}, \quad (10)$$

and by $\nabla_{\mathcal{D}} u$ the function defined a.e. on Ω by $\nabla_{p,y} u$ on $K_{p,y}$.

Let $T > 0$ be given, and $\tau > 0$ such that there exists $N_T \in \mathbb{N}$ with $T = N_T \tau$, We then define $X_{\mathcal{D},\tau} = X_{\mathcal{D}}^{N_T} = \{(u^n)_{n=1,\ldots,N_T}, u^n \in X_{\mathcal{D}}\}$, and we define the mappings $\Pi_{\mathcal{D},\tau} : X_{\mathcal{D},\tau} \to L^2(\Omega)$ and $\nabla_{\mathcal{D},\tau} : X_{\mathcal{D},\tau} \to L^2(\Omega)^d$ by

$$\Pi_{\mathcal{D},\tau} u(x,t) = \Pi_{\mathcal{D}} u^n(x), \text{ for a.e. } x \in \Omega, \forall t \in](n-1)\tau, n\tau], \forall n = 1, \ldots, N_T, \quad (11)$$

$$\nabla_{\mathcal{D},\tau} u(x,t) = \nabla_{\mathcal{D}} u^n(x), \text{ for a.e. } x \in \Omega, \forall t \in](n-1)\tau, n\tau], \forall n = 1, \ldots, N_T. \quad (12)$$

We then define the following gradient scheme approximation [5] for the discretization of Problem (1):

$$u \in X_{\mathcal{D},\tau}, \ D_\tau u(x,t) := \frac{1}{\tau}(\Pi_{\mathcal{D}} u^1(x) - u_{\text{ini}}(x)), \text{ for a.e. } x \in \Omega, \forall t \in]0, \tau],$$

$$D_\tau u(x,t) = \frac{1}{\tau}(\Pi_{\mathcal{D}} u^n(x) - \Pi_{\mathcal{D}} u^{n-1}(x)),$$

$$\text{for a.e. } x \in \Omega, \forall t \in](n-1)\tau, n\tau], \forall n = 2, \ldots, N_T, \quad (13)$$

and

$$\int_0^T \int_\Omega (D_\tau u \, \Pi_{\mathcal{D},\tau} v + G_{\mathcal{D},\tau}(\Pi_{\mathcal{D},\tau} u, x, t) \nabla_{\mathcal{D},\tau} u \cdot \nabla_{\mathcal{D},\tau} v) \, dx dt$$

$$= \int_0^T \int_\Omega r \Pi_{\mathcal{D},\tau} v dx dt, \ \forall v \in X_{\mathcal{D},\tau}, \quad (14)$$

where $G_{\mathcal{D},\tau}(v, x, t)$ is a suitable approximation of $G(v, x, t)$. The mathematical properties of this scheme are studied in [4].

Remark 1. The equations obtained, for a given $y \in \mathcal{V}$, defining $v \in X_{\mathcal{D}}$ for a given $\sigma \in \mathcal{E}_y$ by $v_{\sigma,y} = 1$ and all other degrees of freedom null, constitute a local invertible linear system, allowing for expressing all $(u_{\sigma,y})_{\sigma \in \mathcal{E}_y}$ with respect to all $(u_p)_{p \in \mathcal{M}}$. This leads to a nine-point stencil on rectangular meshes in 2D, 27-point stencil in 3D (this property is the basis of the MPFA O-scheme [1]).

3 Numerical experiments

3.1 Numerical study of the error for the time-delayed Perona-Malik model

We consider equation (1) in case of G defined by (9) and with a right hand side computed such that the function $u(x, y, t) = ((x^2 + y^2)/2 - (x^3 + y^3)/3)t$ is its exact solution. The domain Ω is square $[0, 1] \times [0, 1]$. We consider two cases, first, the time delay $\bar{t} = 0.0625$ and the overal time $T = 0.625$, and then $\bar{t} = 0.625$ and $T = 1.25$. In both cases we used coupling between space and time step $\tau \approx h^2$, where $h = \frac{1}{n}$ is length of the side of finite volume in uniform squared partition of Ω. We observe the second order convergence in L^2 and L^∞ norms of solution (denoted by E_2 and E_∞) and its gradient (denoted by EG_2 and EG_∞) in this special example, see Tables 1 and 2.

Table 1 The errors and EOC for the time-delayed Perona-Malik model, $\bar{t} = 0.0625, T = 0.625$

n	τ	E_2	EOC	E_∞	EOC	EG_2	EOC	EG_∞	EOC
4	0.0625	4.771e-4	-	1.022e-3	-	7.184e-3	-	1.450e-2	-
8	0.015625	1.172e-4	1.429	2.692e-4	1.925	1.707e-3	2.073	3.615e-3	2.004
16	0.00390625	2.913e-5	2.604	6.812e-5	1.982	4.213e-4	2.019	9.031e-4	2.001
32	0.0009765625	7.270e-6	2.002	1.708e-5	1.996	1.050e-4	2.004	2.257e-4	2.000
64	0.000244140625	1.815e-6	2.001	4.273e-6	1.999	2.624e-5	2.000	5.643e-5	1.999

Table 2 The errors and EOC for the time-delayed Perona-Malik model, $\bar{t} = 0.625, T = 1.25$

n	τ	E_2	EOC	E_∞	EOC	EG_2	EOC	EG_∞	EOC
4	0.0625	1.482e-3	-	2.237e-3	-	1.913e-2	-	2.848e-2	-
8	0.015625	3.745e-4	1.985	5.889e-4	1.925	4.651e-3	2.040	7.083e-3	2.007
16	0.00390625	9.379e-5	1.998	1.450e-4	2.022	1.155e-3	2.009	1.768e-3	2.002
32	0.0009765625	2.346e-5	1.999	3.735e-5	1.957	2.881e-4	2.003	4.419e-4	2.000
64	0.000244140625	5.865e-6	2.000	9.343e-6	1.999	7.201e-5	2.003	1.105e-4	2.000

3.2 Image filtering by the time-delayed Perona-Malik model

The example of image filtering by the gradient scheme applied to the time-delayed Perona-Malik equation is presented in Fig. 2. The original clean image can be seen in Fig. 2 left top. It is damaged by 40% additive noise, see Fig. 2 right top. In the bottom raws of Fig. 2 we present 5th, 10th and 20th denoising step which show the reconstruction of the original. In the last step we see the correct shape reconstruction with the keeping of the edge, with only slighly changed intensity values inside and outside quatrefoil due to diffusion. The following parameters were used in computations: $n^{(1)} = n^{(2)} = 200$, $h = 0.0125$, $\tau = 0.01$, $\bar{t} = 0.1$.

3.3 Image filtering by the nonlinear anisotropic tensor diffusion

In this example we present the image denoising by the nonlinear tensor diffusion and show improvement of the coherence of the line structures, which is the basic

Gradient Schemes for Image Processing

Fig. 2 Image filtering by the time-delayed Perona-Malik model: the original image (left top), the noisy image (right top) and the results after 5, 10 and 20 filtering steps

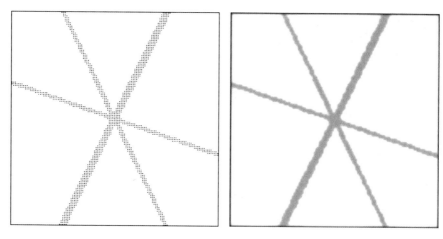

Fig. 3 The enhancement of the coherence by the nonlinear anisotropic tensor diffusion, original image (left) and the result of filtering after 100 time steps (right)

property of such models. Here, in the evaluation of diffusion matrix we use the semi-implicit approach, which means that in (6) we use the solution shifted by one time step backward, $u_{\tilde{t}}(x,t) = (G_{\tilde{t}} * u(\cdot, t-\tau))(x)$, cf. also [6]. The original image with three crackling lines can be seen in Fig. 3 left. On the right, one can see its filtering after 100 time steps which indeed enhance the coherence of those line structures. In this experiment we used the following parameters: $n^{(1)} = n^{(2)} = 250, h = 0.01, \tau = 0.0001, \tilde{t} = 0.0001, \rho = 0.01, \alpha = 0.001, C = 1$.

References

1. I. Aavatsmark, T. Barkve, O. Boe, and T. Mannseth. Discretization on non-orthogonal, quadrilateral grids for inhomogeneous, anisotropic media. *J. Comput. Phys.*, 127(1):2–14, 1996.
2. F. Catté, P.L. Lions, J.M. Morel and T. Coll. Image selective smoothing and edge detection by nonlinear diffusion. *SIAM J. Numer. Anal.* 29:182–193,1992.
3. A. Handlovičová and Z. Krivá. Error estimates for finite volume scheme for Perona - Malik equation. *Acta Math. Univ. Comenianae*,74,(1):79–94, 2005.
4. R. Eymard, A. Handlovičová, R. Herbin, K. Mikula and O. Stašová. Applications of approximate gradient schemes for nonlinear parabolic equations. *in preparation*, 2011.
5. R. Eymard, R. Herbin. Gradient Scheme Approximations for Diffusion Problems. *these proceedings*, 2011.
6. O. Drblíková and K. Mikula. Convergence Analysis of Finite Volume Scheme for Nonlinear Tensor Anisotropic Diffusion in Image Processing. *SIAM Journal on Numerical Analysis*, 46 (1): 37–60,2007.
7. O. Drblíková A. Handlovičová and K. Mikula. Error estimates of the Finite Volume Scheme for the Nonlinear Tensor -Driven Anisotropic Diffusion. *Applied Numerical Mathemtaics*, 59: 2548–2570,2009.

8. K. Mikula and N. Ramarosy. Semi-implicit finite volume scheme for solving nonlinear diffusion equations in image processing. *Numerische Mathematik*, 89, (3):561–590, 2001.
9. P. Perona, J. Malik. Scale space and edge detection using anisotropic diffusion. In: Proc. IEEE Computer Society Workshop on Computer Vision (1987).
10. N. J. Walkington. Algorithms for computing motion by mean curvature. *SIAM J. Numer. Anal.*, 33(6):2215–2238, 1996.
11. J. Weickert. Coherence-enhancing diffusion filtering. *Int. J. Comput. Vision*, 31: 111–127, 1999.

The paper is in final form and no similar paper has been or is being submitted elsewhere.

Gradient Scheme Approximations for Diffusion Problems

Robert Eymard and Raphaèle Herbin

Abstract We propose in this paper the definition and main properties of a family of nonconforming methods, dedicated to the approximation of diffusion problems on general meshes. We give an example of theoretical convergence result in the case of a nonlinear diffusion problem. We then review a few schemes that are part of this family, such as standard conforming and nonconforming finite element schemes, mixed finite element schemes, the SUSHI scheme, the vertex gradient approximation and particular DDFV schemes in 3D.

Keywords Gradient Scheme Approximation, diffusion problems
MSC2010: 65N08

1 Introduction

The 2D [13] and 3D [12] benchmarks for the approximation of heterogeneous and anisotropic diffusion show the large range of schemes which can be used in this setting. The aim of this paper is to propose a simple framework for nonconforming approximation methods, which can include a number of schemes such as some finite volume schemes or nonconforming finite element methods. The interest of this framework is that it provides a simple assessment of the approximation error with respect to some consistency errors. We consider the following problem, posed on an open bounded subset $\Omega \subset \mathbb{R}^d$ (where d is the space dimension), with boundary $\partial \Omega = \overline{\Omega} \setminus \Omega$:

$$\begin{cases} -\mathrm{div}(\Lambda(\overline{u})\nabla \overline{u}) = f \text{ in } \Omega, \\ \overline{u} = 0 \text{ on } \partial \Omega, \end{cases}$$

R. Eymard
Université Paris-Est, France, e-mail: robert.eymard@univ-mlv.fr

R. Herbin
Université Aix-Marseille, France, e-mail: Raphaele.Herbin@latp.univ-mrs.fr

where $\bar{u}$ is an unknown field (temperature, pressure,...), $f \in L^2(\Omega)$ is a volumetric source term, and $\Lambda : L^2(\Omega) \to (L^\infty(\Omega))^{d \times d}$ is a continuous operator with respect to the L^2 norm on both $L^2(\Omega)$ and $(L^\infty(\Omega))^{d \times d}$. Furthermore, we assume that for any $u \in L^2(\Omega)$ and a.e. $x \in \Omega$, the matrix $\Lambda(u)(x)$ is symmetric and the eigenvalues of $\Lambda(u)(x)$ belong to $[\underline{\lambda}, \overline{\lambda}]$, $0 < \underline{\lambda} \le \overline{\lambda}$. Note that, if $\Lambda(u)(x)$ only depends on u through the value $u(x)$ for a.e. $x \in \Omega$, it may be defined by $\Lambda(u)(x) = \tilde{\Lambda}(u(x), x)$ where $\tilde{\Lambda}$ is a Caratheodory function.

We wish to approximate a function $\bar{u}$ solution of the weak form of the problem, that is:

$$\bar{u} \in H_0^1(\Omega) \text{ and } \forall \bar{v} \in H_0^1(\Omega), \int_\Omega \Lambda(\bar{u}) \nabla \bar{u}(x) \cdot \nabla \bar{v}(x) \mathrm{d}x = \int_\Omega f(x) \bar{v}(x) \mathrm{d}x. \quad (1)$$

In order to obtain a consistent approximation of this problem, we define the following nonconforming method, called in this paper a Gradient Scheme Approximation. Defining the set $X_{\mathscr{D},0}$ of all families of discrete unknowns (which may take into account the homogeneous Dirichlet boundary condition if the discrete unknowns include approximate values at the boundary of the domain), we denote for a family of discrete unknowns $u \in X_{\mathscr{D},0}$ by $\Pi_\mathscr{D} u \in L^2(\Omega)$ a reconstruction of a measurable function and by $\nabla_\mathscr{D} u \in L^2(\Omega)^d$ a discrete approximation of its gradient.

Then Problem (1) is naturally approximated by the discrete weak formulation

$$u \in X_{\mathscr{D},0}, \forall v \in X_{\mathscr{D},0}, \int_\Omega \Lambda(\Pi_\mathscr{D} u) \nabla_\mathscr{D} u(x) \cdot \nabla_\mathscr{D} v(x) \mathrm{d}x = \int_\Omega f(x) \Pi_\mathscr{D} v(x) \mathrm{d}x, \quad (2)$$

which yields a numerical scheme once the set $X_{\mathscr{D},0}$ and the operators $\Pi_\mathscr{D}$ and $\nabla_\mathscr{D}$ are defined. In Section 2, we provide the characterisation of the coercivity, compactness, strong and dual approximation properties for given $X_{\mathscr{D},0}$, $\Pi_\mathscr{D}$ and $\nabla_\mathscr{D}$. In the case where Λ does not depend on u and where the Gradient Scheme Approximation checks suitable properties in terms of coercivity, strong and dual approximation, then Scheme (2) may be shown to converge to (1). In the general case of an operator $\Lambda(u)$, a requirement on the compactness property is then needed for proving the convergence of the scheme (2). We then review in Section 3 a few known schemes which can be seen as Gradient Scheme Approximations.

2 Gradient Scheme Approximation

2.1 Definition and properties

Definition 1 (Gradient scheme discretization). Let Ω be an open bounded domain of $\mathbb{R}^d$, with $d \in \mathbb{N}^*$. A gradient scheme discretization $\mathscr{D}$ is defined by $\mathscr{D} = (X_{\mathscr{D},0}, h_\mathscr{D}, \Pi_\mathscr{D}, \nabla_\mathscr{D})$, where:

1. the set of discrete unknowns $X_{\mathcal{D},0}$ is a finite dimensional vector space on $\mathbb{R}$,
2. the space step $h_{\mathcal{D}} \in (0, +\infty)$ is a positive real number,
3. the mapping $\Pi_{\mathcal{D}} : X_{\mathcal{D},0} \to L^2(\Omega)$ is the reconstruction of the approximate function (for any $u \in X_{\mathcal{D},0}$, $\Pi_{\mathcal{D}} u$ is prolonged by 0 outside Ω),
4. the mapping $\nabla_{\mathcal{D}} : X_{\mathcal{D},0} \to L^2(\Omega)^d$ is the reconstruction of the gradient of the function (for any $u \in X_{\mathcal{D},0}$, $\nabla_{\mathcal{D}} u$ is prolonged by 0 outside Ω); accounting for the homogeneous Dirichlet boundary condition, it must be chosen such that $\|\cdot\|_{\mathcal{D}} = \|\nabla_{\mathcal{D}} \cdot\|_{L^2(\Omega)^d}$ is a norm on $X_{\mathcal{D},0}$.

Remark 1. In the case of the homogeneous Neumann boundary condition, one requires that
$$\|\cdot\|_{\mathcal{D}} = \left(\left(\int_\Omega \Pi_{\mathcal{D}} \cdot dx \right)^2 + \|\nabla_{\mathcal{D}} \cdot\|^2_{L^2(\Omega)^d} \right)^{1/2}$$
be a norm on $X_{\mathcal{D},0}$.

Then the **coercivity** of the discretization is measured through the norm $C_{\mathcal{D}}$ of the linear mapping $\Pi_{\mathcal{D}}$, defined by

$$C_{\mathcal{D}} = \max_{v \in X_{\mathcal{D},0} \setminus \{0\}} \frac{\|\Pi_{\mathcal{D}} v\|_{L^2(\Omega)}}{\|v\|_{\mathcal{D}}}. \tag{3}$$

Note that, in the homogeneous Dirichlet boundary condition framework, (3) yields the following "discrete Poincaré" inequality:

$$\|\Pi_{\mathcal{D}} v\|_{L^2(\Omega)} \leq C_{\mathcal{D}} \|\nabla_{\mathcal{D}} v\|_{L^2(\Omega)^d}, \quad \forall v \in X_{\mathcal{D},0}.$$

The **consistency** of the discretization is measured through the interpolation error function $S_{\mathcal{D}} : H^1_0(\Omega) \to [0, +\infty)$, defined by

$$S_{\mathcal{D}}(\varphi) = \min_{v \in X_{\mathcal{D},0}} \left(\|\Pi_{\mathcal{D}} v - \varphi\|^2_{L^2(\Omega)} + \|\nabla_{\mathcal{D}} v - \nabla \varphi\|^2_{L^2(\Omega)^d} \right)^{\frac{1}{2}}, \quad \forall \varphi \in H^1_0(\Omega), \tag{4}$$

The **dual consistency** of the discretization is measured through the conformity error function $W_{\mathcal{D}} : H_{\text{div}}(\Omega) \to [0, +\infty)$, defined by

$$W_{\mathcal{D}}(\varphi) = \max_{v \in X_{\mathcal{D},0} \setminus \{0\}} \frac{\int_\Omega (\nabla_{\mathcal{D}} v(x) \cdot \varphi(x) + \Pi_{\mathcal{D}} v(x) \operatorname{div} \varphi(x)) \, dx}{\|v\|_{\mathcal{D}}}, \tag{5}$$
$$\forall \varphi \in H_{\text{div}}(\Omega).$$

The **compactness** of the discretization is measured through the function $T_{\mathcal{D}} : \mathbb{R}^d \to \mathbb{R}^+$, defined by

$$T_{\mathcal{D}}(\xi) = \max_{v \in X_{\mathcal{D}} \setminus \{0\}} \frac{\|\Pi_{\mathcal{D}} v(\cdot + \xi) - \Pi_{\mathcal{D}} v\|_{L^2(\mathbb{R}^d)}}{\|v\|_{\mathcal{D}}}, \quad \forall \xi \in \mathbb{R}^d. \tag{6}$$

We may remark that the function $\Pi_{\mathcal{D}} u$ lies in a finite dimensional subspace of $L^2(\Omega)$ and therefore $T_{\mathcal{D}}$ is such that $\lim_{|\xi| \to 0} T_{\mathcal{D}}(\xi) = 0$ and that, for any $|\xi| \in \mathbb{R}^d$, $T_{\mathcal{D}}(\xi) \leq 2C_{\mathcal{D}}$, showing the link between compactness and coercivity.

If $\mathcal{D} = (X_{\mathcal{D},0}, h_{\mathcal{D}}, \Pi_{\mathcal{D}}, \nabla_{\mathcal{D}})$ is an approximate gradient discretization, we shall say that (2) is a Gradient Scheme Approximation.

In [11] the following results are proved:

Lemma 1 (**Control of the approximation error, linear case**). *Let Ω be a bounded open domain of $\mathbb{R}^d$, with $d \in \mathbb{N}^*$, let $f \in L^2(\Omega)$ and let $\Lambda \in (L^\infty(\Omega))^{d \times d}$ be such that, for a.e. $x \in \Omega$, the matrix $\Lambda(x)$ is symmetric and the eigenvalues of $\Lambda(x)$ belong to $[\underline{\lambda}, \overline{\lambda}]$, $0 < \underline{\lambda} \leq \overline{\lambda}$. Let $\overline{u} \in H_0^1(\Omega)$ be the solution of (1) (remark that since $f \in L^2(\Omega)$, one has $\Lambda \nabla \overline{u} \in H_{\mathrm{div}}(\Omega)$).*

Let $\mathcal{D}$ be an approximate gradient discretization in the sense of Definition 1. Then there exists one and only one $u_{\mathcal{D}} \in X_{\mathcal{D},0}$, solution to the Gradient Scheme Approximation (2), which moreover satisfies the following inequalities:

$$\|\nabla \overline{u} - \nabla_{\mathcal{D}} u_{\mathcal{D}}\|_{L^2(\Omega)^d} \leq \frac{1}{\underline{\lambda}}(W_{\mathcal{D}}(\Lambda \nabla \overline{u}) + (\overline{\lambda} + \underline{\lambda})S_{\mathcal{D}}(\overline{u})), \tag{7}$$

and

$$\|\overline{u} - \Pi_{\mathcal{D}} u_{\mathcal{D}}\|_{L^2(\Omega)} \leq \frac{1}{\underline{\lambda}}(C_{\mathcal{D}} W_{\mathcal{D}}(\Lambda \nabla \overline{u}) + (C_{\mathcal{D}} \overline{\lambda} + \underline{\lambda})S_{\mathcal{D}}(\overline{u})). \tag{8}$$

Corollary 1 (**Convergence, linear case**). *Under the assumptions of Lemma 1, let $\mathcal{F}$ be a family of gradient discretizations in the sense of Definition 1, which satisfies the following assumptions:*

(P1) there exists $C_P \in \mathbb{R}$ such that $C_{\mathcal{D}} \leq C_P$ for any $\mathcal{D} \in \mathcal{F}$,
(P2) for all $\varphi \in H_0^1(\Omega)$ and $\mathcal{D} \in \mathcal{F}$, $S_{\mathcal{D}}(\varphi)$ tends to 0 as $h_{\mathcal{D}} \to 0$,
(P3) for all $\boldsymbol{\varphi} \in H_{\mathrm{div}}(\Omega)$ and $\mathcal{D} \in \mathcal{F}$, $W_{\mathcal{D}}(\boldsymbol{\varphi})$ tends to 0 as $h_{\mathcal{D}} \to 0$.

For $\mathcal{D} \in \mathcal{F}$, let $u_{\mathcal{D}} \in X_{\mathcal{D},0}$ be the solution to the Gradient Scheme Approximation (2), then $\Pi_{\mathcal{D}} u_{\mathcal{D}}$ converges to $\overline{u}$ in $L^2(\Omega)$ and $\nabla_{\mathcal{D}} u_{\mathcal{D}}$ converges to $\nabla \overline{u}$ in $L^2(\Omega)^d$ as $h_{\mathcal{D}} \to 0$.

Lemma 2. *Let Ω be a bounded open domain of $\mathbb{R}^d$, with $d \in \mathbb{N}^*$. Let $\mathcal{F}$ be a family of approximate gradient discretizations in the sense of Definition 1. Then, for any dense subspace $\mathcal{R}$ of $H_0^1(\Omega)$, the two properties:*

$$\lim_{h_{\mathcal{D}} \to 0} S_{\mathcal{D}}(\varphi) = 0, \ \forall \varphi \in \mathcal{R}, \tag{9}$$

and

$$\lim_{h_{\mathcal{D}} \to 0} S_{\mathcal{D}}(u) = 0, \ \forall u \in H_0^1(\Omega), \tag{10}$$

are equivalent. Furthermore, if there exists $C_P > 0$ such that the following uniform discrete Poincaré inequality holds:

Gradient Scheme Approximations for Diffusion Problems 443

$$C_{\mathcal{D}} \leq C_P, \quad \forall \mathcal{D} \in \mathcal{F}, \tag{11}$$

then for any dense subspace $\mathcal{S}$ of $H_{\text{div}}(\Omega)$, the two properties:

$$\lim_{h_{\mathcal{D}} \to 0} W_{\mathcal{D}}(\varphi) = 0, \quad \forall \varphi \in \mathcal{S}, \tag{12}$$

and

$$\lim_{h_{\mathcal{D}} \to 0} W_{\mathcal{D}}(U) = 0, \quad \forall U \in H_{\text{div}}(\Omega), \tag{13}$$

are equivalent.

Let us now prove a convergence result in the nonlinear case.

Lemma 3 (Convergence of the scheme, nonlinear case). *Let Ω be a bounded open domain of $\mathbb{R}^d$, with $d \in \mathbb{N}^*$, let $f \in L^2(\Omega)$ and let $\Lambda : L^2(\Omega) \to (L^\infty(\Omega))^{d \times d}$ be a continuous operator with respect to the L^2 norm on both $L^2(\Omega)$ and $(L^\infty(\Omega))^{d \times d}$; furthermore, we assume that for any $u \in L^2(\Omega)$ and a.e. $x \in \Omega$, the matrix $\Lambda(u)(x)$ is symmetric and the eigenvalues of $\Lambda(u)(x)$ belong to $[\underline{\lambda}, \overline{\lambda}]$, $0 < \underline{\lambda} \leq \overline{\lambda}$. Let $\mathcal{F}$ be a family of gradient discretizations in the sense of Definition 1, which satisfies properties (P1), (P2) and (P3) of Corollary 1, and moreover satisfies*

(P4) the family of functions $(T_{\mathcal{D}})_{\mathcal{D} \in \mathcal{F}}$ (which is bounded by $2C_P$ thanks to (P1)) is such that

$$\lim_{|\xi| \to 0} \sup_{\mathcal{D} \in \mathcal{F}} T_{\mathcal{D}}(\xi) = 0. \tag{14}$$

Then, for any $\mathcal{D} \in \mathcal{F}$, there exists at least one $u_{\mathcal{D}} \in X_{\mathcal{D},0}$, solution to the Gradient Scheme Approximation (2). Moreover, for a sequence $(\mathcal{D}_n)_{n \in \mathbb{N}}$ of elements of $\mathcal{F}$ such that $h_{\mathcal{D}_n} \to 0$ as $n \to \infty$, there exists $\overline{u} \in H_0^1(\Omega)$, solution to (1) and a subsequence of $(\mathcal{D}_n)_{n \in \mathbb{N}}$, again denoted $(\mathcal{D}_n)_{n \in \mathbb{N}}$, such that then $\Pi_{\mathcal{D}_n} u_{\mathcal{D}_n}$ converges to $\overline{u}$ in $L^2(\Omega)$ and $\nabla_{\mathcal{D}_n} u_{\mathcal{D}_n}$ converges to $\nabla \overline{u}$ in $L^2(\Omega)^d$ as $n \to \infty$.

Remark 2. It is possible to find a family of gradient discretizations in the sense of Definition 1, which only satisfies (P1), (P2), (P3) and not (P4).

Proof. The existence of a solution to (2) is an immediate consequence of the topological degree argument and of the estimate

$$\underline{\lambda} \|\nabla_{\mathcal{D}} u\|_{L^2(\Omega)^d} \leq \|f\|_{L^2(\Omega)} C_{\mathcal{D}}. \tag{15}$$

We then define, for all $n \in \mathbb{N}$, a solution u_n to (2) for $\mathcal{D} = \mathcal{D}_n$. Thanks to properties (P1) and (P4), to (15) and to the Kolmogorov theorem, the family $(\Pi_{\mathcal{D}_n} u_n)_{n \in \mathbb{N}}$ is relatively compact in $L^2(\Omega)$. Then there exists $\overline{u} \in L^2(\Omega)$ and a subsequence of $(\mathcal{D}_n)_{n \in \mathbb{N}}$, again denoted $(\mathcal{D}_n)_{n \in \mathbb{N}}$, such that $\Pi_{\mathcal{D}_n} u_n$ converges to $\overline{u}$ in $L^2(\Omega)$. Extracting again a subsequence, we get that $\nabla_{\mathcal{D}_n} u_n$ converges weakly in $L^2(\mathbb{R}^d)$ to some function $G \in L^2(\mathbb{R}^d)$. Using (P3), we get that $G = \nabla \overline{u}$, hence showing that $\overline{u} \in H_0^1(\Omega)$. Then, for all $v \in H_0^1(\Omega)$, denoting by $v_n \in X_{\mathcal{D}_n,0}$ the element minimising $S_{\mathcal{D}_n}(v)$, we get from (P2) that $\nabla_{\mathcal{D}_n} v_n$ converges in $L^2(\Omega)^d$ to ∇v. It is then possible, by weak/strong limit, to pass to the limit as $n \to \infty$ in (2); thus, $\bar{u}$ is solution to (1). Letting $v = u_n$ in (2) and $\bar{v} = \bar{u}$ in (1) shows that

$$\lim_{n \to \infty} \int_\Omega \Lambda(\Pi_{\mathcal{D}_n} u_n) \nabla_{\mathcal{D}_n} u_n(x) \cdot \nabla_{\mathcal{D}_n} u_n(x) \mathrm{d}x = \int_\Omega \Lambda(\bar{u}) \nabla \bar{u}(x) \cdot \nabla \bar{u}(x) \mathrm{d}x,$$

and therefore:

$$\lim_{n \to \infty} \int_\Omega \Lambda(\Pi_{\mathcal{D}_n} u_n)(\nabla_{\mathcal{D}_n} u_n(x) - \nabla \bar{u}(x)) \cdot (\nabla_{\mathcal{D}_n} u_n(x) - \nabla \bar{u}(x)) \mathrm{d}x = 0,$$

hence proving the convergence of $\nabla_{\mathcal{D}_n} u_{\mathcal{D}_n}$ to $\nabla \bar{u}$ in $L^2(\Omega)^d$ as $n \to \infty$.

3 Application to some schemes

Let us notice that **standard conforming finite element** discretizations may be seen as Gradient Scheme Approximations. If $V_h \subset H_0^1(\Omega)$ is the usual conforming finite element space spanned by the basis functions $\varphi_1, \ldots \varphi_N$, the space $X_{\mathcal{D},0}$ is then $\mathbb{R}^N$ and for $u = (u_1, \ldots, u_N) \in X_{\mathcal{D},0}$, $\Pi_{\mathcal{D}} u = \sum_{i=1}^N u_i \varphi_i$, and $\nabla_{\mathcal{D}} u = \sum_{i=1}^N u_i \nabla \varphi_i = \nabla \Pi_{\mathcal{D}} u$. Hence

$$W_{\mathcal{D}}(\boldsymbol{\varphi}) = 0 \text{ for all } \boldsymbol{\varphi} \in H_{\mathrm{div}}(\Omega). \tag{16}$$

Note that in fact, an approximate gradient discretization is conforming if and only if (16) holds. The compactness property (P4) is satisfied since in this conforming case, we have $T_{\mathcal{D}}(\xi) = |\xi|$.

Let us now turn to the case of the **non conforming P1 finite element discretization** on conforming simplicial meshes. In this case, the basis functions of the finite element space V_h are associated with the N internal faces of the mesh, and V_h is spanned by the basis functions $\varphi_1, \ldots \varphi_N$ which are piecewise affine and continuous at the barycentre of the faces. In this case, the space $X_{\mathcal{D},0}$ is then again $\mathbb{R}^N$ and for $u = (u_1, \ldots, u_N) \in X_{\mathcal{D},0}$, $\Pi_{\mathcal{D}} u = \sum_{i=1}^N u_i \varphi_i$, but $\nabla_{\mathcal{D}} u$ cannot be defined as in the conformal case; it is only piecewise defined as the gradient of $\Pi_{\mathcal{D}} u$. It is possible, under some geometrical conditions on the mesh (see e.g. [9]) to get from classical results that for all $\boldsymbol{\varphi} \in (C^1(\mathbb{R}^d))^d$, $W_{\mathcal{D}}(\boldsymbol{\varphi}) \leq h_{\mathcal{D}} C_{\boldsymbol{\varphi}}$. Property (P4) is also classically shown.

In fact, the **mixed finite element discretization** may also be seen as a Gradient Scheme Approximation. We denote by $(\varphi_i)_{i=1,\ldots,N} \subset L^2(\Omega)$ the basis functions for the approximation of $\bar{u}$, and $(\boldsymbol{\varphi}_i)_{i=1,\ldots,M} \subset H_{\mathrm{div}}(\Omega)$ the basis functions for the approximation of $\Lambda \nabla \bar{u}$. We then define $X_{\mathcal{D},0} \subset \mathbb{R}^{N+M}$ as the set of all families $u = ((u_1, \ldots, u_N), (q_1, \ldots, q_M))$ such that, denoting $\Pi_{\mathcal{D}} u = \sum_{i=1}^N u_i \varphi_i$ and $\nabla_{\mathcal{D}} u = \sum_{i=1}^M q_i \Lambda^{-1} \boldsymbol{\varphi}_i$, the relation $\int_\Omega (\boldsymbol{\varphi}_j(x) \cdot \nabla_{\mathcal{D}} u(x) + \Pi_{\mathcal{D}} u(x) \mathrm{div} \boldsymbol{\varphi}_j(x)) \mathrm{d}x = 0$ holds

Gradient Scheme Approximations for Diffusion Problems

for all $j = 1, \ldots, M$. Then the mixed finite element scheme may be written as (2). The property (P3) is a direct consequence of the imposed relation between $\Pi_{\mathcal{D}} u$ and $\nabla_{\mathcal{D}} u$. The property (P1) is the consequence of the so-called "infsup" condition, and the properties (P2) and (P4) may be shown to be satisfied.

The **SUSHI scheme** [10], as well as the **vertex gradient scheme** [11] are explicitly defined through the space $X_{\mathcal{D},0}$, the reconstruction operator $\Pi_{\mathcal{D}}$ and the discrete gradient $\nabla_{\mathcal{D}}$. The compactness property (P4) is detailed in the appendix of [10]. The study of $S_{\mathcal{D}}$ is also detailed in [10], and that of $W_{\mathcal{D}}$ in [11]. Note that the SUSHI scheme is part of the Mimetic Mixed Hybrid family [8]; however, it does not seem easy to write a general mimetic scheme as a Gradient Scheme Approximation, because the stabilisation term which is needed for the coercivity of the scheme (except in its SUSHI implementation) be included in the gradient term, and therefore the scheme cannot be written under the form (2).

The **DDFV scheme**, see [3, 7, 14] for the two dimensional case and [1, 2, 4–6, 15, 16] for the three dimensional case may also be seen, in some cases, as a Gradient Scheme Approximation. Consider the case where the domain Ω is the union of octahedra which are the so-called diamond cells (such a cell is depicted in Figure 1). Octahedral meshes may be obtained from general hexahedral meshes by introducing an internal point to each hexahedron. We show in Fig. 1 a locally refined face of hexahedral cell where we depict a octahedron constructed with an internal point of the cell and the barycentre of the four points of a face. With such a construction, we can easily take into account boundary conditions and heterogeneous media (each octahedron is homogeneous). The unknown at the centre of the internal faces (point B on the right side of Figure 1), may be easily eliminated. Let us define the space $X_{\mathcal{D},0}$ as $X_{\mathcal{D},0} = \{(u_s)_{s \in \mathcal{V}}, u_s = 0, \forall s \in \mathcal{V}_{\text{ext}}\}$, where $\mathcal{V}$ denotes the set of vertices of the octahedral mesh $\mathcal{M}$ and $\mathcal{V}_{\text{ext}}$ denotes the set of the elements of $\mathcal{V}$ located on the boundary of Ω. Referring to Fig. 1, we define a discrete piecewise constant

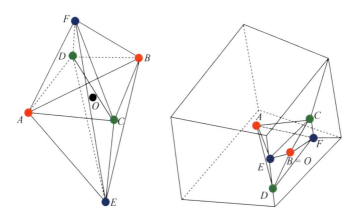

Fig. 1 Left: A generic octahedral cell for the DDFV scheme - Right: An example of construction of an octahedron from a locally refined face of a hexahedron

gradient by its value on the octahedron $K \in \mathscr{M}$:

$$\nabla_{\mathscr{D}} u(x) = \frac{1}{\Delta_K} \Big((u_B - u_A) \overrightarrow{CD} \wedge \overrightarrow{EF} + (u_D - u_C) \overrightarrow{EF} \wedge \overrightarrow{AB} \\ + (u_F - u_E) \overrightarrow{AB} \wedge \overrightarrow{CD} \Big), \qquad \forall x \in K, \tag{17}$$

where $\Delta_K = \text{Det}(\overrightarrow{AB}, \overrightarrow{CD}, \overrightarrow{EF})$. Let O be a well chosen point in $\overline{K}$, for instance the barycentre of the six vertices A, B, C, D, E and F. Taking the example of the vertex F, we denote by σ_{EF} the union of the four triangles OAC, OCB, OBD and ODA, and we denote by $V_{K,F}$ the subset of K of all points which are on the same side of σ_{EF} as F. We proceed similarly for the five other vertices. The reconstruction operator is then defined for $x \in K$ by:

$$\Pi_{\mathscr{D}} u(x) = \tfrac{1}{3} \big(u_A 1_{V_{K,A}}(x) + u_B 1_{V_{K,B}}(x) + u_C 1_{V_{K,C}}(x) \\ + u_D 1_{V_{K,D}}(x) + u_E 1_{V_{K,E}}(x) + u_F 1_{V_{K,F}}(x) \big).$$

With these definitions, (2) is identical to a DDFV scheme [4] formulated on three grids.

Acknowledgements Work supported by Groupement MOMAS and ANR VFSitCom.

References

1. B. Andreianov, M. Bendahmane, and K. Karlsen. A gradient reconstruction formula for finite-volume schemes and discrete duality. In *Finite volumes for complex applications V*, pages 161–168. ISTE, London, 2008.
2. B. Andreianov, M. Bendahmane, K. H. Karlsen, and C. Pierre. Convergence of discrete duality finite volume schemes for the cardiac bidomain model. see http://hal.archives-ouvertes.fr/hal-00526047/PDF/ABKP-submitted.pdf.
3. F. Boyer and F. Hubert. Finite volume method for 2d linear and nonlinear elliptic problems with discontinuities. *SIAM Journal on Numerical Analysis*, 46(6):3032–3070, 2008.
4. Y. Coudière and F. Hubert. A 3D discrete duality finite volume method for nonlinear elliptic equations. 35J65, 65N15, 74S10.
5. Y. Coudière, C. Pierre, O. Rousseau, and R. Turpault. 2D/3D discrete duality finite volume scheme (DDFV) applied to ECG simulation. A DDFV scheme for anisotropic and heterogeneous elliptic equations, application to a bio-mathematics problem: electrocardiogram simulation. In *Finite volumes for complex applications V*, pages 313–320. ISTE, London, 2008.
6. Y. Coudière, C. Pierre, O. Rousseau, and R. Turpault. A 2D/3D discrete duality finite volume scheme. Application to ECG simulation. *Int. J. Finite Vol.*, 6(1):24, 2009.
7. K. Domelevo and P. Omnes. A finite volume method for the laplace equation on almost arbitrary two-dimensional grids. *M2AN Math. Model. Numer. Anal.*, 39(6):1203–1249, 2005.
8. J. Droniou, R. Eymard, T. Gallouët, and R. Herbin. A unified approach to mimetic finite difference, hybrid finite volume and mixed finite volume methods. *Math. Models Methods Appl. Sci.*, 20(2):265–295, 2010.
9. A. Ern and J.-L. Guermond. *Theory and practice of finite elements*, volume 159 of *Applied Mathematical Sciences*. Springer-Verlag, New York, 2004.

10. R. Eymard, T. Gallouët, and R. Herbin. Discretization of heterogeneous and anisotropic diffusion problems on general nonconforming meshes SUSHI: a scheme using stabilization and hybrid interfaces. *IMA J. Numer. Anal.*, 30(4):1009–1043, 2010.
11. R. Eymard, C. Guichard, and R. Herbin. Small-stencil 3d schemes for diffusive flows in porous media. *submitted*.
12. R. Eymard, G. Henry, R. Herbin, F. Hubert, R. Klöfkorn, and G. Manzini. 3d benchmark on discretization schemes for anisotropic diffusion problem on general grids. In *Finite volumes for complex applications VI*. SPringer, 2011.
13. R. Herbin and F. Hubert. Benchmark on discretization schemes for anisotropic diffusion problems on general grids for anisotropic heterogeneous diffusion problems. In R. Eymard and J.-M. Hérard, editors, *Finite Volumes for Complex Applications V*, pages 659–692. Wiley, 2008.
14. F. Hermeline. Approximation of diffusion operators with discontinuous tensor coefficients on distorted meshes. *Comput. Methods Appl. Mech. Engrg.*, 192(16-18):1939–1959, 2003.
15. F. Hermeline. Approximation of 2-D and 3-D diffusion operators with variable full tensor coefficients on arbitrary meshes. *Comput. Methods Appl. Mech. Engrg.*, 196(21-24):2497–2526, 2007.
16. F. Hermeline. A finite volume method for approximating 3D diffusion operators on general meshes. *J. Comput. Phys.*, 228(16):5763–5786, 2009.

The paper is in final form and no similar paper has been or is being submitted elsewhere.

Cartesian Grid Method for the Compressible Euler Equations

M. Asif Farooq and B. Müller

Abstract The accuracy of the Cartesian grid method has been investigated for the 2D compressible Euler equations. We impose wall boundary conditions at ghost points by interpolation or extrapolation at the corresponding mirror points either linearly or quadratically. We find that linear or quadratic interpolation does not affect the accuracy of our node-centered finite volume method. Two different ghost point treatments have been compared.

Keywords Cartesian Grid Method, Ghost Point Treatment, Compressible Euler Equations, Conservation Laws, Oblique Shock Wave
MSC2010: 76J20, 76L05, 35L03, 35L65, 76N15

1 Introduction

The Cartesian grid method has recently become one of the widely used methods in CFD [1–7]. This is due to its simplicity, faster grid generation, simpler programming, lower storage requirements, lower operation count, and easier post processing compared to body fitted structured and unstructured grid methods. The Cartesian grid method is also advantageous in constructing higher order methods. Problems occur at the boundary, when this method is applied to complex domains [8]. When the Cartesian grid method is applied at curved boundaries the cells at the boundaries are not rectangular and these cut-cells create problems for the scheme to be implemented. The time step restriction problem caused by small cut-cells can be solved by merging those cut-cells with neighboring cells [7].

M. Asif Farooq and B. Müller
Department of Energy and Process Engineering, Norwegian University of Science and Technology (NTNU), 7491, Trondheim, Norway, e-mail: asif.m.farooq@ntnu.no, bernhard.muller@ntnu.no

Cut cells are avoided altogether by ghost point treatment at the boundary. In this method symmetry conditions with respect to the boundary are imposed at ghost points in the solid adjacent to the boundary [9]. However, conservativity is lost in this process. Nevertheless, the simplicity of the ghost point treatment has motivated us to use that approach instead of the more complicated cut-cells.

The Cartesian grid method is also called immersed boundary method, in particular when it is applied to the incompressible Navier-Stokes equations. Often the effect of solid boundaries cutting a Cartesian grid has been modelled by a force term in the incompressible momentum equations [10]. Since this approach is not so practical for compressible flow due to the sensitive coupling of all flow variables, it has not been used for compressible flow simulation except for [2, 11]. Instead, the effect of the tangency or slip condition at solid boundaries for inviscid compressible flow is used in the Cartesian grid method to determine the flow variables in ghost cells or at ghost points near solid boundaries [9, 12–16]. In the ghost point treatment we divide our domain into three types of points: fluid, ghost and solid points. For first and second order schemes the methods require one and two ghost points, respectively. Solid and ghost points are flagged inactive.

In this paper we employ the Sjögreen and Petersson ghost point treatment [16], while in [17] we applied a simplified ghost point treatment for the 2D compressible Euler equations. A comparison of these two ghost point treatments at the boundary is also presented in this paper. Sjögreen and Petersson [16] used linear interpolation at the boundary, while we use linear and quadratic interpolation at the boundary. We impose the wall boundary conditions at the ghost points by interpolating the numerical solution at their mirror points with respect to the wall in the fluid domain and mirroring the interpolated values to ensure reflective boundary conditions. If the numerical solution at a mirror point cannot be approximated by interpolation, we employ extrapolation. We employ the local Lax-Friedrichs (lLF) method for the spatial discretization. To increase the accuracy we apply the MUSCL approach with the minmod limiter. For time integration we use the first order explicit Euler and the third order TVD Runge-Kutta (RK3) methods. As a test case, we consider supersonic flow over a wedge and solve the 2D compressible Euler equations by time stepping for the steady state.

The paper is organized as follows. In Section 2 we present the governing equations, i.e. the 2D compressible Euler equations. In Section 3 we outline the boundary conditions. In Section 4 we explain the ghost point treatment at the embedded boundary. In section 5 we present results and discussions. Conclusions are given in section 6.

2 Compressible Euler Equations

The 2D compressible Euler equations serve as a model for a 2D nonlinear hyperbolic system. In conservative form the 2D compressible Euler equations read

$$\frac{\partial U}{\partial t} + \frac{\partial F}{\partial x} + \frac{\partial G}{\partial y} = 0, \tag{1}$$

where

$$U = \begin{bmatrix} \rho \\ \rho u \\ \rho v \\ \rho E \end{bmatrix}, F = \begin{bmatrix} \rho u \\ \rho u^2 + p \\ \rho u v \\ (\rho E + p)u \end{bmatrix}, G = \begin{bmatrix} \rho v \\ \rho u v \\ \rho v^2 + p \\ (\rho E + p)v \end{bmatrix}, \quad (2)$$

with ρ, u, v, E, and p are density, velocity components in x and y-directions, total energy per unit mass and pressure, respectively.

For perfect gas we have the following relation

$$p = (\gamma - 1)(\rho E - \frac{1}{2}\rho(u^2 + v^2)), \quad (3)$$

where γ is the ratio of specific heats. We consider $\gamma = 1.4$ for air.

3 Approximation of Boundary Conditions

The inflow boundary conditions for supersonic flow at $x = 0$ are imposed as $U_{0,j}(t) = g(y_j, t)$. The flow variables at the outlet $x = L_1$ are approximated by $U_{I,j}(t) = U_{I-1,j}(t)$, i.e. by constant extrapolation. This approximation implies that the upwind finite volume method is used to determine the numerical fluxes $F_{I-\frac{1}{2},j}$. The symmetry boundary conditions are implemented by considering an extra line below $y = 0$. There we use $U_{i,1}(t) = diag(1, 1, -1, 1)U_{i,3}(t)$. The boundary conditions at $y = L_2$ are treated as $U_{i,J}(t) = U_{i,J-1}(t)$.

4 Ghost Point Treatment at Embedded Boundary

4.1 Sjögreen and Petersson [16] Ghost Point Treatment for Two Dimensional Embedded Boundary

In Fig. 1 we show the flagging strategy. We flag the ghost and solid points by assigning them 0 and -1 values. The fluid points are assigned values equal to 1. In Fig. 2(a) we show a 2D graphical description of the treatment at the boundary [16]. The distance of ghost point g from the wedge is denoted by b_1. The straight line through g normal to the wedge is intersecting the horizontal lines at three points denoted by vertical lines. At the first intersection point I we obtain the primitive variables V_I by linear interpolation of the values at the neighboring horizontal grid points. And similarly we get V_{II} and V_{III}. We introduce a coordinate s on the line in the direction of the outer unit normal $\mathbf{n}$ of the boundary. Now we proceed as follows. Subtract the distance b_1 from the boundary coordinate s_{wedge} to obtain the mirror point s_m. Then we reach between intersection points I and II on the straight line normal to the boundary. Here we apply either linear interpolation between

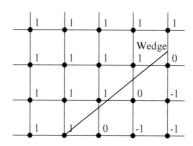

Fig. 1 Flagging strategy for fluid (1), ghost (0) and solid points (-1)

V_I and V_{II} or quadratic interpolation among V_I, V_{II} and V_{III} for normal and tangential components of velocity, pressure p and density ρ. The mathematical description of this strategy is explained as follows.

$$b_1 = s_g - s_{wedge}, \qquad (4)$$

$$s_m = s_{wedge} - b_1, \qquad (5)$$

$$V_m = V_I + \frac{V_{II} - V_I}{\Delta s}(s_I - s_m). \qquad (6)$$

$$V_m = V_{III} + \frac{V_{II} - V_{III}}{s_{II} - s_{III}}(s_m - s_{III}) + \frac{\frac{V_I - V_{II}}{s_I - s_{II}} - \frac{V_{II} - V_{III}}{s_{II} - s_{III}}}{s_I - s_{III}}(s_m - s_{III})(s_m - s_{II}) \qquad (7)$$

where $V = (\rho, u, v, p)$ and $\Delta s = s_I - s_{II}$. Then we use reflection boundary conditions

$$u_{t_g} = u_{t_m}, u_{n_g} = -u_{n_m}, p_g = p_m, \rho_g = \rho_m, \qquad (8)$$

where u_t and u_n denote the tangential and normal components of the velocity vector, respectively.

4.2 Simplified Ghost Point Treatment for Two Dimensional Embedded Boundary

In Fig. 2(b) we show a simplified ghost point treatment at the solid boundary [17]. A ghost point is denoted by G. In the simplified ghost point treatment we consider the fluid point F on the vertical grid line through G adjacent to the boundary as the mirror point. Then, we assume the wedge is in the middle between ghost and fluid points. The mathematical description of this strategy is given as

$$\rho_G = \rho_F, p_G = p_F, u_G = u_F - 2(n_1 u_F + n_2 v_F)n_1, v_G = v_F - 2(n_1 u_F + n_2 v_F)n_2, \qquad (9)$$

where n_1 and n_2 are the x-and y-components of the outer unit normal **n** of the boundary.

Cartesian Grid Method for the Compressible Euler Equations

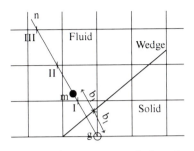 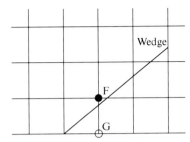

(a) Ghost point treatment at the boundary [16].

(b) Simplified ghost point treatment [17].

Fig. 2 Ghost point treatment

5 Results

5.1 Two Dimensional Compressible Euler Equations

We verify our 2D code of the Cartesian grid method for an oblique shock wave. For the spatial discretization we use the local Lax-Friedrichs (ILF) method, and to increase the order of our method we employ the MUSCL scheme with the minmod limiter. For time integration we use the first order explicit Euler and third order TVD Runge-Kutta (RK3) methods. A supersonic flow moves from left to right and hits a wedge with the wedge angle $\Theta = 15$. The supersonic upstream flow conditions are given as

$$M = 2, p_\infty = 10^5 Pa, \rho_\infty = 1.2 kg/m^3 \qquad (10)$$

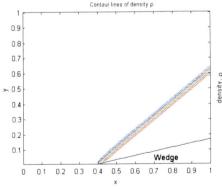 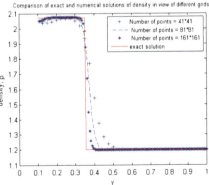

(a) Density contours for supersonic wedge flow ($M_\infty = 2, \Theta = 15$ degrees).

(b) Comparison of exact and numerical solutions for density at different grid levels.

Fig. 3 Left: Density contours. Right: Comparison of exact and numerical solutions for density

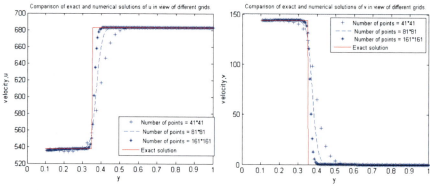

(a) Comparison of exact and numerical solutions for velocity component u at different grid levels.

(b) Comparison of exact and numerical solutions for velocity component v at different grid levels.

Fig. 4 Comparison of exact and numerical solutions for velocity components

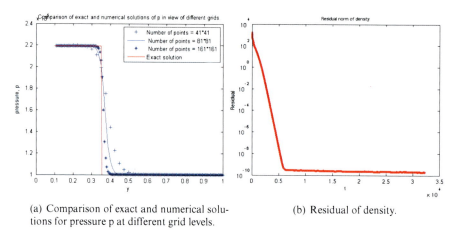

(a) Comparison of exact and numerical solutions for pressure p at different grid levels.

(b) Residual of density.

Fig. 5 Left: Comparison of exact and numerical solutions for pressure. Right: Residual of density

In Fig. 3(a) we present density contours obtained with the TVD RK3 method in time and the local Lax-Friedrichs (lLF) method in space with MUSCL and minmod limiter using the Sjögreen and Petersson ghost point treatment [16]. The apex of the wedge is placed at $x = 0.4$. When the supersonic flow hits the wedge an oblique shock wave is produced which begins at the apex of the wedge.

In Fig. 3(b) and Fig. 4(a) we compare the exact and numerical solutions for density ρ and velocity component u at $x = 0.75m$. We observe that ρ and u are getting closer to the exact solution as we refine the grid. However, there is some discrepancy between the exact and computed solutions near the wall of the wedge. This might be due to the ghost point method not guaranteeing conservativity and

Table 1 Mass flow error for simplified ghost point treatment [17]

	2D Compressible Euler Equations			
	First Order Method		MUSCL with minmod limiter	
Number of points	$\Delta\dot{m}[\frac{kg}{s}]$	$\frac{\Delta\dot{m}}{\dot{m}}$ %	$\Delta\dot{m}[\frac{kg}{s}]$	$\frac{\Delta\dot{m}}{\dot{m}}$ %
41×41	23.7115	2.9797	17.0569	2.1275
81×81	11.5532	1.43	8.3530	1.0298
161×161	5.6317	0.6920	4.0201	0.4930

Table 2 Mass flow error for Sjögreen and Petersson [16] method

	2D Compressible Euler Equations			
	Linear Interpolation		Quadratic Interpolation	
Number of points	$\Delta\dot{m}[\frac{kg}{s}]$	$\frac{\Delta\dot{m}}{\dot{m}}$ %	$\Delta\dot{m}[\frac{kg}{s}]$	$\frac{\Delta\dot{m}}{\dot{m}}$ %
41×41	23.3362	2.9311	23.3219	2.9293
81×81	11.3940	1.41	11.3653	1.4064
161×161	5.5727	0.6847	5.5465	0.6814

due to numerical problems near the apex of the wedge. This apex is acting like a singular point where the flow variables are multivalued.

In Figs. 4(b) and 5(a) we compare the exact and numerical solutions for velocity component v and pressure p. The computed results for v and p are in good agreement with the exact solutions.

In Fig. 5(b) we show the l_2-norm of the residual ($\frac{\rho_{i,j}^{n+1}-\rho_{i,j}^{n}}{\Delta t}$) of the density for the first order method in time and space. We see that the residual has dropped to machine accuracy after 5500 time levels.

In Tables 1 and 2 we present the mass flow error for the simplified ghost point treatment and the Sjögreen and Petersson [16] method. In Table 1 we present results for the first order node-centered finite volume method and the corresponding method with MUSCL and minmod limiter. From Table 1 we observe that by doubling the number of grid points in each direction the percentage of mass flow error is almost halved. In Table 2 we present results for the first order method by using linear and quadratic interpolation using the Sjögreen and Petersson [16] ghost point treatment. We see that linear and quadratic interpolation is not affecting the accuracy of our first order method. The mass flow error in Table 2 obtained with [16] is only slightly lower than the mass flow error of the first order method in Table 1 obtained with the simplified ghost point treatment [17].

6 Conclusions

The Cartesian grid method has been applied to the compressible Euler equations. Local symmetry boundary conditions have been employed at ghost points. The ghost point treatments at the solid boundary are not conservative, and the mass flow error is calculated. We find that linear or quadratic interpolation does not affect the results for the Sjögreen and Petersson method. For supersonic wedge flow, the simplified ghost point treatment on vertical grid lines yields similar results as the ghost point treatment on lines normal to the boundary.

Acknowledgements The current research has been funded by Higher Education Commission (HEC) of Pakistan.

References

1. Almgren, A. S., Bell, J. B., Colella, P., Marthaler, T.: A Cartesian grid projection method for the incompressible Euler equations in complex geometries, SIAM J. Sci. Comput **18**, 1289–1309 (1997).
2. Palma, P. D., de Tullio, M. D., Pascazio, G., Napolitano, M.: An immersed-boundary method for compressible viscous flows, Comput. Fluids **35**, 693–702 (2006).
3. Marshall, D. D., Ruffin, S. M: A new inviscid wall boundary condition treatment for boundary Cartesian grid method, AIAA 2004-0583 42nd AIAA Aerospace Sciences Meeting and Exhibit, Reno, Nevada (2004).
4. Udaykumar, H. S., Krishnann, S. and Marella, S. V. : Adaptively refined parallelisded sharp interface Cartesian grid method for three dimensional moving boundary problem, Int. J. Comput. Fluid Dyn. **23**, 1–24 (2009).
5. Uzgoren, E., Sim, J., Shyy, W.: Marker based 3-D adaptive Cartesian grid method for multiphase flow around irregular geometries, Commun. Comput. Phys. **5**, 1–41 (2009).
6. Wang, Z., Fan, J., Cen, K.: Immersed boundary method for the simulations of 2D viscous flow based on vorticity-velocity formulations., J. Comput. Phys **228**, 1504–1520 (2009).
7. Mittal, R., Iaccarino, G: Immersed boundary method, Annu. Rev. Fluid Mech. **37**, 239–261 (2005).
8. Quirk, J. J.: An alternative to unstructred grids for computing gas dynamic flows around arbitrarily complex two dimensional bodies, Comput. Fluids **23**, 125–142 (1994).
9. Forrer, H., Jeltsch, R: A higher-order boundary treatment for Cartesian grid methods, J. Comput. Phys. **140**, 259–277 (1998).
10. Peskin, C. S. : Flow pattern around heart valves: A numerical method, J. Comput. Phys. **10**, 252–271 (1972).
11. de Tullio, M. D., Palma, P. D. D., Iaccarino, G., Pascazio, G., Napolitano, M.: An immersed boundary method for compressible flows using local grid refinement, J. Comput. Phys. **225**, 2098–2117 (2007).
12. Berger, M. J., Leveque, R. J.: A rotated difference scheme for Cartesian grids in complex geometries, AIAA Paper **CP-91-1602**, 1–9 (1991).
13. Pember, R. B., Bell, J. B., Colella, P., Curtchfield, W. Y., Welcome, M. L.: An adaptive Cartesian grid method for unsteady compressible flow in irregular regions, J. Comput. Phys. **117**, 121-131 (1995).
14. Coirier, W. J., Powell, K. G.: An accuracy assessment of Cartesian-mesh approaches for the Euler equations, J. Comput. Phys. **117**, 121–131 (1995).
15. Colella, P., Graves, D. T., Keen, B. J., Modiano, D.: A Cartesian grid embedded boundary method for hyperbolic conservation laws, J. Comput. Phys. **211**, 347–366 (2006).
16. Sjögreen, B., Petersson, N. A.: A Cartesian embedded boundary method for hyperbolic conservation laws, Commun. Comput. Phys. **2**, 1199–1219 (2007).
17. Farooq, M. A., Müller, B.: Investigation of the accuracy of the Cartesian grid method, Proceedings of International Bhurban Conference on Applied Sciences and Technology Islamabad, Pakistan, January 10-13 (2011).

The paper is in final form and no similar paper has been or is being submitted elsewhere.

Compressible Stokes Problem with General EOS

A. Fettah and T. Gallouët

Abstract In this paper, we propose a discretization for the compressible Stokes problem with an equation of state of the form $p = \varphi(\rho)$ (where p stands for the pressure, ρ for the density and φ is a nondecreasing function belonging to $C^1(\mathbb{R}_+, \mathbb{R}))$. This scheme is based on Crouzeix-Raviart approximation spaces. The discretization of the momentum balance is obtained by the usual finite element technique. The discrete mass balance is obtained by a finite volume scheme, with an upwinding of the density, and two additional terms. We prove existence of a discrete solution and convergence of this approximate solution to a solution of the continuous problem.

Keywords Compressible Stokes, finite element, finite volume
MSC2010: 35Q30,65N12,65N08,65N30

1 Introduction

Let Ω be a bounded open set of $\mathbb{R}^d$, polygonal if $d = 2$ and polyhedral if $d = 3$, and $\mu > 0$. For $M > 0$, $f \in L^2(\Omega)^d$ and $\varphi \in C^1(\mathbb{R}_+, \mathbb{R})$ a nondecreasing function satisfying:

$$\forall s \in \mathbb{R}_+, as^\gamma - b \leq \varphi(s) \leq \tilde{a}s^{2\gamma-1} + \tilde{b}, \tag{1}$$

with $a, \tilde{a}, b, \tilde{b} > 0$ and $\gamma > 1$, we consider the following problem:

A. Fettah and T. Gallouët
Université Aix-Marseille, e-mail: afettah@cmi.univ-mrs.fr, gallouet@cmi.univ-mrs.fr

$$-\mu \Delta u - \frac{\mu}{3}\nabla(div u) + \nabla p = f \text{ in } \Omega, \quad u = 0 \text{ on } \partial\Omega, \tag{2a}$$

$$\text{div}(\rho u) = 0 \text{ in } \Omega, \quad \rho \geq 0 \text{ in } \Omega, \quad \int_\Omega \rho(x)\,dx = M, \tag{2b}$$

$$p = \varphi(\rho) \text{ in } \Omega. \tag{2c}$$

Remark 1. The second inequality in (1) is used only in Section 3, for the passage to the limit in the EOS. It can be replaced by an hypothesis of convexity of φ.

Definition 1. Let $f \in L^2(\Omega)^d$ and $M > 0$. A weak solution of Problem (2) is a function (u, p, ρ) satisfying:

$$(u, p, \rho) \in H_0^1(\Omega)^d \times L^2(\Omega) \times L^{2\gamma}(\Omega), \tag{3a}$$

$$\mu \int_\Omega \nabla u : \nabla v\,dx + \frac{\mu}{3}\int_\Omega \text{div}(u)\text{div}(v)\,dx - \int_\Omega p\,\text{div}(v)\,dx = \int_\Omega f \cdot v\,dx$$

$$\text{for all } v \in (H_0^1(\Omega))^d, \tag{3b}$$

$$\int_\Omega \rho u \cdot \nabla \psi\,dx = 0 \text{ for all } \psi \in W^{1,\infty}(\Omega), \tag{3c}$$

$$\rho \geq 0 \text{ a.e. in } \Omega, \quad \int_\Omega \rho\,dx = M, \quad p = \varphi(\rho) \text{ a.e. in } \Omega. \tag{3d}$$

In Section 2, we give a possible discretization of this problem and we prove the existence of a solution of the discrete problem. In Section 3 we prove the convergence (up to a subsequence, since no uniqueness result of a solution of (3) is avalaible), as the mesh size goes to zero, of this approximate solution to a solution of (3). In particular, we then obtain existence of a solution of (3). The present paper generalizes the results of [3] where convergence was proven if $\varphi(\rho) = \rho^\gamma$. The main additional difficulties of the present paper with respect to [3] are in the proof of the crucial lemma 1 (which yields the estimate on the approximate velocity), in the proof of the estimate of the approximate pressure (Inequality (13)) and in the last proof of the paper, which consists in proving $p = \varphi(\rho)$. The proof of $p = \varphi(\rho)$ cannot be done using a strict monotony argument as in [3] because φ is not necessarily an increasing function. We overcome this difficulty by using the so called "Minty trick" (the drawback of this method is that we do not obtain the a.e. convergence, up to a subsequence, of pressure and density).

2 Discrete spaces and scheme

Let $\mathscr{T}$ be a decomposition of the domain Ω in simplices, which we call hereafter a triangulation of Ω, regardless of the space dimension. By $\mathscr{E}(K)$, we denote the set of the edges ($d = 2$) or faces ($d = 3$) of the element $K \in \mathscr{T}$; for short, each edge or face will be called an edge hereafter. The set of all edges of the mesh is denoted by $\mathcal{E}$; the set of edges included in the boundary of Ω is denoted by $\mathcal{E}_{\text{ext}}$ and the set of internal edges (i.e. $\mathcal{E} \setminus \mathcal{E}_{\text{ext}}$) is denoted by $\mathcal{E}_{\text{int}}$. The decomposition $\mathcal{T}$ is assumed to be regular in the usual sense of the finite element literature. For each internal edge of the mesh $\sigma = K|L$, n_{KL} stands for the normal vector of σ, oriented from K to L (so that $n_{KL} = -n_{LK}$). By $|K|$ and $|\sigma|$ we denote the (d and $d-1$ dimensional) measure, respectively, of an element K and of an edge σ, and h_K and h_σ stand for the diameter of K and σ, respectively. We measure the regularity of the mesh through the parameter θ defined by:

$$\theta = \inf \{\frac{\xi_K}{h_K}, \ K \in \mathcal{T}\} \tag{4}$$

where ξ_K stands for the diameter of the largest ball included in K. Finally, as usual, we denote by h the quantity $\max_{K \in \mathcal{T}} h_K$. The space discretization relies on the Crouzeix-Raviart element (see [1] for the seminal paper and, for instance, [2, pp. 199–201] for a synthetic presentation). The space of approximation for the velocity is the space W_h of vector-valued functions each component of which belongs to V_h: $W_h = (V_h)^d$, where V_h is the discrete space defined as follows:

$$V_h = \{v \in L^2(\Omega) : \forall K \in \mathcal{T}, \ v|_K \in P_1(K); \\ \forall \sigma \in \mathcal{E}_{\text{int}}, \ \sigma = K|L, \ F_\sigma(v|_K) = F_\sigma(v|_L); \ \forall \sigma \in \mathcal{E}_{\text{ext}}, \ F_\sigma(v) = 0\}, \tag{5}$$

where $F_\sigma(v)$ is the mean value of v on σ, denoted hereafter by v_σ. The pressure and the density are approximated in the space L_h of piecewise constant functions, namely $L_h = \{q \in L^2(\Omega) : q|_K = \text{constant}, \ \forall K \in \mathcal{T}\}$. For $u \in W_h$, the discrete gradient and discrete divergence of u are defined by $\nabla_h u = \nabla u$ and $\text{div}_h(u) = \text{div}(u)$ on K, for $K \in \mathcal{T}$. The Crouzeix-Raviart pair of approximation spaces for the velocity and the pressure is *inf-sup* stable, in the sense that there exists $c_i > 0$ only depending on Ω and, in a monotone way, on θ, such that:

$$\forall p \in L_h, \quad \sup_{v \in W_h} \frac{\int_\Omega p \ \text{div}_h(v) \ dx}{\|v\|_{1,b}} \geq c_i \ \|p - m(p)\|_{L^2(\Omega)},$$

where $m(p)$ is the mean value of p over Ω and $\|\cdot\|_{1,b}$ stands for the broken Sobolev H^1 semi-norm, which is defined for scalar as well as for vector-valued functions by:

$$\|v\|_{1,b}^2 = \sum_{K \in \mathcal{T}} \int_K |\nabla v|^2 \ dx = \int_\Omega |\nabla_h v|^2 \ dx.$$

This norm is known to control the L^2 norm by a Poincaré inequality (e.g. [2, lemma 3.31]). We also define a discrete semi-norm on L_h, similar to the usual H^1 semi-norm used in the finite volume context:

$$\forall p \in L_h, \quad |p|_{\mathcal{T}}^2 = \sum_{\substack{\sigma \in \mathcal{E}_{\text{int}}, \\ \sigma = K|L}} \frac{|\sigma|}{h_\sigma} (p_K - p_L)^2.$$

We refer to [1] for the usual properties of the interpolation operator from the Sobolev spaces to the Crouzeix-Raviart spaces. We now describe the numerical scheme. Let ρ^* be the mean density, i.e. $\rho^* = M/|\Omega|$ where $|\Omega|$ stands for the measure of Ω. We consider the following numerical scheme for the discretization of (2):

$$u \in W_h, \; p \in L_h, \; \rho \in L_h, \quad \text{(6a)}$$

$$\forall v \in W_h, \; \mu \int_\Omega \nabla_h u : \nabla_h v \, dx + \frac{\mu}{3} \int_\Omega \text{div}_h(u) \text{div}_h(v) \, dx - \int_\Omega p \, \text{div}_h(v) \, dx$$

$$= \int_\Omega f \cdot v \, dx, \quad \text{(6b)}$$

$$\forall K \in \mathcal{T}, \quad \sum_{\sigma=K|L} v_{\sigma,K}^+ \rho_K - v_{\sigma,K}^- \rho_L + M_K + T_K = 0, \quad \text{(6c)}$$

$$\forall K \in \mathcal{T}, \quad p_K = \varphi(\rho_K), \quad \text{(6d)}$$

where:

- $v_{\sigma,K} = |\sigma| u_\sigma \cdot n_{KL}$, $v_{\sigma,K}^+ = \max(v_{\sigma,K}, 0)$, $v_{\sigma,K}^- = -\min(v_{\sigma,K}, 0)$,
- the terms M_K and T_K read, with $\zeta = \max(0, 2 - \gamma)$, $\alpha \geq 1$ and $0 < \xi < 2$,

$$M_K = h^\alpha |K| (\rho_K - \rho^*), \quad \text{(7a)}$$

$$T_K = \sum_{\sigma=K|L} (h_K + h_L)^\xi \frac{|\sigma|}{h_\sigma} (|\rho_K| + |\rho_L|)^\zeta (\rho_K - \rho_L). \quad \text{(7b)}$$

As it is proven in [3], if $(u, \rho) \in W_h \times L_h$ is solution of (6c), one has necessarily $\rho_K > 0$ for all $K \in \mathcal{T}$, so that (6d) makes sense, and $\sum_{K \in \mathcal{T}} |K| \rho_K = M$. The existence of a solution to the numerical scheme (6) can be proven with the Brouwer fixed point Theorem, using a simple adaptation of the proof of [3], which we therefore omit.

3 Convergence of approximate solutions

We first have to obtain some estimates on the approximate solution. In order to obtain an estimate on the velocity, we will use the following crucial lemma 1.

Lemma 1. *Let $\mathcal{T}$ be a triangulation of the computational domain Ω and $(u, \rho) \in W_h \times L_h$ satisfy Equation (6c). (As above mentioned, this gives $\rho > 0$.) Then:*

$$\int_\Omega \varphi(\rho)\mathrm{div}_h(u)\,dx \le 0.$$

Proof. Let $\psi \in C^1(\mathbb{R}_+^*)$ be a function satisfying $\psi'(s) = \frac{\varphi'(s)}{s}$ (so that ψ is nondecreasing). Multiplying (6c) by $\psi_K = \psi(\rho_K)$ and summing over $K \in \mathcal{T}$ yields $T_1 + T_2 + T_3 = 0$ with:

$$T_1 = \sum_{K \in \mathcal{T}} \psi_K \sum_{\sigma=K|L} |\sigma|\rho_\sigma\, u_\sigma \cdot n_{KL},\quad T_2 = \sum_{K \in \mathcal{T}} h^\alpha |K| \psi(\rho_K)(\rho_K - \rho^*),$$

$$T_3 = \sum_{K \in \mathcal{T}} \psi(\rho_K) \sum_{\sigma=K|L} (h_K + h_L)^\xi \frac{|\sigma|}{h_\sigma}(\rho_K + \rho_L)^\xi (\rho_K - \rho_L).$$

Let $T_4 = \sum_{K \in \mathcal{T}} \int_K \varphi(\rho_K)\mathrm{div}(u) = \sum_{\sigma=K|L} |\sigma| u_\sigma \cdot n_{KL}(\varphi(\rho_K) - \varphi(\rho_L))$. We have $T_4 = T_4 - T_1 - T_2 - T_3$ and then

$$T_4 = \sum_{\sigma=K|L} |\sigma| u_\sigma \cdot n_{KL}[\varphi(\rho_K) - \varphi(\rho_L) - \rho_\sigma(\psi(\rho_K) - \psi(\rho_L))] - T_2 - T_3. \quad (8)$$

The fact that ψ is nondecreasing (and $\sum_{K \in \mathcal{T}} |K|\rho_K = M$) yields:

- $T_2 \ge \sum_{K \in \mathcal{T}} h^\alpha |K|\psi(\rho^*)(\rho_K - \rho^*) = 0$
- $T_3 = \sum_{\sigma=K|L}(h_K + h_L)^\xi \frac{|\sigma|}{h_\sigma}(\rho_K + \rho_L)^\xi (\rho_K - \rho_L)(\psi(\rho_K) - \psi(\rho_L)) \ge 0$.

In order to conclude that $T_4 \le 0$, we now introduce, for $\alpha > 0$, the function Φ defined on $\mathbb{R}_+^*$ by

$$\Phi(s) = \varphi(\alpha) - \varphi(s) - \alpha(\psi(\alpha) - \psi(s)).$$

Since $s\psi'(s) = \varphi'(s)$ (for $s > 0$), one has $\Phi(s) \le 0$ for all $s > 0$ and then:

$$\sum_{\sigma=K|L} |\sigma| u_\sigma \cdot n_{KL}[\varphi(\rho_K) - \varphi(\rho_L) - \rho_\sigma(\psi(\rho_K) - \psi(\rho_L))] \le 0.$$

We then conclude, with (8), that $\int_\Omega \varphi(\rho)\mathrm{div}_h(u)\,dx = T_4 \le 0$.

Theorem 1. *Let $\theta_0 > 0$ and let $\mathcal{T}$ be a triangulation of the computational domain Ω such that $\theta \ge \theta_0$, where θ is defined by (4). Let (u, p, ρ) be a solution of (6). Then, there exist C, only depending on the data of the problem (Ω, μ, f, M and φ) and on θ_0 such that:*

$$\|u\|_{1,b} \le C,\quad \|p\|_{L^2(\Omega)} \le C,\quad \|\rho\|_{L^{2\gamma}(\Omega)} \le C \text{ and } h^{\xi/2} |\rho|_\mathcal{T} \le C. \quad (9)$$

Proof. Let (u, p, ρ) be a solution of (6). Taking u as test function in (6b) yields:

$$\mu \|u\|_{1,b}^2 + \frac{\mu}{3}\int_\Omega \mathrm{div}_h^2(u)\,dx - \int_\Omega p\,\mathrm{div}(u)\,dx = \int_\Omega f \cdot u\,dx. \quad (10)$$

Using Lemma 1, a discrete Poincaré Inequality (as in [3]) and the Hölder inequality, yields the existence of C_1 only depending on Ω, f, μ and θ_0 such that $\|u\|_{1,b} \leq C_1$. Using the *inf-sup* stability of the discretization, we hence get from (10) a control of $\|p - m(p)\|_{L^2(\Omega)}$ (where $m(p)$ stands for the mean value of p over Ω).

In order to obtain an estimate on p, we set (for simplicity) $\bar{\varphi}(s) = s + \varphi(0)$ for $s < 0$ and we define the function Φ from $\mathbb{R}$ to $\mathbb{R}$ by $\Phi(s) = \inf\{t \in \mathbb{R}_+; s = 3\bar{\varphi}(t)\}$. The function Φ satisfies the following properties:

$$s = 3\bar{\varphi}(t) \Rightarrow \Phi(s) \leq t, \tag{11a}$$

$$s = 3\bar{\varphi}(\Phi(s)), \tag{11b}$$

$$\Phi(s) \to +\infty, \text{ as } s \to +\infty, \tag{11c}$$

$$\Phi \text{ is nondecreasing.} \tag{11d}$$

For all $x \in \Omega$ one has $m(p) \leq |m(p) - p(x)| + |p(x)| \leq |m(p) - p(x)| + 2|\varphi(0)| + p(x)$. Then, using (11d),

$$\Phi(m(p)) \leq \Phi(3|m(p) - p(x)|) + \Phi(6|\varphi(0)|) + \Phi(3p(x)).$$

Since $3p(x) = 3\bar{\varphi}(\rho(x))$, (11a) gives $\Phi(m(p)) \leq \Phi(3|m(p) - p(x)|) + \Phi(6|\varphi(0)|) + \rho(x)$. By summing equation (6c) for $K \in \mathcal{T}$, we obtain that the integral of ρ over Ω is M, which yields:

$$\int_\Omega \Phi(m(p))dx \leq \int_\Omega \Phi(3|m(p) - p(x)|)dx + M + \Phi(6|\varphi(0)|)|\Omega|. \tag{12}$$

On the other hand, if $\Phi(s) \geq 0$, one has, with (11b) and the first inequality of (1),

$$\frac{s}{3} = \bar{\varphi}(\Phi(s)) \geq a(\Phi(s))^\gamma - b,$$

and then $\Phi(s) \leq (\frac{|s|}{3a} + \frac{b}{a})^{\frac{1}{\gamma}} \leq (\frac{|s|}{3a} + \frac{b}{a} + 1)^2$. This inequality gives an estimate on $\int_\Omega \Phi(3|m(p) - p(x)|)dx$ from the L^2-estimate on $(p - m(p))$. We hence get, with (12), an estimate on $\Phi(m(p))$. Using (11c) yields an estimate on $m(p)$. Finally, the estimate on $[m(p)]$ and $[p - m(p)]$ gives the existence of C_2 (depending on the data and θ_0) such that

$$\|p\|_{L^2(\Omega)} \leq C_2. \tag{13}$$

Finally, thanks to $p = \bar{\varphi}(\rho)$ and the first inequality of (1), the estimate on ρ follows. For the estimate on $|\rho|_\mathcal{T}$, which comes form the T_K term in (6c), we refer to [3] where the proof is the same.

Let us now state the final convergence result:

Theorem 2. *Let a sequence of triangulations $(\mathcal{T}^{(n)})_{n \in \mathbb{N}}$ of Ω be given. We assume that h_n tends to zero when $n \to \infty$. In addition, we assume that the sequence of* discretizations is regular, in the sense that there exists $\theta_0 > 0$ such that $\theta_n \geq \theta_0$ for all $n \in \mathbb{N}$. For $n \in \mathbb{N}$, we denote by $W_h^{(n)}$ and $L_h^{(n)}$ the discrete spaces associated to $\mathcal{T}^{(n)}$ and b (u_n, p_n, ρ_n) a corresponding solution to the discrete problem (6), with $\alpha \geq 1$ and $0 < \xi < 2$. Then, up to the extraction of a subsequence, when $n \to \infty$:

1. the sequence $(u_n)_{n \in \mathbb{N}}$ (strongly) converges in $L^2(\Omega)^d$ to a limit $u \in H_0^1(\Omega)^d$,
2. the sequence $(p_n)_{n \in \mathbb{N}}$ weakly converges in $L^2(\Omega)$,
3. the sequence $(\rho_n)_{n \in \mathbb{N}}$ weakly converges in $L^{2\gamma}(\Omega)$,
4. (u, p, ρ) is a solution to Problem (3).

Proof. The first item of Theorem 2 (namely the convergence, up to the extraction of a subsequence, of the sequence $(u_n)_{n \in \mathbb{N}}$ and the fact that the limit belongs to $H_0^1(\Omega)^d$) is a consequence of the uniform (with respect to n) estimate of Theorem 1, applying a compactness result wich is proven for instance in [4, Theorem 3.3]. The second and third item of Theorem 2 are trivial consequences of the uniform (with respect to n) estimate of Theorem 1. It remains to prove that (u, p, ρ) is solution to (3b)–(3d).

The proof that the limit satisfies $\rho \geq 0$ a.e. in Ω, $\int_\Omega \rho \, dx = M$ and Equation (3b) is strictly the same as the proof of the same result for a linear equation of state, i.e. Theorem 6.1 in [4]. The fact that (u, ρ) satisfies Equation (3c) follows the proof of [3]. Then, we only need here to prove that the equation of state is satisfied, that is $p = \varphi(\rho)$ a.e. in Ω.

The fact that $\rho \in L^{2\gamma}(\Omega)$, $\rho \geq 0$ a.e. in Ω, $u \in (H_0^1(\Omega))^d$ and that (ρ, u) satisfies (3c) yields, see Lemma 2.1 in [3]:

$$\int_\Omega \rho \operatorname{div}(u) \, dx = 0. \tag{14}$$

Then, using (14), we have, following the proof given in [3]:

$$\lim_{n \to \infty} \int_\Omega (p_n - \operatorname{div}_h(u_n)) \rho_n \, dx - \int_\Omega p \rho \, dx = 0.$$

As in [3], we also have $\limsup_{n \to \infty} \int_\Omega \operatorname{div}_h(u_n) \rho_n \, dx \leq 0$. Hence:

$$\limsup_{n \to \infty} \int_\Omega p_n \rho_n \, dx \leq \int_\Omega p \rho \, dx. \tag{15}$$

We want to deduce from (15) that $p = \varphi(\rho)$. But, since φ in only nondecreasing (and not necessarily increasing), we cannot use the proof given in [3]. We use here the so called Minty trick.

For simplicity, we define φ on $\mathbb{R}^-$ setting $\varphi(s) = \varphi(0)$ if $s < 0$. Let $\bar{\rho} \in L^{2\gamma}$ and, for $n \in \mathbb{N}$, $G_n = (\varphi(\rho_n) - \varphi(\bar{\rho}))(\rho_n - \bar{\rho})$. One has $G_n \geq 0$ a.e. in Ω (since φ is nondecreasing). Thanks to the second inequality of (1) (which is used only in this proof) one has $\varphi(\bar{\rho}) \in L^{2\gamma/(2\gamma-1)}(\Omega)$ and then $\varphi(\bar{\rho})\bar{\rho} \in L^1(\Omega)$. Then, one has

$G_n \in L^1(\Omega)$ and

$$0 \leq \int_\Omega G_n \, dx = \int_\Omega (p_n \rho_n - p_n \bar\rho - \varphi(\bar\rho)\rho_n + \varphi(\bar\rho)\bar\rho) \, dx.$$

Using (15) and the weak convergences of p_n to p and ρ_n to ρ in $L^2(\Omega)$ and $L^{2\gamma}(\Omega)$ respectively, we obtain:

$$0 \leq \limsup_{n \to \infty} \int_\Omega G_n \, dx \leq \int_\Omega (p - \varphi(\bar\rho))(\rho - \bar\rho) \, dx.$$

We have thus proven that

$$\int_\Omega (p - \varphi(\bar\rho))(\rho - \bar\rho) \, dx \geq 0 \text{ for all } \bar\rho \in L^{2\gamma}(\Omega). \tag{16}$$

We now have to choose $\bar\rho$ conveniently to deduce $p = \varphi(\rho)$ from (16). Let $\psi \in C_c^\infty(\Omega, \mathbb{R})$. For $n \in \mathbb{N}^*$, we set $\rho_n = \rho + \frac{1}{n}\psi$. Since $\rho_n \in L^{2\gamma}$, we can choose $\bar\rho = \rho_n$ in (16). We obtain

$$\int_\Omega (p - \varphi(\rho + \frac{1}{n}\psi))\psi \leq 0.$$

We now use the Dominated Convergence Theorem on the sequence $(g_n)_{n \in \mathbb{N}^*}$ with $g_n = (p - \varphi(\rho + \frac{1}{n}\psi))\psi$. The continuity of φ gives $g_n \to (p - \varphi(\rho))\psi$ a.e. in Ω. Since φ is nondecreasing, one has, for all $n \in \mathbb{N}^*$,

$$|g_n| \leq G = |p\psi| + |\varphi(\rho + \|\psi\|_\infty)\psi| + |\varphi(0)\psi| \text{ a.e. in } \Omega.$$

The second inequality of (1) gives $\varphi(\rho + \|\psi\|_\infty) \in L^1(\Omega)$. Then one has $G \in L^1(\Omega)$ and the Dominated Convergence Theorem yields $\int_\Omega (p - \varphi(\rho))\psi \leq 0$. Changing ψ in $-\psi$, we conclude that $\int_\Omega (p - \varphi(\rho))\psi = 0$ for all $\psi \in C_c^\infty(\Omega, \mathbb{R})$. This gives $p = \varphi(\rho)$ a.e. in Ω. The proof of Theorem 2 is now complete.

If φ is increasing, we can prove, as in [3], the a.e. convergence, up to a subsequence, of p and ρ.

Conclusion We gave a scheme for the discretization of the compressible Stokes problem with a quite general EOS and we proved the convergence of the approximate solution to an exact solution (up to a subsequence) as the mesh size goes to zero. The main difficulty of the paper is in the passage to the limit in EOS. This difficulty is due to the nonlinearity of the EOS and the fact that the estimates on pressure and density only lead to weak convergences. It will be now interesting to consider the Navier-Stokes problem along with the evolution problem.

References

1. M. Crouzeix and P.-A. Raviart. Conforming and nonconforming finite element methods for solving the stationary Stokes equations I. *Revue Française d'Automatique, Informatique et Recherche Opérationnelle (R.A.I.R.O.)*, R-3:33–75, 1973.
2. A. Ern and J.-L. Guermond. Theory and practice of finite elements. Number 159 in Applied Mathematical Sciences. Springer, New York, 2004.
3. R. Eymard, T. Gallouët, R. Herbin, and J.-C. Latché. A convergent finite element-finite volume scheme for the compressible Stokes problem. Part II: the isentropic case. *to appear in Mathematics of Computation*, 2009.
4. T. Gallouët, R. Herbin, and J.-C. Latché. A convergent finite element-finite volume scheme for the compressible Stokes problem. Part I: the isothermal case. *Mathematics of Computation*, 267:1333–1352, 2009.

The paper is in final form and no similar paper has been or is being submitted elsewhere.

Asymptotic Preserving Finite Volumes Discretization For Non-Linear Moment Model On Unstructured Meshes

Emmanuel Franck, Christophe Buet, and Bruno Després

Abstract In this work we present a new finite volume discretization of the nonlinear model M_1 [2]. This new method is based on nodal solver for hyperbolic systems [3, 6] and overcomes, on 2-D unstructured meshes, the problem of the inconsistent diffusion limit for schemes based on classical edge formulation. We provide numerical examples to illustrate the properties of the method.

Keywords asymptotic preserving, M_1 model, unstructured, diffusion limit, GLACE scheme

MSC2010: 65M08, 65M08, 65M22, 65Z05, 65M08

1 Introduction

Our physical motivation stems from the discretization of the linear transport equation $\partial_t f(t, \mathbf{x}, \omega) + \frac{1}{\epsilon} \omega \nabla f(t, \mathbf{x}, \omega) = \frac{\sigma}{\epsilon^2} Q(f)$ in diffusive regime. $f(t, \mathbf{x}, \omega) \geq 0$ is the distribution function associated to particles located in $\mathbf{x}$ and having a direction ω. $Q(f)$ is a Lorentz operator for scattering of lights particles. It is well known that on coarse grids, numerical schemes for such hyperbolic systems does not capture the diffusion limit ($\varepsilon << 1$) correctly. Since many years, many Asymptotic Preserving (AP) schemes have been proposed to correct this problem. But extended in 2-D unstructured meshes these methods are not consistent with a diffusion operator in diffusive regimes and coarse grids.

Emmanuel Franck and Christophe Buet
CEA, DAM, DIF, F-91297 Arpajon, France, e-mail: efranck21@gmail.com, christophe.buet@cea.fr

Bruno Després
Laboratoire Jacques Louis Lions, Université Pierre et Marie Curie, 75252 Paris, Cedex 5, France

In this work we present an attempt to overcome this difficulty on a simplified non linear model which is the M_1 model: this model is the first element of a family of angular discretization based on minimization of entropy procedure [2]. It writes

$$\begin{cases} \partial_t E + \dfrac{1}{\epsilon} \nabla . \mathbf{F} = 0 \\ \partial_t \mathbf{F} + \dfrac{1}{\epsilon} \nabla(\hat{P}) = -\dfrac{\sigma}{\epsilon^2} \mathbf{F} \end{cases} \quad (1)$$

E is the energy, $\mathbf{F}$ is the flux and $\hat{P}$ the pressure tensor. The pressure tensor is defined by

$$\hat{P} = \frac{1}{2}((1 - \chi(\mathbf{f}))Id + (3\chi(\mathbf{f}) - 1)\frac{\mathbf{f} \otimes \mathbf{f}}{\|\mathbf{f}\|})E$$

with $\mathbf{f} = \mathbf{F}/E$, $\chi(\mathbf{f}) = \dfrac{3 + 4\mathbf{f}^2}{5 + 2\sqrt{4 - 3\mathbf{f}^2}}$ for M_1 model. When ϵ tends to zero these models tends to the linear diffusion equation $\partial_t E(t, \mathbf{x}) - \frac{1}{3\sigma} \Delta E(t, \mathbf{x}) = 0$.

The original contribution of this work concerns news results for the construction of an AP scheme for the non-linear M_1 models on 2-D unstructured meshes. For this we first rewrite the M_1 model as a compressible gas dynamics like equation as in [4], and second we adapt the linear scheme developed in [5] to the non linear M_1 model. For the P_1 model [5]: one shows that in diffusive regime and on coarse grids the asymptotic limit of the finit volume based scheme is consistent with the right diffusion equation. Numerical results that it is also the case for the discretization of the M_1 model. To finish we present news numerical results for the equilibrium radiative model, with a non-linear coupling between a moment model and a non linear temperature relaxation.

2 Notations on 2-D unstructured meshes

Jin, Levermore in [9] or Gosse, Toscani in [7] proposed methods based on the incorporation to the source term in the fluxes in order to obtain AP schemes. We introduce some notations which are used to define a particular AP scheme on unstructured mesh.

Our idea is to use a nodal scheme like "GLACE" [6] or "CHIC" [3] for the linearized Euler equations, since the hyperbolic heat equation is a special case of them and incorporate the source term in the Riemann solver by Jin-Levermore procedure [9] to obtain an AP scheme. Let us consider the 2D unstructured mesh of Fig. 2. The mesh is defined by the vertices $\mathbf{x}_r$ and the cells Ω_j. We denote by $\mathbf{x}_j$ the gravity center of Ω_j. In each cell j, we define the length and the normal associated to the node of local index r

$$l_{jr} = \frac{1}{2} \|\mathbf{x}_{r+1} - \mathbf{x}_{r-1}\| \text{ and } \mathbf{n}_{jr} = \frac{1}{2l_{jr}} \begin{pmatrix} -y_{r-1} + y_{r+1} \\ x_{r-1} - x_{r+1} \end{pmatrix}. \quad (2)$$

Asymptotic Preserving Finite Volumes Discretization For Non-Linear Moment Model

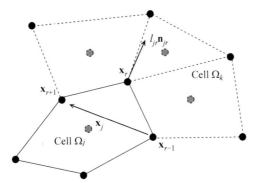

Fig. 1 Notation for the nodal formulation

where (x_r, y_r) are the coordinates of $\mathbf{x}_r$. We use a tensor definition of nodal schemes introduced in [8]. We define the scheme by

$$\begin{cases} |\Omega_j| \partial_t E_j(t) + \dfrac{1}{\varepsilon} \sum_r l_{jr}(\mathbf{F}_r.\mathbf{n}_{jr}) = 0 \\ |\Omega_j| \partial_t \mathbf{F}_j(t) + \dfrac{1}{\varepsilon} \sum_r \mathbf{G}_{jr} = -\dfrac{\sigma}{\varepsilon^2} \sum_r \widehat{\beta}_{jr} \mathbf{F}_r \end{cases} \quad (3)$$

The fluxes associated to these schemes are

$$\begin{cases} \mathbf{G}_{jr} = l_{jr} E_j \mathbf{n}_{jr} + \widehat{\alpha}_{jr}(\mathbf{F}_j - \mathbf{F}_r) - \dfrac{\sigma}{\varepsilon} \widehat{\beta}_{jr} \mathbf{F}_r \\ \left(\sum_j \widehat{\alpha}_{jr} + \dfrac{\sigma}{\varepsilon} \widehat{\beta}_{jr} \right) \mathbf{F}_r = \sum_j l_{jr} E_j \mathbf{n}_{jr} + \widehat{\alpha}_{jr} \mathbf{F}_j \end{cases} \quad (4)$$

$\widehat{\alpha}_{jr}$ and $\widehat{\beta}_{jr}$ are defined by $\widehat{\alpha}_{jr} = l_{jr} \mathbf{n}_{jr} \otimes \mathbf{n}_{jr}$, $\widehat{\beta}_{jr} = l_{jr} \mathbf{n}_{jr} \otimes (\mathbf{x}_r - \mathbf{x}_j)$ For the $\widehat{\alpha}_{jr}$ tensor we can use also the CHIC tensor [5].

The matrix $\sum_j l_{jr} \widehat{\alpha}_{jr}$ is always invertible on non-degenerate meshes. For the matrix $A_r = \sum_j l_{jr} \mathbf{n}_{jr} \otimes (\mathbf{x}_r - \mathbf{x}_j)$ we do not have a complete result. However we give a sufficient condition for positivity of the matrix. We prove that $A_r = V_r \widehat{I}_d + P$, where P is the matrix with $Tr(P) = 0$ and V_r is the control volume around the node r. Studying P we obtain the sufficient condition. For example, on triangular meshes the matrix is positive definite if the angles are superior to 11 degrees. In practice, these matrix are always non singular. Under the condition that the matrix is invertible we prove that the scheme is L^2 stable for the different tensor defined in [5].

The previous scheme tends to a new diffusion scheme on coarse grids

$$\begin{cases} E'_j(t) + \dfrac{1}{|\Omega_j|} \sum_r l_{jr}\left(\mathbf{n}_{jr}, \mathbf{F}_r\right) = 0, \\ A_r \sigma \mathbf{F}_r = \sum_j l_{jr} \mathbf{n}_{jr} E_j, \qquad \text{with } A_r = \left(\sum_j l_{jr} \mathbf{n}_{jr} \otimes (\mathbf{x}_r - \mathbf{x}_j)\right). \end{cases} \quad (5)$$

In [5] we prove the first order convergence to the solution of the linear P_1 model. The numerical results show a convergence at the second order on different unstructured meshes. This scheme may exhibit spurious modes but this problem can be solve by a modification of the normal and length associated to the node.

3 An AP scheme for the M_1 model on 2-D unstructured meshes

We reformulate the M_1 model as gas dynamics equations, see [4] in 1D, and we use a nodal solver as in [3, 6]. The new formulation writes

$$\begin{cases} \partial_t \rho + \dfrac{1}{\epsilon} div(\rho \mathbf{u}) = 0 \\ \partial_t \rho \mathbf{v} + \dfrac{1}{\epsilon} div(\rho \mathbf{u} \otimes \mathbf{v}) + \dfrac{1}{\epsilon} \nabla q = -\dfrac{\sigma}{\epsilon^2} \rho \mathbf{v} \\ \partial_t \rho e + \dfrac{1}{\epsilon} div(\rho u e + q \mathbf{u}) = 0 \\ \partial_t \rho s + \dfrac{1}{\epsilon} div(\rho \mathbf{u} s) = 0 \end{cases} \quad (6)$$

with $S = \rho s$ (S the radiation entropy), $\mathbf{F} = \rho \mathbf{v}$ and $E = \rho e$. We define also the hydrodynamics variables

- $q = \dfrac{1-\chi}{2} E$,
- $\mathbf{u} = \dfrac{3\chi - 1}{2} \dfrac{\mathbf{f}}{|\mathbf{f}|^2}$

with $\mathbf{f} = \dfrac{|\mathbf{F}|}{E}$.

To discretize this model we use a Lagrange+remap scheme. The lagrangian step is solved by a nodal scheme which is a non-linear generalization of (11) coupled with the Jin-Levermore method

Asymptotic Preserving Finite Volumes Discretization For Non-Linear Moment Model

$$\begin{cases} M_j \dfrac{\tau_j^{n+1} - \tau_j^n}{\Delta t} - \dfrac{1}{\epsilon}\sum_r l_{jr}(\mathbf{u}_r, \mathbf{n}_{jr}) = 0 \\ M_j \dfrac{\mathbf{v}_j^{n+1} - \mathbf{v}_j^n}{\Delta t} + \dfrac{1}{\epsilon}\sum_r \mathbf{G}_{jr} = -\dfrac{\sigma}{\epsilon^2}\sum_r \widehat{\beta}_{jr} k_r \mathbf{u}_r \\ M_j \dfrac{e_j^{n+1} - e_j^n}{\Delta t} + \dfrac{1}{\epsilon}\sum_r (\mathbf{u}_r, \mathbf{G}_{jr}) = 0 \end{cases} \quad (7)$$

with the fluxes

$$\begin{cases} \mathbf{G}_{jr} = l_{jr} q_j \mathbf{n}_{jr} + r_{jr}\widehat{\alpha}_{jr}(\mathbf{u_j} - \mathbf{u}_r) - \dfrac{\sigma}{\epsilon} k_r \widehat{\beta}_{jr} \mathbf{u}_r \\ (\sum_j r_{jr}\widehat{\alpha}_{jr} + \dfrac{\sigma}{\epsilon} k_r \widehat{\beta}_{jr})\mathbf{u_r} = \sum_j l_{jr} q_j \mathbf{n}_{jr} + r_{jr}\widehat{\alpha}_{jr}\mathbf{u_j} \end{cases} \quad (8)$$

where $M_j = |\Omega_j|^{n+1} \rho^{n+1} = |\Omega_j|^n \rho^n$, and $k_r = \dfrac{2E_r|\mathbf{f}_r|^2}{(3\chi - 1)}$, $r_{jr} = \dfrac{4}{\sqrt{3}} \dfrac{E_j}{3+|\mathbf{u}|^2}$, r_{jr} is the wave-speed calculated for the one dimensional Riemann solver. In diffusive regime, the previous lagrangian scheme gives the following non-linear positive diffusion scheme.

$$\begin{cases} |\Omega_j| \dfrac{E_j^{n+1} - E_j^n}{\Delta t} + \sum_r \dfrac{1}{12\sigma}((l_{jr} E_j \mathbf{n}_{jr} - \sigma \widehat{\beta}_{jr}\mathbf{u}_r), \dfrac{\mathbf{u}_r}{E_r}) = 0 \\ \sigma \left(\sum_j \widehat{\beta}_{jr}\right)\mathbf{u}_r = \sum_j l_{jr} E_j \mathbf{n}_{jr} \end{cases} \quad (9)$$

This scheme can be seen as the non linear extension to the previous limit scheme. If $\widehat{\beta}_{jr} \simeq l_{jr}\mathbf{n}_{jr} \otimes (\mathbf{x}_r - \mathbf{x}_j)$ and $\mathbf{u}_r$ discretize correctly the gradient then

$$(l_{jr} E_j \mathbf{n}_{jr} - \sigma \widehat{\beta}_{jr}\mathbf{u}_r) \simeq l_{jr} E_r \mathbf{n}_{jr}$$

We obtain a result very close to the linear limit diffusion scheme. For the remap step, we can use any advection scheme. In asymptotic regime the lagrangian step gives a diffusion coefficient $\frac{1}{12\sigma}$. Studying the asymptotic limit we remark that $\mathbf{u}_r$ is homogeneous to $\frac{\nabla E}{4E\sigma}$ when ε tends to zero. Therefore the remap step gives a first order diffusion scheme with the coefficient $\frac{1}{4\sigma}$. The two steps are necessary to obtain the good diffusion coefficient. To obtain a second order scheme, we must use a MUSCL procedure with slop limiter in the remap step.

4 Numerical results

4.1 Numerical results for the limit diffusion scheme

We begin by study he convergence to the limit diffusion scheme. The studied scheme is the sum to (9) and the limit scheme of advection step. The initial condition is the fundamental solution of the heat equation at $t = 0.001$. the final time is $t = 0.01$. We obtain the following order of convergence K the coefficient of deformation for

Mesh	order	negative coef
Cartesian	1.92	0
Rand. quad. mesh	1.9	0
Cartesian trig. mesh	2.23	0
Rand. trig. mesh	2.16	0
Kershaw K=1	1.93	0
Kershaw K=1.5	2.02	0

the Kershaw mesh. This results show that the scheme is a valid second order scheme on unstructured meshes.

Remark: Other test cases show that the scheme is convergent with the second order for th free-streaming regime ($\sigma = 0$) to the M_1 model.

4.2 Numerical results for radiation equilibrium models

The convergence results show that the limit scheme and the hyperbolic for all ϵ scheme are convergent. We solve

$$\begin{cases} \partial_t E + \dfrac{1}{\epsilon} \nabla.\mathbf{F} = \dfrac{\sigma}{\epsilon^2}(aT^4 - E) \\ \partial_t \mathbf{F} + \dfrac{1}{\epsilon} \nabla(\hat{P}) = -\dfrac{\sigma}{\epsilon^2}\mathbf{F} \\ \rho C_v \partial_t T_j = -\dfrac{\sigma}{\epsilon^2}(aT^4 - E) \end{cases} \quad (10)$$

Where T is the material temperature. We define $T_r = (E/a)^{\frac{1}{4}}$ the radiation temperature. To treat this model, we use a splitting strategy. The moment model part is solve with the previous scheme and the temperature relaxation part part is solve with a implicit fixed point procedure.

To obtain this implicit procedure we linearize in time the equation of T. We define $\Theta = aT^4$, consequently we obtain for the relaxation part the scheme

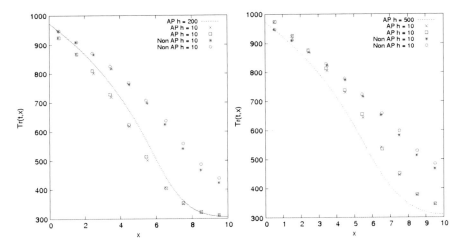

Fig. 2 The curve represent the solution T_r at the final time. The square and cross correspond to the solution with the AP correction on Cartesian and random mesh with 10 cells by direction. The point and circle correspond to the solution without the AP correction on Cartesian and random mesh with 10 cells by direction. At left this is the result for the P_1 model and at right for the M_1 model

$$\begin{cases} \frac{E_j^{q+1}-E_j}{\Delta t} = \frac{\sigma}{\varepsilon^2}(\Theta_j^{q+1} - E_j) \\ \rho C_v \mu_j \frac{E_j^{q+1}-E_j}{\Delta t} = -\frac{\sigma}{\varepsilon^2}(\Theta_j^{q+1} - E_j) \end{cases} \quad (11)$$

with $\mu_j = \frac{T_j^q - T_j}{\Theta_j^q - \Theta^q}$. This method is convergent and preserve the positivity of E and T.

We use the Marshak test describe in [1]. For the test case we consider a material initially cold and at radiative equilibrium. A heat wave enters the domain and we observe this evolution. The calculation is realized on a 2D mesh. We present the results for one line of cells. This test show the AP scheme capture the correct solution on coarse grid contrary to the classical scheme. The Fig. 2 show also that the scheme give the good result on random grid.

5 Conclusion

Starting from an asymptotic preserving scheme on 2-D unstructured meshes obtained in [5] for the linear P_1 model, we propose in this work its extension for the non-linear M_1 model. The scheme is valid on unstructured meshes, and is asymptotic preserving for the non-equilibrium regime (without coupling with matter) and for the equilibrium regime (with the coupling). Future works will be devoted to higher order angular discretization of the linear transport equations such as the discrete ordinates method or P_N equations.

References

1. C. Berthon, P. Charrier and B. Dubroca: An HLLC scheme to solve M_1 model of radiative transfer in two space dimensions, J. Scie. Comput., 31 (2007), pp. 347–389
2. J.L. Feugeas and B. Dubroca, Entropy moment closure hierarchy for the radiative transfer equation, C. R. Acad. Sci., Paris, Sér. I, Math. 329, No.10, 915-920 (1999).
3. P.-H. Maire, R. Abgrall, J. Breil, J. Ovadia *A cell-centered lagragian scheme for two-dimensional compressible flow problems.* SIAM J. Sci. Comput. Vol 29, No. 4, pp. 1781-1824. 2007.
4. C. Buet, B. Després: A gas dynamics scheme for a two moments model of radiative transfer, Mathematical models and numerical methods for radiative transfer, Panorama er synthse 2009.
5. C. Buet, B. Després, E. Franck: Design of asymptotic preserving schemes for hyperbolic heat equation on unstructured meshes. Preprint LJLL UPMC, 2010.
6. G. Carré, S. Del Pino, B. Desprès, E. Labourasse: A Cell-centered lagrangian hydrodynamics scheme on general unstructured meshes in arbitrary dimension, JCP vol. 228 (2009) no14, pp. 5160-518.
7. L. Gosse, G. Toscani: An asymptotic-preserving well-balanced scheme for the hyperbolic heat equations, C. R. Acad. Sci Paris,Ser, I 334 (2002) 337-342.
8. G. Kluth, B. Després: Discretization of hyperelasticity on unstructured mesh with a cell-centered Lagrangian scheme. Journal of Computational Physics, Volume 229, December 2010
9. S. Jin, D. Levermore: Numerical schemes for hyperbolic conservation laws with stiff relaxation terms. JCP 126, 449-467, 1996.

The paper is in final form and no similar paper has been or is being submitted elsewhere.

Mass Conservative Coupling Between Fluid Flow and Solute Transport

Jürgen Fuhrmann, Alexander Linke, and Hartmut Langmach

Abstract We present a coupled discretization approach for species transport in an incompressible fluid. The Navier-Stokes equations for the flow are discretized by the divergence-free Scott-Vogelius element. The convection-diffusion equation for species transport is discretized by the Voronoi finite volume method. The species concentration fulfills discrete global and local maximum principles. We report convergence results for the coupled scheme and an application of the scheme to the interpretation of limiting current measurements in an electrochemical flow cell.

Keywords Incompressible Navier-Stokes Equations, Convection-Diffusion Equation, Finite Element Method, Finite Volume Method, Limiting Current
MSC2010: 76V05 76D05 65N08 65N30

1 Introduction

For the transport of a substance dissolved in a dilute solution in an incompressible fluid characterized by a velocity field $\mathbf{v}$, local mass conservation and maximum principle for the substance concentration c are directly connected to the solenoidal condition $\nabla \cdot \mathbf{v} = 0$ on the velocity field.

The Scott-Vogelius mixed finite element P_k-P_{k-1}^{disc} with order $k \geq 1$ for the Navier-Stokes equations guarantees a point-wise divergence-free discrete velocity field.

Upwinded Voronoi finite volume methods guarantee the desired qualitative properties for the discrete transport problem if the discrete velocity field fulfills a discrete counterpart of the solenoidal condition and if the underlying simplicial mesh fulfills

---

Jürgen Fuhrmann, Alexander Linke, and Hartmut Langmach
Weierstrass Institute, Mohrenstr. 39, 10117 Berlin, Germany,
e-mail: juergen.fuhrmann|alexander.linke|hartmut.langmach@wias-berlin.de

the boundary conforming Delaunay property [1,2]. Recent developments in mesh generation [3,4] allow to consider this approach as a realistic option.

Using exact integration of the normal component of the discrete flow through the faces of the Voronoi volumes, we couple both schemes [5]. We discuss the application to the limiting current problem in a thin layer flow cell [6].

Let $\Omega \subset \mathbb{R}^d$ be a simply connected Lipschitz domain with $d \in \{2,3\}$. We regard the stationary, incompressible Navier-Stokes equations coupled to the equation of stationary transport of a dissolved species The flow is described using the steady, incompressible Navier-Stokes equations:

$$(\mathbf{v} \cdot \nabla)\mathbf{v} + \nabla p - \eta \Delta \mathbf{v} = \mathbf{f}, \qquad \nabla \cdot \mathbf{v} = 0. \qquad (1)$$

Here, $\mathbf{v}$ is the fluid velocity, p is the pressure, η is the fluid viscosity, and $\mathbf{f}$ is a force vector. The steady transport of a species dissolved in the fluid is described by

$$\nabla \cdot \mathbf{q} = s, \qquad \mathbf{q} = -(D\nabla c - c\mathbf{v}) \qquad (2)$$

Here, $\mathbf{q}$ is the species molar flux, c is the species concentration, D is the diffusion coefficient, and s is a given source term.

The boundary conditions correspond to the limiting current problem in a flow cell [6]. Let $\mathscr{I}_\Gamma = \{A, I, O, S, W\}$ be a set of labels for boundary segments. We assume that the boundary $\Gamma = \partial \Omega = \bigcup_{i \in \mathscr{I}_\Gamma} \Gamma_i$ is subdivided into an inlet Γ_I, an outlet Γ_O, an anode Γ_A, and a symmetry boundary on Γ_S. For an illustration, see Fig. 1. The remaining part of Γ is assumed to consist of inert, impermeable walls Γ_W. We further assume that $\Gamma_A, \Gamma_I, \Gamma_O$ are separated from each other by sections belonging either to Γ_W or Γ_S. We impose the following boundary conditions:

Section	c	$(\mathbf{v}, p)$
Inlet Γ_I	$c = c_I(\mathbf{x})$	$\mathbf{v} = \mathbf{v}_I(\mathbf{x})$
Anode Γ_A	$c = 0$	$\mathbf{v} = \mathbf{0}$
Outlet Γ_O	$\frac{\partial c}{\partial \mathbf{n}} = 0$	$\eta \frac{\partial \mathbf{v}}{\partial \mathbf{n}} = p\mathbf{n}$
Symmetry Γ_S	$\frac{\partial c}{\partial \mathbf{n}} = 0$	$\mathbf{v} \cdot \mathbf{n} = 0, \frac{\partial (\mathbf{v} \cdot \mathbf{t})}{\partial \mathbf{n}} = 0$
Wall Γ_W	$\frac{\partial c}{\partial \mathbf{n}} = 0$	$\mathbf{v} = \mathbf{0}.$

(3)

The flow boundary condition at the outlet Γ_O states that the stress $\eta \nabla \mathbf{v} - p \cdot \text{Id}$ projected onto the outward normal direction $\mathbf{n}$ is zero. For the concentration, it states that all solute transported to Γ_O by convection leaves the domain there [2].

Let $\Gamma_D^{NS} = \Gamma_I \cup \Gamma_A \cup \Gamma_W$ denote the Dirichlet boundary for the Navier-Stokes equations. Let $\mathbf{v}_D$ be a vector function on Γ_D^{NS} which is defined by the corresponding boundary values in (3). By applying the differential operator to the extension of $\mathbf{v}_D$ into Ω, and adding the result to the right hand $\mathbf{f}$ side we derive a new right hand side, also denoted by $\mathbf{f}$ which allows to assume that the solution $\mathbf{v}$ is in the space $V = \{\mathbf{v} \in [H^1(\Omega)]^d | \mathbf{v} = 0 \text{ on } \Gamma_D^{NS}, \mathbf{v} \cdot \mathbf{n} = 0 \text{ on } \Gamma_S\}$. The weak formulation of (1) arises as follows: Find $(\mathbf{v}, p) \in V \times L^2(\Omega)$ such that for all $(\mathbf{w}, q) \in V \times L^2(\Omega)$

$$\int_\Omega \eta \nabla \mathbf{v} : \nabla \mathbf{w}\, dx + \int_\Omega ((\mathbf{v} \cdot \nabla)\mathbf{v}) \cdot \mathbf{w}\, dx + \int_\Omega p \nabla \cdot \mathbf{w}\, dx = \int_\Omega \mathbf{f} \cdot \mathbf{w}\, dx \quad (4)$$

$$\int_\Omega q \nabla \cdot \mathbf{v}\, dx = 0.$$

The weak formulation of (2) relies on the particular choice of boundary conditions for **v**. Let $\Gamma_D^T = \Gamma_A \cup \Gamma_I$ be the Dirichlet boundary for the transport equation, and let s be the right hand side containing the Dirichlet boundary conditions.
Let $W = \{c \in H^1(\Omega)|\ c|_{\Gamma_D^T} = 0\}$. Then we look for $c \in W$ such that for all $\phi \in W$,

$$\int_\Omega (D\nabla c - c\mathbf{v}) \cdot \nabla \phi\, dx + \int_{\Gamma_O} \mathbf{v} \cdot \mathbf{n} c \phi\, ds = \int_\Omega s\phi\, dx. \quad (5)$$

2 Scott Vogelius mixed finite elements for fluid flow

Let $\bar{\mathcal{T}}_h$ denote a regular finite element triangulation of the domain Ω in the sense of [7], called macro triangulation. For each simplex $\bar{T} \in \bar{\mathcal{T}}_h$ we connect its barycenter with its vertices, and we thereby get $d+1$ new simplices from each macro simplex. This new triangulation $\mathcal{T}_h$ is called an SV-admissible mesh. We define $V_h := \{\mathbf{v}_h \in [C(\Omega)]^d \cap V : \mathbf{v}_h|_T \in [P_d(T)]^d\ \forall T \in \mathcal{T}_h\}$ as the space of continuous element-wise polynomial vector functions of order d on the triangulation $\mathcal{T}_h$. The pressure space $P_h := \{q \in L^2(\Omega) : q|_T \in P_{d-1}\ \forall T \in \mathcal{T}_h\}$ is defined as the space of element-wise polynomial functions of degree $d-1$ without the constraint of continuity between elements. The derivation of the triangulation $\mathcal{T}_h$ from a macro-triangulation $\bar{\mathcal{T}}_h$ assures that the discrete saddle point problem derived from the Scott-Vogelius element has a unique solution by fulfilling in a stable manner the necessary and sufficient $\inf-\sup$ condition $0 < \beta \leq \beta_h = \inf_{p_h \in P_h, p_h \neq 0} \sup_{\mathbf{v}_h \in V_h} \frac{(\nabla \cdot \mathbf{v}_h, p_h)}{\|p_h\|\|\mathbf{v}_h\|}$ [8–10].

The discretization of of the Navier-Stokes equations is derived in a standard manner from (4): find $(\mathbf{v}_h, p_h) \in V_h \times P_h$ such that $\forall\, (\mathbf{w}_h, q_h) \in V_h \times P_h$,

$$\int_\Omega (\mathbf{v}_h \cdot \nabla)\mathbf{v}_h \cdot \mathbf{w}_h\, dx - \int_\Omega p_h \nabla \cdot \mathbf{w}_h\, dx + \int_\Omega \eta \nabla \mathbf{v}_h : \nabla \mathbf{w}_h\, dx = \int_\Omega \mathbf{f} \cdot \mathbf{w}_h\, dx$$
$$- \int_\Omega q_h \nabla \cdot \mathbf{v}_h\, dx = 0. \quad (6)$$

3 Voronoi Finite Volumes for solute transport

Let $\partial \Omega$ be the union of straight lines resp. planar polygons. Let $\mathcal{P} = \{\mathbf{x}_K\} \subset \bar{\Omega}$ be a set of points which includes all the vertices of the polygons constituting $\partial \Omega$.

A simplicialization of this point set is Delaunay if no circumball of any simplex contains a point x_K of $\mathcal{P}$. Besides the fact that it is related to the same domain Ω, this simplicialization may be completely independent of the triangulations introduced in section 2. For a point $x_K \in \mathcal{P}$, the Voronoi cell $V_K^0 \subset \mathbb{R}^d$ around x_K is defined as the set of points $x \in \mathbb{R}^d$ which are closer to x_K than to any other point x_L of $\mathcal{P}$. We define as the control volume around x_K the Voronoi box V_K associated with x_K as $V_K = V_K^0 \cap \Omega$. The Delaunay simplicialization is boundary conforming [1] if

1. Ω is the union of all simplices;
2. no simplex circumball contains any other discretization vertex;
3. all simplex circumcenters are contained in $\bar{\Omega}$;
4. the boundary sections Γ_i ($i \in \mathcal{I}_\Gamma$) are the unions of simplex faces, and all circumcenters of boundary simplices from Γ_i are contained in $\bar{\Gamma}_i$.

We will use K in order to denote the Voronoi boxes V_K. Let $\mathcal{K}$ denote the set of control volumes K, and $\mathcal{K}_i$ denote the set of control volumes K which share facets with Γ_i. Let $\mathcal{K}_D = (\mathcal{K}_I \cup \mathcal{K}_A)$ denote the set of Dirichlet control volumes and $\mathcal{K}^0 = \mathcal{K} \setminus \mathcal{K}_D$ denote the set of non-Dirichlet control volumes. For two neighboring control volumes K, L, $x_K x_L$ is an edge of the boundary conforming Delaunay simplicialization which is known to be orthogonal to the Voronoi box face $\partial K \cap \partial L$. Let $\mathcal{N}_K$ denote the set of neighbors of K. For $i \in \mathcal{I}_\Gamma$, let $\mathcal{G}_K^i$ be the set of facets of K with nonempty intersection with boundary section Γ_i. Then $\partial K \cap \Gamma_i = \bigcup_{\sigma \in \mathcal{G}_K^i} \sigma$ and

$$\partial K = \left(\bigcup_{L \in \mathcal{N}_K} \partial K \cap \partial L \right) \cup \left(\bigcup_{i \in \mathcal{I}_\Gamma} (\cup_{\sigma \in \mathcal{G}_K^i} \sigma) \right).$$

Let $\mathbf{v} \in [H^1(\Omega)]^d$ fulfill the boundary conditions (3) and $\nabla \cdot \mathbf{v} = 0$. These conditions are fulfilled by every solution $\mathbf{v}$ of (4) and every solution $\mathbf{v}_h$ of (6). For any $K \in \mathcal{K}$, the H^1-regularity of $\mathbf{v}$ allows to define the scaled velocity projections

$$v_{KL} = \frac{1}{|\partial K \cap \partial L|} \int_{\partial K \cap \partial L} \mathbf{v} \cdot (\mathbf{x}_K - \mathbf{x}_L) ds, \qquad L \in \mathcal{N}_K \qquad (7)$$

$$v_\sigma = \frac{1}{|\sigma|} \int_\sigma \mathbf{v} \cdot \mathbf{n}_\sigma ds, \qquad \sigma \in \mathcal{G}_K^i \qquad (8)$$

They are discretely divergence-free in the sense that for all $K \in \mathcal{K}$ holds

$$\sum_{L \in \mathcal{N}_K} \frac{|\partial K \cap \partial L|}{|\mathbf{x}_K - \mathbf{x}_L|} v_{KL} + \sum_{i \in \mathcal{I}} \sum_{\sigma \in \mathcal{G}_K^i} |\sigma| v_\sigma = 0. \qquad (9)$$

We introduce the space of functions $W_h = \{c_h \in L^2(\Omega) : c_h|_K = c_K\}$, consisting of scalar functions which are piecewise constant on each control volume.

For a given upwind function $U(z)$, the average normal flux of $\mathbf{q} = -D\nabla c + c\mathbf{v}$ between two neighboring control volumes K, L is approximated by a flux function $g(c_K, c_L, v_{KL}) = D\left(U\left(\frac{v_{KL}}{D}\right)c_K - U\left(-\frac{v_{KL}}{D}\right)c_L\right)$, depending on the values of the discrete solution in the adjacent control volumes and the velocity projection [2].

Further, we define the discrete right-hand side of the discrete convection-diffusion equation by the average value of the continuous right-hand side over the control volume K $s_K = \frac{1}{|K|}\int_L s(\mathbf{x})\,dx$. Then, the finite volume scheme for the transport equation (5) reads as: we look for $c_h \in W_h$ such that

$$\begin{cases} \sum_{L \in \mathcal{N}_K} \frac{|\partial K \cap \partial L|}{|\mathbf{x}_K - \mathbf{x}_L|} g(c_K, c_L, v_{KL}) + \sum_{\sigma \in \mathcal{G}_K^0} |\sigma| g(c_K, c_K, v_\sigma) = s_K & K \in \mathcal{K}^0 \\ c_K = c_D(\mathbf{x}_K), & K \in \mathcal{K}_D, \end{cases} \qquad (10)$$

where the treatment of the outflow boundary conditions is taken from [2]. For $U(z) = U_{\text{dcd}}(z) = 1 + \frac{z}{2}$ we yield the central difference scheme. The simple upwind discretization is given by the upwind function $U_{\text{dsu}}(z) = 1 + \max\{0, z\}$. Our preferable choice is the Bernoulli function $U_{\exp}(z) = B(z) = \frac{z}{1-e^{-z}}$, leading to the the so-called exponential fitting scheme [11, 12].

4 Convergence of the coupled FVM-FEM scheme

The convergence results are given for homogeneous Dirichlet boundary conditions on $\Gamma = \partial\Omega$: $c|_\Gamma = 0$ $\mathbf{v}|_\Gamma = \mathbf{0}$. As a consequence, in the weak formulations (4) and (5), we assume that $\Gamma = \Gamma_I$ and $\mathbf{v} \in V = [H_0^1(\Omega)]^d$ and $c \in H_0^1(\Omega)$.

We investigate a sequence of mesh pairs $(\mathcal{T}_h, \mathcal{V}_h)$ indexed by the mesh parameter h, and where $\mathcal{T}_h, \mathcal{V}_h$ are SV-admissible and boundary conforming Delaunay, respectively and possess uniform bounds for their respective mesh regularities. The only geometrical assumption relating both sequences is that there are h-independent constants C_1 and C_2 such that $C_1 h_{\text{FEM}}(h) \leq h_{\text{FVM}}(h) \leq C_2 h_{\text{FEM}}(h)$.

Theorem 1 (Finite Element Convergence).

1. Equation (6) has at least one solution $(\mathbf{v}_h, p_h)$ on every SV-admissible grid.
2. For a sequence of Scott-Vogelius solutions $(\mathbf{v}_h)$ in (6) we can extract a subsequence which converges weakly in $V = [H_0^1(\Omega)]^d$ to some $\mathbf{v} \in V$. Moreover, this convergence is strong in $[L^2(\Omega)]^d$ and the limit $\mathbf{v}$ is divergence-free.
3. The limit $\mathbf{v}$ of said subsequence $(\mathbf{v}_h)_h$ is a solution of (4), and $\mathbf{v}_h \xrightarrow{H_0^1} \mathbf{v}$.

Theorem 2 (Convergence of the coupled scheme). We assume that $(\mathbf{v}_h, c_h)$ is a sequence of pairs of discrete solutions of (6) and (10) such that the sequence $(\mathbf{v}_h)$ converges strongly in $[H_0^1(\Omega)]^d$ to a solution $\mathbf{v}$ of (4).

1. From the sequence (c_h) we can extract a subsequence which converges strongly in $L^2(\Omega)$ to some $c \in H_0^1(\Omega)$.
2. The accumulation point $c \in H_0^1(\Omega)$ of said subsequence $(c_h)_h$ is the unique solution of the continuous problem (5), where the solution $\mathbf{v}$ of (4) drives the convection. Therefore, also the entire sequence (c_h) converges strongly in L^2 to the unique c, and not only a subsequence.

The discretization matrix of (10) has the M property. Furthermore, (9) leads to

Lemma 1. *For any solution $(c_K)_{K \in \mathcal{K}}$ of (10) with $(s_K) = 0$, we have:*

1. *Global minimax principle:* $0 \le c_K \le c_I \quad \forall K \in \mathcal{K}$
2. *Local minimax principle:* $\min_{L \in \mathcal{N}_K} c_L \le c_K \le \max_{L \in \mathcal{N}_K} c_L \quad \forall K \in \mathcal{K}^0$

The convergence of the finite volume scheme for an analytically given velocity in the discrete L^2, discrete maximum, and discrete H^1 norms has been investigated [5]. On a mesh obtained from a rectangular mesh by subdividing each rectangle into two triangles, the exponential fitting and the central schemes exhibit second order convergence in all three norms, while the simple upwind scheme is first order. On a genuinely triangular mesh, first order convergence in the discrete H^1-norm for all three schemes has been indicated. Replacing the analytical velocities by velocities obtained using the Scott-Vogelius element resulted in the same asymptotic behavior.

5 Interpretation of a limiting current experiment

We report results from [6]. At the inlet Γ_I, a sulphuric acid (H_2SO_4) based electrolyte with given velocity profile $\mathbf{v}_I(\mathbf{x})$ derived from Poiseuille flow is injected with a concentration c_I of dissolved hydrogen H_2. At a certain potential applied between the anode Γ_A covered with a platinum catalyst, and a counter electrode placed in the electrolyte outside the domain of consideration, the part of the H_2 reaching Γ_A reacts immediately according to $H_2 \to 2H^+ + 2e^-$. The flow containing the unreacted H_2 leaves the cell at the outlet Γ_O. All H_2 reaching the anode Γ_A is consumed by the reaction, so homogeneous Dirichlet boundaries for the concentration are assumed. The source terms $\mathbf{f}, s$ in (1), (2) are zero. The geometry is depicted in Fig. 1. The symmetry of the cell allows to reduce the computational domain to one twelfth of the original problem by applying symmetry boundary conditions at the corresponding cut planes Γ_S. The anode current $I_E = 2F \int_{\Gamma_A} D \frac{\partial c}{\partial n} ds$ is called the *limiting current*.

Figure 2 compares the concentration isosurfaces obtained with the Scott-Vogelius and Taylor-Hood Elements, respectively. We clearly see a striking difference concerning the preservation of the maximum principle.

Figure 3 (left) shows the maximum concentration vs. flow rate for the two finite element discretizations. For the Taylor-Hood element, we are unable to control the violation of the maximum principle. For the Scott-Vogeljus element, we see that the a-priori bound for the concentration given by the inlet velocity is observed.

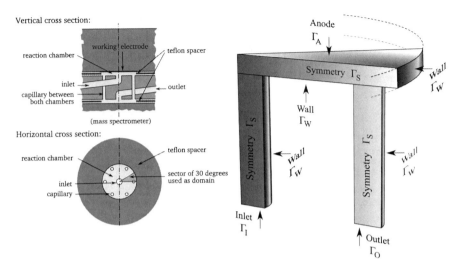

Fig. 1 Left: Schematic of a thin layer flow cell [13]. By symmetry, the problem is reduced to the 30 degrees (gray) circular arc shown. Right: computational domain with boundary segments. Reprinted with permission from [5]

Fig. 2 Concentration profiles for flow rate $80 mm^3/s$ on a coarse grid: Flow calculated using Scott-Vogelius element (left) and Taylor-Hood element (right). Isosurfaces ($c = 1.0, 2.0 \ldots 6.0$) are shown in the interior of the working chamber. Isolines and grayscale color code at surfaces are shown at the inlet, the outlet, and the bottom of the working chamber. The graphical representation has been stretched by a factor around 10 in z direction. Reprinted with permission from [5]

The right plot in Fig. 3 compares the values of the limiting current for different grids and discretizations with those measured in [14]. The grid dependency of this value is well below the accuracy of the experimental data [6]. At the same time one observes that the violation of the maximum principle does not significantly influence the value of the limiting current.

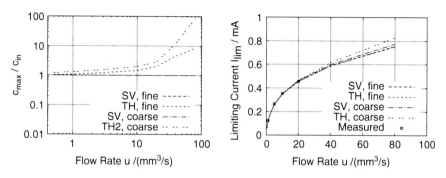

Fig. 3 Maximum concentration vs. flow rate (left). Measured [14] and calculated limiting current for different grids and discretizations (right). Reprinted with permission from [5]

6 Conclusions and Outlook

We presented a new scheme allowing for mass conservative coupling of solute transport and Navier-Stokes flow. It shows the expected convergence properties, and can be used in relevant applications. The approach has some drawbacks. Whereas in the implementation of the scheme (10), only the areas $|\partial K \cap \partial L|$ are used, which can be assembled from simplicial contributions, for the velocity projections (7), the entities $\partial K \cap \partial L$ need to be constructed [5]. The Scott Vogelius element is expensive. Static condensation may allow for more efficient assembly. There may be other routes to the discrete solenoidal condition (9) – for the tangential velocity MAC scheme [15], it exactly corresponds to the discretization of the mass balance for fluid flow.

References

1. H. Si, K. Gärtner, and J. Fuhrmann. Boundary conforming Delaunay mesh generation. *Comput. Math. Math. Phys.*, 50:38–53, 2010.
2. J. Fuhrmann and H. Langmach. Stability and existence of solutions of time-implicit finite volume schemes for viscous nonlinear conservation laws. *Appl. Numer. Math.*, 37(1–2):201–230, 2001.
3. J. R. Shewchuk. triangle version 1.6. URL: http://www.cs.cmu.edu/quake/triangle.html, 2007. Retrieved 2007-09-26.
4. H. Si. TetGen version 1.4.2. URL: http://tetgen.berlios.de/, 2010. Retrieved 2010-09-21.
5. J. Fuhrmann, A. Linke, and H. Langmach. A numerical method for mass conservative coupling between fluid flow and solute transport. *Appl. Numer. Math.*, 61(4):530 – 553, 2011.
6. J. Fuhrmann, A. Linke, H. Langmach, and H. Baltruschat. Numerical calculation of the limiting current for a cylindrical thin layer flow cell. *Electrochimica Acta*, 55(2):430–438, 2009.
7. P. G. Ciarlet. *The Finite Element Method for Elliptic Problems*, volume 4 of *Studies in Mathematics and its Applications*. North-Holland, 1978.
8. J. Qin. *On the convergence of some low order mixed finite elements for incompressible fluids*. PhD thesis, Penn. State Univ., 1994.

9. D. N. Arnold and J. Qin. Quadratic velocity/linear pressure Stokes elements. In R. Vichnevetsky, D. Knight, and G. Richter, editors, *Advances in Computer Methods for Partial Differential Equations VII*, pages 28–34. IMACS, 1992.
10. S. Zhang. A new family of stable mixed finite elements for the 3D Stokes equations. *Math. Comp.*, 74(250):543–554, 2005.
11. D. N. Allen and R. V. Southwell. Relaxation methods applied to determine the motion, in two dimensions, of a viscous fluid past a fixed cylinder. *Quart. J. Mech. and Appl. Math.*, 8:129–145, 1955.
12. A. M. Il'in. A difference scheme for a differential equation with a small parameter multiplying the second derivative. *Mat. zametki*, 6:237–248, 1969.
13. Z. Jusys, H. Massong, and H. Baltruschat. A new approach for simultaneous DEMS and EQCM: Electrooxidation of adsorbed CO on Pt and Pt-Ru. *J. Electrochem. Soc.*, pages 1093–1098, 1999.
14. H. Wang. *Electrocatalytic oxidation of adsorbed CO and methanol on Mo, Ru and Sn modified poly- and mono-crystalline platinum electrodes: A quantitative DEMS study (chinese)*. PhD thesis, Beijing Normal Univ., 2001.
15. R. Eymard, A. Linke, and J. Fuhrmann. MAC schemes on triangular meshes. In *Finite Volumes for Complex Application VI*. Springer, 2011. submitted.

The paper is in final form and no similar paper has been or is being submitted elsewhere.

Large Eddy Simulation of the Stable Boundary Layer

Vladimír Fuka and Josef Brechler

Abstract The model CLMM (Charles University Large-eddy Microscale Model) is a large-eddy simulation model for atmospheric flows. It solves Navier-Stokes equations for incompressible flow using the projection method and the 3rd order Runge-Kutta method in time. The spatial discretization is performed using the finite volume method on a uniform staggered grid.

The capability of the model to compute flows influenced by buoyancy is evaluated in this study in the case of stable stratification of the planetary boundary layer. The results are compared to the results of the project GABLS [2] with a good agreement.

Keywords large eddy simulation, planetary boundary layer, stable stratification, atmosphere
MSC2010: 76F40, 76F45, 76F65

1 Introduction

The large eddy simulation (LES) has been an important tool in boundary layer meteorology for several decades [4]. The presented model CLMM (Charles University Large-eddy Microscale Model) is a nonhydrostatic model for flows in the planetary boundary layer (PBL) and uses LES as it's main framework. It has been extended for the effects of buoyancy (or temperature stratification). The aim of this study is to present the numerical methods used in the dynamical core of the model and evaluate

---

Vladimír Fuka and Josef Brechler
Dep. of Meteorology and Environment Protection, Fac. of Mathematics and Physics, Charles University, V Holešovičkách 2, 18000, Prague 8, Czech Republic, e-mail: vladimir.fuka@mff.cuni.cz, josef.brechler@mff.cuni.cz

it's results in the situations influenced by buoyancy. All computations presented here are performed above a flat homogeneous terrain for simplicity.

2 Numerical methods

The model CLMM solves the Navier-Stokes equations for incompressible flow in the Boussinesq approximation. These equations in the filtered form for the use in LES are as follows

$$\frac{\partial \overline{u_i}}{\partial t} + \frac{\partial \overline{u_i} \overline{u_j}}{\partial x_j} = -\frac{\partial \overline{p}}{\partial x_i} + \frac{\partial \tau_{ij}}{\partial x_j} + \delta_{i3} \frac{g}{\theta_{ref}} (\overline{\theta} - \theta_{ref}) + f \epsilon_{ij3} \overline{u_j} \quad (1)$$

$$\frac{\partial \overline{\theta}}{\partial t} + \frac{\partial \overline{\theta} \overline{u_j}}{\partial x_j} = \frac{\partial q_j}{\partial x_j} \quad (2)$$

$$\tau_{ij} = \overline{u_i} \overline{u_j} - \overline{u_i u_j} \quad (3)$$

$$q_i = \overline{u_i} \overline{\theta} - \overline{u_i \theta}, \quad (4)$$

$$\frac{\partial \overline{u_i}}{\partial x_i} = 0 \quad (5)$$

where θ_{ref} is the reference potential temperature, f the Coriolis parameter and τ_{ij} and q_i are the subgrid stress tensor and the subgrid temperature fluxes respectively. The molecular viscosity and diffusivity is neglected. CLMM uses implicit filtering, i.e. the use of finite grid is considered as a sort of filtering. The overlines will be omitted hereinafter.

The solution of equations (1-5) is based on the method of lines (MOL), i.e. on discretization of time and space separately. The time discretization is based on a projection method [3] and the 3rd order low storage Runge-Kutta method combined with the Crank-Nicolson method [7]. The semi-discretized system can be written as

$$\frac{\hat{u}_i^k - u_i^{k-1}}{\Delta t} = -\gamma_k \left[\frac{\partial u_i u_j}{\partial x_j} \right]^{k-1} - \rho_k \left[\frac{\partial u_i u_j}{\partial x_j} \right]^{k-2} -$$

$$- \alpha_k \frac{\partial p}{\partial x_i} + \frac{\alpha_k}{2} \left(\frac{\partial \tau_{ij}^k}{\partial x_j} + \frac{\partial \tau_{ij}^{k-1}}{\partial x_j} \right) + \alpha_k f \epsilon_{ij3} u_j +$$

$$+ \gamma_k \delta_{i3} \frac{g}{\theta_{ref}} (\theta^{k-1} - \theta_{ref}) + \rho_k \delta_{i3} \frac{g}{\theta_{ref}} (\theta^{k-2} - \theta_{ref}) \quad (6)$$

$$\frac{\partial^2 \varphi}{\partial x_i^2} = \frac{1}{\alpha_k \Delta t} \frac{\partial \hat{u}_i}{\partial x_i} \quad (7)$$

$$u_i^k = \hat{u}_i^k - \alpha_k \Delta t \frac{\partial \varphi}{\partial x_i}, \quad (8)$$

$$p^k = p^{k-1} + \varphi - \frac{\alpha_k \Delta t \, v_t}{2} \frac{\partial^2 \varphi}{\partial x_i^2}, \tag{9}$$

$$\frac{\theta^k - \theta^{k-1}}{\Delta t} = -\gamma_k \left[\frac{\partial \theta u_j}{\partial x_j}\right]^{k-1} - \rho_k \left[\frac{\partial \theta u_j}{\partial x_j}\right]^{k-2} +$$

$$+ \alpha_k \left(\frac{1}{2}\frac{\partial q_j^k}{\partial x_j} + \frac{1}{2}\frac{\partial q_j^{k-1}}{\partial x_j}\right), \tag{10}$$

where

$$k = (1, 2, 3), \tag{11}$$
$$\gamma_k = (8/15, 5/12, 3/4), \tag{12}$$
$$\rho_k = (0, -17/60, -5/12), \tag{13}$$
$$\alpha_k = (8/15, 2/15, 1/3) \tag{14}$$
$$\tag{15}$$

and $\hat{u}_i$ and φ are auxiliary intermediate variables. The $\hat{u}_i$ does not fulfill the continuity equation (4) and is corrected in the latter steps.

The spatial discretization is carried out using the finite volume method on a uniform staggered grid. The standard second order central differences are used for most of the terms. In the case of scalar advection this method is not adequate because negative values and spurious oscillations have to be avoided at the cost of slightly increased numerical diffusion. For this reason CLMM employs a third order non-split semi-discrete advection method [5] which employs a flux limiter. This method is conservative, positive, but is not TVD. It still prevents the spurious oscillations to emerge and for 1D problems TVD and positive schemes may be equivalent [8].

The Poisson equation (7) is solved using a multigrid method with the Gauss-Seidel smoother and the Gaussian elimination solver on the smallest grid.

A crucial part in LES is the evaluation of the subgrid stresses. Many approaches are possible, but the eddy viscosity models are the basic and still widely used ones [6]. In these models it is assumed, that the subgrid stresses and fluxes are correlated to the strain rates and gradients in the same way, as in the case of molecular diffusion. CLMM uses the Vreman [9] algebraic model. It is simple to use, but it's results are claimed to be close to that of a dynamic model. The eddy viscosity is computed using the equation

$$v_t = c \sqrt{\frac{B_\beta}{\alpha_{ij}\alpha_{ij}}}, \tag{16}$$

where

$$\alpha_{ij} = \frac{\partial \bar{u}_j}{\partial x_i}, \qquad (17)$$

$$\beta_{ij} = \sum_m \Delta_m^2 \alpha_{mi}\alpha_{mj} \qquad (18)$$

$$B_\beta = \beta_{11}\beta_{22} - \beta_{12}^2 + \beta_{22}\beta_{33} - \beta_{23}^2 + \beta_{33}\beta_{11} - \beta_{31}^2. \qquad (19)$$

The constant c was set to 0.05 in present computations. The temperature diffusivity is computed using a constant subgrid Prandtl number $\mathrm{Pr}_{sgs} = 0.5$ in all presented cases.

At the surface the flow cannot be accurately resolved and the subgrid terms have to be computed using a wall model. because of the buoyancy effects the Monin-Obukhov similarity theory is employed [1]. The surface stress and the surface temperature flux are computed using the following expressions

$$\frac{U}{u_\star} = \frac{\ln(z/z_0) - \Psi_M(z/L)}{\kappa} \qquad (20)$$

$$\frac{\theta - \theta_0}{\theta_\star} = \frac{\ln(z/z_0) - \Psi_H(z/L)}{\kappa} \qquad (21)$$

where $\kappa = 0.4$ is the von Kármán constant, z_0 is the roughness parameter, θ_0 is the surface potential temperature, $u_\star = \sqrt{\tau_0}$ is the friction velocity and $\theta_\star = -\overline{(w'\theta')}_0/u_\star$ is the friction temperature, τ_0 is the surface stress and

$$L = -\frac{u_\star^3 \theta_0}{\kappa g \overline{(w'\theta')}_0} \qquad (22)$$

is the Obukhov length. The empirical similarity functions are set according to the GABLS [2] recommendation

$$\Psi_M = -4.8\frac{z}{L}, \qquad (23)$$

$$\Psi_H = -7.8\frac{z}{L}. \qquad (24)$$

3 Boundary and initial conditions

The boundary and initial conditions in the present study follow the GABLS intercomparison project. It is based on the results of the Beaufort Sea Arctic Stratus Experiment (BASE) and should approximate a quasi-stationary stable boundary layer over sea ice. The Coriolis parameter was set a value at the latitude of 73° north and the roughness parameter was set to a value of $z_0 = 0.1$ m.

The computational domain measured 400 m in all three dimensions. The computations were been carried out using resolutions $16 \times 16 \times 17$, $32 \times 32 \times 33$, $64 \times 64 \times 65$ and $128 \times 128 \times 129$. The vertical resolution is different from the horizontal one due to a limitation of the multigrid solver in various boundary conditions.

The boundary conditions at the limits of the domain in x and y directions were periodic. At the upper boundary the free-slip condition was used with a sponge layer damping the oscillations in the upper 100 meters.

The initial conditions consists of a neutrally stratified layer in the lowest 100 m and a stable layer with the lapse rate of 0.01 K/m. In the lowest 50 m random fluctuations with amplitude 0.1 K are applied to start-up turbulence. The initial surface temperature is 265 K and drops with a cooling rate of 0.25 K per hour.

4 Results

The model was run with described conditions for 9 hours. The last one hour was used for computing statistics. In the next paragraphs the results of CLMM are combined with the results of the groups participating in the project GABLS. All presented profiles are averaged temporally on the last one hour and spatially on horizontal planes.

4.1 Mean quantities

In Fig. 1 are the profiles of potential temperature at different resolutions. It is obvious, that a proper grid convergence has not been achieved. The GABLS paper [2] suggest the importance of even larger resolution. For the finest computed grid the comparison with GABLS results in Fig. 2 shows noticeable difference in the temperature gradient in the boundary layer. This value is sensitive to the choice of the subgrid model [2] and should be investigated more in the future.

For the wind velocity magnitude (Fig. 3 and the wind direction (Ekman spiral, Fig. 4) the agreement with GABLS results is better. The super-geostrophic jet and the wind turning in the boundary layer are well pronounced.

4.2 Turbulent fluxes

The vertical buoyancy and momentum fluxes are depicted in Figs. 5 and 6 respectively. In both cases the profiles follow the shape and fall within the range

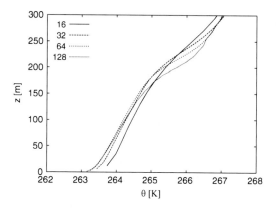

Fig. 1 Grid convergence study for the vertical temperature profile. Other variables yield similar results

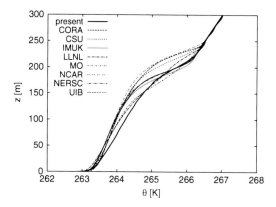

Fig. 2 The vertical temperature profile in comparison with the GABLS results

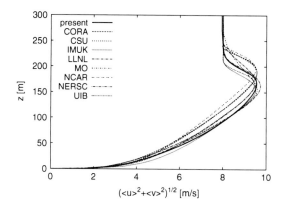

Fig. 3 The vertical profile of the wind velocity in comparison with the GABLS results

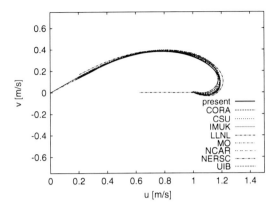

Fig. 4 The Ekman spiral (graph of horizontal velocity components) in comparison with the GABLS results

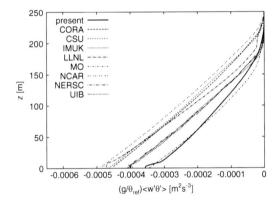

Fig. 5 The vertical buoyancy flux profile in comparison with the GABLS results

of the referenced GABLS simulations. The value of the fluxes is at the lower side of the range.

The gradient Richardson number and the flux Richardson number profiles are in Fig. 7. Their values are almost equal throughout the boundary layer reaching approximately the critical value 0.25 at it's top.

5 Conclusion

The ability of the model CLMM to simulate turbulent flow in the stable boundary layer has been tested. The results agree to those of model intercomparison initiative GABLS. Some inconsistency has been found in the temperature gradient in the boundary layer and will be investigated further. In next development the model will

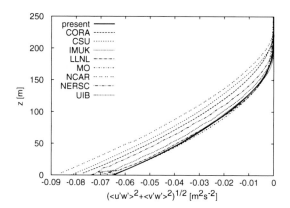

Fig. 6 The vertical momentum flux profile in comparison with the GABLS results

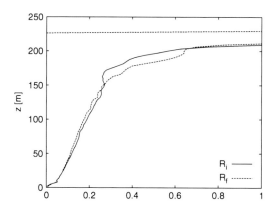

Fig. 7 The vertical profile of the gradient and flux Richardson number

be extended for inhomogeneous or spatially developing flows and for flows over a complex terrain. The model is also aimed to atmospheric dispersion studies.

Acknowledgements The work was supported by grant SVV-2010-261308 and by the Czech Ministry of Education, Youth and Sports under the framework of research plan MSM0021620860. The authors would like to thank Ing. Petr Bauer for valuable help with the implementation of the multigrid method.

References

1. Basu, S., Holtslag, A., van de Wiel, B., Moene, A., Steeneveld, G.J.: An inconvenient truth about using sensible heat flux as a surface boundary condition in models under stably stratified regimes. Acta Geophys. **56**(1), 88–99 (2008)

2. Beare, R.J., Macvean, M.K., Holtslag, A.A.M., Cuxart, J., Esau, I., Golaz, J.C., Jimenez, M.A., Khairoutdinov, M., Kosovic, B., Lewellen, D., Lund, T.S., Lundquist, J.K., Mccabe, A., Moene, A.F., Noh, Y., Raasch, S., Sullivan, P.: An intercomparison of large-eddy simulations of the stable boundary layer. Boundary-Layer Meteorology **118**(2), 247–272 (2006)
3. Brown, D.L., Cortez, R., Minion, M.L.: Accurate projection methods for the incompressible Navier-Stokes equations. J. Comput. Phys. **168**, 464–499 (2001)
4. Deardorff, J.W.: Numerical investigation of neutral and unstable planetary boundary layers. J. Atmos. Sci. **29**, 91–115 (1972)
5. Hundsdorfer, W., Koren, B., van Loon, M., Verwer, J.G.: A Positive Finite-Difference Advection Scheme. J. Comput. Phys. **117**(1), 35–46 (1995)
6. Lilly, D.K.: The representation of small-scale turbulence in numerical simulation experiments. Proc. IBM Sci. Comput. Symp. on Environ. Sci. **29**, 91–115 (1967)
7. Spalart, P.R., Moser, R.D., Rogers, M.M.: Spectral methods for the Navier-Stokes equations with one infinite and two periodic directions. J. Comput. Phys. **96**(2), 297–324 (1991). DOI 10.1016/0021-9991(91)90238-G
8. Thuburn, J.: TVD schemes, positive schemes, and the universal limiter. Monthly Weather Review **125**(8), 1990–1993 (1997)
9. Vreman, A.W.: An eddy-viscosity subgrid-scale model for turbulent shear flow: Algebraic theory and applications. Phys. Fluids **16**(10), 3670–3681 (2004)

The paper is in final form and no similar paper has been or is being submitted elsewhere.

3D Unsteady Flow Simulation with the Use of the ALE Method

Petr Furmánek, Jiří Fürst, and Karel Kozel

Abstract This works deals with three-dimensional numerical simulation of transonic and subsonic inviscid compressible steady and unsteady flow. The problem is solved using finite volume method, namely the so–called Modified Causon's scheme [3] in combination with Arbitrary Lagrangian–Eulerian method [5]. This scheme is based on TVD form of classical MacCormack scheme. Although it is not TVD it retains almost the same precision as the original TVD scheme, but demands approximately 30% less computational memory and power. Both subsonic and transonic regimes of flow over oscillating wings are simulated. The subsonic case (flow over the AS28 wing) is compared with experimental data with a very good agreement. Comparison for the transonic unsteady case (flow over the ONERA M6 wing) is unfortunately not possible, but numerical results show very good properties.

Keywords ALE, FVM, TVD, unsteady flow
MSC2010: 65M08, 65Y20

1 Introduction

Unsteady effects appear in many physical phenomena including flows in external aerodynamics. Their appearance usually entails very unpleasant problems, sometimes even with fatal consequences (e.g. flutter). It is therefore necessary to research unsteady behaviour of the flow - both forced and induced. The authors made a series of numerical experiments featuring subsonic and transonic flow over an oscillating wing using the finite volume method (FVM) [4] in combination with the Arbitrary

Petr Furmánek
VZLÚ a.s., Beranovch 130, 199 05 Praha - Letany, e-mail: petr.furmanek@fs.cvut.cz

Jiří Fürst and Karel Kozel
ÈVUT v Praze, Fakulta strojní, Ústav technick matematiky, Karlovo námstí 13, 12135 Praha, e-mail: jiri.furst@fs.cvut.cz, kozelk@fsik.cvut.cz

Lagrangian–Eulerian method (ALE) in order to study behaviour of the flow field and its development towards unsteady state.

2 Mathematical Model

The flow was considered inviscid and compressible and hence system of the Euler equations was employed as a mathematical model. It can be written down in the following conservative vector form:

$$W_t + F(W)_x + G(W)_y + H(W)_z = 0, \quad (1)$$

where subscripts denote partial derivatives and

$$
\begin{aligned}
W &= (\rho, \rho u, \rho v, \rho w, e)^T, \\
F(W) &= (\rho u, \rho u^2 + p, \rho uv, \rho uw, (e+p)u)^T, \\
G(W) &= (\rho v, \rho uv, \rho v^2 + p, \rho vw, (e+p)v)^T, \\
H(W) &= (\rho w, \rho uw, \rho vw, \rho w^2 + p, (e+p)w)^T.
\end{aligned}
\quad (2)
$$

W is vector of conservative variables with components: ρ - density, $\mathbf{w} = (u, v, w)$ - velocity vector, e - total energy per unit volume and p - static pressure. F, G, H are inviscid fluxes. System (1) is enclosed by the Equation of State:

$$p = (\gamma - 1)\left[e - \frac{1}{2}\rho(u^2 + v^2 + w^2)\right], \quad \gamma = \frac{c_p}{c_v}. \quad (3)$$

where c_p and c_v are specific heat capacities under constant pressure (at constant volume).

3 Numerical Method

When solving (1) by the finite volume method for the case of steady flow the computational domain Ω is divided into a number of quadrilateral cells D_i such that $\Omega = \bigcup_i D_i$ and $i \in \langle 1, N_i \rangle$ where N_i is total number of cells. For each i the following relation must be fulfilled

$$\frac{d}{dt}\int_{D_i} W\, d\Omega_X + \int_{\partial D_i} (F(W)_x, G(W)_y, H(W)_z)\cdot \mathbf{n}\, d\Omega_S = 0, \quad (4)$$

where **n** is unit normal outer vector of D_i. In the unsteady case system (4) is altered in order to meet the needs of the ALE method. Computational cells D_i are now time-dependent and

$$\frac{d}{dt}\int_{D_i(t)} W\, d\Omega_X + \int_{\partial D_i(t)} (\tilde{F}(W,w_1)_x, \tilde{G}(W,w_2)_y, \tilde{H}(W,w_3)_z) \cdot \mathbf{n}\, d\Omega_S = 0 \quad (5)$$

where

$$\tilde{F}(W,w_1)_x = F(W)_x - w_1 W,$$
$$\tilde{G}(W,w_2)_y = G(W)_y - w_2 W, \quad (6)$$
$$\tilde{H}(W,w_3)_z = H(W)_z - w_3 W.$$

(w_1, w_2, w_3) is velocity of mesh vertices during motion [5]. System (5) is now solved by the Modified Causon's scheme in ALE formulation [3]. This cell-centred scheme is derived from TVD form of the MacCormack scheme. It is not TVD but saves approximately 30% of computational time and memory with almost no loss in accuracy. The ALE method uses computation on moving meshes and hence an algorithm for mesh modification is needed. The actual position of mesh vertices x_i is in our case given by the following prescription

$$\mathbf{x}_i(t) = \mathbb{Q}[\phi(t, ||\mathbf{x}_i(0) - \mathbf{x}_{ref}||)](\mathbf{x}_i(0) - \mathbf{x}_{ref}) + \mathbf{x}_{ref}, \quad (7)$$

where

$$\mathbb{Q}(\phi) = \begin{pmatrix} \cos\phi & -\sin\phi \\ \sin\phi & \cos\phi \end{pmatrix}, \quad (8)$$

and

$$\phi(t,r) = \begin{cases} -\alpha_1(t) & \text{for } r < r_1, \\ -\alpha_1(t) f_D(r) & \text{for } r_1 \le r < r_2, \\ 0 & \text{for } r_2 < r. \end{cases} \quad (9)$$

where

$$f_D(r) = \left[2\left(\frac{r-r_1}{r_2-r_1}\right)^3 - 3\left(\frac{r-r_1}{r_2-r_1}\right)^2 + 1 \right] \quad (10)$$

The computational area is divided into three regions by spheres (or hemispheres) with different radius. The hemisphere with centre in $\mathbf{x}_{ref}$ and radius r_1 is rotating according to the prescribed change of pitching angle as a solid body. Outer area of the second hemisphere with radius $r_2 > r_1$ is motion-less and in space between these two hemispheres motion of the mesh is damped by damping function $f_D(\cdot)$. Wing moves according to the following prescription for pitching angle:

$$\alpha_1(t) = \alpha_{init} + A\sin(\omega t) \quad (11)$$

with angular velocity

$$\omega = \frac{2\pi k U_\infty}{c}, \quad (12)$$

$U_\infty = M_\infty$ is the free-stream velocity, c is chord length (in the wing-root) and k is reduced frequency (or $\omega = 2\pi f$ with f being real /dimensional/ frequency). In both simulated cases structured C-mesh was used for discretization of the computational domain. In the case of the AS28 wing the mesh consisted from 396000 cells, in the case of the ONERA M6 wing it was made from 493000 cells.

4 Numerical Results

Forced oscillations of the wing were both in subsonic and transonic regimes given by formally the same relation (11) but with various values of α_{init}, A and ω.

4.1 Unsteady Subsonic Flow over the AS28 wing

The initial conditions for unsteady subsonic flow over the AS28 wing were as follows: inlet Mach number $M_\infty = 0.51$, $\alpha_{init} = -0.5°$, $f = 45$Hz, $A = 3°$. The wing oscillated around reference axis parallel with wing span and going through point $\mathbf{x}_{ref} = [0.25, 0, 0]$. Numerical and experimental results were compared on behaviour of lift coefficient in cuts along the wing (Fig. 2). Development of the periodic state can be observed on Fig. 1.

4.2 Steady Transonic Flow over the ONERA M6 wing

A well-known test case published in AGARD report no. 138 [1] and characterised by inlet Mach number $M_\infty = 0.8395$ and angle of attack $\alpha_{init} = 3.06°$ was chosen as initial condition for unsteady transonic flow computation. Numerical results obtained by the MCS scheme are compared to results of WLSQR scheme [6] with HLLC numerical flux [7] and also to the experimental data (Fig. 3). Agreement between numerical and experimental results is more than satisfactory.

4.3 Unsteady Transonic Flow over the ONERA M6 wing

Simulation of transonic flow over the ONERA M6 wing was based on a test case mentioned Sect. 4.2 with initial conditions: $M_\infty = 0.8395$, $\alpha_{init} = 3.06°$, $f = 10$Hz, $A = 1.5°$. The wing oscillated around reference axis parallel with its span, this time going through point $\mathbf{x}_{ref} = [0.35, 0, 0]$.

As can be seen from Figs. 1 to 4 the scheme delivers very good results for both steady and unsteady flow. Considering subsonic regime, behaviour of c_l coefficient

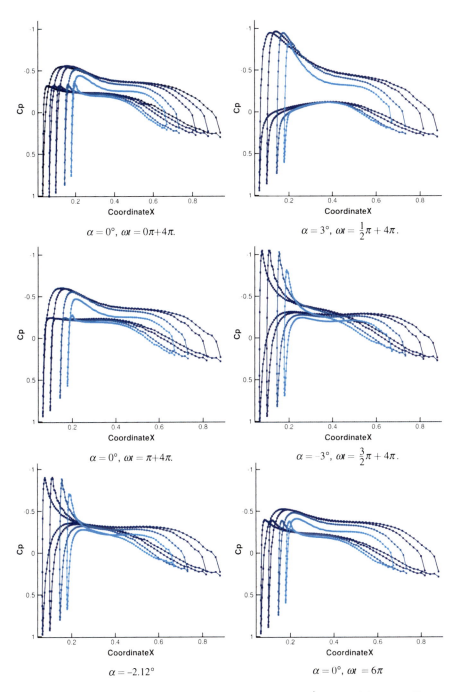

Fig. 1 c_p coefficient in cuts alongside the AS28 wing during 3rd period of forced oscillatory motion. Cuts are placed in 17.05%, 35.38%, 53.72%, 72.05% and 93.38% of the wing span

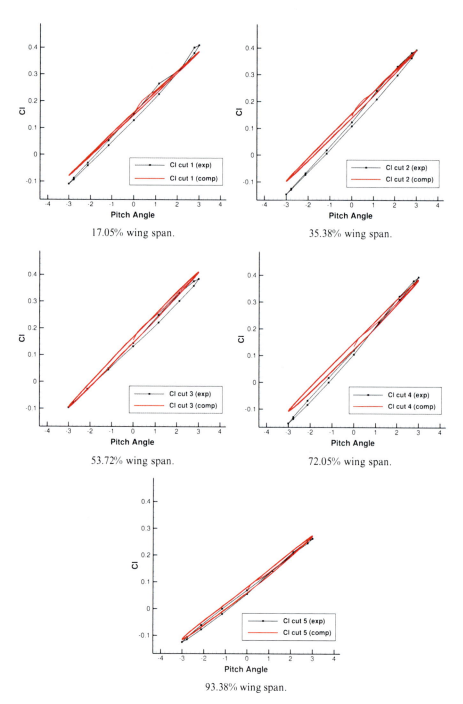

Fig. 2 c_l coefficient in cuts alongside the AS28 wing, forced oscillatory motion. Red line - numerical results, black line - experimental results

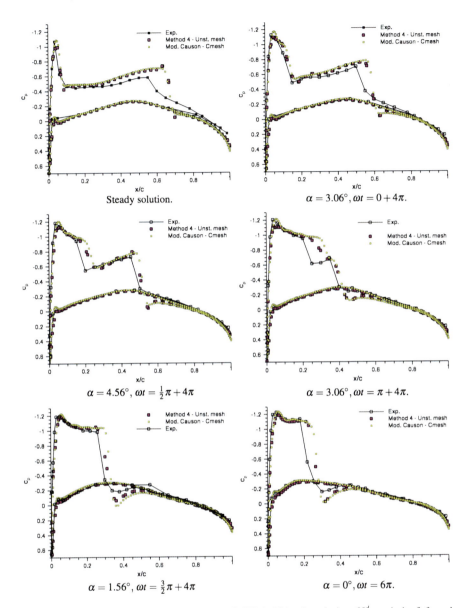

Fig. 3 c_p coefficient in cuts alongside the ONERA M6 wing during 3rd period of forced oscillatory motion

obtained by numerical computation corresponds very well to the experimental observations. Moreover, the results show that fully periodic state has been achieved at least during the 3rd period of oscillatory motion. Pressure coefficient decreases with increasing angle of attack (and vice versa) and the scheme does not produce

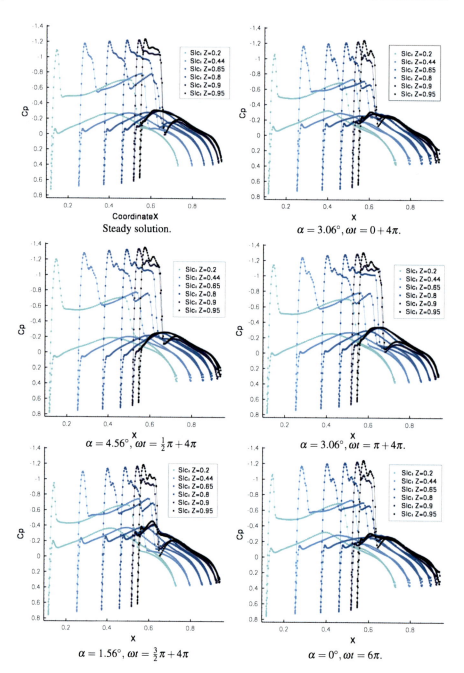

Fig. 4 c_p coefficient in cuts alongside the ONERA M6 wing during 3rd period of forced oscillatory motion

spurious oscillations. Comparison between experimental and numerical data is unfortunately not available in the transonic case, but the numerical results have all the mentioned characteristics as in subsonic flow.

5 Conclusion

The scheme is able to capture important flow characteristics even in the case of inviscid flow and can be used as a reliable numerical simulation of mentioned problems. From Figs. 1 to 4 can be seen that fully periodic state was achieved during at least 3rd period of oscillatory motion. The future steps intended are implementation of implicit version of the scheme and its extension to aero-elastic problems.

Acknowledgements This work was supported by Research Plans MSM 6840770010, MSM 0001066902 and grant GACR No. 201/08/0012 and P101/10/1329.

References

1. Schmitt, V., Charpin, F.: Pressure Distributions on the ONERA-M6-Wing at Transonic Mach Numbers. Experimental Data Base for Computer Program Assessment. Report of the Fluid Dynamics Panel Working Group 04, AGARD AR 138, May 1979.
2. Fürst J.,: A weighted least square scheme for compressible flows. *Flow, Turbulence and Combustion*, 76(4):331–342, June 2006.
3. Furmánek, P., Fürst, J., Kozel, K.: High Order Finite Volume Schemes for Numerical Solution of 2D and 3D Transonic Flows in Kybernetika, Volume 45 no. 4, 567-579, 2009.
4. LeVeque, R., J.: Numerical Methods for Conservation Laws, Basel, 1990, ISBN 3-7643-2464-3.
5. Donea, J.: An arbitrary Lagrangian-Eulerian finite element method for transient fluid- structur interactions. Comput. Methods Apll. Mech. Eng., (1982), 33:689-723.
6. Fürst, J.: A weighted least square scheme for compressible fows. Submitted to "Flow, Turbulence and Combustion", (2005).
7. Batten, P., Leschziner, M. A., Goldberg, U. C.: Average-State Jacobians and Implicit Methods for Compressible Viscous and Turbulent Flows, Journal of computational physics 137, 1997.

The paper is in final form and no similar paper has been or is being submitted elsewhere.

FVM-FEM Coupling and its Application to Turbomachinery

J. Fořt, J. Fürst, J. Halama, K. Kozel, P. Louda, P. Sváček, Z. Šimka, P. Pánek, and M. Hajsman

Abstract The paper deals with the numerical solution of turbulent flows through a 2D turbine cascade considering heat exchange between the gas and the solid blade. The flow field is described by the Favre averaged Navier-Stokes equations, and the temperature field inside the solid blade is given by the Laplace equation. Both parts are coupled in order to achieve continuity of the temperature as well as of the heat flux along the fluid-solid boundary. The analysis of simplified model case is presented and the results obtained with two in-house codes with several two-equation turbulence models are compared to results of commercial software (Fluent).

Keywords finite volumes, finite elements, heat transfer
MSC2010: 65M08, 65M12, 65N30, 76M10, 76M12

1 Introduction

The objective of this paper is to describe the coupled solution of turbulent flows through turbine cascade with heat transfer inside the blade. Due to the geometry of blades and expansion of the compressible fluid there is a temperature jump between suction and pressure side of blade profile, which is overestimated for commonly used adiabatic case compared to case with blade-fluid heat exchange. The correct modeling of heat exchange between blade and fluid is important for blades with high thermal conductivity and of course it is essential when considering some heat source inside the blade.

J. Fürst, J. Fořt, J. Halama, K. Kozel, P. Louda, and P. Sváček
Dept. of Tech. Math., CTU FME Prague, Karlovo nám. 13, CZ-12135 Praha 2, Czech Republic,
e-mail: Jiri.Furst@fs.cvut.cz

Z. Šimka, P. Pánek, and M. Hajsman
ŠKODA POWER a.s., A Doosan company, Tylova 1/57, CZ-30128 Plzeň, Czech Republic,
e-mail: Pavel.Panek@doosanskoda.com

The solution of fluid part is obtained with the help of our in-house code using finite volumes whereas the heat equation is solved with finite elements. In order to achieve the continuity of temperature filed between the fluid and solid parts, the Dirichlet-Neumann coupling is used.

It is well known that the Dirichlet-Neumann coupling is under some conditions divergent in FEM-FEM case (see e.g. [8]). Therefore we do an analysis of simplified 2D problem in our FVM-FEM case in section 2 and we show that the method is under certain conditions stable.

The section 3 describes an application of the coupled algorithm to quite complex case of heat transfer between the turbulent flow field in turbine cascade and the solid blade.

2 Model problem

Our goal is to solve the heat transfer problem in turbomachinery (see section 3). The analysis of full model involving the solution of Navier–Stokes equations with a turbulence model is rather difficult, therefore we will analyze simplified model of Dirichlet–Neumann coupling of temperature field with different heat conductivities. We assume that the domain Ω is divided onto two parts: Ω_f covered by fluid with heat conductivity k_f and thermal capacity C and Ω_s corresponding to solid part with heat conductivity k_s. The solid part is in the interior of Ω (see Fig. 1 a).

We assume that the temperature field is described by parabolic heat equation in fluid part and by elliptic equation in solid part:

$$C\frac{\partial T(\mathbf{x},t)}{\partial t} = k_f \Delta T(\mathbf{x},t), \text{ for } \mathbf{x} \in \Omega_f, \quad (1)$$

$$\Delta \theta(\mathbf{x},t) = 0 \text{ for } \mathbf{x} \in \Omega_s. \quad (2)$$

Here T and θ denote the temperature in fluid and solid parts. The initial-boundary value problem for fluid part is equipped with the initial and boundary condition

$$T(\mathbf{x},0) = T_0(\mathbf{x}) \text{ for } \mathbf{x} \in \Omega_f, \quad (3)$$

$$T(\mathbf{x},t) = g(\mathbf{x},t) \text{ for } \mathbf{x} \in \partial\Omega \text{ and } t > 0. \quad (4)$$

We assume, that the temperature is continuous at the interface $\Gamma = \overline{\Omega_f} \cap \overline{\Omega_s}$. Moreover, the conservation of energy dictates also the continuity of heat fluxes across Γ. Hence

$$\theta(\mathbf{x},t) = T(\mathbf{x},t) \text{ for } \mathbf{x} \in \Gamma, \quad (5)$$

$$k_f \frac{\partial T}{\partial \mathbf{n}} = k_s \frac{\partial \theta}{\partial \mathbf{n}} \text{ for } \mathbf{x} \in \Gamma. \quad (6)$$

Here $\mathbf{n}$ is the outer normal with respect to Ω_f.

The solution is calculated using following semi-discrete time marching algorithm

1. Set $n = 0$ and $T^0 = T_0$.
2. Solve the Laplace equation (2) for θ^n using Dirichlet boundary condition $\theta^n|_\Gamma = T^n|_\Gamma$.
3. Calculate T^{n+1} using the parabolic equation (1) with Dirichlet boundary condition (4) at the $\partial\Omega$ and Neumann boundary condition (6) at the interface Γ calculating the heat flux using θ^n.
4. Increment n and repeat steps 2-4.

2.1 FE solution in Ω_s

We are solving the Laplace equation with Dirichlet boundary condition using standard FEM method with piece-wise linear base functions with triangular mesh in 2D. Let θ_i denotes the solution at FEM mesh node i (we omit the superscript n in this section). The standard discretization then leads to the algebraic system of equations

$$A\theta = \mathbf{b}, \qquad (7)$$

where $b_i = 0$ for internal nodes and $b_j = T_j^n$ for boundary nodes.

It is known, that the Delaunay triangulation in 2D implies for piece-wise linear elements discrete maximum principle, i.e.

$$\min_{j \in \Gamma} T_j^n \leq \theta_i \leq \max_{j \in \Gamma} T_j^n, \qquad (8)$$

where the shorthand notation $j \in \Gamma$ denotes boundary nodes (see e.g. [9] or [3]). This discrete maximum principle is equivalent in our case to the fact, that the solution in internal points i is a convex combination of boundary values T_j^n, i.e.

$$\theta_i = \sum_{j \in \Gamma} \alpha_{ij} T_j^n, \qquad (9)$$

with $\alpha_{ij} \geq 0$ and $\sum_j \alpha_{ij} = 1$.

2.2 FV solution in Ω_f

Assume that the parabolic equation is solved with the explicit cell-centered FV scheme using an unstructured mesh. Assume that the mesh is orthogonal in the sense, that the face between two adjacent cell is orthogonal to the line connecting the cell centers. In that case the flux through the interface between cells i and k is proportional to $T_k - T_i$ and the explicit scheme for internal points reads

$$T_i^{n+1} = T_i^n + \Delta t \sum_k \beta_{ik}(T_k^n - T_i^n), \qquad (10)$$

with $\beta_{ik} \geq 0$ (k goes over cells adjacent to cell i).

Including the Neumann boundary condition $k_f \partial T/\partial \mathbf{n} = \dot{q}$, the scheme for cells adjacent to Γ is

$$T_i^{n+1} = T_i^n + \Delta t \sum_k \beta_{ik}(T_k^n - T_i^n) + \Delta t \beta_i^b \dot{q}_i, \qquad (11)$$

with $\beta_i^b > 0$.

The scheme for internal cells is positive and hence stable for small enough time steps. On the other hand the positivity is not obvious for boundary cells.

2.3 Coupling method

Here we assume that the nodes for FEM correspond to the cell vertices of FVM method at the boundary (see Fig. 1b, quadrilateral FV mesh in the upper part is connected to triangular FE mesh in the lower part).

Before solving the FE problem, we have to obtain a value at the boundary (points A and B at Fig. 1b). We calculate the boundary values using a weighted average of the cell-centered values adjacent to the boundary node with non-negative weights, e.g. $T_A^n = 0.5 T_W^n + 0.5 T_P^n$. Let us note, that this interpolation implies low order of accuracy at the interface. On the other hand we have to use very fine mesh spacing and high aspect ration cells in the fluid part near the interface due to thin boundary layers. Therefore we hope that the low order interpolation doesn't impair the overall accuracy.

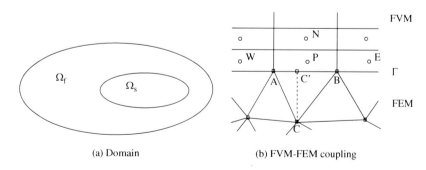

(a) Domain   (b) FVM-FEM coupling

Fig. 1 Domain $\Omega = \Omega_f \cup \Omega_s$, meshes for FVM-FEM coupling

Combining this positive interpolation with the equation (9) we get the solution in the solid part as

$$\theta_i = \sum_{k \in \Gamma} \gamma_{ik} T_k^n \text{ with } \gamma_{ik} \geq 0, \qquad (12)$$

where $k \in \Gamma$ means that k goes over FV cells adjacent to the boundary Γ, i.e. $k = W, P, E, \ldots$ at Fig. 1b.

The "natural" evaluation of the normal derivative of θ in the triangle ABC yields

$$\frac{\partial \theta}{\partial \mathbf{n}}\Big|_{ABC} = f(\theta_A, \theta_B, \theta_C) = \frac{\theta_C - \theta_{C'}}{|CC'|}, \qquad (13)$$

where C' is the orthogonal projection of C onto line AB and $\theta_{C'}$ is the obtained with linear interpolation of θ_A and θ_B. Unfortunately it is very difficult to analyze the scheme with formula. Therefore we propose to use an approximation (see Fig. 1b)

$$\theta_{C'} = T_P^n. \qquad (14)$$

Then the gradient of θ with respect to $\mathbf{n}$ is

$$\frac{\partial \theta}{\partial \mathbf{n}}\Big|_{ABC} = \frac{\theta_C - T_P^n}{|CC'|}, \qquad (15)$$

and taking into account the relation (12) we get

$$\frac{\partial \theta}{\partial \mathbf{n}}\Big|_{ABC} = \frac{\sum_{k \in \Gamma} \gamma_{Ck} T_k^n - T_P^n}{|CC'|}, \qquad (16)$$

Then the final scheme for cell P is

$$T_P^{n+1} = T_P^n + \Delta t \sum_{k \in \{N,W,E\}} \beta_{Pk}(T_k^n - T_P^n) + \Delta t \beta_P^b k_s \frac{\sum_{j \in \Gamma} \gamma_{Cj} T_j^n - T_P^n}{|CC'|} \qquad (17)$$

with β, and γ being non-negative. Therefore we can make the scheme positive using small enough time step.

Note 1: the final scheme for T is under appropriate limit for time step Δt positive

$$T_i^{n+1} = \sum_j b_{ij} T_j^n, \; b_{ij} \geq 0. \qquad (18)$$

Moreover taking $T_j^n = \tau$ a constant, we can easily show that the boundary values $T_A^n, T_B^n, \ldots$ are equal to τ too. Then $\theta_C = \tau$ and finally $T_i^{n+1} = \tau$. Canceling τ yields $\sum_j b_{ij} = 1$. Therefore the T_i^{n+1} is convex combination of values T_j^n and therefore it satisfies the discrete maximum principle (as far as the FE mesh is Delaunay and the FV mesh is orthogonal).

Note 2: the approximation (14) introduces an error in the heat flux. Nevertheless as we stated before, we are using fine near-wall scaling dictated by the turbulence model in the fluid part. In order to eliminate the error caused by "side shift" (i.e. different x coordinate of P and C' at the Fig. 1b) we can use isosceles triangles near the boundary in the FEM part.

3 Fluid-solid heat exchange in turbine cascade

Here we describe the application of this method to the solution of turbulent flows including heat transfer in the solid blade.

The fluid field is described by the set of Favre-averaged Navier-Stokes equations for compressible flows using several two-equations turbulence models. The temperature field inside the blade Ω_s is described by the Laplace equation (2).

The problem was numerically solved using commercial software Fluent at Škoda Plzeň and two different versions of in-house software developed at Czech Technical University.

3.1 Commercial software

The calculation by Škoda was performed in Fluent version 6.2 with the two-dimensional, double-precision, pressure-based solver. The turbulence models tested were the RNG $k-\epsilon$ model and the $k-\omega$ SST model. A second-order discretization scheme was employed. All calculations started on a coarse initial grid generated in Gambit, which was gradually refined at walls using the hanging-node adaption method to achieve adequate mesh resolution for low Reynolds turbulence models. The software does the computation of temperature field in both parts with the same FVM, hence it does not use the above described coupling procedure.

3.2 In-house codes

The coupling algorithm from previous section was used in the combination of two in-house codes developed at the Czech Technical University in Prague. The first one, denoted as solver 1 in later text, uses structured multiblock mesh, AUSM flux by [6] and quasi one-dimensional reconstructions with Van Leer limiter. The time discretization is achieved with backward Euler method for details see [1]. The second solver (solver 2) uses AUSMPW+ flux of [4], unstructured meshes, and multidimensional weighted least squares reconstruction described in [2].

The turbulence is modeled using Low-Reynolds $k-\omega$ model by [10], TNT $k-\omega$ model by [5], and SST model by [7].

The temperature field in solid is solved with FEM and is coupled to fluid part using the algorithm described above with "natural" approximation of heat fluxes, see eq. (13). Moreover, the FV mesh is not orthogonal in the above mentioned sense, therefore the sufficient conditions for positivity of the scheme were not satisfied. Nevertheless we didn't met serious problems with stability in this case.

3.3 Results

Calculations were performed using structured and unstructured hybrid meshes with 10-20 000 cells in fluid part with in-house codes and using an extremely fine mesh with 650 000 cells using Fluent. The flow regime is characterized by the outlet isentropic Mach number $M_{2i} = 0.34$ and with the Reynolds number $Re = 820\,000$ related to the parameters at the outlet and to the blade pitch.

Figure 2 shows the iso-lines of the temperature obtained with solver 2 in the fluid part of the domain (a), the iso-lines of the temperature in the blade (b) calculated

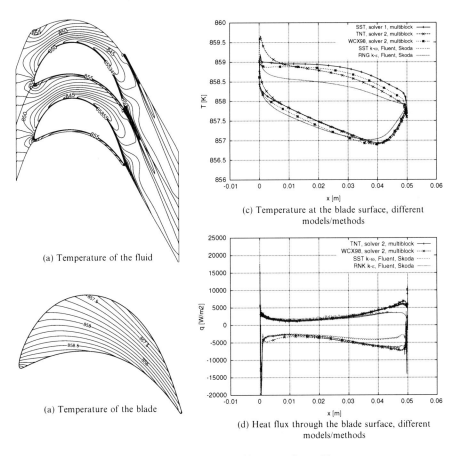

Fig. 2 Temperature and heat flux for the conjugated heat transfer problem

with FEM. Moreover, it compares the distribution of temperature (c) and heat flux (d) along the blade surface obtained with both in-house methods including several turbulence models and the results of calculation made by commercial software. However there are some differences in the temperature and consequently in the heat flux, we can say that the agreement of all methods is satisfactory.

4 Conclusions

The article shows some results concerning the solution of heat transfer between turbulent flow and solid blades. The analysis shows that it is possible to couple FVM with FEM for this kind of problems. The second part shows an application of the procedure to conjugated heat transfer problem. Due to missing experimental data for our case, we were able to compare only our solution to the results obtained with commercial software which uses different approach. However the comparison was quite satisfactory, we have to do a comparison with experimental data in the future.

Acknowledgements The work was supported by the grant No. P101/10/1329 of the Grant Agency of Czech Republic and by the Research Plan MSM No. 6840770010.

References

1. J. Dobeš, J. Fořt, J. Fürst, P. Louda, K. Kozel, and L. Tajč. Numerical methods for transonic flows, application for design of axial and radial stator turbine cascades. In *8th ISAIF Conference Proceedings*, volume 2, pages 569–578. Ecole Centrale de Lyon, 2007.
2. Jiří Fürst. The third order WLSQR scheme on unstructured meshes with curvilinear boundaries. In *Proceedings of the ENUMATH conference*, Graz, 2007. submitted.
3. Antti Hannukainen, Sergey Korotov, and Tom Vejchodsk. On weakening conditions for discrete maximum principles for linear finite element schemes. *Numerical Analysis and Its Applications*, 5434:297–304, 2009.
4. Kyu Hong Kim, Chongam Kim, and Oh-Hyun Rho. Methods for the accurate computations of hypersonic flows I. AUSMPW+ scheme. *Journal of Computational Physics*, (174):38–80, 2001.
5. J. C. Kok. Resolving the dependence on free stream values for the k-omega turbulence model. Technical Report NLR-TP-99295, NLR, 1999.
6. M. S. Liou. A sequel to AUSM: AUSM+. *Journal of Computational Physics*, (129):364–82, 1996.
7. F. R. Menter. Two-equation eddy-viscosity turbulence models for engeenering applications. *AIAA J.*, 8(32):1598–1605, 1994.
8. A. Quarteroni and A. Valli. *Domain Decomposition Methods for Partial Differential Equations*. Oxford University Press, Oxford, 1999.
9. Reiner Vanselow. About delaunay triangulations and discrete maximum principles for the linear conforming FEM applied to the Poisson equation. *Applications of Mathematics*, 46(1):13–28, 2001.
10. David C. Wilcox. *Turbulence Modeling for CFD*. DCW Industries, Inc., second edition edition, 1998.

The paper is in final form and no similar paper has been or is being submitted elsewhere.

Charge Transport in Semiconductors and a Finite Volume Scheme

Klaus Gärtner

Abstract The van Roosbroeck system describes the transport of holes and electrons in semiconductors in a drift-diffusion approximation (a special type of Nernst-Planck-Poisson systems). The classic finite volume scheme used in the field allows to prove the existence of bounded steady state solutions and the uniqueness of the thermodynamic equilibrium solution by using the duality of the boundary conforming Delaunay grid and the Voronoi diagram. The article gives an overview over properties proven for this discrete version. The time dependent problem is dissipative in case of the implicit Euler scheme. The free energy decays exponentially in case of boundary conditions compatible with the thermodynamic equilibrium. The interesting qualitative properties of the analytic problem can be carried over to the discrete case for any h and τ (spatial, time step size respectively).
A weak interpretation of the scheme is helpful: using test functions one to gets estimates, and the weak discrete maximum principle allows to prove the bounds.
An implementation following the theory strictly (Oskar3) is used to solve 3d silicon detector problems, characterized by large volumes, multiple floating regions per detector pixel and extreme charge conservation requirements. An example is discussed to illustrate the problem.

Keywords reaction-diffusion systems, discrete bounded solutions, Delaunay grids, discrete weak maximum principle
MSC2010: 65N08, 65N12

Klaus Gärtner
WIAS, Mohrenstr.39, 10117 Berlin, Germany, e-mail: gaertner@wias-berlin.de

1 The van Roosbroeck system

The continuity equations for the particle densities of electrons and holes are given on a bounded polyhedral domain $\Omega = \cup_i \Omega_i$, Ω_i a subdomain containing one material.

$$\frac{\partial n}{\partial t} - \nabla (D_n n_i \cdot \nabla \frac{n}{n_i} - \mu_n n \cdot \nabla w) + R(n,p) = 0, \tag{1}$$

$$\frac{\partial p}{\partial t} - \nabla (D_p n_i \cdot \nabla \frac{p}{n_i} + \mu_p p \cdot \nabla w) + R(n,p) = 0. \tag{2}$$

The main interaction of electrons and holes is described by the Poisson equation

$$-\nabla \cdot (\varepsilon_r \varepsilon_s \nabla w) = C - n + p. \tag{3}$$

The meaning of the quantities is:
- w - electrostatic potential,
- $n = n_i e^{w-\phi_n}$ - electron density, ϕ_n - quasi-Fermi potential of electrons,
- $p = n_i e^{\phi_p - w}$ - hole density, ϕ_p - quasi-Fermi potential of holes,
- C - density of impurities, n_i intrinsic carrier density,
- $\varepsilon = \varepsilon_r \varepsilon_s$ - dielectric permittivity, ε_r relative permittivity, ε_s scaled permittivity
- R - recombination-generation rate $R = r(x,n,p)(np-1)$, $r(x,n,p) \geq 0$,
- $\mu_{n,p}$ - carrier mobilities $\mu_{n,p} > 0$.

The Einstein relation is supposed to hold (diffusion constant $D_i = \mu_i k_B T/q_e$, k_B Boltzmann constant, T temperature, q_e elementary charge). Hence a natural scaling is to 'measure' all potentials in thermal voltages U_T and densities in a $n_{ref} \approx n_i$ resulting in $\varepsilon_s \approx 1.4 \cdot 10^{-13} \text{m}^2$, $1\text{V} \approx 40 U_T$ at room temperature, and $n+p$ can easily be of the order 10^{10} or 10^{-10} in different parts or states of a device.

The free energy of van Roosbroeck system is (compare [8, 16, 17])

$$F(w,n,p) = \int [n(\ln \frac{n}{n^*} - 1) + n^* + p(\ln \frac{p}{p^*} - 1) + p^*] \, d\Omega + \frac{1}{2} \|w - w^*\|^2, \tag{4}$$

with $\|h\|^2 = \int \varepsilon |\nabla h|^2 \, d\Omega + \int \alpha h^2 \, d\Gamma$, $\nu \cdot \nabla w^* + \alpha(w^* - w_\Gamma) = 0$, ν outer normal, and

$$-\nabla \cdot (\varepsilon \nabla w^*) = C + n_i e^{-w^*} - n_i e^{w^*} \text{ in } \Omega \tag{5}$$

the (w^*, n^*, p^*) (unique weak) thermal equilibrium solution. The dissipation rate is given by

$$d(w,n,p) = \int [n\mu_n |\nabla \phi_n|^2 + p\mu_p |\nabla \phi_p|^2 + r(x,n,p)(np-1)\ln(np)] \, d\Omega \geq 0. \tag{6}$$

The problem is supplemented by boundary conditions on $\Gamma = \Gamma_D \cup \Gamma_N$ describing contacts (Dirichlet boundary conditions for w, n, p on $\Gamma_D = \bar{\Gamma}_D$), gate contacts (third kind boundary condition for w, homogeneous Neumann boundary condition for n, p), and homogeneous Neumann boundary conditions on different parts of the boundary of the domain. The analytic results obtained by different techniques and for different assumptions on data [6, 9, 19–21] can be summarized for the purpose in mind here by: existence of steady state solutions (in $H^1 \cap L^\infty$).

2 Spatial discretization

The aim from this application point of view is to have discretizations, that carry over the analytic properties for classes of grids, hence parameters like step sizes are a question of precision and not of existence of the solution. For boundary conforming Delaunay meshes and the Scharfetter-Gummel scheme together with an implicit Euler time discretization the following table summarizes the proven properties: At a first glance that may look like a pretty comfortable position, but the headroom for improvement by better understanding is large.

A short summary with respect to Delaunay meshes [3] introduces notations and the following part reviews results establishing the lower right part of Table 1, see [10]. Let a vertex $v_k \in I\!R^N$ be denoted by $\mathbf{v}_k = (x_1, \ldots, x_N)^T$, E_l^N is simplex l in the Delaunay grid, $B(E_l^M)$ its circumscribed ball (if $M < N$ the smallest circumscribed ball). Vertex numbers are chosen such that the local coordinate system for each simplex defined by the matrix $P_{l,k} = (\mathbf{v}_{k+1} - \mathbf{v}_k, \ldots, \mathbf{v}_{k+N} - \mathbf{v}_k)$ results in the volume $|P_{l,k}|/N > 0$. Interfaces and Γ are given by $N - 1$ dimensional simplices in the grid. The Delaunay property requires that for all l $\mathbf{v}_j \in\!| B(E_l^N)$, $\forall \mathbf{v}_j \neq P_l$, and $B(E_l^N)$ is the circumscribed ball of E_l^N. The Voronoi volume V_i is the set of all points closer to $\mathbf{v}_i$ than to $\mathbf{v}_j$, $j \neq i$. $\partial V_i = \bar{V}_i \setminus V_i$ denotes the surface of the Voronoi volume and the intersections with the simplex $\mathbf{E}_j^N$ are $V_{ij} = V_i \cap \mathbf{E}_j^N$ and $\partial V_{ij} := \partial(V_i \cap \mathbf{E}_j^N)$. ∂V_{ij} is the union of planar pieces of ∂V_i and those $E_l^{N-1} \in E_j^N$ sharing the vertex i. The Delaunay property guarantees a non negative surface measure per edge in the interior of each subdomain Ω_i, the boundary conformity [5] (per subdomain) requires that all lower dimensional simplices on the boundary have empty smallest circumscribed balls, too. Together with the fact that all interfaces (boundaries) coincide with a set of E_l^{N-1} both types of surface measures (in or orthogonal to each interface) per edge and subdomain are non negative.

Starting with the equation $-\nabla \cdot \varepsilon \nabla w = f$, using Gauss's theorem on V_{ij}, and assuming $\varepsilon = const$ per simplex yields:

Table 1 Proven properties, compare [7, 8, 10–15]

property	analytic	discrete
dissipativity	yes	yes
exponential decay free energy	yes	yes
existence of bounded steady state sol.	yes	yes
uniqueness for small applied voltages	yes	yes

$$\int_{V_{ij}} -\nabla \cdot \varepsilon_l \nabla w \, dV = -\varepsilon_l \int_{\partial V_{ij}} \nabla w \cdot d\mathbf{S_k} = -\varepsilon_l \sum_{k(j)} \int_{\partial V_{i,k(j)}} \nabla w \cdot d\mathbf{S_k} + BI_{V_{ij}}$$

$$\approx -\varepsilon_l \sum_k \frac{\partial V_{i,k(j)}}{|\mathbf{e}_{ik(j)}|} (w_k - w_i) + BI_{V_{ij}} = \varepsilon_l [\gamma_{k(i)}] \tilde{G}_N \mathbf{w}|_{E_j^N} + BI_{V_{ij}},$$

where $BI_{V_{ij}}$ denotes boundary integrals in case of boundary faces. It is compensated in the interior by the neighboring $BI_{V_{ij'}}$, $\mathbf{e}_{i,k(j)}$ is the edge from node i to k in simplex j, and $\tilde{G}_N$ is a difference matrix, mapping from nodes to edges.

$$(\tilde{G}^T \tilde{G})_{ii} > 0, \quad (\tilde{G}^T \tilde{G})_{i>j} < 0, \text{ and } \mathbf{1}^T \tilde{G}^T = \mathbf{0}^T. \tag{7}$$

$$\gamma_{k(i)} = \frac{\partial V_{i,k(i)}}{|\mathbf{e}_{ik(i)}|} \tag{8}$$

denotes the elements of a diagonal matrix of geometric weights per simplex. Functions are approximated by

$$\int_{V_{ij}} f \, dV \approx V_{ij} f(x_i), \quad [V]_i = \sum_j V_{ij}, \tag{9}$$

where $[\cdot]$ denotes a diagonal matrix. Summing over all vertices of the simplex j yields

$$\sum_{V_{ij} \in E_j^N} \int_{V_{ij}} -\nabla \cdot \varepsilon \nabla w \, dV \approx \varepsilon \tilde{G}^T [\gamma] \tilde{G} \mathbf{w}|_{E_j^N} + BI. \tag{10}$$

The explicit form of the boundary integrals (in the generic situation $\xi_1 w + \xi_2 \partial w/\partial v + \xi_3 = 0$, with ξ_i defined on Γ, $\xi_1(x, w, \ldots) \geq 0$, $\xi_2(x, w, \ldots) > 0$) is given by

$$BI_{V_{ij}} = \sum_{i' \neq i, i' \in \mathbf{E}_j^N} \int_{E_{i'}^{N-1} \cap \partial V_{ij}} -\varepsilon \nabla w \cdot d\mathbf{S} \approx \sum_{i' \neq i, i' \in \mathbf{E}_j^N} |E_{i'}^{N-1} \cap \partial V_{ij}| \frac{\varepsilon}{\xi_{2_{i'}}} (\xi_{1_{i'}} w_i + \xi_{3_{i'}}),$$

where $E_{i'}^{N-1}$ denotes the $N-1$ dimensional simplex opposite to $i' \in \mathbf{E}_j^N$, $E_{i'}^{N-1} \in \Gamma$, and $BI = \sum_{i \in \mathbf{E}_j^N} BI_{V_{ij}}$.

Remark 1. A 'discrete weak maximum principle' holds ($\mathbf{w}^+$ pos. part)

$$(\mathbf{w} - \mathbf{w}_0)^{+T} \tilde{G}^T [\gamma] \tilde{G} \mathbf{w} > 0,$$

if $\mathbf{w} > \mathbf{w}_0 = const$ at least for one $\mathbf{x}_i \in \Omega$, as long as the Voronoi faces related to each edge and subdomain fulfill

$$\sum_{E_j^N \ni \mathbf{e}_{ik}, E_j^N \in \Omega_l} \partial V_{ik} \geq 0. \tag{11}$$

Charge Transport in Semiconductors and a Finite Volume Scheme

This is exactly the requirement fulfilled by a 'boundary conforming Delaunay mesh' and has to be preserved for acceptable averages $\bar{\varepsilon}_{ij}$ in case of $\varepsilon = \varepsilon(x, n, p, |\nabla w|, \ldots)$. In the sequel the notation

$$G := [\sqrt{\gamma}]\tilde{G},$$

is used to denote the discrete gradient and summation (elements, edges, nodes) is not indicated any more—the context should indicate a local or global use.

The continuity equations can be transformed by changing variables $n = n_i e^w u$, $p = n_i e^{-w} v$, hence the steady state case reads:

$$-\nabla \cdot n_i D_n e^w \nabla u + R = 0, \tag{12}$$

$$-\nabla \cdot n_i D_p e^{-w} \nabla v + R = 0, \tag{13}$$

with an elliptic main part for bounded electrostatic potentials. Application of the discretization scheme and integration along each edge (the term $w_k - w_i$ in the discrete current expression (7) is just a special case of integrating the equation $w(s)'' = 0$ along the edge from s_i to s_k) using
- $(\bar{\mu} e^{w(x)}(e^{-\phi})')' = 0$, $\bar{\mu}$ edge average,
- $w(x)$ piecewise linear,
- $\text{sh}(s) := \sinh(s)/s$, $b(2s) = e^{-s}/\text{sh}(s) = 2s/(e^{-2s} - 1)$, b Bernoulli function:

$$G^T[\varepsilon]Gw = [V]g(C, \mathbf{n}, \mathbf{p}), \; \mathbf{g} = \mathbf{C} - \mathbf{n} + \mathbf{p}, \; \mathbf{n} = n_i[e^w]\mathbf{u}, \; \mathbf{p} = n_i[e^{-w}]\mathbf{v}, \tag{14}$$

$$A_{S_n}(D_n, \mathbf{w})e^{-\phi_n} = G^T[\bar{D}_n e^{\tilde{w}}/\text{sh}(\tilde{G}\mathbf{w}/2)]G\mathbf{u} = [V][r(\mathbf{x}, \mathbf{n}, \mathbf{p})](\mathbf{1} - [v]\mathbf{u}), \tag{15}$$

$$A_{S_p}(D_p, -\mathbf{w})e^{\phi_p} = G^T[\bar{D}_p e^{-\tilde{w}}/\text{sh}(\tilde{G}\mathbf{w}/2)]G\mathbf{v} = [V][r(\mathbf{x}, \mathbf{n}, \mathbf{p})](\mathbf{1} - [u]\mathbf{v}). \tag{16}$$

The diagonal transformations $\mathbf{n} = [n_i e^w]\mathbf{u}$, $\mathbf{p} = [n_i e^{-w}]\mathbf{v}$ yield the well known Scharfetter-Gummel (Il'in) scheme, dating back to Allen and Southwell ([1, 18, 22], see [4], too), generalized to boundary conforming Delaunay grids and used since the early eighties in semiconductor device simulations (compare [2, 23]).

The thermodynamic equilibrium solution is given by ($\mathbf{w}^*$, $\mathbf{u}^* = \mathbf{1}$, $\mathbf{v}^* = \mathbf{1}$) and $\mathbf{w}^*$ solution of (14) with $\mathbf{u} = \mathbf{u}^*$, $\mathbf{v} = \mathbf{v}^*$.

The proof of bounded steady state solutions for the system (14, 15, 16) proceeds in the following steps (for details see [10]):
a) for $-\infty < \check{w}_0 \leq w \leq \hat{w}_0 < \infty$ matrices A_{S_n}, A_{S_p} are weakly diagonally dominant (positive Dirichlet boundary measure).
b) for some $\check{u} \leq u_i^0 \leq \hat{u}$, $\check{v} \leq v_i^0 \leq \hat{v} \; \forall \; x_i \in \bar{\Omega}$, $\check{C} = \min(C(x))$ and $\hat{C} = \max(C(x))$, the extreme boundary values, and the monotone mapping with respect to w_j, $g_j(C_j, n_j, p_j) = V_j(C_j - n_j + p_j)$ (right hand side of the discrete Poisson equation) allow to bound the solution of (14) by construction contradictions using the weak discrete maximum principle.

c) supposing properly chosen bounds for $\tilde{w} = \max |w_i|$ one proves $e^{-\tilde{w}} \le \mathbf{u}^0$, $\mathbf{v}^0 \le e^{\tilde{w}}$ using the maximum principle and the properties of the recombination-generation term, hence $e^{-\tilde{w}}$, $e^{\tilde{w}}$ is a lower, upper solution respectively for equations (15, 16) with properly frozen $\mathbf{u}^0$, $\mathbf{v}^0$ in $r(x, \mathbf{u}^0, \mathbf{v}^0)$ and $(1 - [v^0]\mathbf{u}^0)$ $(1 - [u^0]\mathbf{v}^0)$ in (15, 16) respectively.

d) Brouwer's fixed point theorem guarantees the existence of at least one fixed point. The bounds are identical with the analytic ones.

The discrete dissipation expression is obtained from testing the discrete equations by the discrete quasi-Fermi potentials.

The uniqueness in the neighborhood of the discrete thermodynamic equilibrium follows by linearization at $(\mathbf{w}^*, \mathbf{u}^*, \mathbf{v}^*)$, resulting in a decoupling of the continuity equations and the Poisson equation, the Schur complement due to elimination of $\mathbf{u}$ is a weakly diagonally dominant M-matrix (compare [10]).

3 Example

Silicon detectors for high energy and astrophysics are nice examples to stress the algorithms used: each new detector is in some sense an extreme design for one special purpose. The new X-ray lasers for instance require high speed, low power, high spatial and energy resolution and the best possible signal to noise ratio. Extreme charge conservation requirements (see Fig. 4) in the interesting parts of the detector, hence in the computations, are typical.

The example shown is a pnCCD for SLAC's Linac Coherent Light Source designed at the MPI HLL, Munich. The Fig. 1 shows the relative simple geometry of two quarter CCD registers, two times two half CCD registers, and again two quarter CCD registers. This is the minimal configuration for testing the charge shift properties of the CCD (Fig. 3). Questions of interest are:

- The maximum number of electrons to be stored in one register (see Fig. 2)?
- Appeare losses to the surface, hence recombination with holes from contacts?
- How fast is shifting by changing boundary conditions on top of the registers?
- Do electrons stay in the start register, reach all the aim register, see Fig. 4?

The computations predicted the possibility to store 5 to 10 times more electrons in an optimized pnCCD. That was verified by experiments just now.

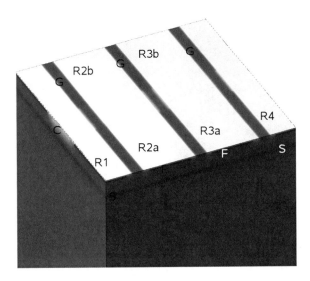

Fig. 1 Doping as equivalent equilibrium potential, white negative, dark positive, R1 quarter of a register, R2a one half register, side a is separated by the 'channel stop' C from side b, R3b one half register, F floating region to create a potential minimum beneath each register in the shift layer S, G gate contact to tune register separation in shift direction; dimensions $x = 75^-$m, $y = 75^-$m, $z = 150^-$m, 958 399 nodes, BACK: -50V, REGISTER 1, 3, 4: -18V, REGISTER 2: -10V, GATE 1, 2, 3: 5V

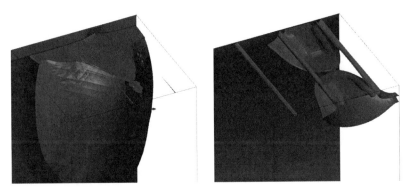

Fig. 2 Overflow of electrons at the arrival of the charge cloud created close to the bottom in the center of register 2 at y_{max}, $10n_i$ iso-surface of $\log(n)$ (left), weakest point in the potential barrier (properly selected electrostatic potential iso-surface, right)

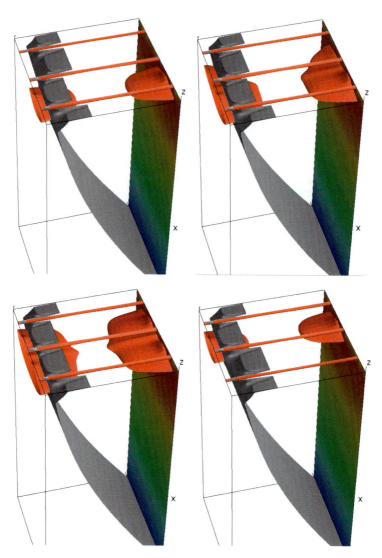

Fig. 3 In computations one can not shift the electrons 1000 times in one direction, hence a minimal configuration is used, to shift them back and forth. Shifting of electrons (not shown) takes place inside the moving potential barriers (iso-surfaces) due to time dependent boundary conditions at register 2 and 3 (compare the plotted electrostatic potential elevation over the $y = y_{max}$ surface). Initial state, the electrons are inside the iso-surface centered at register 2 (top left), the boundary value at register 3 reduces the potential barrier between register 2 and 3 (top right), both registers are 'open' (bottom left), register 2 has pushed out the electrons to register 3 (bottom right). Graphics by `gltools`

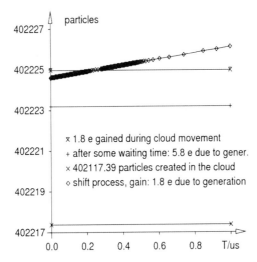

Fig. 4 The electron balance in the volume of interest is one crucial point: after order 1000 shift operations total losses of 0.1% are acceptable in the detector. The charge balance in the computations can be explained up to one third missing electron (out of 402225) after two integrations over 7 orders in time or 668 time steps

Acknowledgements The author has to thank the European XFEL for collaboration and partially funding the work. MPI HLL at Munich is a partner with interesting problems since many years.

References

1. Allen, D.N., Southwell, R.V.: Relaxation methods applied to determine the motion, in two dimensions, of a viscous fluid past a fixed cylinder. Quart. J. Mech. and Appl. Math. **8**, 129–145 (1955)
2. Bank, R., Rose, D., Fichtner, W.: Numerical methods for semiconductor device simulation. SIAM Journal on Scientific and Statistical Computing **4**(3), 416–435 (1983)
3. Delaunay, B.: Sur La Sphére Vide. Izvestia Akademii Nauk SSSR. Otd. Matem. i Estestv. Nauk **7**, 793–800 (1934)
4. Eymard, R., Fuhrmann, J., Gärtner, K.: A finite volume scheme for nonlinear parabolic equations derived from one-dimensionsl local Dirichlet problems. Numer. Math. **102**, 463–495 (2006)
5. Gabriel, K., Sokal, R.: A new statistical approach to geographic analysis. Systematic Zoology **18**, 259–278 (1969)
6. Gajewski, H.: On existence, uniqueness and asymptotic behavior of solutions of the basic equations for carrier transport in semiconductors. Z. Angew. Math. Mech. **65**, 101–108 (1985)
7. Gajewski, H., Gärtner, K.: On the discretization of van Roosbroeck's equations with magnetic field. Z. Angew. Math. Mech. **76**, 247–264 (1996)
8. Gajewski, H., Gröger, K.: On the basic equations for carrier transport in semiconductors. J. Math. Anal. Appl. **113**, 12–35 (1986)

9. Gajewski, H., Gröger, K.: Initial-boundary value problems modelling heterogeneous semiconductor devices. In: Surveys on Analysis, Geometry and Mathematical Physics, *Teubner-Texte Math.*, vol. 117, pp. 4–53. Teubner, Leipzig (1990)
10. Gärtner, K.: Existence of bounded discrete steady state solutions of the van roosbroeck system on boundary conforming delaunay grids. SIAM J. Sci. Comput. **31**, 1347–1362 (2009)
11. Glitzky, A.: Exponential decay of the free energy for discretized electro-reaction-diffusion systems. Nonlinearity **21**, 1989–2009 (2008)
12. Glitzky, A.: Uniform exponential decay of the free energy for Voronoi finite volume discretized reaction-diffusion systems. Preprint 1443, Weierstraß-Institut für Angewandte Analysis und Stochastik, Berlin (2009, to appear in Math. Nachr.)
13. Glitzky, A., Gärtner, K.: Energy estimates for continuous and discretized electro-reaction-diffusion systems. Nonlinear Analysis **70**, 788–805 (2009)
14. Glitzky, A., Gärtner, K.: Existence of bounded steady state solutions to spin-polarized drift-diffusion systems. SIAM J. Math. Anal. **41**, 2489–2513 (2010)
15. Glitzky, A., Hünlich, R.: Energetic estimates and asymptotics for electro–reaction–diffusion systems. Z. Angew. Math. Mech. **77**, 823–832 (1997)
16. Gokhale, B.: Numerical solutions for a one-dimensional sislicon p-n-p transistor. IEEE Trans. Electron Devices **ED-17**, 594–602 (1970)
17. Horn, F., Jackson, R.: General mass action kinetics. Arch. Rat. Mech. Anal. **47**, 81–116 (1972)
18. Il'in, A.M.: A difference scheme for a differential equation with a small parameter multiplying the second derivative. Matematičeskije zametki **6**, 237–248 (1969)
19. Jerome, J.W.: Consistency of Semiconductor Modeling: An Existence/Stability Analysis for the Stationary Van Roosbrook System. SIAM J. Appl. Math. **45**, 565–590 (1985)
20. Markowich, P.A.: The Stationary Semiconductor Device Equations. Springer, Wien (1986)
21. Mock, M.S.: Analysis of Mathematical Models of Semiconductor Devices. Boole Press, Dublin (1983)
22. Scharfetter, D.L., Gummel, H.K.: Large–signal analysis of a silicon read diode oscillator. IEEE Trans. Electr. Dev. **16**, 64 – 77 (1969)
23. Selberherr, S.: Analysis and Simulation of Semiconductor Devices. Springer, Wien-New York (1984)

The paper is in final form and no similar paper has been or is being submitted elsewhere.

Playing with Burgers's Equation

T. Gallouët, R. Herbin, J.-C. Latché, and T.T. Nguyen

Abstract The 1D Burgers equation is used as a toy model to mimick the resulting behaviour of numerical schemes when replacing a conservation law by a form which is equivalent for smooth solutions, such as the total energy by the internal energy balance in the Euler equations. If the initial Burgers equation is replaced by a balance equation for one of its entropies (the square of the unknown) and discretized by a standard scheme, the numerical solution converges, as expected, to a function which is not a weak solution to the initial problem. However, if we first add to Burgers' equation a diffusion term scaled by a small positive parameter ϵ before deriving the entropy balance (this yields a non conservative diffusion term in the resulting equation), and then choose ϵ and the discretization parameters adequately and let them tend to zero, we observe that we recover a convergence to the correct solution.

Keywords Burgers equation, compressible flows, Euler equations, finite volumes
MSC2010: 65M08,76N99

1 Introduction

Computer codes developed for the simulation of inviscid and non heat-conducting compressible flows are in general based on the conservative form of the Euler equations, which read in the one-dimensional case:

T. Gallouët and R. Herbin
Université Aix-Marseille, e-mail: [gallouet,herbin]@cmi.univ-mrs.fr

J.-C. Latché and T.T. Nguyen
Institut de Radioprotection et de Sûreté Nucléaire (IRSN), e-mail: [jean-claude.latche,tan-trung.nguyen]@irsn.fr

$$\partial_t \rho + \partial_x(\rho u) = 0, \tag{1a}$$

$$\partial_t(\rho u) + \partial_x(\rho u^2) + \partial_x p = 0, \tag{1b}$$

$$\partial_t E + \partial_x\big((E+p)u\big) = 0, \tag{1c}$$

where t stands for the time, ρ, u and p are the density, velocity and pressure in the flow, and E stands for the total energy, $E = \rho u^2/2 + \rho e$, with e the internal energy. This system must be complemented by an equation of state, giving for instance the pressure as a function of the density and the internal energy $p = \wp(\rho, e)$.

For physical reasons, the density and internal energy must be non-negative (in usual applications, positive). In addition, for the continuous problem as well as, at the discrete level, for a wide range of schemes (the so-called conservative schemes), the non-negativity of these variables allows a (weak) control on the solution; assuming that ρ and E are known on the parts of the boundary where the flow is entering the computational domain, Equations (1a) and (1c) indeed yield an $L^\infty(0,T;L^1(\Omega))$-estimate (with $\Omega \times (0,T)$ the space-time domain of computation) for the density and the total energy respectively. The positivity of the density at the discrete level is easily obtained from a convenient discretization of (1a). The positivity of the internal energy does not seem easily obtained other than by replacing Equation (1c) by a balance equation for the internal energy in the discrete problem; this balance equation is formally derived (*i.e.* supposing that the solution is regular) from (1b) and (1c) and reads:

$$\partial_t(\rho e) + \partial_x(\rho e u) + p\partial_x u = 0. \tag{2}$$

In this relation, the discrete convection operator may be built so as to respect the positivity of e: provided that the equation of state is such that for any value of ρ, p vanishes for $e = 0$, testing the discrete counterpart of (2) by the negative part of e proves $e \geq 0$ (see [5] for the initial paper, [2, Appendix B] for another proof suitable in this context, and [4] in the framework of the compressible Navier-Stokes equations).

Instead of Equation (1c), one may also prefer to use a conservation equation for the physical entropy s, because this equation (derived for regular solutions) is a simple transport equation:

$$\partial_t(\rho s) + \partial_x(\rho s u) = 0. \tag{3}$$

Let us then consider that, for computational efficiency or robustess reasons, (2) or (3) are prefered to (1c). Since both (2) and (3) are derived from (1c) assuming a regular solution, there is no reason for their discretization to yield the correct weak solutions in the presence of shocks. Nevertheless, we may reasonably expect to recover the correct shock solutions if we use the following strategy:

(i) regularize the problem by adding a small diffusion term,
(ii) derive the counterpart of (2) or (3) taking into account the diffusion terms,

(iii) solve these equations,
(iv) let ϵ tend to zero.

Of course, step (*iii*) is performed numerically, and convergence is monitored by the space and time discretization steps h and k; the question which arises is then to find a convenient way to let ϵ and the numerical parameters h and k tend to zero. The aim of this paper is to perform numerical experiments in order to investigate this issue on a toy problem, namely the inviscid Burgers equation. Note that we only consider explicit schemes in this study.

2 The equations and the numerical schemes

The inviscid Burgers equation reads:

$$\partial_t u + \partial_x (u^2) = 0, \qquad \text{for } x \in \mathbb{R},\ t \in (0, T), \tag{4}$$

which we complement with the initial condition:

$$u(x, 0) = u_0(x), \qquad \text{for } x \in \mathbb{R}. \tag{5}$$

Following the above mentioned strategy (items (i)-(iv)), we first add to (4) a viscous term, to obtain: $\partial_t u + \partial_x (u^2) - \epsilon \partial_{xx} u = 0$. Now, multiplying this relation by $2u$ yields the following perturbed equation:

$$\partial_t u^2 + \frac{4}{3} \partial_x u^3 - 2u\epsilon \partial_{xx} u = 0. \tag{6}$$

For $\varepsilon = 0$, we get the following "Burgers square entropy" equation:

$$\partial_t u^2 + \frac{4}{3} \partial_x u^3 = 0. \tag{7}$$

which also reads, setting $v = u^2$:

$$\partial_t v + \frac{4}{3} \partial_x (v^{\frac{3}{2}}) = 0. \tag{8}$$

We consider the following initial data, chosen such that the entropy solution of (4)-(5) contains a discontinuity:

$$u_0(x) = \begin{cases} 10, & x \leq -0.25 \\ 1, & x > -0.25 \end{cases}. \tag{9}$$

It is well known that for such an initial condition, the entropy weak solutions of equations (4) and (7) differ. Let us then turn to their numerical approximations.

Since the chosen initial data (9) is positive, the celebrated Godunov scheme reduces for both equations to the classical upwind scheme, thanks to the fact that the upwind scheme preserves (for these equations) the sign of the solution; it is well known that it leads to an approximate solution which converges, under a so called CFL condition, to the exact solution as the discretization parameters go to zero [1] (note that this is not the case for the centred finite volume scheme, although it is conservative). For the sake of simplicity, we consider constant time and space steps h and k. For $i \in \mathbb{Z}$, we set $x_i = ih$ and for $n \in \{0, \ldots, M\}$, with $(M-1)k < T \leq Mk$, we set $t_n = nk$. The discrete unknowns are the real numbers $u_i^{(n)}$, with $i \in \mathbb{Z}$ and $n \in \{0, \ldots, M\}$. The values $u_i^{(0)}$ are obtained with the initial condition:

$$u_i^{(0)} = \frac{1}{h}\int_{x_i-\frac{h}{2}}^{x_i+\frac{h}{2}} u_0(x)dx. \tag{10}$$

Since the discrete solution is positive, the upwind scheme for Equation (4) reads:

$$u_i^{(n)} = u_i^{(n-1)} + \frac{k}{h}\left[\left(u_{i-1}^{(n-1)}\right)^2 - \left(u_i^{(n-1)}\right)^2\right]. \tag{11}$$

For this particular problem and scheme, the maximum value for the solution is reached at the initial time step so that the CFL number is the number G such that:

$$k = G\frac{h}{\max\{2s,\ s \in [1, 10]\}} = G\frac{h}{20}. \tag{12}$$

Similarly, the upwind scheme for Equation (8) reads:

$$v_i^{(n)} = v_i^{(n-1)} + \frac{4k}{3h}\left[\left(v_{i-1}^{(n-1)}\right)^{\frac{3}{2}} - \left(v_i^{(n-1)}\right)^{\frac{3}{2}}\right], \tag{13}$$

and the CFL number is the same number G. The numerical solutions obtained with (11) for the Burgers equation (4) and with (13) for the Burgers square entropy equation (7) are depicted in Fig. 1. Both are obtained with CFL equal to 1, for $T = 1/20$ and with various values of N, starting from $N = 200$ and multiplying successively by two the number of cells up to $N = 1600$. As expected, the upwind scheme (13) yields a numerical solution which converges (as the discretization parameters go to zero and under a CFL condition) to a weak solution of (7) (and even to its entropy solution), which is not a weak solution of (4), since the Rankine-Hugoniot conditions differ. At time $T = 1/20$, the shock for the solution of (4) is located at $x = 0.3$, while the shock of the solution of (7) is located at $x > 0.4$.

Remark 1 (Link with a non-conservative diffusion term). For the Burgers equation (4), upwinding may be seen as adding a diffusion, namely discretizing (since $u > 0$):

$$\partial_t u + \partial_x(u^2) - \partial_x((hu - 2ku^2)\partial_x u) = 0.$$

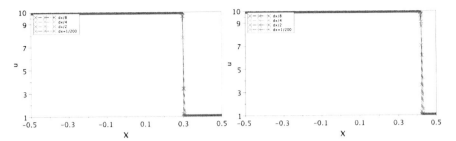

Fig. 1 Upwind Scheme for (4)-(9) (left) and (7)-(9) (right) with different mesh sizes, $CFL = 1$

Note that one has $hu - 2ku^2 \geq 0$ thanks to the CFL condition. For the Burgers square entropy equation (7), upwinding may be seen, formally, as solving the following parabolic equation (since $u > 0$): $\partial_t u^2 + (4/3)\partial_x(u^3) - \partial_x((2hu^2 - 4ku^3)\partial_x u) = 0$. This equation is equivalent to the following parabolic perturbation of the Burgers equation:

$$\partial_t u + \partial_x(u^2) - \frac{1}{u}\partial_x((hu^2 - 2ku^3)\partial_x u) = 0.$$

The third term at the left-hand side may be seen as a numerical diffusion (thanks to the CFL condition) which is not in a conservative form, because of the factor $1/u$. The above numerical results show that such a non conservative diffusion may lead to wrong discontinuities.

3 Numerical solution of the perturbed equation

We then discretize the perturbed equation (6) with $\epsilon = \epsilon_0 h^\alpha$, where $\epsilon_0 > 0$ and $\alpha > 0$ are fixed. Note that, setting $v = u^2$, (6) can also be recast as:

$$\partial_t v + \frac{4}{3}\partial_x(v^{\frac{3}{2}}) - v^{\frac{1}{2}}\epsilon_0 h^\alpha \partial_x(v^{-\frac{1}{2}}\partial_x v) = 0,$$

that is a nonlinear hyperbolic equation augmented with a nonlinear nonconservative diffusion term. The upwind finite volume discretization of this equation reads (in the u variable), with $u_i^{(0)}$ given by (10),

$$\left(u_i^{(n)}\right)^2 = \left(u_i^{(n-1)}\right)^2 + \frac{4k}{3h}\left[\left(u_{i-1}^{(n-1)}\right)^3 - \left(u_i^{(n-1)}\right)^3\right]$$
$$+ \frac{k}{h^2}\epsilon_0 h^\alpha u_i^{(n-1)}\left[u_{i-1}^{(n-1)} - 2u_i^{(n-1)} + u_{i+1}^{(n-1)}\right]. \quad (14)$$

We present in Figs. 2, 3 and 4 the numerical solutions obtained with (14) for $\alpha = 0.5$, $\alpha = 1$ and $\alpha = 2$ respectively, and for the same time $T = 1/20$,

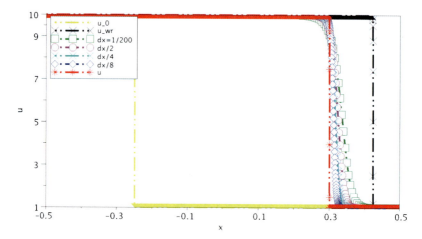

Fig. 2 Upwind Scheme for (6) with non conservative diffusion term, $\alpha = 0.5$

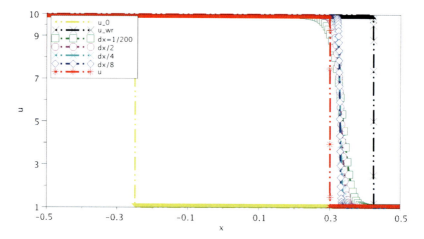

Fig. 3 Upwind Scheme for (6) with non conservative diffusion term, $\alpha = 1$

CFL $= 0.1$ and meshes as in Sect. 2. The parameter ϵ_0 is such that $\epsilon_0 h^\alpha = 0.2$ for $N = 200$ (whatever α may be). Figure 2 shows that for $0 < \alpha < 1$, the sequence of approximate solutions given by (14) converges to a weak solution of the initial Burgers equation (4), as h and k tend to 0, under a stability condition, which, since $\alpha < 1$, becomes more stringent than a CFL condition when h tends to zero. Figure 3 shows that for $\alpha > 1$, we obtain the convergence to the solution of (7); Fig. 4 shows that for $\alpha = 1$, the location of the discontinuity lies in between the discontinuities of the solution to (6) and (7). These results seem to indicate that the convergence to the solution of (7) (resp. (6)) occurs when the added diffusion dominates (resp. is dominated by) the numerical one.

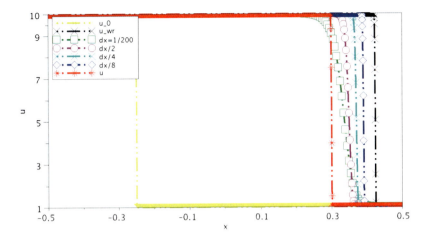

Fig. 4 Upwind Scheme for (6) with non conservative diffusion term, $\alpha = 2$

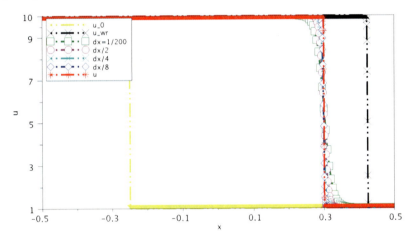

Fig. 5 Centered Scheme for (6) with non conservative diffusion term, $\alpha = 1$

Let us finally study the following finite volume centred scheme for Equation (7), which reads:

$$\left(u_i^{(n)}\right)^2 = \left(u_i^{(n-1)}\right)^2 + \frac{4k}{3h}\left[\left(\frac{u_{i-1}^{(n-1)} + u_i^{(n-1)}}{2}\right)^3 - \left(\frac{u_i^{(n-1)} + u_{i+1}^{(n-1)}}{2}\right)^3\right]$$
$$+ \frac{k}{h^2}\epsilon_0 h^\alpha u_i^{(n-1)}\left[u_{i-1}^{(n-1)} - 2u_i^{(n-1)} + u_{i+1}^{(n-1)}\right]. \quad (15)$$

Results for $\alpha = 1$, $\alpha = 1.5$ and $\alpha = 2$ (and ϵ_0 such that $\epsilon_0 h^\alpha = 0.2$ for $N = 200$, whatever α may be) are reported on Figs. 5, 6 and 7, respectively. The numerical

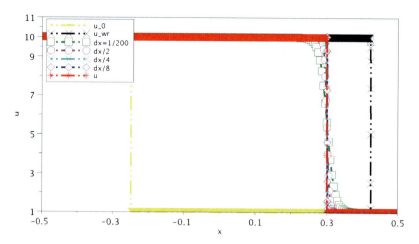

Fig. 6 Centered Scheme for (6) with non conservative diffusion term, $\alpha = 1.5$

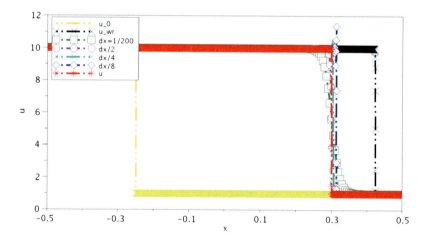

Fig. 7 Centered Scheme for (6) with non conservative diffusion term, $\alpha = 2$

solution now seems to converge to the solution of (7), at least for $\alpha \in (0, 2)$. For the finest mesh and $\alpha = 2$, the diffusion is no longer sufficient to prevent some spurious oscillations near the shock. Last but not least, the additional diffusion which is necessary to recover the right shock location is considerably reduced with respect to the upwind scheme (even if the scheme still appears more diffusive than the standard upwind scheme applied to (4)), which is encouraging in view of practical extensions to Euler equations.

Conclusion We tested two discretizations for the modified equation (6):

- an upwind scheme for which the solution converges to the weak solution of (4) if the viscous term is predominant with respect to the numerical diffusion, that is if $\epsilon = \epsilon_0 h^\alpha$, with $\epsilon_0 > 0$ and $\alpha \in (0, 1)$.
- a centred scheme which yields correct solutions for all values $\alpha \in (0, 2)$.

The extension of this work to Euler equations is under way, and results are encouraging. Indeed, it seems that we are able to build convergent schemes, even in the presence of shocks, using either the entropy or internal energy balance. A next step might be to use a nonlinear viscosity to avoid an excessive smearing of the solutions, following the ideas developed in [3].

References

1. R. Eymard, T. Gallouët, R. Herbin: Finite Volume Methods. Handbook of Numerical Analysis, Volume VII, North Holland (2000).
2. T. Gallouët, A. Larcher, J.-C. Latché: Convergence of a finite volume scheme for the convection-diffusion equation with L^1 data. To appear in Mathematics of Computation (2011).
3. J.-L. Guermond, R. Pasquetti, B. Popov: Entropy viscosity methods for nonlinear conservation laws. To appear in Journal of Computational Physics (2011).
4. R. Herbin, W. Kheriji, J.-C. Latché: An unconditionally stable Finite Element-Finite Volume pressure correction scheme for compressible Navier-Stokes equations. In preparation (2011).
5. B. Larrouturou: How to preserve the mass fractions positivity when computing compressible multi-component flows. Journal of Computational Physics, **95**, 59–84 (1991).

The paper is in final form and no similar paper has been or is being submitted elsewhere.

On Discrete Sobolev–Poincaré Inequalities for Voronoi Finite Volume Approximations

Annegret Glitzky and Jens A. Griepentrog

Abstract We prove a discrete Sobolev–Poincaré inequality for functions with arbitrary boundary values on Voronoi finite volume meshes. We use Sobolev's integral representation and estimate weakly singular integrals in the context of finite volumes. We establish the result for star shaped polyhedral domains and generalize it to the finite union of overlapping star shaped domains.

Keywords Discrete Sobolev inequality, Sobolev integral representation, Voronoi finite volume mesh
MSC2010: 46E35, 46E39, 31B10

1 Introduction and notation

In this paper we study discrete Sobolev inequalities. In the continuous situation the Sobolev embedding estimates

$$\|u\|_{L^q(\Omega)} \leq C_q \|u\|_{H^1(\Omega)} \quad \forall u \in H^1(\Omega) \tag{1}$$

for $q \in [1, \infty)$ in two space dimensions and for $q \in [1, 2n/(n-2)]$ in $n \geq 3$ space dimensions are well known [1, 10, 15].

For the finite volume discretized situation some results can be found in [3, 6]. But these estimates concern only the case of zero boundary values. The two-dimensional case for admissible finite volume meshes (see [6, Definition 9.1]) is treated in [6, Lemma 9.5]. The corresponding three-dimensional result is proved in [3, Lemma 1]. For $p \in [1, 2]$, a discrete Sobolev inequality estimating the L^{p^*}-

---

Annegret Glitzky and Jens A. Griepentrog
WIAS, Mohrenstr. 39, D-10117 Berlin, Germany, e-mail: glitzky@wias-berlin.de, griepent@wias-berlin.de

norm (where $p^* = np/(n-p)$ if $p < n$ and $p^* < \infty$ if $n = p = 2$) by the discrete $W^{1,p}$-norm is presented in [5, Proposition 2.2]. Moreover, for the zero boundary value case and $1 \leq p < \infty$, the discrete embedding of $W_0^{1,p}$ into L^q for some $q > p$, $1 \leq p < \infty$ is established in [7, sect. 5]. A corresponding result for discontinuous Galerkin methods working in the spaces of piecewise polynomial functions on general meshes is obtained in [4, Theorem 6.1]. The idea there is to follow Nirenberg's proof of Sobolev embeddings. Recently in [2], in the context of discontinuous Galerkin finite element methods, broken Sobolev–Poincaré inequalities were proved. There, known classical results in $BV(\Omega)$ and in Sobolev spaces $W^{1,p}(\Omega)$, together with local norm equivalence and global estimates for the reconstruction operator, lead to the desired estimates.

According to our knowledge and to the information of authors of the cited papers concerning finite volume schemes, finding discrete versions of the Sobolev inequality (1) for functions with arbitrary boundary values has been an open question up to now. Only a discrete Poincaré inequality ($q = 2$) is available in [6, Lemmas 10.2, 10.3] and [9, Lemma 4.2]. But in both papers the second step of the proof is done only for two space dimensions.

The aim of the present paper is to establish a discrete Sobolev–Poincaré inequality for functions with nonzero boundary values on Voronoi finite volume meshes. Such results can be applied to more general boundary value problems, for instance, to problems with inhomogeneous Dirichlet, Neumann, or mixed boundary conditions. The technique used here is an adaptation of Sobolev's integral representation and of the treatment of weakly singular integrals in the context of Voronoi finite volume meshes. The Voronoi property of the mesh essentially comes into play in the proofs of the potential theoretical results, Lemmas 1–3.

The plan of the paper is as follows. In the remainder of this section we introduce our notation. In Sect. 2 we formulate our assumptions and our main result, the discrete Sobolev–Poincaré inequality for star shaped domains (see Theorems 1 and 2 for a uniform estimate for a class of Voronoi finite volume meshes having comparable mesh quality). In Sect. 3 we collect three potential theoretical lemmas needed for the proof of our main result. In Sect. 4 we generalize the discrete Sobolev inequality to domains which are a finite union of overlapping star shaped domains (see Theorem 3). The last section contains some remarks concerning applications of discrete Sobolev inequalities.

Let $\Omega \subset B(0, R_0) \subset \mathbb{R}^n$, $n \in \mathbb{N}$, $n \geq 2$, be a bounded, open, polyhedral domain, and let $\partial\Omega$ be its boundary. We work with Voronoi finite volume meshes of Ω, and our notation is basically taken from [3, 6]. Moreover, for set valued arguments we write diam($\cdot$) for the diameter of the corresponding set. By mes($\cdot$) and mes$_d(\cdot)$ we denote the n- and d-dimensional Lebesgue measures, respectively.

A Voronoi finite volume mesh of Ω denoted by $\mathcal{M} = (\mathcal{P}, \mathcal{T}, \mathcal{E})$ is formed by a family of grid points $\mathcal{P}$ in $\overline{\Omega}$, a family $\mathcal{T}$ of Voronoi control volumes, and a family of relatively open parts of hyperplanes in $\mathbb{R}^n$ denoted by $\mathcal{E}$ (which represent the faces of the Voronoi boxes). For a Voronoi mesh we use the following notation:

For each grid point x_K of the set $\mathcal{P}$ the control volume K of the Voronoi mesh belonging to the point x_K is defined by

$$K = \{x \in \Omega : |x - x_K| < |x - x_L| \quad \forall x_L \in \mathcal{P}, \ x_L \neq x_K\}, \quad K \in \mathcal{T}.$$

For $K, L \in \mathcal{T}$ with $K \neq L$ either the $(n-1)$-dimensional Lebesgue measure of $\overline{K} \cap \overline{L}$ is zero or $\overline{K} \cap \overline{L} = \overline{\sigma}$ for some $\sigma \in \mathcal{E}$. In the latter case the symbol $\sigma = K|L$ denotes the Voronoi surface between K and L. We introduce the following subsets of $\mathcal{E}$: The sets of interior and external Voronoi surfaces are denoted by $\mathcal{E}_{int}$ and $\mathcal{E}_{ext}$, respectively. Additionally, for every $K \in \mathcal{T}$ we call $\mathcal{E}_K$ the subset of $\mathcal{E}$ such that $\partial K = \overline{K} \setminus K = \cup_{\sigma \in \mathcal{E}_K} \overline{\sigma}$. Then $\mathcal{E} = \cup_{K \in \mathcal{T}} \mathcal{E}_K$. Moreover, for $\sigma \in \mathcal{E}$ we use the following notation: m_σ represents the $(n-1)$-dimensional measure of the Voronoi surface σ, and x_σ corresponds to the coordinates of the center of gravity of σ.

For $\sigma = K|L \in \mathcal{E}_{int}$ let d_σ be the Euclidean distance between x_K and x_L. For $K \in \mathcal{T}, \sigma \in \mathcal{E}_K$ we define $d_{K,\sigma}$ to be the Euclidean distance between x_K and the hyperplane containing σ. Then, in the case of (isotropic) Voronoi meshes we have $d_{K,\sigma} = \frac{d_\sigma}{2}$ for $\sigma \in \mathcal{E}_{int}$.

We work with half-diamonds $D_{K\sigma} = \{tx_K + (1-t)y : t \in (0,1), \ y \in \sigma\}$, where $n \operatorname{mes}(D_{K\sigma}) = m_\sigma d_{K,\sigma}$. Then due to our definitions,

$$n \operatorname{mes}(K) = \sum_{\sigma \in \mathcal{E}_K} m_\sigma d_{K,\sigma} \quad \forall K \in \mathcal{T}.$$

The mesh size is defined by $\operatorname{size}(\mathcal{M}) = \sup_{K \in \mathcal{T}} \operatorname{diam}(K)$. We denote by $X(\mathcal{M})$ the set of functions from Ω to $\mathbb{R}$ which are constant on each Voronoi box of the mesh. For $u \in X(\mathcal{M})$ the value in the Voronoi box $K \in \mathcal{T}$ is denoted by u_K. Finally, for $u \in X(\mathcal{M})$ the discrete H^1-seminorm $|u|_{1,\mathcal{M}}$ of u is defined by

$$|u|_{1,\mathcal{M}}^2 = \sum_{\sigma \in \mathcal{E}_{int}} \frac{m_\sigma}{d_\sigma} (D_\sigma u)^2,$$

where $D_\sigma u = |u_K - u_L|$, u_K is the value of u in the Voronoi box K, and $\sigma = K|L$.

2 Main result

First we formulate our *assumptions on the geometry and the meshes* as follows:

Assumption 1. We assume that the open, polyhedral domain $\Omega \subset B(0, R_0) \subset \mathbb{R}^n$ is star shaped with respect to some ball $B(0, R)$.

Let the function $\rho : \mathbb{R}^n \to [0, 1]$ be given by

$$\rho(y) = \begin{cases} \exp\left\{-\frac{R^2}{R^2 - |y|^2}\right\} & \text{if } |y| < R, \\ 0 & \text{if } |y| \geq R. \end{cases}$$

We introduce the piecewise constant approximations $\rho^{\mathcal{M}} \in X(\mathcal{M})$ as

$$\rho_K^{\mathcal{M}}(x) = \min_{y \in \overline{K}} \rho(y) \quad \text{for } x \in K. \tag{2}$$

Assumption 2. Let $\mathcal{M} = (\mathcal{P}, \mathcal{T}, \mathcal{E})$ be a Voronoi finite volume mesh of Ω with the property that $\mathcal{E}_K \cap \mathcal{E}_{ext} \neq \emptyset$ implies $x_K \in \partial\Omega$. Moreover, the local mesh size near $B(0, R)$ is assumed to be so small that there exists a constant $\rho_0 > 0$ such that $\int_\Omega \rho^{\mathcal{M}}(x)\,dx \geq \rho_0$.

Let us remark that the constant ρ_0 in Assumption 2 can be fixed, for instance, if we demand that for some $r \leq R/4$ we have $\operatorname{diam}(K) < r$ for all $x_K \in \mathcal{P}$ with $x_K \in B(0, R)$. Then, for almost all $x \in B(0, r)$ we find

$$\rho^{\mathcal{M}}(x) \geq \exp\left\{-\frac{R^2}{R^2 - (2r)^2}\right\} \geq \exp\left\{-\frac{4}{3}\right\}$$

and

$$\int_\Omega \rho^{\mathcal{M}}(x)\,dx \geq \operatorname{mes}(B(0, r)) \exp\left\{-\frac{R^2}{R^2 - (2r)^2}\right\},$$

which can be taken as ρ_0 in Assumption 2.

Under Assumption 2 there exist minimal constants $\kappa_1(\mathcal{M}) > 0$, $\kappa_2(\mathcal{M}) \geq 1$ such that the geometric weights fulfill

$$0 < \operatorname{diam}(\sigma) \leq \kappa_1(\mathcal{M}) d_\sigma \quad \forall \sigma \in \mathcal{E}_{int} \tag{3}$$

and

$$\max_{\sigma \in \mathcal{E}_K \cap \mathcal{E}_{int}} \max_{x \in \overline{\sigma}} |x_K - x| \leq \kappa_2(\mathcal{M}) \min_{\sigma \in \mathcal{E}_K \cap \mathcal{E}_{int}} d_{K,\sigma} \quad \forall x_K \in \mathcal{P}. \tag{4}$$

Having in mind that

$$R_{K,out} := \max_{\sigma \in \mathcal{E}_K \cap \mathcal{E}_{int}} \max_{x \in \overline{\sigma}} |x_K - x|, \quad R_{K,inn} := \min_{\sigma \in \mathcal{E}_K \cap \mathcal{E}_{int}} d_{K,\sigma}$$

are the smallest radius of a circumscribed ball of K centered at x_K and the greatest radius of a ball fully contained in K and centered at x_K, respectively, inequality (4) implies that

$$R_{K,out} \leq \kappa_2(\mathcal{M}) R_{K,inn}.$$

Moreover, inequality (4) implies that

$$\max_{\sigma \in \mathcal{E}_K \cap \mathcal{E}_{int}} |x_K - x_\sigma| \leq \kappa_2(\mathcal{M}) \min_{\sigma \in \mathcal{E}_K \cap \mathcal{E}_{int}} d_{K,\sigma} \quad \forall x_K \in \mathcal{P}. \tag{5}$$

In this prescribed setting of a Voronoi finite volume mesh we establish the discrete Sobolev–Poincaré inequality:

On Discrete Sobolev–Poincaré Inequalities

Theorem 1. *Let Assumptions 1 and 2 be satisfied, and let $q \in [1,\infty)$ for $n = 2$ and $q \in [1, \frac{2n}{n-2})$ for $n \geq 3$, respectively. Then there exists a constant $c_q(\mathcal{M}) > 0$ depending only on n, q, Ω and the constants ρ_0, $\kappa_1(\mathcal{M})$, and $\kappa_2(\mathcal{M})$ such that*

$$\|u - m_\Omega(u)\|_{L^q(\Omega)} \leq c_q(\mathcal{M}) |u|_{1,\mathcal{M}} \quad \forall u \in X(\mathcal{M}),$$

where $m_\Omega(u) = \text{mes}(\Omega)^{-1} \int_\Omega u(x)\,dx$.

For the proof this theorem (see [14, sect. 4]) we adapt techniques used in [16, 17] to the discretized situation using Voronoi diagrams. To do so, we establish some discrete analogue for Sobolev's integral representation (see [16, sect. 116]) and of the treatment of weakly singular integral operators (see [16, sect. 115]).

Note that for $n \geq 3$, the discrete version of the Sobolev embedding $H^1(\Omega) \hookrightarrow L^{2n/(n-2)}(\Omega)$ for the critical Sobolev exponent can not be obtained by using only Sobolev's integral representation. This is exactly the same situation as for the continuous case (see [10, Chap. 7.8], [16, sect. 114–116], [17, sect. 8]).

We generalize our result to a class of Voronoi finite volume meshes having a unified mesh quality. Namely, we additionally assume the following for the meshes:

Assumption 3. There exist constants $\kappa_1 > 0$ and $\kappa_2 \geq 1$ such that the geometric weights fulfill $0 < \text{diam}(\sigma) \leq \kappa_1 d_\sigma$ for all $\sigma \in \mathcal{E}_{int}$ and $\max_{\sigma \in \mathcal{E}_K \cap \mathcal{E}_{int}} |x_K - x_\sigma| \leq \kappa_2 \min_{\sigma \in \mathcal{E}_K \cap \mathcal{E}_{int}} d_{K,\sigma}$ for all $x_K \in \mathcal{P}$.

Now we can formulate the main theorem of our paper, the discrete Sobolev inequality uniformly on a class of Voronoi finite volume meshes $\mathcal{M}$ characterized by Assumptions 2 and 3:

Theorem 2. *Let Ω be an open bounded polyhedral subset of $\mathbb{R}^n$, and let $\mathcal{M}$ be a Voronoi finite volume mesh such that additionally Assumptions 1–3 are fulfilled. Let $q \in [1,\infty)$ for $n = 2$ and $q \in [1, \frac{2n}{n-2})$ for $n \geq 3$, respectively. Then there exists a constant $c_q > 0$ depending only on n, q, Ω and the constants in Assumptions 1–3 such that*

$$\|u - m_\Omega(u)\|_{L^q(\Omega)} \leq c_q |u|_{1,\mathcal{M}} \quad \forall u \in X(\mathcal{M}).$$

Note that the constant c_q in Theorem 2 depends on the fixed constants κ_1, κ_2 from Assumption 3 instead of $\kappa_1(\mathcal{M})$, $\kappa_2(\mathcal{M})$. The dependency on ρ_0 is of the same quality as in Theorem 1.

3 Potential theoretical lemmas

In this section we introduce three potential theoretical lemmas which are essential for the proof of the discrete Sobolev–Poincaré inequality, Theorem 1. The proofs of these lemmas can be found in [14, sect. 5].

Lemma 1. *Let $n \in \mathbb{N}$, $n \geq 2$ and Assumptions 1 and 2 be satisfied. Let $x_{K_0} \in \mathcal{P}$ be a fixed grid point and let $\sigma \in \mathcal{E}_{int}$ be an internal Voronoi surface with gravitational* center x_σ. Then

$$\mathrm{mes}\big(\{x \in B(0, R) : [x_{K_0}, x] \cap \sigma \neq \emptyset\}\big) \leq A_n \frac{m_\sigma}{|x_{K_0} - x_\sigma|^{n-1}},$$

where $A_n := \frac{1}{n} \max\{2, 4\kappa_1(\mathcal{M})\}^{n-1} \mathrm{diam}(\Omega)^n$.

Lemma 2. *Let $n \in \mathbb{N}$, $n \geq 2$ and Assumptions 1 and 2 be satisfied. Let $q \in (2, \infty)$ for $n = 2$ and $q \in (2, \frac{2n}{n-2})$ for $n \geq 3$. Moreover, let $\beta > 0$ be given by $2\beta = \frac{n}{q} - \frac{n}{2} + 1$, and let $x_{K_0} \in \mathscr{P}$ be a fixed grid point. Then*

$$\sum_{K \in \mathscr{T}} \sum_{\sigma \in \mathscr{E}_K} |x_{K_0} - x_\sigma|^{-n+2\beta} m_\sigma d_{K,\sigma} \leq B_n,$$

where $B_n := \frac{n}{2\beta} \max\{1 + 2\kappa_1(\mathcal{M}), 2\}^{n-2\beta} (2R_0)^{2\beta} \mathrm{mes}_{n-1}(\partial B(0, 1))$.

Lemma 3. *Let $n \in \mathbb{N}$, $n \geq 2$ and Assumptions 1 and 2 be satisfied. Let $q \in (2, \infty)$ for $n = 2$ and $q \in (2, \frac{2n}{n-2})$ for $n \geq 3$. Moreover, let $\beta > 0$ be given by $2\beta = \frac{n}{q} - \frac{n}{2} + 1$, let $\sigma \in \mathscr{E}_{int}$ be a fixed inner Voronoi surface, and let x_σ denote its center of gravity. Then*

$$\sum_{K_0 \in \mathscr{T}} \sum_{\sigma_0 \in \mathscr{E}_{K_0}} |x_{K_0} - x_\sigma|^{-n+q\beta} m_{\sigma_0} d_{K_0,\sigma_0} \leq D_n,$$

where $D_n := \frac{n}{q\beta}\big(1 + \kappa_2(\mathcal{M})(1 + 2\kappa_1(\mathcal{M}))\big)^{n-q\beta} (2R_0)^{q\beta} \mathrm{mes}_{n-1}(\partial B(0, 1))$.

4 Sobolev–Poincaré inequalities for more general domains

In this section we discuss how the results of Theorems 1 and 2, which hold true for star shaped domains Ω, can be used to obtain assertions for a more general situation. In the nondiscretized situation the result can be carried over to domains Ω, which are a finite union of star shaped domains Ω_i (see [16, sect. 118], [17, pp. 69–70]). In our discretized situation we assume the following:

Assumption 4. The open, connected, polyhedral domain $\Omega \subset B(0, R_0)$ is a finite union of open, polyhedral sets Ω_i, $i = 1, \ldots, N$, and there are $\delta > 0$, $R > 0$, and points $z^i \in \Omega$ such that Ω_i, as well as the set $\Omega_{i\delta} := \Omega_i \cup \bigcup_{j \neq i} \{x \in \Omega_j : \mathrm{dist}(x, \Omega_i) < \delta\}$, is star shaped with respect to the ball $B(z^i, R)$, $i = 1, \ldots, N$.

We introduce the functions

$$\rho_i : \mathbb{R}^n \to [0, 1], \quad \rho_i(y) = \begin{cases} \exp\left\{-\frac{R^2}{R^2 - |y - z^i|^2}\right\} & \text{if } |y - z^i| < R, \\ 0 & \text{if } |y - z^i| \geq R \end{cases}$$

and their piecewise constant approximations $\rho_i^\mathcal{M} \in X(\mathcal{M})$. Concerning the mesh, we assume the following:

Assumption 5. Let $\mathcal{M} = (\mathcal{P}, \mathcal{T}, \mathcal{E})$ be a Voronoi finite volume mesh of Ω with the property that $\mathcal{E}_K \cap \mathcal{E}_{ext} \neq \emptyset$ implies $x_K \in \partial\Omega$. Moreover, the local mesh size near $B(z^i, R)$, $i = 1, \ldots, N$, is assumed to be so small that there exists a constant $\rho_0 > 0$ such that $\int_\Omega \rho_i^\mathcal{M}(x)\, dx \geq \rho_0$, $i = 1, \ldots, N$.

Then the discrete Sobolev–Poincaré inequalities remain true also for finite unions of δ-overlapping star shaped domains:

Theorem 3. *Let Assumptions 3–5 be satisfied, and $q \in [1, \infty)$ for $n = 2$ and $q \in [1, \frac{2n}{n-2})$ for $n \geq 3$, respectively. Then there exists a constant $C_q > 0$ depending only on n, q, Ω, and the constants in Assumptions 3–5 such that*

$$\|u - m_\Omega(u)\|_{L^q(\Omega)} \leq C_q |u|_{1,\mathcal{M}} \quad \forall u \in X(\mathcal{M}).$$

For a proof we refer to [14, sect. 6]. Since Theorem 2 is a direct consequence of Theorem 1 (with fixed κ_1, κ_2), this statement also remains true for more general domains characterized by Assumption 4.

5 Applications of discrete Sobolev inequalities

A functional analytic tool like a discrete Sobolev–Poincaré inequality enables us to treat discretized boundary value problems similarly to the corresponding continuous boundary value problems. Especially, if the embedding constants hold true for a class of meshes, uniform results with respect to the mesh can be obtained which can be used, for instance, for convergence results, too.

We were forced to prove the discrete Sobolev–Poincaré inequality by the analytical and numerical treatment of (nonlinear) reaction-diffusion systems. For the nondiscretized systems under consideration the free energy decays exponentially to its equilibrium value. We introduced a discretization scheme (Voronoi finite volume in space and fully implicit in time) which has the special property that it preserves the main features of the continuous problem, namely, positivity, dissipativity, and flux conservation (see [13]). For each fixed mesh we proved the exponential decay of the discretized free energy, too (see [11]). This proof works with the finite dimensional quantities.

To obtain uniform decay rates for a class of Voronoi finite volume meshes we had to translate the quantities from the finite dimensional discretized problems into expressions of functions from $X(\mathcal{M})$ being defined on Ω and being constant on Voronoi boxes of the corresponding meshes, and we had to consider limits of such functions belonging to sequences of Voronoi finite volume meshes to find a contradiction in the indirect proof of an estimate of the free energy by the dissipation rate (see [12, Theorem 3.2]). The essential ingredient in that proof is the discrete Sobolev–Poincaré inequality, Theorem 2.

For the application of discrete versions of Sobolev's inequality in the case of homogeneous Dirichlet boundary conditions we refer to [7, sect. 5]. Moreover, this inequality in the discrete $W_0^{1,p}$-setting (where $p \in (1, \infty)$) comes into play in the discretization of nonlinear elliptic problems of the form

$$-\operatorname{div} a(x, \nabla u) = f \quad \text{in } \Omega, \quad u = 0 \quad \text{on } \partial\Omega$$

on general polyhedral meshes in n space dimensions. In [8, sect. 3] it is used to obtain an estimate of the approximate solution. In this setting the Caratheodory function $a : \Omega \times \mathbb{R}^n$ fulfills, with suitable positive constants c_1, c_2, and $d \in L^{p'}(\Omega)$,

$$a(x, \zeta) \cdot \zeta \geq c_1 |\zeta|^p \quad \text{for almost all } x \in \Omega, \forall \zeta \in \mathbb{R}^n,$$

$$(a(x, \zeta) - a(x, \chi)) \cdot (\zeta - \chi) > 0 \quad \text{for almost all } x \in \Omega, \forall \zeta \neq \chi \in \mathbb{R}^n,$$

$$|a(x, \zeta)| \leq d(x) + c_2 |\zeta|^{p-1} \quad \text{for almost all } x \in \Omega, \forall \zeta \in \mathbb{R}^n.$$

Acknowledgements Jens A. Griepentrog gratefully acknowledges financial support by the DFG Research Center Matheon *Mathematics for key technologies* in the framework of the Matheon project C32: *Modeling of phase separation and damage processes in alloys*. Research about the topic of the text has also been carried out under financial support by the European XFEL GmbH in the framework of the XFEL project *Charge cloud simulations*.

References

1. Adams, R.A.: Sobolev Spaces. Academic Press, New York (1975)
2. Buffa, A., Ortner, C.: Compact embeddings of broken Sobolev spaces and applications. IMA J. Numer. Anal. **29**, 827–855 (2009)
3. Coudière, Y., Gallouët, T., Herbin, R.: Discrete Sobolev Inequalities and L^p error estimates for finite volume solutions of convection diffusion equations. M2AN Math. Model. Numer. Anal. **35**, 767–778 (2001)
4. Di Pietro, D., Ern, A.: Discrete functional analysis tools for discontinuous Galerkin methods with applications to the incompressible Navier–Stokes equations. Math. Comp. **79**, 1303–1330 (2010)
5. Droniou, J., Gallouët, T., Herbin, R.: A finite volume scheme for a noncoercive elliptic equation with measure data. SIAM J. Numer. Anal. **41**, 1997–2031 (2003)
6. Eymard, R., Gallouët, T., Herbin, R.: The finite volume method. In: Ciarlet, P., Lions, J.L. (eds.) Handbook of Numerical Analysis VII, pp. 723–1020. North-Holland, Amsterdam (2000)
7. Eymard, R., Gallouët, T., Herbin, R.: Discretization of heterogeneous and anisotropic diffusion problems on general nonconforming meshes SUSHI: A scheme using stabilization and hybrid interfaces. IMA J. Numer. Anal. **30**, 1009–1043 (2010)
8. Eymard, R., Gallouët, T., Herbin, R.: Cell centered discretisation of non linear elliptic problems on general multidimensional polyhedral grids. J. Numer. Math. **17**, 173–193 (2009)
9. Gallouët, T., Herbin, R., Vignal, M.H.: Error estimates on the approximate finite volume solution of convection diffusion equations with general boundary conditions. SIAM J. Numer. Anal. **37**, 1935–1972 (2000)
10. Gilbarg, D., Trudinger, N.S.: Elliptic Partial Differential Equations of Second Order. Springer, Berlin, Heidelberg, New York (1977)

11. Glitzky, A.: Exponential decay of the free energy for discretized electro-reaction-diffusion systems. Nonlinearity **21**, 1989–2009 (2008)
12. Glitzky, A.: Uniform Exponential Decay of the Free Energy for Voronoi Finite Volume Discretized Reaction-Diffusion Systems. WIAS Preprint **1443**, Berlin (2009)
13. Glitzky, A., Gärtner, K.: Energy estimates for continuous and discretized electro-reaction-diffusion systems. Nonlinear Anal. **70**, 788–805 (2009)
14. Glitzky, A., Griepentrog, J.A.: Discrete Sobolev–Poincaré inequalities for Voronoi finite volume approximations. SIAM J. Numer. Anal. **48**, 372–391 (2010)
15. Kufner, A., John, O., Fučik, S.: Function Spaces. Academia, Prague (1977)
16. Smirnow, W.I.: Lehrgang der höheren Mathematik V. Deutscher Verlag der Wissenschaften, Berlin (1962)
17. Sobolew, S.L.: Einige Anwendungen der Funktionalanalysis auf Gleichungen der mathematischen Physik. Akademie-Verlag, Berlin (1964)

The paper is in final form and no similar paper has been or is being submitted elsewhere.

A Simple Second Order Cartesian Scheme for Compressible Flows

Y. Gorsse, A. Iollo, and L. Weynans

Abstract A simple second order scheme for compressible inviscid flows on cartesian meshes is presented. An appropriate Rieman solver is used to impose the impermeability condition. The level set function defines the immersed body and provides some useful geometrical data to increase the scheme accuracy. A modification of the convective fluxes computation for the cells near the solid ensures the boundary condition at second order accuracy. The same procedure is performed in each direction independently. An application to the simulation of a Ringleb flow is presented to demonstrate the accuracy of the method.

Keywords compressible flow, second order scheme, level set method, Riemann solver, cartesian meshes

MSC2010: 65M08, 65M12, 76N15

1 Introduction

The computation of flows in complex unsteady geometries is a crucial issue to perform realistic simulations of physical or biological applications like for instance biolocomotion (fish swimming or insect flight), turbomachines, windmills... To this end several class of methods exist. Here we are concerned with immersed boundary methods, i.e., integration schemes where the grid does not fit the geometry. These methods have been widely developed in the last 15 years, though the first methods were designed earlier (see for example [2,3,10]). The general idea behind immersed boundary methods is to take into account the boundary conditions by a modification of the equations to solve, either at the continuous level or at the discrete one,

---

Y. Gorsse, A. Iollo, and L. Weynans
Institut de Mathematiques de Bordeaux and INRIA Bordeaux Sud-Ouest, Université Bordeaux 1,
e-mail: yannick.gorsse@math.u-bordeaux1.fr, angelo.iollo@math.u-bordeaux1.fr, lisl.weynans@math.u-bordeaux1.fr

rather than by the use of an adapted mesh. The main advantages of using these approaches, compared to methods using body-conforming grids, are that they are easily parallelizable and allow the use of powerful line-iterative techniques. They also avoid to deal with grid generation and grid adaptation, a prohibitive task when the boundaries are moving. A recent through review of immersed boundary methods is provided by Mittal and Iaccarino [6].

In this paper we present a simple globally second order scheme inspired by ghost cell approaches to solve compressible inviscid flows. In the fluid domain, away from the boundary, we use a classical finite-volume method based on an approximate Riemann solver for the convective fluxes and a centered scheme for the diffusive term. At the cells located on the boundary, we solve an *ad hoc* Riemann problem taking into account the relevant boundary condition for the convective fluxes by an appropriate definition of the contact discontinuity speed. These ideas can be adapted to reach higher order accuracy. However, here our objective is to device a method that can easily be implemented in existing codes and that is suitable for massive parallelization.

In section 2 we describe the finite volume scheme used to solve the flow equations in the fluid domain, away from the interface. In section 3 we introduce our method to impose impermeability condition. Finally, in section 4 we present a numerical test in two dimensions to validate the expected order of convergence and to discuss performance compared to others immersed boundary or body fitted methods.

2 Resolution in the fluid domain

We briefly describe how we solve the Euler equations in the fluid domain.

2.1 Governing equations

The compressible Euler equations are:

$$\frac{\partial \rho}{\partial t} + \nabla \cdot \rho \mathbf{u} = 0 \tag{1}$$

$$\frac{\partial \rho \mathbf{u}}{\partial t} + \nabla \cdot (\rho \mathbf{u} \otimes \mathbf{u} + p\mathbf{n}) = 0 \tag{2}$$

$$\frac{\partial E}{\partial t} + \nabla \cdot ((E + p)\mathbf{u}) = 0 \tag{3}$$

where E denotes the total energy per unit volume. For a perfect gas

$$E = \frac{p}{\gamma - 1} + \frac{1}{2}\rho \mathbf{u}^2 \text{ and } p = \rho RT \tag{4}$$

2.2 Discretization

We focus on a two-dimensional setting. Let i and j be integers and consider the rectangular lattice generated by i and j, with spacing h_x and h_y in the x and y direction, respectively.

Let W be the conservative variables, $\mathscr{F}^x(W)$, $\mathscr{F}^y(W)$) the convective flux vectors in the x and y directions, respectively. By averaging the governing equations over any cell of the rectangular lattice we have

$$\frac{dW_{ij}}{dt} + \frac{1}{h_x}(\mathscr{F}^x_{i+1/2\,j} - \mathscr{F}^x_{i-1/2\,j}) + \frac{1}{h_y}(\mathscr{F}^y_{i\,j+1/2} - \mathscr{F}^y_{i\,j-1/2}) = 0 \tag{5}$$

where $W_{i\,j}$ is the average value of the conservative variables on the cell considered, $\mathscr{F}^x_{i+1/2\,j}$ the average flux in the x direction taken on the right cell side, and similarly for the other sides.

The average convective fluxes at cell interfaces are approximated using the Osher numerical flux function [9].

A second order Runge-Kutta scheme is used for the time integration.

3 A second order impermeability condition

For Euler equations, the boundary condition on the interface is the impermeability assumption, i.e., given normal velocity to the boundary (zero for a static wall, but non-zero for a moving solid). We are concerned with recovering second order accuracy on the impermeability condition.

3.1 Level set method

In order to improve accuracy at the solid walls crossing the grid cells we need additional geometric information. This information, mainly the distance from the wall and the wall normal, is provided by the distance function. The level set method, introduced by Osher and Sethian [8], is used to implicitly represent the interface of solid in the computational domain. We refer the interested reader to [11, 12] and [7] for recent reviews of this method. The zero isoline of the level set function represents the boundary Σ of the immersed body. The level set function is defined by:

$$\varphi(x) = \begin{cases} dist_\Sigma(x) & \text{outside of the solid} \\ -dist_\Sigma(x) & \text{inside of the solid} \end{cases} \qquad (6)$$

A useful property of the level set function is:

$$\mathbf{n}(x) = \nabla \varphi(x) \qquad (7)$$

where $\mathbf{n}(x)$ is the outward normal vector of the isoline of φ passing on x. In particular, this allows to compute the values of the normal to the interface, represented by the isoline $\varphi = 0$.

3.2 The impermeability condition in one dimension

To make the ideas clear, let us start from a simple case. The typical situation for a grid that does not fit the body is shown in Fig. 1. The plan is to modify the flux at the cell interface nearest to the boundary of the solid, in order to impose the boundary condition at the actual fluid-solid interface location. For a fixed body, we want to impose $u_b = u(x_b) = 0$ at the boundary point x_b where $\varphi(x_b) = 0$.

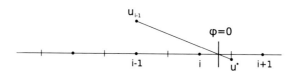

Fig. 1 Mesh near the solid. The interface lies between the center of cell i (fluid) and the center of cell $i + 1$ (solid). The flux in $i + 1/2$ has to be modified in order to account for the boundary conditions

Let u^* be the contact discontinuity speed resulting from the solution of the Riemann problem defined at the interface between cell i and cell $i+1$. The plan is to define a fictitious fluid state in $i + 1$ such that the resulting velocity at the interface u^* takes into account, at the desired degree of accuracy, the boundary condition $u_b = 0$ in x_b. In particular, taking a second order Taylor expansion of the velocity at x_b, we have

$$u^* = u_b + \left(\frac{h_x}{2} - \varphi_i \right) \frac{\partial u}{\partial x}\bigg|_{x_b} + O(h_x^2) \qquad (8)$$

The boundary can be located anywhere between x_i and x_{i+1}, so to ensure a well defined derivative (if $x_b \to x_i$, $\dfrac{u_b - u_i}{x_b - x_i}$ is not numerically well defined), we use x_{i-1} instead of x_i to compute the first order derivative at the interface:

$$\left.\frac{\partial u}{\partial x}\right|_{x_b} = \frac{u_b - u_{i-1}}{h_x + \varphi_i} \qquad (9)$$

To obtain u^* as the contact discontinuity speed of the Riemann problem, having computed the left state of the Riemann problem with the MUSCL reconstruction and slope limiters: $U_- = (u_-, p_-, c_-)$, we create the right state $U_+ = (-u_- + 2u^*, p_-, c_-)$, where c is the speed of sound. The left and right state of the variables p and c are chosen identical to express the continuity of these variables on the interface.

3.3 The impermeability condition in two dimensions

In two dimensions the flow equations are solved by computing independently the flux in each direction, so we want to apply in each direction the same kind of ideas as in one dimension in order to accurately enforce the boundary conditions. When the level set function changes sign between two cells, we need to modify the fluxes at the interface between these cells.

The interface point is the intersection between the interface ($\varphi = 0$) and the segment connecting the two cell centers concerned by the sign change (for example the points A, B and C on Fig. 2). For the flux computation, a fictitious state is created for instance between the cells (i, j) and $(i + 1, j)$ on Fig. 2. The boundary condition that we have to impose now is $\mathbf{u}_b . \mathbf{n}_b = 0$, where $\mathbf{u}_b$ is the speed of the fluid at the boundary, and $\mathbf{n}_b$ the outward normal vector of the body.

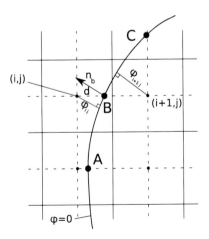

Fig. 2 Example of geometric configuration at the interface. B is the interface point located between (i, j) and $(i + 1, j)$. The flux on cell interface $(i + 1/2, j)$ is modified to enforce the boundary condition on B

With reference to Fig. 2, the level set function changes sign between $x_{i,j}$ and $x_{i+1,j}$ at point B. Let the normal vector point to the fluid side. If we assume that the boundary $\varphi = 0$ is locally rectilinear, using the side splitter theorem, the distance between $x_{i,j}$ and B is

$$d = \frac{h_x |\varphi_{i,j}|}{|\varphi_{i,j}| + |\varphi_{i+1,j}|} \quad (10)$$

and the normal vector $\mathbf{n}_b$ is computed by

$$\mathbf{n}_b = \mathbf{n}_{i,j} + \frac{d}{h_x} \left(\mathbf{n}_{i+1,j} - \mathbf{n}_{i,j} \right) \quad (11)$$

where $\mathbf{n}_{i,j}$ is a second order centered finite-difference approximation of $\nabla \varphi$ at point (i, j). To impose the boundary condition at the interface point B, we determine a value of the contact discontinuity speed $\mathbf{u}^*$, relative to a Riemann problem defined in the direction normal to the cell side through $x_{i+1/2,j}$, consistent at second order accuracy with $\mathbf{u}_b \cdot \mathbf{n}_b = 0$ in B. Figure 3 illustrates graphically the following steps. Let the normal component of $\mathbf{u}^*$ be $u_n^* = \mathbf{u}^* \cdot \mathbf{n}_b$.
u_n^* is computed with a second order Taylor expansion of the normal velocity at the interface point, that is:

$$u_n^* = \mathbf{u}_b \cdot \mathbf{n} + \left(\frac{h_x}{2} - d \right) \frac{\mathbf{u}_b \cdot \mathbf{n} - \mathbf{u}_{i-1} \cdot \mathbf{n}}{h_x + d} + O(h_x^2). \quad (12)$$

Then, let determine u_τ^* the tangential component of $\mathbf{u}^*$ by $u_\tau^* = \mathbf{u}^* \cdot \tau_b$, τ_b being the vector tangential to the interface at point B. We use the continuity property of the tangential component of the velocity to define u_τ^* according to

$$u_\tau^* = \mathbf{u}_- \cdot \tau \quad (13)$$

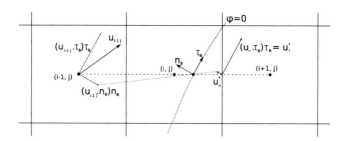

Fig. 3 Graphical illustration of the construction of the $\mathbf{u}^*$ vector

Finally we decompose $\mathbf{u}^*$ in the canonical basis by its horizontal and vertical components u^* and v^*, that is

$$u^* = u_n^* n_x + u_\tau^* \tau_x \tag{14}$$

$$v^* = u_n^* n_y + u_\tau^* \tau_y \tag{15}$$

To obtain $\mathbf{u}^*$ as the contact discontinuity speed of the Riemann problem, the left state resulting from the MUSCL reconstruction with slope limiters being $U_- = (u_-, v^*, p_-, c_-)$, we choose the right state to be $U_+ = (-u_- + 2u^*, v^*, p_-, c_-)$.

4 Numerical illustration: The Ringleb flow

The objective is to ascertain the actual accuracy obtained at the solid interface.

The Ringleb flow refers to an exact solution of Euler equations. The solution is obtained with the hodograph method, see [13]. The streamlines and iso-Mach lines are shown on Fig. 4.

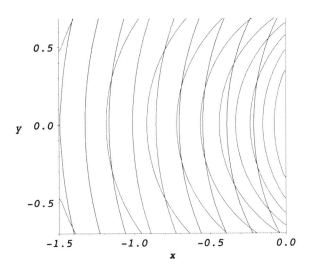

Fig. 4 Streamlines (black) and iso-Mach lines (grey) of the Ringleb flow

The exact solution is formulated in (θ, V) variables with $u = V\cos\theta$, $v = V\sin\theta$ and $V = \sqrt{u^2 + v^2}$.
The stream function is given by $\Psi = \frac{\sin\theta}{V}$.
The streamlines equations are:

$$x = \frac{1}{2\rho}\left(\frac{1}{V^2} - 2\Psi^2\right) + \frac{L}{2}, \qquad y = \frac{\sin\theta\cos\theta}{\rho V^2} \tag{16}$$

with:

$$L = -\left(\frac{1}{2}\ln\frac{1+c}{1-c} - \frac{1}{c} - \frac{1}{3c^3} - \frac{1}{5c^5}\right), \quad c^2 = 1 - \frac{\gamma-1}{2}V^2, \quad \rho = c^5 \quad (17)$$

In our test case, the computational domain is $[-0.5; -0.1] \times [-0.6; 0]$ and we numerically solve the flow between the streamlines $\Psi_1 = 0.8$ and $\Psi_2 = 0.9$. The inlet and outlet boundary condition are supersonic for $y = -0.6$ and $y = 0$ respectively. The convergence orders are calculated for each variable in L_1, L_2, L_∞ norms on four different grids $32 \times 48, 64 \times 96, 128 \times 192$ and 256×384 and presented on Table 1.

Table 1 Global orders of convergence for each variable

Variables	L_1 norm	L_2 norm	L_∞ norm
x-velocity	2.04	1.68	1.28
y-velocity	1.97	1.6	1.13
pressure	2.0	2.02	1.97
sound velocity	1.95	1.58	1.03
entropy	1.9	1.49	1.08

The error for several variables is order 1 for the L_{infty} norm. Colella et al. [5] obtain the same kind of results on other test cases. One argument developed in [5] to explain this order degradation is that the solid wall is characteristic for entropy, and hence the error on this variable accumulates from inlet to outlet. For the same test case, Coirier and Powell [4] observed also a convergence order between one and two in the case of their own cartesian method. In [1], Abgrall et al. obtain a L^2 numerical order of convergence for the density equal to 1.5 with their second order residual distribution scheme.

5 Conclusions

In this paper we have presented a new second order cartesian method to solve compressible flows in complex domains. This method is based on a classical finite volume approach, but the values used to compute the fluxes at the cell interfaces near the solid boundary are determined so to satisfy the boundary conditions with a second order accuracy. A test case for inviscid flows was presented. The order of convergence of the method is similar to those observed in the literature. This method is particularly simple to implement, as it doesn't require any special cell reconstruction at the solid-wall interface. The extension to three-dimensional cases is natural as the same procedure at the boundary is repeated in each direction. Forthcoming work will concern the extension of the present approach to multi-physics problems.

References

1. ABGRALL, R., LARAT, A., AND RICCHIUTO, M. Construction of very high order residual distribution schemes for steady inviscid flow problems on hybrid unstructured meshes, in press, 2010.
2. BERGER, M., AND LEVEQUE, R. An adaptive cartesian mesh algorithm for the euler equations in arbitrary geometries, 1989.
3. BERGER, M., AND LEVEQUE, R. Stable boundary conditions for cartesian grid calculations. *Computing systems in Engineering 1*, 2-4 (1990), 305–311.
4. COIRIER, W., AND POWELL, K. An accuracy assessment of cartesian mesh approaches for the euler equations. *J. Comput. Phys. 117* (1995), 121–131.
5. COLELLA, P., GRAVES, D., KEEN, B., AND MODIANO, D. A cartesian grid embedded boundary method for hyperbolic conservation laws. *J. Comput. Phys. 211*, 1 (2006), 347–366.
6. MITTAL, R., AND IACCARINO, G. Immersed boundary methods, 2005.
7. OSHER, S., AND FEDKIW, R. *Level Set Methods and Dynamic Implicit Surfaces*. Springer, 2003.
8. OSHER, S., AND SETHIAN, J. A. Fronts propagating with curvature-dependent speed: Algorithms based on hamiltonjacobi formulations. *J. Comput. Phys. 79*, 12 (1988).
9. OSHER, S., AND SOLOMAN, F. Upwind difference schemes for hyperbolic systems of conservation laws. *Math. Comp. 38*, 158 (April 1982), 339–374.
10. PESKIN, C. The fluid dynamics of heart valves: experimental, theoretical and computational methods. *Annu. Rev. Fluid Mech. 14* (1981), 235–259.
11. SETHIAN, J. A. *Level Set Methods and Fast Marching Methods*. Cambridge University Press, Cambridge, UK, 1999.
12. SETHIAN, J. A. Evolution, implementation, and application of level set and fast marching methods for advancing fronts. *J. Comput. Phys. 169* (2001), 503–555.
13. SHAPIRO, A. *The Dynamics and Thermodynamics of Compressible Fluid Flow*. Ronald Press, 1953.

The paper is in final form and no similar paper has been or is being submitted elsewhere.

Efficient Implementation of High Order Reconstruction in Finite Volume Methods

Florian Haider, Pierre Brenner, Bernard Courbet, and Jean-Pierre Croisille

Abstract The paper presents a new algorithm for high order piecewise polynomial reconstruction. This algorithm computes a high order approximant in a given cell using data from adjacent cells in several steps, eliminating the need to handle directly large reconstruction stencils. The resulting high order finite volume method is well suited for modern parallel and vector (array) computers.

Keywords High Order Scheme, Unstructured Grid, Finite Volume Method, MUSCL
MSC2010: 65M08,65D15

1 Introduction

The finite volume MUSCL method to solve hyperbolic conservation laws was introduced by B. Van Leer in [7] thirty years ago. The main idea is to increase the accuracy of the first order finite volume scheme by a piecewise linear reconstruction used to evaluate upwinded fluxes at the cell interfaces.

Practical applications for convection dominated flows in complex geometries have motivated many extensions of the MUSCL approach to unstructured grids.

Florian Haider and Bernard Courbet
ONERA 29 avenue de la Division Leclerc 92322 Châtillon France,
e-mail: florian.haider@onera.fr, bernard.courbet@onera.fr

Pierre Brenner
ASTRIUM Space Transportation - Aerodynamics BP3002 - 78133 Les Mureaux France,
e-mail: pierre.brenner@astrium.eads.net

Jean-Pierre Croisille
Université Paul Verlaine-Metz - UFR MIM Département de Mathématiques
Ile du Saulcy 57045 Metz France, e-mail: croisil@poncelet.univ-metz.fr

A typical example is the flow solver CEDRE developed at ONERA. It uses a cell centered finite volume scheme with piecewise linear reconstruction on general polyhedral grids to solve the compressible Navier Stokes equations. A large choice of physical models is available in CEDRE: turbulence, combustion, diphasic flow, radiation etc.

Our experience has shown that second order accuracy becomes insufficient for LES and to capture contact discontinuities. The easiest way to increase the spatial accuracy is to replace the linear interpolation with quadratic or cubic ones. Indeed, the MUSCL scheme with quadratic reconstruction (3rd order) was already discussed by B. Van Leer [7]. The quadratic approach was extended to unstructured grids [1, 2]. The need for large (non compact) stencils seems to have limited the use of cubic reconstructions (4th order), although some practical applications exist [6].

For reasons of performance, the computation of a polynomial reconstruction on a grid cell must be local, using data in neighboring cells only. On the other hand, high order approximation requires sufficiently many data samples, which means that data from cells beyond adjacent cells in the grid must be accessed.

This paper shows how to compute a high order approximant in a given cell using data from adjacent cells in several steps, eliminating the need to handle directly large reconstruction stencils. No additional degrees of freedom are added: the independent variables are the cell averages of the conserved quantity. The resulting high order finite volume method is well suited for modern parallel and vector (array) computers. This aspect is of primary importance for large scale industrial software.

2 Semi-discrete High Order Finite Volume Scheme

- *Geometric notation*: an unstructured grid is a triangulation of a domain $\Omega \subset \mathbb{R}^d$ consisting of N general polyhedra. The cell with number α is denoted $\mathcal{T}_\alpha$, with barycenter $\boldsymbol{x}_\alpha$ and d-volume $|\mathcal{T}_\alpha|$. The face $\mathcal{A}_{\alpha\beta}$, with barycenter $\boldsymbol{x}_{\alpha\beta}$, has a normal vector $\boldsymbol{a}_{\alpha\beta}$ oriented from cell $\mathcal{T}_\alpha$ to $\mathcal{T}_\beta$ and of length $\|\boldsymbol{a}_{\alpha\beta}\|$ equal to the surface $|\mathcal{A}_{\alpha\beta}|$. The oriented normal unit vector of the face $\mathcal{A}_{\alpha\beta}$ is $\boldsymbol{v}_{\alpha\beta}$. Furthermore, define $\boldsymbol{h}_{\alpha\beta} = \boldsymbol{x}_\beta - \boldsymbol{x}_\alpha$ and

$$z_{\alpha\beta}^{(k)} \triangleq \frac{1}{|\mathcal{T}_\beta|} \int_{\mathcal{T}_\beta} (\boldsymbol{x} - \boldsymbol{x}_\alpha)^{\otimes k} \, d\boldsymbol{x} = \frac{1}{|\mathcal{T}_\beta|} \int_{\mathcal{T}_\beta} \underbrace{(\boldsymbol{x} - \boldsymbol{x}_\alpha) \otimes \cdots \otimes (\boldsymbol{x} - \boldsymbol{x}_\alpha)}_{m \text{ factors}} \, d\boldsymbol{x} \quad (1)$$

The kth moment of cell $\mathcal{T}_\alpha$ is then defined as $\boldsymbol{x}_\alpha^{(k)} \triangleq z_{\alpha\alpha}^{(k)}$. Note that $\boldsymbol{x}_\alpha^{(1)} = 0$. For a locally integrable function u, define its average over cell $\mathcal{T}_\alpha$ as $\bar{u}_\alpha$.

- *Semi-discrete* MUSCL *scheme*: consider a hyperbolic conservation law with flux $\boldsymbol{f}$. Its balance equation over a cell $\mathcal{T}_\alpha$ can be written as

$$\frac{d\bar{u}_\alpha(t)}{dt} = -\frac{1}{|\mathcal{T}_\alpha|} \sum_\beta \int_{\mathcal{A}_{\alpha\beta}} \boldsymbol{v}_{\alpha\beta} \cdot \boldsymbol{f}(u(\boldsymbol{x}, t)) \, d\sigma. \quad (2)$$

… Efficient Implementation of High Order Reconstruction …

The semi-discrete MUSCL discretization of such a conservation law gives the finite volume scheme

$$\frac{d\bar{u}_\alpha(t)}{dt} = -\frac{1}{|\mathcal{T}_\alpha|} \sum_\beta \int_{\mathcal{A}_{\alpha\beta}} \widetilde{f}_{\alpha\beta}\left(w_\alpha\left[\bar{u}(t)\right](x), w_\beta\left[\bar{u}(t)\right](x)\right) d\sigma. \quad (3)$$

In (3), $\widetilde{f}_{\alpha\beta} : \mathbb{R} \times \mathbb{R} \to \mathbb{R}$ is a *numerical flux* that is *consistent* with f: $\widetilde{f}_{\alpha\beta}(u,u) = \nu_{\alpha\beta} \cdot f(u)$. The functions w_α and w_β are reconstructed from the cell averages $\bar{u}(t) = (\bar{u}_1(t), \ldots, \bar{u}_N(t))$. The dependence of w_α on the cell averages is denoted by square brackets $w_\alpha[\bar{u}(t)]$ and the dependence on x by $w_\alpha[\bar{u}(t)](x)$.

- *Accuracy*: the piecewise reconstruction operates on each cell so that only the cell averages in a certain neighborhood – the *reconstruction stencil* – of the cell $\mathcal{T}_\alpha$ determine the approximant w_α. Assume that the reconstruction satisfies for all smooth functions u and uniformly in $x \in \mathcal{T}_\alpha$ for all cells $\mathcal{T}_\alpha$

$$\left| w_\alpha[\bar{u}(t)](x) - u(x,t) \right| \le O\left(h^{k+1}\right). \quad (4)$$

Assuming f is Lipschitz continuous, one verifies easily that (3) is k^{th} order accurate.

- *Conservation*: the reconstruction is required to be *conservative*, i.e. the mean value of the function $w_\alpha[\bar{u}(t)]$ over the cell $\mathcal{T}_\alpha$ must always be $\bar{u}_\alpha(t)$.

3 High Order Polynomial Reconstruction

This section gives a short overview of *k-exact reconstruction* along the line of [1,4]. The goal is to reconstruct the functions w_α used in (3) in such a way that they satisfy (4). The time dependency is dropped to simplify the notation.

Let $\mathbb{P}_k(\mathbb{R}^d)$ be the space of polynomials of degree k in $\mathbb{R}^d$. In each cell $\mathcal{T}_\alpha$, the reconstruction procedure is represented by the linear operator

$$\mathfrak{R}_\alpha : \mathbb{R}^N \to \mathbb{P}_k(\mathbb{R}^d) \quad ; \quad \bar{u} \mapsto w_\alpha[\bar{u}]. \quad (5)$$

Define a *neighborhood* of cell $\mathcal{T}_\alpha$ as a set of cells $\mathbb{W}_\alpha \subset \{1 \ldots, N\}$ such that $\alpha \in \mathbb{W}_\alpha$ and associate with $\mathbb{W}_\alpha$ a local cell average operator

$$\mathfrak{P}_{k;\mathbb{W}_\alpha} : \mathbb{P}_k(\mathbb{R}^d) \to \mathbb{R}^N \quad (6)$$

given by $(\mathfrak{P}_{k;\mathbb{W}}(p))_\beta = \bar{p}_\beta$ if $\beta \in \mathbb{W}_\alpha$ and $(\mathfrak{P}_{k;\mathbb{W}}(p))_\beta = 0$ if $\beta \notin \mathbb{W}_\alpha$.

A reconstruction operator $\mathfrak{R}_\alpha : \mathbb{R}^N \to \mathbb{P}_k(\mathbb{R}^d)$ is called *k-exact* if it is a left inverse of (6)

$$\mathfrak{R}_\alpha \mathfrak{P}_{k;\mathbb{W}_\alpha} = \mathrm{Id}_{\mathbb{P}_k(\mathbb{R}^d)}. \tag{7}$$

It can be shown that, under certain conditions, (7) provides an approximation error (4) that is $O\left(h^{k+1}\right)$ [3].

The space of symmetric tensors of rank m in $\mathbb{R}^d$ is denoted $\mathcal{S}^m\left(\mathbb{R}^d\right)$. For all $\boldsymbol{a}, \boldsymbol{b} \in \mathcal{S}^m\left(\mathbb{R}^d\right)$ and $\boldsymbol{c} \in \mathbb{R}^d$ define

$$\boldsymbol{a} \bullet \boldsymbol{b} \triangleq \sum_{i_1=1}^d \cdots \sum_{i_m=1}^d a_{i_1 \cdots i_m} b_{i_1 \cdots i_m} \tag{8}$$

$$(\boldsymbol{a} \cdot \boldsymbol{c})_{j_1 \cdots j_{m-1}} \triangleq \sum_{j_m=1}^d a_{j_1 \cdots j_{m-1} j_m} c_{j_m}. \tag{9}$$

A function u is called k-exact on $\mathbb{W}_\alpha$ if the restriction of u to the cells in $\mathbb{W}_\alpha$ is a polynomial of degree k. Note that the mth derivative of u can be considered as an element of $\mathcal{S}^m\left(\mathbb{R}^d\right)$.

A k-exact m^{th} derivative on the neighborhood $\mathbb{W}_\alpha$ at cell $\mathcal{T}_\alpha$ is defined to be a linear map $\boldsymbol{w}_\alpha^{(m|k)} : \mathbb{R}^N \longrightarrow \mathcal{S}^m\left(\mathbb{R}^d\right)$ such that for all polynomials p of degree k

$$\boldsymbol{w}_\alpha^{(m|k)}\left[\mathfrak{P}_{k;\mathbb{W}_\alpha}(p)\right] = D^{(m)} p\bigg|_{x_\alpha}. \tag{10}$$

Since a polynomial is determined by its cell average and its mth derivatives at a point $\boldsymbol{x}_\alpha$, a k-exact reconstruction operator is equivalent to a set of k-exact m^{th} derivatives $\boldsymbol{w}_\alpha^{(m|k)}$ for $1 \leq m \leq k$. By linearity, (10) can be expressed as

$$\boldsymbol{w}_\alpha^{(m|k)}[\overline{u}] = \sum_{\beta \in \mathbb{W}_\alpha} \boldsymbol{w}_{\alpha\beta}^{(m|k)} \overline{u}_\beta \tag{11}$$

In (11), the symmetric tensors $\boldsymbol{w}_{\alpha\beta}^{(m|k)}$, called the *reconstruction coefficients* of $\boldsymbol{w}_\alpha^{(m|k)}$, depend only on the local cell geometry. In principle, a complete set of $\boldsymbol{w}_{\alpha\beta}^{(m|k)}$ can be computed by applying (10) to a basis of the space $\mathbb{P}_k\left(\mathbb{R}^d\right)$ and solving the resulting linear system in the least squares sense. Since this algorithm computes the $\boldsymbol{w}_\alpha^{(m|k)}$ directly from the cell averages, we will refer to this method in Sect. 5 as *direct least squares reconstruction* (DLS). However, an obvious drawback of this method is that its implementation requires the computation of (11) over large stencils $\mathbb{W}_\alpha$.

Taking into account the constraint of conservation and using k-exact m^{th} derivatives (10), one can write the reconstructed polynomial at cell $\mathcal{T}_\alpha$ in the general form

$$w[\overline{u}](\boldsymbol{x}) = \overline{u}_\alpha + \sum_{m=1}^k \frac{1}{m!} \boldsymbol{w}_\alpha^{(m|k)}[\overline{u}] \bullet \left[(\boldsymbol{x} - \boldsymbol{x}_\alpha)^{\otimes m} - \boldsymbol{x}_\alpha^{(m)}\right]. \tag{12}$$

Efficient Implementation of High Order Reconstruction

In (12), $(x - x_\alpha)^{\otimes m}$ is defined as in (1) and $x_\alpha^{(m)} \triangleq z_{\alpha\alpha}^{(m)}$.

When a k-exact m^{th} derivative (11) is applied to a polynomial p of degree $(k+1)$, the reconstruction error can be expressed as

$$w_\alpha^{(m|k)} \left[\overline{p}\right] - D^{(m)} p \bigg|_{x_\alpha} = \frac{1}{(k+1)!} \sum_{\beta \in \mathbb{W}_\alpha} w_{\alpha\beta}^{(m|k)} \left(z_{\alpha\beta}^{(k+1)} \bullet D^{(k+1)} p \bigg|_{x_\alpha} \right). \quad (13)$$

The interest of (13) is that a $(k+1)$-exact $(k+1)^{\text{th}}$ derivative can be used to compute the right hand side of (13) and to subtract it from $w_\alpha^{(m|k)}$, making $w_\alpha^{(m|k)}$ $(k+1)$-exact.

Finally, we introduce the following smoothing technique: let $\mathbb{V}_\alpha$ be the set of direct neighbors of cell $\mathcal{T}_\alpha$, including $\mathcal{T}_\alpha$ itself. Let $0 \leq \xi_\beta \leq 1$ be such that $\sum_{\beta \in \mathbb{V}_\alpha} \xi_\beta = 1$. If a set of k-exact k^{th} derivatives $w_\beta^{(k|k)}$ is known, one can define a new k-exact k^{th} derivative $\widetilde{w}_\alpha^{(k|k)}$ as a convex combination

$$\widetilde{w}_\alpha^{(k|k)} [\overline{u}] = \sum_{\beta \in \mathbb{V}_\alpha} \xi_\beta w_\beta^{(k|k)} [\overline{u}]. \quad (14)$$

It is natural to choose the weights ξ_β in (14) proportional to the cell volumes $|\mathcal{T}_\beta|$. The stencil of (14) is larger which increases stability, see [3, 5].

4 Efficient Algorithms for High Order Reconstruction

The computation of (11) involves large (non compact) stencils. To avoid this undesirable feature, we compute a $(k+1)$-exact $(k+1)^{\text{th}}$ derivative not directly from the cell averages, but from a family of k-exact k^{th} derivatives $w_\beta^{(k|k)}$ at cells $\mathcal{T}_\beta$ for β in a small neighborhood $\mathbb{W}_\alpha$ of cell $\mathcal{T}_\alpha$. This is done as follows:

Let $\mathbb{W}_\alpha$ be a neighborhood of cell $\mathcal{T}_\alpha$. Let $w_\beta^{(k|k)}$ be a family of k-exact k^{th} derivatives with stencils $\mathbb{W}_\beta^{(k)}$ at cells $\mathcal{T}_\beta$ for $\beta \in \mathbb{W}_\alpha$. Assume that $\bigcup_{\beta \in \mathbb{W}_\alpha} \mathcal{T}_\beta$ is path connected where the paths are piecewise C^1. Let $m_\alpha \triangleq |\mathbb{W}_\alpha| - 1$ and define the linear operator

$$\mathfrak{J}_{\mathbb{W}_\alpha}^{(k+1)} : \mathcal{S}^{(k+1)}\left(\mathbb{R}^d\right) \longrightarrow \left(\mathcal{S}^{(k)}\left(\mathbb{R}^d\right)\right)^{m_\alpha}. \quad (15)$$

The ith component of (15) is defined using $h_{\alpha\beta} = x_\beta - x_\alpha$, (1), (8) and (9) as

$$\left(\mathfrak{J}_{\mathbb{W}_\alpha}^{(k+1)}(b)\right)_i \triangleq b \cdot h_{\alpha\beta_i} + \frac{1}{(k+1)!} \sum_\gamma w_{\beta_i\gamma}^{(k|k)} \left(z_{\beta_i\gamma}^{(k+1)} \bullet b\right) - \\ - \frac{1}{(k+1)!} \sum_\gamma w_{\alpha\gamma}^{(k|k)} \left(z_{\alpha\gamma}^{(k+1)} \bullet b\right). \quad (16)$$

Proposition 1 (Functional Identity for Reconstruction). *Let u be a function that is $(k+1)$-exact on $\bigcup_{\beta \in \mathbb{W}_\alpha} \mathbb{W}_\beta^{(k)}$. Then the following identity holds*

$$\mathfrak{J}_{\mathbb{W}_\alpha}^{(k+1)} \left(D^{(k+1)} u \Big|_{x_\alpha} \right) =$$

$$= \left(w_{\beta_1}^{(k|k)} [\overline{u}] - w_\alpha^{(k|k)} [\overline{u}], \ldots, w_{\beta_{m_\alpha}}^{(k|k)} [\overline{u}] - w_\alpha^{(k|k)} [\overline{u}] \right). \quad (17)$$

The *main result* of this section is

Proposition 2 ($(k+1)$-exact $(k+1)^{\text{th}}$ derivative). *Assume that the operator $\mathfrak{J}_{\mathbb{W}_\alpha}^{(k+1)}$ defined in (16) has a left inverse $\mathfrak{D}_{\mathbb{W}_\alpha}^{(k+1)}$. Then the following expression defines a $(k+1)$-exact $(k+1)^{\text{th}}$ derivative on the neighborhood $\bigcup_{\beta \in \mathbb{W}_\alpha} \mathbb{W}_\beta^{(k)}$:*

$$\widetilde{w}_\alpha^{(k+1|k+1)} [\overline{u}] \triangleq$$

$$\triangleq \mathfrak{D}_{\mathbb{W}_\alpha}^{(k+1)} \left(w_{\beta_1}^{(k|k)} [\overline{u}] - w_\alpha^{(k|k)} [\overline{u}], \ldots, w_{\beta_m}^{(k|k)} [\overline{u}] - w_\alpha^{(k|k)} [\overline{u}] \right) \quad (18)$$

Prop. 2 gives the following algorithm.

Definition 1 (k-exact Coupled Least Squares Algorithm (CLS)).

1. Compute a 1-exact 1st derivative directly from the cell averages on a small stencil.
2. Iterate the following step from $m = 1$ to $m = k - 1$ at each cell:

   a. Compute a $(m+1)$-exact $(m+1)^{\text{th}}$ derivative from a m-exact m^{th} derivative, using the pseudo inverse of (15) in (18).

   b. On tetrahedral grids, apply (14) to the $(m+1)$-exact $(m+1)^{\text{th}}$ derivative.
3. Use (13) to obtain k-exact m^{th} derivatives for $1 \leq m \leq k-1$.

Remark 1. The smoothing step 2b is important on tetrahedral meshes due to stability considerations, see [5].

5 Numerical Results

As a test case, we apply the cell centered finite volume scheme (3) to the linear advection equation with constant velocity $c = \left(\frac{1}{10}, \frac{1}{5}, 1 \right)$ on the unit cube with periodic boundaries. The numerical flux is the classical upwinded flux

$$\widetilde{f}_{\alpha\beta} (u_\alpha, u_\beta) \triangleq (c \cdot v_{\alpha\beta})_+ u_\alpha + (c \cdot v_{\alpha\beta})_- u_\beta.$$

Efficient Implementation of High Order Reconstruction

Table 1 Grid convergence: series of tetrahedral grids (CLS *with* smoothing)

h_{avg}	N	CLS D2(4)	CLS D3(6)	DLS D1(2)	DLS D2(3)	DLS D3(4)
0.042316	5928					
0.037870	8406	2.2661	4.4984	1.9190	1.5232	4.8690
0.032909	12817	2.3654	4.4386	1.9881	1.8080	4.8354
0.027354	22493	2.7521	4.5814	2.2780	2.3162	5.0287
0.022707	39518	2.7832	4.3773	2.3307	2.5409	4.8292
0.018133	77770	2.8736	4.2354	2.3583	2.7250	4.7772
0.013422	192972	2.9989	4.3158	2.3962	2.9245	4.8433

Table 2 Grid convergence: series of cartesian grids (CLS *with* smoothing)

h_{avg}	N	CLS D2(4)	CLS D3(6)	DLS D1(2)	DLS D2(3)	DLS D3(4)
0.045455	10648					
0.035714	21952	2.4243	4.4868	1.9355	1.9313	4.4325
0.029412	39304	2.6872	4.5062	2.1330	2.3892	4.4078
0.025000	64000	2.8119	4.4633	2.1923	2.6240	4.3497
0.021739	97336	2.8791	4.4349	2.2001	2.7541	4.3267
0.019231	140608	2.9175	4.3359	2.1894	2.8310	4.2191
0.017241	195112	2.9406	4.3225	2.1720	2.8779	4.2215

Table 3 Grid convergence: series of polyhedral grids (CLS *without* smoothing)

h_{avg}	N	CLS D2(2)	CLS D3(3)	DLS D1(2)	DLS D2(3)	DLS D3(4)
0.044784	13819					
0.041544	17933	3.1771	5.4782	1.3943	1.0037	4.9159
0.038507	22983	2.9794	4.9878	1.4959	1.2227	4.5057
0.033027	35595	2.4432	3.3753	1.4035	1.3847	3.9068
0.029400	52487	3.3681	4.8044	2.1503	2.2422	5.1493
0.025212	80995	2.5399	2.3076	1.6894	1.9848	4.0666
0.021547	135609	3.5112	5.4666	2.4056	2.8601	5.3057

The scheme has been tested with the CLS reconstruction of Def. 1 for $k = 2$ – named CLS D2 – and $k = 3$ – named CLS D3. The direct least squares reconstruction mentioned in Sect. 3 serves as comparison, named DLS Dk for $k = 1, 2, 3$. Tables 1, 2 and 3 display the convergence rate for the ℓ_2 error at $t = 10$ as a function of the average cell diameter h_{avg} on three different shapes of grids for the initial value $u_0(x, y, z) = \sin(2\pi x) \sin(2\pi y) \sin(2\pi z)$. The column N displays the number of cells. The number (n) indicates that the effective stencil is the nth neighborhood: The first neighborhood of the cell $\mathcal{T}_\alpha$ consists of the cell $\mathcal{T}_\alpha$ itself and its adjacent cells $\mathcal{T}_\beta$. The second neighborhood of the cell $\mathcal{T}_\alpha$ is the union of the first neighbors of the first neighbors, etc.

Observe that the algorithm CLS in Def. 1 gives the desired convergence rates for quadratic (3rd order) and cubic reconstruction (4th order). The rates are comparable to those of the direct method DLS.

6 Conclusion

The algorithm of Def. 1 avoids large stencils in implementing high order finite volume schemes (3), without introducing additional degrees of freedom. The integration of the CLS algorithm in the CEDRE software is an ongoing work. This requires appropriate limiting techniques to deal with monotonicity.

References

1. Barth, T.J., Frederickson, P.O.: Higher order solution of the Euler equation on unstructured grids using quadratic reconstruction. In: AIAA 90, AIAA-90-0013, pp. 1–12. AIAA, Reno Nevada (1990)
2. Delanaye, M., Essers, J.A.: Quadratic-reconstruction finite volume scheme for compressible flows on unstructured adaptive grids. AIAA Journal **35**(4), 631 – 639 (1997)
3. Haider, F.: Discrétisation en maillage non structuré et applications les. Ph.D. thesis, Université Pierre et Marie Curie Paris VI (2009)
4. Haider, F., Brenner, P., Courbet, B., Croisille, J.P.: High order approximation on unstructured grids: Theory and implementation. Preprint (2011)
5. Haider, F., Croisille, J.P., Courbet, B.: Stability analysis of the cell centered finite-volume MUSCL method on unstructured grids. Numer. Math. **113**, 555 – 600 (2009). DOI 10.1007/s00211-009-0242-6
6. Khosla, S., Dionne, P., Lee, M., Smith, C.: Using fourth order spatial integration on unstructured meshes to reduce LES run time. AIAA 2008-782. 46th AIAA Aerospace Sciences Meeting and Exhibit, AIAA (2008)
7. van Leer, B.: Towards the ultimate conservative difference scheme. IV. A new approach to numerical convection. Journal of Computational Physics **23**(3), 276 – 299 (1977). DOI 10.1016/0021-9991(77)90095-X. URL http://www.sciencedirect.com/science/article/B6WHY-4DD1MM2-4J/2/61bfce9111ba17f514bbf0fbdb2f2ee4

The paper is in final form and no similar paper has been or is being submitted elsewhere.

A Well-Balanced Scheme For Two-Fluid Flows In Variable Cross-Section ducts

Philippe Helluy and Jonathan Jung

Abstract We propose a finite volume scheme for computing two-fluid flows in variable cross-section ducts. Our scheme satisfies a well-balanced property. It is based on the VFRoe approach. The VFRoe variables are the Riemann invariants of the stationnary wave and the cross-section. In order to avoid spurious pressure oscillations, the well-balanced approach is coupled with an ALE (Arbitrary Lagrangian Eulerian) technique at the interface and a random sampling remap.

Keywords Well-balanced scheme, Glimm scheme, Lagrange projection, two-fluid flows.
MSC2010: 65M08, 76M12, 76T10, 35Q31

Introduction

Classical finite volume solvers generally have a bad precision for solving two-fluid interfaces or flows in varying cross-section ducts. Several cures have been developed for improving the precision.

- For cross-section ducts, the well-balanced approach of Greenberg and Leroux [4] (see also [7] and [5]) is an efficient tool to improve the precision.
- For two-fluid flows the pressure oscillations phenomenon (see [6] and [2] for instance) can be cured by a recent tool developed in [3] and [1]. It is based on an ALE (Arbitrary Lagrangian Eulerian) scheme followed by a random sampling projection step.

In this paper, we show that is is possible to mix the two approaches in order to design an efficient scheme for computing two-fluid flows in variable cross-section ducts.

Philippe Helluy, Jonathan Jung
IRMA, Université de Strasbourg, 7 rue Descartes 67084 Strasbourg,
e-mail: jonathan.jung@math.unistra.fr

1 A well-balanced two-fluid ALE solver

1.1 Model

We consider the flow of a mixture of two compressible fluids (a gas (1) and a liquid (2), for instance) in a cross-section duct. The time variable is noted t and the space variable along the duct is x. We denote by $A(x)$ the cross-section at position x. The unknowns are the density $\rho(x,t)$, the velocity $u(x,t)$, the internal energy $e(x,t)$ and the fraction of gas $\varphi(x,t)$. Following Greenberg and Leroux [4] it is now classical to consider the cross-section A as an artificial unknown. The equations are the Euler equations in a duct, which read

$$\partial_t(A\rho) + \partial_x(A\rho u) = 0, \tag{1}$$

$$\partial_t(A\rho u) + \partial_x(A(\rho u^2 + p)) = p\partial_x A, \tag{2}$$

$$\partial_t(A\rho E) + \partial_x(A(\rho E + p)u) = 0, \tag{3}$$

$$\partial_t(A\rho\varphi) + \partial_x(A\rho\varphi u) = 0, \tag{4}$$

$$\partial_t A = 0, \tag{5}$$

with

$$p = p(\rho, e, \varphi), \tag{6}$$

$$E = e + \frac{u^2}{2}. \tag{7}$$

Without loss of generality, in this paper we consider a stiffened gas pressure law (see [8] and included references)

$$p(\rho, e, \varphi) = (\gamma(\varphi) - 1)\rho e - \gamma(\varphi)\pi(\varphi). \tag{8}$$

The mixture pressure law parameters $\gamma(\varphi)$ and $\pi(\varphi)$ are obtained from the pure fluid parameters $\gamma_i > 1, \pi_i, i = 1, 2$ thanks to the following interpolation, which is justified in [2]

$$\frac{1}{\gamma(\varphi) - 1} = \varphi \frac{1}{\gamma_1 - 1} + (1 - \varphi)\frac{1}{\gamma_2 - 1}, \tag{9}$$

$$\frac{\gamma(\varphi)\pi(\varphi)}{\gamma(\varphi) - 1} = \varphi \frac{\gamma_1 \pi_1}{\gamma_1 - 1} + (1 - \varphi)\frac{\gamma_2 \pi_2}{\gamma_2 - 1}. \tag{10}$$

We define the vector of conservative variables

$$W = (A\rho, A\rho u, A\rho E, A\rho\varphi, A)^T. \tag{11}$$

The conservative flux is

$$F(W) = (A\rho u, A(\rho u^2 + p), A(\rho E + p)u, A\rho\varphi u, 0)^T, \qquad (12)$$

and the non-conservative source term is

$$S = (0, p\partial_x A, 0, 0, 0), \qquad (13)$$

such that the system (1)-(5) becomes

$$\partial_t W + \partial_x F(W) = S(W). \qquad (14)$$

We define the vector of primitive variables

$$Y = (\rho, u, p, \varphi, A)^T. \qquad (15)$$

We define also the following quantities

$$Q = \text{mass flow rate} = \rho A u, \qquad (16)$$
$$s = \text{entropy} = (p + \pi(\varphi))\rho^{-\gamma(\varphi)}, \qquad (17)$$
$$h = \text{enthalpy} = e + \frac{p}{\rho}, \qquad (18)$$
$$H = \text{total enthalpy} = h + \frac{u^2}{2}. \qquad (19)$$

The entropy is solution of the partial differential equation

$$Tds = de - \frac{p}{\rho^2}d\rho + \lambda d\varphi. \qquad (20)$$

It is useful to express also the pressure p and the enthalpy h as functions of (ρ, s, φ)

$$p = p(\rho, s, \varphi), \quad h = h(\rho, s, \varphi). \qquad (21)$$

Then in these variables the sound speed c satisfies

$$c^2 = p_\rho = \rho h_\rho. \qquad (22)$$

The jacobian matrix $F'(W)$ in system (14) admits real eigenvalues

$$\lambda_0 = 0, \quad \lambda_1 = u - c, \quad \lambda_2 = \lambda_3 = u, \quad \lambda_4 = u + c. \qquad (23)$$

However, the system may be resonant (when $\lambda_0 = \lambda_1$ or $\lambda_0 = \lambda_4$.) The quantities φ, s, Q and H are independant Riemann invariants of the stationnary wave λ_0. In the sequel, the vector of "stationary" variables Z will play a particular role

$$Z = (A, \varphi, s, Q, H)^T. \tag{24}$$

1.2 VFRoe ALE numerical flux

We recall now the principles of the VFRoe solver. We first consider a arbitrary change of variables $U = U(W)$. In practice, we will take the set of primitive variables $U = Y$ (15) or the set of stationnary variables $U = Z$ (24). The vector U satisfies a non-conservative set of equations

$$\partial_t U + C(U)\partial_x U = 0. \tag{25}$$

The system (1)-(5) is approximated by a finite volume scheme with cells $]x_{i-1/2}, x_{i+1/2}[$, $i \in \mathbb{Z}$. We denote by τ the time step and by $\Delta x_i = x_{i+1/2} - x_{i-1/2}$ the size of cell i. We denote by W_i^n the conservative variables in cell i at time step n. The cross-section A is approximated by a piecewise constant function, $A = A_i$ in cell i.

We consider first a very general scheme where the boundary of the cell $x_{i+1/2}$ moves at the velocity $v_{i+1/2}^n$ between time steps n and $n+1$, thus we have

$$x_{i+1/2}^{n+1} = x_{i+1/2}^n + \tau v_{i+1/2}^n. \tag{26}$$

In a VFRoe-type scheme, we have to define linearized Riemann problems at interface $i + 1/2$ between the state $W_L = W_i^n$ and $W_R = W_{i+1}^n$, we introduce

$$\overline{U} = \frac{1}{2}(U_L + U_R). \tag{27}$$

In this way, it is possibe to define

$$\overline{W} = W(\overline{U}), \quad \overline{C} = C(\overline{U}). \tag{28}$$

We then consider the linearized Riemann problem

$$\partial_t U + \overline{C}\partial_x U = 0, \tag{29}$$

$$U(x, 0) = \begin{cases} U_L \text{ if } x < 0, \\ U_R \text{ if } x > 0. \end{cases} \tag{30}$$

We denote its solution by

$$U(U_L, U_R, \frac{x}{t}) = U(x, t). \tag{31}$$

Because of the stationary wave, $U(U_L, U_R, \frac{x}{t})$ is generally discontinuous at $x/t = 0$. We are then able to define a discontinuous Arbitrary Lagrangian Eulerian (ALE) numerical flux

$$F(W_L, W_R, v^{\pm}) := F(W(U(U_L, U_R, v^{\pm}))) - vW(U(U_L, U_R, v^{\pm})). \qquad (32)$$

The sizes of the cells evolve as

$$\Delta x_i^{n+1} = \Delta x_i^n + \tau(v_{i+1/2}^n - v_{i-1/2}^n). \qquad (33)$$

If $v_{i+1/2}^n \leq 0$ and $v_{i-1/2}^n \geq 0$, the ALE scheme is

$$\Delta x_i^{n+1} W_i^{n+1,-} - \Delta x_i^n W_i^n +$$
$$\tau \left(F(W_i^n, W_{i+1}^n, v_{i+1/2}^{n,-}) - F(W_{i-1}^n, W_i^n, v_{i-1/2}^{n,+}) \right) = 0. \qquad (34)$$

If $v_{i+1/2}^n > 0$ then we have to add the following term to the left of the previous equation

$$\tau \left(F(W_i^n, W_{i+1}^n, 0^-) - F(W_i^n, W_{i+1}^n, 0^+) \right). \qquad (35)$$

If $v_{i-1/2}^n < 0$ then we have to add also the following term

$$\tau \left(F(W_{i-1}^n, W_i^n, 0^-) - F(W_{i-1}^n, W_i^n, 0^+) \right). \qquad (36)$$

1.3 ALE velocity

We have now to detail the choice of the variable U and the velocity v according to the data W_L and W_R. The idea is to use the classical well-balanced scheme everywhere but at the interface between the two fluids, where we use the Lagrange flux. When our initial data satisfy $\varphi \in \{0, 1\}$, the algorithm reads

- If we are not at the interface, i.e. if $\varphi_L = \varphi_R$, we take $U = Z$ and $v = 0$. This choice corresponds to the VFRoe well-balanced scheme described in [5].
- If we are at the interface, i.e. if $\varphi_L \neq \varphi_R$ then we choose $U = Y$. This choice ensures that the linearized Riemann solver presents no jump of pressure and velocity at the contact discontinuity. We thus denote by $u^*(W_L, W_R)$ and $p^*(W_L, W_R)$ the velocity and the pressure at the contact. We take $v = u^*(W_L, W_R)$, $A^* = A_L$ if $v < 0$ and $A^* = A_R$ if $v > 0$. The lagrangian numerical flux then takes the form

$$F(W_L, W_R, v^{\pm}) = (0, A^* p^*, A^* u^* p^*, 0, -A^* u^*)^T. \qquad (37)$$

1.4 Glimm remap

We go back to the original Euler grid by the Glimm procedure.

We construct a sequence of pseudo-random numbers $\omega_n \in [0, 1[$. In practice, we consider the $(5, 3)$ van der Corput sequence [1]. According to this number we take

$$W_i^{n+1} = W_{i-1}^{n+1,-} \text{ if } \omega_n < \frac{\tau_n}{\Delta x_i} \max(v_{i-1/2}^n, 0), \tag{38}$$

$$W_i^{n+1} = W_{i+1}^{n+1,-} \text{ if } \omega_n > 1 + \frac{\tau_n}{\Delta x_i} \min(v_{i+1/2}^n, 0), \tag{39}$$

$$W_i^n = W_i^{n+1,-} \text{ if } \frac{\tau_n}{\Delta x_i} \max(v_{i-1/2}^n, 0) \le \omega_n \le 1 + \frac{\tau_n}{\Delta x_i} \min(v_{i+1/2}^n, 0). \tag{40}$$

1.5 Properties of the scheme

The constructed scheme has many interesting properties:

- it is well-balanced in the sense that it preserves exactly all stationary states (i.e. initial data for which the quantities φ, s, Q, H are constant);
- for constant cross-section ducts, it computes exactly the contact discontinuities, with no smearing of the density and the mass fraction;
- if at the initial time the mass fraction is in $\{0, 1\}$, then this property is exactly preserved at any time.

For detailed proofs, we refer to [5] and [1]. Some other subtleties are given in the same references. For instance, the change of variables $Z = Z(W)$ is not always invertible. This implies to define a special procedure for constructing completely rigorously the well-balanced VFRoe solver.

2 Numerical results

In order to test our algorithm, we consider a Riemann problem for which we know the exact solution. The initial data are discontinuous at $x = 1$. The data of the problem are given in Table 1.

The pressure law parameters are $\gamma_1 = 1.4$, $\pi_1 = 0$, $\gamma_2 = 1.6$ and $\pi_2 = 2$. We compute the solution on the domain $[0.4; 1.6]$ with approximately 2000 cells. The final time is $T = 0.2$ and the CFL number is 0.6. The density, the velocity and the pressure are represented on Figs. 1, 2 and 3. We observe an excellent agreement between the exact and the approximate solution. The mass fraction is not represented: it is not smeared at all and perfectly matches the exact solution.

quantity	Left	Right
ρ	2	3.230672602
u	0.5	-0.4442565900
p	1	12
φ	1	0
A	1.5	1

Table 1 Numerical results. Data of the Riemann problem

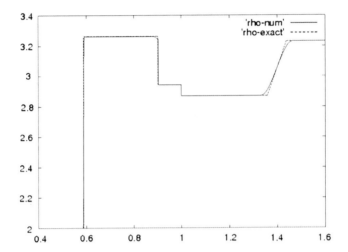

Fig. 1 Two-fluid, discontinuous cross-section Riemann problem. Density plot. Comparison of the exact solution (dotted line) and the approximate one (continuous line)

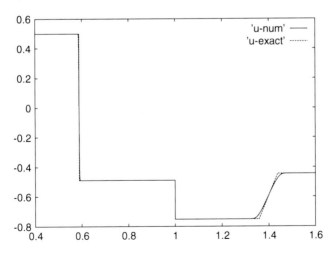

Fig. 2 Two-fluid, discontinuous cross-section Riemann problem. Pressure plot. Comparison of the exact solution (dotted line) and the approximate one (continuous line)

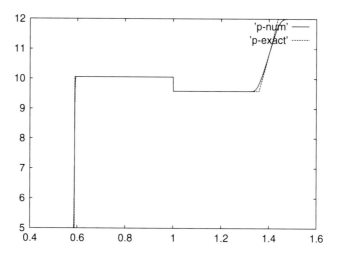

Fig. 3 Two-fluid, discontinuous cross-section Riemann problem. Velocity plot. Comparison of the exact solution (dotted line) and the approximate one (continuous line)

3 Conclusion

We have constructed and validated a new scheme for computing two-fluid flows in variable cross-section ducts. Our scheme relies on two ingredients:

- a well-balanced approach for dealing with the varying cross-section;
- a Lagrange plus remap technique in order to avoid pressure oscillations at the interface. The random sampling remap ensures that the interface is not diffused at all.

On preliminary test cases, our approach gives very satisfactory results. We intend to apply it to the computation of the oscillations of cavitation bubbles. More results will be presented at the conference.

The authors wish to thank Jean-Marc Hérard for many fruitful discussions.

References

1. M. Bachmann, P. Helluy, H. Mathis, S. Mueller. Random sampling remap for compressible two-phase flows. Preprint HAL http://hal.archives-ouvertes.fr/hal-00546919/fr/
2. T. Barberon, P. Helluy, S. Rouy. Practical computation of axisymmetrical multifluid flows. Int. J. Finite Vol. 1 (2004), no. 1, 34 pp. http://ijfv.org
3. C. Chalons, F. Coquel. Computing material fronts with a Lagrange-Projection approach. HYP2010 Proc. http://hal.archives-ouvertes.fr/hal-00548938/fr/
4. J.-M. Greenberg, A.Y., Leroux. A well balanced scheme for the numerical processing of source terms in hyperbolic equations", SIAM J. Num. Anal., vol. 33 (1), pp. 1–16, 1996.

5. P. Helluy, J.-M. Hérard, H. Mathis. A Well- Balanced Approximate Riemann Solver for Variable Cross- Section Compressible Flows. AIAA-2009-3540. 19th AIAA Computational Fluid Dynamics. June 2009.
6. S. Karni. Multicomponent flow calculations by a consistent primitive algorithm. J. Comput. Phys. 112 (1994), no. 1, 31–43.
7. D. Kroner, M.-D. Thanh. Numerical solution to compressible flows in a nozzle with variable cross-section", SIAM J. Numer. Anal., vol. 43(2), pp. 796–824, 2006.
8. R. Saurel, R. Abgrall. A simple method for compressible multifluid flows. SIAM J. Sci. Comput. 21 (1999), no. 3, 11151145.

The paper is in final form and no similar paper has been or is being submitted elsewhere.

Discretization of the viscous dissipation term with the MAC scheme

F. Babik, R. Herbin, W. Kheriji, and J.-C. Latché

Abstract We propose a discretization for the MAC scheme of the viscous dissipation term $\tau(u) : \nabla u$ (where $\tau(u)$ stands for the shear stress tensor associated to the velocity field u), which is suitable to obtain an unconditionally stable scheme for the compressible Navier-Stokes equations. It is also shown, in some model cases, to ensure the strong convergence in L^1 of the dissipation term.

Keywords compressible Navier-Stokes, MAC scheme
MSC2010: 65M12

1 Introduction

Let us consider the compressible Navier-Stokes equations, which may be written as:

$$\partial_t \rho + \mathrm{div}(\rho u) = 0, \tag{1a}$$

$$\partial_t (\rho u) + \mathrm{div}(\rho u \otimes u) + \nabla p - \mathrm{div}(\tau(u)) = 0, \tag{1b}$$

$$\partial_t (\rho e) + \mathrm{div}(\rho e u) + p\, \mathrm{div}\, u + \mathrm{div}(q) = \tau(u) : \nabla u, \tag{1c}$$

$$\rho = \wp(p, e), \tag{1d}$$

where t stands for the time, ρ, u, p and e are the density, velocity, pressure and internal energy in the flow, $\tau(u)$ stands for the shear stress tensor, q for the energy

---

F. Babik, W. Kheriji, and J.-C. Latché
Institut de Radioprotection et Sûreté Nucléaire (IRSN),
e-mail: [fabrice.babik,walid.kheriji,jean-claude.latche]@irsn.fr

R. Herbin
Université de Provence, e-mail: herbin@cmi.univ-mrs.fr

diffusion flux, and the function $\wp$ is the equation of state. This system of equations is posed over $\Omega \times (0, T)$, where Ω is a domain of $\mathbb{R}^d$, $d \leq 3$. It must be supplemented by a closure relation for $\tau(u)$ and for q, assumed to be:

$$\tau(u) = \mu(\nabla u + \nabla^t u) - \frac{2\mu}{3} \operatorname{div} u \, I, \quad q = -\lambda \nabla e, \tag{2}$$

where μ and λ stand for two (possibly depending on x) positive parameters.

Let us suppose, for the sake of simplicity, that u is prescribed to zero on the whole boundary, and that the system is adiabatic, i.e. $q \cdot n = 0$ on $\partial \Omega$. Then, formally, taking the inner product of (1b) with u and integrating over Ω, integrating (1c) over Ω, and, finally, summing both relations yields the stability estimate:

$$\frac{d}{dt} \int_\Omega \left[\frac{1}{2} \rho |u|^2 + \rho e\right] dx \leq 0. \tag{3}$$

If we suppose that the equation of state may be set under the form $p = f(\rho, e)$ with $f(\cdot, 0) = 0$ and $f(0, \cdot) = 0$, Equation (1c) implies that e remains positive (still at least formally), and so (3) yields a control on the unknown. Mimicking this computation at the discrete level necessitates to check some arguments, among them:

(i) to have available a discrete counterpart to the relation:

$$\int_\Omega \left[\partial_t (\rho u) + \operatorname{div}(\rho u \otimes u)\right] \cdot u \, dx = \frac{d}{dt} \int_\Omega \frac{1}{2} \rho |u|^2 \, dx.$$

(ii) to identify the integral of the dissipation term at the right-hand side of the discrete counterpart of (1c) with the (discrete) L^2 inner product between the velocity and the diffusion term in the discrete momentum balance equation (1b).

(iii) to be able to prove that the right-hand side of (1c) is non-negative, in order to preserve the positivity of the internal energy.

The point (i) is extensively discussed in [5] (see also [6]), and is not treated here. Indeed, we focus here on a discretization technique which allows to obtain (ii) and (iii) with the usual Marker and Cell (MAC) discretization [3, 4], and which is implemented in the ISIS free software developed at IRSN [8] on the basis of the software component library PELICANS [10]. We complete the presentation by showing how (ii) may also be used, in some model problems, to prove the convergence in L^1 of the dissipation term.

2 Discretization of the dissipation term

2.1 The two-dimensional case

Let us begin with a two-dimensional case. The first step is to propose a discretization for the diffusion term in the momentum equation. We begin with the x-component of the velocity, for which we write a balance equation on $K^x_{i-\frac{1}{2},j} = (x_{i-1}, x_i) \times (y_{j-\frac{1}{2}}, y_{j+\frac{1}{2}})$ (see Figs. 1 and 2 for the notations). Integrating the x component of the momentum balance equation over $K^x_{i-\frac{1}{2},j}$, we get for the diffusion term:

$$\bar{T}^{\text{dif}}_{i-\frac{1}{2},j} = -\left[\int_{K^x_{i-\frac{1}{2},j}} \text{div}[\tau(u)]\,\text{d}x\right]\cdot e^{(x)} = -\left[\int_{\partial K^x_{i-\frac{1}{2},j}} \tau(u)\,n\,\text{d}\gamma\right]\cdot e^{(x)}, \quad (4)$$

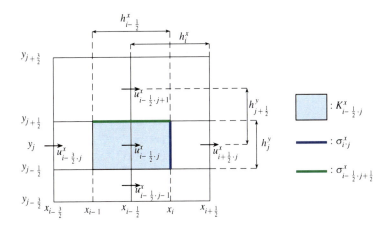

Fig. 1 Dual cell for the x-component of the velocity

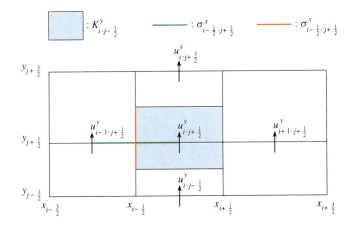

Fig. 2 Dual cell for the y-component of the velocity

where $e^{(x)}$ stands for the first vector of the canonical basis of $\mathbb{R}^2$. We denote by $\sigma_{i,j}^x$ the right face of $K_{i-\frac{1}{2},j}^x$, i.e. $\sigma_{i,j}^x = \{x_i\} \times (y_{j-\frac{1}{2}}, y_{j+\frac{1}{2}})$. Splitting the boundary integral in (4), the part of $\bar{T}_{i-\frac{1}{2},j}^{\mathrm{dif}}$ associated to $\sigma_{i,j}^x$, also referred to as the viscous flux through $\sigma_{i,j}^x$, reads:

$$-\left[\int_{\sigma_{i,j}^x} \tau(u)\, n\, d\gamma\right] \cdot e^{(x)} = -2\int_{\sigma_{i,j}^x} \mu\, \partial_x u^x\, d\gamma + \frac{2}{3}\int_{\sigma_{i,j}^x} \mu\, (\partial_x u^x + \partial_y u^y)\, d\gamma,$$

and the usual finite difference technique yields the following approximation for this term:

$$-\frac{4}{3}\int_{\sigma_{i,j}^x} \mu\, \partial_x u^x\, d\gamma + \frac{2}{3}\int_{\sigma_{i,j}^x} \mu\, \partial_y u^y\, d\gamma$$

$$\approx -\frac{4}{3}\mu_{i,j}\frac{h_j^y}{h_i^x}(u_{i+\frac{1}{2},j}^x - u_{i-\frac{1}{2},j}^x) + \frac{2}{3}\mu_{i,j}\frac{h_j^y}{h_j^y}(u_{i,j+\frac{1}{2}}^y - u_{i,j-\frac{1}{2}}^y), \quad (5)$$

where $\mu_{i,j}$ is an approximation of the viscosity at the face $\sigma_{i,j}^x$. Similarly, let $\sigma_{i-\frac{1}{2},j+\frac{1}{2}}^x = (x_{i-1}, x_i) \times \{y_{j+\frac{1}{2}}\}$ be the top edge of the cell. Then:

$$-\left[\int_{\sigma_{i-\frac{1}{2},j+\frac{1}{2}}^x} \tau(u)\, n\, d\gamma\right] \cdot e^{(x)} = -\int_{\sigma_{i-\frac{1}{2},j+\frac{1}{2}}^x} \mu\, (\partial_y u^x + \partial_x u^y)\, d\gamma$$

$$\approx -\mu_{i-\frac{1}{2},j+\frac{1}{2}}\left[\frac{h_{i-\frac{1}{2}}^x}{h_{j+\frac{1}{2}}^y}(u_{i-\frac{1}{2},j+1}^x - u_{i-\frac{1}{2},j}^x) + \frac{h_{i-\frac{1}{2}}^x}{h_{i-\frac{1}{2}}^x}(u_{i,j+\frac{1}{2}}^y - u_{i-1,j+\frac{1}{2}}^y)\right],$$

where $\mu_{i-\frac{1}{2},j+\frac{1}{2}}$ stands for an approximation of the viscosity at the edge $\sigma_{i-\frac{1}{2},j+\frac{1}{2}}^x$.

Let us now multiply each discrete equation for u^x by the corresponding degree of freedom of a velocity field v (i.e. the balance over $K_{i-\frac{1}{2},j}^x$ by $v_{i-\frac{1}{2},j}^x$) and sum over i and j. The viscous flux at the face $\sigma_{i,j}^x$ appears twice in the sum, once multiplied by $v_{i-\frac{1}{2},j}^x$ and the second one by $-v_{i+\frac{1}{2},j}^x$, and the corresponding term reads:

$$T_{i,j}^{\mathrm{dis}}(u,v) =$$

$$\mu_{i,j}\left[-\frac{4}{3}\frac{h_j^y}{h_i^x}(u_{i+\frac{1}{2},j}^x - u_{i-\frac{1}{2},j}^x) + \frac{2}{3}\frac{h_j^y}{h_j^y}(u_{i,j+\frac{1}{2}}^y - u_{i,j-\frac{1}{2}}^y)\right](v_{i-\frac{1}{2},j}^x - v_{i+\frac{1}{2},j}^x)$$

$$= \mu_{i,j}\, h_j^y h_i^x \left[\frac{4}{3}\frac{u_{i+\frac{1}{2},j}^x - u_{i-\frac{1}{2},j}^x}{h_i^x} - \frac{2}{3}\frac{u_{i,j+\frac{1}{2}}^y - u_{i,j-\frac{1}{2}}^y}{h_j^y}\right]\frac{v_{i+\frac{1}{2},j}^x - v_{i-\frac{1}{2},j}^x}{h_i^x}. \quad (6)$$

Similarly, the term associated to $\sigma^x_{i-\frac{1}{2},j+\frac{1}{2}}$ appears multiplied by $v^x_{i-\frac{1}{2},j}$ and by $-v^x_{i-\frac{1}{2},j+1}$, and we get:

$$T^{\text{dis}}_{i-\frac{1}{2},j+\frac{1}{2}}(u,v) = \mu_{i-\frac{1}{2},j+\frac{1}{2}} \, h^x_{i-\frac{1}{2}} h^y_{j+\frac{1}{2}}$$

$$\left[\frac{u^x_{i-\frac{1}{2},j+1} - u^x_{i-\frac{1}{2},j}}{h^y_{j+\frac{1}{2}}} + \frac{u^y_{i,j+\frac{1}{2}} - u^y_{i-1,j+\frac{1}{2}}}{h^x_{i-\frac{1}{2}}} \right] \frac{v^x_{i-\frac{1}{2},j+1} - v^x_{i-\frac{1}{2},j}}{h^y_{j+\frac{1}{2}}}. \quad (7)$$

Let us now define the discrete gradient of the velocity as follows:
- The derivatives involved in the divergence, $\partial^{\mathcal{M}}_x u^x$ and $\partial^{\mathcal{M}}_y u^y$, are defined over the primal cells by:

$$\partial^{\mathcal{M}}_x u^x(x) = \frac{u^x_{i+\frac{1}{2},j} - u^x_{i-\frac{1}{2},j}}{h^x_i}, \quad \partial^{\mathcal{M}}_y u^y(x) = \frac{u^y_{i,j+\frac{1}{2}} - u^y_{i,j-\frac{1}{2}}}{h^y_j}, \quad \forall x \in K_{i,j}. \quad (8)$$

- For the other derivatives, we introduce another mesh which is vertex-centred, and we denote by K^{xy} the generic cell of this new mesh, with $K^{xy}_{i+\frac{1}{2},j+\frac{1}{2}} = (x_i, x_{i+1}) \times (y_j, y_{j+1})$. Then, $\forall x \in K^{xy}_{i+\frac{1}{2},j+\frac{1}{2}}$:

$$\partial^{\mathcal{M}}_y u^x(x) = \frac{u^x_{i+\frac{1}{2},j+1} - u^x_{i+\frac{1}{2},j}}{h^y_{j+\frac{1}{2}}}, \quad \partial^{\mathcal{M}}_x u^y(x) = \frac{u^y_{i+1,j+\frac{1}{2}} - u^y_{i,j+\frac{1}{2}}}{h^x_{i+\frac{1}{2}}}. \quad (9)$$

With this definition, we get:

$$T^{\text{dis}}_{i,j}(u,v) = \mu_{i,j} \int_{K_{i,j}} \left[\frac{4}{3} \partial^{\mathcal{M}}_x u^x - \frac{2}{3} \partial^{\mathcal{M}}_y u^y \right] \partial^{\mathcal{M}}_x v^x \, dx,$$

and:

$$T^{\text{dis}}_{i-\frac{1}{2},j+\frac{1}{2}}(u,v) = \mu_{i-\frac{1}{2},j+\frac{1}{2}} \int_{K^{xy}_{i-\frac{1}{2},j+\frac{1}{2}}} (\partial^{\mathcal{M}}_y u^x + \partial^{\mathcal{M}}_x u^y) \, \partial^{\mathcal{M}}_y v^x \, dx.$$

Let us now perform the same operations for the y-component of the velocity. Doing so, we are lead to introduce an approximation of the viscosity at the edge $\sigma^y_{i-\frac{1}{2},j+\frac{1}{2}} = \{x_{i-\frac{1}{2}}\} \times (y_j, y_{j+1})$ (see Fig. 2). Let us suppose that we take the same approximation as on $\sigma^x_{i-\frac{1}{2},j+\frac{1}{2}}$. Then, the same argument yields that multiplying each discrete equation for u^x and for u^y by the corresponding degree of freedom of a velocity field v, we obtain a dissipation term which reads:

$$T^{\text{dis}}(u,v) = \int_\Omega \tau^{\mathcal{M}}(u) : \nabla^{\mathcal{M}} v \, dx, \quad (10)$$

where $\nabla^{\mathcal{M}}$ is the discrete gradient defined by (8)-(9) and $\tau^{\mathcal{M}}$ the discrete tensor:

$$\tau^{\mathcal{M}}(u) = \begin{bmatrix} 2\mu\,\partial_x^{\mathcal{M}} u_x & \mu^{xy}\,(\partial_y^{\mathcal{M}} u_x + \partial_x^{\mathcal{M}} u_y) \\ \mu^{xy}\,(\partial_y^{\mathcal{M}} u_x + \partial_x^{\mathcal{M}} u_y) & 2\mu\,\partial_y^{\mathcal{M}} u_y \end{bmatrix} - \frac{2}{3}\mu\,(\partial_x^{\mathcal{M}} u_x + \partial_y^{\mathcal{M}} u_y)\,I, \quad (11)$$

where μ is the viscosity defined on the primal mesh by $\mu(x) = \mu_{i,j}$, $\forall x \in K_{i,j}$ and μ^{xy} is the viscosity defined on the vertex-centred mesh, by $\mu(x) = \mu_{i+\frac{1}{2},j+\frac{1}{2}}$, $\forall x \in K_{i+\frac{1}{2},j+\frac{1}{2}}^{xy}$.

Now the form (10) suggests a natural to discretize the viscous dissipation term in the internal energy balance in order for the consistency property *(ii)* to hold. Indeed, if we simply set on each primal cell $K_{i,j}$:

$$(\tau(u) : \nabla u)_{i,j} = \frac{1}{|K_{i,j}|} \int_{K_{i,j}} \tau^{\mathcal{M}}(u) : \nabla^{\mathcal{M}} u \, dx, \quad (12)$$

then, thanks to (10), the property *(ii)* which reads:

$$T^{\text{dis}}(u,u) = \sum_{i,j} |K_{i,j}|\,(\tau(u) : \nabla u)_{i,j}.$$

holds. Furthermore, we get from Definition (11) that $\tau^{\mathcal{M}}(u)(x)$ is a symmetrical tensor, for any i, j and $x \in K_{i,j}$, and therefore an elementary algebraic argument yields:

$$(\tau(u) : \nabla u)_{i,j} = \frac{1}{|K_{i,j}|} \int_{K_{i,j}} \tau^{\mathcal{M}}(u) : \nabla^{\mathcal{M}} u \, dx$$

$$= \frac{1}{2|K_{i,j}|} \int_{K_{i,j}} \tau^{\mathcal{M}}(u) : \left[\nabla^{\mathcal{M}} u + (\nabla^{\mathcal{M}} u)^t\right] dx \geq 0.$$

Remark 1 (Approximation of the viscosity). Note that, for the symmetry of $\tau^{\mathcal{M}}(u)$ to hold, the choice of the same viscosity at the edges $\sigma_{i-\frac{1}{2},j+\frac{1}{2}}^x$ and $\sigma_{i-\frac{1}{2},j+\frac{1}{2}}^y$ is crucial even though other choices may appear natural. Assuming for instance the viscosity to be a function of an additional variable defined on the primal mesh, the following construction seems reasonable:

1. define a constant value for μ on each primal cell,
2. associate a value of μ to the primal edges, by taking the average between the value at the adjacent cells,

3. finally, split the integral of the shear stress over $\sigma^x_{i-\frac{1}{2},j+\frac{1}{2}}$ in two parts, one for the part included in the (top) boundary of $K_{i-1,j}$ and the second one in the boundary of $K_{i,j}$.

Then the viscosities on $\sigma^x_{i-\frac{1}{2},j+\frac{1}{2}}$ and $\sigma^y_{i-\frac{1}{2},j+\frac{1}{2}}$ coincide only for uniform meshes, and, in the general case, the symmetry of $\tau^{\mathcal{M}}(u)$ is lost.

2.2 Extension to the three-dimensional case

Extending the computations of the preceding section to three space dimensions yields the following construction.

- First, define three new meshes, which are "edge-centred": $K^{xy}_{i+\frac{1}{2},j+\frac{1}{2},k} = (x_i, x_{i+1}) \times (y_i, y_{j+1}) \times (z_{k-\frac{1}{2}}, z_{k+\frac{1}{2}})$ is staggered from the primal mesh $K_{i,j,k}$ in the x and y direction, $K^{xz}_{i+\frac{1}{2},j,k+\frac{1}{2}}$ in the x and z direction, and $K^{yz}_{i,j+\frac{1}{2},k+\frac{1}{2}}$ in the y and z direction.
- The partial derivatives of the velocity components are then defined as piecewise constant functions, the value of which is obtained by natural finite differences:
    - for $\partial^{\mathcal{M}}_x u^x$, $\partial^{\mathcal{M}}_y u^y$ and $\partial^{\mathcal{M}}_z u^z$, on the primal mesh,
    - for $\partial^{\mathcal{M}}_y u^x$ and $\partial^{\mathcal{M}}_x u^y$ on the cells ($K^{xy}_{i+\frac{1}{2},j+\frac{1}{2},k}$),
    - for $\partial^{\mathcal{M}}_z u^x$ and $\partial^{\mathcal{M}}_x u^z$ on the cells ($K^{xz}_{i+\frac{1}{2},j,k+\frac{1}{2}}$),
    - for $\partial^{\mathcal{M}}_y u^z$ and $\partial^{\mathcal{M}}_z u^y$ on the cells ($K^{yz}_{i,j+\frac{1}{2},k+\frac{1}{2}}$).
- Then, define four families of values for the viscosity field, μ, μ^{xy}, μ^{xz} and μ^{yz}, associated to the primal and the three edge-centred meshes respectively.
- The shear stress tensor is obtained by the extension of (11) to $d = 3$.
- And, finally, the dissipation term is given by (12).

3 A strong convergence result

We conclude this paper by showing how the consistency property *(ii)* may be used, in some particular cases, to obtain the strong convergence of the dissipation term, and then pass to the limit in a coupled equation having the dissipation term as right-hand side. To this purpose, let us just address the model problem:

$$-\Delta \underline{u} = \underline{f} \text{ in } \Omega = (0,1) \times (0,1), \qquad \underline{u} = 0 \text{ on } \partial\Omega, \tag{13}$$

with $\underline{u}$ and $\underline{f}$ two scalar functions, $f \in L^2(\Omega)$. Let us suppose that this problem is discretized by the usual finite volume technique, with the uniform MAC mesh associated to the x-component of the velocity. We define a discrete function as a piecewise constant function, vanishing on the left and right sides of the domain (so on the left and right stripes of staggered (half-)meshes adjacent to these boundaries), and we define the discrete H^1-norm of a discrete function v by:

$$\|v\|_1^2 = \int_\Omega (\partial_x^\mathcal{M} v)^2 + (\partial_y^\mathcal{M} v)^2 \, dx.$$

Let $(\mathcal{M}^{(n)})_{n \in \mathbb{N}}$ be a sequence of such meshes, with a step h^n tending to zero, and let $(u^{(n)})_{n \in \mathbb{N}}$ be the corresponding sequence of discrete solutions. Then, with the variational technique employed in the preceding section, we get, with the usual discretization of the right-hand side:

$$\|u^{(n)}\|_1^2 = \int_\Omega (\partial_x^\mathcal{M} u^{(n)})^2 + (\partial_y^\mathcal{M} u^{(n)})^2 \, dx = \int_\Omega \underline{f} u^{(n)} \, dx. \qquad (14)$$

Since the discrete H^1-norm controls the L^2-norm (*i.e.* a discrete Poincaré inequality holds [2]), this yields a uniform bound for the sequence $(u^{(n)})_{n \in \mathbb{N}}$ in discrete H^1-norm. Hence the sequence $(u^{(n)})_{n \in \mathbb{N}}$ converges in $L^2(\Omega)$ to a function $\bar{u} \in H_0^1(\Omega)$, possibly up to the extraction of a subsequence [2], and he discrete derivatives $(\partial_x^\mathcal{M} u^{(n)})_{n \in \mathbb{N}}$ and $(\partial_y^\mathcal{M} u^{(n)})_{n \in \mathbb{N}}$ weakly converge in $L^2(\Omega)$ to $\partial_x \bar{u}$ and $\partial_y \bar{u}$ respectively. This allows to pass to the limit in the scheme, and we obtain that $\bar{u}$ satisfies the continuous equation (13). Thus, taking $\bar{u}$ as a test function in the variational form of (13):

$$\int_\Omega (\partial_x \bar{u})^2 + (\partial_y \bar{u})^2 \, dx = \int_\Omega \underline{f} \bar{u} \, dx.$$

But, passing to the limit in (14), we get:

$$\lim_{n \to \infty} \int_\Omega (\partial_x^\mathcal{M} u^{(n)})^2 + (\partial_y^\mathcal{M} u^{(n)})^2 \, dx = \lim_{n \to \infty} \int_\Omega \underline{f} u^{(n)} \, dx = \int_\Omega \underline{f} \bar{u} \, dx,$$

which, comparing to the preceding relation, yields:

$$\lim_{n \to \infty} \int_\Omega (\partial_x^\mathcal{M} u^{(n)})^2 + (\partial_y^\mathcal{M} u^{(n)})^2 \, dx = \int_\Omega (\partial_x \bar{u})^2 + (\partial_y \bar{u})^2 \, dx.$$

Since the discrete gradient weakly converges and its norm converges to the norm of the limit, the discrete gradient strongly converges in $L^2(\Omega)^2$ to the gradient of the solution. Let us now imagine that Equation (13) is coupled to a balance equation for another variable, the right-hand side of which is $|\nabla \underline{u}|^2$; this situation occurs for instance in models involving ohmic losses [1], or RANS turbulence models [9]. The discretization (12) of the dissipation term in the cell K, which reads here:

$$\left(|\nabla u^{(n)}|^2\right)_K = \frac{1}{|K|} \int_K (\partial_x^{\mathcal{M}} u^{(n)})^2 + (\partial_y^{\mathcal{M}} u^{(n)})^2 \, dx,$$

yields a convergent right-hand side, in the sense that, for any regular function $\varphi \in C_c^\infty(\Omega)$, we have:

$$\lim_{n \to \infty} \sum_K \int_K \left(|\nabla u^{(n)}|^2\right)_K \varphi \, dx = \int_\Omega |\nabla \underline{u}|^2 \varphi \, dx.$$

(A declination of) this argument has been used to prove the convergence of numerical schemes in [1,9].

References

1. A. Bradji, R. Herbin: Discretization of the coupled heat and electrical diffusion problems by the finite element and the finite volume methods. IMA Journal of Numerical Analysis, **28**, 469–495 (2008).
2. R. Eymard, T. Gallouët, R. Herbin, Finite Volume Methods. Handbook of Numerical Analysis, Volume VII, 713–1020, North Holland (2000).
3. F.H. Harlow, J.E. Welsh: Numerical calculation of time-dependent viscous incompressible flow of fluid with free surface. Physics of Fluids, **8**, 2182–2189 (1965).
4. F.H. Harlow, A.A. Amsden: A numerical fluid dynamics calculation method for all flow speeds. Journal of Computational Physics, **8**, 197–213 (1971).
5. L. Gastaldo, R. Herbin, W. Kheriji, C. Lapuerta, J.-C. Latché: Staggered discretizations, pressure correction schemes and all speed barotropic flows. Finite Volumes for Complex Applications VI (FVCA VI), Prague, Czech Republic, June 2011.
6. R. Herbin, J.-C. Latché: A kinetic energy control in the MAC discretization of compressible Navier-Stokes equations. International Journal of Finite Volumes **2** (2010).
7. R. Herbin, W. Kheriji, J.-C. Latché: An unconditionally stable Finite Element-Finite Volume pressure correction scheme for compressible Navier-Stokes equations. In preparation (2011).
8. ISIS: a CFD computer code for the simulation of reactive turbulent flows,
   https://gforge.irsn.fr/gf/project/isis.
9. A. Larcher, J.-C. Latché: Convergence analysis of a finite element-finite volume scheme for a RANS turbulence model. Submitted (2011).
10. PELICANS: Collaborative Development Environment.
    https://gforge.irsn.fr/gf/project/pelicans.

The paper is in final form and no similar paper has been or is being submitted elsewhere.

A Sharp Contact Discontinuity Scheme for Multimaterial Models

Angelo Iollo, Thomas Milcent, and Haysam Telib

Abstract We present a method to capture the evolution of a contact discontinuity separating two different materials. This method builds on the ghost-fluid idea: a locally non-conservative scheme allows an accurate and stable simulation of problems involving non-miscible media that have significantly different physical properties. Compared to the ghost-fluid approach, the main difference is that with the present method no ghost fluid is introduced. Numerical illustrations involving one-dimensional interfaces show that with this scheme the contact discontinuity stays sharp and oscillation free.

Keywords sharp contact discontinuity, multimaterial, compressible elasticity
MSC2010: 65N08, 65Z05, 74S10, 76M12

1 Introduction

Physical and engineering problems that involve several materials are ubiquitous in nature and in applications: multi-phase flows, fluid-structure interaction, particle flows, to cite just a few examples. The main contributions in the direction of simulating these phenomena go back to [7] and [8] for the model and to [1] for a consistent and stable discretization. The idea is to model the eulerian stress tensor through a constitutive law reproducing the mechanical characteristics of the

---

Angelo Iollo, Thomas Milcent
Institut de Mathématiques de Bordeaux UMR 5251 Université Bordeaux 1 and INRIA Bordeaux-Sud Ouest, équipe-projet MC2, 351, Cours de la Libération, 33405 Talence, France, e-mail: angelo.iollo@math.u-bordeaux1.fr, thomas.milcent@math.u-bordeaux1.fr

Haysam Telib
Dipartimento di Ingegneria Aeronautica e Spaziale, Politecnico di Torino, C.so Duca degli Abruzzi 24, 10129 Torino, Italy, e-mail: haysam.telib@polito.it

medium under consideration. Hence, for example, an elastic material or a gas will be modeled by the same set of equations except for the constitutive law relating the deformation and the stress tensor. The system of conservation laws thus obtained can be cast in the framework of quasi-linear hyperbolic partial differential equations (PDEs). From the numerical view point this is convenient since classical integration schemes can be employed in each material. However, it turns out that the evolution of the interface, which is represented in this model by a contact discontinuity, is particularly delicate because standard Godunov schemes fail. In [1] it was shown that a simple and effective remedy to this problem is the definition of a ghost fluid across the interface. A remarkable application based on this approach is presented in [6]. This method, however, has the disadvantage that the interface is diffused over a certain number of grid points. From a practical view point this can be a serious drawback if one is interested in the geometric properties of the evolving interface, as, for example, in the case of surface tension or when the interface itself is elastic. In this paper we propose a simple first-order accurate method to recover a sharp interface description keeping the solution stable and non-oscillating.

2 The model

This approach was discussed in [4, 6, 7], and [8]. We develop here the principal elements of the formulation. The starting point is classical continuum mechanics. Let Ω_0 be the reference or initial configuration of a single material and Ω_t the deformed configuration at time t. We define $X(\xi, t)$ as the image at time t of a material point ξ belonging to the initial configuration, in the deformed configuration, i.e., $X : \Omega_0 \times [0, T] \longrightarrow \Omega_t$, $(\xi, t) \mapsto X(\xi, t)$, and the corresponding velocity field u as $u : \Omega_t \times [0, T] \longrightarrow \mathbb{R}^3$, $(x, t) \mapsto u(x, t)$ where $X_t(\xi, t) = u(X(\xi, t), t)$ completed by the initial condition $X(\xi, 0) = \xi$. Also we introduce the backward characteristics $Y(x, t)$ that for a time t and a point x in the deformed configuration, gives the corresponding initial point ξ in the initial configuration, i.e., $Y : \Omega_t \times [0, T] \longrightarrow \Omega_0$, $(x, t) \mapsto Y(x, t)$ with the initial condition $Y(x, 0) = x$. Of course, we have $[\nabla_\xi X(\xi, t)] = [\nabla_x Y(x, t)]^{-1}$ and $Y_t + (u \cdot \nabla)Y = 0$.

In elasticity, the internal energy is a function of the strain tensor $\nabla_\xi X$ and the entropy s: $W = W(\nabla_\xi X, s)$. The potential W has to be Galilean invariant and, eventually, isotropic. It can be proven that (Rivlin-Eriksen theorem [3]) this is the case if, and only if, $\mathcal{E}$, the energy, is expressed as a function of $s(\xi, t)$, the entropy, and of the invariants of $C(\xi, t) = [\nabla_\xi X]^T [\nabla_\xi X]$, the right Cauchy-Green tensor. The invariants often considered in the literature are $J(\xi, t) = \det([\nabla_\xi X])$, $\mathrm{Tr}(C(\xi, t))$ and $\mathrm{Tr}(\mathrm{Cof}(C(\xi, t)))$. We assume that the internal energy per unit volume is the sum of a term depending on volume variation and entropy, and a term accounting for isochoric deformation. In general the term relative to an isochoric transformation will also depend on entropy. Here, we will limit the discussion to materials where shear forces are conservative. The governing equations derived from the above formulation in the deformed configuration are:

$$\begin{cases} \rho_t + \text{div}_x(\rho u) = 0 \\ \rho(u_t + (u \cdot \nabla_x)u) - \text{div}_x \sigma = 0 \\ \rho(\varepsilon_t + (u \cdot \nabla_x)\varepsilon) - \sigma : \nabla_x u = 0 \\ Y_t + (u \cdot \nabla_x)Y = 0 \end{cases} \quad (1)$$

where $\sigma(x,t)$ is the Cauchy stress tensor in the physical domain. The unknowns are the backward characteristics of the problem $Y(x,t)$, the velocity $u(x,t)$, the internal energy per unit mass $\varepsilon(x,t) = W/\rho_0$ and the density $\rho(x,t)$. The initial velocity $u(x,0)$, the initial internal energy $\varepsilon(x,0)$ and $Y(x,0) = x$ are given. If the initial density $\rho_0(\xi)$ is known, the equation of mass conservation is actually redundant because $\rho(x,t) = \det(\nabla_x Y(x,t))\rho_0(Y(x,t))$.

To close the system, a constitutive law which connects σ to Y is necessary. In the deformed domain energy can be written

$$\mathcal{E} = \int_{\Omega_t} \left(W_{\text{vol}}(J,s) + W_{\text{iso}}(\text{Tr}(\overline{B}), \text{Tr}(\text{Cof}(\overline{B})) \right) J^{-1} dx \quad (2)$$

where

$$\overline{B}(x,t) = \frac{B(x,t)}{\det(B(x,t))^{\frac{1}{3}}} \qquad B(x,t) = [\nabla_x Y(x,t)]^{-1}[\nabla_x Y(x,t)]^{-T} \quad (3)$$

with $B(x,t)$ the left Cauchy-Green tensor and

$$J(x,t) = \det([\nabla_x Y(x,t)])^{-1} = \det(B(x,t))^{\frac{1}{2}}. \quad (4)$$

It can be shown that

$$\sigma(x,t) = W'_{\text{vol}}(J,s)I + 2J^{-1}\left(\overline{\sigma}_{\text{iso}} - \frac{1}{3}I(\overline{\sigma}_{\text{iso}} : I)\right) \quad (5)$$

with

$$\overline{\sigma}_{\text{iso}} = \frac{\partial W_{\text{iso}}}{\partial a}\overline{B} - \frac{\partial W_{\text{iso}}}{\partial b}\overline{B}^{-1}. \quad (6)$$

By definition pressure is given by $p = -\frac{1}{3}\text{Tr}(\sigma) = -W'_{\text{vol}}(J,s)$.

For the objectives of this paper, we restrict our investigation to an elastic one-dimensional isoentropic case with non-zero transverse velocity. Let x_i, $i = 1\ldots 2$ be the coordinates in the canonical basis of $\mathbb{R}^2$, u_i the velocity components, Y^i the components of Y and σ^{ij} the components of the stress tensor. Also, let us denote by $,i$ differentiation with respect to x_i. We consider the governing equations in two space dimensions and we assume that ∇Y is a function only of one direction (x_1), as well as u_1 and u_2. In this case we have that $(Y^1_{,2})_t = (Y^2_{,2})_t = 0$. Since $Y(x,0) = x$, we have also that $Y_{,2}^1 = 0$, $Y_{,2}^2 = 1$ and hence

$$[\nabla Y] = \begin{pmatrix} Y_{,1}^1 & 0 \\ Y_{,1}^2 & 1 \end{pmatrix}. \tag{7}$$

The governing equations in conservative form become

$$\Psi_t + (F(\Psi))_{,1} = 0$$

with

$$\Psi = \begin{pmatrix} \rho \\ \phi_1 \\ \phi_2 \\ Y_{,1}^1 \\ Y_{,1}^2 \end{pmatrix} \qquad F(\Psi) = \begin{pmatrix} \phi_1 \\ \dfrac{(\phi_1)^2}{\rho} - \sigma^{11} \\ \dfrac{\phi_1 \phi_2}{\rho} - \sigma^{21} \\ \dfrac{\phi_1 Y_{,1}^1}{\rho} \\ \dfrac{\phi_1 Y_{,1}^2 + \phi_2}{\rho} \end{pmatrix}$$

In two dimensions we define $\overline{B} = \dfrac{B}{\det(B)^{\frac{1}{2}}}$, so that $\det(\overline{B}) = 1$. Let now

$$\varepsilon = \dfrac{W_{vol} + W_{iso}}{\rho_0} = \dfrac{\exp\left(\dfrac{s - s_0}{c_v}\right) \rho^{\gamma - 1}}{\gamma - 1} + \dfrac{p_\infty}{\rho} + \chi(\text{Tr}(\overline{B}) - 2) \tag{8}$$

where s_0 is the reference entropy, $\phi_i = \rho u_i$ and γ, p_∞, $\chi \in \mathbb{R}^+$ are constants that characterize a given material. This model accounts for elastic deformations in the transverse direction, i.e., $\sigma^{21} \neq 0$. Here the term W_{vol} represents a stiff gas, the term W_{iso} a Mooney-Rivlin solid.

3 Multimaterial solver

We assume that the initial condition at time t_n, the n-th time step, is known. Let $\Psi_k^n = \Psi(x_k, t_n)$, with x_k the spatial coordinate x of grid point k. The discretization points are $N + 1$ and let $I = \{1, \cdots, N\}$. Consider two non-miscible materials separated by a physical interface located, at time t_n, in x_f^n and let $\iota = i$ such that $x_i \leq x_f < x_{i+1}$, $i \in I$. The space and time discretization $\forall k \in I$ and $k \neq \iota, \iota + 1$ is as follows

$$\dfrac{\Psi_k^{n+1} - \Psi_k^n}{\Delta t} = -\dfrac{\mathscr{F}_{k+1/2}^n(\Psi_k^n, \Psi_{k+1}^n) - \mathscr{F}_{k-1/2}^n(\Psi_{k-1}^n, \Psi_k^n)}{\Delta x} \tag{9}$$

where $\Delta t = t^{n+1} - t^n$, $\Delta x = x_{k+1/2} - x_{k-1/2}$ and $\mathscr{F}^n_{k\pm 1/2}$ are the numerical fluxes evaluated at the cell interface located at $x_{k\pm 1/2}$. For consistency $\mathscr{F}$ is a regular enough function of both arguments and $\mathscr{F}(\Psi, \Psi) = F(\Psi)$. Numerical conservation requires that $\mathscr{F}(\Psi', \Psi) = \mathscr{F}(\Psi, \Psi')$. The numerical flux function $\mathscr{F}_{k+1/2}(\Psi^n_k, \Psi^n_{k+1})$ is computed by an approximate Riemann solver. In the following we use the HLLC [9] approximate solvers.

In any case, we assume that Riemann solver employed defines at least two intermediate states Ψ^n_- and Ψ^n_+, in addition to Ψ^n_k and Ψ^{n+1}_{k+1} and a contact discontinuity of speed u^n_*. The fluid speed is continuous across the states Ψ^n_- and Ψ^n_+. These states are defined so that mechanical equilibrium is ensured at the contact discontinuity.

Let us assume that Ψ^n_- is the state to the left of the contact discontinuity and Ψ^n_+ to the right. The main idea is to use a standard numerical flux function $\mathscr{F}(\Psi_k, \Psi_{k+1})$, $\forall k \in I, k \neq \iota, \iota + 1$ and from (9) to deduce Ψ^{n+1}_k. In contrast, for Ψ^{n+1}_ι and $\Psi^{n+1}_{\iota+1}$ we have

$$\begin{cases} \dfrac{\Psi^{n+1}_\iota - \Psi^n_\iota}{\Delta t} = -\dfrac{\mathscr{F}^n_-(\Psi^n_-) - \mathscr{F}^n_{\iota-1/2}(\Psi^n_{\iota-1}, \Psi^n_\iota)}{\Delta x} \\ \dfrac{\Psi^{n+1}_{\iota+1} - \Psi^n_{\iota+1}}{\Delta t} = -\dfrac{\mathscr{F}^n_{\iota+3/2}(\Psi^n_{\iota+1}, \Psi^n_{\iota+2}) - \mathscr{F}^n_+(\Psi^n_+)}{\Delta x} \end{cases} \quad (10)$$

where $\mathscr{F}^n_\pm = F(\Psi^n_\pm)$. The scheme is locally non conservative since $\mathscr{F}^n_+ \neq \mathscr{F}^n_-$. However, the effect on the approximation of shocks is negligible. The interface position is updated in time using u^n_*, i.e., $x^{n+1}_f = x^n_f + u^n_* \Delta t$. For numerical stability, the integration step is limited by the fastest of the characteristics over the grid points. Hence, the interface position will belong to the same interval between two grid points for more than one time step. When the physical interface overcomes a grid point, i.e., $x^{n+1}_f \geq x_{i+1}$ or $x^{n+1}_f < x_i$ then $\iota = i \pm 1$ accordingly. In other words, the above integration scheme is simply shifted of one point to the right or to the left.

When the interface crosses a grid point, however, the corresponding conservative variables Ψ^{n+1}_ι do not correspond anymore to the material present at that grid point before the integration step. When $\iota = i + 1$, i.e., the physical interface moves to the right of $i + 1$, then we take $\Psi^{n+1}_\iota = \Psi^n_-$, whereas if $\iota = i - 1$, $\Psi^{n+1}_\iota = \Psi^n_+$. The scheme proposed in [2] can be recast in a form similar to what we presented here.

4 Results

As a first test case we show the results (Fig. 1) of a computation on 200 grid points of the classical perfect gas ($\gamma = 1.4$) shock tube with conditions $\rho = 1$, $u_1 = 1.0$ and $p = 0.75$ to the left and $\rho = 0.125$, $u_1 = 0$ and $p = 0.1$ to the right. The governing equations are in this case the usual one-dimensional compressible Euler equations. The multimaterial solver is applied across the contact discontinuity, that stays sharp during the transient.

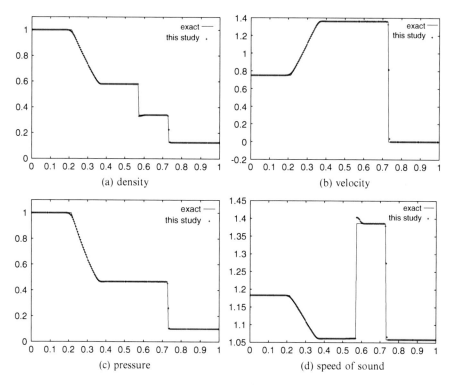

Fig. 1 Shock tube

The subsequent test case is relevant to elastic materials with different physical properties separated by an interface. It represents a one-dimensional configuration with non-zero transverse velocity. This velocity is constant in the transverse direction but may be variable in the longitudinal direction. This configuration is similar to that presented in [5].

The test case concerns a copper-air interface with discontinuous initial conditions in the copper. The copper-air interface is at $x_1 = 0.4$. To left of the interface there is copper with $p_\infty = 342 \cdot 10^8$, $\gamma = 4.22$, $\chi = 9.2 \cdot 10^{10}$ and $\rho_0 = 8.9 \cdot 10^3$. To the right there is air with $p_\infty = 0$, $\gamma = 1.4$, $\chi = 0$ and $\rho_0 = 1$. The initial conditions are uniform static pressure (10^5) and uniform horizontal velocity (0) across the materials. Inside copper the vertical velocity is 10^3 between $x_1 = 0$ and $x_1 = 0.15$ and 0 elsewhere. The vertical velocity in air is 0. The left boundary conditions are such that the solution is symmetric.

The results obtained on 2000 grid points are presented in Fig. 2 and Fig. 3. The initial discontinuity in vertical velocity at time 0 breaks down in two waves travelling in opposite directions. These waves are reflected on the left border and on the copper-air interface, giving rise to subsequent wave interactions. The sharp contact discontinuity is between copper and air. When the transverse wave hits this interface, since $\sigma^{21} = 0$, the transverse speed is discontinuous. The results are in good accordance with those presented in [5].

A Sharp Contact Discontinuity Scheme for Multimaterial Models

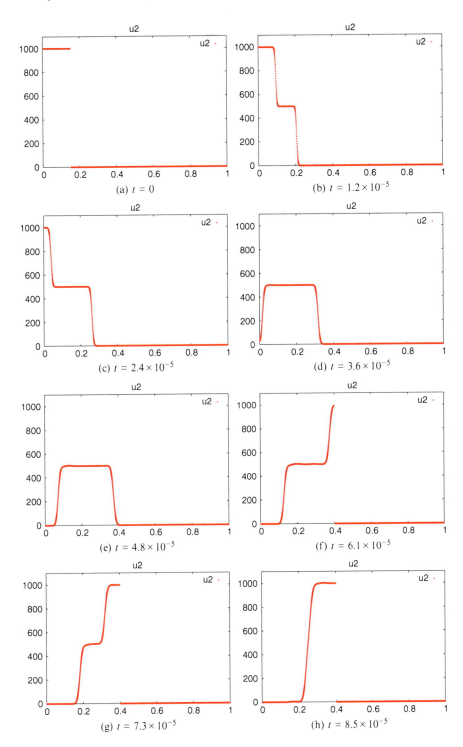

Fig. 2 Copper-air. Vertical velocity u_2

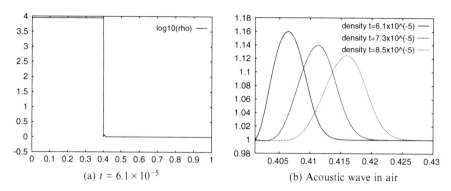

Fig. 3 Copper-air. (a) Density in logarithmic scale: copper to the left of the interface, air to the right. The time snapshot shown correspond to the first shear wave interaction with the copper-air interface ((f) in Fig. 2). Copper density is nearly constant during the transient. In (b) a zoom of the acoustic wave in air for several time steps ((f), (g) and (h) in Fig. 2)

References

1. R. Abgrall and S. Karni. Computations of compressible multifluids. *Journal of computational physics*, 169(2):594–623, 2001.
2. A. Chertock, S. Karni, and A. Kurganov. Interface tracking method for compressible multifluids. *ESAIM: Mathematical Modelling and Numerical Analysis*, 42:991–1019, 2008.
3. P.G. Ciarlet. *Mathematical elasticity Vol I, Three dimensional elasticity*. Volume 20 of Studies in Mathematics and its Applications, 1994.
4. G.-H. Cottet, E. Maitre, and T. Milcent. Eulerian formulation and level set models for incompressible fluid-structure interaction. *M2AN*, 42:471–492, 2008.
5. N. Favrie, S.L. Gavrilyuk, and R. Saurel. Solidfluid diffuse interface model in cases of extreme deformations. *Journal of computational physics*, 228(16):6037–6077, 2009.
6. S.L. Gavrilyuk, N. Favrie, and R. Saurel. Modelling wave dynamics of compressible elastic materials. *Journal of computational physics*, 227(5):2941–2969, 2008.
7. S.K. Godunov. Elements of continuum mechanics. *Nauka Moscow*, 1978.
8. G.H. Miller and P. Colella. A high-order eulerian godunov method for elasticplastic flow in solids. *Journal of computational physics*, 167(1):131–176, 2001.
9. E.F. Toro, M. Spruce, and W. Speares. Restoration of the contact surface in the hll-riemann solver. *Shock Waves*, 4:25–34, 1994.

The paper is in final form and no similar paper has been or is being submitted elsewhere.

Numerical Simulation of Viscous and Viscoelastic Fluids Flow by Finite Volume Method

Radka Keslerová and Karel Kozel

Abstract This paper deals with the numerical modeling of steady incompressible laminar flows of viscous and viscoelastic fluids. The governing system of the equations is based on the system of balance laws for mass and momentum for incompressible fluid. Two models for the stress tensor are tested. The models used in this study are generalized Newtonian model with power-law viscosity model and Oldroyd-B model with constant viscosity. The numerical results for these models are presented.

Keywords viscous and viscoelastic stress tensor, generalized Newtonian and Oldroyd-B model, Runge–Kutta method, finite volume method
MSC2010: 65N08, 76D05, 65L06, 76A05, 76A10

1 Introduction

Generalized Newtonian fluids can be subdivided according to the viscosity behavior. For Newtonian fluids the viscosity is constant and is independent of the applied shear stress. Shear thinning fluids are characterized by decreasing viscosity with increasing shear rate. Shear thickening fluids are characterized by increasing viscosity with increasing shear rate. The Fig. 1 shows the dependence of viscosity on the shear rate, see e.g. [2].

R. Keslerová and K. Kozel
CTU in Prague, Karlovo nám. 13, Praha, Czech Republic, e-mail: keslerov@marian.fsik.cvut.cz, kozelk@fsik.cvut.cz

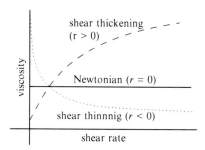

Fig. 1 Viscosity generalized Newtonian fluid as a function of shear rate for power-law fluid

2 Mathematical Model

The governing system of equations is the system of balance laws of mass and momentum for incompressible fluids [1], [7]:

$$\text{div } \boldsymbol{u} = 0 \tag{1}$$

$$\rho \frac{\partial \boldsymbol{u}}{\partial t} + \rho(\boldsymbol{u}.\nabla)\boldsymbol{u} = -\nabla P + \text{div } \mathsf{T} \tag{2}$$

where P is the pressure, ρ is the constant density, $\boldsymbol{u}$ is the velocity vector. The symbol T represents the stress tensor.

2.1 Stress tensor

In this work the different choices of the definition of the stress tensor are used.
a) Viscous fluids
The simple viscous model is *Newtonian model*:

$$\mathsf{T} = 2\mu \mathsf{D} \tag{3}$$

where μ is the dynamic viscosity and tensor D is the symmetric part of the velocity gradient.

This model could be generalized by extending Newtonian model for shear thinning and thickening fluids flow. For this case the viscosity μ is no more constant, but is defined as the viscosity function by the power-law model [4]

$$\mu = \mu(\dot{\gamma}) = \mu_\epsilon \left(\sqrt{\text{tr}\mathsf{D}^2}\right)^r, \tag{4}$$

where μ_ϵ is a constant, e.g. the dynamic viscosity for Newtonian fluid. The symbol $\text{tr } \mathsf{D}^2$ denotes the trace of the tensor D^2. The exponent r is the power-law index. This model includes Newtonian fluids as a special case ($r = 0$). For $r > 0$ the power-law fluid is shear thickening, while for $r < 0$ it is shear thinning, (see Fig. 1).

b) Viscoelastic fluids

The behavior of the mixture of viscous and viscoelastic fluids can be described by *Oldroyd-B model* and it has the form

$$\mathsf{T} + \lambda_1 \frac{\delta \mathsf{T}}{\delta t} = 2\mu \left(\mathsf{D} + \lambda_2 \frac{\delta \mathsf{D}}{\delta t} \right). \tag{5}$$

The parameters λ_1, λ_2 are *relaxation* and *retardation time*.

The stress tensor T is decomposed to the Newtonian part T_s and viscoelastic part T_e ($\mathsf{T} = \mathsf{T}_s + \mathsf{T}_e$) and

$$\mathsf{T}_s = 2\mu_s \mathsf{D}, \qquad \mathsf{T}_e + \lambda_1 \frac{\delta \mathsf{T}_e}{\delta t} = 2\mu_e \mathsf{D}, \tag{6}$$

where

$$\frac{\lambda_2}{\lambda_1} = \frac{\mu_s}{\mu_s + \mu_e}, \qquad \mu = \mu_s + \mu_e. \tag{7}$$

The *upper convected derivative* $\frac{\delta}{\delta t}$ is defined (for general tensor) by the relation (see [7])

$$\frac{\delta \mathsf{M}}{\delta t} = \frac{\partial \mathsf{M}}{\partial t} + (\boldsymbol{u}.\nabla)\mathsf{M} - (\mathsf{W}\mathsf{M} - \mathsf{M}\mathsf{W}) - (\mathsf{D}\mathsf{M} + \mathsf{M}\mathsf{D}) \tag{8}$$

where D is the symmetric part of the velocity gradient $\mathsf{D} = \frac{1}{2}(\nabla \boldsymbol{u} + \nabla \boldsymbol{u}^T)$ and W is the antisymmetric part of the velocity gradient $\mathsf{W} = \frac{1}{2}(\nabla \boldsymbol{u} - \nabla \boldsymbol{u}^T)$.

The governing system (1), (2) of equations is completed by the equation for the viscoelastic part of the stress tensor

$$\frac{\partial \mathsf{T}_e}{\partial t} + (\boldsymbol{u}.\nabla)\mathsf{T}_e = \frac{2\mu_e}{\lambda_1}\mathsf{D} - \frac{1}{\lambda_1}\mathsf{T}_e + (\mathsf{W}\mathsf{T}_e - \mathsf{T}_e\mathsf{W}) + (\mathsf{D}\mathsf{T}_e + \mathsf{T}_e\mathsf{D}). \tag{9}$$

3 Numerical Solution

Numerical solution of the described models is based on cell-centered finite volume method using explicit Runge–Kutta time integration. The unsteady system of equations with steady boundary conditions is solved by finite volume method. Steady state solution is achieved for $t \to \infty$. In this case the artificial compressibility method can be applied. It means that the continuity equation is completed by the time derivative of the pressure in the form (for more details see e.g. [8]):

$$\frac{1}{\beta^2}\frac{\partial p}{\partial t} + \operatorname{div} \boldsymbol{u} = 0, \quad \beta \in \mathbb{R}^+. \tag{10}$$

The system of equations (including the modified continuity equation) could be rewritten in the vector form.

$$\tilde{R}_\beta W_t + F_x^c + G_y^c = F_x^v + G_y^v + S, \quad \tilde{R}_\beta = \operatorname{diag}(\frac{1}{\beta^2}, 1, 1, 1, 1, 1). \tag{11}$$

where W is the vector of unknowns, F^c, G^c are inviscid fluxes, F^v, G^v are viscous fluxes defined as

$$W = \begin{pmatrix} p \\ u \\ v \\ t_{11} \\ t_{12} \\ t_{22} \end{pmatrix}, \quad F^c = \begin{pmatrix} u \\ u^2 + p \\ uv \\ ut_{11} \\ ut_{12} \\ ut_{22} \end{pmatrix}, \quad G^c = \begin{pmatrix} v \\ uv \\ v^2 + p \\ vt_{11} \\ vt_{12} \\ vt_{22} \end{pmatrix}, \tag{12}$$

$$F^v = \begin{pmatrix} 0 \\ 2\mu(\dot{\gamma})u_x \\ \mu(\dot{\gamma})(u_y + v_x) \\ 0 \\ 0 \\ 0 \end{pmatrix}, \quad G^v = \begin{pmatrix} 0 \\ \mu(\dot{\gamma})(u_y + v_x) \\ 2\mu(\dot{\gamma})v_y \\ 0 \\ 0 \\ 0 \end{pmatrix} \tag{13}$$

and the source term S is defined as where t_{ij} are components of the symmetric tensor T_e

$$S = \begin{pmatrix} 0 \\ t_{11x} + t_{12y} \\ t_{12x} + t_{22y} \\ 2\frac{\mu_e}{\lambda_1}u_x - \frac{t_{11}}{\lambda_1} + 2(u_x t_{11} + u_y t_{12}) \\ \frac{\mu_e}{\lambda_1}(u_y + v_x) - \frac{t_{12}}{\lambda_1} + (u_x t_{12} + u_y t_{22} + v_x t_{11} + v_y t_{12}) \\ 2\frac{\mu_e}{\lambda_1}v_y - \frac{t_{22}}{\lambda_1} + 2(v_x t_{12} + v_y t_{22}) \end{pmatrix} \tag{14}$$

The following special parameters are used:

Newtonian	$\mu(\dot{\gamma}) = \mu_s = const.$	$\mathsf{T}_e \equiv 0$
Generalized Newtonian	$\mu(\dot{\gamma})$	$\mathsf{T}_e \equiv 0$
Oldroyd-B	$\mu(\dot{\gamma}) = \mu_s = const.$	T_e

The eq. (11) is discretized in space by the cell-centered finite volume method (see [3]) and the arising system of ODEs is integrated in time by the explicit multistage Runge–Kutta scheme (see [4], [6], [9]):

$$W_i^n = W_i^{(0)}$$
$$W_i^{(s)} = W_i^{(0)} - \alpha_{s-1} \Delta t \mathcal{R}(W)_i^{(s-1)} \qquad (15)$$
$$W_i^{n+1} = W_i^{(M)} \qquad s = 1, \ldots, M,$$

where $M = 3$, $\alpha_0 = \alpha_1 = 0.5$, $\alpha_2 = 1.0$, the steady residual $\mathcal{R}(W)_i$ is defined by finite volume method as

$$\mathcal{R}(W)_i = \frac{1}{\sigma_i} \sum_{k=1}^{4} \left[\left(\overline{F}_k^c - \overline{F}_k^v \right) \Delta y_k - \left(\overline{G}_k^c - \overline{G}_k^v \right) \Delta x_k \right] + \overline{S}, \qquad (16)$$

where σ_i is the volume of the cell, $\sigma_i = \int \int_{C_i} dx \, dy$. The symbols $\overline{F}_k^c, \overline{G}_k^c$ and $\overline{F}_k^v, \overline{G}_k^v$ denote the numerical approximation of the inviscid and viscous fluxes, for more details see [5], symbol $\overline{S}$ represents the numerical approximation of the source term with central approximation of derivatives.

4 Numerical results

The steady numerical results in the branching channel for two dimensional generalized Newtonian fluids are shown in the Sect. 4.1. In the Sect. 4.2 the comparison of Newtonian and Oldroyd-B fluids is presented for simple 2D channel.

4.1 Two Dimensional Case

In this section the steady numerical results are presented. The comparison of Newtonian and non-Newtonian shear thickening and shear thinning fluids for Re = 400 in the geometry of the branching channel in the form of the velocity isolines is shown in the Fig. 2.

The following choices of the power-law index were used: for Newtonian fluid $r = 0$, for shear thickening and shear thinning fluid values $r = 0.5$ and $r = -0.5$. In the inlet the velocity is prescribed by the parabolic profile. The histories of the convergence are also presented in the Fig. 2. One can observe some differences between tested fluids in the size of the separation region.

The nondimensional axial velocity profile for steady fully developed flow of considered fluids is shown in the Fig. 3. In these figure the small channel is sketched. The line (inside the domain) marks the position where the cuts for the velocity profile were done.

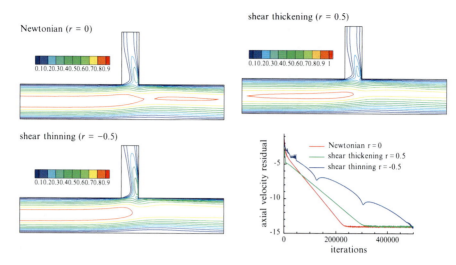

Fig. 2 Velocity isolines and history of the convergence of steady flows for generalized Newtonian fluids

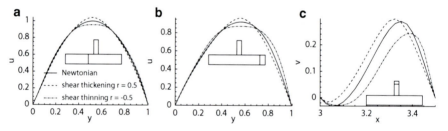

Fig. 3 Nondimensional velocity profile for steady fully developed flow of generalized Newtonian fluids in the branching channel (the line legend in the a) is the same for all figures)

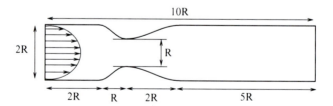

Fig. 4 Structure of the computational domain

4.2 Viscous and Visoelastic Model

This section deals with the comparison of the numerical results of Newtonian and Oldroyd-B fluids. Fig. 4 shows the shape of the tested domain.

The following model parameters are:

$$\mu_e = 4.0 \cdot 10^{-4} Pa \cdot s \quad \mu_s = 3.6 \cdot 10^{-3} Pa \cdot s$$
$$\lambda_1 = 0.06s \quad \lambda_2 = 0.054s$$
$$U_0 = 0.0615 m \cdot s^{-1} \quad L_0 = 2R = 0.0062m$$
$$\mu_0 = \mu = \mu_s + \mu_e \quad \rho = 1050 kg \cdot m^{-3}$$

In the Figs. 5 and 6 the comparison of the axial velocity isolines and the pressure distributions is presented.

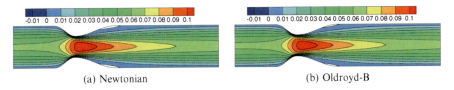

(a) Newtonian   (b) Oldroyd-B

Fig. 5 Axial velocity isolines for Newtonian and Oldroyd-B fluids

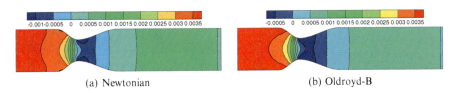

(a) Newtonian   (b) Oldroyd-B

Fig. 6 Pressure distribution for Newtonian and Oldroyd-B fluids

Pressure and velocity distribution along the axis for both tested fluids models is shown in the Fig. 7. By simple observation one can conclude that the main effect of the Oldroyd-B fluids behavior is visible mainly in the recirculation zone.

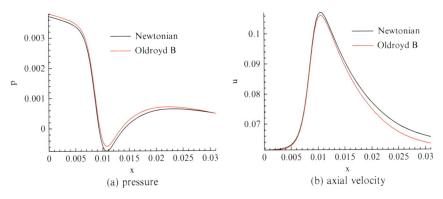

(a) pressure   (b) axial velocity

Fig. 7 Pressure and axial velocity distribution along the central axis of the channel

5 Conclusions

Newtonian model with its generalized modification and Oldroyd-B model have been considered for numerical simulation of fluids flow in the branching channel and in the idealized axisymmetric stenosis. The cell-centered finite volume solver for incompressible laminar viscous and viscoelastic fluids flow has been described. Generalized Newtonian model was used for testing of different choices of the power-law index r in the branching channel. For time integration the explicit Runge–Kutta method was considered. The numerical results obtained by this method are presented. We can conclude that the numerical results of the tested fluids agrees with well-known non-Newtonian behavior. The differences between these three fluids are given mainly in the separation region.

In the idealized stenosis we tested the Newtonian and Oldroyd-B fluids models. Here the two definitions of the stress tensor were used. Based on the above numerical results we can conclude that the difference between the viscous and viscoelastic fluids is visible in the recirculation zone.

Acknowledgements This work was partly supported by the grant GACR 101/09/1539 and GACR 201/09/0917, Research Plan MSM 684 077 0003 and Research Plan MSM 684 077 0010.

References

1. Dvořák, R., Kozel, K.: Mathematical Modelling in Aerodynamics (in Czech). CTU, Prague, Czech Republic (1996).
2. Robertson, A.M., Sequeira, A., Kameneva, M.V.: Hemorheology. Birkhäuser Verlag Basel, Switzerland (2008).
3. LeVeque, R.: Finite-Volume Methods for Hyperbolic Problems. Cambridge University Press, (2004).
4. Keslerová, R., Kozel, K.: Numerical modelling of incompressible flows for Newtonian and non-Newtonian fluids, Mathematics and Computers in Simulation, **80**, 1783–1794 (2010).
5. Keslerová, R., Kozel, K.: Numerical solution of laminar incompressible generalized Newtonian fluids flow, Applied Mathematics and Computation, **217**, 5125–5133 (2011).
6. Jameson, A., Schmidt, W., Turkel, E.: Numerical solution of the Euler equations by finite volume methods using Runge-Kutta time-stepping schemes, AIAA 14th Fluid and Plasma Dynamic Conference California (1981).
7. Bodnar, T., Sequeira, A.: Numerical study of the significance of the non-Newtonian nature of blood in steady flow through stenosed vessel (Editor: R. Ranacher, A. Sequeira), Advances in Mathematical Fluid Mechanics, 83–104 (2010).
8. Chorin, A.J.: A numerical method for solving incompressible viscous flow problem, Journal of Computational Physics, **135**, 118–125 (1967).
9. Vimmr, J., Jonášová, A.: Non-Newtonian effects of blood flow in complete coronary and femoral bypasses, Mathematics and Computers in Simulation, **80**, 1324–1336 (2010).

The paper is in final form and no similar paper has been or is being submitted elsewhere.

An Aggregation Based Algebraic Multigrid Method Applied to Convection-Diffusion Operators

Sana Khelifi, Namane Méchitoua, Frank Hülsemann, and Frédéric Magoulès

Abstract The paper focuses on an aggregation-based algebraic multigrid method applied to convection/diffusion problems. We show that for an unstructured finite volume approach on arbitrary shaped cells, the separation of the two operators associated with suitable smoothers improves the aggregation-based multigrid. While the convection is treated by a piecewise constant prolongation, the off-diagonals entries of the diffusion P_0 Galerkin operator are scaled by a parameter representative of the mesh spacing ratio between the fine and coarse mesh in the vicinity of the coarse mesh cell boundaries. Some numerical examples are shown to assess the rate of convergence and the robustness of the proposed approach.

Keywords finite volumes, algebraic multigrid, convection-diffusion
MSC2010: 76M12

Introduction

The ongoing increase in computing power renders the solving of ever larger linear systems possible, so that the use of algorithms with optimal algebraic complexity becomes necessary. From this point of view, multigrid methods (MG) represent a

---

Sana Khelifi
EDF R&D, MFEE, 6 Quai Wattier F-78401 Chatou Cedex and Ecole Centrale de Paris, Grande voie des Vignes, F-92295 Chatenay-Malabry, e-mail: sana.khelifi@edf.fr

Namane Méchitoua
EDF R&D, MFEE, 6 Quai Wattier F-78401 Chatou, Cedex, e-mail: namane.mechitoua@edf.fr

Frank Hülsemann
EDF R&D, SINETICS, 1, avenue du Général de Gaulle, F-92140 Clamart Cedex, e-mail: frank.hulsemann@edf.fr

Frédéric Magoulès
Ecole Centrale de Paris, Grande voie des Vignes, F-92295 Chatenay-Malabry, e-mail: frederic.magoules@hotmail.com

viable alternative to other solution strategies since their theoretical computational complexities scale quasi-linearly with the problem size, especially for elliptic dominated problems [11]. The key ingredient of the multigrid technique relies on the use of a hierarchy of grids for solving a linear set of equations. The fast convergence of the multigrid scheme is based on the fact that, for each component of the error, there exists a grid level on which the error component in question is efficiently reduced. Compared to the standard geometric procedure, the algebraic multigrid has become very competitive. The hierarchy of grids is created automatically, taking into account the matrix entries of the discretized operator. This procedure allows the effective solution of a large class of linear systems arising from highly non homogeneous PDE discretized with unstructured meshes. Among the different interpolation schemes used for the restriction and prolongation operators involved in the coarse level matrix calculation, the piecewise constant interpolation is a limiting case of the "Ruge-Stüben" multigrid procedure [11]. Consisting of agglomerating the fine mesh points for creating coarse mesh points, this approach is widely used in finite volume based CFD solvers [8]. In order to recover the theoretical convergence for elliptic second order operators [6], the rescaling of the coarse level P_0 Galerkin operator is a judicious and quite simple mean. The trivial method, consisting of rescaling by a global number is very simple to implement but it is limited to problems that can be solved effectively with a geometric multigrid procedure [9]. A second approach called "smoothed aggregation" [12] improves the MG convergence, but it increases the number of off-diagonal entries of the coarse level matrix and therefore destroys the simplicity of the original approach. A third approach, although only studied with a finite volume scheme on fully unstructured meshes, greatly improves the scaled P_0 Galerkin multigrid procedure, thanks to an original face based rescaling [10]. It maintains the simplicity of P_0 interpolation. However, for equations mixing convection and diffusion operators, acting preferentially at different grid scales, the optimal use of multilevel techniques is less evident, with the presence of several different strategies for overcoming these difficulties [2], [4], [5], [7]. The aim of the paper is to present a strategy for solving such systems, in the framework of the face based rescaling algebraic multigrid procedure.

1 Finite Volume procedure

The scalar convection/diffusion equation $div(\vec{Q}C - \kappa\vec{\nabla}C) = b$, where $\vec{Q}, \kappa$ and C represent respectively a divergence free velocity field, a diffusion coefficient and the unknown scalar to be solved, is representative of the transport/diffusion terms of the momentum, energy or stationary mass fraction equations used in CFD solvers [1]. The integration over a discrete cell Ω_I is written as:

$$\int_{\Omega_I} div(\vec{Q}C - \kappa\vec{\nabla}C) = \sum_{J \in V_I}(\vec{Q}_{IJ}.\vec{N}_{IJ})C_{IJ} + \sum_{J \in V_I}(-\kappa\vec{\nabla}C)_{IJ}.\vec{N}_{IJ} = b_I|\Omega_I|, \quad (1)$$

where V_I represents the neighbourhood of the cell I, i.e the set of cells J sharing a non-zero area surface IJ with the cell I and $\vec{N}_{IJ}$ designs the normal vector to the face IJ pointing from Ω_I to Ω_J. Numerical consistency and precision for diffusive and convective fluxes for non-orthogonal cells are taken into account using a gradient reconstruction technique. This technique is useful for increasing the order of some numerical schemes, when applied to complex situations as unstructured meshes for instance. It concerns both first order (convection) and second order (diffusion) differential equations, discretized with finite volume methods. Among the various numerical fluxes, assuming regular coefficient κ, the following one is used for the diffusion:

$$(-\kappa \vec{\nabla} C)_{IJ} . \vec{N}_{IJ} = \frac{\kappa_{IJ}(C_I - C_J) + (\vec{II'} - \vec{JJ'}).(\kappa \vec{\nabla} C)_{IJ}^c}{I'J'} . |\vec{N}_{IJ}|, \qquad (2)$$

where $(\kappa \vec{\nabla} C)_{IJ}^c \approx 0.5(\kappa_I \vec{\nabla} C_I + \kappa_J \vec{\nabla} C_J)$ represents the cell gradient projected at the face centre F, either evaluated with a finite volume or a least square formulation and κ_{IJ} represents the face interpolation of the diffusion coefficient (with arithmetic, harmonic or geometric interpolation). For the convective part of the fluxes, the simplest upwind scheme remains consistent for non orthogonal smooth meshes, but the order can be less than one. Among the various numerical higher order convective fluxes, the following one is used for a centred interpolation of the convected variable C at the face centre F (also named IJ) of the interface separating two cells I and J (see Fig.1):

$$\begin{aligned} C_{IJ} &= C_F = C_O + \vec{OF}.\vec{\nabla} C_{IJ}^c \\ C_O &= \tfrac{OJ}{IJ} C_I + \tfrac{OI}{IJ} C_J \end{aligned} \qquad (3)$$

In order to avoid instabilities, the convective schemes for all variables, except the pressure, are non-linear centred or second order upwind schemes. The switch between first order upwind and higher order interpolation is triggered if the non-monotony of the variable in the neighbourhood of the interpolation point is detected. The linear set of equations arising from discrete formulations (1) with the higher order diffusive flux (2) and convective flux (3), is not solved directly with GMRES or BICG-STAB method, because the resolution can be too expensive

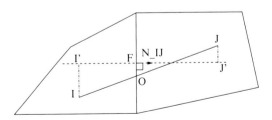

Fig. 1 Geometrical parameters at the face separating cells I and J. I' (res. J') is the orthogonal projection of I (res. J) on the normal through the face centre F

or even impossible without robust pre-conditioners. The computation of such systems is made through a defect correction technique. The explicit (or initial) convective/diffusive flux (named old) computed with the scheme (2) and (3) is taken into account in the right hand side. The iteration matrix for solving the correction, making use only of the original finite volume neighbourhood, is a positive M matrix, which can be solved by suitable iterative methods, such as conjugate gradient or multigrid solvers.

$$\int_{\Omega_I} div(\vec{Q}\phi - \kappa \vec{\nabla}\phi) \approx \sum_{J \in V_I} (\vec{Q}_{IJ}.\vec{N}_{IJ})\phi_{IJ}^{upwind} + \frac{\kappa_{IJ}(\phi_I - \phi_J)}{I'J'}|\vec{N}_{IJ}|$$
$$= b_I \Omega_I - \sum_{J \in V_I} (\vec{Q}_{IJ}.\vec{N}_{IJ}\phi_{IJ} - \kappa \vec{\nabla} C_{IJ}^{old}.\vec{N}_{IJ}) \quad (4)$$

$$C^{new} = C^{old} + \phi$$

In (4), $\phi_{IJ}^{upwind} = \phi_I$ if $(\vec{Q}_{IJ}.\vec{N}_{IJ}) > 0$ and $\phi_{IJ}^{upwind} = \phi_J$ otherwise. The defect correction procedure (4) is repeated, by replacing C^{old} by C^{new}, until the residual tends towards a user defined value.

2 Multigrid procedure

Many of the multigrid approaches for CFD solvers are based on the idea of separating the elliptic part from the non-elliptic one in the PDE [3]. System (4) can be written as $(C + D)\phi = f$, where C is a non-symmetric M-matrix obtained by upwind-biased discretization of the convection and D is a symmetric M-matrix corresponding to a low order discrete scheme (on non orthogonal meshes) of diffusion. The piecewise constant interpolation for restriction and prolongation operators involved in the coarse level matrix construction preserves the M-matrix properties of the 2 operators, and hence the smoothing properties of simple Gauss-Seidel or Jacobi-type relaxation schemes. The coarse convection operator is constructed based on the Galerkin product with a piecewise constant interpolation operator [11]. The off-diagonal entries XC^0 of the coarse convection operator read as:

$$XC_{I_C J_C}^0 = \sum_{(I_k, J_k)} Min((\vec{Q}_{I_k J_k} \vec{N}_{I_k J_k}), 0)$$
$$XC_{J_C I_C}^0 = -\sum_{(I_k, J_k)} Max((\vec{Q}_{I_k J_k} \vec{N}_{I_k J_k}), 0)$$

The upwind character of the finest level discretization is propagated to all coarse levels. This property, associated with an algebraic multigrid procedure allowing the aggregation along the streamlines and associated with an appropriate smoother ensures a physical coarse grid correction for the convective part. The piecewise constant interpolation for the elliptic part is not optimal from a theoretical viewpoint

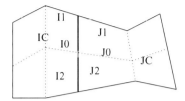

Fig. 2 Sketch of an aggregate and coarse mesh boundary where I_C and J_C are the gravity centres of the coarse cells, (I_k, J_k) are the fine mesh cells situated on both sides of the coarse mesh interface and I_0 (resp. J_0) is an average value of I_1, I_2 (resp. J_1, J_2)

[6]. The optimal multigrid convergence of the elliptic part is obtained in a quite simple manner, considering a finite volume discretization on fully unstructured meshes [10]. A geometrical face-based rescaling of the off-diagonal entries XD^0 of the P_0 coarse mesh matrix takes into account the mesh spacing ratio between the fine and coarse level in the vicinity of the coarse mesh cell boundaries, as represented in Fig.2. In our notation, the rescaling reads as follows:

$$XD_{I_C J_C} = \frac{I'_0 J'_0}{I'_C J'_C} XD^0_{I_C J_C}, \text{ with } XD^0_{I_C J_C} = -\sum_{(I_k, J_k)} \frac{\kappa_{IJ}}{I'_k J'_k} |\vec{N}_{I_k J_k}|$$

The detailed derivation of the rescaling method for the elliptic part is given in [10]. The singularly perturbed character of convection complicates the use of multigrid techniques, because of the possible poor coarse grid approximation of the convective part. The typical approach is to use a smoother which eliminates the singular perturbation errors on the finest level, so that the coarse grid correction can handle efficiently the remaining elliptic part of the errors. Gauss-Seidel like methods are quite fast iterative solvers for convection operators discretized with upwind biased techniques. Downwind numbering w.r.t constant characteristics [2] or, more generally, curved characteristics following the vortices in the flow [5] renders the Gauss-Seidel iteration sufficiently robust. Nevertheless, these formulations cannot be easily applied to complex flows inside complex geometries, in terms of implementation and set up phase computing time. The symmetric Gauss-Seidel (SSOR) relaxation scheme, for which upwind and downwind directions are swept, represents a viable alternative for complex situations. Previous numerical assessments have shown that it is nearly as robust as circular ordering [7].

The simplest cycle in the multigrid resolution, the V-cycle, is used, in combination with SSOR methods acting as smoothers. The algebraic multigrid procedure is based upon the strength of the matrix connectivity, defined in a symmetric way for an aggregation-based procedure. Two cells numbered i and j are merged if $max(A_{ij}^2, A_{ji}^2)/A_{ii}A_{jj}$ is greater than a threshold value, progressively relaxed until the targeted coarsening is reached. Each coarse grid has approximately one third of the number of cells of the previous fine grid, representing a good trade off between the efficiency of the smoothing with few iterations (3 for a coarsening ratio of about 3) and grid complexity.

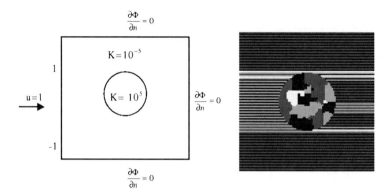

Fig. 3 The square test case: initial parameters and the coarsest level with a finest mesh of 100×100

3 Numerical examples

In the following examples, we tested the based piecewise constant multigrid scheme (referred as P_0) and the new scheme with the rescaled diffusion operator (referred as P_1). The multigrid scheme is used as a stand alone solver. Our unit of measure is the equivalent of the number of matrix-vector products performed on the finest level. It is a representative measure of the arithmetic and memory access operations performed during the resolution that does not depend on the computer.

3.1 The square test

In this example, the diffusion coefficient is piecewise constant with a strong jump and the convection velocity is horizontal and equal to 1, as shown on Fig. 3. The right hand side is homogeneous and equal to 0. The initial solution is zero on the whole domain with the boundary conditions shown on Fig. 3. The stopping criterion is based on a threshold on a normalised residual. The number of iterations presented in Table 1 stands for the equivalent of the number of matrix-vector products performed on the initial mesh (the finest grid). The number between brackets represents the number of cycles performed. For Gauss-Seidel as a stand alone solver, one iteration counts for 3 matrix-vector products (2 sweeps and the computing of the residual). Observing the different results, we notice a stability of the P_1 multigrid solver while the P_0 one exhibits a mesh dependency. The mesh dependency of the aggregation P_0 solver is a well known fact. The rescaling of this operator yields much better results. For the SSOR stand alone solver, it is clear that it is not competitive. It is not efficient because of diffusive part.

Table 1 The number of matrix-vector products (MG cycles), for different solvers on a sequence of 2D meshes for the square problem

Number of cells in one direction	10	100	500	1,000
Symmetric Gauss-Seidel	63	3,372	74,409	284,301
P_0 V-cycle	71 (5)	547 (31)	1,588 (94)	5,353 (298)
P_1 V-cycle	55 (4)	126 (8)	135 (9)	135 (9)

3.2 The 600 MW corner fired boiler

The second example concerns the steady transport/diffusion source term of NO_x polluant inside a 600 MW corner fired boiler, see the Fig. 4. The boiler is fitted with 24 burners displayed at 3 levels. The velocity field and the diffusivity are relatively complex [13]. The hierarchy of grids obtained is summarised in Table 2. A reduction by a factor around 3 is noticed between all the levels. The agglomeration is stopped when the coarse level size drops below 1% of the cells number in the finest level.

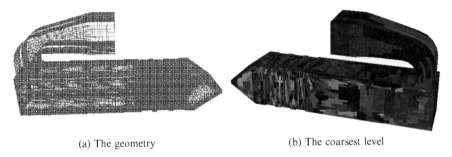

(a) The geometry   (b) The coarsest level

Fig. 4 The geometry, the initial mesh and the coarsest level of the boiler test case

Table 2 Hierarchy of grids obtained for the boiler test case

Grid level	0	1	2	3	4	5
$Ncel_k$	462,784	158,169	52,631	17,403	5,739	1,898
$Nfac_k$	1,363,784	629,626	278,518	113,296	42,760	15,106

The different results obtained with the multigrid scheme, the stand alone symmetric Gauss-Seidel solver and the stand alone BICG-Stab solver are summarised in Table 3. As we recompute the residual at the end of each BICG-Stab iteration, each iteration counts for 3 matrix-vector products. The number between brackets represents the number of MG cycles performed. It is clear that symmetric Gauss-Seidel is not competitive because of the presence of diffusive dominated regions and recirculation zones. The results obtained with BICG-Stab are reasonable but far away from those obtained with the proposed scheme.

Conclusion: Based on a finite volume approach on arbitrary cell shapes, the rescaling of the piecewise constant elliptic part of the convection/diffusion equation

Table 3 Number of matrix-vector products (and multigrid cycles) for the boiler test case

solver	P_0 V-cycle	P_1 V-cycle	SSOR	BICG-Stab
number of iterations	919 (49)	387(20)	115,806	4227

was successfully accomplished. The separation of the convection and the diffusion operators in order to enable the use of the basic piecewise constant interpolation operator for the convection while rescaling the Galerkin coarse grid operator of the diffusion, associated with suitable smoothers, yields better results than using the same interpolation for both operators. The numerical examples show the robustness and the convergence rate of the proposed technique.

References

1. Archambeau F., Mechitoua N., Sakiz M.: Code_Saturne: a finite volume code for the computation of turbulent incompressible flows- Industrial Applications. Int. J. on Finite Volumes, **1** (2004). http://www.latp.univ-mrs.fr/IJFV/.
2. Bey J., Wittum G.: Downwind numbering: robust multigrid for convection-diffusion problems. Applied Numerical Mathematics, **23**,177–192 (1997).
3. Brandt A., Yavneh I.: On multigrid solution of high Reynolds incompressible entering flow. J. Comp. Phys., **101**, 151–164 (1992).
4. Guillard H., Vanek P.: An Aggregation Multigrid Solver for Convection-diffusion Problems on Unstructured Meshes. University of Colorado at Denver, technical report(1998).
5. Hackbusch W., Probst T.: Downwind Gauss-Seidel Smoothing for Convection Dominated Problems. Numerical Linear Algebra with Applications, **4**, 85–102 (1997).
6. Hemker P.W.: On the order of prolongations and restrictions in multigrid procedures. J. Comp. Applied Math., 423–429 (1990).
7. Kanschat G.: Robust smoothers for high-order discontinuous Galerkin discretizations of advection-diffusion problems. J. of Comp. and App. Math., **218**, 53–60 (2008).
8. Lonsdale R.D.: An algebraic multigrid solver for the Navier Stokes equations on unstructured meshes. Int. J. Num. Meth. Heat and FluidFlow, **3**, 3–14 (1993).
9. Mavriplis D.J., Venkatakrishnan V.: Agglomeration Multigrid for 2 Dimensional Viscous Flows. Computers and Fluids, **24**, 553–570, (1995).
10. Mechitoua N., Hülsemann F., Fournier Y.: Improvement of a Finite Volume Based Multigrid Method Applied to Elliptic Problems. Int. Conf. on Math., Comp. Methods & Reactor Physics (M&C09), Saratoga Springs, New York, May 3-7, (2009).
11. Trottenberg U., Oosterlee C., Schuller A.: Multigrid. Elsevier Academic Press. ISBN 0-12-701070-X, (2001).
12. Vanek P., Mandel J., Brezina M.: Algebraic Multigrid by smoothed aggregation for 2nd order and 4th order elliptic problems. Computing, **56**, 179–196, (1996).
13. Dal Secco S., Schuck Y.: Using three dimensional simulation of pulverized coal combustion in a 600 MW corner fired boiler to identify low NO_x operating conditions. VGB conference in Potsdam, "Power plants in competition, operation, technology and environment", (2005).

The paper is in final form and no similar paper has been or is being submitted elsewhere.

Stabilized DDFV Schemes For The Incompressible Navier-Stokes Equations

Stella Krell

Abstract "Discrete Duality Finite Volume" schemes (DDFV for short) on general meshes are studied here for the Navier-Stokes problem with Dirichlet boundary conditions. The DDFV method falls in the class of the so-called staggered scheme: the discrete unknowns, the components of the velocity and the pressure, are located on different nodes. The scheme is stabilized using a finite volume analogue to Brezzi-Pitkäranta techniques. We prove the wellposedness of the scheme for general meshes and we derive the first energy estimates. Finally, we illustrate the convergence properties with numerical experimentations.

Keywords Finite-volume methods, Navier-Stokes problem, DDFV scheme.
MSC2010: 65N08, 76D03, 76D05.

1 Introduction

We restrict here the presentation to the Navier-Stokes problem with homogeneous Dirichlet boundary conditions and a smooth viscosity which depends on the spatial variable. The system reads as follows:

$$\begin{cases} \partial_t \mathbf{u} + \mathrm{div}\,(-2\eta(x)\mathrm{D}\mathbf{u} + p\mathrm{Id}) + (\mathbf{u}\cdot\nabla)\mathbf{u} = \mathbf{f}, \text{ in }]0,T[\times\Omega, \\ \mathrm{div}(\mathbf{u}) = 0, \text{ in }]0,T[\times\Omega, \end{cases} \quad (1)$$

where the unknowns are the velocity $\mathbf{u}:]0,T[\times\Omega \to \mathbb{R}^2$ and the pressure $p:]0,T[\times\Omega \to \mathbb{R}$ such that $\int_\Omega p(t,x)\mathrm{d}x = 0$, for all $t \in]0,T[$, Ω is a polygonal open bounded connected subset of $\mathbb{R}^2$, $T > 0$. We recall that $\mathrm{D}\mathbf{u} = \frac{1}{2}(\nabla\mathbf{u} + {}^t\nabla\mathbf{u})$.

Stella Krell
INRIA, Lille, France, e-mail: stella.krell@inria.fr

and $(\mathbf{u} \cdot \nabla)\mathbf{u} = \sum_{i=1}^{2} \mathbf{u}_i \partial_i \mathbf{u}$ for $\mathbf{u} = (\mathbf{u}_1, \mathbf{u}_2)$. We supplement the system (1) with the following boundary and initial conditions:

$$\begin{cases} \mathbf{u} = 0, \text{ on }]0, T[\times \partial \Omega, \\ \mathbf{u}(0,.) = \mathbf{u}_{ini}, \text{ in } \Omega. \end{cases}$$

We assume that $\mathbf{f}$ is a function in $(L^2(]0, T[\times\Omega))^2$, $\mathbf{u}_{ini}$ is a function in $(L^\infty(\Omega))^2$ and the viscosity η is a function in $W^{1,\infty}(\Omega)$ with $\underset{\Omega}{\text{Inf }} \eta > 0$.

Finite volume approximation of Navier-Stokes problem is a current research topic, we refer to [5, 6, 9–11, 17] for the description and the analysis of the main available schemes up to now. We consider here the class of finite volume schemes called DDFV, which have been first introduced and studied in [7,12] to approximate the solution of the Laplace equation on a large class of 2D meshes including non-conformal and distorted meshes and without "orthogonality" assumptions as for classical finite volume methods. This strategy has been extended to a wide class of PDE problems [1–4, 13, 15] and gives a staggered method for the Navier-Stokes equations: the approximate velocity is located at the centers and at the vertices of the mesh and the approximate pressure at the edges of the mesh. In a previous work [15], we proposed a stabilized DDFV scheme for the Stokes problem with variable viscosity, which is equivalent (except on the boundary) to two uncoupled MAC schemes [14, 16] when the grid is cartesian and the viscosity is constant. We will use here the same discretization for the viscous part of momentum conservation and the mass conservation equations of (1).

This paper is organized as follows. In Sect. 2, we construct the approximation of the non-linear convective term. In Sect. 3, we introduce the DDFV stabilized scheme for the Navier-Stokes problem (1), we begin with existence and uniqueness of the approximate solution, then we present the first energy estimates. Finally, in Sect. 4, we illustrate the convergence with numerical results.

2 The DDFV framework

We use the same notation as in [15] for the Stokes problem. We do not recall here the complete description of meshes and operators:

- the DDFV meshes $(\mathfrak{T}, \mathfrak{D})$: $\mathfrak{T}$ is constituted by the primal mesh $\mathfrak{M} \cup \partial \mathfrak{M}$, which is the initial mesh, and the dual mesh $\mathfrak{M}^* \cup \partial \mathfrak{M}^*$, whose cells κ^* are built around the vertices of the primal mesh (Fig. 1), and $\mathfrak{D}$ is the diamond mesh, whose cells D are built around the edges of the primal mesh.
- a discrete gradient $\nabla^\mathfrak{D} : (\mathbb{R}^2)^\mathfrak{T} \to (M_2(\mathbb{R}))^\mathfrak{D}$, its discrete dual operator $\mathbf{div}^\mathfrak{T} : (M_2(\mathbb{R}))^\mathfrak{D} \to (\mathbb{R}^2)^\mathfrak{T}$, its trace $\text{div}^\mathfrak{D} : (\mathbb{R}^2)^\mathfrak{T} \to \mathbb{R}^\mathfrak{D}$, a discrete strain rate tensor $D^\mathfrak{D} : (\mathbb{R}^2)^\mathfrak{T} \to (M_2(\mathbb{R}))^\mathfrak{D}$ and a stabilization term $\Delta^\mathfrak{D} : \mathbb{R}^\mathfrak{D} \to \mathbb{R}^\mathfrak{D}$.

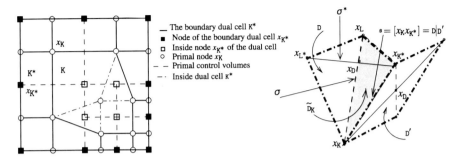

Fig. 1 The mesh $\mathcal{T}$ (left). A diamond D with a neighbour diamond D'(right)

Concerning the discretization of the nonlinear term $(\mathbf{u}^n \cdot \nabla)\mathbf{u}^{n+1}$, a conflict appears when writing the DDFV scheme because $\mathbf{u}^n$ is defined on centers and vertices of the mesh whereas $\nabla \mathbf{u}^{n+1}$ is defined on the diamond mesh. We have to approach $\int_V (\mathbf{u}^n \cdot \nabla)\mathbf{u}^{n+1} dx$ on both the primal and dual cells. Using a Stokes formula, we get $\sum_{\sigma \in \partial V} \int_\sigma (\mathbf{u}^n \cdot \mathbf{n}_{\sigma,V})\mathbf{u}^{n+1} ds$, this quantity will be approached using a scheme of the form: $\sum_{\sigma \in \partial V} F_{\sigma,V} \mathbf{u}_{\mathfrak{E}}^{n+1}$. In the following section, we explain how to define $F_{\sigma,V}$ and in Sect. 2.2, how to define $\mathbf{u}_{\mathfrak{E}}^{n+1}$.

2.1 Approximation of the normal flux

The major difficulty in the construction of the scheme lies in the approximation of

$$\int_\sigma (\mathbf{u}^n \cdot \mathbf{n}_{\sigma,V}) ds. \qquad (2)$$

We use the idea already presented in [8, 11] that is to define discrete mass fluxes taking into account the stabilization term. This allows to ensure the convenient property given below in Proposition 2.

Expression of discrete mass fluxes through a diamond edge. At the continuous level, the Stokes formula gives:

$$\int_D \text{div}(\mathbf{u}^n) dx = \sum_{s \in \partial D} \int_s \mathbf{u}^n \cdot \mathbf{n}_{sD} ds.$$

The discrete counterpart of this equality is:

$$|D| \text{div}^D(\mathbf{u}^n) - \lambda d_D^2 |D| \Delta^D p^n = \sum_{s \in \partial D} G_{s,D}(\mathbf{u}^n, p^n),$$

where d_D is the diameter of D and for $s = [x_K, x_{K^*}] = D|D'$ (see Fig. 1), we have:

$$G_{s,D}(\mathbf{u}^{\mathfrak{T}}, p^{\mathfrak{D}}) = m_s \frac{\mathbf{u}_K + \mathbf{u}_{K^*}}{2} \cdot \mathbf{n}_{sD} - \lambda(d_D^2 + d_{D'}^2)(p^{D'} - p^D).$$

We can approach the mass fluxes $\int_s \mathbf{u}^n \cdot \mathbf{n}_{sD} ds$ by using $G_{s,D}(\mathbf{u}^n, p^n)$.

Link between the integral (2) and the mass conservation equation. Noting $\widetilde{D}_K$ the triangle whose vertices are x_K, x_{K^*} and x_{L^*} (see Fig. 1), we remark that:

$$0 = \int_{\widetilde{D}_K} \text{div}(\mathbf{u}^n) dx = \int_\sigma \mathbf{u}^n \cdot \mathbf{n}_{\sigma K} ds + \sum_{s \in \mathfrak{S}_K \cap \partial D} \int_s \mathbf{u}^n \cdot \mathbf{n}_{sD} ds.$$

where $\mathfrak{S}_K = \{s \in \mathfrak{S}, \text{s. t. } s \subset K\}$ for all $K \in \mathfrak{M}$ and $\mathfrak{S}_{K^*} = \{s \in \mathfrak{S}, \text{s. t. } s \subset K^*\}$ for all $K^* \in \mathfrak{M}^*$, recalling that $\mathfrak{S}$ is the set of interior diamond sides. Thus, we define $F_{\sigma,K}$ and F_{σ^*,K^*}, the approximation of mass fluxes (2), as follows:

$$F_{\sigma,K}(\mathbf{u}^{\mathfrak{T}}, p^{\mathfrak{D}}) = -\sum_{s \in \mathfrak{S}_K \cap \partial D} G_{s,D}(\mathbf{u}^{\mathfrak{T}}, p^{\mathfrak{D}}), \quad \text{where } \sigma \subset D, \forall K \in \mathfrak{M},$$

$$F_{\sigma^*,K^*}(\mathbf{u}^{\mathfrak{T}}, p^{\mathfrak{D}}) = -\sum_{s \in \mathfrak{S}_{K^*} \cap \partial D} G_{s,D}(\mathbf{u}^{\mathfrak{T}}, p^{\mathfrak{D}}), \quad \text{where } \sigma^* \subset D, \forall K^* \in \mathfrak{M}^*. \quad (3)$$

We remark that if $(\mathbf{u}^{\mathfrak{T}}, p^{\mathfrak{D}})$ satisfies $\text{div}^{\mathfrak{D}}(\mathbf{u}^{\mathfrak{T}}) - \lambda d_{\mathfrak{D}}^2 \Delta^{\mathfrak{D}} p^{\mathfrak{D}} = 0$, we have the conservativity of the fluxes:

$$F_{\sigma,K}(\mathbf{u}^{\mathfrak{T}}, p^{\mathfrak{D}}) = -F_{\sigma,L}(\mathbf{u}^{\mathfrak{T}}, p^{\mathfrak{D}}), \quad \forall \sigma = K|L,$$

$$F_{\sigma^*,K^*}(\mathbf{u}^{\mathfrak{T}}, p^{\mathfrak{D}}) = -F_{\sigma^*,L^*}(\mathbf{u}^{\mathfrak{T}}, p^{\mathfrak{D}}), \quad \forall \sigma^* = K^*|L^*.$$

With our choice, we obtain that the approximation of the integral of the velocity divergence on the primal and dual cells vanishes:

Proposition 1. *Let $\mathfrak{T}$ be a DDFV mesh. For all $(\mathbf{u}^{\mathfrak{T}}, p^{\mathfrak{D}}) \in (\mathbb{R}^2)^{\mathfrak{T}} \times \mathbb{R}^{\mathfrak{D}}$, we have*

$$\forall K \in \mathfrak{M}, \sum_{\sigma \in \partial K} F_{\sigma,K}(\mathbf{u}^{\mathfrak{T}}, p^{\mathfrak{D}}) = 0 \text{ and } \forall K^* \in \mathfrak{M}^*, \sum_{\sigma^* \in \partial K^*} F_{\sigma^*,K^*}(\mathbf{u}^{\mathfrak{T}}, p^{\mathfrak{D}}) = 0.$$

2.2 Discretization of the non-linear term

Using the definition of the mass fluxes given by (3), we can define the discretization of the non-linear term with an upwind method.

Definition 1. We define $\mathbf{b}^{\mathfrak{T}} : (\mathbb{R}^2)^{\mathfrak{T}} \times \mathbb{R}^{\mathfrak{D}} \times (\mathbb{R}^2)^{\mathfrak{T}} \to (\mathbb{R}^2)^{\mathfrak{T}}$, as follows:

$$\mathbf{b}_K(\mathbf{u}^{\mathfrak{T}}, p^{\mathfrak{D}}, \mathbf{v}^{\mathfrak{T}}) = \frac{1}{|K|} \sum_{\sigma \in \partial K} F_{\sigma,K}(\mathbf{u}^{\mathfrak{T}}, p^{\mathfrak{D}}) \mathbf{v}_{\sigma+}, \quad \forall K \in \mathfrak{M},$$

$$b_{\kappa^*}(\mathbf{u}^{\mathfrak{T}}, p^{\mathfrak{D}}, \mathbf{v}^{\mathfrak{T}}) = \frac{1}{|\kappa^*|} \sum_{\sigma^* \in \partial \kappa^*} F_{\sigma^*, \kappa^*}(\mathbf{u}^{\mathfrak{T}}, p^{\mathfrak{D}}) \mathbf{v}_{\sigma^*+}, \quad \forall \kappa^* \in \mathfrak{M}^*,$$

where

$$\mathbf{v}_{\sigma+} = \begin{cases} \mathbf{v}_{\kappa} & \text{if } F_{\sigma,\kappa}(\mathbf{u}^{\mathfrak{T}}, p^{\mathfrak{D}}) \geq 0, \\ \mathbf{v}_{\mathbf{L}} & \text{elsewhere.} \end{cases} \qquad \mathbf{v}_{\sigma^*+} = \begin{cases} \mathbf{v}_{\kappa^*} & \text{if } F_{\sigma^*, \kappa^*}(\mathbf{u}^{\mathfrak{T}}, p^{\mathfrak{D}}) \geq 0, \\ \mathbf{v}_{\mathbf{L}^*} & \text{elsewhere.} \end{cases}$$

The unconditional stability of the scheme is ensured by the crucial result:

Proposition 2. *Let $\mathfrak{T}$ be a DDFV mesh. For all $(\mathbf{u}^{\mathfrak{T}}, p^{\mathfrak{D}}, \mathbf{v}^{\mathfrak{T}}) \in \mathbb{E}_0 \times \mathbb{R}^{\mathfrak{D}} \times \mathbb{E}_0$ such that $\operatorname{div}^{\mathfrak{D}}(\mathbf{u}^{\mathfrak{T}}) - \lambda d_{\mathfrak{D}}^2 \Delta^{\mathfrak{D}} p^{\mathfrak{D}} = 0$, we have*

$$\sum_{\kappa \in \mathfrak{M}} |\kappa| b_{\kappa}(\mathbf{u}^{\mathfrak{T}}, p^{\mathfrak{D}}, \mathbf{v}^{\mathfrak{T}}) \cdot \mathbf{v}_{\kappa} + \sum_{\kappa^* \in (\mathfrak{M}^* \cup \partial \mathfrak{M}^*)} |\kappa^*| b_{\kappa^*}(\mathbf{u}^{\mathfrak{T}}, p^{\mathfrak{D}}, \mathbf{v}^{\mathfrak{T}}) \cdot \mathbf{v}_{\kappa^*} \geq 0.$$

3 DDFV schemes for the Navier-Stokes equation

Let $N \in \mathbb{N}^*$. We note $\delta t = \frac{T}{N}$ and $t_n = n \delta t$ for $n \in \{0, \cdots, N\}$. We use an implicit Euler time discretization except for the non-linear term which is linearized thanks to a standard semi-implicit approximation $(\mathbf{u}^n \cdot \nabla) \mathbf{u}^{n+1}$. The DDFV scheme for the problem (1) reads as follows:

- **Initialization:** we define $\mathbf{u}^0 \in \mathbb{E}_0$ and $p^0 \in \mathbb{R}^{\mathfrak{D}}$ as follows:

$$\begin{cases} \mathbf{u}^0 = \mathbb{P}_m^{\mathfrak{T}} \mathbf{u}_{\text{ini}} \in \mathbb{E}_0, \\ p^0 \in \mathbb{R}^{\mathfrak{D}}, \text{ s. t. } \Delta^{\mathfrak{D}} p^0 = \frac{1}{\lambda d_{\mathfrak{D}}^2} \operatorname{div}^{\mathfrak{D}}(\mathbb{P}_m^{\mathfrak{T}} \mathbf{u}_{\text{ini}}) \text{ with } \sum_{D \in \mathfrak{D}} |D| p_D^0 = 0. \end{cases} \quad (4)$$

Note that with this choice of $(\mathbf{u}^0, p^0)$, we have $\operatorname{div}^{\mathfrak{D}}(\mathbf{u}^0) - \lambda d_{\mathfrak{D}}^2 \Delta^{\mathfrak{D}} p^0 = 0$.

- **Time stepping:** assume that $(\mathbf{u}^n, p^n) \in \mathbb{E}_0 \times \mathbb{R}^{\mathfrak{D}}$ are given ($n \in \{0, \cdots, N-1\}$). We have to find $\mathbf{u}^{n+1} \in \mathbb{E}_0$ and $p^{n+1} \in \mathbb{R}^{\mathfrak{D}}$ such that:

$$\begin{cases} \forall \kappa \in \mathfrak{M}, \dfrac{\mathbf{u}_{\kappa}^{n+1} - \mathbf{u}_{\kappa}^n}{\delta t} + \operatorname{div}^{\kappa}(-2\eta^{\mathfrak{D}} D^{\mathfrak{D}} \mathbf{u}^{n+1} + p^{n+1} \mathrm{Id}) + b_{\kappa}(\mathbf{u}^n, p^n, \mathbf{u}^{n+1}) = \mathbf{f}_{\kappa}^{n+1}, \\ \forall \kappa^* \in \mathfrak{M}^*, \dfrac{\mathbf{u}_{\kappa^*}^{n+1} - \mathbf{u}_{\kappa^*}^n}{\delta t} + \operatorname{div}^{\kappa^*}(-2\eta^{\mathfrak{D}} D^{\mathfrak{D}} \mathbf{u}^{n+1} + p^{n+1} \mathrm{Id}) + b_{\kappa^*}(\mathbf{u}^n, p^n, \mathbf{u}^{n+1}) = \mathbf{f}_{\kappa^*}^{n+1}, \\ \operatorname{div}^{\mathfrak{D}}(\mathbf{u}^{n+1}) - \lambda d_{\mathfrak{D}}^2 \Delta^{\mathfrak{D}} p^{n+1} = 0, \\ \sum_{D \in \mathfrak{D}} |D| p_D^{n+1} = 0, \end{cases}$$

(5)

where $\eta^{\mathfrak{D}} = (\eta(x_{\scriptscriptstyle D}))_{D \in \mathfrak{D}}$, $\lambda > 0$ given, $\mathbf{f}_{\kappa}^{n+1} = \dfrac{1}{\delta t \, |\kappa|} \displaystyle\int_{t_n}^{t_{n+1}} \int_{\kappa} \mathbf{f}(t,x) \mathrm{d}x \mathrm{d}t$ for all $\kappa \in \mathfrak{M}$ and $\mathbf{f}_{\kappa^*}^{n+1} = \dfrac{1}{\delta t \, |\kappa^*|} \displaystyle\int_{t_n}^{t_{n+1}} \int_{\kappa^*} \mathbf{f}(t,x) \mathrm{d}x \mathrm{d}t$ for all $\kappa^* \in \mathfrak{M}^*$.

Note that, in order to be able to apply Proposition 2, we have to ensure the property $\mathrm{div}^{\mathfrak{D}}(\mathbf{u}^n) - \lambda d_{\mathfrak{D}}^2 \Delta^{\mathfrak{D}} p^n = 0$ even for the initial time (i.e. for $n \in \{0, \cdots, N\}$). This permits to prove the following stability proposition.

Proposition 3 (**Discrete energy estimates**). *Let $\mathfrak{T}$ be a DDFV mesh. The finite volume scheme (4)-(5) with $\lambda > 0$ admits a unique solution $(\mathbf{u}^n, p^n)_{n \in \{0,\cdots,N\}}$. For $N > 1$, there exists a constant $C > 0$, depending only on Ω, λ, η, $\mathbf{u}_{ini}$ and $\mathbf{f}$, such that:*

$$\sum_{n=1}^{N} \delta t \|\nabla^{\mathfrak{D}} \mathbf{u}^n\|_2^2 \leq C, \ \sum_{n=1}^{N} \delta t |p^n|_h^2 \leq C, \ \sum_{n=0}^{N-1} \|\mathbf{u}^{n+1} - \mathbf{u}^n\|_2^2 \leq C \text{ and } \|\mathbf{u}^N\|_2^2 \leq C.$$

4 Numerical results

We show here some numerical results obtained on a domain $\Omega =]0,1[^2$ with $T = 1$ and $\delta t = 10^{-2}$. Error estimates are given on a test with a stabilization coefficient chosen to be $\lambda = 10^{-3}$. In order to illustrate error estimates, the family of meshes (see Fig. 2) are obtained by successive global refinement of the original mesh.

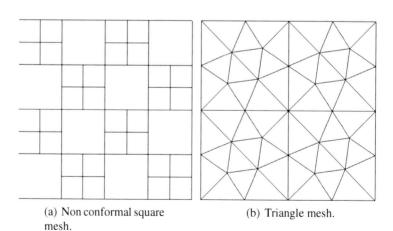

(a) Non conformal square mesh.

(b) Triangle mesh.

Fig. 2 Family of meshes

The exact solution is the Green-Taylor vortex:

$$\mathbf{u} = \begin{pmatrix} -\cos(2\pi x)\sin(2\pi y)e^{-2t\eta} \\ \sin(2\pi x)\cos(2\pi y)e^{-2t\eta} \end{pmatrix}, \quad p = -\frac{1}{4}(\cos(4\pi x) + \cos(4\pi y))e^{-4t\eta}.$$

The viscosity η being chosen, we define the source term **f** and the boundary data **g** in such a way that (1) is satisfied.

We compare the relative $L^2(\Omega \times]0,T[)$-norm of the error obtained with the DDFV scheme, for the pressure (denoted Erpre), for the velocity gradient (denoted Ergradvel) and for the velocity (denoted Ervel) respectively. On the two tables, we give the number of primal cells (denoted NbCell) and the convergence rates (denoted Ratio).

Table 1 $\eta = 1$ on the non conformal square mesh Fig. 2(a)

NbCell	Ervel	Ratio	Ergradvel	Ratio	Erpre	Ratio
208	2.804E-02	-	8.508E-02	-	1.526E+00	-
736	6.761E-03	2.052	4.309E-02	0.9815	6.574E-01	1.215
2752	1.803E-03	1.907	2.158E-02	0.9973	3.237E-01	1.022
10624	6.045E-04	1.577	1.079E-02	1.001	1.633E-01	0.9874

When the viscosity is equal to 1 (Table 1), we observe a first order convergence for the L^2-norm of the velocity gradient and of the pressure, which seems to be optimal. We obtain a super-convergence for the L^2-norm of the velocity. Furthermore, let us emphasize that the convergence rate is not sensitive to the presence of non conformal control volumes.

Table 2 $\eta = 10^{-3}$ on the triangle mesh Fig. 2(b)

NbCell	Ervel	Ratio	Ergradvel	Ratio	Erpre	Ratio
256	2.952E-01	-	4.403E-01	-	5.181E-01	-
960	2.080E-01	0.5049	3.551E-01	0.3105	3.718E-01	0.4788
3712	1.292E-01	0.6871	2.465E-01	0.5262	2.420E-01	0.6195
14592	7.432E-02	0.7975	1.643E-01	0.5858	1.432E-01	0.7573

When the viscosity is equal to 10^{-3} (Table 2), the convective term is dominant and we observe that the scheme is still convergent even if the convergence of the velocity gradient deteriorates.

5 Conclusion

In this paper, we proposed a stabilized DDFV scheme for the Navier-Stokes problem. This scheme is well-posed on 2D general meshes. Its convergence properties were illustrated with numerical experimentations. In a work in progress, we provide a proof of this result.

References

1. B. Andreianov, F. Boyer, and F. Hubert. Discrete duality finite volume schemes for Leray-Lions type elliptic problems on general 2D-meshes. *Numer. Methods PDE*, (2007), **23**(1):145–195.
2. F. Boyer and F. Hubert. Finite volume method for 2D linear and nonlinear elliptic problems with discontinuities. *SIAM J. Num. Anal.*, (2008), **46**(6):3032–3070.
3. Y. Coudière and G. Manzini. The Discrete Duality Finite Volume Method for Convection-diffusion Problems. *SIAM J. Numer. Anal. Volume*, (2010), **47**(6):4163–4192.
4. S. Delcourte, K. Domelevo, and P. Omnes. A discrete duality finite volume approach to Hodge decomposition and div-curl problems on almost arbitrary two-dimensional meshes. *SIAM J. Numer. Anal.*, (2007), **45**(3):1142–1174.
5. J. Droniou and R. Eymard. Study of the mixed finite volume method for Stokes and Navier-Stokes equations. *Num. Meth. PDEs*, (2009), **25**(1):137–171.
6. S. Delcourte. *Développement de méthodes de volumes finis pour la mécanique de fluides.* Ph.D. thesis, Univ. Paul Sabatier, Toulouse, 2007.
7. K. Domelevo and P. Omnès. A finite volume method for the Laplace equation on almost arbitrary two-dimensional grids. *Math. Model. Numer. Anal.*, (2005), **39**(6):1203–1249.
8. R. Eymard, T. Gallouët, R. Herbin, and J.-C. Latché. Analysis tools for finite volume schemes. *Acta Math. Univ. Comenian. (N.S.)*, (2007) **76**(1):111–136.
9. R. Eymard and R. Herbin. A new colocated finite volume scheme for the incompressible Navier-Stokes equations on general non matching grids. *C. R. Math. Acad. Sci. Paris*, (2007) **344**(10):659–662.
10. R. Eymard, R. Herbin, and J.-C. Latché. Convergence analysis of a colocated finite volume scheme for the incompressible Navier-Stokes equations on general 2D or 3D meshes. *SIAM J. Numer. Anal.*, (2007) **45**(1):1–36.
11. R. Eymard, R. Herbin, J.-C. Latché, and B. Piar. Convergence analysis of a locally stabilized collocated finite volume scheme for incompressible flows. *Math. Model. Numer. Anal.*, (2009), **43**(5):889–927.
12. F. Hermeline. A finite volume method for the approximation of diffusion operators on distorted meshes. *J. Comput. Phys.*, (2000), **160**(2):481–499.
13. F. Hermeline. Approximation of diffusion operators with discontinuous tensor coefficients on distorted meshes. *Comput. Methods Appl. Mech. Engrg.*, (2003), **192**(16-18):1939–1959.
14. F. Harlow and J. Welch. Numerical calculation of time-dependent viscous incompressible flow of fluid with free surface. *The physics of fluids*, (1965), **8**(12):2182–2189.
15. S. Krell. Stabilized DDFV schemes for Stokes problem with variable viscosity on general 2D meshes. *Num. Meth. PDEs*, (2011), available on-line: http://dx.doi.org/10.1002/num.20603.
16. R. A. Nicolaides. Analysis and convergence of the MAC scheme. I. The linear problem. *SIAM J. Numer. Anal.*, (1992), **29**(6):1579–1591.
17. S. Perron, S. Boivin, and J.-M. Hérard. A finite volume method to solve the 3D Navier-Stokes equations on unstructured collocated meshes. *Comput. & Fluids*, (2004), **33**(10):1305–1333.

The paper is in final form and no similar paper has been or is being submitted elsewhere.

Higher-Order Reconstruction: From Finite Volumes to Discontinuous Galerkin

Václav Kučera

Abstract This work is concerned with the introduction of a new numerical scheme based on the discontinuous Galerkin (DG) method. We follow the methodology of higher order finite volume (FV) and spectral volume (SV) schemes and introduce a reconstruction operator into the discontinuous Galerkin (DG) method. This operator constructs higher order piecewise polynomial reconstructions from the lower order DG scheme. We present two variants, the generalization of standard FV schemes, already proposed by Dumbser et al. (2008) and the generalization of the SV method. Theoretical aspects are discussed and numerical experiments are carried out.

Keywords Discontinuous Galerkin, finite volumes, reconstruction
MSC2010: 65M15, 65M60, 65M12

1 Problem formulation and notation

For simplicity, we shall be concerned with a scalar hyperbolic equation, although the same arguments basically hold for any time-dependent PDE. We treat a nonlinear nonstationary scalar hyperbolic equation in a bounded domain $\Omega \subset I\!R^d$ with a Lipschitz-continuous boundary $\partial \Omega$. We seek $u : \Omega \times [0,T] \to I\!R$ such that

$$\frac{\partial u}{\partial t} + \mathrm{div}\mathbf{f}(u) = 0 \quad \text{in } \Omega \times (0,T) \tag{1}$$

along with an appropriate initial and boundary condition. Here $\mathbf{f} = (f_1, \cdots, f_d)$ and $f_s, s = 1, \ldots, d$ are Lipschitz continuous fluxes in the direction $x_s, s = 1, \ldots, d$.

Václav Kučera
Charles University in Prague, Faculty of Mathematics and Physics, Sokolovská 83, Praha 8, 186 75, Czech Republic, e-mail: vaclav.kucera@email.cz

Let $\mathcal{T}_h$ be a partition (triangulation) of the closure $\overline{\Omega}$ into a finite number of closed simplices $K \in \mathcal{T}_h$. In general we do not require the standard conforming properties of $\mathcal{T}_h$ used in the finite element method (i.e. we admit the so-called hanging nodes). We shall use the following notation. By ∂K we denote the boundary of an element $K \in \mathcal{T}_h$ and set $h_K = \mathrm{diam}(K)$, $h = \max_{K \in \mathcal{T}_h} h_K$.

Let $K, K' \in \mathcal{T}_h$. We say that K and K' are *neighbours*, if they share a common face $\Gamma \subset \partial K$. By $\mathcal{F}_h$ we denote the system of all faces of all elements $K \in \mathcal{T}_h$.

For each $\Gamma \in \mathcal{F}_h$ we define a unit normal vector $\mathbf{n}_\Gamma$, such that for $\Gamma \in \mathcal{F}_h^B$ the normal $\mathbf{n}_\Gamma$ has the same orientation as the outer normal to $\partial \Omega$.

Over a triangulation $\mathcal{T}_h$ we define the *broken Sobolev spaces*

$$H^k(\Omega, \mathcal{T}_h) = \{v; v|_K \in H^k(K), \forall K \in \mathcal{T}_h\}.$$

For each face $\Gamma \in \mathcal{F}_h^I$ there exist two neighbours $K_\Gamma^{(L)}, K_\Gamma^{(R)} \in \mathcal{T}_h$ such that $\Gamma \subset K_\Gamma^{(L)} \cap K_\Gamma^{(R)}$. We use the convention that $\mathbf{n}_\Gamma$ is the outer normal to $K_\Gamma^{(L)}$. For $v \in H^1(\Omega, \mathcal{T}_h)$ and $\Gamma \in \mathcal{F}_h^I$ we introduce the following notation:

$$v|_\Gamma^{(L)} = \text{trace of } v|_{K_\Gamma^{(L)}} \text{ on } \Gamma, \quad v|_\Gamma^{(R)} = \text{trace of } v|_{K_\Gamma^{(R)}} \text{ on } \Gamma, \quad [v]_\Gamma = v|_\Gamma^{(L)} - v|_\Gamma^{(R)}.$$

On boundary edges we define $v|_\Gamma^{(R)} = [v]_\Gamma := v|_\Gamma^{(L)}$.

Let $n \geq 0$ be an integer. We define the space of discontinuous piecewise polynomial functions

$$S_h^n = \{v; v|_K \in P^n(K), \forall K \in \mathcal{T}_h\},$$

where $P^n(K)$ is the space of all polynomials on K of degree $\leq n$. Specifically,

- S_h^0: is the space of piecewise constant functions as known from the FV method,
- S_h^n, $n \geq 0$: the DG solution lies in this space of piecewise nth degree polynomials,
- S_h^N, $N > n$: the higher order reconstructed DG solution will lie in this space.

2 Discontinuous Galerkin (DG) formulation

We multiply (1) by an arbitrary $\varphi_h^n \in S_h^n$, integrate over an element $K \in \mathcal{T}_h$ and apply Green's theorem. By summing over all $K \in \mathcal{T}_h$ and rearranging, we get

$$\frac{d}{dt}\int_\Omega u(t)\, \varphi_h^n\, dx + \sum_{\Gamma \in \mathcal{F}_h} \int_\Gamma \mathbf{f}(u) \cdot \mathbf{n}\, [\varphi_h^n]\, dS - \sum_{K \in \mathcal{T}_h} \int_K \mathbf{f}(u) \cdot \nabla \varphi_h^n\, dx = 0. \quad (2)$$

The boundary convective terms will be treated similarly as in the finite volume method, i.e. with the aid of a numerical flux $H(u, v, \mathbf{n})$:

$$\int_\Gamma \mathbf{f}(u) \cdot \mathbf{n} \, [\varphi_h^n] \, dS \approx \int_\Gamma H(u^{(L)}, u^{(R)}, \mathbf{n}) [\varphi_h^n] \, dS. \qquad (3)$$

We assume that H is *Lipschitz continuous*, *consistent* and *conservative*, cf. [4]. Finally, we define the *convective form* $b_h(\cdot, \cdot)$ defined for $v, \varphi \in H^1(\Omega, \mathcal{T}_h)$:

$$b_h(v, \varphi) = \int_{\mathcal{F}_h} H(v^{(L)}, v^{(R)}, \mathbf{n})[\varphi] \, dS - \sum_{K \in \mathcal{T}_h} \int_K \mathbf{f}(v) \cdot \nabla \varphi \, dx.$$

Definition 1 (Standard DG scheme). We seek $u : [0, T] \to S_h^n$ such that

$$\frac{d}{dt}\left(u_h(t), \varphi_h^n\right) + b_h\left(u_h(t), \varphi_h^n\right) = 0, \quad \forall \varphi_h^n \in S_h^n, \, \forall t \in (0, T). \qquad (4)$$

We note that if we take $n = 0$, i.e. $u_h : (0, T) \to S_h^0$, then from the definition of b_h, we see that the DG scheme (4) is equivalent to the standard FV method.

3 Reconstructed discontinuous Galerkin (RDG) formulation

For $v \in L^2(\Omega)$, we denote by $\Pi_h^n v$ the $L^2(\Omega)$-projection of v on S_h^n:

$$\Pi_h^n v \in S_h^n, \quad \left(\Pi_h^n v - v, \varphi_h^n\right) = 0, \quad \forall \varphi_h^n \in S_h^n. \qquad (5)$$

The basis of the proposed method lies in the observation that (2) can be viewed as an equation for the evolution of $\Pi_h^n u(t)$, where u is the exact solution of (1). In other words, due to (5), $\Pi_h^n u(t) \in S_h^n$ satisfies the following equation for all $\varphi_h^n \in S_h^n$:

$$\frac{d}{dt} \int_\Omega \Pi_h^n u(t) \, \varphi_h^n \, dx + \int_{\mathcal{F}_h} \mathbf{f}(u) \cdot \mathbf{n} \, [\varphi_h^n] \, dS - \sum_{K \in \mathcal{T}_h} \int_K \mathbf{f}(u) \cdot \nabla \varphi_h^n \, dx = 0. \qquad (6)$$

Now, let $N > n$ be an integer. We assume that there exists a piecewise polynomial function $U_h^N(t) \in S_h^N$, which is an approximation of $u(t)$ of order $N + 1$, i.e.

$$U_h^N(x, t) = u(x, t) + O(h^{N+1}), \quad \forall x \in \Omega, \, \forall t \in [0, T]. \qquad (7)$$

This is possible, if u is sufficiently regular in space, e.g. $u(t) \in W^{N+1,\infty}(\Omega)$, cf.[1]. Now we incorporate the approximation $U_h^N(t)$ into (6): the exact solution u satisfies

$$\frac{d}{dt}\left(\Pi_h^n u(t), \varphi_h^n\right) + b_h\left(U_h^N(t), \varphi_h^n\right) = E(\varphi_h^n, t), \quad \forall \varphi_h^n \in S_h^n, \, \forall t \in (0, T), \qquad (8)$$

where $E(\varphi_h^n)$ is an error term defined as

$$E(\varphi_h^n, t) = b_h\left(U_h^N(t), \varphi_h^n\right) - b_h\left(u(t), \varphi_h^n\right). \qquad (9)$$

Lemma 1. *The following estimate holds for all $t \in [0, T]$:*

$$E(\varphi_h^n, t) = O(h^N) \|\varphi_h^n\|_{L^2(\Omega)}. \tag{10}$$

Proof: Due to the consistency and Lipschitz continuity of H, we have on $\Gamma \in \mathcal{F}_h$

$$\mathbf{f}(u) \cdot \mathbf{n} - H(U_h^{N,(L)}, U_h^{N,(R)}, \mathbf{n}) = H(u, u, \mathbf{n}) - H(U_h^{N,(L)}, U_h^{N,(R)}, \mathbf{n}) = O(h^{N+1}).$$

Furthermore, due to the Lipschitz-continuity of $\mathbf{f}$, we have on element $K \in \mathcal{T}_h$

$$\mathbf{f}(u) - \mathbf{f}(U_h^N) = O(h^{N+1}).$$

Estimate (10) follows from these results and the application of the *inverse* and *multiplicative trace inequalities*, cf [4]. □

It remains to construct a sufficiently accurate approximation $U_h^N(t) \in S_h^N$ to $u(t)$, such that (7) is satisfied. This leads to the following problem.

Definition 2 (**Reconstruction problem**). Let $v : \Omega \to \mathbb{R}$ be sufficiently regular. Given $\Pi_h^n v \in S_h^n$, find $v_h^N \in S_h^N$ such that $v - v_h^N = O(h^{N+1})$ in Ω. We define the corresponding reconstruction operator $R : S_h^n \to S_h^N$ by $R \Pi_h^n v := v_h^N$.

By setting $U_h^N(t) := R \Pi_h^n u(t)$ in (8), we obtain the following semidiscrete, formally Nth order scheme for the $L^2(\Omega)$-projections of the exact solution u onto S_h^n:

$$\frac{d}{dt}(\Pi_h^n u(t), \varphi_h^n) + b_h(R \Pi_h^n u(t), \varphi_h^n) = O(h^N) \|\varphi_h^n\|_{L^2(\Omega)}, \quad \forall \varphi_h^n \in S_h^n. \tag{11}$$

By neglecting the right-hand side and approximating $u_h^n(t) \approx \Pi_h^n u(t)$, we arrive at the following definition of the *reconstructed discontinuous Galerkin* (RDG) scheme.

Definition 3 (**Reconstructed DG scheme**). We seek $u_h^n : [0, T] \to S_h^n$ such that

$$\frac{d}{dt}(u_h^n(t), \varphi_h^n) + b_h(R u_h^n(t), \varphi_h^n) = 0, \quad \forall \varphi_h^n \in S_h^n, \ \forall t \in (0, T). \tag{12}$$

There are several points worth mentioning.
- The derivation of the RDG scheme follows the methodology of higher order FV and SV schemes, cf. [7]. The basis of these schemes is an equation for the evolution of averages of the exact solution on individual elements (i.e. an equation for $\Pi_h^0 u(t)$). Equation (11) is a direct generalization for the case of higher order $L^2(\Omega)$-projections $\Pi_h^n u(t)$, $n \geq 0$.
- Both $u_h^n(t)$ and φ_h^n lie in S_h^n. Only $R u_h^n(t)$, lies in the higher dimensional space S_h^N. Despite this fact, equation (11) indicates that we may expect $u - R u_h^n = O(h^{N+1})$, although $u - u_h^n = O(h^{n+1})$.

- Numerical quadrature must be employed to evaluate surface and volume integrals in (12). Since test functions are in S_h^n, as compared to S_h^N in the corresponding Nth order standard DG scheme, we may use lower order (i.e. more efficient) quadrature formulae as compared to standard DG.

As in the case of higher order FVM, we use an explicit time stepping method. For simplicity, we formulate the forward Euler method, which is only first order accurate, however in Section 5, higher order Adams-Bashforth methods are used.

Let us construct a partition $0 = t_0 < t_1 < t_2 \ldots$ of the time interval $[0, T]$ and define the time step $\tau_k = t_{k+1} - t_k$. We use the approximation $u_h^n(t_k) \approx u_h^{n,k} \in S_h^n$. The forward Euler scheme is given by:

Definition 4 (**Explicit RDG scheme**). We seek $u_h^{n,k} \in S_h^n$, $k = 0, 1, \ldots$ such that

$$\left(\frac{u_h^{n,k+1} - u_h^{n,k}}{\tau_k}, \varphi_h^n \right) + b_h\left(R u_h^{n,k}, \varphi_h^n\right) = 0, \quad \forall \varphi_h^n \in S_h^n, \; k = 0, 1, \ldots, \qquad (13)$$

where $u_h^{n,0} = u_{h,0}$ is an S_h^n approximation of the initial condition u^0.

The upper limit on stable time steps, given by a CFL-like condition, is more restrictive with growing N. However, in the RDG scheme, stability properties are inherited from the lower order scheme, therefore a larger time step is possible as compared to the corresponding Nth order standard DG scheme.

3.1 Construction of the reconstruction operator

3.1.1 'Standard' approach

In the *standard approach*, a stencil (a group of neighboring elements and the element under consideration) is used to build an Nth-degree polynomial approximation to u on the element under consideration ([5] [6]). In the FV method, the von Neumann neighborhood of an element is used as a stencil to obtain a piecewise linear reconstruction, cf. Fig. 1, 1). However, for higher order reconstructions, the size of the stencil increases dramatically, cf. Fig. 1, 2), rendering higher degrees than quadratic very time consuming. In the case of the RDG scheme, we need not increase the stencil size to obtain higher order accuracy, it suffices to take the von Neumann neighborhood and increase the order of the underlying DG scheme.

In analogy to the FV method, the reconstruction operator R is constructed on each stencil independently and satisfies that $R\Pi_h^n$ is in some sense *polynomial preserving*. Specifically, for each element K and its corresponding stencil S, we require that for all $p \in P^N(S)$

$$\left((R\Pi_h^n)\big|_S p \right)\Big|_K = p\big|_K. \qquad (14)$$

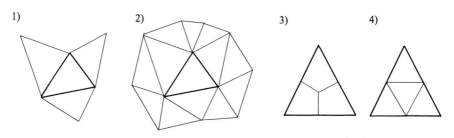

Fig. 1 1) FV stencil for linear reconstruction, 2) FV stencil for quadratic reconstruction, 3) Control volumes in a spectral volume for linear reconstruction, 4) Analogy to the SV approach for DG - partition of triangle into control volumes, e.g. cubic reconstruction from linear data

This requirement allows us to study approximation properties of R using the Bramble–Hilbert technique as in the standard finite element method, [1]. The disadvantage of this approach is that for unstructured meshes, the coefficients of the reconstruction operator must be stored for each individual stencil.

In the FV method, different conditions on R than (14) are often used, e.g. continuous or discrete least squares. Special care must be taken in the vicinity of steep gradients and discontinuities, where the Gibbs phenomenon may occur. In this case different strategies are employed, e.g. limiting, ENO and WENO schemes, TVD etc. The generalization of these concepts to the RDG method is left for future work.

3.1.2 Spectral volume approach

In the *spectral volume approach*, we start with a partition of Ω into so-called *spectral volumes* S, for example triangles in 2D. The triangulation $\mathcal{T}_h$ is formed by subdividing each spectral volume S into sub-cells K, called *control volumes*, cf. [7]. In the FV method, the order of accuracy of the reconstruction determines the number of control volumes to be generated in each spectral volume. For example, for a linear reconstruction on a triangle, the triangle is divided into three control volumes, Fig. 1, 3). Again, in the RDG scheme, we may use only the smallest available partition into control volumes, and increase the accuracy by increasing the order of the underlying scheme, cf. Fig. 1, 4).

The reconstruction operator is constructed on each spectral volume independently such that it is in some sense polynomial preserving, i.e. for each stencil S, we require that for all $p \in P^N(S)$

$$\left(R\Pi_h^n\right)\big|_S p = p. \tag{15}$$

The advantage of this approach is that all spectral volumes are affine equivalent, we construct the reconstruction operator R only on one reference spectral volume.

4 Relation between RDG and standard DG

The only difference between the DG scheme (4) and RDG scheme (12) is the presence of the reconstruction operator R in the first variable of $b_h(\cdot,\cdot)$. While the error analysis of (4) is well understood (at least for convection-diffusion problems [4]), the analysis of (12) or (13) poses a new challenge. The problem lies in the fact that we cannot test (12) with $\varphi_h^n := Ru_h^{n,k}$ or something similar, since $Ru_h^{n,k} \notin S_h^n$. Therefore, we need to establish a relation between (12) and Nth order DG, instead of only nth order DG.

Definition 5 (Auxiliary problem). We seek $\tilde{u}_h^{N,k} \in S_h^N$ such that

$$\left(\frac{\tilde{u}_h^{N,k+1} - \tilde{u}_h^{N,k}}{\tau_k}, \varphi_h^N\right) + b_h\left(R\Pi_h^n \tilde{u}_h^{N,k}, \varphi_h^N\right) = 0, \quad \forall \varphi_h^N \in S_h^N, \ k = 0, 1, \ldots, \tag{16}$$

where $\tilde{u}_h^{N,0}$ is an S_h^N approximation of the initial condition u^0.

Lemma 2. Let $u_h^{n,0} = \Pi_h^n \tilde{u}_h^{N,0}$. Then $u_h^{n,k} \in S_h^n$, the solution of (13) and the solution $\tilde{u}_h^{N,k} \in S_h^N$ of (16) satisfy

$$u_h^{n,k} = \Pi_h^n \tilde{u}_h^{N,k}, \quad \forall k = 0, 1, \cdots. \tag{17}$$

Proof: We prove (17) by induction:
$k = 1$: Since $u_h^{n,0} = \Pi_h^n \tilde{u}_h^{N,0}$, we have for all $\varphi_h^n \in S_h^n$

$$(\Pi_h^n \tilde{u}_h^{N,1}, \varphi_h^n) = (\tilde{u}_h^{N,1}, \varphi_h^n) = (\tilde{u}_h^{N,0}, \varphi_h^n) - \tau_k b_h(R\Pi_h^n \tilde{u}_h^{N,0}, \varphi_h^n)$$
$$= (u_h^{n,0}, \varphi_h^n) - \tau_k b_h(Ru_h^{n,0}, \varphi_h^n) = (u_h^{n,1}, \varphi_h^n),$$

hence $(\Pi_h^n \tilde{u}_h^{N,1} - u_h^{n,1}, \varphi_h^n) = 0$ for all $\varphi_h^n \in S_h^n$. Therefore $\Pi_h^n \tilde{u}_h^{N,1} = u_h^{n,1}$.
$k > 1$: Assume (17) holds for some $k > 1$. Then for all $\varphi_h^n \in S_h^n$

$$(\Pi_h^n \tilde{u}_h^{N,k+1}, \varphi_h^n) = (\tilde{u}_h^{N,k+1}, \varphi_h^n) = (\tilde{u}_h^{N,k}, \varphi_h^n) - \tau_k b_h(R\Pi_h^n \tilde{u}_h^{N,k}, \varphi_h^n)$$
$$= (u_h^{n,k}, \varphi_h^n) - \tau_k b_h(Ru_h^{n,k}, \varphi_h^n) = (u_h^{n,k+1}, \varphi_h^n),$$

therefore $\Pi_h^n \tilde{u}_h^{N,k+1} = u_h^{n,k+1}$. This completes the induction step $k \to k+1$. □

As a corollary, error estimates for the auxiliary problem imply error estimates for the RDG scheme (12). Problem (16) is basically the standard Nth order DG scheme with the operator $R\Pi_h^n$ in the first variable of $b_h(\cdot,\cdot)$. Therefore, sufficient knowledge of the properties of $R\Pi_h^n$ (which is polynomial preserving) and standard DG error estimates would imply the estimates for the RDG scheme.

5 Numerical experiments

We present numerical experiments for the periodic advection of a 1D sine wave on uniform meshes. Experimental orders of accuracy α in various norms on meshes with N elements are given in Tables 1 and 2. Here $e_h = u - Ru_h^n$ at t corresponding to ten periods. The increase in accuracy due to reconstruction is clearly visible.

Table 1 1D advection of sine wave, P^2 RDG scheme with P^8 reconstruction

N	$\|e_h\|_{L^\infty(\Omega)}$	α	$\|e_h\|_{L^2(\Omega)}$	α	$|e_h|_{H^1(\Omega,\mathcal{T}_h)}$	α
4	5.82E-03	–	3.49E-03	–	3.65E-02	–
8	7.53E-05	6.27	4.43E-05	6,30	1.06E-03	5,11
16	9.07E-07	6.38	5.95E-07	6,22	3.58E-05	4,89
32	1.82E-08	5.64	8.70E-09	6,10	1.16E-06	4,95
64	3.41E-10	5.74	1.33E-10	6,03	3.67E-08	4,98

Table 2 1D advection of sine wave, P^2 RDG scheme with P^8 reconstruction

N	$\|e_h\|_{L^\infty(\Omega)}$	α	$\|e_h\|_{L^2(\Omega)}$	α	$|e_h|_{H^1(\Omega,\mathcal{T}_h)}$	α
4	2.90E-03	–	1.85E-03	–	1.63E-02	–
8	7.75E-06	8.55	3.56E-06	9.02	1.03E-04	7.30
16	2.10E-08	8.53	6.64E-09	9.07	4.34E-07	7.89
32	7.21E-11	8.18	4.02E-11	7.37	1.76E-09	7.94

6 Conclusions

We have presented a possible generalization of higher-order reconstruction operators as used in the FV method to the DG method. Two constructions of the reconstruction operator R are presented, the first analogous to the standard FV case (already treated in [2]) and the construction analogous to the SV method. The resulting scheme has many advantages over standard DG, FV and SV schemes:

- To increase the order of the scheme, the reconstruction stencil need not be enlarged, we may simply increase the order of the underlying DG scheme.
- Test functions are from the lower order space, hence more efficient quadratures may be used than in the corresponding higher order DG scheme.
- Since the RDG scheme is basically a lower order DG scheme with higher order reconstruction, the CFL condition is less restrictive than for the corresponding higher order DG scheme.

Acknowledgements This work is a part of the research project MSM 0021620839 of the Ministry of Education of the Czech Republic. The research was also supported by the project P201/11/P414 of the Czech Science Foundation.

References

1. P.G. Ciarlet: *The Finite Elements Method for Elliptic Problems*, North-Holland, Amsterdam, New York, Oxford, 1979.
2. M. Dumbser, D. Balsara, E.F. Toro, C.D. Munz: *A unified framework for the construction of one-step finite-volume and discontinuous Galerkin schemes*, J. Comput. Phys. 227 (2008), pp. 8209–8253.
3. M. Feistauer, J. Felcman, I. Straškraba: *Mathematical and Computational Methods for Compressible Flow*, Oxford University Press, Oxford, 2003.
4. M. Feistauer, V. Kučera: *Analysis of the DGFEM for nonlinear convection-diffusion problems*, Electronic Transactions on Numerical Analysis, Vol. 32, No.1, (2008), pp. 33–48.
5. D. Kröner: *Numerical Schemes for Conservation Laws*, Wiley und Teubner, 1996.
6. R.J. LeVeque: *Finite Volume Methods for Hyperbolic Problems*, Cambridge University Press, Cambridge, 2002.
7. Z. J. Wang: *Spectral (Finite) Volume Method for Conservation Laws on Unstructured Grids. Basic Formulation*, J. Comput. Phys. 178 (2002), pp. 210 – 251.

The paper is in final form and no similar paper has been or is being submitted elsewhere.

Flux-Based Approach for Conservative Remap of Multi-Material Quantities in 2D Arbitrary Lagrangian-Eulerian Simulations

Milan Kucharik and Mikhail Shashkov

Abstract Remapping is one of the essential parts of most Arbitrary Lagrangian-Eulerian (ALE) methods. It conservatively interpolates all fluid quantities from the original (Lagrangian) computational mesh to the new (rezoned) one. This paper focuses on the situation when more materials are present in the computational domain – the multi-material remap. We present a new remapping method based on the computation of the material exchange integrals (using intersections), and construction of the inter-cell fluxes of all quantities from them. As we are interested in the staggered ALE, we also briefly discuss the remap of nodal mass and velocity. Properties of the method are demonstrated on a selected numerical example.

Keywords Multi-material remap, conservative interpolation, staggered arbitrary Lagrangian-Eulerian methods
MSC2010: 35L65, 41A45, 65D05, 76T99

1 Introduction

Traditionally, there have been two families of numerical method for computational fluid dynamics, utilizing the Lagrangian or the Eulerian framework, each with its own advantages and disadvantages. In the pioneering paper [1], Hirt et al. developed the formalism combining both frameworks, and showed that this general framework could be used to combine the best properties of the Lagrangian and Eulerian methods. This class of methods has been termed Arbitrary Lagrangian-Eulerian or

---

Milan Kucharik
Faculty of Nuclear Sciences and Physical Engineering, Czech Technical University in Prague, Brehova 7, Praha 1, 115 19, Czech Republic, e-mail: kucharik@newton.fjfi.cvut.cz

Mikhail Shashkov
XCP-4 Group, MS-B284, Los Alamos National Laboratory, Los Alamos, NM, 87545, USA, e-mail: shashkov@lanl.gov

---

J. Fořt et al. (eds.), *Finite Volumes for Complex Applications VI – Problems & Perspectives*, Springer Proceedings in Mathematics 4,
DOI 10.1007/978-3-642-20671-9_66, © Springer-Verlag Berlin Heidelberg 2011

ALE. This methodology has become very popular in recent years and many authors contributed to this topic, see for example [2–6].

For multi-material flows, the initial mesh is usually aligned with the material interfaces – each cell of the mesh contains only one material. For simple flows, it is possible to rezone the mesh in each material separately and keep the interfaces aligned with the mesh that is, do not move nodes on the interface at all or move them along the interface. Unfortunately for realistic simulations, the material interfaces get often distorted and their rezoning leads to the appearance of mixed cells containing two or more materials. We focus on the explicit material representation in form of pure material sub-polygons in each mixed cell, constructed by the modern moment-of-fluid (MOF) [7] method, which appears to be most optimal for this kind of application [8].

The ALE algorithm is usually separated into three distinct stages: 1) a Lagrangian stage in which the solution and the computational mesh are updated; 2) a rezoning stage in which the nodes of the computational mesh are moved to more optimal positions; and 3) a remapping stage in which the Lagrangian solution is interpolated onto the rezoned mesh. Here, we focus on the last stage of the ALE algorithm – the multi-material remapping. In the multi-material case, the fast and simple swept region method [9] cannot be used and one must switch to an intersection-based method.

In this paper, we present a new remapping method for multi-material quantities in the staggered discretization. The remapping algorithm is based on the computation of the exchange integrals between the Lagrangian and rezoned meshes, which represent fluxes of the basic geometry integrals through the computational cell boundaries, and are computed using intersections (overlays) of the original and rezoned meshes. These exchange integrals can be pre-computed at the beginning of the remapping step, and fluxes of all quantities are composed from these integrals. Due to the flux form of the remapper, this method is best suitable for continuous remap, where the original and rezoned meshes are similar.

This paper is organized as follows. In Section 2, we describe the construction of the exchange integrals. In the following two Sections 3 and 4, we describe the construction of material/average quantity fluxes and remapping all cell-centered and nodal fluid quantities. In Section 5, we demonstrate the properties of the method on a selected numerical example. The whole paper is concluded in Section 6.

2 Construction of Exchange Integrals

Our approach is based on expressing the standard overlay formula in the equivalent flux form [10],

$$\tilde{c} = c \cup \left(\bigcup_{c' \in C'(c)} c' \cap \tilde{c} \right) \setminus \left(\bigcup_{\tilde{c}' \in C'(\tilde{c})} c \cap \tilde{c}' \right), \qquad (1)$$

where ˜ denotes a particular cell in the new mesh, and $C'(c)$ is the set of all cells neighboring with c (including the corner neighbors). For the construction of the intersection polygons, we intersect the original cell with the halfplanes defined by the edges of the rezoned cell [11]. This robust approach works well for intersection of the generally non-convex polygons (Lagrangian cells) with the convex polygons (rezoned cells). The situation is demonstrated in Fig. 1. The first term in parentheses represents the outward part of the flux, while the second term is the inward part of the flux. Both inward and outward parts can be seen as two light triangles in image (c) of the Figure. The same expression can be written for each pure material polygon of cell c. An example of original material polygons is shown in images (a) and (b) of the Figure, the fluxes of different materials are shown in different shades in images (d-f) of the Figure. As we can see, the flux between c and a particular neighbor can have non-zero values of both inward and outward components of the flux, and each component can include fluxes of several materials (including the corner fluxes).

Now, suppose that we want to remap volume of a particular materials k of cell c to the new cell $\tilde{c}$. The new material volume can be written as

$$V_{\tilde{c},k} = \int_{\tilde{c}_k} 1\, dV, \qquad (2)$$

and after employing formula (1), we can rewrite it as

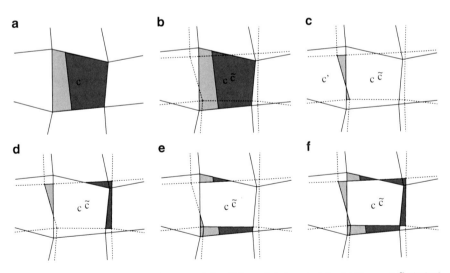

Fig. 1 (a) Original cell c (solid line) containing light and dark materials. (b) New cell $\tilde{c}$ (dashed line). (c) Fluxes between cell c and its left neighbor c'. (d) All outward fluxes around c. (e) All inward fluxes around c. (f) All fluxes around c

$$V_{\tilde{c},k} = V_{c,k} + \sum_{c' \in C'(c)} F^V_{c,c',k}, \qquad (3)$$

where the material volume fluxes are defined as

$$F^V_{c,c',k} = I^1_{c'_k \cap \tilde{c}} - I^1_{c_k \cap \tilde{c}'}, \qquad I^f_P = \int_P f\, dV. \qquad (4)$$

Here c_k is the polygon of pure material k in cell c, and the total material volume flux has its outward and inward components. The exchange integral of function f over polygon P is denoted by the I^f_P symbol. The exchange integrals can be pre-computed at the beginning of the remapping step from the mesh geometry, and can be used for the construction of fluxes of all fluid quantities. Later, we will need the exchange integrals for polynomials up to the second order, i.e. $f = 1, x, y, x^2, x\,y, y^2$. Let us note that these are all integrals of polynomials over polygons ($c'_k \cap \tilde{c}$ or $c_k \cap \tilde{c}'$), which can be evaluated analytically.

3 Remap of Cell-Centered Quantities

In this Section, we demonstrate the remap of the cell-centered and material-centered quantities. All material quantities will be remapped in the same form as we have shown in equation (3) for material volumes. Material centroids $x_{c,k}$ (needed as reference centroids for MOF) are remapped as

$$x_{\tilde{c},k} V_{\tilde{c},k} = x_{c,k} V_{c,k} + \sum_{c' \in C'(c)} F^x_{c,c',k}, \qquad F^x_{c,c',k} = I^x_{c'_k \cap \tilde{c}} - I^x_{c_k \cap \tilde{c}'} \qquad (5)$$

and similarly for $y_{c,k}$.

Material density is reconstructed in a piece-wise linear way

$$\rho_{c,k}(x, y) = \rho_{c,k} + S^x_{c,k}(x - x_{c,k}) + S^y_{c,k}(y - y_{c,k}), \qquad (6)$$

where the material density mean values $\rho_{c,k}$ are known, and the slopes $S^{x,y}_{c,k}$ are determined by minimization of the discrepancy between the reconstructed values in the centroids of the same material polygons in the neighboring cells from the mean values there [12]. Limiting by the Barth-Jespersen limiter [13] guarantees preservation of the local density extrema, while in the mixed cells the 0 and $+\infty$ limits are used for limiting to avoid excessive slope degradation in case of thin material filaments with only few neighbors containing the same material. This approach implies second order of accuracy of the remapper. The material mass remap is then performed in a similar form,

$$m_{\tilde{c},k} = m_{c,k} + \sum_{c' \in C'(c)} F^m_{c,c',k}, \qquad F^m_{c,c',k} = F^m_{c'_k \cap \tilde{c}} - F^m_{c_k \cap \tilde{c}'}, \qquad (7)$$

where the inward and outward mass fluxes are obtained by the integration of the reconstructed density over the intersection, which can be composed from the pre-computed exchange integrals, for example

$$F^m_{c_k \cap \tilde{c}'} = \int_{c_k \cap \tilde{c}'} \rho_{c,k}(x,y) \, dV = \left(\rho_{c,k} - S^x_{c,k} x_{c,k} - S^y_{c,k} y_{c,k} \right) I^1_{c_k \cap \tilde{c}'} + S^x_{c,k} I^x_{c_k \cap \tilde{c}'} + S^y_{c,k} I^y_{c_k \cap \tilde{c}'}. \qquad (8)$$

For the material internal energy, same approach

$$\varepsilon_{\tilde{c},k} m_{\tilde{c},k} = \varepsilon_{c,k} m_{c,k} + \sum_{c' \in C'(c)} F^\varepsilon_{c,c',k}, \qquad F^\varepsilon_{c,c',k} = F^\varepsilon_{c'_k \cap \tilde{c}} - F^\varepsilon_{c_k \cap \tilde{c}'} \qquad (9)$$

and (8) with the material specific internal energy ε instead of the density ρ can be used. However, this approach does not guarantee satisfaction of the local-bound conservation condition, so a more advanced approach described in [14] must be used, which constructs the energy fluxes by integration of the reconstructed density multiplied by the reconstructed specific internal energy, for example

$$F^\varepsilon_{c_k \cap \tilde{c}'} = \int_{c_k \cap \tilde{c}'} \rho_{c,k}(x,y) \, \varepsilon_{c,k}(x,y) \, dV. \qquad (10)$$

The reconstruction of the specific internal energy cannot be done the same way as we did for density in (6), it must be centered in material centers of mass $x^m_{c,k} = (\int_{c_k} \rho_{c,k}(x,y) x \, dV)/m_{c,k}$ instead of material centroids,

$$\varepsilon_{c,k}(x,y) = \varepsilon_{c,k} + S^{x,\varepsilon}_{c,k} \left(x - x^m_{c,k} \right) + S^{y,\varepsilon}_{c,k} \left(y - y^m_{c,k} \right). \qquad (11)$$

Both centers of mass and energy fluxes (10) can be composed from the pre-computed exchange integrals as we did for mass (8), however, integrals of the second order polynomials are needed now.

The last cell-centered quantity we need to remap is the average cell pressure needed for the next Lagrangian step (the material pressures are updated from the remapped material energies using the equation of state). We suggest to remap the average pressure in the following form

$$p_{\tilde{c}} V_{\tilde{c}} = p_c V_c + \sum_{c' \in C'(c)} F^p_{c,c'}, \qquad F^p_{c,c'} = F^p_{c' \cap \tilde{c}} - F^p_{c \cap \tilde{c}'}, \qquad (12)$$

where the pressure fluxes are obtained as the exchange volumes multiplied by the pressure reconstructed by same formula as (6) in the centroid of the intersection polygon, for example

$$F^p_{c\cap\tilde{c}'} = I^q_{c\cap\tilde{c}'}\, p_c(x_{c\cap\tilde{c}'}, y_{c\cap\tilde{c}'}), \qquad \{x,y\}_{c\cap\tilde{c}'} = \frac{I^{\{x,y\}}_{c\cap\tilde{c}'}}{I^1_{c\cap\tilde{c}'}}. \qquad (13)$$

All terms here can be composed from the pre-computed exchange integrals again.

4 Remap of Nodal Quantities

Nodal mass is tied with the total cell mass through the sub-zonal masses [15]. Our approach is to remap the nodal mass in a similar flux form,

$$m_{\tilde{n}} = m_n + \sum_{n' \in N'(n)} F^m_{n,n'} \qquad (14)$$

where $N'(n)$ is the set of nodes neighboring with n and $F^m_{n,n'}$ are the inter-nodal mass fluxes, which can be defined either by interpolation from the inter-cell fluxes [16], or by minimizing of their difference from given reference fluxes [17]. All remaining nodal quantities are remapped in the same form as (14) by attaching the particular nodal quantity to the inter-nodal mass flux, for example

$$u_{\tilde{n}} m_{\tilde{n}} = u_n m_n + \sum_{n' \in N'(n)} u_{n,n'} F^m_{n,n'} \qquad (15)$$

for nodal velocity, where the value of $u_{n,n'}$ is the reconstructed velocity inside the inter-nodal swept region, for example a simple bilinear interpolation or the kinetic-energy conservative approach [18]. Similarly, the kinetic energy can be remapped in order to perform the standard energy fix [2] ensuring total energy conservation.

5 Numerical Example

To demonstrate the properties of the described remapping method, we present simulation of the triple point problem described in [19]. The initial data are shown in image (a) of Fig. 2. This problem contains three materials, the interfaces are initially aligned with the mesh edges. The simulation was performed in the context of our 2D staggered research multi-material ALE (RMALE) code on the orthogonal 140×60 mesh. To stress the influence of the remapper, we run the simulation in the Eulerian manner – remapping to the initial mesh is done after every single Lagrangian step. The light material generates a shock wave propagating in different speeds into the gray and dark materials, causing development of a vertex.

The material distribution in the final time $t = 5$ can be seen in image (b) of Fig. 2. We can see the thin filament of the dark material which stays compact and does not break apart even though the width of its tail is smaller than 1 cell size. The profiles of material density, specific internal energy, and pressure can be seen in images (c-e) of Fig. 2. As we can see, all fields are smooth without any numerical problems. Finally, in image (f) of Fig. 2, the material velocity field is shown displaying the vortex around the triple point. Again, the velocity field is smooth and does not contain any numerical artifacts.

6 Conclusion

We have briefly described a new method for remapping of all fluid quantities between similar meshes in the context of staggered multi-material 2D ALE. This method is flux based, and fluxes of all fluid quantities are constructed from the pre-computed exchange integrals. As these integrals are computed just once, at the beginning of the remapping step, computational cost of this method is not excessive

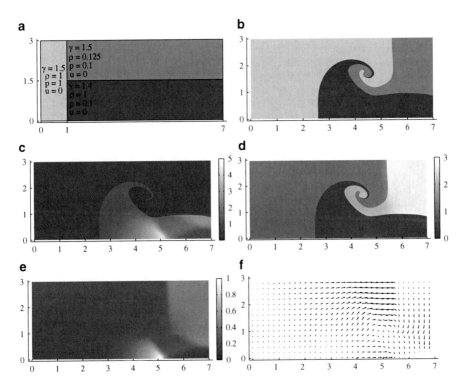

Fig. 2 Triple point problem: (a) materials in $t = 0$; (b) materials in $t = 5$; (c) material density in $t = 5$; (d) material energy in $t = 5$; (e) material pressure in $t = 5$; (f) velocity field in $t = 5$

although it involves intersections. Due to the flux nature of the method, this method is conservative for all quantities (total energy conservation is assured by the energy fix). If high order reconstructions are used for fluxes of all quantities, this method is second order accurate. We have demonstrated that this method can be used as a remapper in the framework of a full hydrodynamic code.

Acknowledgements This work was performed under the auspices of the National Nuclear Security Administration of the US Department of Energy at Los Alamos National Laboratory under Contract No. DE-AC52-06NA25396 and supported by the DOE Advanced Simulation and Computing (ASC) program. The authors acknowledge the partial support of the DOE Office of Science ASCR Program. Milan Kucharik was supported by the Czech Ministry of Education grants MSM 6840770022, MSM 6840770010, and LC528, and the Czech Science Foundation project P201/10/P086. The authors thank H. Ahn, D. Bailey, A. Barlow, K. Lipnikov, R. Loubere, P.-H. Maire, and M. Owen for fruitful and stimulating discussions over many years.

References

1. C. W. Hirt, A. A. Amsden, and J. L. Cook. An arbitrary Lagrangian-Eulerian computing method for all flow speeds. *Journal of Computational Physics*, 14(3):227–253, 1974.
2. D. J. Benson. Computational methods in Lagrangian and Eulerian hydrocodes. *Computer Methods in Applied Mechanics and Engineering*, 99(2-3):235–394, 1992.
3. L. G. Margolin. Introduction to "An arbitrary Lagrangian-Eulerian computing method for all flow speeds". *Journal of Computational Physics*, 135(2):198–202, 1997.
4. J. S. Peery and D. E. Carroll. Multi-material ALE methods in unstructured grids. *Computer Methods in Applied Mechanics and Engineering*, 187(3-4):591–619, 2000.
5. R. W. Anderson, N. S. Elliott, and R. B. Pember. An arbitrary Lagrangian-Eulerian method with adaptive mesh refinement for the solution of the Euler equations. *Journal of Computational Physics*, 199(2):598–617, 2004.
6. R. Loubere, P.-H. Maire, M. Shashkov, J. Breil, and S. Galera. ReALE: A reconnection-based arbitrary-LagrangianEulerian method. *Journal of Computational Physics*, 229(12):4724–4761, 2010.
7. V. Dyadechko and M. Shashkov. Reconstruction of multi-material interfaces from moment data. *Journal of Computational Physics*, 227(11):5361–5384, 2008.
8. M. Kucharik, R.V. Garimella, S.P. Schofield, and M.J. Shashkov. A comparative study of interface reconstruction methods for multi-material ALE simulations. *Journal of Computational Physics*, 229(7):2432–2452, 2010.
9. M. Kucharik, M. Shashkov, and B. Wendroff. An efficient linearity-and-bound-preserving remapping method. *Journal of Computational Physics*, 188(2):462–471, 2003.
10. L. G. Margolin and M. Shashkov. Second-order sign-preserving conservative interpolation (remapping) on general grids. *Journal of Computational Physics*, 184(1):266–298, 2003.
11. M. Kucharik and M. Shashkov. Conservative multi-material remap for staggered discretization. 2011. In prep.
12. D. J. Mavriplis. Revisiting the least-squares procedure for gradient reconstruction on unstructured meshes. In *AIAA 2003-3986*, 2003. 16th AIAA Computational Fluid Dynamics Conference, June 23-26, Orlando, Florida.
13. T. J. Barth. Numerical methods for gasdynamic systems on unstructured meshes. In C. Rohde D. Kroner, M. Ohlberger, editor, *An introduction to Recent Developments in Theory and Numerics for Conservation Laws, Proceedings of the International School on Theory and Numerics for Conservation Laws*, Berlin, 1997. Lecture Notes in Computational Science and Engineering, Springer. ISBN 3-540-65081-4.

14. J. K. Dukowicz and J. R. Baumgardner. Incremental remapping as a transport/advection algorithm. *Journal of Computational Physics*, 160(1):318–335, 2000.
15. R. Loubere and M. Shashkov. A subcell remapping method on staggered polygonal grids for arbitrary-Lagrangian-Eulerian methods. *Journal of Computational Physics*, 209(1):105–138, 2005.
16. R. B. Pember and R. W. Anderson. A comparison of staggered-mesh Lagrange plus remap and cell-centered direct Eulerian Godunov schemes for Eulerian shock hydrodynamics. Technical report, LLNL, 2000. UCRL-JC-139820.
17. J. M. Owen and M. J. Shashkov. Arbitrary Lagrangian Eulerian remap treatments consistent with corner based compatible total energy conserving Lagrangian methods. In prep., 2010.
18. D. Bailey, M. Berndt, M. Kucharik, and M. Shashkov. Reduced-dissipation remapping of velocity in staggered arbitrary Lagrangian-Eulerian methods. *Journal of Computational and Applied Mathematics*, 233(12):3148–3156, 2010.
19. S. Galera, J. Breil, and P.-H. Maire. A 2D unstructured multi-material cell-centered arbitrary lagrangianeulerian (CCALE) scheme using MOF interface reconstruction. *Computers & Fluids*, 2010. In press. doi:10.1016/j.compfluid.2010.09.038.

The paper is in final form and no similar paper has been or is being submitted elsewhere.

Optimized Riemann Solver to Compute the Drift-Flux Model

Anela Kumbaro and Michaël Ndjinga

Abstract This paper discusses the development of an approximated optimized Riemann solver applied to the two-phase flow drift-flux model. The solver makes use of a partial eigenstructure information while maintaining the Roe solver accuracy. Moreover, it allows to take into account the contribution of the dynamic and thermal non-equilibrium in the upwinding matrix. A further optimization of the solver is realized by scaling the global matrix which results in better preconditioning. Both the partial eigenstructure decomposition and the scaling of the matrix are inspired from the eigenstructure of the two-phase flow model. A number of physical benchmarks are presented to illustrate this method. Comparison between the computational results obtained with the optimized solver and the conventional Roe-type solver demonstrates the efficiency of the new methodology.

Keywords Riemann solver, eigenstructure decomposition, drift-flux model
MSC2010: 76T10, 76M12, 35L65, 76E19, 65F35

1 Introduction

The drift-flux is commonly used to simulate water-vapor flows in nuclear power plants. Various industrial codes within the nuclear community, for example FLICA4 code of CEA, or THYC of EDF, both dedicated to design and safety studies of nuclear reactors, rely on this model. When compared with codes that use more advanced two-phase models, such as the two-fluid or the multifield model, their strong point is the code-efficiency. Reducing furthermore the CPU time cost is crucial for the survival of these type of codes. Our work is done within the FLICA-OVAP code [3], which is a new platform dedicated to core thermal-hydraulic studies,

---

Anela Kumbaro and Michaël Ndjinga
CEA-Saclay, DEN/DM2S/SFME/LETR, F-91191 Gif-sur-Yvette, France, e-mail: anela.kumbaro@cea.fr, michael.ndjinga@cea.fr

funded by the Thermal-hydraulics Simulation project of CEA. To provide a relevant response to different core concepts and multiple industrial applications, several models coexist in FLICA-OVAP platform: the Homogeneous Equilibrium model, the drift flux model which is directly derived from the previous CEA core code FLICA-4 [1]-[2], the two-fluid model, and finally, a general multifield model [4], with a variable number of fields for both vapor and liquid phases. We present in this paper two techniques to reduce the execution time and improve the code's performance while using the drift-flux model. Our starting point solver is the weak formulation of the Roe's approximate Riemann solver, adapted to low Mach number [6]. Based on the eigenstructure of the drift-flux model we propose to rewrite the solver in a more optimized form.

On the other hand, to go forward in time, a fully implicit integrating step is used that provides fast running steady state calculations. We introduce a scaling of the implication matrix so that the matrix coefficients have the same order of magnitude. This allows for a much better preconditioning of the matrix and significantly reduces the global CPU time.

This paper is organized as follows: to begin with, Sect. 2 briefly describes the standard two-phase flow drift-flux model we deal with. Next, we introduce an evaluation of its eigenstructure. In Sect. 4 we present the numerical solver based on the specific eigenstructure of the drift-flux model and discuss its accuracy. Section 5 introduces the scaling of the matrix. We show that the coefficients of the upwinding matrix which have different orders of magnitude, have the same magnitude after the scaling. Some numerical results are presented in Sect. 6 to illustrate the behavior of the numerical solver. Finally, some conclusions are presented in the last section.

2 Drift-flux two-phase flow model

We introduce here the FLICA-OVAP drift-flux model. For the sake of simplicity this model is represented without taking into account the porosity variable and the viscous term. The balance equations for the drift-flux model read:

$$\frac{\partial}{\partial t}\rho + \nabla \cdot (\rho \mathbf{u}) = 0, \tag{1}$$

$$\frac{\partial}{\partial t}(\rho \mathbf{u}) + \nabla \cdot (\rho \mathbf{u} \otimes \mathbf{u} + \rho c(1-c)\mathbf{u}_r \otimes \mathbf{u}_r) + \nabla p = \mathbf{F}_{ext} + \mathbf{F}^w \tag{2}$$

$$\frac{\partial}{\partial t}(\rho E) + \nabla \cdot (\rho H \mathbf{u} + \rho c(1-c)\mathbf{u}_r(L + \frac{\mathbf{u}_v^2 - \mathbf{u}_l^2}{2})) = Q_{tot} + \mathbf{F}_{ext} \cdot \mathbf{u} \tag{3}$$

$$\frac{\partial}{\partial t}(\rho c) + \nabla \cdot (\rho c \mathbf{u} + \rho c(1-c)\mathbf{u}_r) = \Gamma, \tag{4}$$

where c is the vapor concentration, $\mathbf{u}$, ρ, p, E, H are the mixture velocity, mixture density, pressure, total energy and enthalpy, respectively, $\mathbf{u}_r$ is the relative velocity between vapor and liquid phases given by a drift-flux model, Γ is the mass transfer term, $\mathbf{F}_{ext}$ is the external forces term, F^w is the wall friction term, and L is the latent heat.

The model is closed by a general equation of state $\rho = \rho(p, h, c)$, and by the assumption that the vapor is saturated in presence of liquid: $h_v = h_v^{sat}(P)$, where h_g is the vapor enthalpy. Closure laws (wall transfer, mass exchange, diffusion, ...) for this model come from FLICA-4 code and have been described in [1].

3 Eigenstructure of the drift-flux model

If we introduce an orthonormal basis $(\mathbf{n}, \tau_1, \tau_2)$ of the three dimensional space R^3, the one-dimensional formulation of the above drift-flux system (1-4) is:

$$\frac{\partial \mathbf{V}}{\partial t} + \frac{\partial \mathbf{F}_n}{\partial n} = \mathbf{S} \tag{5}$$

with the conservative vector $\mathbf{V} = (\rho, \rho\mathbf{u}, \rho E, \rho c) \in \mathbb{R}^m$, where $m = d + 3$ and d is the space dimension, and S the source term vector. The expression of the flux is separated into two part, the zero order relative velocity part, $\mathbf{F}_{n0}$, and the relative velocity dependent part, $\mathbf{F}_{nr}$:

$$\mathbf{F}_{n0} = \begin{pmatrix} \rho \mathbf{u} \cdot \mathbf{n} \\ \rho (\mathbf{u} \cdot \mathbf{n})\mathbf{u} + p\mathbf{n} \\ \rho H \mathbf{u} \cdot \mathbf{n} \\ \rho c \mathbf{u} \cdot \mathbf{n} \end{pmatrix}, \quad \mathbf{F}_{nr} = \begin{pmatrix} 0 \\ \rho c(1-c)\mathbf{u}_r \cdot \mathbf{n} \mathbf{u}_r \\ \rho c(1-c)(L + \frac{u_v^2 - u_l^2}{2})(\mathbf{u}_r \cdot \mathbf{n}) \\ \rho c(1-c)\mathbf{u}_r \cdot \mathbf{n} \end{pmatrix} \tag{6}$$

Let first consider only the part without relative velocity contribution. The eigenvalues are $\lambda^- = \mathbf{u} \cdot \mathbf{n} - a$, $\lambda_u = \mathbf{u} \cdot \mathbf{n}$ (multiplicity $d + 1$), and $\lambda^+ = \mathbf{u} \cdot \mathbf{n} + a$, where a is the mixture sound velocity. The right and left eigenvectors associated to the sound waves are

$$\mathbf{r}^\pm = \begin{bmatrix} 1 \\ \mathbf{u} - (\mathbf{u} \cdot \mathbf{n} - \lambda^\pm)\mathbf{n} \\ H - (\mathbf{u} \cdot \mathbf{n} - \lambda^\pm)\mathbf{u} \cdot \mathbf{n} \\ c \end{bmatrix} \quad \mathbf{l}^\pm = \frac{1}{2a^2}\begin{bmatrix} \chi \mp a\mathbf{u} \cdot \mathbf{n} \\ -\kappa \mathbf{u} \pm a\mathbf{n} \\ \kappa \\ \xi \end{bmatrix} \tag{7}$$

where $\chi = \frac{\partial P}{\partial \rho}$, $\kappa = \frac{\partial P}{\partial \rho E}$, and $\xi = \frac{\partial P}{\partial \rho c}$. If we consider the relative velocity dependent part, the problem becomes very complex. Indeed, the relative velocity is drift flux model dependent, and the drift flux model depends on the flow configuration. We will not represent here any analytical expression about the eigenvalues but we will only assume that the drift-flux model, like the general multifield model [4], has two fast eigenvalues of $\mathbf{u} \pm a$ order of magnitude, while the other eigenvalues are between $\mathbf{u}_v \cdot \mathbf{n}$ and $\mathbf{u}_l \cdot \mathbf{n}$. Hence, these so called intermediate eigenvalues have the same order of magnitude as the mixture velocity.

4 Simplified eigenstructure decomposition solver (SEDES)

Let τ be a meshing of Ω defined as the union of control volumes K. The discrete unknowns are denoted by $\mathbf{V}_K^n$ and represent the approximation of a mean value of $\mathbf{V}$ on the control volume K at time t^n.

The Roe-type approximate Riemann solver is the current solver in the CEA industrial code FLICA-4 [2] and it will be used as the reference solver for this study. This solver requires the solution of a one-dimensional Riemann problem at cell interfaces on a non-staggered grid, to define backward and forward differences used to approximate the spatial derivatives. The numerical flux is the following:

$$\Phi_{K,L}^{n+1} = \frac{\mathbf{F}_n(\mathbf{V}_K^{n+1}) + \mathbf{F}_n(\mathbf{V}_L^{n+1})}{2} - \frac{|\mathbf{A}_n(\mathbf{V}_K^n, \mathbf{V}_L^n)|}{2}(\mathbf{V}_L^{n+1} - \mathbf{V}_K^{n+1}) \qquad (8)$$

So, at the heart of such scheme is the so-called Roe matrix, first introduced in [5] for the single-phase flow equations, taken at the Roe average state on the interface between K and L.

To construct this matrix the FLICA-4 standard Roe-type solver makes use of a complete eigenstructure decomposition:

$$|\mathbf{A}_n(\mathbf{V}_K^n, \mathbf{V}_L^n)| = \sum_{k=1}^{m} |\lambda_k| l_k \otimes r_k \qquad (9)$$

with λ_k, l_k and r_k the eigenvalues, the left and right eigenvectors of the system matrix.

First case: If the relative velocity contribution is not considered into the system matrix, the eigenvalues are $\lambda^- = \mathbf{u} \cdot \mathbf{n} - a$, $\lambda_u = \mathbf{u} \cdot \mathbf{n}$ (multiplicity d), and $\lambda^+ = \mathbf{u} \cdot \mathbf{n} + a$, and the eigenvectors are easily obtained. Therefore, we propose to rewrite Equation (9)

$$|\mathbf{A}_n| = (|\lambda^-| - |u_n|)l^- \otimes r^- + (|\lambda^+| - |u_n|)l^+ \otimes r^+ + |u_n|\mathbb{I}. \qquad (10)$$

with $l^{\pm}$ and $r^{\pm}$ given by (7). Eq. (10), while corresponding exactly to the Roe upwinding matrix, provides a more efficient way to calculate this matrix.

Second case: The relative velocity contribution is considered into the system matrix. In this case Eq. (9) requires the computation of all the eigenstructure of the system matrix. This computation has to be done using a numerical algorithm and this means a consistent increase in CPU time. For this reason the FLICA-4 standard Roe-type solver does not take into account the relative velocity in the upwind part of the numerical fluxes.

On the other hand, based on the structure of the complete matrix eigenvalues, we remark that to compute the absolute value of the upwinding matrix it is essential to take into account the contributions of the fastest eigenvalues, while the remaining eigenvalues which have more or less the same order of magnitude, can be represented by an unique candidate, for instance, the fastest one of this group, that we will denote simply by $\tilde{\lambda}$.

We can rewrite Eq. (10) using $\tilde{\lambda}$ instead of $|u_n|$:

$$|\mathbf{A}_n^{SEDES}| = (|\lambda^-| - \tilde{\lambda})l^- \otimes r^- + (|\lambda^+| - \tilde{\lambda})l^+ \otimes r^+ + \tilde{\lambda}\mathbb{I}. \qquad (11)$$

Eq. (11) corresponds to a simplified eigenstructure decomposition, hence the name SEDES of the solver, as it uses only the fastest waves contributions. To construct the SEDES flux we need the eigenvalues and eigenvectors associated to the sound waves, which are determined using a shifted power method, and $\tilde{\lambda}$ calculated as $\tilde{\lambda} = \mathbf{u} \cdot \mathbf{n} + |\mathbf{u}_r \cdot \mathbf{n}|$.

The spectrum of the approximated upwind matrix $|\mathbf{A}_n^{SEDES}|$ is very close to the spectrum of the standard complete Roe decomposition (9), since the spectrum of the following matrix is close to zero

$$|\mathbf{A}_n^{SEDES}| - |\mathbf{A}_n^{Roe}| = \sum_{k=2}^{m-1} l_k \otimes r_k (\tilde{\lambda} - |\lambda_k|). \qquad (12)$$

The general structure of the drift-flux system eigenvalues ensures that $\tilde{\lambda} - |\lambda_k| \leq |\mathbf{u}_r \cdot \mathbf{n}|$, so the truncation error in upwinding matrix remains small, especially when compared with the two first terms on the right hand side of Eq. (11), as the velocities are small compared to the sound speeds.

5 Matrix scaling for better preconditioning

We are interested in using a fully implicit method for transient calculations. To this end we rewrite the numerical flux (8) that gives its contribution on the right hand side of the disretized system, in either of the two equivalent forms:

$$\Phi_{K,L}^{n+1} = \mathbf{F}_n(\mathbf{V}_K^{n+1}) + \mathbf{A}_n^-(\mathbf{V}_K^n, \mathbf{V}_L^n)(\mathbf{V}_L^{n+1} - \mathbf{V}_K^{n+1}) \qquad (13)$$

or

$$\Phi_{K,L}^{n+1} = \mathbf{F}_n(\mathbf{V}_L^{n+1}) - \mathbf{A}_n^+(\mathbf{V}_K^n, \mathbf{V}_L^n)(\mathbf{V}_L^{n+1} - \mathbf{V}_K^{n+1}) \qquad (14)$$

where $\mathbf{A}_n^\pm(\mathbf{V}_K^n, \mathbf{V}_L^n)$ are the negative/positive part of the upwind matrix. The derivatives of these fluxes give contribution on the implicitation matrix that depends only on the matrices $\mathbf{A}_n^\pm(\mathbf{V}_K^n, \mathbf{V}_L^n)$. We remark that both vapor and liquid velocity projections on the normal at the cells interface have the same sign in most kind of two-phase flow configurations, as the relative velocity is smaller compared with the mixture velocity. Hence, we expect the eigenvalues of the two-phase flow system to have rather the same sign, except one. We choose in the code to compute first whichever matrix $\mathbf{A}_n^+(\mathbf{V}_K^n, \mathbf{V}_L^n)$ or $\mathbf{A}_n^-(\mathbf{V}_K^n, \mathbf{V}_L^n)$ corresponding to a minimal number of eigenvalues of the same sign. Then, we obtain the other one using the relation $\mathbf{A}_n^+(\mathbf{V}_K^n, \mathbf{V}_L^n) - \mathbf{A}_n^-(\mathbf{V}_K^n, \mathbf{V}_L^n) = |\mathbf{A}_n(\mathbf{V}_K^n, \mathbf{V}_L^n)|$, with the absolute value matrix obtained using the partial eigenstructure decomposition method explained in Sect. 5.

The implicit numerical method finally leads to the solving of the system

$$MX = b. \qquad (15)$$

n In order to solve efficiently (15) using an iterative solver [7], one needs to find an approximation of M^{-1}. This is usually done through an approximate factorization $M \approx LU$ where U is upper triangular and L is lower triangular. The error made in the approximate factorisation using an incomplete Gauss factorisation depends on the size of off-diagonal coefficients of the matrix. Hence one may benefit from working with matrix having off-diagonal coefficient of smallest possible magnitude.

When looking at the coefficients of the system eigenvectors (7), one sees that they have very different magnitudes. Indeed in the particular case where $\mathbf{u} = 0$, the Roe matrix has only two non zero eigenvalues, $\pm a$, with the respective eigenvectors

$$\mathbf{r}^\pm = [1, \pm a\mathbf{n}, h, c] \qquad \mathbf{l}^\pm = \tfrac{1}{2a^2}[\chi, \pm a\mathbf{n}, \kappa, \xi] \qquad (16)$$

For better readability, the rest of the analysis is presented in the 1D case ($\mathbf{n} = 1$) and derivatives χ and ξ having the same order as κh will be replaced by κh. One has $A^\pm = \pm a \, (\mathbf{l}_\pm \otimes \mathbf{r}_\pm)$:

$$A^\pm = \frac{1}{2a^2} \begin{pmatrix} h\kappa & \pm ah\kappa & h^2\kappa & ch\kappa \\ \pm a & a^2 & \pm ah & \pm ac \\ \kappa & \pm a\kappa & h\kappa & c\kappa \\ h\kappa & \pm ah\kappa & h\kappa & ch\kappa \end{pmatrix}. \qquad (17)$$

We remark that $h\kappa$ has the same order of magnitude as a^2. One can see that the disequilibrium in $A^\pm$ coefficients comes from the difference in magnitude of the left and right eigenvectors of A. Multiplying $A^\pm$ to the left (respectively to the right) by a diagonal matrix with coefficients $d_{scale} = diag(1, a, \frac{a^2}{\kappa}, 1))$ (respectively $d_{scale}^{-1} = diag(1, \frac{1}{a}, \frac{\kappa}{a^2}, 1))$ one obtains a new matrix with better balanced coefficients

$$\tilde{A} = d_{scale} A^{\pm} d_{scale}^{-1} = \frac{1}{2a^2} \begin{pmatrix} h\kappa & \pm h\kappa & \frac{h^2\kappa^2}{a^2} & ch\kappa \\ \pm a^2 & a^2 & \pm h\kappa & \pm a^2 c \\ a^2 & \pm a^2 & h\kappa & a^2 c \\ h\kappa & \pm h\kappa & \frac{h^2\kappa^2}{a^2} & ch\kappa \end{pmatrix} \sim \frac{1}{2} \begin{pmatrix} 1 & \pm 1 & 1 & c \\ \pm 1 & 1 & \pm 1 & \pm c \\ 1 & \pm 1 & 1 & c \\ 1 & \pm 1 & 1 & c \end{pmatrix}$$
(18)

We propose to build two diagonal matrices D_{scale} and D_{scale}^{-1} having the size of the mesh and containing the successive coefficients of the local matrices d_{scale} and d_{scale}^{-1}. Instead of solving system (15) it is equivalent to solve

$$\tilde{M} Y = \tilde{b}$$
(19)

where $\tilde{M} = D_{scale} M D_{scale}^{-1}$, $Y = D_{scale} X$ and $\tilde{b} = D_{scale} b$. System (19) can be more easily resolved using an ILU preconditioner. Once the solution Y is obtained we compute $D_{scale}^{-1} Y$ to obtain the original unknown vector X.

6 Numerical investigation

We present here three applications to demonstrate the overall efficiency of the new solver. All the simulations are realized without taking into account the relative velocity into the upwinding matrix. In this case the SEDES solver gives the identical results with Roe solver and we can concentrate our attention only to the solver efficiency. The first two test cases correspond to 1D configurations. The first test case is a boiling flow in a 1D heated channel and the second one corresponds to a flow in a PWR reactor core. We have used a 50-cells mesh and a CFL number of 30 and 833, respectively. Table 1 represents the dimensionless CPU time for the simplified eigenstructure decomposition solver (SEDES) and the old solver (Roe solver). The last test corresponds to a steady state computation of a full charge 3D PWR reactor core configuration. The simulation is run using a 157 assemblies and 32 cells in the axial direction. The CFL number is equal to 2000. We have realized two runs with and without the scaling using the SEDES solver and compared the CPU time with the standard Roe solver time results. For this simulation we use the standard steady state algorithm of FLICA-OVAP code such as described in [2] which saves the matrix of the linear system at the first time step and uses it for the whole steady-state calculation. Nevertheless, the scaling decreases both the number

Table 1 Dimensionless CPU time: 1D test cases

CPU Time	Boiling	PWR
SEDES	0.63	0.55
Roe	1	1

Table 2 Dimensionless CPU time: 3D PWR

Solver	CPU Time
SEDES without scaling	0.89
SEDES with scaling	0.70
Roe	1

of iterations and the cost of an iteration during the resolution of the non linear system and the new solver is still more efficient as the old one as shown by the results represented in Table 2. In this last case, the result is nearly the same when relative velocity is taken into account in the upwind matrix, as the relative velocity is smaller than mixture velocity which is about 4m/s and both, are much smaller than the sound velocity.

7 Conclusions

This paper has presented how a simplified account for the system's eigenvalues can be considered in order to build a more efficient Riemann solver for the resolution of two-phase flow drift-flux model considered in industrial codes to assess the safety of nuclear plants or to support research on two phase thermal-hydraulics which conserves the accuracy of a Roe solver. Moreover, a procedure is presented to scale the coefficient of the upwinding matrix in order to obtain a better preconditioning, which is particularly efficient in complex geometry. Various test cases have shown that the methodology greatly improves the code efficiency during the simulation of two-phase flows with realistic state equations in mono-dimensional and multi-dimensional settings.

References

1. E. Royer, S. Aniel, A. Bergeron, P. Fillion, D. Gallo, F. Gaudier, O. Grgoire, M. Martin, E. Richebois, P. Salvadore, S. Zimmer, T. Chataing, P. Clment, and F. Franois, FLICA4: Status of numerical and physical models and overview of applications, in Proceedings of NURETH-11, (Avignon, France), October 2-6 (2005)
2. I. Toumi, A. Bergeron, D. Gallo, E. Royer, and D. Caruge, FLICA-4: a three-dimensional two-phase flow computer code with advanced numerical methods for nuclear application, Nucl. Eng. Design, vol. 200, pp. 139-155 (2000)
3. Fillion P, Chanoine A, Dellacherie S, Kumbaro A, FLICA-OVAP: a New Platform for Core Thermal-hydraulic Studies, NURETH-13, Japan, September 27-October 2, (2009)
4. Kumbaro A., Application of the Simplified Eigenstructure Decomposition Solver to the Simulation of General Multifield Models, 7th International Topical Meeting on Nuclear Reactor Thermal Hydraulics, Operation and Safety, Seoul, Korea, October 5-9, (2008)

5. Roe P.L., Approximate Riemann solvers, parameter vectors, and difference schemes, J. Comp. Phys. 43(2) (1981)357-372.
6. Toumi I, A weak formulation of Roe's approximate Riemann solver , *J. Comput. Phys.*, 102 (1992) 360-373.
7. Michele Benzi, Preconditioning Techniques for Large Linear Systems: A Survey *J. Comput. Phys.*, 182 (2002), 418-477.

The paper is in final form and no similar paper has been or is being submitted elsewhere.

Finite Volume Schemes for Solving Nonlinear Partial Differential Equations in Financial Mathematics

Pavol Kútik and Karol Mikula

Abstract In order to estimate a fair value of financial derivatives, various generalizations of the classical linear Black–Scholes parabolic equation have been made by adjusting the constant volatility to be a function of the option price itself. We present a second order numerical scheme, based on the finite volume method discretization, for solving the so–called Gamma equation of the Risk Adjusted Pricing Methodology (RAPM) model. Our new approach is based on combination of the fully implicit and explicit schemes where we solve the system of nonlinear equations by iterative application of the semi–implicit approach. Presented numerical experiments show its second order accuracy for the RAPM model as well as for the test with exact Barenblatt solution of the porous–medium equation which has a similar character as the Gamma equation.

Keywords finite volume method, second-order scheme, financial mathematics
MSC2010: 65A05, 35K20, 35K55, 65M08

1 Motivation from Financial Mathematics

Black–Scholes linear model In 1973 Black and Scholes in [4] and independently Merton in [9] derived a simple model for pricing financial derivatives based on the solution of a linear PDE. To obtain the governing equation, they assumed that the underlying asset S follows a geometric Brownian motion $dS = (\mu - q)S dt + \hat{\sigma} S dW$, where $\mu > 0$ is a constant drift, $\hat{\sigma} > 0$ is a constant volatility, $q > 0$ is a dividend yield rate and W is a standard Wiener process. Denoting the price of an option as $V(S,t)$ and applying Itō's lemma to obtain the stochastic differential dV,

---

Pavol Kútik and Karol Mikula
Department of Mathematics, Radlinského 11, 813 68, Bratislava, Slovak University of Technology, e-mail: kutik@math.sk, mikula@math.sk

the equation takes the following form [7]:

$$\frac{\partial V}{\partial t} + \frac{\hat{\sigma}^2}{2} S^2 \frac{\partial^2 V}{\partial S^2} + (r-q) S \frac{\partial V}{\partial S} - rV = 0, \qquad (1)$$

where r represents the riskless interest rate. In the case of an European call option the terminal pay–off condition in time $t = T$ for the strike price E looks as follows:

$$V(S, T) = \max(S - E, 0). \qquad (2)$$

For plain vanilla options, an exact solution to (1)–(2) is known (see [7]).

Nonlinear extensions If we assume the volatility parameter to be non-constant, it can be defined by a function $\sigma = \sigma(\partial_S^2 V, S, T - t)$, where $\partial_S^2 V$ is the so–called Γ of an option. In financial theory and practice various nonlinear generalizations of Black–Scholes linear model exist with such defined volatility function. For instance, Leland in [8] proposed a model which takes transaction costs into account. Avellaneda et al. in [1] described option pricing in incomplete markets. Barles and Soner in [3] adjusted the volatility depending on investor's preferences. Illiquid market effects were studied by Schönbucher and Wilmott in [11]. Another model which we deal with in this paper is the so–called **Risk Adjusted Pricing Methodology (RAPM) model** derived by Kratka in [6] and further generalized by Jandačka and Ševčovič in [5]. Notice that the numerical scheme presented in the next section can be applied to all the above mentioned models since they can be represented by a PDE in the general form (5). Interestingly, the nonlinear porous–medium equation (13) which we deal with in the last section is also a special case of the Gamma equation (5).

The RAPM model assumes that the portfolio is rehedged only at discrete times, since continuous rehedging would lead to infinite costs. The more often the portfolio is being rehedged, the higher the risk associated with transaction costs becomes. On the other hand, seldom rehedging implies higher risk arising from its weak protection against the movement of the assets's price. Hence, there exists an optimal time step, representing the hedge interval, for which the sum of both risks is minimal. Using such ideas, the governing PDE in the following form is obtained [5]:

$$\frac{\partial V}{\partial t} + \frac{1}{2} \hat{\sigma}^2 S^2 \left[1 + \mu \left(S \frac{\partial^2 V}{\partial S^2} \right)^{\frac{1}{3}} \right] \frac{\partial^2 V}{\partial S^2} + (r-q) S \frac{\partial V}{\partial S} - rV = 0, \qquad (3)$$

where $\mu = 3 \left(\frac{C^2 R}{2\pi} \right)^{\frac{1}{3}}$, $C \geq 0$ represents the relative transaction costs for buying or selling one stock and $R \geq 0$ is the marginal value of investor's exposure to risk. Since $1 + \mu (S\Gamma)^{\frac{1}{3}} \geq 1$, the option price computed by this equation is slightly above that from the linear Black–Scholes model, i.e. we obtain a so–called Ask price. On the contrary, if $1 - \mu (S\Gamma)^{\frac{1}{3}} \leq 1$, then we get the lower Bid price of an option.

Gamma equation Let us define a function $\beta(H) := \frac{1}{2}\hat{\sigma}^2\left(1+\mu H^{\frac{1}{3}}\right)H$. Since the equation (3) contains the term $S\Gamma$ we introduce the following transformation:

$$H(x,\tau) = S\Gamma = S\partial_S^2 V(S,t), \qquad (4)$$

where $x = \ln\left(\frac{S}{E}\right)$, $x \in R$ and $\tau = T - t$, $\tau \in (0,T)$. Moreover, if we take the second derivative of equation (3) with respect to x it turns out that the function $H(x,\tau)$ is a solution to the following nonlinear PDE, the so-called **Gamma equation** [12]:

$$\frac{\partial H(x,\tau)}{\partial \tau} = \frac{\partial^2 \beta(H)}{\partial x^2} + \frac{\partial \beta(H)}{\partial x} + (r-q)\frac{\partial H(x,\tau)}{\partial x} - qH(x,\tau). \qquad (5)$$

Notice that unlike in equation (3), all terms containing spatial derivatives in the Gamma equation (5) are in divergent form, thus it is suitable to use finite volume method discretization which follows. Furthermore, since $\partial_S^2 V$ tends asymptotically to zero as $S \to 0$, respectively $S \to \infty$, from (4) it follows that the transformed Dirichlet boundary conditions are $H(-\infty,\tau) = H(\infty,\tau) = 0$.

2 Finite Volume Approximation Schemes

The most general form of the Gamma equation is as follows:

$$\frac{\partial H(x,\tau)}{\partial \tau} = \frac{\partial^2 \beta(H,x,\tau)}{\partial x^2} + \frac{\partial \beta(H,x,\tau)}{\partial x} + f(x)\frac{\partial H(x,\tau)}{\partial x} + g(x)H(x,\tau), \qquad (6)$$

Notice that

$$\frac{\partial^2 \beta(H(x,\tau),x,\tau)}{\partial x^2} = \frac{\partial}{\partial x}\left(\beta'_H(H,x,\tau)\frac{\partial H(x,\tau)}{\partial x} + \beta'_x(H,x,\tau)\right), \qquad (7)$$

where $\beta'_H(H,x,\tau)$ and $\beta'_x(H,x,\tau)$ are partial derivatives of the function $\beta(H(x,\tau),x,\tau)$ by H and x, respectively. Moreover,

$$f(x)\frac{\partial H(x,\tau)}{\partial x} = \frac{\partial}{\partial x}(f(x)H(x,\tau)) - H(x,\tau)f'_x(x). \qquad (8)$$

Inserting (7) and (8) into (6) and integrating over the finite volume $\left(x_{i-\frac{1}{2}}, x_{i+\frac{1}{2}}\right)$, with center point denoted by x_i, we get

$$\int_{x_{i-\frac{1}{2}}}^{x_{i+\frac{1}{2}}} \frac{\partial H}{\partial \tau} dx = \int_{x_{i-\frac{1}{2}}}^{x_{i+\frac{1}{2}}} \frac{\partial}{\partial x} \left(\beta'_H \frac{\partial H}{\partial x} + \beta'_x + \beta + f(x)H \right) dx$$
$$+ \int_{x_{i-\frac{1}{2}}}^{x_{i+\frac{1}{2}}} (g(x) - f'_x(x)) H \, dx. \qquad (9)$$

Using central spatial differences, Newton–Leibniz formula and notations
$\beta^*_{i+\frac{1}{2}} = \beta(H^*_{i+\frac{1}{2}}, x_{i+\frac{1}{2}}, \tau_*)$, $\beta'^*_{x\,i+\frac{1}{2}} = \beta'_x(H^*_{i+\frac{1}{2}}, x_{i+\frac{1}{2}}, \tau_*)$, $\beta'^*_{H\,i+\frac{1}{2}} = \beta'_H(H^*_{i+\frac{1}{2}}, x_{i+\frac{1}{2}}, \tau_*)$,

we obtain the following *general numerical scheme* for solving (6):

$$h \frac{H_i^{j+1} - H_i^j}{k} = \beta'^*_{H\,i+\frac{1}{2}} \frac{H^*_{i+1} - H^*_i}{h} - \beta'^*_{H\,i-\frac{1}{2}} \frac{H^*_i - H^*_{i-1}}{h} + \beta'^*_{x\,i+\frac{1}{2}} - \beta'^*_{x\,i-\frac{1}{2}} + \beta^*_{i+\frac{1}{2}}$$
$$- \beta^*_{i-\frac{1}{2}} + f(x_{i+\frac{1}{2}}) \frac{H^*_{i+1} + H^*_i}{2} - f(x_{i-\frac{1}{2}}) \frac{H^*_i + H^*_{i-1}}{2} + h H^*_i \left(g(x_i) - f'_x(x_i) \right), \qquad (10)$$

where H_i^j represents the approximate value of the solution in point x_i at time τ_j and $\star \in \{j, j+1\}$ represents the chosen time layer. Depending on in which time we evaluate the terms on the right–hand side in (10) we obtain three distinct first–order schemes.

Explicit scheme is obtained by taking all terms from the old time layer, i.e. $\star = j$.

Semi–implicit scheme is obtained by taking all linear terms from the old time layer, i.e. $\star = j$, and all nonlinear terms from the new time layer, i.e. $\star = j+1$. The solution is found by solving a tridiagonal system of linear equations by the Thomas algorithm.

Fully–implicit scheme is obtained if all terms are taken from the new time layer, i.e. $\star = j+1$. We get a system of nonlinear equations. The algorithm for solving such a system is based on iterative solution of the semi–implicit scheme. We start the iterative process by assigning the old time step solution vector to the starting iteration solution vector for the new time step. Then, in each iteration, we insert the solution vector into the nonlinear terms, to get their actual iteration. If we collect all unknowns from the solution vector, i.e. the linear terms from the new layer, on the left–hand side and all remaining terms, i.e. the nonlinear terms and the linear term from the old layer, on the right–hand side we obtain a linear tridiagonal system for determining next iteration of the solution vector. The whole process is terminated when the successive solution vectors are close enough [2].

New second–order scheme is of the **Crank–Nicolson type** and is obtained by the arithmetic average of the explicit and the fully–implicit scheme. The system of nonlinear equations has a similar structure to that from the fully–implicit scheme, thus we solve it using the same principles.

Stability As noticed above, the linear systems arising in our schemes are solved by the Thomas algorithm. Its numerical stability is guaranteed by the strict diagonal dominance of the system's matrix which can be always achieved by a suitable choice of time step k in (10). Another important issue is the study of stability which is usually related to the approximation of diffusion and advection terms. Inspecting the Gamma equation (5), one can see that the diffusion coefficient is given by β'_H while the speed of the advection is proportional to $\beta'_H + r - q$ and thus they are comparable ($\hat{\sigma}^2 \approx r - q$). The fully explicit scheme gives oscillations for the coupling $k \approx h$ due to violating the CFL condition in approximation of the diffusion term. On the other hand, all other schemes are implicit and we did not observe any oscillations, mainly due to the fact that the advection does not dominate the diffusion.

3 Numerical Experiments

Three different numerical experiments were made. The first two are concerned with the approximate solution to the RAPM Gamma equation and the last one deals with the numerical solution to a nonlinear porous–medium PDE.

RAPM Gamma equation experiments As no comparative exact solution to such an equation is known, a natural choice is to take the exact solution of the linear Black–Scholes model. Clearly, to maintain the equality in the Gamma equation we have to add a residual term $Res(x, \tau)$ into (5) which balances the difference between the Black–Scholes solution and the higher Ask price of the RAPM model:

$$\frac{\partial H}{\partial \tau} = \frac{\partial^2 \beta}{\partial x^2} + \frac{\partial \beta}{\partial x} + (r - q)\frac{\partial H}{\partial x} - qH + Res, \tag{11}$$

where $\beta(H) = \frac{\hat{\sigma}^2}{2}(1 + \mu H^{\frac{1}{3}})H$. The first two experiments differ from each other in two main aspects: the coefficient μ and the initial condition. Following parameters were set for both cases the same: $\hat{\sigma} = 0.30$, $r = 0.03$, $q = 0.01$, $E = 25$. In all numerical experiments we impose boundary conditions $H(x_L, \tau) = H(x_R, \tau) = 0$, where x_L and x_R are boundaries of the space interval.

The intention of *the first experiment* is to show how well the proposed numerical schemes can handle the nonlinearity in the Gamma equation (11). We put the coefficient $\mu = 0.2$, hence the function $\beta(H)$ is nonlinear. As the initial condition $H(x, \tau_0)$ we consider Black–Scholes solution $V(S, T - \tau_0)$ transformed by (4), in time $\tau_0 = 1$. Measurements of the estimated error $||e_n^m||_{L_2}$ are done by comparison with the exact solution $H(x, \tau)$ to (11) for $\tau > \tau_0$. Since all first-order schemes exhibited very similar features, we show here outputs just for the semi-implicit scheme. The reason for exclusion of the explicit scheme was its instability using coupling $k = h$. Regarding the fully-implicit scheme, experiments show that the accuracy of the semi-implicit scheme is very close to the fully-implicit scheme, thus it is sufficient to use just the former one which is less time consuming. The

Table 1 Outputs obtained by solving the RAPM Gamma equation (11) ($\tau_0 = 1, k = h$) using the semi–implicit scheme: estimated error $||e_n^m||_{L_2}$, CPU–time and EOC with respect to $||e_n^m||_{L_2}$

n	h	$||e_n^m||_{L_2}$	CPU	$EOC_{k\sim h}$
20	0.1	0.00777657	2.231	–
40	0.05	0.00333385	9.126	1.22194
80	0.025	0.00153036	36.614	1.12332
160	0.0125	0.00073141	147.078	1.06512
320	0.00625	0.00035733	582.929	1.03343

Table 2 Outputs obtained by solving the RAPM Gamma equation (11) ($\tau_0 = 1, k = h$) using the Crank–Nicolson type scheme: estimated error $||e_n^m||_{L_2}$, CPU–time and EOC with respect to $||e_n^m||_{L_2}$

n	h	$||e_n^m||_{L_2}$	CPU	$EOC_{k\sim h}$
20	0.1	0.00272286	4.383	–
40	0.05	0.000666762	17.785	2.02988
80	0.025	0.000165182	71.136	2.01311
160	0.0125 5	0.0000412598	294.062	2.00125
320	0.00625	0.0000108204	1206.53	1.93099

experiment was done on the time–space domain $(x, \tau) = [-2, 2] \times [1, 2]$. Tables 1 and 2 indicate that for this type of problem the semi–implicit scheme is first order accurate while the Crank–Nicolson type scheme is second order accurate.

In the *the second experiment* we set $\mu = 0$ and we show how the regularization of the transformed initial condition and the backward transformation of the Gamma equation solution affects the total accuracy of the method. In this case the solution of the Gamma equation coincides with the transformed solution $H(x, \tau)$ of the linear Black–Scholes equation (1) which implies that the residual term in (11) is zero. The initial condition $H(x, \tau_0)$ is considered for $\tau_0 = 0$. Hence the transformed payoff function, see (2) and (4), is the Dirac delta function, $H(x, 0) = \delta(x)$, $x \in R$. In order to get a suitable initial condition for our computation, we consider its regularization given by the function $H(x, 0) = \frac{N'(d)}{\hat{\sigma}\sqrt{\tau^*}}$, where $\tau^* > 0$ is sufficiently small, $N(d)$ is the cumulative distribution function of the normal distribution and $d = \frac{x+(r-q-\hat{\sigma}^2/2)\tau^*}{\hat{\sigma}\sqrt{\tau^*}}$ [12]. The backward transformation of numerical solution is done by using formula

$$V(S_k, T - \tau_j) = h \sum_{i=-n}^{n} \max(S_k - Ee^{x_i}, 0) H_i^j = h \sum_{i=-n}^{k} (S_k - Ee^{x_i}) H_i^j$$

$$= h S_k \sum_{i=-n}^{k} H_i^j - h E \sum_{i=-n}^{k} e^{x_i} H_i^j = h S_k F_k - h E G_k, \quad (12)$$

Table 3 Outputs obtained by solving numerically Gamma equation (11) ($\tau_0 = 0, k = h/4$) using the Crank–Nicolson type scheme and using formula (12) for backward transformation

n	h	τ^*	$||e_n^m||_{L_2}$	$EOC_{k \sim h}$	CPU Gamma	CPU Transform	CPU Total
5	0.4	0.46765	4.0644	–	0.047	0.011	0.058
10	0.2	0.14602	1.4586	1.4784	0.141	0.016	0.157
20	0.1	0.04371	0.4617	1.6595	0.624	0.047	0.671
40	0.05	0.01269	0.1379	1.7432	2.372	0.187	2.559
80	0.025	0.00361	0.0399	1.787	9.173	0.843	10.016
160	0.0125	0.00101	0.0113	1.816	41.091	3.323	44.414
320	0.00625	0.00028	0.0031	1.8270	150.525	12.87	163.396

where $F_k = F_{k-1} + H_k^j$, $G_k = G_{k-1} + e^{x_k} H_k^j$ and $S_k = Ee^{x_k}$. Formula (12) is obtained by integration of (4). Measurements of the estimated error $||e_n^m||_{L_2}$ are done by comparison with the Black–Scholes solution $V(S, t)$. However, in practice, doing computations with such an initial condition is not as straightforward task as in the first experiment. The problem is that we do not know a priori the optimal value of τ^* for a given time–space mesh. We consider the optimal value of τ^* as a value for which the estimated error of the numerical solution is minimized. Numerical outputs for the discretized time–space domain $(x, \tau) = [-2, 2] \times [0, 1]$ are summarized in the Table 3. Since the total error is influenced not only by the discretization error, but also by the error related to the regularization and backward transformation, the Crank–Nicolson method exhibits EOC slightly below the second order. Finally, in Fig. 1 we present the numerical solution of the RAPM model for a call option using parameter τ^* obtained by the above described strategy but considering nonzero μ. Such an experiment is of particular interest also for practical applications.

Experiment with an exact (Barenblatt) solution The goal of the *third experiment* was to investigate the accuracy of the proposed Crank–Nicolson type scheme using exact solution of the following (porous–medium type) equation [10]:

$$\partial_t v = \partial_x^2 (v^\omega), \quad x \in R, \ t > 0, \ \omega > 1 \qquad (13)$$

which is a special case of the Gamma equation (5). The exact solution has the form $v(x, t) = \frac{1}{\omega(t)} \max\left[0, 1 - \left(\frac{x}{\omega(t)}\right)^2\right]^{\frac{1}{\omega-1}}$, where $\lambda(t) = \left[\frac{2\omega(\omega+1)}{\omega-1}(t+1)\right]^{\frac{1}{\omega+1}}$ represents a sharp interface of the solution's finite support. EOC of the Crank–Nicolson type scheme in L_1-norm which is used due to the singularity in the exact solution, is equal to 2, see Table 4.

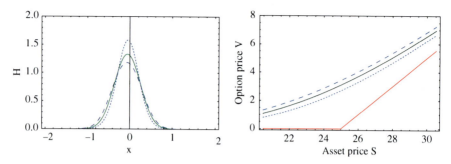

Fig. 1 A comparison of Bid and Ask option prices computed by means of the RAPM model for a call option in time $T - t = 1$. Left (right) figure presents the results before (after) the backward transformation. The dashed (fine–dashed) curve indicates the Ask (Bid) price of a call option. The solid curve represents the option prices computed by the linear Black–Scholes model and the solid broken line is the payoff function. Parameters: $n = 80$, $h = 0.025$, $m = 160$, $k = 0.00625$, $\tau^* = 0.00391$, $\hat{\sigma} = 0.30$, $\mu = \pm 0.2$, $r = 0.011$, $q = 0.0$, $X = 25$

Table 4 Numerical approximation of the Barenblatt exact solution using Crank–Nicolson type scheme

n	h	$\|\|e_n^m\|\|_{L_1}$	CPU	$EOC_{k \sim h}$
25	0.1	0.000629	0.312	–
50	0.05	0.000173	1.139	1.8584
100	0.025	0.000048	4.258	1.8543
200	0.0125	0.000012	17.036	1.9161
400	0.00625	$3.31 \cdot 10^{-6}$	67.798	1.9399
800	0.003125	$8.52 \cdot 10^{-7}$	250.475	1.9597
1600	0.0015625	$2.16 \cdot 10^{-7}$	881.905	1.97824

4 Conclusions

In this paper we proposed a new nonlinear second order Crank–Nicolson type numerical scheme based on the finite volume method. Our main goal was to provide an efficient and precise numerical solution to nonlinear PDEs arising in financial mathematics. Various experiments have shown such properties of the new scheme.

Acknowledgements This work was supported by grants APVV–0351–07 and VEGA 1/0733/10.

References

1. Avellaneda, M., Levy, A., and Paras, A.: Pricing and hedging derivative securities in markets with uncertain volatilities. Applied Mathematical Finance 2, (1995) 73-88
2. Balažovjech M., Mikula K.: A Higher Order Scheme for the Curve Shortening Flow of Plane Curves. Proceedings of ALGORITMY, STU Bratislava, (2009) 165–175

3. Barles, G., and Soner, H. M.: Option pricing with transaction costs and a nonlinear Black-Scholes equation. Finance Stoch. 2, 4 (1998) 369-397
4. Black, F., and Scholes, M.: The pricing of options and corporate liabilities. The Journal of Political Economy 81, (1973) 637-654
5. Jandačka, M., Ševčovič, D.: On the risk-adjusted pricing-methodology-based valuation of vanilla options and explanation of the volatility smile. J. Appl. Math. 3, (2005) 235-258
6. Kratka, M.: No mystery behind the smile. Risk 9, (1998) 67-71
7. Kwok, Y. K.: Mathematical models of financial derivatives. Springer-Verlag, Singapore (1998)
8. Leland, H. E.: Option pricing and replication with transaction costs. Journal of Finance 40, (1985) 1283-1301
9. Merton, R.: Theory of rational option pricing. The Bell Journal of Economics and Management Science, (1973) 141-183
10. Mimura, M., Tomoeda, K.: Numerical approximations to interface curves for a porous media equation. Hiroshima Math. J. 13, Hiroshima University, (1983) 273–294
11. Schönbucher, P. J., and Wilmott, P.: The feedback effect of hedging in illiquid markets. SIAM J. Appl. Math. 61, 1 (2000) 232-272
12. Ševčovič, D., Stehlíkova, B., Mikula, M.: Analytical and numerical methods for pricing financial derivatives. Nova Science Publishers, Hauppauge NY (2011)

The paper is in final form and no similar paper has been or is being submitted elsewhere.

Monotonicity Conditions in the Mimetic Finite Difference Method

Konstantin Lipnikov, Gianmarco Manzini, and Daniil Svyatskiy

Abstract The maximum principle is a major property of solutions of partial differential equations. In this work, we analyze a few constructive algorithms that allow one to embed this property into a mimetic finite difference (MFD) method. The algorithms search in the parametric family of MFD methods for a member that guarantees the discrete maximum principle (DMP). A set of sufficient conditions for the DMP is derived for a few types of meshes. For general meshes, a numerical optimization procedure is proposed and studied numerically.

Keywords Mimetic method, discrete maximum principle, M-matrix
MSC2010: 35B50, 65N06, 65N50

1 Mimetic finite difference method with parameters

The maximum principle is one of the most important properties of solutions of partial differential equations [3, 4]. Its numerical analog, the discrete maximum principle (DMP), is one of the most difficult properties to achieve in numerical methods, especially when the computational mesh is distorted to adapt and conform to the physical domain or the problem coefficients are highly heterogeneous and anisotropic. In this work, we investigate sufficient conditions to ensure the DMP in the mimetic finite difference (MFD) method [2]. We extend the analysis proposed in [5] by considering an optimization procedure as a way to achieve the DMP.

K. Lipnikov · D. Svyatskiy
Los Alamos National Laboratory, USA, e-mail: lipnikov@lanl.gov,dasvyat@lanl.gov

G. Manzini
IMATI-CNR and CESNA-IUSS, Pavia, Italy, e-mail: marco.manzini@imati.cnr.it

We consider the mimetic discretization of the steady diffusion problem for the scalar and vector solution fields, p and $\mathbf{u}$, given by

$$\mathbf{u} + \mathsf{K}\nabla p = 0 \quad \text{in } \Omega, \tag{1}$$

$$\text{div}(\mathbf{u}) = q \quad \text{in } \Omega, \tag{2}$$

$$p = g^D \quad \text{on } \Gamma. \tag{3}$$

Here, Ω is an open bounded polygonal subset of R^d with Lipshitz boundary Γ; K is a $d \times d$ bounded, strongly elliptic and symmetric diffusion tensor; $q \in L^2(\Omega)$ is the forcing term and $g^D \in H^{1/2}(\Gamma)$ is the given boundary function.

Hopf's lemmas in weak and strong form can be summarized as follows [4]. Let $-\text{div}(\mathsf{K}\nabla p) \leq 0$ in Ω. Then, the weak maximum principle holds:

$$\max_{\mathbf{x} \in \overline{\Omega}} p(\mathbf{x}) \leq \max\left(0, \max_{\mathbf{x} \in \Gamma} p(\mathbf{x})\right).$$

This implies immediately that $p \geq 0$ if f and g^D are non-negative functions. Finally, the strong maximum principle says that if p attains a nonnegative maximum $\widehat{p}$ at an interior point of Ω, then $p = \widehat{p}$ in $\overline{\Omega}$.

Let Ω_h denote a conforming and *face-connected* partition of Ω into control volumes P, which are general polyhedra in 3-D and polygons in 2-D. The degrees of freedom for the scalar variable p are p_P and p_f. They approximate the average of p over elements P and faces f, respectively. The degrees of freedom of the vector variable $\mathbf{u}$ are $U_{\mathsf{P},\mathsf{f}}$. They approximate the normal component of $\mathbf{u}$ over mesh faces f. Any internal face f shared by two elements P' and P'' is characterized by two flux unknowns $U_{\mathsf{P}',\mathsf{f}}$ and $U_{\mathsf{P}'',\mathsf{f}}$ that must satisfy the flux conservation condition:

$$U_{\mathsf{P}',\mathsf{f}} + U_{\mathsf{P}'',\mathsf{f}} = 0. \tag{4}$$

Let the boundary of P be formed by the m faces f_i, $i = 1, \ldots, m$, with measure $|\mathsf{f}_i|$ (length in 2-D, surface area in 3-D). We consider the numerical discretization of (1) that reads

$$\begin{pmatrix} U_{\mathsf{P},\mathsf{f}_1} \\ \vdots \\ U_{\mathsf{P},\mathsf{f}_m} \end{pmatrix} = \mathbb{W}_\mathsf{P} \begin{pmatrix} |\mathsf{f}_1|(p_{\mathsf{f}_1} - p_\mathsf{P}) \\ \vdots \\ |\mathsf{f}_m|(p_{\mathsf{f}_m} - p_\mathsf{P}) \end{pmatrix}, \tag{5}$$

where $\mathbb{W}_\mathsf{P}$ is a symmetric and positive definite (SPD) matrix.

Let $\mathbf{U}_\mathsf{P} = (U_{\mathsf{P},\mathsf{f}_1}, U_{\mathsf{P},\mathsf{f}_2}, \ldots, U_{\mathsf{P},\mathsf{f}_m})^T$ be the m-sized vector of numerical fluxes across faces f_i of P. We write the numerical approximation of (2) as

$$\text{div}_\mathsf{P} \mathbf{U}_\mathsf{P} = \overline{q}_\mathsf{P}, \quad \text{and} \quad \text{div}_\mathsf{P} \mathbf{U}_\mathsf{P} = \frac{1}{|\mathsf{P}|} \sum_{i=1}^{m} |\mathsf{f}_i| U_{\mathsf{P},\mathsf{f}_i}. \tag{6}$$

where $\bar{q}_P$ is the average of q over P, and div$_P$ is the primary mimetic divergence operator. The MFD method is given by equations (4), (5), (6). The Dirichlet boundary conditions are imposed by assigning average values of g^D on boundary faces f to corresponding unknowns p_f.

2 Construction of monotone mimetic methods

In the MFD method, the SPD matrix $\mathbb{W}_P$ is is built in accordance with a *stability* and a *consistency* conditions [2]. A rich family of matrices satisfies these conditions. To achieve the DMP, we will impose additional constraints on this family.

The *stability condition* states that

$$\frac{\sigma_*}{|P|} \mathbf{V}_P^T \mathbf{V}_P \leq \mathbf{V}_P^T \mathbb{W}_P \mathbf{V}_P \leq \frac{\sigma^*}{|P|} \mathbf{V}_P^T \mathbf{V}_P \quad \forall \mathbf{V}_P, \tag{7}$$

where σ_* and σ^* are two constants independent of P and of the mesh Ω_h. This condition states that matrix $\mathbb{W}_P$ is spectrally equivalent to the scalar matrix $|P|^{-1} \mathbb{I}_m$.

Let $\mathbf{x}_P$ and $\mathbf{x}_f$ be centers of gravity of element P and face f, respectively. Let $\mathbf{n}_f$ be the external unit normal vector to f. We introduce the following two matrices:

$$\mathbb{R}_P = \begin{pmatrix} |f_1|(\mathbf{x}_{f_1} - \mathbf{x}_P)^T \\ \vdots \\ |f_m|(\mathbf{x}_{f_m} - \mathbf{x}_P)^T \end{pmatrix} \quad \text{and} \quad \mathbb{N}_P = \begin{pmatrix} \mathbf{n}_{f_1}^T \\ \vdots \\ \mathbf{n}_{f_m}^T \end{pmatrix} \mathbb{K}. \tag{8}$$

The *consistency condition* takes the form $\mathbb{W}_P \mathbb{R}_P = \mathbb{N}_P$.

A straightforward calculation shows that $\mathbb{N}_P^T \mathbb{R}_P = |P| \mathbb{K}$. It is proved in [2] that matrix $\mathbb{W}_P$ is given by

$$\mathbb{W}_P = \mathbb{N}_P (\mathbb{N}_P^T \mathbb{R}_P)^{-1} \mathbb{N}_P^T + \mathbb{D}_P \mathbb{U}_P \mathbb{D}_P^T, \tag{9}$$

where $\mathbb{D}_P$ is a maximum rank $d \times (m-d)$-sized matrix such that $\mathbb{R}_P^T \mathbb{D}_P = 0$, and $\mathbb{U}_P$ is a $(m-d) \times (m-d)$-sized SPD matrix of parameters.

An effective way to ensure that the monotonicity property holds is to construct a numerical method such that the final discretization matrix is an M-matrix [1]. In the MFD method, this occurs when $\mathbb{W}_P$ satisfies two geometric conditions formulated below. Let $\mathbb{W}_P = \{w_{ij}\}_{i,j=1}^m$. Since this is an SPD matrix, we obtain that $w_{ii} > 0$. We assume that

(A1) The matrix $\mathbb{W}_P$ satisfies the geometric constraint:

$$w_{ii}|f_i| + \sum_{j \neq i} w_{ij}|f_j| \geq 0 \quad \forall i,$$

and the inequality is strict for at least one matrix row.

(A2) The matrix $\mathbb{W}_P$ is a Z-matrix, i.e., $w_{ij} \leq 0$ for $i \neq j$.

Sufficient conditions (A1) and (A2) together with positive definiteness of matrix $\mathbb{U}_P$ result in a set of inequalities for every element P. These local optimization problems can be solved analytically on special meshes or numerically to provide an MFD method for which the following theorem holds.

Theorem 1 (Discrete Maximum Principle). *Let p_P, p_f and $U_{P,f}$ be the solutions of the MFD method under assumptions (A1) and (A2). Let q and g^D be nonnegative functions. Then, $p_P \geq 0$ for any $P \in \Omega_h$. Furthermore, if $q = 0$, then the values of p_P are bounded by the maximum and minimum values of p_f on boundary faces* f.

3 Oblique parallelepipeds

Let us consider a mesh Ω_h consisting of regular oblique parallelepipeds. We assume that parallelepiped faces are planar. To construct matrices $\mathbb{N}_P$ and $\mathbb{R}_P$, we refer to the numbering order shown in Fig. 1. Let $\mathbf{n}_1 = \mathbf{n}_{BCGF}, \mathbf{n}_2 = \mathbf{n}_{DCGH}, \mathbf{n}_3 = \mathbf{n}_{EFGH}$ and $\alpha := |f_{BCGF}| = |f_{ADHE}|$, $\beta := |f_{DCGH}| = |f_{ABFE}|$, $\gamma := |f_{EFGH}| = |f_{ABCD}|$. We define the *rotated diffusion tensor*:

$$\mathsf{K}^\theta = (\mathsf{K}^\theta_{ij})_{i,j=1}^3 \qquad \mathsf{K}^\theta_{ij} = \mathbf{n}_i^T \mathsf{K} \mathbf{n}_j.$$

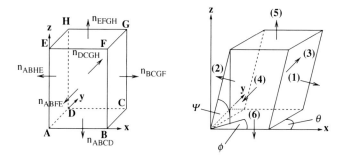

Fig. 1 Geometry of an orthogonal (left) and oblique (right) parallelepiped

Let us choose matrix $\mathbb{D}_P^T$ as follows:

$$\mathbb{D}_P^T = \begin{pmatrix} 1 & 1 & 0 & 0 & 0 & 0 \\ 0 & 0 & 1 & 1 & 0 & 0 \\ 0 & 0 & 0 & 0 & 1 & 1 \end{pmatrix}. \tag{10}$$

This choice of $\mathbb{D}_P$ allows us to simplify analysis of the MFD method and to prove the following results.

Lemma 1. *Assumptions (A1) and (A2) imply that*

$$0 < |P|\mathbb{U}_P \begin{pmatrix} \alpha \\ \beta \\ \gamma \end{pmatrix} \leq \widetilde{K}^\theta \begin{pmatrix} \alpha \\ \beta \\ \gamma \end{pmatrix},$$

where $\widetilde{K}^\theta = \{\widetilde{K}^\theta_{ij}\}_{i,j=1}^3$ *with* $\widetilde{K}^\theta_{ii} = K^\theta_{ii}$ *and* $\widetilde{K}^\theta_{ij} = -|K^\theta_{ij}|$ *for* $i \neq j$.

From this and assumptions on $\mathbb{U}_P$, we derive two *necessary conditions* for existence of a monotone MFD method:

$$\widetilde{K}^\theta \begin{pmatrix} \alpha \\ \beta \\ \gamma \end{pmatrix} > 0 \quad \text{and} \quad \widetilde{K}^\theta \text{ is SPD}. \tag{11}$$

Conditions (11) impose constraints on the range of values that the coefficients in K and the face areas $|f_i|$ may attain in order to have a monotone mimetic discretization. If $\widetilde{K}^\theta$ is an SPD matrix, a possible choice for $\mathbb{U}_P$, which maximizes the sparsity its structure, is $\mathbb{U}_P = |P|^{-1} \widetilde{K}^\theta$.

Let Ω be the unit cube a hole $]0.6; 0.8[\times]0.438; 0.563[\times]0.5; 0.6[$. The computational mesh is 10×16×20 with a 2×2×2 hole. A tilted domain and the corresponding mesh are obtained through the linear transformation $x := x + z \cos(\theta)$, $y := y + z \cos(\theta)$, $z := z \sin(\theta)$. The diffusion tensor is

$$K = \begin{pmatrix} 100 & 0.25 & 0.15 \\ 0.25 & 1 & 0.25 \\ 0.15 & 0.25 & 1 \end{pmatrix}. \tag{12}$$

We set $f = 0$, $g^D = 2$ on the interior boundary (surface of the hole) and $g^D = 0$ on the exterior boundary. The exact solution is not known but must vary between 0 and 2 due to the maximum principle.

In Table 1, we consider a range of parameters θ. The significant violation of the minimum principle is clearly observed in the MFD method [2] where $\mathbb{U}_P$ is set to a scalar matrix. It is due to huge number of negative entries in the inverse of the stiffness matrix (second column). Fig. 2 shows cuts through the element-based

Table 1 Original and monotone MFD methods on tilted parallelepiped meshes

θ	%	Original MFD				Monotone MFD	
		$\min(p_t)$	$\max(p_t)$	$\min(p_P)$	$\max(p_P)$	$\min(p_P)$	$\max(p_P)$
90	36.0	$-3.651\ 10^{-1}$	2.083	$-9.371\ 10^{-2}$	1.669	$6.304\ 10^{-12}$	1.700
80	35.9	$-3.478\ 10^{-1}$	2.089	$-9.039\ 10^{-2}$	1.659	$5.224\ 10^{-11}$	1.694
70	35.6	$-3.657\ 10^{-1}$	2.092	$-9.464\ 10^{-2}$	1.665	$5.403\ 10^{-11}$	1.695
60	35.2	$-4.019\ 10^{-1}$	2.084	$-1.028\ 10^{-1}$	1.679	$9.995\ 10^{-12}$	1.699

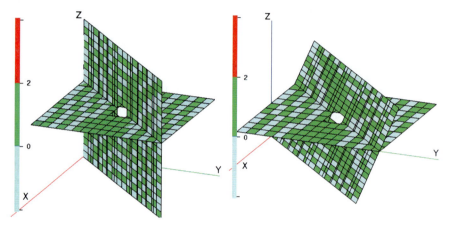

Fig. 2 Original MFD method: undershoots in the element-based numerical solution on the tilted domains with angles $\theta = 90°$ (left) and $61°$ (right)

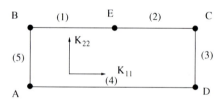

Fig. 3 A rectangular element $ABCD$ with a handing node E

discrete solution for $\theta = 90°$ and $60°$ in the original MFD method. For visualization clarity only two colors are used and the lighter color corresponds to negative solution values. The monotone MFD method satisfies the DMP.

4 Locally refined rectangular meshes

Let us consider a locally refined rectangular meshes (see Fig. 4). In the MFD framework, these meshes are considered as general polygonal meshes; thus, no special treatment of handing nodes is required (see Fig. 3).

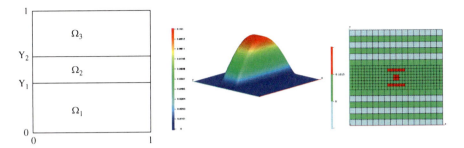

Fig. 4 The computational domain (left), discrete solution calculated with the monotone MFD method (middle), and the locally refined mesh (right)

We assume that the diffusion tensor is diagonal, $\mathsf{K} = \text{diag}\{\mathsf{K}_{11}, \mathsf{K}_{22}\}$. Let $r^2 = \frac{|f_{AD}|}{|f_{AB}|}$ be the aspect ratio of pentagon $ABECD$. Analysis of the MFD method leads to the following results.

Lemma 2. *A matrix $\mathbb{W}_P$ satisfying assumptions (A1) and (A2) exists when*

$$r^4 < 4\frac{\mathsf{K}_{11}}{\mathsf{K}_{22}}. \tag{13}$$

For each aspect ratio r satisfying (13), we obtain a family of monotone mimetic methods. The closer the aspect ratio to the limiting value $4\mathsf{K}_{11}/\mathsf{K}_{22}$, the narrower this family. Among many of possible choices, we present a member which reduces the number of nonzero elements in the matrix $\mathbb{W}_P$:

$$\mathbb{U}_P = \frac{1}{|P|}\begin{pmatrix} \frac{\mathsf{K}_{11}}{r^4} + \frac{\mathsf{K}_{22}}{4} & -\frac{\mathsf{K}_{22}}{2} & \frac{\mathsf{K}_{11}}{r^4} \\ -\frac{\mathsf{K}_{22}}{2} & \mathsf{K}_{22} & -\frac{\mathsf{K}_{22}}{2} \\ \frac{\mathsf{K}_{11}}{r^4} & -\frac{\mathsf{K}_{22}}{2} & \frac{\mathsf{K}_{22}}{r^4} + \frac{\mathsf{K}_{11}}{4} \end{pmatrix}, \quad \mathbb{D}_P^T = \begin{pmatrix} 2 & -2\,r^2 & 0 & 0 \\ 1 & 1 & 0 & 1 & 0 \\ -2 & 2 & 0 & 0 & r^2 \end{pmatrix}. \tag{14}$$

This choice imposes a stronger condition on the aspect ration, $r^4 < 8\mathsf{K}_{22}/(3\mathsf{K}_{11})$.

Let Ω be the unit square divided into three subdomains $\Omega_1 = (0,1) \times (0, Y_1)$, $\Omega_2 = (0,1) \times (Y_1, Y_2)$, and $\Omega_3 = (0,1) \times (Y_2, 1)$ as shown in Fig. 4. We set the forcing term and the diffusion tensor as follows:

$$f(x,y) = \begin{cases} 0, & (x,y) \in \Omega_1 \cup \Omega_3, \\ 10^3 \sin(\pi x) & (x,y) \in \Omega_2, \end{cases} \quad \mathsf{K} = \begin{pmatrix} 10^3 & 0 \\ 0 & 1 \end{pmatrix}.$$

In our experiments $Y_1 = 3/8$ and $Y_2 = 5/8$. The exact solution to this problem can be calculated using the separation of variables. It is shown in Fig. 4.

The solution profile has sharp gradients around interfaces between subdomains. Therefore, we refine the subdomain Ω_2 and obtain a set of meshes similar to that shown in Fig. 4. According to the DMP, the solution has to be strictly positive inside the computational domain. The numerical results show that the original MFD method produces numerical solutions that violate the DMP and have large subdomains with overshoots and undershoots. A cell-centered solution is plotted in Fig. 4 using three pseudo-colors. The lighter color represents negative solution, the darker color represents solution overshoot.

The numerical solutions obtained with the original MFD method still violates the DMP after one mesh refined. With one additional refinement, the undershoots become comparable with the solver tolerance. The monotone MFD method uses the parameter matrix in (14) and provides a monotone solution which is bounded by the minimum and maximum of the analytical solution.

5 Monotone MFD methods based on numerical optimization

On more general meshes, analysis of the MFD family of methods becomes too complicated. Therefore, we reformulate the problem of constructing an M-matrix $\mathbb{W}_\mathsf{P}$ as a constrained optimization problem:

$$\min_{\mathbb{U}_\mathsf{P} \in \mathscr{U}_\mathsf{P}} \Phi(\mathbb{W}_\mathsf{P}(\mathbb{U}_\mathsf{P})),$$

where $\mathscr{U}_\mathsf{P}$ is a set of SPD matrices with the smallest eigenvalue bounded from below by $\lambda_{\min}(\mathsf{K}_\mathsf{P})/2$ and the functional Φ penalizes positive off-diagonal entries in $\mathbb{W}_\mathsf{P}$ as well as violation of the assumption (**A1**):

$$\Phi(\mathbb{W}_\mathsf{P}) = \sum_{i \neq j} (w_{ij} + |w_{ij}|)^2 + \sum_i (s_i - |s_i|)^2, \qquad s_i = w_{ii}|f_i| + \sum_{i \neq j} w_{ij}|f_j|.$$

The functional achieves its minimal value when $w_{ij} \leq 0$ for $i \neq j$ and $s_i \geq 0$. Restriction imposed on the minimal eigenvalue of $\mathbb{U}_\mathsf{P}$ guarantees that the matrix $\mathbb{W}_\mathsf{P}$ is SPD.

We implemented a simple minimization algorithm based on numerical calculation of the gradient of Φ and functional minimization along this direction. Let us consider again the problem from Sec. 3. Table 2 shows minimal and maximal solution values for two MFD methods. The first method uses a scalar matrix $\mathbb{U}_\mathsf{P} = a_\mathsf{P} \mathbb{I}_\mathsf{P}$, where $a_\mathsf{P} = \mathrm{trace}(\mathsf{K}_\mathsf{P})/3$ lies in a middle of spectrum of K_P. The second method uses this matrix as the initial guess for the minimization algorithm. Since for every P, the number of parameters is six, we terminate the algorithm after six steps.

A simple optimization procedure is sensitive to an initial guess. Therefore, in the future, we plan to analyze more advanced optimization strategies as well as different functionals.

Table 2 Original and optimized MFD methods on tilted parallelepiped meshes

θ	Original MFD ($\mathbb{U}_P = a_P \mathbb{I}_P$)		Optimized MFD	
	$\min(p_P)$	$\max(p_P)$	$\min(p_P)$	$\max(p_P)$
70	$-7.267\ 10^{-2}$	1.577	$5.855\ 10^{-11}$	1.641
60	$-8.320\ 10^{-2}$	1.602	$9.801\ 10^{-12}$	1.648
50	$-8.998\ 10^{-2}$	1.628	$2.378\ 10^{-14}$	1.641

6 Conclusions

In this paper, we present a new methodology for the construction of mimetic discretizations which satisfy the discrete maximum principle. A set of sufficient conditions is derived to ensure that such monotone subfamily exists.

Acknowledgements The work of the first (K.L.) and third author (D.S.) was supported by the Department of Energy (DOE) Advanced Scientific Computing Research (ASCR) program in Applied Mathematics. The work of the second author (G.M.) was partially supported by the Italian MIUR through the program PRIN2008.

References

1. A. Berman and R. J. Plemmons. *Nonnegative matrices in the mathematical sciences*. Academic Press, New York, 1979. Computer Science and Applied Mathematics.
2. F. Brezzi, K. Lipnikov, and V. Simoncini. A family of mimetic finite difference methods on polygonal and polyhedral meshes. *Math. Mod. Meth. Appl. Sci.*, 15(10):1533–1551, 2005.
3. P. Grisvard. *Elliptic problems in nonsmooth domains*, volume 24 of *Monographs and Studies in Mathematics*. Pitman (Advanced Publishing Program), Boston, MA, 1985.
4. E. Hopf. Elementare Bemerkungen uber die L osungen partieller Differentialgleichungen zweiter Ordnung vom elliptischen Typus. *Sitzungsber. Preuss. Akad. Wiss.*, 19, 1927.
5. K. Lipnikov, G. Manzini, and D. Svyatskiy. Analysis of the monotonicity conditions in the mimetic finite difference method for elliptic problems *J. Comput. Phys.*, 230(7): 2620–2642, 2011.

The paper is in final form and no similar paper has been or is being submitted elsewhere.

Discrete Duality Finite Volume Method Applied to Linear Elasticity

Benjamin Martin and Frédéric Pascal

Abstract We present the Discrete Duality Finite Volume method (DDFV) for solving the linear elasticity problem on unstructured mesh applied to solids undergoing mechanical loads. The procedure is described in detail for three dimensional problems and some theoretical results are provided: the discrete problem is well-posed, stable and convergent. A number of numerical test problems demonstrates the ability of this finite volume scheme to approach the solution and some comparisons with the conventional finite element method are provided.

Keywords Finite volume methods, linear elasticity, stability and convergence
MSC2010: 65N08, 73C02, 65N12

1 Motivation

The finite volume method is extensively used in computational fluid dynamics, on its part the finite element method is the conventional tool for solving solid mechanics. However there is a multitude of physical problems combining fluid and solid mechanics where finite volume methods appear to be a pertinent alternative. Let us quote for instance fluid-structure interaction, deformation of geomechanical reservoir, or even the frost heave problem in freezing soils where the moving frozen fringe introduces a discontinuity in the physical parameters. The finite volume approach for elasticity problems has already been discussed and published in [3],

Benjamin Martin and Frédéric Pascal
CMLA, ENS de Cachan, CNRS, 61 Avenue du Pt Wilson, 94235 Cachan, France,
e-mail: benjamin.martin@cmla.ens-cachan.fr, frederic.pascal@cmla.ens-cachan.fr

[15], [17], [18] for cell-vertex formulations, in [12] for cell centered formulation with a decoupled strategy for each component, in [6] and [7] for a coupled cell-center version. In this study, we address the DDFV implementation for solving linear elasticity. Let us recall that the principle of the DDFV discretization consists in integrating the system both over a given primal mesh and a dual mesh built from the primal one. Presentation, convergence analysis and numerical tests of DDFV for diffusion, convection-diffusion and Stokes problems are available in [1], [2], [4], [8], [10], [11], [13], [14].

We limit ourselves to the simplest mathematical model of a linear elastic solid which consists in finding the displacement $\mathbf{u} \in \mathbb{R}^3$ such that

$$-\operatorname{div} \sigma(\mathbf{u}) = \mathbf{f} \text{ on } \Omega, \quad \mathbf{u} = \mathbf{g} \text{ on } \Gamma_D, \quad \sigma(\mathbf{u}) \cdot \mathbf{n} = \mathbf{h} \text{ on } \Gamma_N \quad (1)$$

where $\mathbf{n}$ is the outward normal, $\partial \Omega = \Gamma_D \cup \Gamma_N$ and where the stress tensor σ depends on $\mathbf{u}$ by the Hooke relation that links the strain tensor and the trace of the gradient

$$\sigma(\mathbf{u}) = \lambda \mathbb{D}\mathrm{iv}\mathbf{u} + 2\mu \mathbb{D}\mathbf{u} \quad \text{with} \quad \mathbb{D}\mathbf{u} = \frac{\nabla \mathbf{u} + (\nabla \mathbf{u})^T}{2} \quad \text{and} \quad \mathbb{D}\mathrm{iv}\mathbf{u} = \operatorname{div} \mathbf{u} \text{ Id}. \quad (2)$$

For sake of clarity, we assume that Ω is a bounded polyhedral subset of $\mathbb{R}^3$ and that the Lamé coefficients λ and μ are constant.

2 Finite volume discretization

A mesh of Ω is defined by the three sets $\{\mathfrak{M}, \mathfrak{M}^*, \mathfrak{D}\}$, corresponding to the primal, dual and diamond mesh. They form a non overlapping partition of Ω, so that

$$\overline{\Omega} = \bigcup_{D \in \mathfrak{D}} D = \bigcup_{K \in \mathfrak{M}} K = \frac{1}{2} \bigcup_{K^* \in \mathfrak{M}^*} K^*.$$

The set $\mathfrak{M}$ is a conforming triangulation of tetraedra. Each element K in $\mathfrak{M}$ is supplied with a center $\mathbf{x}_K$, in practice the barycenter of K and $\partial \mathfrak{M}$ denotes the set of faces on the boundary of the domain. The elements of $\mathfrak{M}^*$ are polygons K^* corresponding to the primal mesh vertices $\mathbf{x}_{K^*}$. These polygons are the union of all tetrahedra spanned, for each faces $s = K \cap L$ or $s = K \cap \partial \Omega$ having $\mathbf{x}_{K^*}$ as vertex, by $\mathbf{x}_{K^*}$ himself, $\mathbf{x}_K$ or $\mathbf{x}_L$ if it exists, $\mathbf{x}_s$ the center of the face s, and one of the other vertices of the face s. In order to take into account the boundary conditions, the dual mesh is splitted into the internal volumes and the boundary ones corresponding to vertices on the boundary: $\mathfrak{M}^* = \mathfrak{M}^{*i} \cup \mathfrak{M}^{*b}$. On its side, diamond cell D in $\mathfrak{D}$ associated to the internal face $s = K \cap L$ is the union of the two tetrahedra $D_{K,s}$ and $D_{L,s}$ spanned by the face s and respectively by the centers $\mathbf{x}_K$ and $\mathbf{x}_L$ (see Fig. 1a). For the boundary face $s = K \cap \partial \Omega$, the corresponding diamond cell is reduced to

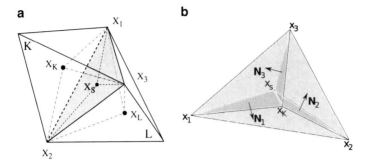

Fig. 1 (a) Primal and diamond cell - (b) Normal orientations in the diamond cell

the tetrahedron $D_{K,s}$. The number of primal and dual cells is denoted by τ and the number of diamond cell by δ.

2.1 Discrete operators

The idea of the DDFV discretization is to construct gradient and divergence operators that are under discrete duality relation by a formula that mimics the Green fomula for continuous functions (see for instance [5] for a detailed construction). A discrete unknown $\mathbf{u}_K$ (resp. $\mathbf{u}_{K^*}$) is associated to each volume K (resp. K^*) of the primal mesh (resp. dual mesh). They are gathered and denoted by

$$\mathbf{u}^\tau = (\mathbf{u}_K, \mathbf{u}_{K^*})_{K \in \mathfrak{M}, K^* \in \mathfrak{M}^*}.$$

For a vector field $\mathbf{u}^\tau$ in $(\mathbb{R}^d)^\tau$, we define on each diamond cell a consistent discrete gradient operator $\nabla^{\mathcal{D}} \mathbf{u}^\tau = (\nabla^D \mathbf{u}^\tau)_{D \in \mathcal{D}}$ in $(\mathcal{M}_d(\mathbb{R}))^\delta$ and a consistent discrete divergence operator $\text{div}^{\mathcal{D}} \mathbf{u}^\tau = (\text{div}^D \mathbf{u}^\tau)_{D \in \mathcal{D}}$ in $\mathbb{R}^\delta$ such that on the internal face $s = K \cap L$ and for the associated diamond cell $D = D_{K,s} \cup D_{L,s}$, the gradient is given by $\nabla^D \mathbf{u}^\tau = \frac{|D_{K,s}|}{|D|} \nabla^{D_{K,s}} \mathbf{u}^\tau + \frac{|D_{L,s}|}{|D|} \nabla^{D_{L,s}} \mathbf{u}^\tau$ and the divergence by $\text{div}^D \mathbf{u}^\tau = \frac{|D_{K,s}|}{|D|} \text{div}^{D_{K,s}} \mathbf{u}^\tau + \frac{|D_{L,s}|}{|D|} \text{div}^{D_{L,s}} \mathbf{u}^\tau$ where for K, we take

$$\nabla^{D_{K,s}} \mathbf{u}^\tau = \frac{1}{3 \, |D_{K,s}|} (\mathbf{u}_s - \mathbf{u}_K) \otimes \mathbf{N}_{Ks} + \frac{1}{3 \, |D_{K,s}|} \sum_{i=1}^{d} \mathbf{u}_i \otimes (\mathbf{N}_{i-1} - \mathbf{N}_{i+1}) \quad (3)$$

$$\text{div}^{D_{K,s}} \mathbf{u}^\tau = \frac{1}{3 \, |D_{K,s}|} (\mathbf{u}_s - \mathbf{u}_K) \cdot \mathbf{N}_{Ks} + \frac{1}{3 \, |D_{K,s}|} \sum_{i=1}^{d} \mathbf{u}_i \cdot (\mathbf{N}_{i-1} - \mathbf{N}_{i+1}). \quad (4)$$

Here $|\cdot|$ denotes the measure and $(\mathbf{x}_i)_{i=1}^d$, respectively $(\mathbf{u}_i)_{i=1}^d$, the vertices of the face s, respectively the corresponding unknowns, with the local numbering convention $\mathbf{x}_0 = \mathbf{x}_d$. The outward normals are defined by (see Fig. 1b)

$$\mathbf{N}_{Ks} = \sum_{i=1}^{d} \mathbf{N}_{s,i-1,i} \quad \text{with} \quad \mathbf{N}_{s,i-1,i} = \frac{1}{2}(\mathbf{x}_i - \mathbf{x}_s) \wedge (\mathbf{x}_{i-1} - \mathbf{x}_s)$$
$$\mathbf{N}_i = \frac{1}{2}(\mathbf{x}_K - \mathbf{x}_s) \wedge (\mathbf{x}_i - \mathbf{x}_s) \tag{5}$$

and $\mathbf{u}_s$ is chosen in order to satisfy the continuity of fluxes (see 7). Otherwise, on a boundary face $s \in \partial\mathfrak{M}$ and for the corresponding diamond cell $D = D_{K,s}$, the gradient and the divergence are simply $\nabla^D \mathbf{u}^\tau = \nabla^{D_{K,s}} \mathbf{u}^\tau$ and $\text{div}^D \mathbf{u}^\tau = \text{div}^{D_{K,s}} \mathbf{u}^\tau$ but $\mathbf{u}_s$ depending on the boundary datas is explicited in (8).

2.2 The DDFV scheme

For $\mathbf{u}^\tau$ in $(\mathbb{R}^d)^\tau$, we are now able to define the discrete strain tensor $\mathbb{D}^\mathfrak{D} \mathbf{u}^\tau = (\mathbb{D}^D \mathbf{u}^\tau)_{D \in \mathfrak{D}}$ and the divergence one $\text{Div}^\mathfrak{D} \mathbf{u}^\tau = (\text{Div}^D \mathbf{u}^\tau)_{D \in \mathfrak{D}}$ by

$$\mathbb{D}^D \mathbf{u}^\tau = \frac{\nabla^D \mathbf{u}^\tau + (\nabla^D \mathbf{u}^\tau)^T}{2}, \quad \text{Div}^D \mathbf{u}^\tau = \text{div}^D \mathbf{u}^\tau \, \text{Id} \quad \forall D \in \mathfrak{D}. \tag{6}$$

After extending this definition to each tetrahedron that composes the diamond cell, we can specify that the displacement $\mathbf{u}_s$ at an internal face $s = K \cap L$ has to satisfy the continuity of the fluxes

$$(\lambda \text{Div}^{D_{K,s}} \mathbf{u}^\tau + 2\mu \mathbb{D}^{D_{K,s}} \mathbf{u}^\tau)\mathbf{N}_{Ks} = -(\lambda \text{Div}^{D_{L,s}} \mathbf{u}^\tau + 2\mu \mathbb{D}^{D_{L,s}} \mathbf{u}^\tau)\mathbf{N}_{Ls}. \tag{7}$$

Now for a tensor field $\xi^\mathfrak{D}$ in $(\mathcal{M}_d(\mathbb{R}))^\delta$, we define a consistent approximation of the discrete divergence operator equal to

$$(\text{div}^\mathfrak{M} \xi^\mathfrak{D}, \text{div}^{\mathfrak{M}^*} \xi^\mathfrak{D}) = \left((\text{div}^K \xi^\mathfrak{D})_{K \in \mathfrak{M}}, (\text{div}^{K^*} \xi^\mathfrak{D})_{K^* \in \mathfrak{M}^*}\right)$$

with

$$\text{div}^K \xi^\mathfrak{D} = \frac{1}{|K|} \sum_{s \in \partial K} \xi^D \mathbf{N}_{Ks} \quad \text{and} \quad \text{div}^{K^*} \xi^\mathfrak{D} = \frac{1}{|K^*|} \sum_{s \ni \mathbf{x}_{K^*}} \xi^D \mathbf{N}_{K^* s}$$

and where D is the diamond cell associated to the face s. $\mathbf{N}_{K^* s}$ is the normal to ∂K^* pointing outward K^* and it can be explicited using local numbering and applying formula (5):

$$\mathbf{N}_{K^* s} = \begin{cases} \mathbf{N}_{i+1}^K - \mathbf{N}_{i-1}^K + \mathbf{N}_{i+1}^L - \mathbf{N}_{i-1}^L & \text{for an internal face } s = K \cap L \\ \mathbf{N}_{i+1} - \mathbf{N}_{i-1} + \mathbf{N}_{s,i-1,i} + \mathbf{N}_{s,i,i+1} & \text{for a boundary face } s = K \cap \partial \Omega \end{cases}$$

where we assume that $x_{K^*} = x_i^K$ (resp. $x_{K^*} = x_i^L$) in the volume K (resp. L).

Let us now denote $\mathbf{f}^{\mathfrak{M}} = (\mathbf{f}^K)_{K \in \mathfrak{M}}$ and $\mathbf{f}^{\mathfrak{M}^{*i}} = (\mathbf{f}^{K^*})_{K^* \in \mathfrak{M}^{*i}}$, where $\mathbf{f}^K$ and $\mathbf{f}^{K^*}$ are the average of the external force $\mathbf{f}$ on primal and dual cells. Then the DDFV scheme, written here, for sake of simplicity, only for displacement boundary conditions, consists in finding $\mathbf{u}^\tau \in (\mathbb{R}^d)^\tau$ such that

$$\begin{cases} -\mathbf{div}^{\mathfrak{M}}(\lambda \mathbb{D}\mathrm{iv}^{\mathcal{D}}\mathbf{u}^\tau + 2\mu \mathbb{D}^{\mathcal{D}}\mathbf{u}^\tau) = \mathbf{f}^{\mathfrak{M}} \\ -\mathbf{div}^{\mathfrak{M}^{*i}}(\lambda \mathbb{D}\mathrm{iv}^{\mathcal{D}}\mathbf{u}^\tau + 2\mu \mathbb{D}^{\mathcal{D}}\mathbf{u}^\tau) = \mathbf{f}^{\mathfrak{M}^{*i}} \\ \mathbf{u}_s = \mathbf{g}(\mathbf{x}_s), \quad \forall s \in \partial \mathfrak{M} \\ \mathbf{u}_{K^*} = \mathbf{g}(\mathbf{x}_{K^*}), \quad \forall K^* \in \mathfrak{M}^{*b}. \end{cases} \quad (8)$$

2.3 Existence, stability and convergence results

Applying discrete Green formula, Korn and Poincaré inequalities, divergence equality and approximation results on the center value projection operator (see [14]), we prove that the numerical scheme is well-posed, stable and convergent:

Theorem 1. *Under the assumption that* $\mathrm{mes}(\Gamma_D) \neq 0$, *the DDFV scheme for linear elasticity* (8) *yields to a symmetric positive definite system of linear equations. So it admits exactly one solution* $\mathbf{u}^\tau \in (\mathbb{R}^d)^\tau$

Theorem 2. *Let* $\mathbf{u}^\tau \in (\mathbb{R}^d)^\tau$ *be the solution of the discrete problem* (8). *Then there exists a constant C depending only on the regularity of the mesh such that*

$$\mu \| \nabla^{\mathcal{D}} \mathbf{u}^\tau \|_2^2 + \frac{\lambda}{3} \| \mathbb{D}iv^{\mathcal{D}} \mathbf{u}^\tau \|_2^2 \leq C \| \mathbf{f}^\tau \|_2^2 \quad (9)$$

Theorem 3. *Assuming that the exact solution of the continuous problem* (1) *is regular enough then there exists a constant C depending only on the regularity of the mesh, such that*

$$\| \mathbf{u} - \mathbf{u}^\tau \|_2 + \| \nabla \mathbf{u} - \nabla^{\mathcal{D}} \mathbf{u}^\tau \|_2 \leq C \, \mathrm{size}(\mathfrak{M}) \quad (10)$$

3 Numerical experiments

The DDFV method has been implemented in two and three dimensions. Free and imposed traction conditions (described in [16]) are also taken into account. Both homogeneous and non homogeneous test cases are considered. Comparisons are made with the analytical solution or with the clasical finite element one.

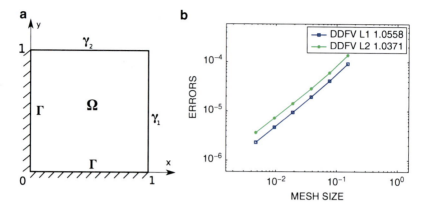

Fig. 2 (a) Geometry and test setup - (b) L^1 and L^2 nors of the error between the analytical and the numerical displacement.

3.1 Two dimensional examples

Following a study of [9], we apply the code to a simple test case with analytical solution in order to study the convergence properties. The geometry of the homogeneous square plate and the specified boundary conditions are shown in Fig. 2(a). Lamé coefficients $(\lambda, \mu) = (2.9\,10^9, 1.9\,10^9)$ correspond to Young modulus and Poisson ratio $(E, \nu) = (5\,10^9, 0.3)$. The displacement $\mathbf{g}$ is null on Γ boundary and the traction is imposed on γ_1 and γ_2 boundaries:

$$\mathbf{g}_{|\gamma_1} = \begin{pmatrix} ((2\mu + \lambda)y - 2\lambda)10^{-2} \\ \mu(1 - 2y)10^{-2} \end{pmatrix} \qquad \mathbf{g}_{|\gamma_2} = \begin{pmatrix} \mu(x - 2)10^{-2} \\ (-2(2\mu + \lambda)x + \lambda)10^{-2} \end{pmatrix}.$$

The external force is equal to $\mathbf{f} = (\mu + \lambda)10^{-2}(2, -1)$ and the corresponding exact displacement is $\mathbf{u} = xy10^{-2}(1, -2)$. The comparison between the analytical and numerical displacement obtained for various primal meshes are plotted in Fig. 2(b) with an order of convergence of one.

The second example concerns a domain with non homogeneous material properties. The plate (without deformation) is composed of the part $[0, 3] \times [0, 1]$ with a hole inside and $(\lambda, \mu) = (5.6, 2.6)$ and the part $[3, 4] \times [0, 1]$ with $(\lambda, \mu) = (10, 8)$. Null displacement is imposed on the left side of the domain, a load of 1 (resp. a displacement of 1) is imposed on the right side for Fig. 3 (resp. for Fig. 4). There is a free traction elsewhere. The deformed domain obtained with the present scheme (above) and with the conventional finite element method (below) are plotted. In both case, solution are similar, the largest differences are observed in the load one.

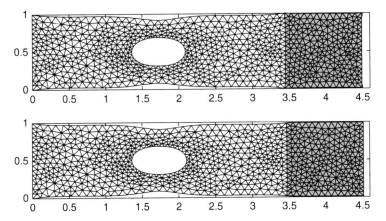

Fig. 3 Deformed domain for the non homogeneous case with an imposed load on the right. DDFV above and FE below

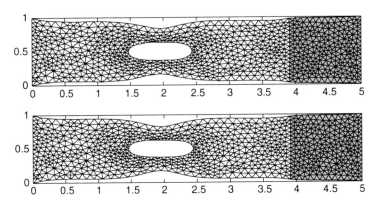

Fig. 4 Deformed domain for the non homogeneous case with an imposed displacement on the right. DDFV above and FE below

3.2 Three dimensional test

The domain is the unit cube with an embedding condition on the bottom ($z = 0$), imposed displacement $(0, 0, -0.5)$ on the top ($z = 1$) simulating a compression of the domain (see Fig. 5b) and free traction conditions on the vertical sides of the cube. For Lamé coefficients $(\lambda, \mu) = (28.8, 19.2)$, the solution is compared with the P1 finite element one on a series of meshes: Figure 5a shows the behavior of the error in L^2 and L^1 norms and reveals that the DDFV solution of the linear elasticity problem converges as we expect.

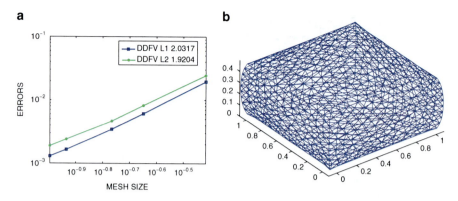

Fig. 5 (a) Differences with the finite element solution - (b) DDFV deformed domain

References

1. Andreianov, B., Bendahmane, M., Karlsen, K., Pierre, C.: Convergence of discrete duality finite volume schemes for the cardiac bidomain model. Arxiv preprint arXiv:1010.2718 (2010)
2. Andreianov, B., Boyer, F., Hubert, F.: Discrete duality finite volume schemes for Leray-Lions-type elliptic problems on general 2D meshes. Numer. Methods Partial Differential Equations **23**(1), 145–195 (2007)
3. Bailey, C., Cross, M.: A finite volume procedure to solve elastic solid mechanics problems in three dimensions on an unstructured mesh. Int. J. Numer. Methods Eng. **38**(10), 1757–1776 (1995)
4. Coudière, Y., Manzini, G.: The discrete duality finite volume method for convection-diffusion problems. SIAM J. Numer. Anal. **47**(6), 4163–4192 (2010)
5. Coudière, Y., Pierre, C., Rousseau, O., Turpault, R.: A 2D/3D discrete duality finite volume scheme. Application to ECG simulation. Int. J. Finite Vol. **6**(1), 24 (2009)
6. Demirdžić, I., Muzaferija, S.: Finite volume method for stress analysis in complex domains. Int. J. Numer. Methods Eng. **37**(21), 3751–3766 (1994)
7. Demirdžić, I., Muzaferija, S., Perić, M.: Benchmark solutions of some structural analysis problems using finite-volume method and multigrid acceleration. Int. J. Numer. Methods Eng. **40**(10), 1893–1908 (1997)
8. Domelevo, K., Omnes, P.: A finite volume method for the laplace equation on almost arbitrary two-dimensional grids. M2AN Math. Model. Numer. Anal. **39**, 1203–1249 (2005)
9. Figueiredo, J., Viano, J.: Finite elements q1-lagrange for the linear elasticity problem. Tech. rep., Universidade de Santiago de Compostela (2005)
10. Herbin, R., Hubert, F.: Benchmark on discretization schemes for anisotropic diffusion problems on general grids. In: Finite volumes for complex applications V, pp. 659–692. ISTE, London (2008)
11. Hermeline, F.: A finite volume method for the approximation of diffusion operators on distorted meshes. J. Comput. Phys. **160**(2), 481–499 (2000)
12. Jasak, H., Weller, H.: Application of the finite volume method and unstructured meshes to linear elasticity. Int. J. Numer. Methods Eng. **48**(2), 267–287 (2000)
13. Krell, S.: Stabilized DDFV schemes for stokes problem with variable viscosity on general 2d schemes. Numer. Methods Partial Differential Equations (2010)
14. Krell, S., Manzini, G.: The discrete duality finite volume method for the stokes equations on 3-d polyhedral meshes (2010). URL http://hal.archives-ouvertes.fr/hal-00448465/en/

15. Maitre, J.F., Rezgui, A., Souhail, H., Zine, A.M.: High order finite volume schemes. Application to non-linear elasticity problems. In: Finite volumes for complex applications, III (Porquerolles, 2002), pp. 391–398. Hermes Sci. Publ., Paris (2002)
16. Martin, B.: Résolution du problème de l'élasticité linéaire en volumes finis. Ph.D. thesis, ENS de Cachan (2011)
17. Souhail, H.: Schéma volumes finis : Estimation d'erreur a posteriori hiérarchique par éléments finis mixtes. Résolution de problèmes d'élasticité non-linéaire. Ph.D. thesis, Ecole Centrale de Lyon (2004). URL http://tel.archives-ouvertes.fr/tel-00005418/en/
18. Wenke, P., Wheel, M.: A finite volume method for solid mechanics incorporating rotational degrees of freedom. Computers & Structures **81**(5), 321–329 (2003)

The paper is in final form and no similar paper has been or is being submitted elsewhere.

Model Adaptation for Hyperbolic Systems with Relaxation

Hélène Mathis and Nicolas Seguin

Abstract We address the numerical coupling of two hyperbolic systems, a relaxation model and the associated equilibrium model, separated by spatial interfaces that automatically evolve in time, the whole being approximated by finite volume schemes. The criterion to choose where each model has to be used results of the Chapman–Enskog expansion of the relaxed model, both on a continuous and a discrete view point. Numerical tests illustrate the good behavior of the algorithm.

Keywords Hyperbolic system, adaptation, relaxation
MSC2010: 65M08, 65M15, 35L45, 35L65

1 Introduction

In the framework of the modeling of problems coming from complex phenomena, it is common to handle different scales of modeling depending on the accuracy we need. It leads to the use of a hierarchy of models based on the different scales brought into play both in the spatial and time sides. The problem of spatial coupling of different hyperbolic models has been the topic of numerous papers, see for instance [1]. So far the theoretical and numerical techniques developed lied on the hypothesis that the spatial domains where each model has to be applied is initially prescribed and fixed in time. We aim at developing analytical and numerical tools to determine automatically the space–time domains in which each model has to be used, taking into account the local accuracy and the characteristics of the flow.

Hélène Mathis and Nicolas Seguin
UPMC Univ Paris 06, UMR 7598, LJLL, F-75005, Paris, France; CNRS, UMR 7598, LJLL, F-75005, Paris, France. e-mail: mathis@ann.jussieu.fr, nicolas.seguin@upmc.fr

The whole procedure must allow:

- to increase the accuracy in the domain where the phenomenon scales are small, by means of using a fine model,
- to improve execution time, by the use of a coarse model elsewhere.

It consists thus in constructing local error estimates of modeling and developing adapted numerical schemes. In the sequel we concentrate on an academic problem, involving: a fine model by means of a *hyperbolic system with relaxation* (the information being contained in the relaxation source term) and a coarse model corresponding to the *associated equilibrium* model. To automatically handle the dynamical decomposition of the computational domain into two sub-domains, we have to provide a criterion. It consists in using an intermediate model from which an error estimate is deduced. Since we are dealing with hyperbolic models with relaxation, we propose to consider the first order corrector resulting from the *Chapman–Enskog expansion* of the relaxation model around an equilibrium state. At the interfaces between fine and coarse models, the coupling strategies we use are based on techniques for thin coupling interfaces developed in [1] (see also references therein).

Section 2 is devoted to the structural study of the hyperbolic relaxation system and its associated equilibrium system of conservation laws. The Chapman–Enskog expansion is recalled for such systems. Section 3 addresses the finite volume schemes we use, the derivation of the discrete estimator and the global algorithm for adaptation. Numerical tests are provided in Section 4 and Section 5 is the conclusion.

Let us emphasize that this work is still under development. Most of our attention is paid in this note to the relevance of the developed error estimator but since our framework is still academic (rather simple 1D models), the CPU time saving is not significant (see Sections 4 and 5). More results will be provided at the conference.

2 Hyperbolic systems with relaxation

In order to simplify the presentation, we consider systems of hyperbolic equations which comply with the form

$$\partial_t U + \partial_x f(U, v) = 0, \tag{1}$$

$$\partial_t v + \partial_x g(U, v) = \frac{1}{\varepsilon}(h(U) - v), \tag{2}$$

where the fluxes $f: \mathbb{R}^n \times \mathbb{R} \to \mathbb{R}^n$, $g: \mathbb{R}^n \times \mathbb{R} \to \mathbb{R}$ and $h: \mathbb{R}^n \to \mathbb{R}$ are smooth functions, defining the evolution of the state vector $U: \mathbb{R} \times \mathbb{R}^+ \to \mathbb{R}^n$ and the variable $v: \mathbb{R} \times \mathbb{R}^+ \to \mathbb{R}$. In the following, we will also use the condensed notations $W = (U, v)^T$, $\mathscr{F} = (f, g)^T$ and $R = (0, h(U) - v)^T$. Note that we only focus on scalar-valued functions v in order to make the notations clearer in the following.

When the relaxation parameter ε tends to 0, the relaxation model (1)–(2) reduces to the following system of n conservation laws:

$$\partial_t U + \partial_x f_e(U) = 0, \quad \text{where } f_e(U) = (U, h(U)). \tag{3}$$

The relaxation process thus determines a local equilibrium state $W_e(U) = (U, h(U))$. Several stability conditions exist to justify the asymptotics $\varepsilon \to 0$ [2,4]. Let $(\lambda_{e,k})_{1 \leq k \leq n}$ be the (ordered) eigenvalues of $\nabla_U f_e(U)$ (i.e. of the equilibrium system (3)) and $(\lambda_k)_{1 \leq k \leq n+1}$ be the (ordered) eigenvalues of $\nabla_W \mathscr{F}(W_e)$ (i.e. of the relaxation system (1)–(2) restricted to equilibrium states). Here, we assume that the so-called subcharacteristic condition is satisfied:

$$\lambda_k \leq \lambda_{e,k} \leq \lambda_{k+1}, \quad \forall 1 \leq k \leq n. \tag{4}$$

For more details on the different stability conditions for hyperbolic systems with relaxation, see [2,4].

2.1 Chapman–Enskog expansion

With the aim of coupling the relaxed model and the equilibrium one in an adaptive way, we want to determine a criterion to move from one model to the other. A natural choice is to consider the first order error resulting from the Chapman–Enskog expansion of the relaxed model around an equilibrium state. The Chapman–Enskog method amounts to considering *smooth* solutions of (1)–(2) near equilibrium which we assume to satisfy

$$v = h(U) + \varepsilon v_1 + \mathscr{O}(\varepsilon^2). \tag{5}$$

Plugging (5) into (1)–(2) leads to

$$\partial_t U + \partial_x (f(U, h(U))) + \varepsilon \partial_x (\nabla_2 f(U, h(U)) v_1) = \mathscr{O}(\varepsilon^2), \tag{6}$$

$$\partial_t h(U) + \partial_x g(U, h(U)) = -v_1 + \mathscr{O}(\varepsilon), \tag{7}$$

where $\nabla_\alpha q$ denotes the derivative of the vector field q w.r.t. its α-th variable, $\alpha = 1, 2$. Multiplying (6) by $\nabla h(U)^T$ and combining with (7) leads to

$$v_1 = \nabla h(U)^T \partial_x (f(U, h(U))) - \partial_x g(U, h(U)) + \mathscr{O}(\varepsilon). \tag{8}$$

Finally, dropping second order terms with respect to ε yields:

$$\partial_t U + \partial_x f(U, h(U)) = -\varepsilon \partial_x \big[\nabla_2 f(U, h(U)) \\ \big(\nabla h(U)^T \partial_x f(U, h(U)) - \partial_x g(U, h(U))\big)\big]. \tag{9}$$

This parabolic system can be interpreted as an intermediate model between the relaxation model (1)–(2) and the equilibrium model (3). Indeed, smooth solutions of (1)–(2) solve (9) up to $\mathcal{O}(\epsilon^2)$ and if the right-hand side in (9) vanishes, one recover (3).

Remark 1. Note that the equivalence between the subcharacteristic condition (4) and the dissipativity of the second order term in (9) only holds in rather classical cases. We refer once again to [2,4] for more details.

2.2 Finite volume methods

We depict now the finite volume schemes used to approximate the equilibrium and the relaxed models. We first consider the equilibrium model (3). We introduce the equilibrium state vector $W_{e,i}^n = (U_i^n, h(U_i^n))$ within each cell $C_i = [x_{i-1/2}, x_{i+1/2}]$. The classical finite volume formulation reads

$$\frac{1}{\Delta x}(U_i^{n+1} - U_i^n) + \frac{1}{\Delta t}(\varphi(U_i^n, U_{i+1}^n) - \varphi(U_{i-1}^n, U_i^n)) = 0. \tag{10}$$

The 2-point numerical flux $\varphi : \mathbb{R}^n \times \mathbb{R}^n \to \mathbb{R}^n$ is consistent with the flux f_e in the sense of finite volume methods, i.e. $\varphi(U, U) = f_e(U)$. We now address the numerical scheme to approximate the relaxation system (1)–(2), for which a splitting strategy between the convective part and the source term has been adopted. We introduce the two numerical fluxes F and G respectively consistent with the fluxes f and g. In a first step the convective part is approximated by

$$\frac{1}{\Delta x}(U_i^{n+1,-} - U_i^n) + \frac{1}{\Delta t}(F(W_i^n, W_{i+1}^n) - F(W_{i-1}^n, W_i^n)) = 0. \tag{11}$$

$$\frac{1}{\Delta x}(v_i^{n+1,-} - v_i^n) + \frac{1}{\Delta t}(G(W_i^n, W_{i+1}^n) - G(W_{i-1}^n, W_i^n)) = 0. \tag{12}$$

Then the value $U_i^{n+1,-}$ is taken as the initial data for solving the source term:

$$U_i^{n+1} = U_i^{n+1,-}, \tag{13}$$

$$v_i^{n+1} = v_i^{n+1,-} + \frac{\Delta t}{\varepsilon}(h(U_i^{n+1}) - v_i^{n+1}). \tag{14}$$

Here, the classical implicit Euler scheme has been chosen in order to ensure the unconditional stability of the second step.

Note that it is natural, at least from the academic viewpoint, to impose the compatibility condition between the numerical fluxes $\varphi(U_l, U_r) = F(W_e(U_l), W_e(U_r))$ and it has been done for the numerical results of Section 4.

3 Model adaptation

This section is devoted to the description of the adaptive coupling procedure from the numerical point of view. First we detail the dynamical cutting of the space–time computational domain. Following Section 2.1 the criterion we choose to realize the cutting corresponds to the first order error coming from the discrete Chapman–Enskog expansion. The global algorithm is described in 3.3.

3.1 Basic principles

We want to provide a criterion which enables to automatically determine the space domains $\mathscr{D}^R(t)$ and $\mathscr{D}^E(t)$ where the fine (that is the relaxation system (1–2)) and the coarse (that is the equilibrium system (3)) models have to be used respectively. These two domains evolve as time t increases, without overlapping in such way that their intersection corresponds to the interfaces where the coupling is performed.

We propose to make the cutting of the space into the sub-domains $\mathscr{D}^E(t)$ and $\mathscr{D}^R(t)$ depend on the first order error ϵv_1 which results from the Chapman–Enskog expansion (9). Let θ being a threshold arbitrarily chosen. We then have:

- The region where $|\epsilon v_1| \leq \theta$ is chosen to be $\mathscr{D}^E(t)$. In that domain the error between the equilibrium model and the relaxed one is assumed to be negligible, so that the coarse model (3) is applied.
- The domain $\mathscr{D}^R(t)$ corresponds to the region where $|\varepsilon v_1| \geq \theta$ and the relaxation model (1)–(2) is solved inside.
- At the interfaces separating the sub-domains $\mathscr{D}^E(t)$ and $\mathscr{D}^R(t)$, a numerical coupling method as those developed in [1, 3] is used.

Let us note that several strategies can be applied; in the sequel we give a preference to the state coupling, that is only the value v is transmitted through the interface at each time step. Since the interfaces of coupling are always located in a region where the two models are very close to each other, the different methods of coupling should provide very similar results (see [1, 3]).

It is important to note that thanks to (8), the estimator ϵv_1 can be computed from the solution of each model, (1)–(2) and (3), since it only depends on U.

3.2 Estimators for adaptation

We now give the discrete estimator we use to perform the adaptive coupling, following the strategy of the Chapman–Enskog method. First, we take the ansatz

$$v_i = h(U_i) + \varepsilon v_{1,i} + \mathscr{O}(\varepsilon^2)$$

and denote $W_{e,i}^n = W_e(U_i^n)$. Plugging the ansatz in (11)–(13) and (12)–(14) and dropping high order terms yields

$$U_i^{n+1} - U_i^n + \frac{\Delta t}{\Delta x}(F(W_{e,i}^n, W_{e,i+1}^n) - F(W_{e,i-1}^n, W_{e,i}^n))$$
$$+ \frac{\varepsilon \Delta t}{\Delta x}\bigl[\nabla_1 F(W_{e,i}^n, W_{e,i+1}^n) \cdot (0, v_{1,i}^n) + \nabla_2 F(W_{e,i}^n, W_{e,i+1}^n) \cdot (0, v_{1,i+1}^n)$$
$$- \nabla_1 F(W_{e,i-1}^n, W_{e,i}^n) \cdot (0, v_{1,i-1}^n) + \nabla_2 F(W_{e,i}^n, W_{e,i+1}^n) \cdot (0, v_{1,i}^n)\bigr] = 0. \quad (15)$$

$$h(U_i^{n+1}) - h(U_i^n) + \frac{\Delta t}{\Delta x}(G(W_{e,i}^n, W_{e,i+1}^n) - G(W_{e,i-1}^n, W_{e,i}^n)) = -\Delta t\, v_{1,i}^{n+1}, \quad (16)$$

Since h is smooth, there exists $\overline{U}(.,.)$ such that $\nabla h(\overline{U}(U^n, U^{n+1}))^T (U^{n+1} - U^n) = h(U^{n+1}) - h(U^n)$. Multiplying (16) by $\nabla h(\overline{U}(U_i^n, U_i^{n+1}))^T$ leads to

$$h(U_i^{n+1}) - h(U_i^n) + \frac{\Delta t}{\Delta x}\nabla h(\overline{U})^T (F(W_{e,i}^n, W_{e,i+1}^n) - F(W_{e,i-1}^n, W_{e,i}^n)) = 0.$$

Combining with (16) provides the following expression of $v_{1,i}^{n+1}$:

$$v_{1,i}^{n+1} = \nabla h(\overline{U}) \frac{1}{\Delta x}(F(W_{e,i}^n, W_{e,i+1}^n) - F(W_{e,i-1}^n, W_{e,i}^n))$$
$$- \frac{1}{\Delta x}(G(W_{e,i}^n, W_{e,i+1}^n) - G(W_{e,i-1}^n, W_{e,i}^n)) \quad (17)$$

which is the discretisation of (8). Replacing the terms $v_{1,i}^n$ and $v_{1,i-1}^n$ into (15) allows us to determine the discrete counterpart of (9). Note that in practice, the term $\nabla h(\overline{U}(U_i^n, U_i^{n+1}))$ can be approximated by $\nabla h(U_i^n)$ so that the estimator $\varepsilon v_{1,i}^{n+1}$, at time t^{n+1}, is an explicit function of the discrete solution $(U_i^n)_{i \in \mathbb{Z}}$, at time t^n.

3.3 The general algorithm

We now detail the general algorithm of the dynamical coupling between the fine and the coarse models. Let $(W_i^n)_{i \in \mathbb{Z}}$ be a sequence known at time t^n to be advanced to time t^{n+1}. The algorithm follows the steps:

- For all $i \in \mathbb{Z}$, compute in cell C_i the numerical error $e_i^{n+1} := \varepsilon v_{1,i}^{n+1}$ using (17)
- For all $i \in \mathbb{Z}$, if $[|e_i^{n+1}| > \theta]$ then
  $C_i \in \mathscr{D}^R(t^n)$

Else
$C_i \in \mathscr{D}^E(t^n)$
- At this stage, $\mathscr{D}^R(t^n) \cup \mathscr{D}^E(t^n) = \mathbb{R}$. For all $i \in \mathbb{Z}$:
  - If $[C_{i_1}, C_i, C_{i+1} \in \mathscr{D}^R(t^n)]$ (resp. $\in \mathscr{D}^E(t^n)$) then

    Compute W_i^{n+1} using the numerical scheme (11–14) (resp. compute $W_i^{n+1} = W_e(U_i^{n+1})$ using the numerical scheme (10)
  - Else

    Compute W_i^{n+1} using the state coupling method described in [1, 3]

Besides let us note that the estimator we use is not exactly an *a posteriori* error estimate since the adaptation process add a numerical error. The study of this error estimate is an ongoing work.

4 Numerical experiments

We now present some numerical results in order to illustrate the reliability of the coupling procedure, using the Rusanov scheme for the approximation of each model. The problem we address corresponds to a fluid flow problem governed by the relaxation system

$$\partial_t \tau - \partial_x u = 0, \tag{18}$$

$$\partial_t u + \partial_x \Pi = 0, \tag{19}$$

$$\partial_t \mathscr{T} = \frac{1}{\varepsilon}(\tau - \mathscr{T}) \tag{20}$$

which is derived from the works of Suliciu [7] but also corresponds to Chaplygin gas (see for instance [6]). The state variable τ and u stand for the specific volume and the velocity while $\mathscr{T}$ is a perturbed specific volume. The extended pressure law Π is defined by $\Pi(\tau, \mathscr{T}) = p(\mathscr{T}) + a^2(\mathscr{T} - \tau)$ where p follows a perfect gas law $p = p(\tau) = p^{-\gamma}$, $\gamma > 1$. The associated equilibrium system is obtained setting $\mathscr{T} = \tau$ and is the so-called p-system

$$\partial_t \tau - \partial_x u = 0, \tag{21}$$

$$\partial_t u + \partial_x p = 0. \tag{22}$$

The constant a is assumed to satisfy the Whitham's condition $a^2 > \max_s(-p'(s))$, which ensures that the subcharacteristic condition (4) is satisfied. Plugging the expansion $\mathscr{T} = \tau + \varepsilon \mathscr{T}_1 + \mathscr{O}(\varepsilon^2)$ into (19) yields the parabolic equation, which corresponds to (9),

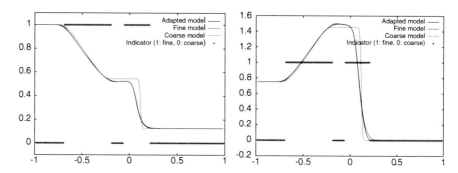

Fig. 1 Density $1/\tau$ (left) and velocity (right) with 200 cells at time $T = 0.5$. The indicator corresponds to the characteristic function of $\mathscr{D}^R(T)$

$$\partial_t u + \partial_x p(\tau) = \varepsilon \partial_x \big(\partial_x u(p'(\tau) + a^2)\big). \tag{23}$$

Figure 1 presents the solution of test case with $\gamma = 1.4$ and $a = 1.5$. The initial data are $\tau_L = 1$, $u_L = 0.75$, $\tau_R = 8$, $u_R = 0$ and the discontinuity is applied at $x = 0$. The relaxation parameter is $\varepsilon = 10^{-6}$. The mesh contains 200 cells and threshold for the adaptation is $\theta = 0.5$. One may check that our method of adaptation only uses the fine model in the regions of large variations of the solution. The results are very close to those with the fine model. One may also note that the results of coarse model are sensibly different: less diffusion and a different intermediate state.

5 Conclusion

In the frame of hyperbolic systems with relaxation, we have proposed a new algorithm for dynamical adaptation of models, based on the Chapman–Enskog expansion. It enables to quantify the difference between a fine model and a coarse model, from the continuous and the discrete points of view. The global algorithm of adaptation is based on a series of works on interface coupling of hyperbolic models and is easy to implement. This is a preliminary work but the first results are encouraging. We are aware that the presented test case is very academic, but it only aims at illustrating the relevance of our estimator. Besides, the fine model (18–20) and its numerical resolution are rather classical and since the space domain is 1D, comparison of CPU times between a computation with the fine model in the whole domain and a computation with our algorithm of adaptation would be meaningless. More complex models (such that nonlinear relaxation terms coming from models of phase transition) and 2D computations will be presented during the conference.

Acknowledgements The first author is supported by the LRC Manon (Modélisation et Approximation Numérique Orientées pour l'énergie Nucléaire — CEA/DM2S-LJLL). The authors would like to thank C. Cancès, F. Coquel and E. Godlewski for their useful comments. The authors also thank the referees who gave them the opportunity to deeply improve the quality of this paper.

References

1. Ambroso, A., Chalons, C., Coquel, F., Godlewski, E., Lagoutière, F., Raviart, P.A., Seguin, N.: The coupling of homogeneous models for two-phase flows. Int. J. Finite Volumes **4**(1), 1–39 (2007)
2. Bouchut, F.: A reduced stability condition for nonlinear relaxation to conservation laws. J. Hyperbolic Differ. Equ. **1**(1), 149–170 (2004)
3. Caetano, F.: Sur certains problèmes de linéarisation et de couplage pour les systèmes hyperboliques non linéaires. Ph.D. thesis, Université Pierre et Marie Curie-Paris6, France (2006)
4. Chen, G.Q., Levermore, C.D., Liu, T.P.: Hyperbolic conservation laws with stiff relaxation terms and entropy. Comm. Pure Appl. Math. **47**(6), 787–830 (1994)
5. Rusanov, V.V.: The calculation of the interaction of non-stationary shock waves with barriers. Ž. Vyčisl. Mat. i Mat. Fiz. **1**, 267–279 (1961)
6. Serre, D.: Multidimensional shock interaction for a Chaplygin gas. Arch. Ration. Mech. Anal. **191**(3), 539–577 (2009)
7. Suliciu, I.: On the thermodynamics of rate-type fluids and phase transitions. I. Rate-type fluids. Internat. J. Engrg. Sci. **36**(9), 921–947 (1998)

The paper is in final form and no similar paper has been or is being submitted elsewhere.

Inflow-Implicit/Outflow-Explicit Scheme for Solving Advection Equations

Karol Mikula and Mario Ohlberger

Abstract We present new method for solving non-stationary advection equations based on the finite volume space discretization and the semi-implicit discretization in time. Its basic idea is that outflow from a cell is treated explicitly while inflow is treated implicitly. Since the matrix of the system in this new I^2OE method is determined by the inflow fluxes it is an M-matrix yielding favourable solvability and stability properties. The method allows large time steps at a fixed spatial grid without losing stability and not deteriorating precision which makes it attractive for practical applications. Our new method is exact for any choice of a discrete time step on uniform rectangular grids in the case of constant velocity transport of quadratic functions in any dimension. We show that it is formally second order accurate in space and time for 1D advection problems with variable velocity and numerical experiments indicates its second order accuracy for smooth solutions in general.

Keywords Advection equation, semi-implicit scheme, finite volume method
MSC2010: 35L04, 65M08, 65M12

1 Introduction

In this paper we present the inflow-implicit/outflow-explicit (I^2OE) method for solving variable velocity advection equations of the form

$$u_t + \mathbf{v} \cdot \nabla u = 0 \qquad (1)$$

Karol Mikula
Department of Mathematics, Faculty of Civil Engineering, Slovak University of Technology, Radlinského 11, 81368 Bratislava, Slovakia, e-mail: mikula@math.sk

Mario Ohlberger
Institut für Numerische und Angewandte Mathematik, Universität Münster, Einsteinstr. 62, D-48149 Münster, Germany, e-mail: mario.ohlberger@uni-muenster.de

where $u \in \mathbb{R}^d \times [0, T]$ is the unknown function and $\mathbf{v}(x)$ is a vector field. The basic idea of our new method is that outflow from a cell is treated explicitly while inflow is treated implicitly. Such an approach is natural, since we know what is flowing out from a cell at an old time step $n - 1$ but we leave the method to resolve a system of equations determined by the inflows to obtain a new value in the cell at time step n. Since the matrix of the system is determined by the inflow fluxes it is an M-matrix for Voronoi like grids and thus it has favourable discrete minimum-maximum properties. Consequently, the method allows large time steps at a fixed spatial grid without losing stability. Interestingly, the new I^{2}OE scheme is exact on rectangular grids for constant velocity transport of quadratic polynomials in any dimension and for any length of a time step. In general, it is second order accurate for smooth solutions, both for variable velocity and nonlinear advection problems [5]. A comparison with the second order Lax-Wendroff method for variable velocity shows good properties of the new scheme with respect to precision and CPU times. In [5], the I^{2}OE method was introduced in more general settings where $\mathbf{v} = \mathbf{v}(x, u, \nabla u)$. The semi-implicit forward-backward diffusion level set approach for motion in normal direction [4] is its special case. The variable and nonlinear velocity fields to which our method can be successfully applied arise in many applications, e.g. in level set methods and other transports with non-divergence free velocities and nonlinear conservation laws or in image segmentation by the active contours.

2 The inflow-implicit/outflow-explicit scheme

Let us consider equation (1) in a bounded polygonal domain $\Omega \subset \mathbb{R}^d$, $d = 2, 3$, and time interval $[0, T]$. Let $\mathcal{Q}_h$ denote a primal polygonal partition of Ω. Let p be a finite volume (cell) of a corresponding dual Voronoi tessellation $\mathcal{T}_h$ with measure m_p and let e_{pq} be an edge between p and q, $q \in N(p)$, where $N(p)$ is a set of neighbouring finite volumes (i.e. $\bar{p} \cap \bar{q}$ has nonzero $(d - 1)$-dimensional measure). Let c_{pq} be the length of e_{pq} and n_{pq} be the unit outer normal vector to e_{pq} with respect to p. We shall consider $\mathcal{T}_h$ to be an admissible mesh in the sense of [1], i.e., there exists a representative point x_p in the interior of every finite volume p such that the joining line between x_p and x_q, $q \in N(p)$, is orthogonal to e_{pq}. We denote by x_{pq} the intersection of this line segment with the edge e_{pq}. The length of this line segment is denoted by d_{pq}, i.e. $d_{pq} := |x_q - x_p|$. As we have build $\mathcal{T}_h$ based on the primal mesh $\mathcal{Q}_h$, we assume that the points x_p coincide with the vertices of $\mathcal{Q}_h$. Let us denote by u_p a (constant) value of the solution in a finite volume p computed by the scheme. For the solution representation inside the finite volume p we use either this value u_p or a reconstructed (but again constant) value denoted by $\bar{u}_p$. A constant value of the solution assigned to the edge e_{pq} (given again by a reconstruction) is denoted by $\bar{u}_{pq}$. Let us rewrite (1) in the formally equivalent form with conserving and non-conserving parts [2]

$$u_t + \nabla \cdot (\mathbf{v} u) - u \nabla \cdot \mathbf{v} = 0. \tag{2}$$

Integrating (2) over a finite volume p then yields

$$\int_p u_t \, dx + \int_p \nabla \cdot (vu) \, dx - \int_p u \nabla \cdot \mathbf{v} \, dx = 0.$$

Applying the divergence theorem and using constant representations of the solution on the cell p, denoted by $\bar{u}_p$, and on the cell interfaces e_{pq}, denoted by $\bar{u}_{pq}$, we get

$$\int_p u_t \, dx + \sum_{q \in N(p)} \bar{u}_{pq} \int_{e_{pq}} \mathbf{v} \cdot n_{pq} \, ds - \bar{u}_p \sum_{q \in N(p)} \int_{e_{pq}} \mathbf{v} \cdot n_{pq} \, ds = 0.$$

If we denote the fluxes in the inward normal direction to the finite volume p by

$$\bar{v}_{pq} = -\int_{e_{pq}} \mathbf{v} \cdot n_{pq} \, ds, \tag{3}$$

we finally arrive at the equation

$$\int_p u_t \, dx + \sum_{q \in N(p)} \bar{v}_{pq} (\bar{u}_p - \bar{u}_{pq}) = 0. \tag{4}$$

The novelty of our scheme is to split the resulting fluxes into the corresponding inflow and outflow parts to the cell p. This is done by defining

$$a_{pq}^{in} = \max(\bar{v}_{pq}, 0), \quad a_{pq}^{out} = \min(\bar{v}_{pq}, 0). \tag{5}$$

We then approximate u_t by the time difference $\frac{u_p^n - u_p^{n-1}}{\tau}$, where τ is a uniform time step size, and take the inflow parts implicitly and the outflow parts explicitly in (4). This yields the following system of equations for the finite volume solution u_p^n, $p \in \mathcal{T}_h$ at the n-th discrete time step, representing the general I^{2}OE scheme:

$$m_p u_p^n + \tau \sum_{q \in N(p)} a_{pq}^{in} (\bar{u}_p^n - \bar{u}_{pq}^n) = m_p u_p^{n-1} - \tau \sum_{q \in N(p)} a_{pq}^{out} (\bar{u}_p^{n-1} - \bar{u}_{pq}^{n-1}). \tag{6}$$

The most natural choice for reconstructions $\bar{u}_p^n$ and $\bar{u}_{pq}^n$ at any time step n (i.e. old and new time steps) is given by $\bar{u}_p^n = u_p^n$, $\bar{u}_{pq}^n = \frac{1}{2}(u_p^n + u_q^n)$ and leads to the basic I^{2}OE scheme:

$$m_p u_p^n + \frac{\tau}{2} \sum_{q \in N(p)} a_{pq}^{in} (u_p^n - u_q^n) = m_p u_p^{n-1} - \frac{\tau}{2} \sum_{q \in N(p)} a_{pq}^{out} (u_p^{n-1} - u_q^{n-1}). \tag{7}$$

The equation (4) has the form of a discretization of a diffusion equation, where $\bar{v}_{pq}$ would represent the so-called transmissive coefficients (integrated diffusion fluxes divided by distances between cell centers). In standard forward diffusion all these coefficients are strictly positive which leads to a weighted averaging of the solution and the implicit schemes are natural in this case. On the other hand the negative coefficients would correspond to backward diffusion in which case information propagates outside the cell and explicit schemes are thus natural. In our case the sign of the coefficients is given by the inflow or outflow character of the cell boundary and the inflow-implicit/outflow-explicit approach is thus natural. It is also well-known that in the second order schemes for solving advection problems one can identify the "forward diffusion" part (like the first order upwinding) and the "backward diffusion" part given by the additional sharpening terms coming (sometimes surprisingly) from the second order Taylor's expansions, cf. the Lax-Wendroff scheme [3]. In our method this splitting arises naturally, gives second order accuracy and when treating it semi-implicitly it brings significant improvements in stability of computations.

Let us present the I^{2}OE scheme for 1D variable velocity equation $u_t + v(x) u_x = 0$, which will be used in numerical computations of Section 4. Let p_i be the cell with the spatial index i, length h, center point x_i, left border $x_{i-\frac{1}{2}}$ and right border $x_{i+\frac{1}{2}}$. Let us denote u_i^n the value of the numerical solution at time step n and $\bar{u}_i^n, \bar{u}_{i-\frac{1}{2}}^n$ the reconstructed values. We define

$$a_{i-\frac{1}{2}}^{in} = \max(v(x_{i-\frac{1}{2}}),0), \quad a_{i-\frac{1}{2}}^{out} = \min(v(x_{i-\frac{1}{2}}),0),$$

$$a_{i+\frac{1}{2}}^{in} = \max(-v(x_{i+\frac{1}{2}}),0), \quad a_{i+\frac{1}{2}}^{out} = \min(-v(x_{i+\frac{1}{2}}),0),$$

and if we use the reconstructions $\bar{u}_i^n = u_i^n$, $\bar{u}_{i-\frac{1}{2}}^n = \frac{1}{2}(u_i^n + u_{i-1}^n)$ in both new and old time steps, the basic one-dimensional I^{2}OE scheme has the following form

$$u_i^n + \frac{\tau}{2h} a_{i-\frac{1}{2}}^{in}(u_i^n - u_{i-1}^n) + \frac{\tau}{2h} a_{i+\frac{1}{2}}^{in}(u_i^n - u_{i+1}^n) = u_i^{n-1} \qquad (8)$$
$$- \frac{\tau}{2h} a_{i-\frac{1}{2}}^{out}(u_i^{n-1} - u_{i-1}^{n-1}) - \frac{\tau}{2h} a_{i+\frac{1}{2}}^{out}(u_i^{n-1} - u_{i+1}^{n-1}).$$

The scheme (8) requires to solve a tridiagonal system in every time step which is done by using the standard tridiagonal solver (also called the Thomas algorithm). In practice, the I^{2}OE scheme allows to use much larger time steps without losing L_∞-stability than given by a standard CFL condition for explicit schemes, cf. Section 3. However, the "backward diffusion" (outflow) explicit part is not necessarily always dominated by the implicit part in the basic form of the scheme (8). Some oscillations (not unboundedly growing in time) may arise e.g. on coarse grids or in solutions tending to a shock. One possibility is to leave the method with oscillations and remove them at the end of computations using e.g. some edge preserving filters. Another approach is to supress the oscillations during the computation. In our scheme, one can use an averaging (by a larger stencil) in the reconstruction of $\bar{u}_p^{n-1}$, similarly to the FBD schemes from [4], or to modify the "backward diffusion" part on the right hand side of (8) by using the standard limiters, for details see [5].

Theorem 1. *Let us consider the equation (1) in 1D with constant velocity v and I^2OE scheme (8) on uniform grid. If the initial condition is given by a second order polynomial, then the scheme gives the exact solution for any choice of time step.*

Proof. The initial condition has the form $u_0(x) = ax^2 + bx + c$ and the exact solution is given by $u(x, \tau) = u^0(x - v\tau)$. For $v > 0$ the scheme (8) takes the form

$$u_i^n + \frac{\tau v}{2h}(u_i^n - u_{i-1}^n) = u_i^{n-1} - \frac{\tau(-v)}{2h}(u_i^{n-1} - u_{i+1}^{n-1}) \tag{9}$$

One can easily check that if we plug the exact values in grid points x_i, x_{i-1}, x_{i+1} at time steps $n = 1$ and $n - 1 = 0$, namely

$$u_i^{n-1} = ax_i^2 + bx_i + c, \quad u_{i+1}^{n-1} = a(x_i + h)^2 + b(x_i + h) + c, \tag{10}$$

$$u_i^n = a(x_i - v\tau)^2 + b(x_i - v\tau) + c, \quad u_{i-1}^n = a(x_i - h - v\tau)^2 + b(x_i - h - v\tau) + c,$$

into the scheme (9), we get true identity, and the same we obtain for $v < 0$. □

It is also possible to make similar considerations as above in higher dimensional case for uniform rectangular grids and constant velocity vector field. One can plug a general 2D or 3D quadratic polynomial as initial condition and the corresponding exact solution at time τ into the I^2OE scheme (7), use a symbolic computational software like the Mathematica, and check that the scheme is exact in such situations.

Theorem 2. *Let us consider the equation (1) in 1D with variable velocity $v(x) \geq 0$ (or $v(x) \leq 0$) and the I^2OE scheme (8) on a uniform grid. Then the scheme is formally second order and the consistency error is of order $\mathcal{O}(h^2) + \mathcal{O}(\tau h) + \mathcal{O}(\tau^2)$.*

Proof. We write our transport equation as $\partial_t u + f(v, \partial_x u) = 0$ with $f(v, \partial_x u) := v(x)\partial_x u$ and let $v(x) \geq 0$. We will use notations $u^n := u(t^n)$, $f^n := f(v, \partial_t u^n)$. The Taylor expansion in time yields

$$u^n = u^{n-1} + \tau \partial_t u^{n-1} + \frac{\tau^2}{2}\partial_t^2 u^{n-1} + \mathcal{O}(\tau^3), \quad u^{n-1} = u^n - \tau \partial_t u^n + \frac{\tau^2}{2}\partial_t^2 u^n + \mathcal{O}(\tau^3).$$

Subtracting these two equations we derive relation

$$u^n - u^{n-1} = \frac{\tau}{2}(\partial_t u^n + \partial_t u^{n-1}) + \frac{\tau^2}{4}(\partial_t^2 u^{n-1} - \partial_t^2 u^n) + \mathcal{O}(\tau^3). \tag{11}$$

We can see that the second term on the right hand side is also $\mathcal{O}(\tau^3)$ and using the equation $\partial_t u + f(v, \partial_x u) = 0$, we get for the first term of the right hand side

$$I = \frac{\tau}{2}(\partial_t u^n + \partial_t u^{n-1}) = -\frac{\tau}{2}(f^n + f^{n-1}). \tag{12}$$

Using the notation $f_i := f(x_i) = v(x_i)\partial_x u(x_i)$, by the Taylor expansion in space we have (for $v(x) \geq 0$)

$$f^n_{i-1/2} = f^n_i - \frac{h}{2}\partial_x f^n_i + \mathcal{O}(h^2), \quad f^{n-1}_{i+1/2} = f^{n-1}_i + \frac{h}{2}\partial_x f^{n-1}_i + \mathcal{O}(h^2) \quad (13)$$

or (for $v(x) \leq 0$)

$$f^{n-1}_{i-1/2} = f^{n-1}_i - \frac{h}{2}\partial_x f^{n-1}_i + \mathcal{O}(h^2), \quad f^n_{i+1/2} = f^n_i + \frac{h}{2}\partial_x f^n_i + \mathcal{O}(h^2). \quad (14)$$

We continue (for $v(x) \geq 0$) and using (12)-(13) we derive

$$I_i = -\frac{\tau}{2}(f^n_i + f^{n-1}_i) = -\frac{\tau}{2}\left(f^n_{i-1/2} + f^{n-1}_{i+1/2} + \frac{h}{2}(\partial_x f^n_i - \partial_x f^{n-1}_i) + \mathcal{O}(h^2)\right).$$

The second term in the brackets on the right hand side is of order $\mathcal{O}(\tau h)$ and we shall analyse the first one. We know that

$$\partial_x u^n_{i-1/2} = \frac{1}{h}(u^n_i - u^n_{i-1}) + \mathcal{O}(h^2), \quad \partial_x u^{n-1}_{i+1/2} = \frac{1}{h}(u^{n-1}_{i+1} - u^{n-1}_i) + \mathcal{O}(h^2)$$

and resubstituting for $f^n_{i-1/2} = v_{i-1/2}\partial_x u^n_{i-1/2}$ and $f^{n-1}_{i+1/2} = v_{i+1/2}\partial_x u^{n-1}_{i+1/2}$ we get

$$I_i = -\frac{\tau}{2}\left(v_{i-1/2}\frac{1}{h}(u^n_i - u_{i-1}) + v_{i+1/2}\frac{1}{h}(u^{n-1}_{i+1} - u^{n-1}_i)\right) + \mathcal{O}(\tau^2 h) + \mathcal{O}(\tau h^2). \quad (15)$$

From (11) and (15) we finally get

$$u^n_i - u^{n-1}_i = -\frac{\tau}{2}\left(\frac{v_{i-1/2}}{h}(u^n_i - u^n_{i-1}) + \frac{v_{i+1/2}}{h}(u^{n-1}_{i+1} - u^{n-1}_i)\right)$$
$$+ \mathcal{O}(\tau^2 h) + \mathcal{O}(\tau h^2) + \mathcal{O}(\tau^3)$$

where we recognize the scheme (8) for $v(x) \geq 0$, cf. also (9), and dividing by τ we get the consistency error of the I^{2}OE scheme stated in the theorem. □

3 Numerical experiments

First, let us consider 1D equation (1) with $v(x) \equiv 1$ in interval $\Omega = (-1, 1)$ and time interval $I = (0, T)$, $T = 1$. Let the initial condition u_0 be given by a quadratic polynomial $u_0(x) = 1 - \frac{1}{2}(x^2 - x)$. The exact solution is given $u(x, t) = u_0(x - vt)$. We solve this problem numerically using the exact Dirichlet boundary conditions and compare the results of the I^{2}OE method (8), the standard Lax-Wendroff and explicit up-wind schemes [3] with the exact solution. In all experiments we used

Inflow-Implicit/Outflow-Explicit Scheme

Table 1 Report on the $L_2(I, L_2)$ errors of the I^{2}OE method, the Lax-Wendroff scheme, and the explicit up-wind scheme for the initial quadratic polynomial and for various choices of time step. We note that all the methods are exact for $\tau = h$

n	$\tau = h/2$	NTS	I^{2}OE	Lax-Wendroff	Up-wind
20	0.05	20	$3.7\ 10^{-16}$	$5.1\ 10^{-17}$	$1.83\ 10^{-2}$
40	0.025	40	$8.0\ 10^{-16}$	$7.5\ 10^{-17}$	$8.99\ 10^{-3}$
80	0.0125	80	$1.1\ 10^{-15}$	$8.3\ 10^{-17}$	$4.45\ 10^{-3}$
160	0.00625	160	$2.4\ 10^{-15}$	$9.9\ 10^{-17}$	$2.22\ 10^{-3}$
n	$\tau = 2h$	NTS	I^{2}OE	Lax-Wendroff	Up-wind
20	0.2	5	$2.1\ 10^{-16}$	$1.1\ 10^{-11}$	$5.02\ 10^{-2}$
40	0.1	10	$2.1\ 10^{-16}$	$1.4\ 10^{-9}$	0.641
80	0.05	20	$3.9\ 10^{-16}$	0.466	$3.8\ 10^{+3}$
160	0.025	40	$5.7\ 10^{-16}$	$1.6\ 10^{+16}$	$1.3\ 10^{+12}$
160	$\tau = 10h = 0.125$	8	$2.5\ 10^{-15}$	–	–
160	$\tau = 40h = 0.5$	2	$1.7\ 10^{-15}$	–	–
160	$\tau = 80h = 1$	1	$2.6\ 10^{-15}$	–	–

increasing number n of finite volumes discretizing Ω, $h = 2/n$, and we consider various choices of time step τ and corresponding number of time steps NTS. In Table 1 we report the errors in $L_2(I, L_2)$ norm for all the methods. As one can see, the I^{2}OE method is exact for any relation between space and time step, see Theorem 1, and one can use extremely large (e.g. just one time step $\tau = T$) without any deterioration of the numerical result. Here the errors are comparable to machine precision, they are not exact zeros because we have to solve a tridiagonal system in every time step yielding some rounding errors which, however, do not propagate even in a long run. The Lax-Wendroff method, as the second order, is exact for any quadratic initial function whenever it is stable, i.e. $\tau \leq h$. For Courant numbers larger than 1, one can see instabilities in the third and 4th rows of Table 1, when $\tau = 2h$ and grid is refined. The explicit upwind scheme is the first order and exact for any initial data only if the relation $\tau = h$ is fulfilled. Its first order accuracy can be seen for $\tau = h/2$, and oscillations occur soon for $\tau > h$ as documented in Table 1.

Next, let us consider an example with variable velocity field $v(x) = -\sin(x)$ and let the initial profile be given by $u_0(x) = \sin(x)$, $\Omega = (-1, 1)$ and $I = (0, T)$, $T = 1$. The exact solution can be derived by the method of characteristics and is given as $u(x, t) = u_0(\frac{2}{\pi}\arctg(e^{\pi t}\tg(\frac{\pi x}{2})))$. We compare the precision and CPU-time of the I^{2}OE and the Lax-Wendroff scheme [3]. In the solutions a strong peak is formed at $T = 1$, see Fig. 1. Both schemes are stable with slight overshoot and undershoot in the result by the Lax-Wendroff scheme on coarser grids. No overshoot or undershoot is observed for the I^{2}OE scheme, cf. Fig. 1. Figure 2 shows log-log plots of CPU time versus error of the schemes. We can see superior behavior of the I^{2}OE scheme in this example with considerable speed-up when using larger time steps up to 4-8 times exceeding the CFL condition, which must be respected in the

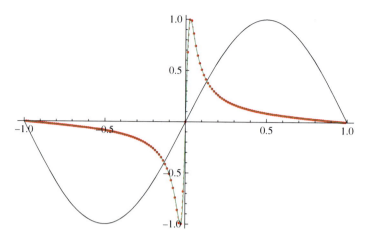

Fig. 1 The result of the I²OE scheme (up, red points) at time $T = 1$, computed with $n = 160$ and $\tau = h$. By green line we plot the exact solution at T and by black line the initial condition

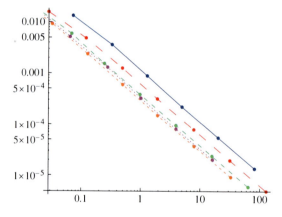

Fig. 2 CPU versus $L_2(I,L_2)$-error for the Lax-Wendroff method (blue solid line) and for the I²OE scheme with CFL=1 (red large dashing, $\tau = h$), CFL=2 (green medium dashing, $\tau = 2h$), CFL=4 (orange small dashing, $\tau = 4h$) and CFL=8 (magenta tiny dashing, $\tau = 8h$) for the experiment from Fig. 1. The plots indicate that I²OE scheme is about 4–times faster in order to get the same $L_2(I,L_2)$-error

Lax-Wendroff scheme. In this case both schemes are second order accurate which holds true for any time step size of the I²OE scheme.

Further 1D and 2D numerical experiments are reported in [5] showing the second order convergence of the I²OE method for any choice of the time steps. This is the main advantage of the new scheme when comparing with standard explicit second order methods, or, when using limiters, in comparison with the so-called high resolution methods for solving advection equations.

Acknowledgements The work of the first author was supported by the grants APVV-0351-07 and VEGA 1/0269/09.

References

1. Eymard, R., Gallouet, T., & Herbin R.: The finite volume methods, Handbook for Numerical Analysis, 2000, Vol. 7 (Ph. Ciarlet, J. L. Lions, eds.), Elsevier.
2. Frolkovič, P., Mikula, K.: Flux-based level set method: a finite volume method for evolving interfaces, Applied Numerical Mathematics, Vol. 57, No. 4 (2007) pp. 436-454.
3. LeVeque R.J., Finite Volume Methods for Hyperbolic Problems, Cambridge Texts in Applied Mathematics. Cambridge University Press, 2002.
4. Mikula, K., Ohlberger, M.: A new level set method for motion in normal direction based on a semi-implicit forward-backward diffusion approach, SIAM J. Scientific Computing, Vol. 32 , No. 3 (2010) pp. 1527-1544.
5. Mikula, K., Ohlberger, M.: A New Inflow-Implicit/Outflow-Explicit Finite Volume Method for Solving Variable Velocity Advection Equations, Preprint 01/10 - N, Angewandte Mathematik und Informatik, Universität Münster, June 2010, pp.1-20

The paper is in final form and no similar paper has been or is being submitted elsewhere.

4D Numerical Schemes for Cell Image Segmentation and Tracking

K. Mikula, N. Peyriéras, M. Remešíková, and M. Smíšek

Abstract The paper introduces new techniques for 4D (space-time) segmentation and tracking in time sequences of 3D images of zebrafish embryogenesis. Instead of treating each 3D image individually, we consider the whole time sequence as a single 4D image and perform the extraction of objects and cell tracking in four dimensions. The segmentation of the spatiotemporal objects corresponding to the time evolution of the individual cells is realized by using the generalized subjective surface model [1], that is discretized by a new 4D finite volume scheme. Afterwards, we use the distance functions to the borders of the segmented spatiotemporal objects and to the initial cell center positions in order to backtrack the cell trajectories that can be understood as 4D parametrized curves. The distance functions are obtained by numerical solution of the time relaxed eikonal equation.

Keywords Cell tracking, image segmentation, finite volume method
MSC2010: 65M08, 65-06, 92B05

1 Introduction

Cell tracking, i.e. finding the space-time trajectories and moments of divisions of the cells of a developing organism, is one of the most interesting challenges of modern biology. A reliable backward tracking can answer a lot of questions concerning the origin and formation of cell structures and organs, the global and local movement

---

Karol Mikula, Mariana Remešíková, and Michal Smíšek
Slovak University of Technology, Radlinského 11, 81368 Bratislava, Slovakia
e-mail: mikula@math.sk, remesikova@math.sk, michal.smisek@gmail.com

Nadine Peyriéras
CNRS-DEPSN, Avenue de la Terasse, 91198, Gif sur Yvette, France
e-mail: nadine.peyrieras@inaf.cnrs-gif.fr

of the cells, the cell division rate and localization etc. They all are fundamental questions of developmental biology.

In this paper, we introduce the basic concepts of a novel technique that can be used for the cell tracking from time sequences of 3D images of embryogenesis. A cell can be represented by the surface of its nucleus or by its membrane, depending on the type of images we have at disposal. The time evolution of a cell can be seen as spatiotemporal tube whose cross-section by a chosen time hyperplane corresponds to the 3D representation of the cell at the selected time. This 4D tube is bifurcated in the time moments when the cell undergoes division. Thus, we get a tree-like object corresponding to any cell present at the beginning of the time sequence. In order to track a cell, we need to descend from its current position to the root of the tree in which it is situated. This implies that the tracking procedure consists in solving the following two problems:

1. Segmentation of the 4D cell evolution trees from the spatiotemporal image.
2. Finding the way to the root of a tree from any of its inner points.

In our paper, we discuss the solution of both of these problems. We test our methods on artificial data and on time sequences of 3D images corresponding to the zebrafish embryonic development obtained by a confocal microscope. In order to be able to apply the described methods, we need to have at disposal the approximate positions of cell centers for all cells visible in the images. For the artificial data, these points are known by construction and for the zebrafish images, the approximate cell centers are computed by a level set object detection technique [1].

In order to solve the first problem, we apply the generalized subjective surface model [1, 6]

$$u_t - w_a \nabla g \cdot \nabla u - w_c g |\nabla u| \nabla \cdot \left(\frac{\nabla u}{|\nabla u|} \right) = 0, \qquad (1)$$

solved in the domain $[0, T_S] \times \Omega$ where $\Omega \subset R^4$ is the spatiotemporal image domain, i.e. the whole time sequence of 3D images. We set $u(0, x) = u_0(x)$ and we consider the zero Dirichlet boundary condition on $\partial \Omega$. The edge detector function $g = g(|\nabla G_\sigma * I_0|)$, I_0 being the 4D image intensity function, and w_a and w_c are the advection and curvature parameters of the model that determine the way the function u is evolving [1]. The desired cell evolution tree segmentation is represented by a selected isosurface of the function $u(x, T_S)$. We would like to point out the importance of performing this segmentation in 4D. Although the cell evolution tree object could be more easily composed of less time and memory consuming 3D cell segmentations, this could lead to spurious interruptions of the cell trajectories in the points where the cell center and consequently the corresponding cell segmentation is missing for some reason. Looking for a whole spatiotemporal structure rather than a composition of 3D objects makes the procedure more robust and resistant to the possible errors of the center detection technique.

Having segmented the tree object, we now want to find a way down to its root from any of its inner points. Since the root can be represented by the center of the root cell, a reasonable descend direction indicator could be the gradient of the distance function d_1 to this center computed inside the segmented 4D object. However, this might not be sufficient. In real data containing a large number of cells, we can observe that the trees corresponding to different root cells are not always perfectly isolated. In order to prevent dropping into a wrong tree, it is desirable to descend along the center line of the tree branches. For this purpose, we compute the distance function to the border of the 4D tree, denoted by d_2, whose negative gradient leads us towards the center line that we want to follow. The distance function to a set Ω_0 can be computed by solving the time relaxed eikonal equation

$$d_t + |\nabla d| = 1 \qquad (2)$$

in the domain $[0, T_D] \times \Omega_D$. In our case, Ω_D is the inner part of the segmented tree object, i.e. the part where $u(x, T_S) > V$, V being the isosurface value chosen to represent the segmentation result. The equation (2) has to be coupled with a Dirichlet type condition

$$d(x, t) = 0, \quad x \in \Omega_0 \qquad (3)$$

where Ω_0 can represent the root point of the tree or its boundary, i.e. the set of points where $u(x, T_S) = V$.

The descend to the root of the tree is performed as follows. Given an arbitrary point (doxel center) $[x_1, x_2, x_3, x_4]$ inside the tree, we move to the center of the nearest doxel in the direction given by ∇d_1. Supposing that x_4 represents the time dimension of the 4D data, we repeat this step until we drop to the level $x_4 - 1$. After, we move in the direction of $-\nabla d_2$ until we find the nearest ridge point of d_2. Thus we are situated on the center line of the current branch of the tree. From there we repeat the whole procedure until we descend to the level $x_4 = 0$, resp. to the root of the tree.

2 Discretization of the models

The time discretization of the generalized subjective surface model (1) is semi-implicit since in this way we can guarantee unconditional stability of the curvature term. Let τ_S be the time discretization step, $\tau_S = T_S/N_S$. Then for any $n = 1 \ldots N_S$ we get

$$\frac{u^n - u^{n-1}}{\tau_S} - w_a \nabla g \cdot \nabla u^{n-1} - w_c g |\nabla u^{n-1}| \nabla \cdot \frac{\nabla u^n}{|\nabla u^{n-1}|} = 0. \qquad (4)$$

where u^n represents the numerical solution on the nth time level.

The space discretization is realized by applying the finite volume strategy where one doxel of the 4D image corresponds to one volume of the discretization. Let us suppose that the volumes are 4D cubes of side length h and let V_i denote the volume with index vector $\mathbf{i} = (i,j,k,l)$ and u_i^n the value of the numerical solution u^n in the center c_i of this volume. Further, let $\mathbf{e}_p$, $p = 1\ldots 4$ represent the standard basis vectors in R^4, F_i^{+p} and F_i^{-p} the two faces of V_i orthogonal to $\mathbf{e}_p$, $v_i^{\pm p}$ the normal of the face $F_i^{\pm p}$ and $m(F_i^{\pm p})$ its measure.

Now let us integrate (4) over V_i. We get

$$\int_{V_i} \frac{u^n - u^{n-1}}{\tau_S} dx - \int_{V_i} w_a \nabla g \cdot \nabla u^{n-1} dx - \int_{V_i} w_c g |\nabla u^{n-1}| \nabla \cdot \frac{\nabla u^n}{|\nabla u^{n-1}|} dx = 0. \quad (5)$$

The time derivative term is approximated by

$$\int_{V_i} \frac{u^n - u^{n-1}}{\tau_S} dx \approx m(V_i) \frac{u_i^n - u_i^{n-1}}{\tau_S}. \quad (6)$$

The advection term is approximated by the upwind approach, i.e.

$$\int_{V_i} (-w_a \nabla g \cdot \nabla u) dx \approx \quad (7)$$

$$w_a m(V_i) \sum_{p=1}^{4} \left(\max(-D_i^p g, 0) \frac{u_i^{n-1} - u_{i-e_p}^{n-1}}{h} + \min(-D_i^p g, 0) \frac{u_{i+e_p}^{n-1} - u_i^{n-1}}{h} \right)$$

where $D_i^p g = (g_{i+e_p} - g_{i-e_p})/(2h)$ and g_i is the average value of g in V_i. For the curvature term we get the approximation

$$\int_{V_i} w_c g |\nabla u^{n-1}| \nabla \cdot \frac{\nabla u^n}{|\nabla u^{n-1}|} dx = w_c g_i \bar{Q}_i^{n-1} \sum_{p=1}^{4} \sum_{q=-p,+p} \int_{F_i^q} \frac{\nabla u^n}{|\nabla u^{n-1}|} \cdot v_i^q \, d\gamma, \quad (8)$$

where $\bar{Q}_i^{n-1}$ is the average value of $|\nabla u^{n-1}|$ in V_i. Further

$$\int_{F_i^{\pm p}} \frac{\nabla u^n}{|\nabla u^{n-1}|} \cdot v_i^{\pm p} \, d\gamma \approx \frac{m(F_i^{\pm p})}{Q_i^{\pm p; n-1}} \frac{u_{i\pm e_p}^n - u_i^n}{h} \quad (9)$$

where $Q_i^{\pm p; n-1}$ is the average value of $|\nabla u^{n-1}|$ on the face $F_i^{\pm p}$.

As we can see, in order to properly perform the approximations indicated in (7) and (8), we need to find an appropriate approximation of the average value of $|\nabla u^{n-1}|$ in both V_i and on the faces $F^{\pm p}$ and the average modulus of $g(|\nabla I_\sigma|)$, $I_\sigma = G_\sigma * I_0$, in V_i. There are various possibilities how to do that [4].

Let us first consider the approximation of ∇u^{n-1} in the barycenter $c_{i\pm\frac{1}{2}e_p}$ of the doxel face $F_i^{\pm p}$. The component corresponding to the direction of e_p is simply approximated by

$$D^{\pm p} u_i^{n-1} = \pm \frac{u_{i\pm e_p}^{n-1} - u_i^{n-1}}{h}. \tag{10}$$

The other components corresponding to the directions of e_q, $q = 1\ldots 4$, $q \neq p$, can be approximated as follows. The doxel face $F_i^{\pm p}$ is a 3D cube with faces denoted by $F_i^{\pm p, \pm q}$. The barycenter of $F_i^{\pm p, \pm q}$ can be expressed as $c_{i\pm\frac{1}{2}e_p\pm\frac{1}{2}e_q}$. Thus, the value of u^{n-1} at this point can be approximated as

$$u_{i\pm\frac{1}{2}e_p\pm\frac{1}{2}e_q}^{n-1} = \frac{1}{4}(u_i^{n-1} + u_{i\pm e_p}^{n-1} + u_{i\pm e_q}^{n-1} + u_{i\pm e_p\pm e_q}^{n-1}). \tag{11}$$

The partial derivatives of u^{n-1} at $c_{i\pm\frac{1}{2}e_p}$ are then approximated as

$$D^{\pm p,q} u_i^{n-1} = \frac{u_{i\pm\frac{1}{2}e_p+\frac{1}{2}e_q}^{n-1} - u_{i\pm\frac{1}{2}e_p-\frac{1}{2}e_q}^{n-1}}{h} \tag{12}$$

Finally, we can define the required approximations

$$Q_i^{\pm p; n-1} = \sqrt{(D^{\pm p} u_i^{n-1})^2 + \sum_{q \neq p}(D^{\pm p, q} u_i^{n-1})^2}, \quad \bar{Q}_i^{n-1} = \frac{1}{8}\sum_{p=1}^{4}\sum_{q=-p,+p} Q_i^{q;n-1} \tag{13}$$

$$G_i^{\pm p} = \sqrt{(D^{\pm p} I_{\sigma; i})^2 + \sum_{q \neq p}(D^{\pm p, q} I_{\sigma; i})^2}, \quad g_i = \frac{1}{8}\sum_{p=1}^{4}\left(G_i^{-p} + G_i^{+p}\right) \tag{14}$$

Combining (6)–(14), we get the finite volume scheme for solving the problem (1).

The eikonal equation (2) is discretized by the Rouy-Tourin scheme [5]. Let $\tau_D = T_D/N_D$ be the time discretization step and d_i^n the value of the numerical solution in the barycenter of the doxel V_i on the nth time level. Let us define for $p = 1\ldots 4$

$$D_i^{\pm p} = \left(\min\left(d_{i\pm e_p}^{n-1} - d_i^{n-1}\right)\right)^2, \quad M_i^p = \max\left(D_i^{-p}, D_i^{+p}\right)$$

Then the numerical scheme is written as follows

$$d_i^n = d_i^{n-1} + \tau_D - \frac{\tau_D}{h}\sqrt{\sum_{p=1}^{4} M_i^p} \tag{15}$$

This scheme is stable for $\tau_D \leq h/4$ and it produces monotonically increasing updates that gradually approach a steady state. This leads to an efficient implementation of the scheme that uses a fixing strategy [2].

3 Experiments

Before we proceed to the experiments concerning the actual segmentation and tracking, we test the experimental order of convergence of the finite volume scheme presented above on a simple regularized mean curvature flow equation

$$\partial_t u = |\nabla u| \nabla \cdot \left(\frac{\nabla u}{|\nabla u|}\right) \tag{16}$$

with the exact solution $u(x_1, x_2, x_3, x_4, t) = \frac{x_1^2 + x_2^2 + x_3^2 + x_4^2 - 1}{6} + t$. We use the Dirichlet boundary condition and the initial condition given by this analytical solution. The problem was solved in the domain $[-1.25, 1.25]^4 \times [0, 0.08]$. The spatial domain consisted of n^4 doxels with $h = 2.5/n$ and $\tau \sim h^2$. The error of the numerical solution was measured in $L_\infty(I, L_2(\Omega))$ norm. The result of this test is displayed in Table 1.

Table 1 The experimental order of convergence of the finite volume scheme described in Sec. 2

n	τ	error	EOC
10	0.04	5.531426e-3	
20	0.01	7.276024e-4	2.926
40	0.0025	1.407815e-4	2.370
80	0.000625	3.264185e-5	2.109

The second experiment illustrates the segmentation of artificial 4D data. The 4D image was constructed as an analogy of the cell nuclei evolution. The cell nuclei are more or less spherical objects, so we started with two spheres. In each time slice of the 4D image, these two spheres are situated at different positions but not far from their positions in the previous time slice. We construct 25 time slices. At time $x_4 = 9$, one of the spheres divides and from then on, we have 3 spheres in the image. To make the situation more general, the radii of the spheres change in time. The centers of these spheres are used to construct the initial segmentation function. We place a 4D ellipsoid with radii a, b, c, d in each of these centers and we set $u_0(x) = 1$ inside these ellipsoids and $u_0(x) = 0$ outside. The model parameters were set as follows: $g(|\nabla I_0|) = 1/(1 + K|\nabla I_0|^2)$, $K = 1.0$, $h = 1.0$, $\tau_S = 0.1$, $w_a = 5.0$, $w_c = 0.1$, $T_D = 30$. Instead of $|\nabla u|$ we use its regularization $\sqrt{\varepsilon + |\nabla u|^2}$ with $\varepsilon = 10^{-6}$. The procedure is illustrated in Fig. 1. In order to visualize a 4D discrete function $u(x_1, x_2, x_3, x_4)$ with m slices in x_4-coordinate, we construct its 3D representation by setting the value in each 3D voxel (x_1, x_2, x_3) to $\max_{x_4=1...m} u(x_1, x_2, x_3, x_4)$. Then we visualize an isosurface of this representation.

Another experiment shows the segmentation of the zebrafish embryogenesis data. We segmented a sequence of 20 3D cell nuclei images preprocessed (denoised) by the geodesic mean curvature flow filter [3]. The initial segmentation function was constructed in the same way as in the case of the artificial data. Further, we set $g(|\nabla I_\sigma|) = 1/(1 + K|\nabla I_\sigma|^2)$, $I_\sigma = G_\sigma * I_0$, $K = 100.0$, $\sigma = 0.01$, $h = 1.0$,

Fig. 1 Segmentation of artificial 4D data. Left, the isosurface $V = 128$ of the 3D representation of the data. Middle, the isosurface $V = 15$ of the 3D representation of the initial segmentation function. Right, the isosurface $V = 15$ of the 3D representation of the segmentation result. This isosurface was chosen as the best representation of the segmented object

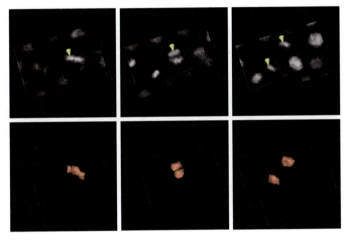

Fig. 2 Segmentation of the zebrafish embryogenesis data. On the top, we display 2D slices of the 4D image corresponding to different x_4 (time) values with indication of the position of the segmented object. On the bottom, we provide the corresponding segmentation result in the form of isosurface $V = 128$ of x_4-slices of the 4D segmentation function

$\tau_S = 0.1$, $w_a = 10.0$, $w_c = 1.0$, $T_D = 50$, $\varepsilon = 10^{-6}$. Fig. 2 displays 2D slices of the 4D data (more precisely, 2D slices of x_4-slices of the 4D data). The object that we tried to segment was a simple cell evolution tree containing one cell division. Together with the image slices, we provide the segmentation result, now displayed as isosurfaces of x_4 (time) slices of the segmentation function.

Fig. 3 shows the result of the cell tracking performed on the artificial data described above. We backtrack the cells (spheres) from the positions of their centers at the end of the time sequence. Both distance functions d_1 and d_2 were computed by setting $h = 1.0$, $\tau_D = 0.25$. The result of the tracking is a set of 4D points characterizing the cell position on the individual time levels. At each time level, we get one point that represents the intersection of the time hyperplane with the ridge of the 4D distance function d_2 (note that these points in general do not correspond

Fig. 3 The result of the cell tracking performed on artificial 4D data. We can see the points characterizing the positions of the cells at each time level visualized by neglecting their x_4 coordinate. The starting points for the tracking are situated on the top of the point sequences

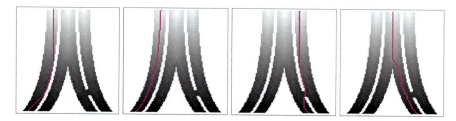

Fig. 4 The effect of using the distance function d_2. From the left: first, the tracking line in an isolated branch obtained by using only d_1, second, the tracking line in the same branch when using d_1 and d_2, third, the tracking line in a branch interconnected with a neighboring branch drops into a wrong branch if only d_1 and not d_2 is applied, fourth, by applying both d_1 and d_2, the line remains in the correct branch. The grey level shading of the branches represents the values of d_1

to the geometrical centers of the individual 3D spheres). The points are visualized by neglecting their x_4 coordinate.

Finally, we present a test illustrating the effect of using the distance function d_2. In Fig. 4, we can see four branches of 2D cell evolution trees. As we can observe, if using only the distance function d_1, the tracking lines tend to go along the borders of their

4 Conclusions

To conclude, we presented the main ideas of a new cell tracking technique and we illustrated the validity of the procedure on several test examples. The method is now prepared to be applied to long time sequences of biological data.

Acknowledgements This work was supported by the grant VEGA 1/0733/10.

References

1. Bourgine, P., Čunderlík, R., Drblíková-Stašová, O., Mikula, O., Peyriéras, N., Remešíková, M., Rizzi, M., Sarti, A.: 4D embryogenesis image analysis using PDE methods of image processing. Kybernetika **46 (2)**, 226–259 (2010).
2. Bourgine, P., Frolkovič, P., Mikula, K., Peyriéras, N., Remešíková, M.: Extraction of the intercellular skeleton from 2D microscope images of early embryogenesis. In Lecture Notes in Computer Science **5567** (Proceeding of the 2nd International Conference on Scale Space and Variational Methods in Computer Vision, Voss, Norway, June 2009) (Springer, 2009), p. 38–49.
3. Krivá, Z., Mikula, K., Peyriéras, N., Rizzi, B., Sarti, A., Stašová, O.: Zebrafish early embryogenesis 3D image filtering by nonlinear partial differential equations. Medical Image Analysis, **14 (4)**, 510–526 (2010).
4. Mikula, K., Remešíková, M.: Finite volume schemes for the generalized subjective surface equation in image segmentation. Kybernetika **45 (4)**, 646–656 (2009).
5. Rouy, E., Tourin, A.: Viscosity solutions approach to shape-from-shading. SIAM Journal on Numerical Analysis **29 (3)**, 867–884 (1992).
6. Zanella, C., Campana, M., Rizzi, B., Melani, C., Sanguinetti, G., Bourgine, P., Mikula, K., Peyriéras, N., Sarti, A.: Cells Segmentation from 3-D Confocal Images Of Early Zebrafish Embryogenesis. IEEE Transactions on Image Processing **19 (2)**, (2010).

The paper is in final form and no similar paper has been or is being submitted elsewhere.

Rhie-Chow interpolation for low Mach number flow computation allowing small time steps

Yann Moguen, Tarik Kousksou, Pascal Bruel, Jan Vierendeels, and Erik Dick

Abstract Low Mach number flow computation in co-located grid arrangement requires pressure-velocity coupling in order to prevent the checkerboard phenomenon. A Rhie-Chow interpolation technique can be formulated with such a coupling involving an explicit time step dependence, suitable for unsteady computations. Following this approach, it is observed that unphysical pressure oscillations arise again for sufficiently small time steps. Some remedies have been proposed for incompressible flows. A simple adaptation of these remedies for low Mach number flow computation is numerically investigated. A slight departure from the original approach appears to be suitable.

Keywords Pressure correction, pressure-velocity coupling, co-located grid
MSC2010: 65M08, 65N22, 76G25, 76N15

Yann Moguen, Jan Vierendeels, and Erik Dick
Ghent University - Department of Flow, Heat and Combustion Mechanics,
Sint-Pietersnieuwstraat, 41 - 9 000 Gent, Belgium, e-mail: yann.moguen@free.fr, jan. vierendeels@ugent.be, erik.dick@ugent.be

Tarik Kousksou
Université de Pau et des Pays de l'Adour - Laboratoire des Sciences de l'Ingénieur Appliquées à la Mécanique et au Génie Electrique, ENSGTI, rue Jules Ferry - 64 075 Pau, France,
e-mail: tarik.kousksou@univ-pau.fr

Pascal Bruel
CNRS and Université de Pau et des Pays de l'Adour - Laboratoire de Mathématiques et de leurs Applications, UMR 5142 CNRS-UPPA, avenue de l'Université, BP 1155 - 64 013 Pau, France,
e-mail: pascal.bruel@univ-pau.fr

1 Introduction

Coupling between velocity and pressure difference on cell faces is necessary in low Mach number flow computations on grids with co-located arrangement. This allows to avoid the checkerboard phenomenon, which means unphysical pressure oscillations, increasing as a Mach number representative of the flow goes to zero. A pressure-velocity coupling that involves an explicit time step dependence, which appears to be advantageous with unsteady computations, can be obtained by Rhie-Chow interpolation method [5]. Unfortunately, with this choice, some pressure oscillations arise again for 'small' time steps. This may lead to useless computations. Some remedies have been proposed, but they concern incompressible flows [6]. In the present contribution, we investigate numerically this issue in the case of low Mach number flow computations.

For simplicity, a one-dimensional flow of a perfect and ideal gas in a nozzle with a variable section is considered. In the following, x denotes the coordinate in the flow direction. The flow model is given by the Euler equations:

$$\partial_t(\rho S) + \partial_x(\rho v S) = 0 \tag{1a}$$

$$\partial_t(\rho v S) + \partial_x((\rho v^2 + p)S) = p \mathrm{d}_x S \tag{1b}$$

$$\partial_t(\rho E S) + \partial_x(\rho v H S) = 0 \tag{1c}$$

$$E = e + \frac{1}{2}v^2 \tag{1d}$$

$$\rho H = \rho E + p \tag{1e}$$

$$\rho e = \frac{p}{\gamma - 1} \tag{1f}$$

where t, ρ, p, v, e, E and H represent time, density, pressure, velocity, internal energy, total energy and total enthalpy per unit mass, respectively. Furthermore, γ denotes the specific heats ratio and S the cross-section area of the nozzle.

The x axis along the nozzle is divided into a number N of cells of length Δx. A finite volume formulation in co-located arrangement is applied.

2 Pressure correction algorithm

To solve the set (1) of equations, the energy-based pressure correction algorithm that we consider takes the following form, where the superscripts $\star$ and $\prime$ denote estimated and correction quantities of each iteration (first iteration: $k = n$), respectively:

1. Pre-estimation step: Generate a transporting velocity v^T at the cell-faces, that will be used in the following two steps.
2. Estimation step: With $p_i^* = p_i^k$, calculate ρ_i^* and $(\rho v)_i^*$ using

$$\frac{1}{2}(3\rho_i^* - 4\rho_i^n + \rho_i^{n-1}) + \frac{\tau}{S_i}\{[\rho_i^* + \frac{1}{2}\psi_i^k(\rho)(\rho_i^k - \rho_{i-1}^k)]v_{i+1/2}^T S_{i+1/2}$$
$$- [\rho_{i-1}^* + \frac{1}{2}\psi_{i-1}^k(\rho)(\rho_{i-1}^k - \rho_{i-2}^k)]v_{i-1/2}^T S_{i-1/2}\} = 0 \quad (2)$$

where τ is formally defined as $\Delta t / \Delta x$ and practically calculated as $\mathrm{CFL}_v/v_{\max}$, and v^T is positive. A similar equation holds for the momentum. Here ψ denotes a slope limiter, for instance MinMod, allowing to reach second-order accuracy in space, while the same order of accuracy in time is obtained by using the second-order backward discretization. From the estimated density, momentum and pressure, calculate the estimated total energy and total enthalpy.
3. Correction step: Calculate the pressure correction p' by solving the energy equation in second-order accurate backward discretization form in time. Flux terms are expanded as

$$(\rho v H)_{i+1/2}^{k+1} = (\rho H)_{i+1/2}^* v_{i+1/2}^* + H_{i+1/2}^*(\rho v)_{i+1/2}' + (\rho H)_{i+1/2}' v_{i+1/2}^* \quad (3)$$

where $H_{i+1/2}^*$ and $(\rho H)_{i+1/2}^*$ are upwinded in second-order accurate form, as convected quantities. Furthermore, neglecting the kinetic energy contribution, $(\rho H)_{i+1/2}' = \frac{\gamma}{\gamma-1} p_{i+1/2}'$ and the calculation of $(\rho v)_{i+1/2}'$ is derived from the momentum equation.
4. Updates: $p_i^{k+1} = p_i^* + p_i'$, $\rho_i^{k+1} = \rho_i^* + (\partial_p \rho)_i^* p_i'$, $(\rho v)_i^{k+1} = (\rho v)_i^* + (\rho v)_i'$, where $(\rho v)_i'$ is derived from the momentum equation in accordance with the derivation of $(\rho v)_{i+1/2}'$. The total energy and the cell-face pressure and velocity are finally updated.

3 Calculation of cell-face quantities

Let us provide some details on the AUSM/Rhie-Chow combination, which we consider as suitable for future unsteady computations.

3.1 AUSM interpolation

For explanations on the AUSM$^+$ and AUSM$^+$-up schemes, we refer to [2] and focus only on a low Mach number adaptation of AUSM$^+$, by using a simple scaling function suggested in this reference for the construction of the AUSM$^+$-up scheme.

The notation L or R, which refers to the left or right side of the face $i + 1/2$, is adopted since an extrapolation technique will be used in the following. Thus, a Mach number on the side S is defined as

$$M_S = \frac{v_S}{c_{i+1/2}}, \quad S = L, R \tag{4}$$

where $c_{i+1/2}$ is the speed of sound at the face $i + 1/2$ (cf. Ref. [2]). A mean Mach number at the face $i + 1/2$ is also defined,

$$\overline{M}_{i+1/2} = \sqrt{\frac{(v_L)^2 + (v_R)^2}{2c_{i+1/2}^2}} \tag{5}$$

and a reference Mach number $M_{0,i+1/2}$ by

$$M_{0,i+1/2}^2 = \min\{1, \max\{\overline{M}_{i+1/2}^2, \mathrm{Ma}_{co}^2\}\} \tag{6}$$

where Ma_{co} is a cut-off Mach number value such that: $\mathrm{Ma}_{co} = \mathcal{O}(\mathrm{Ma}_\infty)$. A scaling function suggested in Ref. [2] is

$$f(M) = M(2 - M) \tag{7}$$

The use of this function permits the proper asymptotic behaviour of the pressure dissipation term for $M \searrow 0$ in the face velocity (see Ref. [2]), defined by the following construction:

$$M_{(1)}^\pm(M) = \frac{1}{2}(M \pm |M|) \tag{8}$$

$$M_{(4)}^\pm(M) = \pm\frac{1}{4}(M \pm 1)^2 \pm \frac{1}{8}(M^2 - 1)^2 \tag{9}$$

$$P_{(0)}^\pm(M) = M_{(1)}^\pm(M)/M \tag{10}$$

$$P_{(5)}^\pm(M) = \frac{1}{4}(M \pm 1)^2(2 \mp M) \pm \frac{3}{16}(5(f(M_0))^2 - 4)M(M^2 - 1)^2 \tag{11}$$

$$\mathcal{M}^\pm(M) = \begin{cases} M_{(1)}^\pm(M), & |M| \geq 1 \\ M_{(4)}^\pm(M), & |M| < 1 \end{cases} \tag{12}$$

$$\mathcal{P}^\pm(M) = \begin{cases} P_{(0)}^\pm(M), & |M| \geq 1 \\ P_{(5)}^\pm(M), & |M| < 1 \end{cases} \tag{13}$$

$$p_{i+1/2} = \mathcal{P}^+(M_L)p_L + \mathcal{P}^-(M_R)p_R \tag{14}$$

$$M_{i+1/2} = \mathcal{M}^+(M_L) + \mathcal{M}^-(M_R) \tag{15}$$

$$v_{i+1/2} = c_{i+1/2} \, M_{i+1/2} \tag{16}$$

To reach second-order accuracy in space, the primitive variables p, ρ and v, which are used in the AUSM$^+$ scheme, are extrapolated at the face $i + 1/2$ according to

$$\phi_L = \phi_i + \frac{1}{2}\psi_i(\phi)(\phi_i - \phi_{i-1}) \quad , \quad \phi_R = \phi_{i+1} - \frac{1}{2}\psi_{i+1}(\phi)(\phi_{i+1} - \phi_i)$$

where ψ denotes a slope limiter. Practically, we choose

$$\psi_i(\theta) = \mathrm{MinMod}(\theta_i - \theta_{i-1}, \theta_{i+1} - \theta_i) / (\theta_i - \theta_{i-1})$$

where

$$\mathrm{MinMod}(a, b) = \frac{\mathrm{sign}(a) + \mathrm{sign}(b)}{2} \min\{|a|, |b|\}$$

3.2 Rhie-Chow interpolation

The pressure-velocity coupling, especially needed at low Mach number, is achieved through the construction of a transporting velocity with a Rhie-Chow interpolation technique. According to the 'classical' Rhie-Chow approach (see e.g. Ref. [1]), the preceding face velocities are interpolated when assembling the current transporting velocity. In this case, steady results which are not time step dependent can be ascertained when a certain interpolation practice is satisfied (see Ref. [4]). However, the numerical dissipation associated with the pressure-velocity coupling arising with this choice can lead to unphysical unsteady computations [3]. An alternative way to allow 'small' time step computations consists to avoid interpolation by directly using the preceding transporting velocities. This approach has been proposed in Ref. [6] for incompressible flows. Its application for low Mach number flow will be addressed in the rest of this paper.

First, an auxiliary density, that can be thought as a 'pre-predicted' one, is defined by solving the continuity equation, as

$$\frac{1}{2}(3\rho_i^{**} - 4\rho_i^n + \rho_i^{n-1}) + \frac{\tau}{S_i}\{[\rho_i^{**} + \frac{1}{2}\psi_i^k(\rho)(\rho_i^k - \rho_{i-1}^k)]v_{i+1/2}^k S_{i+1/2}$$

$$- [\rho_{i-1}^k + \frac{1}{2}\psi_{i-1}^k(\rho)(\rho_{i-1}^k - \rho_{i-2}^k)]v_{i-1/2}^k S_{i-1/2}\} = 0 \tag{17}$$

where the cell-face velocity, which is positive, is given at the last known iteration, and calculated by the scheme described in subsection 3.1. Then, similarly, an auxiliary pre-predicted velocity v^{**} is defined from the momentum equation in which the pressure gradient influence has been removed, by

$$a_i \rho_i^{**} v_i^{**} = -\frac{\tau}{S_i}\left[\frac{1}{2}\psi_i^k(\rho v)[(\rho v)_i^k - (\rho v)_{i-1}^k] v_{i+1/2}^k S_{i+1/2}\right.$$

$$\left. - \{(\rho v)_{i-1}^k + \frac{1}{2}\psi_{i-1}^k(\rho v)[(\rho v)_{i-1}^k - (\rho v)_{i-2}^k]\} v_{i-1/2}^k S_{i-1/2}\right]$$

$$+ (1-\varepsilon)[2(\rho v)_i^n - \frac{1}{2}(\rho v)_i^{n-1}] \quad (18)$$

where $0 \le \varepsilon \le 1$ and $a_i = \frac{3}{2} + \frac{\tau}{S_i} v_{i+1/2}^k S_{i+1/2}$. With a similar equation for v_{i+1}^{**}, a transporting velocity is defined according to a Rhie-Chow interpolation, as

$$v_{i+1/2}^T = \frac{1}{2}(v_i^{**} + v_{i+1}^{**}) - \frac{\tau}{2}\left(\frac{1}{a_i \rho_i^{**}} + \frac{1}{a_{i+1} \rho_{i+1}^{**}}\right)(p_{i+1}^k - p_i^k)$$

$$+ \frac{\varepsilon}{2}\left(\frac{1}{a_i} + \frac{1}{a_{i+1}}\right)[2(v_{i+1/2}^T)^n - \frac{1}{2}(v_{i+1/2}^T)^{n-1}] \quad (19)$$

Now, the pressure correction-mass flux correction coupling is defined accordingly, as

$$(\rho v)'_{i+1/2} = -\frac{\tau}{2}\left(\frac{1}{a_i} + \frac{1}{a_{i+1}}\right)(p'_{i+1} - p'_i) \quad (20)$$

and then, the momentum correction is written as

$$(\rho v)'_i = -\frac{\tau}{a_i}(p'_{i+1/2} - p'_{i-1/2}) \quad (21)$$

Taking $\varepsilon = 1$ in expressions (18) and (19) corresponds to the remedy suggested in Ref. [6] to the non-physical pressure oscillations that occur when Rhie-Chow interpolation is used with small time steps for incompressible flows computation. As a preliminary discussion on the suitable choice of ε, the rest of this paper is devoted to numerical experiments illustrating some numerical difficulties encountered and remedies.

4 Numerical experiments

A low Mach number flow in a converging-diverging nozzle is considered. The prescribed inlet velocity is such that the throat Mach number is 10^{-3} at convergence. The cut-off Mach number Ma_{co} in expression (6) is set as 10^{-3}. At the inlet, a constant target value of the density is fixed as $\rho_{in} = 1.2086$ kg/m^3. At the outlet, a constant target value of the pressure is fixed as $p_{out} = 101\,300$ Pa. Target values are mentioned since a non-reflecting treatment of the boundaries is applied, but this does not relate to the discussion in the present paper. Let us point out that a well-known complete analytical solution is available for the steady flow under consideration.

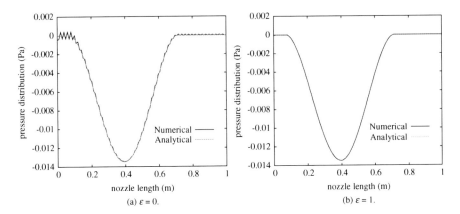

Fig. 1 Pressure distribution (Pa) along the nozzle. $\Delta t = 1.75 \; 10^{-5}$ s ; $\varepsilon = 1$: Ref. [6]. Number of cells: 100

First, unphysical pressure oscillations arising with $\varepsilon = 0$ are shown in Fig. 1, left. The oscillations are more pronounced on the left side of the nozzle, which is the less constraining for the pressure variable – in the sense that no target value is imposed at the inlet –, but they are present more or less on the totality of the nozzle. As shown in Fig. 1, right, a simple remedy consists in the choice of $\varepsilon = 1$ in Eqs. (18) and (19), as suggested in Ref. [6].

Between 0 and 1, an intermediate value for ε is also possible, that can be thought as a parameter that manages the deferred treatment of the temporal terms in the momentum equation at the 'pre-prediction' stage of the algorithm (see section 3.2). In Fig. 2, the plot of the error in the pressure versus the parameter ε reveals that the optimal value of ε is not 1, as far as the accuracy is the criterion. The optimal value ε_{opt} that minimizes the pressure error is slightly less than 1. Let us notice that, as $\Delta t \searrow 0$, the transporting velocity is such that

$$v^T_{i+1/2} \to \frac{1}{3}(1-\varepsilon)\left[2\left(\frac{(\rho v)^n_i}{\rho^{**}_i} + \frac{(\rho v)^n_{i+1}}{\rho^{**}_{i+1}}\right) - \frac{1}{2}\left(\frac{(\rho v)^{n-1}_i}{\rho^{**}_i} + \frac{(\rho v)^{n-1}_{i+1}}{\rho^{**}_{i+1}}\right)\right]$$
$$+ \frac{2}{3}\varepsilon\left[2(v^T_{i+1/2})^n - \frac{1}{2}(v^T_{i+1/2})^{n-1}\right] \quad (22)$$

With $\varepsilon \neq 1$ but close to 1, Eq. (22) corresponds to a slight departure from the original approach of Ref. [6]. The observed sensitivity of ε_{opt} to Δt suggests that a comprehensive study of the ε_{opt} dependency on CFL_v should be carried out.

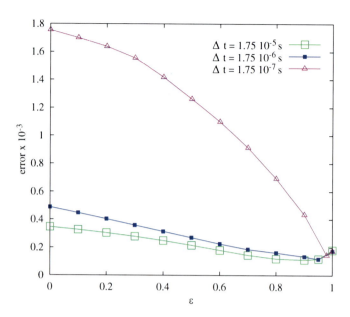

Fig. 2 Error in pressure (L2 norm) vs. ε of Eqs. (18) and (19). $\varepsilon = 1$: Ref. [6]. Number of cells: 100

5 Conclusion

The Rhie-Chow stabilisation method with a pressure-velocity coupling that involves an explicit time step dependence does not avoid unphysical oscillations when the time step is sufficiently small. According to Ref. [6], these oscillations originate from the interpolation of the previous velocities in the transporting velocity construction. However, in the steady case considered here, a slight departure from the approach of Ref. [6] is suitable concerning the deferred treatment of the temporal terms in the momentum equation.

Since the problem of oscillations also occurs for unsteady computations, the next step of this study will be to examine how the considered correction of the Rhie-Chow interpolation works in the case of unsteady low Mach number flow computations. Last but not least, the appropriate choice of the parameter ε if the exact solution is unknown is also an issue to be investigated.

References

1. Lien F.S. and Leschziner M.A.: A general non-orthogonal collocated finite volume algorithm for turbulent flow at all speeds incorporating second-moment turbulence-transport closure, Part 1: Computational implementation. Comput. Methods Appl. Mech. Engrg. **114**, 123–148 (1994)

2. Liou M.-S.: A Sequel to AUSM, part II: AUSM$^+$-up for all speeds. J. Comp. Phys. **214**, 137–170 (2006)
3. Moguen Y., Kousksou T., Dick E. and Bruel P.: On the role of numerical dissipation in unsteady low Mach number flow computations. Proceedings of the Sixth International Conference on Computational Fluid Dynamics. Springer (2011). To appear.
4. Pascau A.: Cell face velocity alternatives in a structured colocated grid for the unsteady Navier-Stokes equations. Int. J. Numer. Meth. Fluids **65**, 812-833 (2011)
5. Rhie C.M. and Chow W.L.: Numerical Study of the Turbulent Flow Past an Airfoil with Trailing Edge Separation. AIAA J. **21**(11), 1525–1532 (1983)
6. Shen W.Z., Michelsen J.A. and Sørensen J.N.: Improved Rhie-Chow Interpolation for Unsteady Flow Computations. AIAA J. **39**(12), 2406–2409 (2001)

The paper is in final form and no similar paper has been or is being submitted elsewhere.

Study and Approximation of IMPES Stability: the CFL Criteria

C. Preux and F. McKee

Abstract Whether it is for the recovery of hydrocarbons or the injection and storage of CO2, the industry uses numerical simulation. The first stage of study consists in the construction of a geologic model. In view of the field size, the fine model thus built contains several tens of million cells. Numerical fluid flow simulations may require a lot of CPU time and are generally impossible to achieve. To reduce the simulation cost, reservoir engineers use an upscaling of the fine mesh in a coarse one. If the upscaling of absolute permeabilities was already the object of detailed research, it is not the case for polyphasics flows. These flows are described by relative permeabilities and capillary pressure curves defined by limit points and normalized forms. The curves are different according to the facies of the model and in this way, according to the cells. The subject of this paper is to study the IMPES scheme (Implicit Pressure, Explicit Saturation [1], [2], [3]) and to simulate a diphasic flow in order to prepare the upscaling step. A stability analysis of this scheme can highlight a CFL condition [4]. This paper proposes a study of this CFL number for the case of reservoir simulation.

Keywords Diphasics, reservoir simulation, IMPES scheme, stability, CFL
MSC2010: 76T99, 76S05, 65N08, 65N12, 65N12

C. Preux and F. McKee
IFP Energies Nouvelles, 1 et 4 avenue de Bois-Préau - 92852 Rueil-Malmaison Cedex - France,
e-mail: christophe.preux@ifpenergiesnouvelles.fr, francois.mckee@ifpenergiesnouvelles.fr

1 Introduction

The acronym IMPES was used in 1968 in a description of a numerical model for simulating black oil reservoir behavior. The IMPES method was generalised in 1980 to apply to simulation models involving any number n of conservation equations. The basic principle is the elimination of differences in non-pressure variables from the model's set of conservation equations to obtain a single pressure equation. Stone [1], Sheldon, Harris, Bavly [5], and Martin [6] used the same principle in deriving the total compressibility of multiphase black oil systems. For Coats [7], the first black oil IMPES model was published by Fagin and Stewart in 1966 [8]. In this work, we deal with the flow of two immiscible and incompressible fluids. We present the system gouverning a two phase flow and proceed to simplifications. We consider the case where there is no gravity and no capillary pressure. This system is based on Darcy's Law generalized for two-phase flow.

1.1 Fully coupled formulation

The most commonly used description for macro-scale two-phase flow in porous media uses a phenomenological extension of Darcy's law introducing the saturation-dependent parameter: relative permeability. This extension was proposed, in the thirties, by Leverett [9] and Wyckoff and Botset [10] and the authors have validated this generalization experimentally.

$$v_i = -\frac{k_{ri}}{\mu_i} K(\nabla p_i - \rho_i g) \tag{1}$$

i represents the considered phase. This equation must be formulated for each phase (for details, see [11–14]). We note k_{ri} the relative permeability, μ_i the dynamic fluid viscosity, p_i the pressure, ρ_i the density of the phase i, and g is the gravity vector. We can define $\lambda_i = \frac{k_{ri}}{\mu_i}$ the mobility of the phase i. We can now write the mass balance equation for the phase i of a multiphase system:

$$\partial_t(\Phi \rho_i S_i) + \nabla.(\rho_i v_i) - \rho_i q_i = 0 \tag{2}$$

where Φ is the porosity, S_i the saturation of the phase i, q_i the source/sink term. Inserting the generalised Darcy law (1) into the mass balance equation (2) and considering the two phases flow system leads to this system of equations:

$$\partial_t(\Phi \rho_w S_w) - \nabla.(\rho_w \lambda_w K(\nabla p_w - \rho_w g)) - \rho_w q_w = 0 \tag{3}$$

$$\partial_t(\Phi \rho_n S_n) - \nabla.(\rho_n \lambda_n K(\nabla p_n - \rho_n g)) - \rho_n q_n = 0 \tag{4}$$

where w represents the wetting phase and n the not-wetting phase. In our case, we consider that Φ is constant in time. Moreover, we suppose that we don't have mass transfer between the two phases. The supplementary constraints to close the system of equations are: the sum of saturations is equal to one ($S_w + S_n = 1$) and the capillary pressure between the two phases (dependent on the saturations) is defined as $p_c(S) = p_n - p_w$. So, if we add the two equations and neglect the source/sink terms, we can present this system with the primary variables p_n et S_w:

$$\phi \partial_t (S_w) - \nabla.(\lambda_w K(\nabla p_n - \nabla p_c - \rho_w g)) = 0 \qquad (5)$$

$$-\nabla.(\lambda_w K(\nabla p_n - \nabla p_c - \rho_w g)) - \nabla (\lambda_n K(\nabla p_n - \rho_n g)) = 0 \qquad (6)$$

1.2 Simplifications: no capillary pressure, no gravity

In this paper, the problem is simplified and the effects of gravity and capillary pressure are not considered. This may be the case in large homogeneous porous media at high capillary number: viscous forces dominate capillary forces. So, $p_n = p_w$ and the indice n for pressure is neglected. Furthermore, if the total mobility $\lambda_T = \lambda_w + \lambda_n$ is introduced, we can define the total velocity v_T with a form similar to the one of the Darcy law:

$$v_T = -\lambda_T K(\nabla p) \qquad (7)$$

In this step the fractional flow function $f_i = \frac{\lambda_i}{\lambda_T}$ is introduced (notice that the fractional flow function depends on the saturation: $f_i = f_i(S_w)$) and if we write the system in term of v_T, the equations (5) and (6) become:

$$\phi \partial_t (S_w) + \nabla.(f_w v_T)) = 0 \qquad (8)$$

$$\nabla v_T = 0 \qquad (9)$$

2 Discretisation of the fully coupled formulation

In this section, the system is discretized in time on a mesh. We introduce the flux $F(x, t, S_w) = f_w(S_w) v_T$ and we note:

$$F_{w_{I/J}}(S^*_{w_I}, S^*_{w_J}) \approx \frac{1}{\Delta t} \int_{t^m}^{t^{m+1}} \int_{I/J} f_{w_{I/J}}(S_w, x, t) v_T(x, t) n_{I/J}(s) ds dt \qquad (10)$$

where $n_{I/J}(s)$ is the normal vector to the I/J cell interface and where $*$ is an indecision: if $* = m$ then an explicit scheme is obtained and if $* = m + 1$ then the scheme is implicit. We get:

$$\phi \frac{S_w^{m+1} - S_w^m}{\Delta t} + \sum_{J \in Neighbor(I)} F_{w_{I/J}}(S_{w_I}^*, S_{w_J}^*) = 0$$

Moreover, a monotonic flux is needed, i.e: $F_{w_{I/J}}(X, Y)$ increasing with X et decreasing with Y and $F_{w_{I/J}}(X, Y) = -F_{w_{I/J}}(Y, X)$. Also, since a fully upwind scheme is used, the mobility at the cell interface is determined in the following way (we note $\mathbf{V_T} = \int_{t^m}^{t^{m+1}} \int_{I/J} \lambda_T(S_w) K(\nabla p) \mathbf{n}(s)_{I/J} ds dt$):

$$f_{w_{I/J}}(S_w) = f_w(S_{w_I}) \quad if \quad \mathbf{V_T} > 0 \quad (11)$$

$$f_{w_{I/J}}(S_w) = f_w(S_{w_J}) \quad if \quad \mathbf{V_T} < 0 \quad (12)$$

The IMPES scheme is obtained for a two phase flow: an implicit treatment for the pressure terms and an explicit model for the saturation:

$$V\phi \frac{S_{w_I}^{m+1} - S_{w_I}^m}{t^{m+1} - t^m} + \sum_{J \in Neighbor(I)} f_w(S_{w_I}) \left(\mathbf{V_T^{m+1}}\right)^+ + f_w(S_{w_J}) \left(\mathbf{V_T^{m+1}}\right)^- = 0 \ (13)$$

$$\sum_{J \in Neighbor(I)} \left(\mathbf{V_T^{m+1}}\right)^+ + \left(\mathbf{V_T^{m+1}}\right)^- = 0 \ (14)$$

3 The CFL criteria

We present here how the CFL criteria of this scheme can be computed using the boundary conditions.

3.1 Consideration of the boundary conditions

In this paper, a waterflooding is considered: the porous media is subjected to a difference of pressure, as shown on Fig. 1.

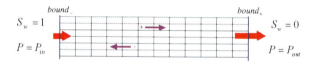

Fig. 1 Pressure imposed

Study and Approximation of IMPES Stability: the CFL Criteria

On the left side $P = P_{in}$ and $S_w = 1$. On the right side, we have $P = P_{out}$ and $S_w = 0$. We must now differentiate the surfaces σ between the cells (interior of the media) or between a cell and the edge. Taking into account these conditions and multiplying by $f_w(S_{w_I})$, the equation 14 becomes:

$$\sum_{\sigma=I|J\in\partial I\cap\Sigma_{int}} f_w(S_{w_I})\left(\mathbf{V_{T,I|J}}^{int,m+1}\right)^+ + f_w(S_{w_J})\left(\mathbf{V_{T,I|J}}^{int,m+1}\right)^-$$

$$+ \sum_{\sigma=I\in\partial I\cap\Sigma_{bound}} f_w(S_{w_I})\left(\mathbf{V}_{T,I|\sigma}^{bound,m+1}\right) = 0 \quad (15)$$

where:

$$\mathbf{V_{T,I|J}}^{int,m+1} = -\mathbf{V_{T,J|I}}^{int,m+1} = \lambda_{T,\sigma}^{int,m} T_\sigma^{int}(p_I^{m+1} - p_J^{m+1}), \sigma = I|J \in \Sigma_{int}, \quad (16)$$

$$\mathbf{V}_{T,I|\sigma}^{bound,m+1} = \lambda_{T,\sigma}^{bound,m} T_\sigma^{bound}(p_I^{m+1} - p_\sigma^{m+1}), \sigma = \Sigma_{bound}. \quad (17)$$

with $T_\sigma^{int} = \frac{S_\sigma K_{I|J}}{D_{I|J}}$ and $T_\sigma^{bound} = \frac{S_\sigma K_I}{D_{I(\sigma)}}$. Taking into account the boundary conditions in the right and left side, we obtain for the equation 13:

$$V_I\phi\frac{S_{w_I}^{m+1}-S_{w_I}^m}{\Delta t} + \sum_{\sigma=I|J\in\partial I\cap\Sigma_{int}} f_w(S_{w_I}^m)\left(\mathbf{V_{T,I|J}}^{int,m+1}\right)^+ + f_w(S_{w_J}^m)\left(\mathbf{V_{T,I|J}}^{int,m+1}\right)^-$$

$$+ \sum_{\sigma=I|J\in\partial I\cap\Sigma_{bound+}} f_w(S_{w_I}^m)\left(\mathbf{V_{T,I|\sigma}}^{bound,m+1}\right)^+ + f_w(0)\left(\mathbf{V}_{T,I|\sigma}^{bound,m+1}\right)^- \quad (18)$$

$$+ \sum_{\sigma=I|J\in\partial I\cap\Sigma_{bound-}} f_w(1)\left(\mathbf{V}_{T,I|\sigma}^{bound,m+1}\right)^+ + f_w(S_{w_I}^m)\left(\mathbf{V}_{T,I|\sigma}^{bound,m+1}\right)^- = 0$$

The last term corresponds to the velocity at the left boundary, where $S_w = 1$ is imposed. Thus we introduce the rates of change and we note:

$$d_{I|J} = \frac{f_w(S_{w_I}) - f_w(S_{w_J})}{S_{w_I} - S_{w_J}}; \quad d_{I|1} = \frac{f_w(S_{w_I}) - f_w(1)}{S_{w_I} - 1}; \quad d_{I|0} = \frac{f_w(S_{w_I}) - f_w(0)}{S_{w_I}} \quad (19)$$

The subtraction of these two equations (15, 18) gives:

$$S_{w_I}^{m+1} = S_{w_I}^m \left(1 + \frac{\Delta t}{V_I\phi}\left[\sum_{\sigma=I|J\in\partial I\cap\Sigma_{int}} \left(\mathbf{V_{T,I|J}}^{int,m+1}\right)^- d_{I|J} \right.\right. \quad (20)$$

$$\left.\left. + \sum_{\sigma=I\in\partial I\cap\Sigma_{bound-}} \left(\mathbf{V}_{T,I|\sigma}^{bound,m+1}\right)^- d_{I|1} + \sum_{\sigma=I\in\partial I\cap\Sigma_{bound+}} \left(\mathbf{V}_{T,I|\sigma}^{bound,m+1}\right)^- d_{I|0}\right]\right)$$

$$-\frac{\Delta t}{V_I \phi} \sum_{\sigma=IJ \in \partial I \cap \Sigma_{int}} \left(\mathbf{V}_{\mathrm{T,I|J}}^{int,m+1}\right)^{-} d_{I|J} S_{wJ} - \frac{\Delta t}{V_I \phi} \sum_{\sigma=I \in \partial I \cap \Sigma_{bound-}} \left(\mathbf{V}_{\mathrm{T,I}|\sigma}^{bound,m+1}\right)^{-} d_{I|1}$$

If we introduce the notation:

$$V_{CFL} = \sum_{\sigma=IJ \in \partial I \cap \Sigma_{int}} -\left(\mathbf{V}_{\mathrm{T,I|J}}^{int,m+1}\right)^{-} d_{I|J} + \sum_{\sigma=I \in \partial I \cap \Sigma_{bound-}} -\left(\mathbf{V}_{\mathrm{T,I}|\sigma}^{bound,m+1}\right)^{-} d_{I|1}$$

$$+ \sum_{\sigma=I \in \partial I \cap \Sigma_{bound+}} -\left(\mathbf{V}_{\mathrm{T,I}|\sigma}^{bound,m+1}\right)^{-} d_{I|0} \tag{21}$$

To ensure the scheme stability, we must satisfy:

$$\Delta t \leq \frac{\inf_I (V_I) \phi}{\sup [V_{CFL}] \sup_{0 \leq S_w \leq 1} f'_w(S_w)} \tag{22}$$

So, the CFL number is thereby defined: $C = \frac{\inf_I (V_I) \phi}{\sup [V_{CFL}] \sup_{0 \leq S_w \leq 1} f'_w(S_w)}$

3.2 The fractional flow function: Brooks and Corey [15] [16]

The goal of this section is to describe the terms used in the CFL condition formula. Classically, in reservoir multiphase simulations, models are used to define relative permeability. The most known formula was developed by Brooks and Corey and is based on a power law:

$$k_{rw}(S_w) = k_{rw_{max}} \left(\frac{S_w - S_{wi}}{1 - S_{wi} - S_{nr}}\right)^{a_w} \quad ; \quad k_{rn}(S_w) = k_{rn_{max}} \left(\frac{1 - S_w - S_{nr}}{1 - S_{wi} - S_{nr}}\right)^{a_n} \tag{23}$$

where a_w and a_n are the Corey exponent. Note that S_w varies between S_{wi} and $1 - S_{or}$. In the inequation 22, the fractional flow function $f_w(S_w)$ is present:

$$f_w(S_w) = \frac{\lambda_w}{\lambda_w + \lambda_n} = \frac{\frac{k_{rw}}{\mu_w}}{\frac{k_{rw}}{\mu_w} + \frac{k_{rn}}{\mu_n}} = \mu_n \frac{k_{rw}}{\mu_n k_{rw} + \mu_w k_{rn}}. \tag{24}$$

Its derivative is computed:

$$f'_w(S_w) = \mu_n \frac{k'_{rw}[\mu_n k_{rw} + \mu_w k_{rn}] - k_{rw}[\mu_n k'_{rw} + \mu_w k'_{rn}]}{[\mu_n k_{rw} + \mu_w k_{rn}]^2} \tag{25}$$

3.3 Numerical test case

The aim of this test is to compare the theoretical value C_{Th} of the CFL condition (inequation 22) and C_{Ma} the one found experimentally by slowly raising Δt until the results stop being physically acceptable. Water is injected in a oil-saturated pipe. Its dimensions are: $[L_x, L_y, L_z] = [100, 10, 10]$ m. The built mesh has 100 elements. Each cell I has therefore a size of $1 \times 10 \times 10$ m and $V_I = 100$ m^3. We choose these parameters for the test case:

- Independent intrinsic permeability K$= 100 \times 10^{-15}$ m^2 and porosity $\Phi = 0.2$
- Residual oil saturation $S_{nr} = 0.5$ and critical water saturation $S_{wi} = 0.25$
- Oil relative permeability at connate water saturation $k_{rn_{max}} = 1$
- Water relative permeability at the residual oil saturation $k_{rw_{max}} = 0.5$
- Exponents of relative permeability curves $a_n = 2$ and $a_w = 2$
- Entry pressure $P_{in} = 20 \times 10^5$ Pa and exit pressure $P_{out} = 10 \times 10^5$ Pa
- Non-wetting fluid (oil) viscosity $\mu_n = 0.01$ Pa.s
- Wetting fluid (water) viscosity $\mu_w = 0.001$ Pa.s
- Time spent ($\Delta t \times$ iterations number) $T = 173.61$ days

The result of this simulation is drawn in Fig. 2: the water saturation front is apparent.

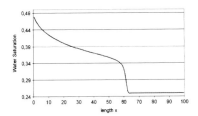

Fig. 2 Water Saturation in the porous media after simulation $T=173$ days

In the first step, we take care of the theoretical value of C. Regarding the values of the parameters, $\inf_I V_I = 100 m^3$ and $\Phi = 0.2$. The derivative of f_w is drawn in Fig. 3 and shows a single maximum: resolving $f_w'' = 0$ for these parameters, we find easily $\sup_{0.25 \leq S_w \leq 0.5} f_w'(S_w) \simeq 9.81$

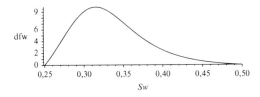

Fig. 3 Derivative of f_w in response of S_w

During a simulation, the value of V_{CFL} is calculated at each iteration and we find $\sup [V_{CFL}] \simeq 1.53 \times 10^{-5}$ m^3.s^{-1}. With the equation 22, we can then compute the theorical value of C:

$$C_{Th} \simeq 1.33 \times 10^5 \text{ s.}$$

In a second step, the experimental value of C is computed. Many simulations are performed while Δt is increased manually. For any time during one simulation, if the water saturation curve (Fig. 2) stops being monotonic, Δt has reached its maximum and the result is $C_{Ma} = \Delta t_{max}$. We find:

$$C_{Ma} \simeq 1.33 \times 10^5 \text{ s.}$$

C_{Th} and C_{Ma} are very similar, so we can conclude that if Δt doesn't exceed C_{Th}, the CFL condition won't be violated and the results will be physically acceptable.

3.4 Conclusion

In this paper, we have presented the IMPES scheme for two immiscible and incompressible fluids. For this two-phase flow system, gravity and capillary pressure are neglected. We also consider that there is neither structural evolution nor mass transfer. This CFL condition is a necessary condition for convergence. The approximation was applied to a short example in order to confront the theoretical value with the one found experimentally. More intricate cases must be studied by adding spatial dimensions: a complex mesh with numerous cells will set a significant limit for a heterogeneous test case. Besides, the contribution of gravity and capillary pressure can be studied.

References

1. H.L.Stone, A.O. Jr. Garder, Analysis of gas-Cap or dissolved gas drive reservoirs, Trans., AIME, (1961) 222.
2. L.C.Young, R.E. Stephenson, A generalized compositionnal approach for reservoir simulation, SPEJ (October 1983) 727.
3. K. H. Coats, A note of IMPES and some IMPES-based simulation models, SPEJ (September 2000) 245.
4. K. H. Coats, IMPES stability, the CFL limit, SPEJ, Vol. 8, No. 3, (September 2003).
5. J. W. Sheldon, C. D. Harris, D Bavly, A method for general reservoir behavior simulation on digital computers, paper SPE 1521-G, 1960 SPE annual fall meeting, Denver, Colorado, 2-5 October (1960).
6. J. C. Martin, Simplified equations of flow in gas drive reservoirs and the theoretical foundation of multiphase pressure buildup analyses, Trans. AIME 216 (1959).
7. K. H. Coats, Computer simulation of three-phase flow in reservoir, U. Of Austin, Texas (1968).
8. R. G. Fagin, C. H. Stewart, A new approach to the two-dimensional multiphase reservoir simulator, SPEJ 175, Trans., AIME, 237 (June 1966).

9. M. C. Leverett, Flow of oil-water mixture through unconsolidated sands, Trans. AIME 132, 149, (1938).
10. R. D. Wyckoff, H. G. Botset, The flow of gas-liquid mixtures through unconsolidated sands, Physics 7, 325 (1936).
11. J. Wolf, Comparison of mathematical and numerical models for twophase flow in porous media. Diplomarbeit, Institut für Wasserbau, Universität Stuttgart, (2008).
12. R. Helmig, Multiphase flow and transport processes in the subsurface. Springer, (1997).
13. A. Scheidegger, The physics of flow through porous media. In: University of Toronto Press. Toronto and Buffalo, 3rd edition, (1974).
14. J. Niessner, S. Berg, S. Majid Hassanizadeh, Comparison of two-phase Darcy's law with Thermodynamically consistent approach. Transp. Porous Med, DOI 10.1007/s11242-011-9730-0, (2011).
15. R. J. Brooks, A. T. Corey, Hydraulic properties of porous media. In Hydrol. Pap. 3, Colo. State Univ., Fort Collins, (1964).
16. A. Corey, Mechanics of heterogeneous fluids in porous media. In Water Resour., 150 pp., Publ., Fort Collins, Colo., (1977).

The paper is in final form and no similar paper has been or is being submitted elsewhere.

Numerical Solution of 2D and 3D Atmospheric Boundary Layer Stratified Flows

Jiří Šimonek, Karel Kozel, and Zbyněk Jaňour

Abstract The work deals with the numerical solution of the 3D turbulent stratified flows in atmospheric boundary layer over the "sinus hill". Mathematical model for the turbulent stratified flows in atmospheric boundary layer is the Boussinesq model - Reynolds averaged Navier-Stokes equations (RANS) for incompressible turbulent flows with addition of the density change equation. The artificial compressibility method and the finite volume method have been used in all computed cases and Lax-Wendroff scheme (MacCormack form) has been used together with the Cebecci-Smith algebraic turbulence model. Computations have been performed with Reynold's number $10^8 \approx u_\infty = 1.5 \frac{m}{s}$ and with density range $\rho \in [1.2; 1.1] \frac{kg}{m^3}$.

Keywords CFD, Finite Volume Method, Variable density Flows, Atmospheric Boundary Layer Flows

MSC2010: 65N08, 65N40

1 Mathematical model

Reynolds averaged Navier-Stokes equations for incompressible flows with addition of the equation of density change (Boussinesq model) have been used as a mathematical model for flows in atmospheric boundary layer:

Jiří Šimonek and Karel Kozel
Czech Technical University, Faculty of Mechanical Engineering (Department of Technical Mathematics), Karlovo nám. 13, 121 35 Praha 2, Czech Rep., e-mail: jiri.simonek@fs.cvut.cz, karel.kozel@fs.cvut.cz

Zbyněk Jaňour
Institute of Thermomechanics - Academy of Sciences of the Czech Republic Dolejškova 1402/5, 182 00 Praha 8, Czech Rep. e-mail: janour@it.cas.cz

$$u_x + v_y + w_z = 0 \tag{1}$$
$$u_t + (u^2 + p)_x + (u \cdot v)_y + (u \cdot w)_z = (v_e u_x)_x + (v_e u_y)_y + (v_e u_z)_z \tag{2}$$
$$v_t + (u \cdot v)_x + (v^2 + p)_y + (w \cdot v)_z = (v_e v_x)_x + (v_e v_y)_y + (v_e v_z)_z \tag{3}$$
$$w_t + (u \cdot w)_x + (v \cdot w)_y + (w^2 + p)_z = (v_e w_x)_x + (v_e w_y)_y + (v_e w_z)_z - \frac{\rho}{\rho_0} g \tag{4}$$
$$\rho_t + u \cdot \rho_x + v \cdot \rho_y + w \cdot \rho_z = 0, \tag{5}$$

where (u, v, w) is a velocity vector, $p = \frac{P}{\rho_0}$ (P- static pressure, ρ_0 - initial maximal density), ρ - density, $v_e = v + v_t$, v - laminar kinematic viscosity, v_t - turbulent kinematic viscosity computed by the Cebecci-Smith algebraic turbulence model and g - gravity acceleration. Upstream density and pressure are changing depending on height (z-axis) according to the hydrostatic equilibrium pressure function:

$$\rho_\infty(z) = -\frac{\rho_0 - \rho_h}{h} \cdot z + \rho_0 \tag{6}$$
$$\frac{\partial p_\infty}{\partial z} = -\frac{\rho_\infty(z)}{\rho_0} \cdot g \tag{7}$$

The (6) is the linear decreasing function of density and the (7) is the hydrostatic pressure function.

It is possible to separate $p = p_\infty + p'$, where the term p_∞ is the initial state of pressure, the term p' is the pressure disturbance. Using this substitution and $\rho = \rho_\infty + \rho'$ in the system (1) - (5) leads to:

$$u_x + v_y + w_z = 0 \tag{8}$$
$$u_t + (u^2 + p')_x + (u \cdot v)_y + (u \cdot w)_z = (v_e u_x)_x + (v_e u_y)_y + (v_e u_z)_z \tag{9}$$
$$v_t + (u \cdot v)_x + (v^2 + p')_y + (w \cdot v)_z = (v_e v_x)_x + (v_e v_y)_y + (v_e v_z)_z \tag{10}$$
$$w_t + (u \cdot w)_x + (v \cdot w)_y + (w^2 + p')_z = (v_e w_x)_x + (v_e w_y)_y + (v_e w_z)_z - \frac{\rho'}{\rho_0} g \tag{11}$$
$$\rho_t + u \cdot \rho_x + v \cdot \rho_y + w \cdot \rho_z = 0, \tag{12}$$

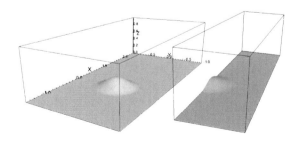

Fig. 1 Computational domains

2 Boundary conditions for 3D computations

Inlet boundary conditions $u = u_\infty = 1.5$, $v = v_\infty = 0$, $w = w_\infty = 0$, $\rho = \rho_\infty(z)$, where $\rho_\infty(z)$ is a linear function which is decreasing with increasing z:

$$\rho_\infty(z) = -\frac{\rho_0 - \rho_h}{h} \cdot z + \rho_0,$$

where $\rho_0 = 1.2 \frac{kg}{m^3}$ is a lower (maximal) density and $\rho_h = 1.1 \frac{kg}{m^3}$ is a upper (minimal) density.
Outlet boundary conditions: $p' = 0$
Boundary conditions on the wall: $u = 0$, $v = 0$, $w = 0$, $\rho = \rho_0$.
Boundary conditions on the upper domain boundary: $p' = 0$, $\frac{\partial u}{\partial n} = 0$, $\frac{\partial v}{\partial n} = 0$, $\frac{\partial w}{\partial n} = 0$, $\rho = \rho_h$
Boundary conditions on side-walls of the domain: symmetry boundary conditions.

3 Boundary conditions for 2D computations

Inlet boundary conditions $u = u_\infty = 1.0$, $w = w_\infty = 0$, $\rho = \rho_\infty(z)$, where $\rho_\infty(z)$ is a linear function which is decreasing with increasing z:

$$\rho_\infty(z) = -\frac{\rho_0 - \rho_h}{h} \cdot z + \rho_0,$$

where $\rho_0 = 1.2 \frac{kg}{m^3}$ is a lower (maximal) density and $\rho_h = 0.6 \frac{kg}{m^3}$ is a upper (minimal) density.
Outlet boundary conditions: $p' = 0$
Boundary conditions on the wall: $u = 0$, $w = 0$, $\rho = \rho_0$
Boundary conditions on the upper domain boundary: $p' = 0$, $u = 1.0$, $\frac{\partial w}{\partial n} = 0$, $\rho = \rho_h$

4 Numerical solution

In all cases the artificial compressibility method has been used, i.e. continuity equation is completed by term $\frac{p'_t}{\beta^2}$, $\beta^2 \in R^+$ - then the modified RANS system is valid only for steady state solutions in which $\frac{p'_t}{\beta^2} = 0$. Modified equations can be expressed in a vector form as follows:

$$W_t + F_x + G_y + H_z = R_x + S_y + T_z + K \tag{13}$$

$$W = \begin{Vmatrix} \frac{p'}{\beta^2} \\ u \\ v \\ w \\ \rho \end{Vmatrix} \quad F = \begin{Vmatrix} u \\ u^2 + p' \\ u \cdot v \\ u \cdot w \\ u \cdot \rho \end{Vmatrix} \quad G = \begin{Vmatrix} v \\ v \cdot u \\ v^2 + p' \\ v \cdot w \\ v \cdot \rho \end{Vmatrix} \quad H = \begin{Vmatrix} w \\ w \cdot u \\ w \cdot v \\ w^2 + p' \\ w \cdot \rho \end{Vmatrix} \quad (14)$$

$$R = v_e \begin{Vmatrix} 0 \\ u_x \\ v_x \\ w_x \\ 0 \end{Vmatrix} \quad S = v_e \begin{Vmatrix} 0 \\ u_y \\ v_y \\ w_y \\ 0 \end{Vmatrix} \quad T = v_e \begin{Vmatrix} 0 \\ u_z \\ v_z \\ w_z \\ 0 \end{Vmatrix} \quad K = -\frac{\rho - \rho_\infty}{\rho_0} \begin{Vmatrix} 0 \\ 0 \\ 0 \\ g \\ 0 \end{Vmatrix} \quad (15)$$

Where W is the vector of conservative variables, F, G, H are convective fluxes, R, S, T are diffusive fluxes and K is the source term, $v_e = v_{laminar} + v_{turbulent}$.

The finite volume method has been used on structured grid of hexahedral cells (uniform in x and y direction, refined near walls in z direction up to $\Delta z = 10^{-5}$, 200x100x80 cells) in 3D and the grid of quadrilateral cells (uniform in x and refined near walls in z direction $\Delta z = 10^{-5}$, 100x40 cells) in 2D.

Lax-Wendroff predictor-corrector scheme (MacCormack form) has been used in a following form:

$$W_i^{n+\frac{1}{2}} = W_i^n - \frac{\Delta t}{V_i} \left(\sum_{k=1}^{6} (\tilde{F} - \tilde{R}, \tilde{G} - \tilde{S}, \tilde{H} - \tilde{T})_{i,k}^n \mathbf{n}_{i,k}^0 \Delta S_{i,k} \right) + \Delta t K_i^n \quad (16)$$

$$W_i^{n+1} = \frac{1}{2}(W_i^{n+\frac{1}{2}} + W_i^n) - \frac{\Delta t}{2V_i} \left(\sum_{k=1}^{6} (\tilde{F} - \tilde{R}, \tilde{G} - \tilde{S}, \tilde{H} - \tilde{T})_{i,k}^{n+\frac{1}{2}} \mathbf{n}_{i,k}^0 \Delta S_{i,k} \right) + \frac{\Delta t}{2} K_i^{n+\frac{1}{2}}$$
(17)

Convective fluxes have been taken in a forward direction in a predictor step and in a backward direction in a corrector step (see Fig. (2)). Viscous fluxes have been computed centrally (see Fig. (3)).

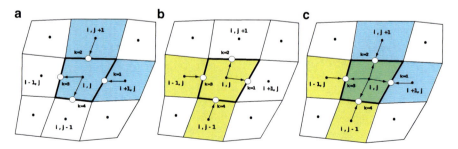

Fig. 2 Stencil for inviscid fluxes computation, (a) predictor step, (b) corrector step, (c) predictor + corrector

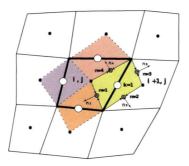

Fig. 3 Stencil for viscous fluxes computation (dual cells)

Jameson's artificial dissipation has been used to stabilize numerical solution.

Cebecci-Smith algebraic turbulence model has been used to compute the turbulent viscosity ν_t. Domain Ω is divided into two subdomains. In the inner subdomain (near walls) the inner turbulent viscosity ν_{Ti} is computed. In the outer subdomain the outer turbulent viscosity ν_{To} is computed. Most common procedure is to compute both turbulent viscosities and use the minimal one:

$$\nu_T = \min(\nu_{Ti}, \nu_{To}). \tag{18}$$

For turbulent viscosity computing is necessary to use local systems of coordinates (X, Y), where X is parallel with the profile and Y is normal of the profile. In inner subdomain the turbulent viscosity is defined as follows:

$$\nu_{Ti} = \rho l^2 \left| \frac{\partial U}{\partial Y} \right|, \tag{19}$$

where ρ is the density of fluid, (U, V) are components of velocity vector in direction of (X, Y) and l is given by equation:

$$l = \kappa Y F_D, \tag{20}$$

where:

$$F_D = 1 - \exp\left(-\frac{1}{A^+} u_r Y Re\right), \tag{21}$$

$$u_r = \left(\nu \left|\frac{\partial U}{\partial Y}\right|\right)_\omega^{\frac{1}{2}}. \tag{22}$$

In outer subdomain the turbulent viscosity is defined by Clauser's equation:

$$\nu_{To} = \rho \alpha \delta^* U_e F_k, \tag{23}$$

$$F_k = \left[1 + 5.5 \left(\frac{Y}{\delta}\right)^6\right]^{-1}, \quad U_e = U(\delta) \qquad (24)$$

where δ is the thickness of boundary layer and

$$\delta^* = \int_0^\delta \left(1 - \frac{U}{U_e}\right) dY. \qquad (25)$$

Following values of the constants were used:

$$\kappa = 0.4, \ \alpha = 0.0168, \ A^+ = 26. \qquad (26)$$

5 3D Numerical results

Following cases of stratified turbulent flows in atmospheric boundary layer have been computed (see Figs. 7 - 10). Authors consider flows over a geometry with the "sinus hill" with the height 10% of its basis length - half domain symmetrical case and the general 3D geometry and the "hill" with the height 15% of its basis length - general 3D geometry (see Fig. (1)). All the computations have been solved with $Re = 10^8 \approx u_\infty = 1.5 \frac{m}{s}$ and with density change $\rho_\infty \in [1.2; 1.1]$.

6 2D Numerical results

One case with $Re = 6.67 \cdot 10^7 \approx u_\infty = 1.0 \frac{m}{s}$ and with density change $\rho_\infty \in [1.2; 0.6]$ has been solved. One can see in the Figs. (4) (5) (6) the waving character of the flow field. These waves are so called Lee waves which should be seen in the results of the stratified computations. Lee waves were only computed in the 2D case with a coarser mesh (uniform in x and refined near walls in z direction up to $\Delta z = 10^{-5}$, 100x40 cells).

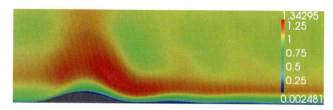

Fig. 4 2D case - Velocity magnitude $\left[\frac{m}{s}\right]$

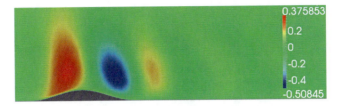

Fig. 5 2D case - Z-velocity $\left[\frac{m}{s}\right]$

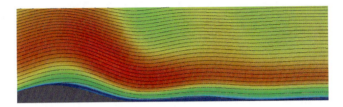

Fig. 6 2D case - stream lines

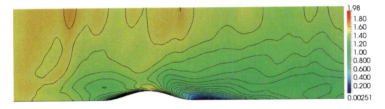

Fig. 7 Half domain symmetrical solution - y-slice in the middle; Velocity mag. $\left[\frac{m}{s}\right]$

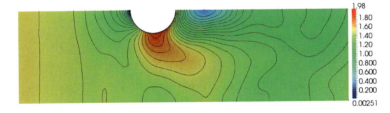

Fig. 8 Half domain symmetrical solution - z-slice in the middle of the hill; Velocity mag. $\left[\frac{m}{s}\right]$

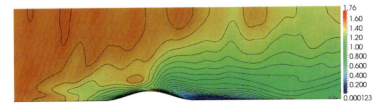

Fig. 9 Full domain solution - y-slice in the middle of the hill; Velocity mag. $\left[\frac{m}{s}\right]$

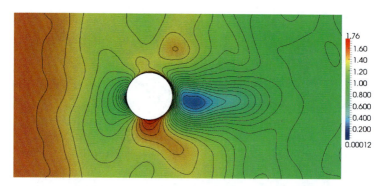

Fig. 10 Full domain solution - z-slice in the middle of the hill; Velocity mag. [$\frac{m}{s}$]

7 Conclusions

Three results of the 3D incompressible turbulent stratified flows in atmospheric boundary layer over the "sinus hill" with Reynolds number $Re = 10^8 \approx u_\infty = 1.5\,\frac{m}{s}$ and with range of density change $\rho \in [1.2; 1.1]\,\frac{kg}{m^3}$ have been presented. As one can see in the Figs. (8) and (10) the solution is not symmetrical and therefore it is necessary to perform only the full domain computations in the future. Lee waves were only computed in the 2D case with a coarser mesh.

The future work will be to extend this model for more complex geometries in 3D and to make a comparison with other numerical methods and mathematical models for variable density flows.

Acknowledgements This work was partially supported by Research Plan MSM 6840070010 and MSM 6840770003 and grant GA ČR no. 201/08/0012.

References

1. Eidsvik, K., Utnes, T.: Flow separation and hydrostatic transition over hills modeled by the Reynolds equations, Journal of Wind Engineering and Industrial Aerodynamics, Issues 67 - 68 (1997), p. 408–413
2. Feistauer, M., Felcman, J., Straškraba, I. Mathematical and Computational Methods for Compressible Flow, Clarendon Press, Oxford 2003.
3. Hirsh, C.: Numerical Computation of Internal and External Flows I and II, John Willey and Sons, New York 1991.
4. Fletcher, C., A., J.: Computational Techniques for Fluid Dynamics I and II, Springer Verlag, Berlin 1996.
5. Šimonek, J., Kozel, K., Fraunié, Ph., Jaňour, Z.: Numerical Solution of 2D Stratified Flows in Atmospheric Boundary Layer, Topical Problems of Fluid Mechanics 2008, Prague 2008.
6. Uchida, T., Ohya, Y.: Numerical Study of Stably Stratified Flows over a Two Dimensional Hill in a Channel of Finite Depth, Fluid Dynamics Research 29/2001 (p. 227 - 250).

The paper is in final form and no similar paper has been or is being submitted elsewhere.

On The Numerical Validation Study of Stratified Flow Over 2D–Hill Test Case

Sládek Ivo, Kozel Karel, and Janour Zbynek

Abstract The paper deals with flow validation study performed using our in–house 3D–code which implements mathematical and numerical model capable to compute stratified atmospheric boundary layer flows over hills or terrain obstacles. The objectives of the paper are at first to formulate the applied mathematical/numerical model and at second to present some results from the validation study of thermally stratified flow over an isolated 2D–hill test case. The mathematical model is based on system of RANS equations closed by a two–equation high–Reynolds number k–ε turbulence model. The finite volume method and the explicit Runge–Kutta time integration method are utilized for numerical procedure.

Keywords Turbulent Boundary Layers, Stratification effects, k–ε modeling, Finite Volume Method, Runge–Kutta method
MSC2010: 76F40, 76F45, 76F60, 65N08, 65L06

1 Mathematical model

The flow itself is assumed to be turbulent, viscous, incompressible, stationary and neutrally/stably stratified in general. The mathematical model is based on the Reynolds–averaged Navier–Stokes equations (RANS) modified by the Boussinesq approximation according to which the following decomposition is utilized for pressure p, density ρ and potential temperature Θ

Sládek Ivo and Janour Zbynek
Institute of Thermomechanics, Dolejskova 5, ZIP 182 00, Prague 8, CZ, e-mail: islad@tiscali.cz, janour@it.cas.cz

Kozel Karel
Faculty of Mechanical Engineering, U12101, Karlovo náměstí 13, ZIP 121 35, Prague 2, CZ
e-mail: karel.kozel@fs.cvut.cz

$$p = p_0 + p', \quad \rho = \rho_0 + \rho', \quad \Theta = \Theta_0 + \Theta'$$

where $_0$ denotes synoptic large scale part and $'$ concerns the small scale deviation from the synoptic part due to local conditions. The potential temperature Θ is defined as temperature of the atmospheric air after adiabatic compression or expansion to the reference pressure $p_{ref} = 1\,bar$, so $\Theta = T(p_{ref}/p)^\kappa$.

The governing equations can be re-casted in the conservative and vector form as follows, [3], [4]

$$(\mathbf{F})_x + (\mathbf{G})_y + (\mathbf{H})_z = (\mathbf{R})_x + (\mathbf{S})_y + (\mathbf{T})_z + \mathbf{f}, \qquad (1)$$

where the terms $\mathbf{F}$, $\mathbf{G}$, $\mathbf{H}$ represent the physical inviscid fluxes and $\mathbf{R}$, $\mathbf{S}$, $\mathbf{T}$ denote the viscous fluxes. The system (1) is then modified in order to be solved by the artificial compressibility method

$$\mathbf{W}_t + \begin{pmatrix} u \\ u^2 + \frac{p'}{\rho_0} \\ uv \\ uw \\ u\Theta' \end{pmatrix}_x + \begin{pmatrix} v \\ vu \\ v^2 + \frac{p'}{\rho_0} \\ vw \\ v\Theta' \end{pmatrix}_y + \begin{pmatrix} w \\ wu \\ wv \\ w^2 + \frac{p'}{\rho_0} \\ w\Theta' \end{pmatrix}_z = \begin{pmatrix} 0 \\ Ku_x \\ Kv_x \\ Kw_x \\ \tilde{K}\Theta'_x \end{pmatrix}_x + \begin{pmatrix} 0 \\ Ku_y \\ Kv_y \\ Kw_y \\ \tilde{K}\Theta'_y \end{pmatrix}_y + \begin{pmatrix} 0 \\ Ku_z \\ Kv_z \\ Kw_z \\ \tilde{K}\Theta'_z \end{pmatrix}_z + \mathbf{f} \qquad (2)$$

where

$$\mathbf{W} = (p'/\beta^2, u, v, w, \Theta')^T, \quad \mathbf{f} = (0, 0, 0, +g\frac{\Theta'}{\Theta_0}, -w\Theta'_z)^T \qquad (3)$$

where $\mathbf{W}$ is vector of unknown variables and $\mathbf{f}$ is the buoyancy force due to the thermal stratification.

The velocity vector components read u, v, w, the term g is the gravitational acceleration, the parameters K, $\tilde{K}$ refer to the turbulent diffusion coefficients for the velocity components and for the potential temperature deviation and β is related to the artificial sound speed.

The synoptic scale part of the potential temperature is taken as $\Theta_0 = \Theta_w + \gamma z$ where Θ_w is the wall potential temperature and γ refers to the wall–normal gradient to be > 0 for the stable stratification and $= 0$ for the neutral stratification.

The system (2) is solved in the computational domain Ω under a stationary boundary conditions for $t \to \infty$ (t is an artificial time variable) to obtain the expected steady–state solution for all the unknown variables involved in the vector $\mathbf{W}$.

2 Turbulence model

Closure of the system of governing equations (2) is achieved by a standard k–ε turbulence model without damping functions [7], [1]. Two additional transport equations are added to the system (2), one for the turbulent kinetic energy abbreviated by

k and one for the rate of dissipation of turbulent kinetic energy denoted by ε. The thermal stratification is taken into account

$$(ku)_x + (kv)_y + (kw)_z = \left(K^{(k)} k_x\right)_x + \left(K^{(k)} k_y\right)_y + \left(K^{(k)} k_z\right)_z + P + G - \varepsilon \quad (4)$$

$$(\varepsilon u)_x + (\varepsilon v)_y + (\varepsilon w)_z = \left(K^{(\varepsilon)} \varepsilon_x\right)_x + \left(K^{(\varepsilon)} \varepsilon_y\right)_y + \left(K^{(\varepsilon)} \varepsilon_z\right)_z +$$
$$C_{\varepsilon 1}(1 + C_{\varepsilon 3} R_f)\frac{\varepsilon}{k}(P + G) - C_{\varepsilon 2}\frac{\varepsilon^2}{k} \quad (5)$$

where $G = \beta_\Theta g \frac{\nu_T}{\sigma_\Theta} \frac{\partial \Theta}{\partial z}$ abbreviates the buoyancy term, $R_f = -\frac{G}{P}$ where $P = \tau_{ij} \frac{\partial v_i}{\partial x_j}$ denotes the turbulent production term for the Reynolds stress written as

$$\tau_{ij} = -\frac{2}{3} k \delta_{ij} + \nu_T \left(\frac{\partial v_i}{\partial x_j} + \frac{\partial v_j}{\partial x_i}\right) \quad (6)$$

and the terms $K^{(k)}$, $K^{(\varepsilon)}$, $\tilde{K}$ stand for the diffusion coefficients and ν_T for the turbulence viscosity

$$K^{(k)} = \nu + \frac{\nu_T}{\sigma_k}, \quad K^{(\varepsilon)} = \nu + \frac{\nu_T}{\sigma_\varepsilon}, \quad \tilde{K} = \nu + \frac{\nu_T}{\sigma_\Theta}, \quad \nu_T = C_\mu \frac{k^2}{\varepsilon}. \quad (7)$$

The model constants are as follows

$$C_\mu = 0.09, \quad \sigma_k = 1.0, \quad \sigma_\varepsilon = 1.3, \quad C_{\varepsilon 1} = 1.44, \quad C_{\varepsilon 2} = 1.92, \quad C_{\varepsilon 3} = 0.7. \quad (8)$$

Note that the buoyancy term $G = 0$ in case of neutral stratification.

3 Numerical model

The cell–centered type of finite volume method is applied on structured non–orthogonal grid made of hexahedral control cells Ω_{ijk}. The system of equations (2)+(4)+(5) is integrated over each control cell using the divergence theorem and the mean value theorem, [2]

$$\mathbf{W}_t\big|_{ijk} = -\frac{1}{\mu_{ijk}} \oint_{\partial \Omega_{ijk}} \left[(\mathbf{F} - K \cdot \mathbf{R}) \, dS_1 + (\mathbf{G} - K \cdot \mathbf{S}) \, dS_2 + (\mathbf{H} - K \cdot \mathbf{T}) \, dS_3\right],$$
(9)

where $\mathbf{W}_t\big|_{ijk}$ is the mean value of $\mathbf{W}_t$ over the control cell and $\mu_{ijk} = \int_{\Omega_{ijk}} dV$. The right hand side of (9) is approximated by

$$\mathbf{W}_t\big|_{ijk} \approx -\frac{1}{\mu_{ijk}} \sum_{l=1}^{6} \Big[(\tilde{\mathbf{F}}_l - K_l \cdot \tilde{\mathbf{R}}_l)\Delta S_1^l + (\tilde{\mathbf{G}}_l - K_l \cdot \tilde{\mathbf{S}}_l)\Delta S_2^l + (\tilde{\mathbf{H}}_l - K_l \cdot \tilde{\mathbf{T}}_l)\Delta S_3^l\Big]. \tag{10}$$

Space discretization of the convective terms in (10) is performed using central differencing while the dual control volumes of octahedral shape is utilized for computation of the viscous terms in (10) at each face of Ω_{ijk}. The resulting semi-discrete system of ordinary differential equations is then integrated in time using the (3)–stage explicit Runge–Kutta method, [6], [9], [5].

The numerical method is second order accurate both in time and space on orthogonal grids. Also the artificial viscosity term of the 4th order is applied due to central differencing of the convective terms in (10) which effectively removes a spurious, high frequency oscillations generated in the computed flow–field.

4 Boundary conditions

The system (2)+(4)+(5) is solved with the following boundary conditions [1], [7]

- Inlet: $u = \frac{u^*}{\kappa}\ln\left(\frac{z}{z_0}\right)$, $v = 0$, $w = 0$, $k = \frac{u^{*2}}{\sqrt{C_\mu}}\left(1 - \frac{z}{D}\right)^2$, $\varepsilon = \frac{C_\mu^{3/4} \cdot k^{3/2}}{\kappa \cdot z}$, $\Theta' = 0$

  where the expression for u velocity component is used to cover the boundary layer depth D while constant value $u = U_0$ is prescribed above the boundary layer depth up to the top of computational domain.
- Outlet: homogeneous Neumann conditions for all quantities
- Top: $u = U_0$, $v = 0$, $\frac{\partial w}{\partial z} = 0$, $\frac{\partial k}{\partial z} = 0$, $\frac{\partial \varepsilon}{\partial z} = 0$, $\frac{\partial C}{\partial z} = 0$, $\frac{\partial \Theta'}{\partial z} = 0$
- Wall: standard wall functions are applied and $\frac{\partial C}{\partial n} = 0$ for the concentration and $\Theta' = 0$ for the potential temperature deviation which is equivalent to $\Theta_0 = 300\,K$

where U_0 represents the free–stream velocity magnitude, u^* is the friction velocity, $\kappa = 0.40$ denotes the von Karman constant, z_0 represents the roughness parameter and the parameter D refers to the boundary layer depth.

The wall–function approach enables to apply a wall–coarser grid where near–wall profiles of computed quantities are reconstructed using the algebraic relations, [8].

5 Validation case

The reference numerical results due to Eidsvik and Utnes [10] have been used for comparison. The computational domain extended distance $-15H$ up and $+25H$ downwind of the hill summit and to vertical height $10H$, where $H = 1000\,m$ is the

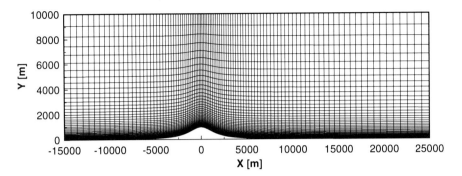

Fig. 1 The whole computational domain and grid 100x40 cells

hill height, see Fig. 1. The integration was performed on grid having 100x40 cells non–uniformly expanding upwind, downwind from the hill summit and vertically from wall using the expansion ratio parameters $ax = 1.04$ and $ay = 1.10$ leading to minimum space increments $\Delta x_{min} = 165\,m$ and $\Delta y_{min} = 20\,m$. Details regarding the grid spacing used by Eidsvik are not available in the reference paper [10].

The flow–field input data used in [10]: the free–stream air velocity $U_0 = 10.5\,m/s$, boundary layer depth of $D = 100\,m$, the friction velocity $u^* = 0.406\,m/s$, the roughness parameters $z_0 = 5\,mm$ and the Reynolds number based on U_0, hill height H and the air kinematic viscosity $\nu = 1.5 \cdot 10^{-5}\,m^2/s$ is $Re = 6.7 \cdot 10^8$.

The inlet profiles for velocity vector components u, v, w, turbulence quantities k, ε as well as for potential temperature deviation Θ' were constructed as described in the Sect. 4.

Totally four different computations have been performed using the same labeling as in [10], N0, N1, N2 and N3. Specifically, the following thermal stratifications of the atmospheric boundary layer were tested

- N0–case: neutral stratification conditions $\gamma = \frac{\partial \Theta_0}{\partial z} = 0\,K/m$
- N1–case: weak stable stratification conditions $\gamma = \frac{\partial \Theta_0}{\partial z} = 3.09 \cdot 10^{-3}\,K/m$
- N2–case: middle stable stratification conditions $\gamma = \frac{\partial \Theta_0}{\partial z} = 12.36 \cdot 10^{-3}\,K/m$
- N3–case: strong stable stratification conditions $\gamma = \frac{\partial \Theta_0}{\partial z} = 27.80 \cdot 10^{-3}\,K/m$

5.1 Numerical results

Separation zone behind hill was found in N0–case under neutral stratification conditions having separation point at $x_1 = 0.9H$ and reattachment point at $x_2 = 3.3H$ downstream from the hill top, see Fig. 2. The recirculation zone in our case is smaller compared to Eidsvik [10] under the same flow conditions where his

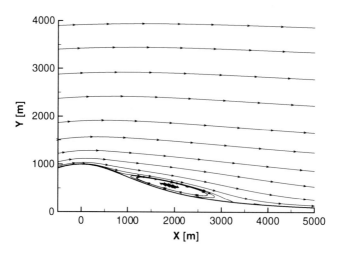

Fig. 2 Zoom to separation zone in N0–case under neutral stratification conditions

separation, reattachment points are $x_1 = 0.8H$, $x_2 = 5.3H$, respectively. Contours of the wall–normal velocity component w are shown in the following four Figs. 3–6 corresponding to N0–, N1–, N2– and N3–case under neutral, weak, middle and strong stratification conditions, respectively. All contours are labeled using levels of w wall–normal velocity component in $[m/s]$. The lee–waves in cases N1, N2 and N3 are well captured as closed isolines of the wall–normal w velocity component changing sign from "+" zone where the wave has an increasing slope to "-" zones where it has a decreasing slope.

According to theory of the internal gravitational waves [11], it is possible to estimate the wavelength of the lee–waves depending on selected stratification conditions. The relation can be written as

$$\lambda_{theoretical} = 2\pi U_0 \left(\frac{g}{\Theta_0} \frac{\partial \Theta_0}{\partial z} \right)^{-1/2} \quad (11)$$

Our prediction of the wavelength is compared to the theory and also to the predictions by Eidsvik [10]

- N0–case: no lee-waves present
- N1–case: $\lambda_{computed} = 6.5\,km$, $\lambda_{Eidsvik} = 6.5\,km$, $\lambda_{theoretical} = 6.3\,km$
- N2–case: $\lambda_{computed} = 4.0\,km$, $\lambda_{Eidsvik} = 3.7\,km$, $\lambda_{theoretical} = 3.1\,km$
- N3–case: $\lambda_{computed} = 2.8\,km$, $\lambda_{Eidsvik} = 2.5\,km$, $\lambda_{theoretical} = 2.1\,km$.

There is a good matching between $\lambda_{computed}$ and $\lambda_{Eidsvik}$ in the N1–case, however there is a difference about $0.3\,km$ in the other two stratification cases N2 and N3. The reason is not clearly known for different wavelength predictions in N2 and N3 cases. It can be attributed to a stretched nature of the computational grid applied

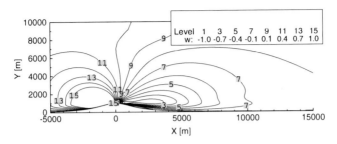

Fig. 3 Contours of the wall–normal w velocity component in N0–case under neutral stratification conditions

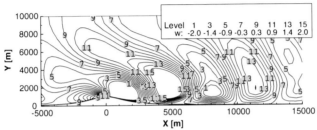

Fig. 4 Contours of the wall–normal w velocity component in N1–case under weak stratification conditions

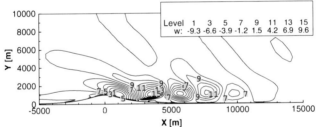

Fig. 5 Contours of the wall–normal w velocity component in N2–case under middle stratification conditions

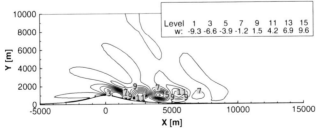

Fig. 6 Contours of the wall–normal w velocity component in N3–case under strong stratification conditions

mainly in the vertical direction along the wall and also to the turbulence model. It will be further investigated.

It is also possible to observe a decreasing tendency of the lee–wave amplitude as moving further downstream from the hill summit due to a viscous nature of the flow. Significantly increased flow velocity magnitude was found close to wall on lee–side of the hill mainly for N2 and N3 cases.

6 Conclusion

The above formulated mathematical/numerical model is capable to simulate the atmospheric boundary layer flow problems under different thermal stratification conditions.

The presented validation test case was related to the thermal stratification 2D study where the reference numerical data are due to Eidsvik [10]. The computed lee–waves were observed in all thermally stratified cases. The wavelength was found to be decreasing for increasing thermal stratification conditions. Matching between our predictions of lee wavelength and the reference numerical data is quite good for the weak stratification N1–case. However, there is difference about 0.3 km in the other two cases N2 and N3. Further numerical tests will follow to clarify the differences. The presented work was supported by the Research Plan VZ6840770010.

References

1. Janour Z. (2006): On the mathematical modelling of stratified atmosphere, Institute of Thermodynamics, Report T-470/06, Prague. (in Czech)
2. Sládek I., Kozel K., Janour Z., Gulíková E. (2004): On the Mathematical and Numerical Investigation of the Atmospheric Boundary Layer Flow with Pollution Dispersion, In: COST Action C14 "Impact of Wind and Storm on City Life and Built Environment", von Karman Institute for Fluid Dynamics, p. 233–242, ISBN 2-930389-11-7.
3. Benes L., Sládek I., Janour Z. (2004): On the Numerical Modelling of 3D – Atmospheric Boundary Layer Flow, In: "Harmonization within Atmospheric Dispersion Modelling for Regulatory Purposes", Garmisch–Parten Kirchen, Germany, p. 340–344, Vol. 1, ISBN 3-923704-44-5.
4. Bodnár T., Kozel K., Sládek I., Fraunié Ph.: Numerical Simulation of Complex Atmospheric Boundary Layer Flows Problems, In: ERCOFTAC bulletin No. 60: Geophysical and Environmental Turbulence Modeling, p. 5–12, 2004.
5. Sládek I., Bodnár T., Kozel K. (2007): On a numerical study of atmospheric 2D- and 3D-flows over a complex topography with forest including pollution dispersion, Journal of Wind Engineering and Industrial Aerodynamics, Vol. 95, Issues 9–11, p. 1422-1444.
6. Sládek, I. (2005) Mathematical modelling and numerical solution of some 2D– and 3D–cases of atmospheric boundary layer flow, PhD thesis, Czech Technical University, Prague.
7. Castro I.P., Apsley P.P. (1996): Flow and dispersion over topography: A comparison between numerical and laboratory data for two-dimensional flows, Atmospheric Environment, Vol.31, No.6, p.839–850.
8. Wilcox D.C. (1993): Turbulence modeling for CFD, DCW Industries, Inc.

9. Sládek I., Kozel K., Janour Z. (2008): On the 2D-validation study of the atmospheric boundary layer flow model including pollution dispersion, Engineering Mechanics, Vol.16, No.5, p. 323-333, 2009.
10. Eidsvik K.J., Utnes T.: Flow separation and hydraulic transitions over hills modelled by the Reynolds equations, Journal of wind engineering and industrial aerodynamics, Vol. 67 & 68, p.403–413, 1997
11. Holton James: An introduction to dynamic meteorology, Academic Press INC., ISBN 0-12-354360-6, 1979.

The paper is in final form and no similar paper has been or is being submitted elsewhere.

A Multipoint Flux Approximation Finite Volume Scheme for Solving Anisotropic Reaction–Diffusion Systems in 3D

Pavel Strachota and Michal Beneš

Abstract In [15], our DT–MRI visualization algorithm based on anisotropic texture diffusion is introduced. The diffusion is modeled mathematically by the problem for the Allen–Cahn equation with a space–dependent anisotropic diffusion operator. To preserve its anisotropic properties in the discretized version of the problem, an appropriate numerical treatment is necessary, reducing the isotropic numerical diffusion. The first part of this contribution is concerned with the design and investigation of the finite volume scheme with multipoint flux approximation. Its desirable properties are investigated by means of our technique based on total variation measurement. The second part presents the recent achievements in applying the same scheme to the phase field model of dendritic crystal growth.

Keywords Anisotropic diffusion, finite volume method, multipoint flux approximation, phase field model, microstructure growth
MSC2010: 35K55, 65M08, 65Z05

1 Introduction

The phase field formulation of the Stefan problem [11] describing phase interface evolution during material solidification involves the Allen–Cahn equation [1]. Besides its original purpose, this equation can also be applied in image processing and mathematical visualization [6, 14]. In particular, in order to visualize the streamlines of a given tensor field in 3D, an initial boundary value problem for the modified Allen–Cahn equation with incorporated anisotropy can be used [15], yielding similar results to the LIC method [9]. We begin with the problem

---

P. Strachota and M. Beneš
Department of Mathematics, Faculty of Nuclear Sciences and Physical Engineering, Czech Technical University in Prague, e-mail: pavel.strachota@fjfi.cvut.cz, michal.benes@fjfi.cvut.cz

formulation and describe its numerical solution using several flux approximation schemes on a rectangular grid. The schemes suffer from an undesired numerical dissipation effect which demonstrates itself as an additional isotropic diffusion of the solution. Hence, we proceed with the development of a measurement technique that would provide for assessing the amount of the numerical diffusion produced by the schemes. A quantitative scheme comparison criterion is thereby created, indicating a clear advantage of our *multipoint flux approximation* (MPFA) scheme. This scheme is then used for the discretization of the complete phase field model of dendritic crystal growth in 3D.

2 Allen–Cahn Equation in Tensor Field Visualization

2.1 Formulation

Assume there is a symmetric positive definite tensor field $\mathbf{D} : \bar{\Omega} \mapsto \mathbb{R}^{3\times 3}$ where $\Omega \subset \mathbb{R}^3$ is a block shaped domain. On the time interval $\mathscr{I} = (0, T)$, the initial boundary value problem for the anisotropic Allen–Cahn equation reads

$$\xi \frac{\partial p}{\partial t} = \xi \nabla \cdot \mathbf{D} \nabla p + \frac{1}{\xi} f_0(p) \qquad \text{in } \mathscr{I} \times \Omega, \qquad (1)$$

$$\left. \frac{\partial p}{\partial n} \right|_{\partial \Omega} = 0 \qquad \text{on } \bar{\mathscr{I}} \times \partial \Omega, \qquad (2)$$

$$p|_{t=0} = I \qquad \text{in } \Omega \qquad (3)$$

where $f_0(p) = p(1-p)\left(p - \frac{1}{2}\right)$. Let $x \in \Omega$. Thanks to $\mathbf{D}(x)$ in the diffusion term on the right hand side of (1), the diffusion of p at x has a directional distribution described by the ellipsoid $\left\{ \eta \in \mathbb{R}^3 \,\middle|\, \eta^T \mathbf{D}(x)^{-1} \eta = 1 \right\}$. In terms of tensor field visualization, we choose the initial condition I in (3) as a noisy texture, preferably an impulse noise. Due to the anisotropic diffusion process carried out by solving (1)–(3), the solution p changes in time from noise to an organized structure. Streamlines of the field of principal eigenvectors of $\mathbf{D}$ can be recognized there as parts with locally similar value of p. The term f_0 efficiently increases contrast of the resulting 3D image provided that the parameter ξ and the final time T are chosen appropriately (in our case by experiment). In order to actually view the resulting 3D image $p(\cdot, T)$, 2D slices through Ω can be helpful.

2.2 Numerical Solution

For numerical solution, the *method of lines* is utilized. Applying a finite volume discretization scheme in space, the problem (1)–(3) is converted to a semidiscrete scheme in the form

$$\xi \frac{d}{dt} p_K(t) = \xi \sum_{\sigma \in \mathscr{E}_K} F_{K,\sigma}(t) + \frac{1}{\xi} f_{0,K}(t) \qquad \forall K \in \mathscr{T} \qquad (4)$$

where $\mathscr{T}$ is an admissible finite volume mesh [7], $K \in \mathscr{T}$ is one particular control volume (cell) and $\mathscr{E}_K$ is the set of all faces of the cell K. $F_{K,\sigma}(t)$ represent the respective numerical fluxes at the time t, which contain difference quotients approximating the derivatives $\partial_x p, \partial_y p, \partial_z p$ at the center of the face σ. To solve (4), we employ the 4th order Runge–Kutta–Merson solver with adaptive time stepping.

2.3 Numerical Diffusion and Finite Volume Scheme Design

As indicated in the introduction, all schemes introduce a certain amount of *numerical isotropic diffusion* depending on the exact form of $F_{K,\sigma}$. This phenomenon caused by high frequency structures in the solution deteriorates the visual quality of the resulting images by *blurring*. It needs to be suppressed as much as possible e.g. by using difference operators of a sufficient order in $F_{K,\sigma}$ [10].

We have assembled and investigated numerical schemes using the following approximations of the derivatives in the flux term:

- second order central difference approximation with linear interpolation of the missing points in the difference stencil;
- fourth order *multipoint flux approximation* (MPFA) central difference scheme with linear interpolation;
- fourth order MPFA central difference scheme with *cubic* interpolation.

Thereto, a classical forward–backward first order finite difference (FD) scheme has been added for comparison. In the MPFA scheme, the numerical flux $F_{K,\sigma}$ is obtained using the rules below:

- The difference quotient approximating the derivative in the direction perpendicular to the face σ uses a non–equidistant point distribution in order to avoid redundant interpolation (Fig. 1a). Its 1–dimensional analog for a function $u \in C^1(\mathbb{R})$ can be represented by the formula

$$\left. \frac{du}{dx} \right|_{x_{i+\frac{1}{2}}} \approx \frac{1}{24h} (u_{i-1} - 27u_i + 27u_{i+1} - u_{i+2}) \qquad (5)$$

where $x_j = j \cdot h, u_j = u(x_j)$ for $j \in \mathbb{Z}, h > 0$.
- The remaining derivatives are approximated using a uniform 5–point stencil. Again, its 1D analog can be written as

$$\left. \frac{du}{dx} \right|_{x_i} \approx \frac{1}{12h} (u_{i-2} - 8u_{i-1} + 8u_{i+1} - u_{i+2}). \qquad (6)$$

Moreover, the stencil points (the crosses along the dashed line in Fig. 1b) are interpolated from the neighboring grid nodes using 1–dimensional cubic interpolation.

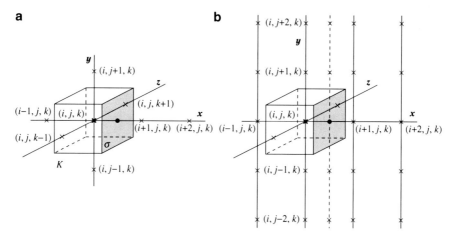

Fig. 1 Point stencils of difference quotients for derivative approximations in the MPFA finite volume scheme

3 Numerical Diffusion Measurement

Having the results available obtained by using different schemes but based on identical input settings, one can try to compare them visually to decide on the scheme with the least artificial diffusion. In Fig. 2, an example of such comparison is demonstrated on a real–data DT–MRI neural tract visualization. In the center part of the images, a major neural tract in the shape of U is displayed in the form of streamlines. It can be observed that the FD scheme produces undesired isotropic diffusion greatly dependent on the prescribed direction of diffusion. This is related to the asymmetry of the difference stencil. The 2nd order central difference flux approximation used in the FV scheme is already symmetric. However, it is clearly outperformed by the scheme based on MPFA which causes significantly weaker blurring.

3.1 Scheme Assessment by Total Variation

In this part we introduce a quantitative measure of the artificial diffusion in the schemes. For this purpose, the total variation of the numerical solution $p^h = p^h(t)$

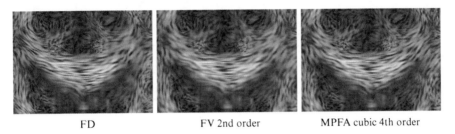

Fig. 2 Artificial diffusion in different numerical schemes. Crops from colorized DT–MRI visualizations based on real data, transverse plane slice *(Input data: Courtesy of IKEM, Prague)*

finds its rather unusual application. It is defined as

$$TV\left(p^h\right) = \sum_{K \in \mathcal{T}} \left|\nabla_h p_K^h\right| m\left(K\right) \qquad (7)$$

where $\nabla^h p_K^h$ represents the discrete approximation of the gradient and $m(K)$ is the measure of the cell K. From the image processing point of view, the value of TV is proportional to both the number of edges in the image p^h and its contrast. Both these quantities assume their maxima for the noisy initial condition and change in time along with the diffuse evolution of the numerical solution. Performing two computations with identical settings except for the choice of the numerical scheme, it is possible to directly compare the TV values of the results. The scheme producing an image with a greater value of TV exhibits less artificial diffusion as it maintains more edges, more contrast, or both.

3.2 Scheme Comparison Methodology

We have performed extensive testing with phantom input tensor fields to investigate the behavior of the schemes depending on the prescribed direction of diffusion. For each triple of spherical coordinates $(r = 1, \varphi, \theta)$ where $\varphi \in [0, 360°]$, $\theta \in [-90°, 90°]$, let a unit vector $\mathbf{v}_1(\varphi, \theta) = (\cos\varphi \cos\theta, \sin\varphi \cos\theta, \sin\theta)$ represent the principal eigenvector of a uniform tensor field $\mathbf{D}(\varphi, \theta)$, corresponding to the eigenvalue $\lambda_1 = 100$. The remaining eigenvalues are $\lambda_2 = \lambda_3 = 1$ and the eigenvectors $\mathbf{v}_2, \mathbf{v}_3$ complete the orthonormal basis of R^3. Afterwards, a computation is carried out using $\mathbf{D}(\varphi, \theta)$ as input data and subsequently, TV is evaluated from the resulting datasets. The TV values alone are not of particular interest since they depend on both the grid dimensions and the size of the domain Ω. However, the relative differences of TV between schemes provide the desired information.

The results of the procedure described above performed for all the four schemes in several time levels are shown in Fig. 3. In all graphs, TV is normalized so that

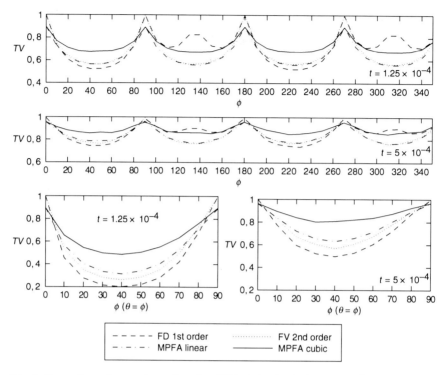

Fig. 3 Comparison of num. schemes based on TV in 2 time levels, $\xi = 5 \times 10^{-3}$. The investigated angles are $\theta = 0$, $\varphi \in [0°, 350°]$ in the upper two graphs and $\theta \in [0°, 90°]$, $\varphi = \theta$ in the lower two

the maximum in each chart is 1. In the upper two graphs, the latitude θ is fixed to 0 and the longitude φ traverses the angles from 0° to 350° with the step 10°. The lower two graphs depict the "diagonal" cut through the space (φ, θ) in the range from 0° to 90°, including the worst situation for all schemes where $\varphi = \theta = 45°$. Observations from Fig. 3 can be summed up as follows:

- Artificial diffusion clearly depends on $\mathbf{v}_1$ and occurs least when the direction $\mathbf{v}_1$ is aligned with coordinate axes. (In the degenerate case $\lambda_2 = \lambda_3 \to 0$, the equation systems for different rows of grid nodes along $\mathbf{v}_1$ become independent.)
- The performance of all schemes improves (i.e. TV rises) with growing time as the ongoing diffusion gradually limits the frequency spectrum of the solution.
- The FD scheme exhibits an asymmetric behavior; FV schemes are symmetric.
- The FV scheme with MPFA and cubic interpolation outperforms all other schemes in the comparison.

4 The full Phase Field Model for Crystal Growth

We apply the MPFA scheme to the phase field formulation of the simplified Stefan Problem, as studied in [4] and extended to the anisotropic case in [3]. Given a domain Ω and time interval $\mathscr{J}$ as in Sect. 2, the full system of phase field equations reads

$$\frac{\partial u}{\partial t} = \Delta u + L \frac{\partial p}{\partial t} \qquad \text{in } \mathscr{J} \times \Omega, \qquad (8)$$

$$\alpha \xi^2 \frac{\partial p}{\partial t} = \xi^2 \nabla \cdot T^0 (\nabla p) + a f_0 (p) - \beta \xi^2 \phi^0 (\nabla p) (u - u^*) \quad \text{in } \mathscr{J} \times \Omega, \qquad (9)$$

$$u|_{t=0} = u_{ini}, \quad p|_{t=0} = p_{ini} \qquad \text{in } \Omega, \qquad (10)$$

with either Dirichlet or Neumann boundary conditions. u represents the temperature field and p the phase field implicitly determining the phase interface Γ by the relation $\Gamma(t) = \{ \mathbf{x} \in \mathbb{R}^3 | \, p(\mathbf{x}, t) = \frac{1}{2} \}$. The model parameters involve the melting point of the material u^*, the latent heat L, the attachment kinetics coefficient α, a positive constant a and the parameter ξ controlling the recovery of the sharp interface model [5]. The anisotropic operator T^0 (see [2]) is derived from the dual Finsler metric $\phi^0(\eta^*)$ as $T^0(\eta^*) = \phi^0(\eta^*) \phi^0_\eta(\eta^*)$ where $\phi^0_\eta = \left(\partial_{\eta_1^*} \phi^0, \partial_{\eta_2^*} \phi^0, \partial_{\eta_3^*} \phi^0 \right)^T$. Putting $\phi^0(\eta^*) = |\eta^*| \psi \left(-\frac{\eta^*}{|\eta^*|} \right)$, ψ has the meaning of the anisotropic surface energy [8, 12] and assumes different forms depending on the degree of anisotropy.

The modifications of the MPFA numerical scheme compared to (4) consist in:

1. discretizing the components of ∇p in the last term of (9) by the equidistant stencil (6) at the cell centers,
2. expressing the term $\frac{\partial p}{\partial t}$ in (8) from the equation (9) and using the already computed discretization of its right hand side.

As seen in Fig. 4, the solution of the model can form nontrivial dendritic shapes. It has been confirmed by early comparisons with the standard 2nd order flux

Fig. 4 Sample simulations of dendritic crystal growth with (from left to right) 4–fold, 6–fold *crystalline* anisotropy and a cut through an 8–fold crystal

approximation that the MPFA scheme inclines to the development of dendritic structures more easily, capturing the shape complexity even on lower resolution meshes. This feature was expected due to its low numerical diffusion.

5 Conclusion and Further Research

We have developed an antidiffusive finite volume scheme based on MPFA combined with higher order interpolation. Its properties are demonstrated by our method for measuring the amount of numerical isotropic diffusion. Thorough computational studies based on phantom input data confirm that this technique fulfills the given objective and produces results in agreement with an intuitive notion of blurring observable in images obtained by solving (4). The experimental order of convergence [13] of the MPFA scheme has also been measured and found to be equal to 2. However, the details are beyond the scope of this contribution. Recently, we have finished a MPFA–based parallel numerical algorithm solving the phase field model for crystal growth. Despite the promising results, further investigation of the advantages and verification of the convergence of the numerical solution need to be performed.

Acknowledgements This work was partially supported by the following projects: The project of the Ministry of Education of the Czech Republic MSM6840770010 "Applied Mathematics in Technical and Physical Sciences". The Grant Agency of the Czech Technical University in Prague, grant No. SGS10/086/OHK4/1T/14. The project "Jindřich Nečas Center for Mathematical Modeling", No. LC06052.

References

1. Allen, S., Cahn, J.W.: A microscopic theory for antiphase boundary motion and its application to antiphase domain coarsening. Acta Metall. **27**, 1084–1095 (1979)
2. Bellettini, G., Paolini, M.: Anisotropic motion by mean curvature in the context of Finsler geometry. Hokkaido Math. J. **25**(3), 537–566 (1996)
3. Beneš, M.: Anisotropic phase-field model with focused latent-heat release. In: FREE BOUNDARY PROBLEMS: Theory and Applications II, *GAKUTO International Series in Mathematical Sciences and Applications*, vol. 14, pp. 18–30 (2000)
4. Beneš, M.: Mathematical and computational aspects of solidification of pure substances. Acta Math. Univ. Comenianae **70**(1), 123–151 (2001)
5. Beneš, M.: Diffuse-interface treatment of the anisotropic mean-curvature flow. Appl. Math-Czech. **48**(6), 437–453 (2003)
6. Beneš, M., Chalupecký, V., Mikula, K.: Geometrical image segmentation by the Allen-Cahn equation. Appl. Numer. Math. **51**(2), 187–205 (2004)
7. Eymard, R., Gallouët, T., Herbin, R.: Finite volume methods. In: P.G. Ciarlet, J.L. Lions (eds.) Handbook of Numerical Analysis, vol. 7, pp. 715–1022. Elsevier (2000)
8. Gurtin, M.E.: Thermomechanics of Evolving Phase Boundaries in the Plane. Oxford Mathematical Monographs. Oxford University Press (1993)

9. Hsu, E.: Generalized line integral convolution rendering of diffusion tensor fields. In: Proc. Intl. Soc. Mag. Reson. Med, vol. 9, p. 790 (2001)
10. Lomax, H., Pulliam, T.H., Zingg, D.W.: Fundamentals of Computational Fluid Dynamics. Springer (2001)
11. Meirmanov, A.M.: The Stefan Problem. De Gruyter Expositions in Mathematics. Walter de Gruyter (1992)
12. R. E. Napolitano, S.L.: Three-dimensional crystal-melt Wulff-shape and interfacial stiffness in the Al-Sn binary system. Phys. Rev. B **70**(21), 214,103 (2004)
13. Rice, J.R., Mu, M.: An experimental performance analysis for the rate of convergence of 5-point star on general domains. Tech. rep., Department of Computer Sciences, Purdue University (1988)
14. Strachota, P.: Vector field visualization by means of anisotropic diffusion. In: M. Beneš, M. Kimura, T. Nakaki (eds.) Proceedings of Czech Japanese Seminar in Applied Mathematics 2006, *COE Lecture Note*, vol. 6, pp. 193–205. Faculty of Mathematics, Kyushu University Fukuoka (2007)
15. Strachota, P.: Implementation of the MR tractography visualization kit based on the anisotropic Allen-Cahn equation. Kybernetika **45**(4), 657–669 (2009)

The paper is in final form and no similar paper has been or is being submitted elsewhere.

Higher Order Chimera Grid Interface for Transonic Turbomachinery Applications

Petr Straka

Abstract In this paper a higher-order accuracy chimera mesh interface for transonic flow in linear turbine blade cascades is described. Proposed method for calculation of the flow in a transonic blade cascade is applied. Conservation of mass flux through the blade cascade is evaluated. Results of calculation are compared with experimental data.

Keywords Finite-volume method, chimera grid
MSC2010: 35L65, 76H05, 76F40, 65N08

1 Introduction

It is possible to cover a computational domain with structured mesh, even in cases of complex geometry, using the structured chimera mesh. In transonic turbomachinery applications, shock waves structures are formed. The interface between overlapped meshes must operate correctly even if the shock wave intersects. Using of standard interpolation methods (linear, bilinear, polynomial) as well as conservative interpolation methods proposed for subsonic flow [1–3] leads to non-physical reflections of the shock waves at the chimera mesh interface in the case of supersonic flow. A gradient limiting technique is used in this contribution for suppression of the non-physical reflection of shock wave at the chimera mesh interface.

Petr Straka
Aeronautical Research and Test Institute, Plc, Beranových 130, 199 05 Prague - Letňany, Czech Republic, e-mail: straka@vzlu.cz

2 Govering equation

The linear blade cascade is a simple model of an axial turbine stator or rotor wheel. Flow in the linear turbine blade cascade is modeled as 2D compressible viscous turbulent flow of perfect gas. This model is described by the Favre-averaged Navier–Stokes equation

$$\frac{\partial \mathbf{W}}{\partial t} + \frac{\partial \mathbf{F}_i}{\partial x_i} = \mathbf{Q}, \quad (1)$$

where $\mathbf{W} = [\rho, \rho u_1, \rho u_2, e]^T$ is conservative variable vector, ρ is density, u_1 and u_2 are velocity vector components, e is total energy per unit volume, $\mathbf{F}_i = \mathbf{F}_i^{inv} - \mathbf{F}_i^{vis}$ ($i = 1, 2$) stands for flux vectors and $\mathbf{Q}$ is source term. In our case is $\mathbf{Q} = \mathbf{0}$. System (1) is closed by two-equation TNT $k - \omega$ turbulence model [4] which can be formulated in vector form as follows:

$$\frac{\partial \mathbf{W}_t}{\partial t} + \frac{\partial \mathbf{F}_{t,i}}{\partial x_i} = \mathbf{Q}_t, \quad (2)$$

where $\mathbf{W}_t = [\rho k, \rho \omega]^T$ is vector of turbulent quantities, k is turbulent kinetic energy, ω is specific dissipation rate, $\mathbf{F}_{t,i} = \mathbf{F}_{t,i}^{inv} - \mathbf{F}_{t,i}^{vis}$ ($i = 1, 2$) are flux vectors of turbulent quantities and $\mathbf{Q}_t$ is production and dissipation source term for turbulent quantities.

3 Numerical solution method

The govering equations are discretized on the structured multiblock mesh with quadrilateral elements using a cell-centered finite-volume technique and solved through a time-marching scheme. Both, the mean flow and the turbulence equations are integrated over a control volume D_i and some area integrals are transformed into line integrals along its boundary ∂D_i by the Green-Gauss theorem. Thus

$$\int\int_{D_i} \frac{\partial \mathbf{W}}{\partial t} dx_1 dx_2 + \oint_{\partial D_i} \mathbf{F}_n ds = \int\int_{D_i} \mathbf{Q} dx_1 dx_2, \quad (3)$$

where $\mathbf{F}_n = n_i \mathbf{F}_i / |\mathbf{n}|$ ($\mathbf{n} = [n_1, n_2]$ is the boundary outward normal vector). The line integrals take the following discrete forms

$$\oint_{\partial D_i} \mathbf{F}_n^{inv} ds = \sum_{j=1}^{4} \Phi^{inv}(\mathbf{W}_j^L, \mathbf{W}_j^R, \mathbf{n}_j) s_j, \quad (4)$$

$$\oint_{\partial D_i} \mathbf{F}_n^{vis} ds = \sum_{j=1}^{4} \Phi^{vis}(\mathbf{W}_j^C, (\nabla \mathbf{W})_{D_j^{dual}}, \mathbf{n}_j) s_j. \quad (5)$$

In the mean flow equations, the inviscid numerical fluxes Φ^{inv} are computed by means of the Osher-Solomon flux splitting scheme [5]. Higher order accuracy is achieved through the 2D linear reconstruction method which will be discussed later. In eq. (4) $\mathbf{W}_j^L$ and $\mathbf{W}_j^R$ denote the *left* and *right* states in the corresponding Riemann problem [6]. In the turbulence equations, the inviscid numerical fluxes are computed by the first-order upwind flux-splitting scheme, based on the local convective velocity normal to the cell boundary. The numerical viscous fluxes Φ^{vis} are computed using second-order central scheme, where $\nabla \mathbf{W}$ is approximated through the Green-Gauss formula on a dual cell (Fig. 1) and the local conservative variables vector at the cell boundary is computed as $\mathbf{W}_j^C = (\mathbf{W}_j^L + \mathbf{W}_j^R)/2$.

The time integration is performed using first-order backward Euler scheme

$$\left(\mathbf{I} + \Delta t \, \frac{\partial \mathbf{R}_i^{low}}{\partial \mathbf{W}_i} \right) \Delta \mathbf{W}_i^{n+1/2} + \Delta t \sum_{j=1}^{4} \frac{\partial \mathbf{R}_i^{low}}{\partial \mathbf{W}_j} \Delta \mathbf{W}_j^{n+1/2} = -\Delta t \, \mathbf{R}_i^n \,, \qquad (6)$$

where $\Delta \mathbf{W}^{n+1/2} = \mathbf{W}^{n+1} - \mathbf{W}^n$, and residual approximation $\mathbf{R}_i$ is given as

$$\mathbf{R}_i = \frac{1}{|D_i|} \sum_{j=1}^{4} [\Phi^{inv}(\mathbf{W}_j^L, \mathbf{W}_j^R, \mathbf{n}_j) - \Phi^{vis}(\mathbf{W}_j^C, (\nabla \mathbf{W})_{D_j^{dual}}, \mathbf{n}_j)] s_j \,. \qquad (7)$$

$\mathbf{R}_i^{low}$ denotes first-order approximation in eq. (6).

3.1 Linear reconstruction technique

As mentioned above, higher order accuracy is achived through the 2D linear reconstruction method, which is used for extrapolation of state vector $\mathbf{W}$ at the cell boundary. A piecewise linear function is used for reconstruction of the components

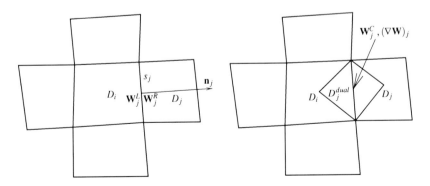

Fig. 1 Scheme of the structured quadrilateral mesh with dual volume

w_k ($k = 1, \ldots 4$) of vector $\mathbf{W}$

$$w_k = f_k(x_1, x_2) = a_k + b_k x_1 + c_k x_2. \tag{8}$$

Coefficients a_k, b_k and c_k are given by supposition

$$w_{k,l} = \frac{1}{|D_l|} \int\int_{D_l} f_k(x_1, x_2)\,dx_1\,dx_2, \tag{9}$$

where l denotes index of three neighbouring cells. We define four linear functions $f_k^1, \ldots, f_k^4$ for reconstruction of state vector components $w_{k\,C}$ ($k = 1, \ldots 4$) from centre of cell denoted C (Fig. 2) to centre of boundary with cell denoted R, where index l in eq. (9) is for function f_k^1: $l = $ C, R T, for function f_k^2: $l = $ C, T, L, for function f_k^3: $l = $ C, L, B and for function f_k^4: $l = $ C, B, R (R, T, L, B are designations of cells adjacent to cell C - Fig. 2). Components $w_{k,\,CR}^L$ of reconstructed state vector $\mathbf{W}_{CR}^L$ (where L means left side in outward normal direction) are given as

$$w_{k,\,CR}^L = w_{k\,C} + \delta_{k\,CR}, \tag{10}$$

where $w_{k\,C}$ are components of state vector in centre of cell C and $\delta_{k\,CR}$ is defined as

$$\delta_{k\,CR} = \delta_{k\,CR}^{min} \psi(r(\delta_{k\,CR}^{min}, \delta_{k\,CR}^{max})), \tag{11}$$

$$\delta_{k\,CR}^{min} = \min_m \{f_k^m(x_{CR,1}, x_{CR,2}) - w_{k\,C}\}, \quad m = 1, \ldots, 4, \tag{12}$$

$$\delta_{k\,CR}^{max} = \max_m \{f_k^m(x_{CR,1}, x_{CR,2}) - w_{k\,C}\}, \quad m = 1, \ldots, 4. \tag{13}$$

There is $r(\delta^{min}, \delta^{max}) = \delta^{max}/\delta^{min}$ and ψ stands for the limiting function enforcing monotonicity to the solution in eq. (11). The limiters of Van Albada, Van Leer and the super-bee limiter are used in this work.

$$\psi_{VA}(r) = \begin{cases} 0, & r \leq 0, \\ (r^2 + r)/(r^2 + 1), & r > 0, \end{cases} \tag{14}$$

$$\psi_{VL}(r) = \begin{cases} 0, & r \leq 0, \\ 2r/(r+1), & r > 0, \end{cases} \tag{15}$$

$$\psi_{SB}(r) = \begin{cases} 0, & r \leq 0, \\ 2r, & 0 \leq r \leq 1/2, \\ 1, & 1/2 \leq r \leq 1, \\ r, & 1 \leq r \leq 2, \\ 2, & r \geq 2. \end{cases} \tag{16}$$

3.2 Chimera grid interface

For numerical solution of flow in the turbine blade cascade, O-type mesh is used around the blade profile, H-type mesh covers the chanel between blades as shown in Fig. 3. For simple implementation of the chimera mesh the cells of mesh are classified into three categories: category C0 refers to the regular cell in which the conservative variables vector **W** is solved, category C1 refers to the hidden cell which is skiped during the solution procedure, category C2 refers to the interpolation cell in which the vector **W** is interpolated from overlapped mesh. The same method, as described in paragraph 3.1, is used for interpolation of state vector **W** into the centre of the cell type C2. We need to have state vector $\mathbf{W}^R$ for calculation of flux through boundary between cells type C0 and C2 (Fig. 4 left), which is obtained using the linear reconstruction in cell type C2. One can see, that for correct reconstruction of vector $\mathbf{W}^R$ at the boundary between cells type C0 and C2, we need to have two layers of interpolation cells type C2 (Fig. 4 right).

Proposed chimera mesh interface is very simple for implementation, is higher-order of accuracy and is robust for transonic flow calculation. The mass flux conservation error will be discussed later.

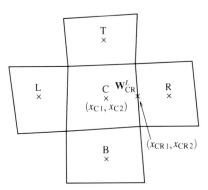

Fig. 2 Scheme of the linear reconstruction on the structured quadrilateral mesh

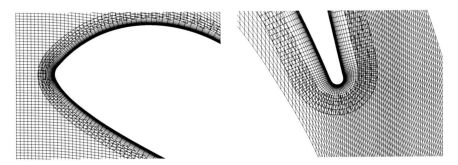

Fig. 3 Detail of chimera mesh around leading and trailing edge

4 Application

The numerical method described in sec. 3 is used for solution of flow in the linear transonic turbine blade cascade VS33R. The computational domain with set boundary conditions types is shown in Fig. 5 (left). The solution was calculated for isentropic output Mach number $0.5 < M_{is,out} < 1.3$, isentropic output Reynolds number $Re_{is,out} = 8.5 \times 10^5$, zero angle of attack and 2 % of inlet turbulence intensity. Transonic flow field in Mach number isolines form is shown in Fig. 5 right. Error of conservation of the mass flux through the blade cascade given as $(1 - q_{in}/q_{out}) \cdot 100$ (where q is the mass flux) is shown in Fig. 7. Further distribution of the total pressure loss coefficient $\eta = 1 - p_{tot,out}/p_{tot,in}$ is compared with the experimental data [7] in Fig. 6.

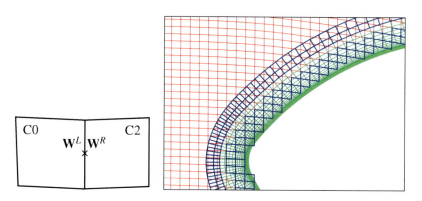

Fig. 4 Left: detail of interface between regular and interpolation cell. Right: two layers of interpolation cells of C2 type (blue)

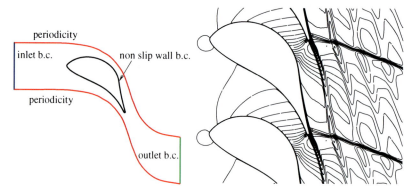

Fig. 5 Left: scheme of computational domain in linear blade cascade. Right: Mach number isolines ($M_{is,out} = 1.3$)

Higher Order Chimera Grid Interface for Transonic Turbomachinery Applications

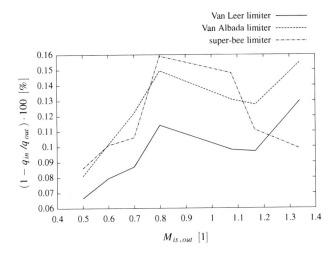

Fig. 6 Distribution of the mass flux conservation error

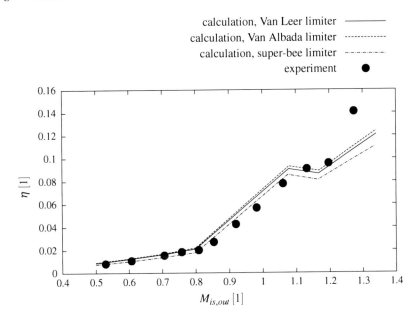

Fig. 7 Distribution of the total pressure loss coefficient

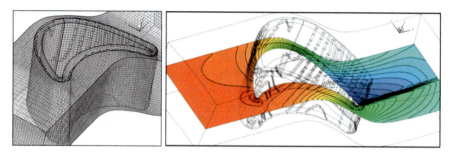

Fig. 8 Chimera mesh and the pressure distribution in 3D problem

4.1 Extension for 3D problem

It is simple to extend the method described in sec. 3 for 3D problems. Two-blocks structured chimera mesh with the hexahedral elements was used for solution of 3D transonic inviscid flow (described by Euler equation: $\frac{\partial \mathbf{W}}{\partial t} + \frac{\partial \mathbf{F}_i^{inv}}{\partial x_i}$, where $\mathbf{W} = [\rho, \rho u_1, \rho u_2, \rho u_3, e]^T$ and $\mathbf{F}_i^{inv}$ ($i = 1, 2, 3$) stands for inviscid flux vector) in an axial tubine cascade ST6 [8] for isentropic output Mach number $M_{is,out} = 1.3$, isentropic output Reynolds number $Re_{is,out} = 7.5 \times 10^5$ and angle of attack $\alpha = 45°$. Distribution of pressure is shown in Fig. 8.

5 Conclusion

The chimera mesh interface described in this contribution is simple for the implementation and robust for the transonic turbomachinery applications. Proposed method was applied for the calculation of transonic flow through the linear blade cascade VS33R. The results are in good agreement with the experimental data. Although the condition of conservation is not directly included in the chimera mesh interface, evaluation of the mass flux conservation error (Fig. 6) shows reasonably good conservation.

References

1. Kangle, X., Gang, S.: Assessment of an interface conservative algorithm MFBI in a chimera grid flow solver for multi-element airfoils. Proceedings of the World Congress on Engineering 2009 Vol II, London (2009)
2. Emmert, T., Lafon, P., Bailly, C.: Numerical study of self-induced transonic ow oscillations behind a sudden duct enlargement. Physics of Fluids, **21**, 106105 (2009)
3. Tang, H.S.: Chimera grid method for incompressible flows and its applications in actual problems. 10th Symposium on Overset Composite Grids and Solution Technology, NASA Ames Research Center, CA (2010)

4. Kok, J.C.: Resolving the dependence on freestream values for the $k - \omega$ turbulence model. AIAA Journal. **38**, 1292–1295 (2000)
5. Osher, S., Solomon, F.: Upwind diference schemes for hyperbolic system of conservation laws. Mathematics of Computation, **38**, 339-374 (1982)
6. Toro, E.F.: Riemann solvers and numerical methods for fluid dynamics, A practical introduction, 2nd edn. (Springer, Berlin, 1999)
7. Benetka, J., Kladrubský, M., Valenta, R., Vích, K.: Measurement of turbine blade cascade VS33R. Research report of Aeronautical research and test institute, R-3435/02, (Prague, 2002) (in Czech)
8. Straka, P.: Calculation of 3D unsteady inviscid flow in turbine stage ST6. Research report of Aeronautical research and test institute, R-4910, (Prague, 2010) (in Czech)

The paper is in final form and no similar paper has been or is being submitted elsewhere.

Application of Nonlinear Monotone Finite Volume Schemes to Advection-Diffusion Problems

Yuri Vassilevski, Alexander Danilov, Ivan Kapyrin, and Kirill Nikitin

Abstract Two conservative schemes for the nonstationary advection-diffusion equation featuring nonlinear monotone finite volume methods (FVMON) are considered. The first one is an operator-splitting scheme which uses discontinuous finite elements for the advection operator discretization and FVMON for the diffusion operator. The second one introduces another type of FVMON and is implicit second-order BDF in time. A brief description of the schemes and their properties is given. A numerical study is conducted in order to check their convergence and to compare them with conventional methods.

Keywords Monotone finite volumes, advection-diffusion.
MSC2010:65M08

1 Formulation of the methods

1.1 Model Problem

Let Ω be a bounded polyhedral domain in $\mathbb{R}^3$ with a boundary $\partial\Omega$. Consider the following model advection-diffusion problem (for simplicity, with homogeneous Dirichlet boundary conditions):

Yuri Vassilevski, Alexander Danilov, Ivan Kapyrin, and Kirill Nikitin
Institute of Numerical Mathematics RAS, 8 Gubkina, Moscow 119333, Russia, e-mail:
vasilevs@dodo.inm.ras.ru, danilov@dodo.inm.ras.ru, kapyrin@dodo.inm.ras.ru, nikitink@dodo.inm.ras.ru

$$\frac{\partial C}{\partial t} - \nabla \cdot D\nabla C + \mathbf{b} \cdot \nabla C = F \quad \text{in } \Omega \times (0, T], \tag{1a}$$

$$C = 0 \quad \text{on } \partial\Omega \times (0, T], \tag{1b}$$

$$C = C_0(x) \text{ in } \Omega \text{ at } t = 0. \tag{1c}$$

Here, C is the contaminant concentration, $\mathbf{b} = \mathbf{b}(x)$ is a conservative convective flux field, $F = F(x)$ is the function of sources or sinks, and $D = D(x)$ is a symmetric positive definite 3×3 diffusion tensor.

1.2 Operator-splitting scheme: DFEM+FVMON

The operator-splitting scheme is designed for tetrahedral grids. It is explained in details in [9], here we give only a brief description. Let a conformal tetrahedral mesh ε_h be introduced in the computational domain Ω. Denote the mesh cells by E_i, $i = 1, \ldots, N_E$, the nodes by $O_i = (x_i, y_i, z_i)$, $i = 1, \ldots, N_P$. We define the space of discontinuous piecewise linear functions on ε_h

$$W_h = \{v \in L_2(\Omega), v_{|E} \in P_1(E), v_{|\partial E \cap \partial \Omega} = 0 \, \forall E \in \varepsilon_h\}.$$

The concentration is approximated by piecewise linear discontinuous functions from W_h. The scheme involves splitting over physical components, and the diffusion and convection operators are handled at different substeps. More specifically, at each substep, we solve the incomplete equation (see [8]). The time step of the scheme is defined as follows:

$$I. \int_E \frac{C_h^{n+\frac{1}{2}} - C_h^n}{\Delta t/2} w_h dx - \int_E \mathbf{b} C_h^n \cdot \nabla w_h dx + \int_{\partial E} \mathbf{b} C_{h,in}^n \cdot \mathbf{n} w_h ds = \int_E F^n w_h dx$$

$$\forall w_h \in W_h(E), \, \forall E \in \varepsilon_h, \tag{2a}$$

$$II. \int_E \frac{C_h^{*,ad} - C_h^n}{\Delta t} w_h dx - \int_E \mathbf{b} C_h^{n+\frac{1}{2}} \cdot \nabla w_h dx + \int_{\partial E} \mathbf{b} C_{h,in/out}^{n+\frac{1}{2}} \cdot \mathbf{n} w_h ds =$$

$$= \int_E F^{n+\frac{1}{2}} w_h dx \quad \forall w_h \in W_h(E), \, \forall E \in \varepsilon_h, \tag{2b}$$

$III.$ Slope limiter: $C_h^{*,ad} \longrightarrow C_h^{n+1,ad}$,

$$IV. \int_E \frac{\bar{C}_{h,E}^{n+1} - \bar{C}_{h,E}^{n+1,ad}}{\Delta t} dx = -\sum_{i=1}^{4} r_{E,i}^{n+1} = -\int_{\partial E} \mathbf{r}_E^{n+1} \cdot \mathbf{n} ds \quad \forall E \in \varepsilon_h, \tag{2c}$$

$$V. \, C_h^{n+1} = C_h^{n+1,ad} + (\bar{C}_h^{n+1} - \bar{C}_h^{n+1,ad}). \tag{2d}$$

Application of Nonlinear Monotone Finite Volume Schemes

The convection operator is approximated by an explicit predictor-corrector scheme with an upwind regularization in the corrector. The intermediate concentration $C_h^{n+\frac{1}{2}}$ is calculated in predictor (2a), while $C_h^{\star,ad}$ in the corrector is calculated from the convective fluxes at the intermediate time level. In the integral over the boundary, $C_{h,in/out}^{n+\frac{1}{2}}$ is taken on the tetrahedron lying upstream. The slope-limiting procedure (2c) is applied to $C_h^{\star,ad}$. Next, implicit scheme (2d) is used to calculate the addition to the mean concentration, $\bar{C}_h$ due to the diffusive fluxes $r_{E,i}^{n+1}$ through the i-th faces of E. The values of $\bar{C}_h^{n+1}$ and $r_{E,i}^{n+1}$ are determined by the nonlinear monotone finite-volume method (FVMON) [3]. Its goal is to derive an as sparse as possible monotone approximation matrix by forming two-point diffusive flux approximations. Then, the solution $\bar{C}_h^{n+1}$ remains nonnegative for nonnegative $\bar{C}_h^{n+1,ad}$. The idea of the two-dimensional FVMON for diffusion problems was set forth in [4]. The details of the present scheme formulation can be found in [9]. The main idea of the scheme construction is based on the following steps:

1. Define the collocation points bearing the degrees of freedom inside each tetrahedron. For cell E we define the point X_E.
2. For two neighbouring tetrahedra E_+, E_- and the corresponding degrees of freedom C_{X_+}, C_{X_-} we define the diffusion flux through the common face e:

$$\mathbf{r}_e \cdot \mathbf{n}_e = K_+(\mathbf{C_X})C_{X_+} - K_-(\mathbf{C_X})C_{X_-}. \qquad (3)$$

Here $\mathbf{C_X}$ is the global vector of unknowns, $\mathbf{n}_e$ is the normal vector to face e. The flux defined in (3) has a two-point approximation stencil with coefficients $K_+(\mathbf{C_X}), K_-(\mathbf{C_X})$ depending on the vector of unknown concentrations. The algorithm of their calculation guarantees positivity of coefficients in case of non-negative vector $\mathbf{C_X}$.

3. Assemble the global nonlinear system and solve it.

To implement step (2c), we find the projection $\hat{c}$ of the solution $C_h^{n+1,ad}$ onto the set $\mathbb{B}$ of the collocation points in cells and solve the FVMON problem for the desired concentrations $\hat{c}^{diff}$ at the points of $\mathbb{B}$:

$$\left(\mathbf{V} + \mathbf{A}(\hat{c}^{diff})\Delta t\right)\hat{c}^{diff} = \mathbf{V}\hat{c}. \qquad (4)$$

Here, V is a diagonal matrix of element volumes and $\mathbf{A}(\hat{c}^{diff})$ is an asymmetric matrix whose elements depend on $\hat{c}^{diff}$. All the off-diagonal and diagonal nonzero elements of $\mathbf{A}(\hat{c}^{diff})$ are negative and positive, respectively, for nonnegative $\hat{c}^{diff}$. Moreover, the transpose $(\mathbf{A}(\hat{c}^{diff}))^T$ is row diagonally dominant. Therefore, $(\mathbf{A}(\hat{c}^{diff}))^T$ is an M-matrix and $([(\mathbf{A}(\hat{c}^{diff}))^T]^{-1})_{ij} \geq 0$. Since $(\mathbf{A}^{-1})^T = (\mathbf{A}^T)^{-1}$, the matrix $\mathbf{A}(\hat{c}^{diff})$ is monotone. Nonlinear system (4) is solved by the Picard iteration algorithm

$$\left(\mathbf{V} + \mathbf{A}(\hat{c}^{diff,k})\Delta t\right)\hat{c}^{diff,k+1} = \mathbf{V}\hat{c}$$

with the initial approximation $\hat{c}^{diff,0} = \hat{c} \geq 0$. Since the matrix $\mathbf{V} + \mathbf{A}(\hat{c}^{diff,k})\Delta t$ is monotone for any nonnegative $\hat{c}^{diff,k}$, all the iterative approximations $\hat{c}^{diff,k+1}$ are nonnegative as well; i.e., scheme (4) is monotone.

After $\hat{c}^{diff}$ is determined, we use the formula

$$\bar{C}_{h,E}^{n+1} - \bar{C}_{h,E}^{n+1,ad} = \hat{c}_E^{diff} - \hat{c}_E \quad \forall E \in \varepsilon_h$$

and find the addition to the mean concentrations due to diffusive fluxes, as required in (2e).

1.3 Implicit FVMON scheme

The idea of the implicit nonlinear monotone finite volume scheme is to derive a discretization for the total advective-diffusive flux $\mathbf{r} = -D\nabla C + C\mathbf{b}$ and use the implicit second-order BDF discretization in time. The method is applicable to arbitrary conformal meshes with polyhedral cells and jumping full anisotropic diffusion tensors as well as variable convection fields.

For each cell E, we assign one degree of freedom, C_E, for concentration C. If two cells E_+ and E_- have a common face f and the normal $\mathbf{n}_f$ is exterior to E_+, the two-point flux approximation is as follows:

$$\mathbf{r}_f \cdot \mathbf{n}_f = M_f^+ C_{E_+} - M_f^- C_{E_-}, \qquad (5)$$

where M_f^+ and M_f^- are some coefficients. In a linear FV method, these coefficients are equal and fixed. In the nonlinear FV method, they may be different and depend on concentrations in surrounding cells.

Diffusive flux $\mathbf{r}_d = -D\nabla C$ is discretized using the nonlinear two-point flux approximation [1, 5] with non-negative coefficients $K_f^{\pm}(C) \geq 0$:

$$(-D\nabla C)_f \cdot \mathbf{n}_f = K_f^+(C) C_{T_+} - K_f^-(C) C_{T_-}. \qquad (6)$$

Advective flux $\mathbf{r}_a = C\mathbf{b}$ is approximated via an upwinded linear reconstruction $\mathscr{R}_T$ of the concentration over cell T [6, 7]:

$$\mathbf{r}_{f,a} \cdot \mathbf{n}_f = b_f^+ \mathscr{R}_{E_+}(\mathbf{x}_f) + b_f^- \mathscr{R}_{E_-}(\mathbf{x}_f), \qquad (7)$$

where

$$b_f^+ = \frac{1}{2}(b_f + |b_f|), \qquad b_f^- = \frac{1}{2}(b_f - |b_f|), \qquad b_f = \frac{1}{|f|}\int_f \mathbf{b} \cdot \mathbf{n}_f \, ds.$$

We define the reconstruction $\mathscr{R}_E$ as a linear function

$$\mathscr{R}_E(\mathbf{x}) = C_E + \mathbf{g}_E \cdot (\mathbf{x} - \mathbf{x}_E), \qquad \forall \mathbf{x} \in E, \qquad (8)$$

with a gradient vector $\mathbf{g}_E$. Since C_E is collocated at the barycenter of E, this reconstruction preserves the mean value of the concentration for any choice of $\mathbf{g}_E$.

The gradient vector $\mathbf{g}_E$ is the solution to the following constrained minimization problem:

$$\mathbf{g}_E = \arg\min_{\tilde{\mathbf{g}}_E \in \mathscr{G}_E} \mathscr{J}_E(\tilde{\mathbf{g}}_E), \tag{9}$$

where the functional

$$\mathscr{J}_E(\tilde{\mathbf{g}}_E) = \frac{1}{2} \sum_{\mathbf{x}_k \in \Sigma_E} [C_E + \tilde{\mathbf{g}}_E \cdot (\mathbf{x}_k - \mathbf{x}_E) - C_k]^2$$

measures deviation of the reconstructed function from the targeted values C_k collocated at points $\mathbf{x}_k$ from a set Σ_E of the neighbouring collocation points.

The set of admissible gradients $\mathscr{G}_E$ is defined via three constraints surpressing non-physical oscillations (see [7] for more details).

As the result, we represent the advective flux as the sum of a linear part (the first-order approximation) and a nonlinear part (the second-order correction):

$$\mathbf{r}_{f,a} \cdot \mathbf{n}_f = A_f^+(C)C_+ - A_f^-(C)C_-, \tag{10}$$

where

$$A_f^{\pm}(C) = \pm b_f^{\pm}(1 + \mathbf{g}_{\pm} \cdot (\mathbf{x}_f - \mathbf{x}_{\pm})C_{\pm}^{-1}) \geq 0, \tag{11}$$

subscript $\pm$ stands for $E_{\pm}$ and $\mathbf{g}_{\pm} = \mathbf{g}_{E_{\pm}}$.

The resulting nonlinear system is solved using the Picard iterations method. The matrix is monotone on each iteration (see [7]) providing a nonnegative solution.

2 Results of numerical experiments

2.1 Smooth analytical solution

In the first test case the computational domain Ω is a unit cube $[0;1]^3$, the advection field $\mathbf{b} = (0.1; z/10; y/10)$. Two diffusion tensors and the corresponding analitycal solutions are considered:

1. $D = I$, $\quad C(x,y,z,t) = (1-x^2)\sin(y)e^{-z}\sin(t)$
2. $D = 10^{-5}I$ $\quad C(x,y,z,t) = x^2\sin(y)e^{-z}\sin(t)$

The first test case features dominating diffusion, the second - dominating advection. The choice of analytical solutions is explained by the desire to obtain nonnegative right-hand sides in the discretization of Eq. (1) in order to verify the monotonicity of the schemes. Recall that only the FVMON guarantees the absence of negative concentrations in this case (although it is unsuitable for problems admitting negative concentrations). Three uniform structured tetrahedral meshes were used in the

Table 1 Solution and flux L_2-errors for DFEM+FVMON scheme

Mesh	case $D = I$		case $D = 10^{-5}I$	
	e_C	e_r	e_C	e_r
1	$1.4 \cdot 10^{-3}$	$2.8 \cdot 10^{-2}$	$4 \cdot 10^{-4}$	$3.8 \cdot 10^{-7}$
2	$4 \cdot 10^{-4}$	$1.3 \cdot 10^{-2}$	$1 \cdot 10^{-4}$	$1.8 \cdot 10^{-7}$
3	$1.2 \cdot 10^{-4}$	$6 \cdot 10^{-3}$	$2.5 \cdot 10^{-5}$	$8.3 \cdot 10^{-8}$

Table 2 Solution L_2-error (e_C) for the BDF FVMON scheme

Mesh	case $D = I$	case $D = 10^{-5}I$
1	$2.2 \cdot 10^{-3}$	$5.3 \cdot 10^{-3}$
2	$5.7 \cdot 10^{-4}$	$1.5 \cdot 10^{-3}$
3	$1.4 \cdot 10^{-4}$	$4.0 \cdot 10^{-4}$

computations. The coarsest of them consisted of 3072 tetrahedra (mesh 1). The other two were obtained by uniformly refining the first and contained 24 576 (mesh 2) and 196 608 (mesh 3) elements, respectively (the mesh size was halved in each refinement procedure). In all the schemes, the time steps used in the tests were 0.025 for mesh 1, 0.0125 for mesh 2, and 0.00625 for mesh 3. The errors were calculated for the solution at the time $T = 1$ and can be seen in tables 1 and 2.

For both schemes we calculate the discrete L_2-error for the solution e_C. For the splitting scheme the diffusion flux L_2-error e_r is computed as well (not implemented yet for the implicit scheme). In both cases we observe second order convergence for the solution, the splitting scheme shows first order convergence for the diffusion fluxes.

2.2 Sharp front resolution

Consider the front of concentration propagating from a constant source occupying a section on the boundary of the domain $\Omega = (0; 1) \times (-0.5; 0.5) \times (-0.5; 0.5)$. More specifically, the following inhomogeneous boundary conditions are set at $x = 0$:

$$C(0, y, z) = \begin{cases} 1 & \text{if } |y| < \frac{1}{4}, |z| < \frac{1}{4}, \\ 0 & \text{elsewhere.} \end{cases}$$

The initial concentration is zero in the entire domain Ω, and the convective flux is $\mathbf{b} = (1, 0, 0)$. For the solution to have a sharp front, the diffusion tensor is chosen to be small with respect to convection: $D = 10^{-4}I$.

The analytical solution to this problem in the half-space $x \geq 0$ was found in [2]. Passing to the bounded domain Ω, we set Dirichlet conditions on all its boundaries. A non-uniform tetrahedral grid is used for the domain discretization. The numerical solutions are compared with the analytical one at the time $T = 0.5$ and with the

Application of Nonlinear Monotone Finite Volume Schemes 767

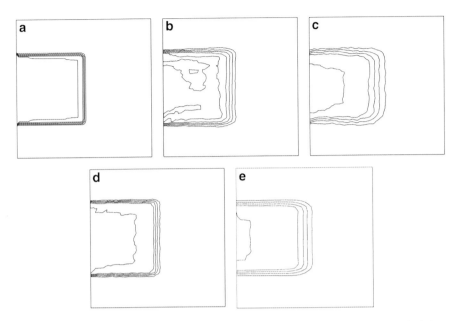

Fig. 1 Analytical and numerical solutions of the front propagation problem: a—analytical; b—implicit BDF P_1-FEM with SUPG; c— BDF implicit HMFEM scheme; d— operator-splitting scheme DFEM+FVMON; e— BDF implicit FVMON scheme.

Table 3 Minima of mean cell concentrations

P_1-FEM	MFEM	DFEM+FVMON	impl. FVMON
$-1.8 \cdot 10^{-1}$	$-6.4 \cdot 10^{-2}$	0	0

solutions obtained by conventional methods: BDF implicit schemes of P_1-FEM with SUPG and HMFEM with upwinding.

Figure 1 displays the exact (a) and approximate (b-e) solutions at $T = 0.5$ in the plane $y = 0$. The contour lines correspond to the concentration values 0.2, 0.4, 0.6, 0.8, and 1. The conventional methods are nonmonotone, so the solution takes negative values (Table 3), whereas the considered monotone schemes guarantee non-negativity of the solution. Figure 1b shows that FEM with SUPG exhibits strong oscillations. Since the FEM is strongly dispersive, a concentration contour line corresponding to 1 appears in Fig. 1b in the area where the solution must be the identical unit. Hybrid MFEM demonstrates high numerical dissipation in Fig. 1c. The operator-splitting scheme shows the lowest numerical diffusion (rf.Fig.1d). The implicit FVMON scheme exhibits numerical diffusion comparable to that of FEM with SUPG method.

Conclusions

The two schemes featuring the nonlinear monotone finite volumes prove to be a good alternative to conventional methods especially in cases when monotonicity (in the sense of non-negative concentrations) is important. The operator-splitting scheme makes use of discontinuous finite elements applied for the advection operator discretization. It produces low numerical diffusion. In this scheme diffusion is treated implicitly, and advection explicitly. Thus the time step of the scheme depends on the CFL number. An efficient solution to accelerate its performance is to use different time steps for advection and diffusion. The extra computational burden due to nonlinearity seems to be admissible: the scheme is approximately 20% slower than a linear similar splitting scheme [9].

The BDF implicit scheme of FVMON also shows second order convergence on analytical solutions both for advection and diffusion dominated problems. While suffering from higher numerical diffusion, the scheme has no time step restriction and thus can be more suitable in terms of computational efficiency. Also the scheme is applicable to arbitrary polyhedral cells. Both schemes guarantee non-negativity of the solution in case of non-negative source terms and proper boundary conditions.

Acknowledgements This work has been supported in part by RFBR grants 09-01-00115-a, 11-01-00971-a and the federal program "Scientific and scientific-pedagogical personnel of innovative Russia"

References

1. Danilov A., Vassilevski Yu. A monotone nonlinear finite volume method for diffusion equations on conformal polyhedral meshes. *Russian J. Numer. Anal. Math. Modelling*, No.24, pp.207–227, 2009.
2. Feike J.L., Dane J.H. Analitical solutions of the one-dimensional advection equation and two- or three-dimensional dispersion equation. *Water Resources Research*, vol.26, No.7, pp.1475–1482, 1990.
3. Kapyrin I.V. A Family of Monotone Methods for the Numerical Solution of Three-Dimensional Diffusion Problems on Unstructured Tetrahedral Meshes. *Doklady Mathematics*, Vol. 76, No. 2, pp. 734–738, 2007.
4. Le Potier C. Schema volumes finis monotone pour des operateurs de diffusion fortement anisotropes sur des maillages de triangle non structures. *C.R.Acad. Sci. Paris*, Ser. I 341, pp.787–792, 2005.
5. Lipnikov K., Svyatskiy D., Vassilevski Yu. Interpolation-free monotone finite volume method for diffusion equations on polygonal meshes. *J. Comp. Phys.* Vol.228, No.3, pp.703–716, 2009.
6. Lipnikov K., Svyatskiy D., Vassilevski Yu. A monotone finite volume method for advection-diffusion equations on unstructured polygonal meshes. *J. Comp. Phys.* Vol.229, pp.4017–4032, 2010.
7. Nikitin K., Vassilevski Yu. A monotone nonlinear finite volume method for advection-diffusion equations on unstructured polyhedral meshes in 3D. *Russian J. Numer. Anal. Math. Modelling*, Vol.25, pp.335–358, 2010.

8. Siegel P., Mose R., Ackerer Ph. and Jaffre J. Solution of the advection-diffusion equation using a combination of discontinuous and mixed finite elements. *International Journal for Numerical Methods in Fluids*, Vol.24, p.595–613, 1997.
9. Vassilevski Yu.V., Kapyrin I.V. Two Splitting Schemes for Nonstationary Convection-Diffusion Problems on Tetrahedral Meshes. *Computational Mathematics and Mathematical Physics*, Vol.48, No. 8, pp. 1349-1366, 2008.

The paper is in final form and no similar paper has been or is being submitted elsewhere.

Scale-selective Time Integration for Long-Wave Linear Acoustics

Stefan Vater, Rupert Klein, and Omar M. Knio

Abstract In this note, we present a new method for the numerical integration of one dimensional linear acoustics with long time steps. It is based on a scale-wise decomposition of the data using standard multigrid ideas and a scale-dependent blending of basic time integrators with different principal features. This enables us to accurately compute balanced solutions with slowly varying short-wave source terms. At the same time, the method effectively filters freely propagating compressible short-wave modes. The selection of the basic time integrators is guided by their discrete-dispersion relation. Furthermore, the ability of the schemes to reproduce balanced solutions is shortly investigated. The method is meant to be used in semi-implicit finite volume methods for weakly compressible flows.

Keywords linear acoustics, implicit time discretization, large time steps, balanced modes, multiscale time integration
MSC2010: 35L05, 65M06, 86-08, 86A10

1 Introduction

General circulation models (GCMs) currently used for planetary flow simulations, are based on the Hydrostatic Primitive Equations. This approximation of the full compressible flow equations suppresses vertically propagating sound waves, but it still admits horizontally traveling long wave acoustics, so called "Lamb waves". These and other effects of compressibility are increasingly considered to be non-negligible for planetary-scale dynamics [1, 6]. On the other hand, modern

---

Stefan Vater and Rupert Klein
Institut für Mathematik, Freie Universität Berlin, Berlin, Germany, e-mail: stefan.vater@math.fu-berlin.de, rupert.klein@math.fu-berlin.de

Omar M. Knio
Dept. of Mechanical Eng., Johns Hopkins University, Baltimore, USA, e-mail: knio@jhu.edu

high-performance computing hardware is beginning to allow the usage of grids with horizontal spacings of merely a few kilometers in such applications (see e.g. [5]). This development introduces considerable numerical difficulties. For explicit time integration schemes, the propagation of sound perturbations introduced by compressibility require very small time steps $\Delta t \sim \Delta x/c$, where Δx is the typical computational grid size, and c a characteristic sound speed. Alternatively, the application of implicit time discretizations solves the problem of the severe time step restriction, but it introduces potentially undesirable numerical dispersion: Most – if not all – existing implicit schemes slow down modes with high wave numbers. Furthermore, there are quite popular schemes, such as the implicit trapezoidal scheme, which preserve the amplitude for all wave numbers. Being a desirable feature at the first glance, it is a potential source of nonlinear instabilities in practice.

In the present work, a new discretization of the linearized acoustic equations is introduced, which overcomes some of the disadvantages of standard implicit discretizations with respect to the representation of compressibility. This means that the scheme should represent the "slaved" dynamics of short-wave solution components induced by slow forcing or arising in the form of high-order corrections to long-wave modes with second-order accuracy. Furthermore, it should eliminate freely propagating compressible short-wave modes that are under-resolved in time, while minimizing dispersion for resolved modes. Here, we describe first successful steps to achieve our goals.

Governing equations. The equations for one dimensional linear acoustics are given by the system

$$\begin{aligned} m_t + p_x &= 0 \\ p_t + c^2 m_x &= q(t, \tfrac{x}{\varepsilon}) \,, \end{aligned} \quad (1)$$

where $p = p(t, x)$ and $m = m(t, x)$ are the pressure and momentum fields. The speed of sound is specified by c, and the source term $q(t, \tfrac{x}{\varepsilon})$, $\varepsilon \ll 1$, is assumed to be slowly varying in time with small scale variations in space. This source term could simulate the release of latent heat from localized condensation, for example.

For traveling waves $(m, p)(t, x) = (m_0, p_0) \exp(i(\omega t - \kappa x))$, the dispersion relation of (1) is

$$\omega^2 - \kappa^2 c^2 = 0 \,. \quad (2)$$

Thus $\omega(\kappa) = \pm c\kappa$, so that in the continuous system all waves travel with the same velocity, $c = \pm \omega/\kappa$, without dispersion. Also, one can show, that the system preserves a global pseudo energy.

2 Implicit second-order staggered grid schemes

Before the new time integration scheme is introduced, we investigate two standard implicit second-order discretizations. These are the implicit trapezoidal rule and the BDF(2) scheme, which are commonly used in meteorological applications [3].

Their ability to compute reliable approximations to solutions of (1) is discussed with respect to the *discrete-dispersion relations* of these schemes (see [3,9] for details).

Furthermore, the capability of the schemes to reproduce balanced modes is discussed. In the case of slow, short-wave forcing the balance is described by

$$c^2 m_x = q\left(t, \frac{x}{\varepsilon}\right) \quad \text{and} \quad p \equiv 0 \tag{3}$$

up to small perturbations introduced by the variation in time of the source term. The schemes should be able to essentially keep this balance. Furthermore, they should reproduce the balanced state in one time step by letting the step going to infinity.

Considering a semi-discretization in time, we leave the choice of a spatial discretization open for the moment. In the subsequent numerical experiments we choose a staggered grid with central differences for simplicity only.

Implicit trapezoidal rule. The implicit trapezoidal rule is derived by integrating the differential equation from t^n to t^{n+1}. The time integral on the right-hand side is then approximated by the trapezoidal quadrature rule. For the system of linear acoustics (1) this results into a Helmholtz problem for p^{n+1}, which is given by

$$p^{n+1} - \frac{c^2 \Delta t^2}{4} \frac{\partial^2 p^{n+1}}{\partial x^2} = p^n - c^2 \Delta t \frac{\partial m^n}{\partial x} + \frac{c^2 \Delta t^2}{4} \frac{\partial^2 p^n}{\partial x^2} + \Delta t \, q^{n+1/2}. \tag{4}$$

The update for m is then obtained by

$$m^{n+1} = m^n - \frac{\Delta t}{2}\left(\frac{\partial p^n}{\partial x} + \frac{\partial p^{n+1}}{\partial x}\right). \tag{5}$$

The method is symplectic and A-stable [4]. The discrete-dispersion relation results in a frequency-wave number relationship of the form

$$\omega_r = \pm \frac{2}{\Delta t} \arctan\left(\text{cfl} \cdot \sin\left(\frac{k \Delta x}{2}\right)\right) \tag{6}$$

where $\text{cfl} = \frac{c \Delta t}{\Delta x}$ is the Courant–Friedrichs–Lewy (CFL) number, and the amplification factor per time step is given by $|A| \equiv 1$. Thus, essentially, the frequency ω_r depends not only on the wave number k, as in the continuous case, but it is also a function of the CFL number.

Figure 1 shows the discrete-dispersion relation for the trapezoidal rule (dashed line) applied to the linear acoustic equations for a CFL number $\text{cfl} = 1$. The scheme slows down modes at almost all wave numbers, and this behavior is amplified the higher the wave number and the higher the CFL number are. Additionally, the trapezoidal rule is free of numerical dissipation. By letting $\Delta t \to \infty$, one obtains the relations

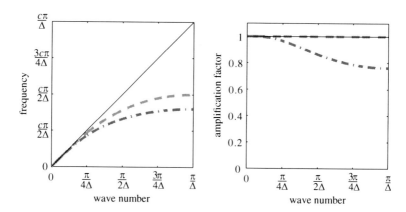

Fig. 1 Discrete-dispersion relations for the trapezoidal (dashed) and the BDF(2) rules (dot-dashed) applied to the linear acoustic equations using cfl = 1. Dispersion relation for continuous system is displayed as black line

$$\frac{c^2}{2}\left(\frac{\partial m^n}{\partial x} + \frac{\partial m^{n+1}}{\partial x}\right) = q^{n+1/2} \quad \text{and} \quad \frac{\partial p^{n+1}}{\partial x} = -\frac{\partial p^n}{\partial x}. \tag{7}$$

This reflects the inability to reproduce balanced modes of the trapezoidal rule, and any perturbation of the system cannot dissipate.

BDF(2) scheme. The BDF(2) scheme is a two-step method from the family of the so called *Backward Differentiation Formulas (BDF)*. Here, the left-hand side is approximated by the derivative of a parabola at t^{n+1}, which interpolates the solution at times t^{n-1}, t^n and t^{n+1}. For the acoustic system, this discretization results again in a Helmholtz problem for p^{n+1}, which is

$$p^{n+1} - \frac{4c^2 \Delta t^2}{9} \frac{\partial^2 p^{n+1}}{\partial x^2} = \frac{4}{3} p^n - \frac{1}{3} p^{n-1} - c^2 \Delta t \left(\frac{8}{9} \frac{\partial m^n}{\partial x} - \frac{2}{9} \frac{\partial m^{n-1}}{\partial x}\right) + \frac{2}{3} \Delta t \, q^{n+1}. \tag{8}$$

The update for m is obtained by

$$m^{n+1} = \frac{4}{3} m^n - \frac{1}{3} m^{n-1} - \frac{2}{3} \Delta t \frac{\partial p^{n+1}}{\partial x}. \tag{9}$$

The method is A- and L-stable [4].

The discrete-dispersion relation for the BDF(2) scheme is given again in Fig. 1 (dot-dashed line). Concerning the phase error, it shows the same behavior as the trapezoidal rule, although it is considerably amplified. On the other hand, the scheme introduces dissipation for almost all modes. The damping is amplified for high wave and CFL numbers. In the limit $\Delta t \to \infty$ one obtains

$$c^2 \frac{\partial m^{n+1}}{\partial x} = q^{n+1} \quad \text{and} \quad \frac{\partial p^{n+1}}{\partial x} = 0,$$

and the scheme achieves balance in a single, sufficiently large, time step. This behavior is characteristic to backward differences formulas by construction [2].

This analysis reveals the dichotomy a practitioner is faced with when having to choose between the two time integrators: Either he could choose to minimize dispersion and preserve the amplitude of well resolved modes by using the trapezoidal rule, or he could ensure that the solution rapidly relaxes to the balanced mode in case of short wave number forcing by the application of the BDF(2) scheme.

3 Multilevel method for long-wave linear acoustics

As described above, the ultimate goal is to filter out all acoustic short wave modes, which are not resolved in time, while sufficiently long wave data is integrated as accurate as possible. Here, we present a strategy for combining the two aspects into one single, scale-dependent numerical time integrator. It is exemplified by using the implicit trapezoidal rule and the BDF(2) scheme as base schemes. One could also use other time integrators (see [9] for a more general presentation), the only restriction is that they are linear in p^{n+1} and m^{n+1}.

Assume that we have scale dependent splittings of the pressure and momentum fields, i.e.

$$p = \sum_{\nu=0}^{\nu_m} p^{(\nu)} \quad \text{and} \quad m = \sum_{\nu=0}^{\nu_m} m^{(\nu)} \qquad (10)$$

which could be a quasi-spectral or wavelet decomposition, splitting p and m into (local) high and low wave number components. The idea is to use for each scale component $(p^{(\nu)}, m^{(\nu)})$ a scale dependent blending of the two time integrators. Taking the μ-dependent convex combination with $\mu \in [0, 1]$ of the two equations (4) and (8), and summing over the scales results in the Helmholtz problem

$$p^{n+1} - c^2 \Delta t^2 \sum_{\nu=0}^{\nu_M} \left(\frac{\mu_\nu}{4} + \frac{4(1-\mu_\nu)}{9} \right) p_{xx}^{(\nu),n+1} = \sum_{\nu=0}^{\nu_M} \left(\mu_\nu R_{\text{TRA}}^{p,(\nu)} + (1-\mu_\nu) R_{\text{BDF2}}^{p,(\nu)} \right), \qquad (11)$$

where

$$R_{\text{TRA}}^{p,(\nu)} = p^{(\nu),n} - c^2 \Delta t\, m_x^{(\nu),n} + \frac{c^2 \Delta t^2}{4} p_{xx}^{(\nu),n} + \Delta t\, q^{(\nu),n+1/2},$$

$$R_{\text{BDF2}}^{p,(\nu)} = \frac{4}{3} p^{(\nu),n} - \frac{1}{3} p^{(\nu),n-1} - c^2 \Delta t \left(\frac{8}{9} m_x^{(\nu),n} - \frac{2}{9} m_x^{(\nu),n-1} \right) + \frac{2}{3} \Delta t\, q^{(\nu),n+1}. \qquad (12)$$

The momentum update is derived from the blending of (5) and (9), which is

$$m^{n+1} = \sum_{\nu=0}^{\nu_M} \mu_\nu \left[m^{(\nu),n} - \frac{\Delta t}{2} \left(p_x^{(\nu),n} + p_x^{(\nu),n+1} \right) \right] + $$
$$(1 - \mu_\nu) \left[\frac{4}{3} m^{(\nu),n} - \frac{1}{3} m^{(\nu),n-1} - \frac{2}{3} \Delta t \, p_x^{(\nu),n+1} \right]. \qquad (13)$$

The scale splitting is obtained by the application of restriction and prolongation operators used in standard multigrid algorithms. Let $\varphi = \sum \varphi^{(\nu)}$ be a grid function, which is decomposed into parts $\varphi^{(\nu)}$ living on the associated grid levels. Then, the grid function on the coarsest level is obtained by the operation

$$\varphi^{(0)} = \left(R^{(0)} \circ R^{(1)} \circ \cdots \circ R^{(\nu_M - 1)} \right) \varphi \qquad (14)$$

and the grid functions on finer levels are computed by

$$\varphi^{(\nu)} = \left(I - P^{(\nu-1)} \circ R^{(\nu-1)} \right) \circ \left(R^{(\nu)} \circ R^{(\nu+1)} \circ \cdots \circ R^{(\nu_M - 1)} \right) \varphi. \qquad (15)$$

In our current approach the pressure is decomposed using the *full weighting* (restriction) and the *linear interpolation* (prolongation) operators [7]. They can be defined by their stencil, which are

$$R^{(\nu)} = \frac{1}{4} \begin{bmatrix} 1 & 2 & 1 \end{bmatrix} \quad \text{and} \quad P^{(\nu)} = \frac{1}{2} \begin{bmatrix} 1 & 2 & 1 \end{bmatrix}. \qquad (16)$$

On a staggered grid the matching splitting in the momentum field is then defined by (for further details, see [9])

$$R^{(\nu)} = \frac{1}{8} \begin{bmatrix} 1 & 3 & 3 & 1 \end{bmatrix} \quad \text{and} \quad P^{(\nu)} = \begin{bmatrix} 1 & 1 \end{bmatrix}. \qquad (17)$$

The description of the scheme is completed by the definition of the weighting function $\mu(\nu)$. In the subsequent tests, it is chosen such that the scheme in (11) and (13) associates the standard implicit trapezoidal scheme with all pressure modes corresponding to coarse grids with grid-CFL number cfl $\leq$ 1, while we nudge the discretization towards BDF(2) for pressure modes living on grids with cfl > 1.

4 Numerical Results

Here, we shortly describe a test case with "multiscale" initial data in a periodic domain $x \in [0, 1]$. Pressure and momentum fields are chosen in such a way that one obtains a right running acoustic simple wave with a sound speed of $c = 1$. The initial conditions are displayed in Fig. 2 (top row). No source term is present

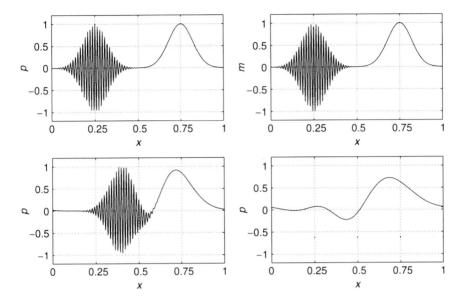

Fig. 2 Top row: "Multiscale" initial data. Bottom row: Numerical solution (pressure) with cfl = 10 at time $t_{\text{end}} = 3$ using the trapezoidal rule (left) and the BDF(2) scheme (right). Grid with 512 cells

(for further details see [9]). We use a grid with 512 cells (i.e. $\Delta x = 1/512$) and a CFL number cfl $= 10$. The results are compared at a final time $t_{\text{end}} = 3.0$, which is equivalent to 154 time steps. At this time the exact solution is identical to the initial data, and the wave has traveled three times across the domain.

The implicit trapezoidal rule produces the results in Fig. 2 (bottom left). Here, and in the following, only pressure is displayed, since the momentum field is essentially the same. The results show what has already been revealed by the discrete-dispersion relation, i.e., the scheme achieves large-CFL stability by slowing down the short wave components of the solution. While the long-wave pulse has traveled at nearly correct speed, the short-wave oscillations have essentially stayed in place. Furthermore, their amplitude has not diminished.

A different behavior is displayed by the BDF(2) scheme (Fig. 2, bottom right). It has considerably more dispersion than the trapezoidal rule, and the damping of the scheme results in a smaller final amplitude, even for the long wave data. On the short scales, the diffusion is so high that at the final time this part of the solution has essentially vanished. Thus, the scheme is able to balance the short-wave modes that are not resolved in time, but it pays the price of simultaneously damping and dispersing the long scales.

The result of the simulation using the new blended scheme with five grid levels is displayed in Fig. 3. For comparison, the result of the trapezoidal rule applied only to long wave data is also shown (dashed line). As one can see, the two results are nearly identical: The short wave data is filtered in such a way that only the long wave

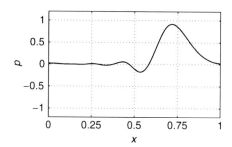

Fig. 3 Numerical solution (pressure) using the blended scheme on a grid with 512 cells and cfl = 10 at time $t_{end} = 3$ (black line). For comparison, the result of trapezoidal rule obtained with only long wave initial data is plotted as gray dashed line

data is left after some time. On the other hand, the long wave data is integrated as well as one could hope when using a second-order method.

5 Conclusion

The presented scheme effectively filters freely propagating compressible short-wave components, which cannot be accurately represented at long time steps. At the same time, dispersion and the amplitude errors for long-wave modes are minimized. Further tests show that in the presence of a source term, which slowly varies in time but has rapid spatial variations, solutions relax to an asymptotic balanced state (see [9]).

One of the next goals is to apply this scheme into a semi-implicit scheme for weakly compressible flows. The latter is an extension of a second-order projection method for incompressible flows as described in [8]. By using the trapezoidal rule in the implicit part of the scheme, one is faced with instabilities near shocks. This can partly be cured by so called off-centering. However, it also decreases the order of the scheme to one. The authors hope to obtain a second-order version by using the new scheme described in this note.

Acknowledgements R.K. thanks Piotr Smolarkiewicz for stimulating and challenging discussions on the theme of this paper. The authors thank Deutsche Forschungsgemeinschaft for its partial support of this work through grants KL 611/14 and KL 611/22 (MetStröm), and the Leibniz Association for partial support through its PAKT program.

References

1. Davies, T., Staniforth, A., Wood, N., Thuburn, J.: Validity of anelastic and other equation sets as inferred from normal-mode analysis. Q. J. R. Meteorolog. Soc. **129**(593), 2761–2775 (2003)
2. Deuflhard, P., Bornemann, F.: Scientific Computing with Ordinary Differential Equations, Texts in Applied Mathematics, vol. 42. Springer (2002)

3. Durran, D.R.: Numerical Methods for Fluid Dynamics: With Applications to Geophysics, 2 edn. No. 32 in Texts in Applied Mathematics. Springer (2010)
4. Hairer, E., Lubich, C., Wanner, G.: Geometric Numerical Integration: Structure-Preserving Algorithms for Ordinary Differential Equations. Springer (2006)
5. Ohfuchi, W., Nakamura, H., Yoshioka, M., Enomoto, T., Takaya, K., Peng, X., Yamane, S., Nishimura, T., Kurihara, Y., Ninomiya, K.: 10-km mesh meso-scale resolving simulations of the global atmosphere on the Earth Simulator: Preliminary outcomes of AFES. J. Earth Simulator **1**, 8–34 (2004)
6. Smolarkiewicz, P.K., Dörnbrack, A.: Conservative integrals of adiabatic Durran's equations. Int. J. Numer. Methods Fluids **56**(8), 1513–1519 (2008)
7. Trottenberg, U., Oosterlee, C., Schüller, A.: Multigrid. Academic Press (2001)
8. Vater, S., Klein, R.: Stability of a Cartesian grid projection method for zero Froude number shallow water flows. Numer. Math. **113**(1), 123–161 (2009)
9. Vater, S., Klein, R., Knio, O.M.: A scale-selective multilevel method for long-wave linear acoustics. Acta Geophysica (2011). Submitted

The paper is in final form and no similar paper has been or is being submitted elsewhere.

Nonlocal Second Order Vehicular Traffic Flow Models And Lagrange-Remap Finite Volumes

Florian De Vuyst, Valeria Ricci, and Francesco Salvarani

Abstract In this paper a second order vehicular macroscopic model is derived from a microscopic car–following type model and it is analyzed. The source term includes nonlocal anticipation terms. A Finite Volume Lagrange–remap scheme is proposed.

Keywords Vehicular traffic flow, modeling, car–following, microscopic, macroscopic, finite volumes, Lagrange–remap
MSC2010: 65M08, 65M22, 65P40, 65Y20, 65Z05

1 Motivation and introduction

There are many ways to describe and model a vehicular traffic flow. Microscopic models e.g. [3] describe the interaction between two successive vehicles. It is known that car–following models may have a complex dynamics (see for example [8, 9]) and are able to reproduce all the flow regimes. In the macroscopic models, conservation laws and balance equations on mean quantities are searched. Since the pioneer works by Lighthill, Whitham and Richards (LWR model), numerous improvements and contributions have been proposed. In 2000, Aw and Rascle [1] derived an interesting second order model that fixed the drawbacks of Payne's

Florian De Vuyst
Centre de Mathématiques et de leurs Applications, École Normale Supérieure de Cachan, 61 avenue du Président Wilson, 94235 Cachan France, e-mail: devuyst@cmla.ens-cachan.fr

Valeria Ricci
Dipartimento di Metodi e Modelli Matematici Universita' di Palermo, Viale delle Scienze, Edificio 8, 90128 Palermo, Italy, e-mail: valeria.ricci@unipa.it

Francesco Salvarani
Dipartimento di Matematica, Università degli Studi di Pavia, Via Ferrata 1 - I-27100 Pavia, Italy, e-mail: francesco.salvarani@unipv.it

model, emphasized by Daganzo [4]. More recently, Aw et al. [2] derived the Aw–Rascle model from microscopic follow–the–leader models. Illner et al. [7] were also able to retrieve the Aw–Rascle model from a kinetic Vlasov description. For related works, see for example [5,6]. In this paper, a continuum traffic flow model is derived from a more complex car–following model.

2 Car-following rule and microscopic model

Let us consider a vehicular traffic flow made of N vehicles, indexed by i, $i = 1, \ldots, N$. For simplicity, we will assume that all the vehicles are identical, of length ℓ. The car indexed by i follows the car $(i + 1)$. At time t, the vehicle i is located at position $x_i(t)$ with speed $\dot{x}_i = v_i$. The spatial gap between the two vehicles i and $(i + 1)$ is then given by $x_{i+1}(t) - x_i(t) - \ell$ (see Fig. 1). The maximum (permitted) speed will be denoted by v_M. Let us also denote by $g^s(v_i, v_{i+1})$ the safety spatial gap for the vehicle i, depending on the vehicle speeds i and $(i + 1)$. A simple relaxation rule for the spatial gap is

$$\frac{d}{dt}(x_{i+1} - x_i - \ell) = \frac{g^s(v_i, v_{i+1}) - (x_{i+1} - x_i - \ell)}{x_{i+1} - x_i - \ell} a_{i,i+1} \quad (1)$$

where $a_{i,i+1}$ is a local characteristic speed. The denominator forbids the collision between the two vehicles. Then we get a target speed v_i^{target} equal to

$$v_i^{target} = v_{i+1} + \left(1 - \frac{g^s(v_i, v_{i+1})}{x_{i+1} - x_i - \ell}\right) a_{i,i+1}. \quad (2)$$

A simple acceleration rule toward the target speed is given by the relaxation scheme

$$\frac{dv_i}{dt} = \frac{v_i^{target} - v_i}{\lambda} = \frac{v_{i+1} - v_i}{\lambda} + \left(1 - \frac{g^s(v_i, v_{i+1})}{x_{i+1} - x_i - \ell}\right) \frac{a_{i,i+1}}{\lambda} \quad (3)$$

using a characteristic relaxation time $\lambda > 0$. Let us comment three interesting cases. If $0 < x_{i+1} - x_i - \ell \ll 1$, then there is a strong breaking in order not to collide. If $x_{i+1} - x_i - \ell \equiv g^s(v_i, v_{i+1})$, the vehicle i is at the right safe distance, and in that case we have the simple car–following rule $\dot{v}_i = (v_{i+1} - v_i)/\lambda$. If $x_{i+1} - x_i \gg 1$,

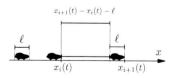

Fig. 1 Microscopic description of the vehicular traffic

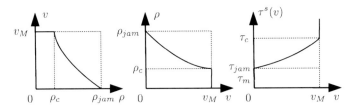

Fig. 2 Fundamental diagram of traffic flow and link with the spatial safety gap

the vehicle's driver i should not be worried about vehicle $(i + 1)$ because it is too far from him. In that case, the driver i should accelerate up to the limit speed v_M according to the rule $\dot{v}_i = (v_M - v_i)/\lambda$. This suggests us to choose $a_{i,i+1} = v_M - v_{i+1}$. To summarize, we get the microscopic model

$$\frac{dv_i}{dt} = \frac{v_{i+1} - v_i}{\lambda} + \left(1 - \frac{g^s(v_i, v_{i+1})}{x_{i+1} - x_i - \ell}\right) \frac{v_M - v_{i+1}}{\lambda}. \quad (4)$$

3 Macroscopic quantities and spatial safety gap

From the microscopic quantities, one can define some macroscopic ones. The specific volume $\tau_{i+1/2}(t) := x_{i+1}(t) - x_i(t)$ has the dimension of a length. The density $\rho_{i+1/2}(t) = (\tau_{i+1/2}(t))^{-1}$ returns the local number of vehicles per unit length. The quantity $\rho_M = \ell^{-1}$ represents the maximum density (nose-to-tail vehicles) and $\tau_m = \ell$ is the minimum specific volume. Now in (4), we need a closure for the safety gap function g^s. From g^s one can define a safety specific volume τ^s such that $g^s = \tau^s - \ell = \tau^s - \tau_m$ and a safety density $\rho^s = (\tau^s)^{-1}$. The density ρ^s can be identified to the fundamental diagram of traffic flow which gives a relation between the density and the equilibrium (safe) speed (see Fig. 2). We shall here consider

$$g^s(v_i, v_{i+1}) = \tau^s\left(\frac{v_i + v_{i+1}}{2}\right) - \tau_m.$$

4 Macroscopic model

In order to derive a macroscopic model, let us introduce some interpolation functions $v(x,t)$ and $\tau(x,t)$ such that

$$v(x_i(t), t) = v_i(t), \quad \tau(x_{i+1/2}(t), t) = x_{i+1}(t) - x_i(t) \quad \forall i = 1, \ldots, N.$$

A Taylor expansion allows us to write

$$v_{i+1}(t) - v_i(t) = v(x_{i+1}(t),t) - v(x_i(t),t) = \left(\tau \frac{\partial v}{\partial x}\right)(x_{i+1/2}(t),t) + o((x_{i+1} - x_i)^2).$$

From the motion equation $\dot{x}_i = v_i$, one can write $\frac{d}{dt}(x_{i+1}(t) - x_i(t)) = v_{i+1}(t) - v_i(t)$. Then we have

$$\frac{D\tau}{Dt}(x_{i+1/2}(t),t)) = \frac{\partial v}{\partial x}(x_{i+1/2}(t),t)\,\tau(x_{i+1/2}(t),t) + o((x_{i+1}(t) - x_i(t))^2).$$

We omit the remaining term and consider that the expression holds almost everywhere, then we get the continuity equation

$$\rho \frac{D\tau}{Dt} - \frac{\partial v}{\partial x} = 0 \quad \Leftrightarrow \quad \frac{\partial \rho}{\partial t} + \frac{\partial}{\partial x}(\rho v) = 0. \tag{5}$$

Consider now the acceleration equation. First remark that

$$\left(\frac{\partial v}{\partial x}\tau\right)(x_{i+1/2}(t),t) = \left(\frac{\partial v}{\partial x}\tau\right)(x_i(t),t) + \frac{\tau}{2}\frac{\partial}{\partial x}\left(\tau \frac{\partial v}{\partial x}\right)(x_i(t),t) + o(x_{i+1} - x_i).$$

One can also write $v_{i+1}(t) = v(x_{i+1}(t),t) = v\left(x_i(t) + \tau(x_{i+1/2}(t),t), t\right)$, which allows us to derive the balance equation in Lagrangian form

$$\rho \frac{Dv}{Dt} = \frac{1}{\lambda}\frac{\partial v}{\partial x} + \frac{1}{2\lambda}\frac{\partial}{\partial x}\left(\tau \frac{\partial v}{\partial x}\right) + \left(1 - \frac{g^s(v(x + \tau/2, t))}{\tau - \tau_m}\right)\rho \frac{v_M - v(x + \tau, t)}{\lambda}. \tag{6}$$

i.e. in Eulerian form

$$\frac{\partial}{\partial t}(\rho v) + \frac{\partial}{\partial x}\left(\rho v^2 - \frac{1}{\lambda}v\right) - \frac{1}{2\lambda}\frac{\partial}{\partial x}\left(\tau \frac{\partial v}{\partial x}\right) = \left(1 - \frac{g^s(v(x + \tau/2, t))}{\tau - \tau_m}\right)\rho \frac{v_M - v(x + \tau, t)}{\lambda}. \tag{7}$$

By multiplying formally equation (6) by v and using the continuity equation we get

$$\frac{\partial}{\partial t}(\rho v^2/2) + \frac{\partial}{\partial x}\left(\rho v^3/2 - \frac{1}{\lambda}v^2/2\right) - \frac{1}{2\lambda}\frac{\partial}{\partial x}\left(\tau \frac{\partial(v^2/2)}{\partial x}\right)$$

$$- \left(1 - \frac{g^s(v(x + \tau/2, t))}{\tau - \tau_m}\right)\rho v \frac{v_M - v(x + \tau, t)}{\lambda} = -\frac{1}{2\lambda}\tau \left(\frac{\partial v}{\partial x}\right)^2. \tag{8}$$

This shows that $S = \rho v^2/2$ is an entropy for the system. It is easy to show that S is convex with respect to the conservative variables $(\rho, \rho v)$ (but not strictly convex). More generally, for any $\mathscr{C}^2$ strictly convex function $h : \mathbb{R}^+ \to \mathbb{R}^+$, the function $S = \rho h(v)$ is a (non strictly) convex entropy for the system.

Properties Let us consider the first order homogeneous part of the system, i.e.

$$\partial_t \rho + \partial_x(\rho v) = 0, \quad \partial_t(\rho v) + \partial_x(\rho v^2 - \frac{1}{\lambda} v) = 0. \qquad (9)$$

In primitive variables (τ, v) we get

$$\partial_t (\tau, v)^T + \begin{pmatrix} v & -\tau \\ 0 & (v - \frac{\tau}{\lambda}) \end{pmatrix} \partial_x (\tau, v)^T = 0.$$

The system is strictly hyperbolic in the admissible space $\Omega_\varepsilon^{ad} = \{(\rho, v), \rho \in [\varepsilon, \rho_M], v \in [0, v_M]\}$ for any $\varepsilon > 0$. The characteristic speeds are $\lambda_1 = v$ and $\lambda_2 = v - \tau/\lambda$. It easy to check that the two characteristic fields are both linearly degenerate (LD) so that the eigenvalues of the system λ_i, $i = 1, 2$ are the Riemann invariants. One gets a straightforward structure of the solutions of the Riemann problem made of two contact discontinuities.

5 Finite volume scheme

For the sake of simplicity, we shall only deal with the inviscid part of the system above. Let us consider a uniform discretization of the spatial domain (with constant mesh step h) made of discrete points $(x_j)_{j \in \mathbb{Z}}$, $x_{j+1} = x_j + h$ and cells $I_j = (x_{j-1/2}, x_{j+1/2})$, $x_{j+1/2} = (x_j + x_{j+1})/2$. From time t^n, the time advance is performed using a time step Δt^n subject to stability constraints that will be detailed later on. The numerical discretization here follows ideas from Billot et al. [5].

Homogeneous Part Because of the structure of the eigenwaves in (9), a Lagrange–remap conservative FV approach is particularly well suited. Initially the discrete solution is piecewise constant on each control volume I_j with density ρ_j^n, specific volume $\tau_j^n = (\rho_j^n)^{-1}$, and speed v_j^n. In the Lagrange step, the computational grid moves according to the flow; the states into each cell evolve according to the Lagrangian equations. For an initial volume $I_j = (x_{j-1/2}, x_{j+1/2})$, the interface points $x_{j-1/2}$ are moved according to the motion equations $\dot{x}_{j+1/2} = v_{j+1/2}^n$ over a time step Δt^n: this gives $x_j^{n+1,-} = x_j^n + \Delta t^n v_{j+1/2}^n$. The choice $v_{j+1/2}^n = v_{j+1}^n$ is compatible with the structure of the solutions of the local Riemann problems, leading to a stable upwind process. After a time step, the cell sizes $h_j^{n+1,-}$ become

$$h_j^{n+1,-} = h + \Delta t^n \left(v_{j+1}^n - v_j^n \right). \qquad (10)$$

The continuity equation shows that the number of vehicles m_j into each Lagrangian cell I_j is conserved, i.e. $m_j^n = \rho_j^n h = \rho_j^{n+1,-} h_j^{n+1,-} = m_j^{n+1,-}$. Combining (10) and mass conservation, we get the equivalent script

$$\tau_j^{n+1,-} = \tau_j^n + \frac{\Delta t^n}{h}\left(v_{j+1}^n - v_j^n\right). \tag{11}$$

The CFL–like condition forbids the 1–waves to interact with the moving interfaces:

$$\frac{\Delta t^n}{h} \sup_{j \in \mathbb{Z}} \left[v_j^n - \min(0, v_j^n - \frac{\tau_j^n}{\lambda})\right] \leq 1. \tag{12}$$

By defining

$$v_j^{n+1,-} = \frac{\int_{I_j^{n+1,-}} \rho^{n+1,-}(x) v^{n+1,-}(x)\, dx}{\int_{I_j^{n+1,-}} \rho^{n+1,-}(x)\, dx} = \frac{\int_{I_j^{n+1,-}} \rho^{n+1,-}(x) v^{n+1,-}(x)\, dx}{m_j^n}$$

the speed in the cell $I_j^{n+1,-}$ before projection, we get the following scheme

$$v_j^{n+1,-} = v_j^n + \frac{\Delta t^n}{m_j^n \lambda}\left(v_{j+1}^n - v_j^n\right). \tag{13}$$

The Lagrange phase is followed by a conservative projection onto the initial uniform mesh. Denoting by $\alpha_{j+1/2}^n = v_{j+1}^n \Delta t^n / h$ the local Courant number related to the flow speed, the projection of the density in the cell I_j reads

$$\rho_j^{n+1} = \alpha_{j-1/2}^n \rho_{j-1}^{n+1,-} + \left(1 - \alpha_{j-1/2}^{n+1,-}\right) \rho_j^{n+1,-} \tag{14}$$

as soon as the time step Δt^n satisfies the additional CFL condition $\frac{\Delta t^n}{h} \sup_{j \in \mathbb{Z}} v_j^n \leq 1$. Similarly, the projection of the conservative quantity (ρv) writes

$$(\rho v)_j^{n+1} = \alpha_{j-1/2}^n (\rho v)_{j-1}^{n+1,-} + \left(1 - \alpha_{j-1/2}^{n+1,-}\right)(\rho v)_j^{n+1,-} \tag{15}$$

and gives v_j^{n+1}. It is easy to prove that this numerical scheme fulfills a discrete entropy inequality for the family of entropy functions $S = \rho h(v)$.

Source Term Integration The second equation has a source term that acts as a speed relaxation toward the maximum speed v_m in the case a free flow regime. The differential problem to solve is

$$\frac{dv}{dt} = \left(1 - \frac{g^s(v(x + \tau/2, t))}{\tau - \tau_m}\right) \frac{v_M - v(x + \tau, t)}{\lambda}. \tag{16}$$

When spatially discretized, we have to solve the differential problem

$$\frac{dv_j}{dt} = \left(1 - \frac{g^s(v(x_j + \tau_j/2, t))}{\tau_j - \tau_m}\right) \frac{v_M - v(x_j + \tau_j, t)}{\lambda}, \quad v_j(0) = v_j^0. \tag{17}$$

The problem (17) is nonstandard because of the presence of delays, nonlocal terms (due to the anticipation by the drivers) but also the coupling between the space variable x and the specific volume τ. A computational approach for (17) requires an interpolation of the function v, such as piecewise linear interpolation for example. If, from the discrete point of view, one expects a local influence of the anticipation, we have to assume that h is "large enough" to fulfill the inequality

$$\inf_{j \in \mathbb{Z}} \rho_j^n \geq h^{-1}. \tag{18}$$

The condition (18) may appear surprising, but actually it expresses that the spatial discretization must be compatible with the maximum space headway. As example, consider a road section of length $L = 200$ km and a uniform mesh made of $M = 1000$ points. Then $h = L/M = 0.2$ km and $h^{-1} = 5$ km^{-1}. The discretization of the source term may be local as soon as the vehicle density does not go below 5 veh/km.

Whole Fractional Step Method A consistent second–order accurate time splitting of the full inhomogeneous system may be achieved using the Strang fractional step approach. Each time iteration is made of three substeps: (i) a time integration of the source term over a time step $\Delta t^n/2$; (ii) a time advance of the homogeneous system over a time step Δt^n, (iii) a time integration of the source term over Δt^n as in (i).

6 Numerical experiments

In this experiment we use $v_M = 130$, a section length $L = 200$, a uniform mesh composed of 500 points, $\lambda = 4/3600$ and $\rho_M = 260$. We use periodic boundary conditions. The safety density is chosen as

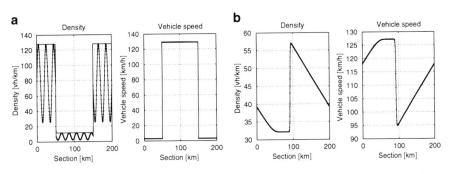

Fig. 3 (a) Initial condition: density (left) and speed (right). (b) Discrete solution at final time: density (left) and speed (right)

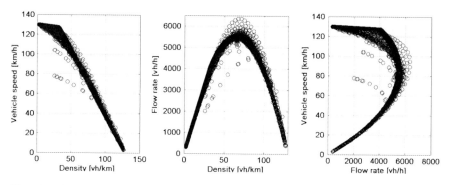

Fig. 4 Discrete solution in the phase space. From left to right: (ρ, v), $(\rho, \rho v)$ and $(\rho v, v)$ diagram

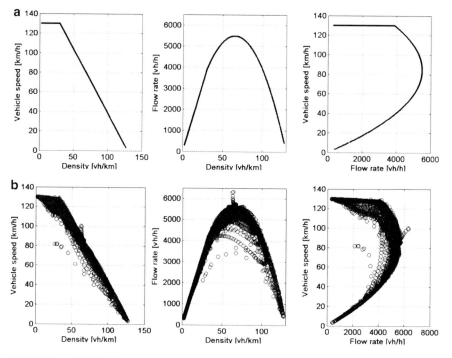

Fig. 5 Numerical Fundamental Diagram computed with: (a) LWR model, (b) Aw–Rascle model

$$\rho^s(v) = \min\left(1500\,\frac{v_M - v}{v_M},\; \rho_{jam} + (\rho_c - \rho_{jam})\frac{v}{v_M}\right)$$

with $\rho_c = 30$ and $\rho_{jam} = 130$. The initial velocity field is a piecewise constant function equal to 3 on $[0, L/4] \cup [3L/4, L]$ and equal to 129 on the interval $(L/4, 3L/4)$. The initial density profile $\rho^0(x) = (0.6 + 0.4\,\sin(20\pi x/L))\,\rho^s(v)$ mimics some nonequilibrium and traffic instabilities (see Fig. 3 (a)). On Fig. 3 (b), the discrete solution at final simulation time $t = 2.22$ is plotted and shows a very good behaviour of the numerical scheme with strong numerical stability, particularly through shock waves. Figure 4 shows the discrete solution for all discrete times in the phase space. One can observe a very good agreement with what is physically expected. The computed numerical discrete fundamental diagram is compared to those obtained with the LWR and Aw–Rascle models, respectively. For the Aw–Rascle model $\partial_t \rho + \partial_x (\rho v) = 0$, $\partial_t v + (v - \rho p'(\rho)) \partial_x v = \frac{A}{T}(v^{eq}(\rho) - v)$, we used $v^{eq}(\rho) = \min\left((v_M (1 - \frac{\rho}{1500}), v_M - v_M \frac{\rho - \rho_c}{\rho_{jam} - \rho_c}\right)$, $p(\rho) = v_M - v^{eq}(\rho)$, $A = 1$, $T = \lambda$.

References

1. A. Aw and M. Rascle, *Resurrection of "second order" models of traffic flow*. SIAM J. Appl. Math., Vol. 60 **(3)**,(2000), 916-938.
2. A. Aw, A. Klar, T. Materne, and M. Rascle. Derivation of continuum traffic flow models from microscopic follow-the-leader models, SIAM J. Applied Math., 63 **(1)**, 259–278 (2002).
3. M. Bando, K. Hasebe, A. Nakayama, A. Shibata, Y. Sugiyama, Phys. Rev. E 51, 1035 (1995).
4. C. F. Daganzo, *Requiem for second order fluid approximations of traffic flow*, Transp. Research B, 29, (1995), 277–286.
5. R. Billot, C. Chalons, F. De Vuyst, N. E. El Faouzi, J. Sau, A conditionally linearly stable second-order traffic model derived from a Vlasov kinetic description, Comptes Rendus Mécanique, Volume 338 **(9)** (2010), 529–537.
6. D. Helbing and A. Johansson, *On the controversy around Daganzos requiem for and Aw-Rascle's resurrection of 2nd-order traffic flow models*, Eur. Phys. J. B 69(4), (2009), 549–562.
7. R. Illner, C. Kirchner and R. Pinnau, *A Derivation of the Aw-Rascle traffic models from the Fokker-Planck type kinetic models*, Quart. Appl. Math. 67, (2009), 39–45.
8. B.S. Kerner, Springer, Berlin, New York (2009).
9. E. Tomer, L. Safonov and S. Havlin, Presence of Many Stable Nonhomogeneous States in an Inertial Car-Following Model, Phys. Rev. Lett. 84 **(2)**, 382385 (2000).

The paper is in final form and no similar paper has been or is being submitted elsewhere.

Unsteady Numerical Simulation of the Turbulent Flow around an Exhaust Valve

M. Žaloudek, H. Deconinck, and J. Fořt

Abstract The article presents numerical results of the flow which is exhausted from the combustion chamber of a four-stroke engine. The unsteady simulations shown correspond to one working cycle of an exhaust valve.

The flow has been described by the set of Reynolds–averaged Navier–Stokes equations. The working medium has been assumed an ideal gas. The numerical solution has been acquired with an in-house numerical code, *COOLFluiD*, based on a finite volume method (FVM). The numerical code is being developed by the team of engineers with wide range of specialization. Our major contribution has been connected to the implementation of the advanced turbulence models for both steady and unsteady simulations on moving grids.

The current work focuses on the turbulence modelling and on the simulation of the real valve movement. The flow structure and the mass flow rate are observed.

Due to a lack of experimental data, the computations are performed in a stepwise manner, validating each implementation step on the testcases known, before being applied to the valve geometry. The results presented therefore correspond to a planar model. The article focuses on the implementation of turbulence models and their application to complex geometry problems, rather than exploring new numerical methods.

Keywords numerical simulation, exhaust systems, turbulence
MSC2010: 76N15, 76F55, 76H05

Milan Žaloudek and Jaroslav Fořt
Dept. of Technical Mathematics, Czech Technical University, Karlovo nám. 13, CZ-12135 Praha 2, e-mail: Milan.Zaloudek@fs.cvut.cz, Jaroslav.Fort@fs.cvut.cz

Herman Deconinck
von Kármán Institute for Fluid Dynamics, Chaussée de Waterloo 72, B-1640 Rhode-Saint-Genèse, e-mail: deconinck@vki.ac.be

1 RANS Equations

The flow is governed by conservation laws of mass, momentum and energy and two transport equations of the turbulence model.

$$\frac{\partial \mathbf{W}}{\partial t} + \frac{\partial \mathbf{F}_i^I}{\partial x_i} = \frac{\partial \mathbf{F}_i^V}{\partial x_i} + \mathbf{Q}, \qquad (1)$$

with t representing time, $\mathbf{x}$ the Cartesian coordinates, $\mathbf{W}$ the vector of conservative unknowns, $\mathbf{F}^I/\mathbf{F}^V$ the convective/viscous fluxes and $\mathbf{Q}$ the source term.

$$\mathbf{W} = |\rho, \rho w_1, \rho w_2, e, \rho k, \rho \omega|^T \qquad (2)$$

$$\mathbf{F}_i^I = w_i |\rho, \rho w_1 + \tilde{p}\delta_{i1}, \rho w_2 + \tilde{p}\delta_{i2}, e + \tilde{p}, \rho k, \rho \omega|^T$$

$$\mathbf{F}_i^V = \left|0, \tau_{i1}, \tau_{i2}, \tau_{ij} w_j - q_i - q_i^t, (\mu + \sigma_k \mu_t)\frac{\partial k}{\partial x_i}, (\mu + \sigma_\omega \mu_t)\frac{\partial \omega}{\partial x_i}\right|^T$$

$$\mathbf{Q} = \left|0,0,0,0, P - \beta^* \rho k \omega, \frac{\gamma}{\nu_t} P - \beta \rho \omega^2 + (1 - F_1) \rho \frac{2\sigma_2}{\omega} \frac{\partial k}{\partial x_j} \frac{\partial \omega}{\partial x_j}\right|^T$$

The unknowns ρ, $\mathbf{w} = (w_1; w_2)$, e, p, T, k, ω denote in turns the density, the velocity components, the total energy, the pressure, the temperature, the turbulent kinetic energy and the specific dissipation rate. The stress tensor τ_{ij} is expressed as

$$\tau_{ij} = (\mu + \mu_t) S_{ij}, \quad S_{ij} = \frac{1}{2}\left(\frac{\partial w_i}{\partial x_j} + \frac{\partial w_j}{\partial x_i}\right) - \frac{2}{3} \cdot \delta_{ij} \cdot \frac{\partial w_k}{\partial x_k}, \qquad (3)$$

with δ_{ij} the Kronecker delta and μ, μ_t the molecular and turbulent dynamic viscosity

$$\mu = \frac{C_1 T^{3/2}}{T + S}, \quad \mu_t = \gamma^* \rho \frac{k}{\omega}. \qquad (4)$$

The heat flux and the production term read

$$q_i = -\frac{\lambda}{\Pr}\frac{\partial T}{\partial x_i}, \quad q_i^t = q_i \frac{\mu_t}{\mu}\frac{\Pr}{\Pr_t}, \quad P = \mu_t S_{ij} S_{ij}. \qquad (5)$$

Unknowns $\sigma_k, \sigma_\omega, \beta, \beta^*, \gamma, \gamma^*, \sigma_2, C_1, S, \lambda$ represent various constants to be found in the literature [7] and Pr stands for the Prandtl number. The function F_1 provides a blending between the $k-\epsilon$ model in freestream regions and the $k-\omega$ model near the wall surfaces. The system is completed with the state equation. The next turbulence models presented, have used a similar formulation as (1) and their specifics have been published in [10] (EARSM model) and [11] (Wilcox $k-\omega$, rev. 2008).

ALE Formulation. For unsteady simulations with a moving valve the arbitrary Lagrangian–Eulerian formulation of the RANS equations has been used, see [9]. The relative velocity $\mathbf{w}_R$ is defined as

$$\mathbf{w}_R = \mathbf{w} - \mathbf{w}_V, \tag{6}$$

with $\mathbf{w}$ the flow velocity and $\mathbf{w}_V$ the velocity of the valve (given by the movement imposed, see fig. 6). The convective flux $\mathbf{F}^I$ is then updated to a form

$$\mathbf{F}_i^{I,ALE} = w_{iR} \, |\rho, \rho w_1 + \tilde{p}\delta_{i1}, \rho w_2 + \tilde{p}\delta_{i2}, e + 2\tilde{p}, \rho k, \rho \omega|^T \tag{7}$$

2 Mathematical Formulation

The system (1) is solved upon the computational domain, see the Fig. 1. Although the real configuration is fully 3D, the computational domain has been considered symmetric with respect to the valve axis. Hence, only a half of the domain has been solved. A mathematic solution fulfils the equation (1) upon the domain interior, the *initial condition* at $t = 0$ and the following *boundary conditions* on the domain borders:

inlet total pressure, total temperature, incidence angle, turbulent variables according to the paper [8]:

$$\omega^{in} = \frac{|\mathbf{w}^{in}|}{L_{ref}}, \quad k^{in} = \omega^{in} \cdot \frac{\mu_\infty}{100}. \tag{8}$$

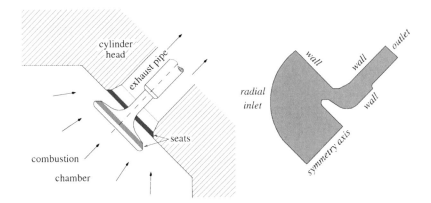

Fig. 1 Detail of the exhaust valve (left), scheme of the computational domain (right)

outlet pressure, velocity, temperature and turbulent variables

$$p = p^{out}, \quad \frac{\partial w_i}{\partial n} = \frac{\partial T}{\partial n} = \frac{\partial k}{\partial n} = \frac{\partial \omega}{\partial n} = 0 \quad (9)$$

wall the adiabatic no-slip condition. The turbulent variables use the expressions suggested at [8]

$$\mathbf{w} \equiv 0, \quad \frac{\partial T}{\partial n} = 0, \; k^w = 0, \; \omega^w = \frac{60\nu}{\beta_1 y_0^2} \quad (10)$$

3 Discretization and Numerical Method

The computational domain has been discretized by a structured triangular grid, see the Fig. 2. The steady flow computations were achieved with the time marching method based on a finite volume method, discretizing the equations (1) as

$$\frac{W_i^{n+1} - W_i^n}{\Delta t} = \frac{1}{\mu_i} \sum_{k=1}^{\#faces} \left(-\tilde{F}_k^I \cdot \mathbf{n}_k + \tilde{F}_k^V \cdot \mathbf{n}_k \right), \quad (11)$$

with μ_i the area of the i-th volume, $\tilde{F}_k^I / \tilde{F}_k^V$ the numerical approximation of the advection/viscous fluxes and $\mathbf{n}_k$ the unit outward normal vector to the k-th face of the volume i. The fully implicit time integration has been used

$$W_i^{n+1} = W_i^n + \frac{\Delta t}{\mu_i} \sum_{k=1}^{\#faces} \left(-\tilde{F}^I \left(W^n, W^{n+1} \right)_k \cdot \mathbf{n}_k + \tilde{F}^V \left(W^n, W^{n+1} \right)_k \cdot \mathbf{n}_k \right), \quad (12)$$

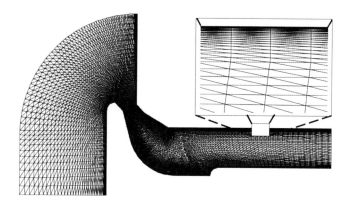

Fig. 2 Overview of the computational grid with the detail of its structure

as it is described in [5]. The linear system has been solved numerically by the GMRES iterative solver, provided by the PETSc library. As the flowfield contains both regions with the gas of negligible velocity (inside the chamber, Mach number ≈ 0.05) and regions with the gas of supersonic velocity (between the seats, $M \approx 2.0$) the numerical scheme $AUSM^{+up}$ able to capture all the velocity scales has been used. The algorithm is based on a solution of the Riemann problem (flux over a discontinuous step between two states) and thanks to the pressure and Mach number correction terms it improves the convergence also for the low velocity regions. The scheme has been published in [6]. The viscous fluxes have been computed as a central approximation, using a diamond dual cell approach.

The spatial accuracy has been improved by a piecewise linear reconstruction that has been built by the least squares interpolation method, complemented with the Barth limiter [2].

Unsteady Flow. The computational domain (and grid) changes with the advancing time. The solution is therefore based on the ALE formulation for moving grids. In order to avoid the situation of two disjunct subdomains with no flow between them for the closed valve a minimal valve opening (treshold) has always been used. Later, due to the significant grid deformation the domain has been remeshed and the current solutions interpolated in a conservative way. The series of three meshes (initial valve lift: 0.5 mm, 2.5 mm, 7.0 mm) have been used to resolve one working cycle of the exhaust valve.

The steady computations algorithm is modified to ensure the accuracy and consistency also for the unsteady flow. The time accurate solution has been obtained with the dual time stepping technique, consisting of an *outer time stepping loop* for a real time-accurate time step Δt and an *inner time stepping loop* with a fictitious time step τ to solve the system at each real time step. For the initiation phase the single step Crank–Nicholson method has been used, followed by the backward differentiation formula BDF2

$$\frac{W^{n+1,\alpha+1} - W^{n+1,\alpha}}{\tau} + \frac{3W^{n+1,\alpha+1} - 4W^n + W^{n-1}}{2\Delta t} = \qquad (13)$$

$$\frac{1}{\mu_i} \sum_{k=1}^{\#faces} \left[\tilde{F}^V \left(W^n, W^{n+1,\alpha}, , W^{n+1,\alpha+1} \right)_k \cdot \mathbf{n}_k - \tilde{F}^I \left(W^n, W^{n+1,\alpha}, , W^{n+1,\alpha+1} \right)_k \cdot \mathbf{n}_k \right]$$

4 Steady Flow Numerical Results

The Fig. 3 reveals the steady solutions with different turbulence models. The computations have been stated by the parameters: valve opening 4 mm, temperature 500 K, pressure ratio $\frac{p_{inlet}}{p_{outlet}} = 2.5$, with the outlet pressure 100 kPa, corresponding to the exhaust to the atmosphere. The flow topology is similar for all models, consisting of a main beam (in approximately same position), surrounded by separation zones on both sides. The differences are visible on the pressure distribution along a streamline that passes the middle of a channel throat, see the Fig. 4. The BSL and

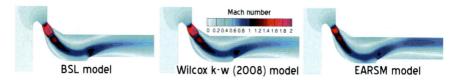

Fig. 3 Contours of Mach number for various turbulence models

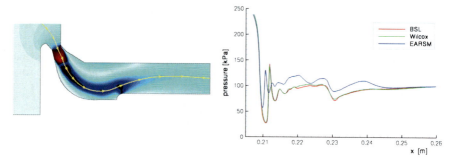

Fig. 4 The streamline for extracting the flow characteristics (left), comparison of the pressure through the exhaust pipe (right)

	lower wall [mm]			upper wall [mm]		
	start	end	length	start	end	length
BSL	2.8231	21.237	18.414	2.8284	35.730	32.901
Wilcox	2.1243	22.024	19.899	2.8284	35.774	32.945
EARSM	0.3643	18.091	17.727	0.0000	34.873	34.873

Fig. 5 Position of separation zones meassured from the channel throat

Wilcox models have a similar nature, which justifies similar results achieved by these models. By the contrary, the EARSM model allows anisotropic turbulence, see [10], leading to the milder peaks predicted and higher outlet velocity. However, the qualitative agreement is observed across all the models. Similar behaviour can be seen also on the comparison of the separation zone positions in the Table 5.

The next expansion is allowed due to the separations which form an artificial nozzle-like channel inside the exhaust pipe. These separations are described in the Fig. 5.

5 Unsteady Flow Numerical Results

The movement of the exhaust valve is shown in the Fig. 6a (valve lift vs. time). The next graph, Fig. 6b, shows the time evolution of the inlet pressure for a spark-ignition (SI), a compression-ignition (CI) engine and the outlet pressure. Values are taken from [4] and represent the boundary conditions for the unsteady simulations.

Unsteady Numerical Simulation of the Turbulent Flow around an Exhaust Valve 797

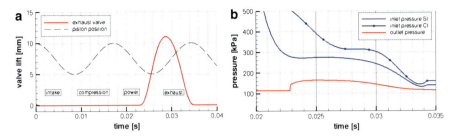

Fig. 6 The movement of the exhaust valve (left), the operating conditions at the inlet and outlet (right)

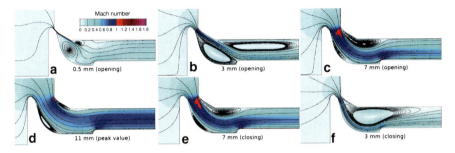

Fig. 7 SI engine. Contours of Mach number, velocity streamlines

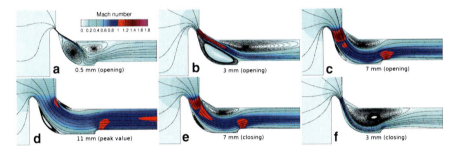

Fig. 8 CI engine. Contours of Mach number, velocity streamlines

The computations start at $t = 0.022$ (see Fig. 6) and the exhaust valve cycle lasts approximately 0.015 seconds. This interval has been resolved with the timestep $\Delta t = 10^{-6}$ s and with a valve lift treshold 0.5 mm. The results of the unsteady computations correspond to the valve lifts: $0.5 \rightarrow 3 \rightarrow 7 \rightarrow 11 \rightarrow 7 \rightarrow 3$ mm for both SI and CI inlet pressure evolutions.

The lift 7 mm has also been supplied by a pair of steady computations at boundary condition of the CI engine for the given lift, see the Fig. 9. The last Fig. 10 shows the mass flow rate over the valve cycle for the SI and CI engines, the steady solutions are mapped by two points. The last graph compares the pressure along the streamline (see Fig. 4) for the unsteady (Fig. 8e) and steady (Fig. 9b) computations at the same valve lift 7 mm.

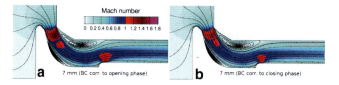

Fig. 9 Steady results. Boundary conditions correspond to CI engine at valve lift 7 mm in the opening (left) and the closing phase. Contours of Mach number, velocity streamlines

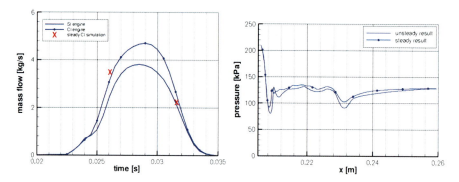

Fig. 10 Comparison of the SI and CI engine model: mass flow rate (left). Comparison of the unsteady and steady model: pressure development in the exhaust pipe (right)

6 Conclusions

The steady results have shown similar behaviour for all the turbulence models tested.

The flowfield of unsteady results are in qualitative agreement with equivalent steady solutions, however, the mass flow rate can differ up to approximately 10% (see Fig. 10). Also the pressure development along the mean streamline behind the channel throat differs from the steady state.

In case of the CI engine (due to higher inlet pressure) one observes the aerodynamical choking and larger supersonic regions, compared to the SI engine. The SI model is choking-free in the dominant time of the valve cycle. The negligible mass flow in the early and late stages of the valve cycle also justifies the use of grids with minimal (non-zero) valve opening. The oncoming work will be aimed at more advanced turbulence models for the unsteady simulations, flow characteristics at different rpm and mainly on 3D unsteady simulations.

Acknowledgements This work has been supported by the grant of the Czech Science Foundation No. P101/10/1329 and by the project 1M6840770002 Josef Božek Research Center of the Ministry of Education of the Czech Republic.

References

1. COOLFluiD homepage [on-line], http://coolfluidsrv.vki.ac.be, Cited 17 Feb 2011
2. Barth, T. J., Jesperson, D. C.: The design and application of upwind schemes on unstructured meshes. AIAA Paper **89(0366)** (1989)
3. Favre, A.: Equations des gaz turbulents compressibles. J. de Mecanique **4**, 361–421 (1965)
4. Heywood, J. B.: Internal Combustion Engine Fundamentals. McGraw-Hill, Inc. USA (1988)
5. Lani, A.: An Object Oriented and High Performance Platform for Aerothermodynamics Simulation. Doctoral thesis, VKI, Belgium (2009)
6. Liou M. S.: A sequel to AUSM, Part II: $AUSM^+$ up for all speeds, J. of Computational Physics **214**, 137–170 (2006)
7. Menter, F. R., Rumsey, C. L.: Assessment of Two-Equation Turbulence Models for Transonic Flows. AIAA 25th Fluid Dynamics Conference, Colorado Springs, USA (1994)
8. Menter, F. R.: Two-Equation Eddy-Viscosity Turbulence Models for Engineering Applications, AIAA Journal **32-8** (1994)
9. Michler, C.: Development of an ALE Formulation for Unsteady Flow Computations on Moving Meshes using RD Schemes, Project Report **2000-13**, VKI, Belgium (2000)
10. S. Wallin: Engineering turbulence modelling for CFD with focus on explicit algebraic Reynolds stress models. Dissertation Thesis, Norsteds Tryckeri AB, Sweden (2000)
11. D. C. Wilcox, Formulation of the $k - \omega$ Turbulence Model Revisited, AIAA J. **46-11** (2008)

The paper is in final form and no similar paper has been or is being submitted elsewhere.